Means
Estimating
Handbook

Means
Estimating
Handbook

SENIOR EDITOR
Jeffrey M Goldman

CONTRIBUTING EDITORS
Allan B. Cleveland
John Chiang
Donald D. Denzer
Mary P. Greene
Melville J. Mossman
John J. Moylan
Jeannene D. Murphy
Kornelis Smit
Phillip Waier
Edward B. Wetherill

TECHNICAL COORDINATOR
Marion Schofield

COMPOSITION COORDINATOR
Joan C. Marshman

TECHNICAL ASSISTANTS
Wayne Anderson
Eleanor J. Morris

COMPOSITION
Denise M. Fitzgerald
Dianne J. Messier
Mary S. Mowrey
Diane E. Silva

Book and jacket designed by Norman R. Forgit
Illustrations by Carl W. Linde

R.S. MEANS COMPANY, INC.
CONSTRUCTION CONSULTANTS & PUBLISHERS
100 Construction Plaza
P.O. Box 800
Kingston, MA 02364-0800
(617) 585-7880
© 1990

Printed in the United States of America

10   9   8   7   6   5   4   3

Library of Congress Catalog Number 91-115713

ISBN 0-87629-177-9

Table of Contents

Foreword
How This Book Is Arranged

This book is organized according to the sixteen divisions of the MASTERFORMAT System of classification as developed by the Construction Specifications Institute (CSI) and Construction Specifications Canada (CSC). This system, the most widely accepted in the industry, is used extensively by architects and engineers for construction specifications, by contractors for estimating and record keeping, and by manufacturers and suppliers for the categorization of materials and products.

The CSI MASTERFORMAT Divisions:

Division 01—General Requirements
Division 02—Site Work
Division 03—Concrete
Division 04—Masonry
Division 05—Metals
Division 06—Wood and Plastics
Division 07—Moisture-Thermal Control
Division 08—Doors, Windows, and Glass
Division 09—Finishes
Division 10—Specialties
Division 11—Equipment
Division 12—Furnishings
Division 13—Special Construction
Division 14—Conveying Systems
Division 15—Mechanical
Division 16—Electrical

Each division of this book is divided into four sections: an Introduction, an Estimating Data section, a Checklist and Tips.

- The Introduction provides an overview of what is included in each division.
- The Estimating Data, the main section of each division, includes numerous charts and tables with information and assistance for all aspects of construction estimating.
- The Checklists will help to make sure that all items are included in an estimate.
- The Tips provide shortcuts, hints on how to avoid pitfalls, and other handy information.

The Appendix includes abbreviations, conversion tables, equivalents, formulas, and other general information useful when preparing construction estimates. The comprehensive index can be used to pinpoint any specific topic.

Acknowledgments

We would like to express our appreciation to the following organizations, which have granted permission to reproduce specific charts or text from their publications.

American Society of Heating, Refrigeration, and
Air-Conditioning Engineers
1791 Tullie Circle, N.E.
Atlanta, GA 30329

Asphalt Roofing Manufacturers Association
6288 Montrose Road
Rockville, MD 20852

Brick Institute of America
11490 Commerce Park Drive
Reston, VA 22091

Caterpillar Tractor Company
North American Commercial Division
100 NE Adams Street
Peoria, IL 61629

Craftsman Book Company
6058 Corte del Cedro
P.O. Box 6550
Carlsbad, CA 92008;
Estimating Tables for Home Building, Paul I. Thomas

Peckham Industries
50 Haarlem Road
White Plains, N.Y.

Prentice Hall, Inc.
Rte. 9W
Englewood Cliffs, NJ
How to Estimate Building Losses and Construction Costs, Paul I. Thomas

Steel Door Institute
712 Lakewood Center, N.
14600 Detroit Ave.
Cleveland, OH 44107

Introduction
Estimating
Review

Estimating Review

The purpose of this book is to provide the estimator with information to assist in estimating quantities of material as well as labor required for construction projects. Construction estimating can be separated into two basic components: how many units are required, and reasonable costs for those units. The determination of quantities is commonly called the *quantity takeoff*. It is, simply, the counting of the physical units of materials needed for the construction of the project. While this may sound like a simple enough task, any estimator knows that it can be quite an involved process. For example, when estimating a concrete footing (excavation and backfill not included here), the items necessary include formwork (forms, ties or spreaders, keyway, inserts, form oil, delivery, erecting, stripping, and clearing); reinforcing (delivery, cutting, bendin, ties, splices, chairs, other accessories; and the concrete itself placing, compacting, finishing, curing, protecting, patching). A working knowledge of construction materials and methods is a must for a successful estimator. This knowledge helps to ensure that all items are accounted for and tabulated correctly—a sound basis for a good estimate.

The determination of "reasonable" costs for tabulated quantities is one of the main reasons why estimates vary. What may be reasonable for one job may be unrealistic for another. Labor rates (and productivity) will vary from region to region. Costs for the same materials also vary—from city to city and even supplier to supplier in the same town. The experienced estimator evaluates each project individually, investigates fluctuations in costs, and uses that knowledge to advantage.

Estimating for building construction is certainly not as simple a task as many people believe. Before one can obtain a quantity and apply a cost per unit to that quantity, many long hours, days, or even weeks of hard work may be put into a detailed project before arriving at that one "magic number."

Types of Estimates

Several different levels of estimates are used to project construction costs. Each has a different purpose. The various types may be referred to by different names and some may not be recognized by all as necessary or definitive, though most estimators will agree to several basic levels, each of which has its place in the construction estimating process. In this text, the levels of estimates are broken down as follows. (See Figure 0.1 for time vs. accuracy.)

1. **Order of Magnitude Estimates:** The order of magnitude estimate could be loosely described as an educated guess. It can be completed in a matter of minutes. Accuracy is plus or minus 20%.

Figure 0.1 Estimating Time vs. Accuracy
(Based on a $2,000,000 Building)

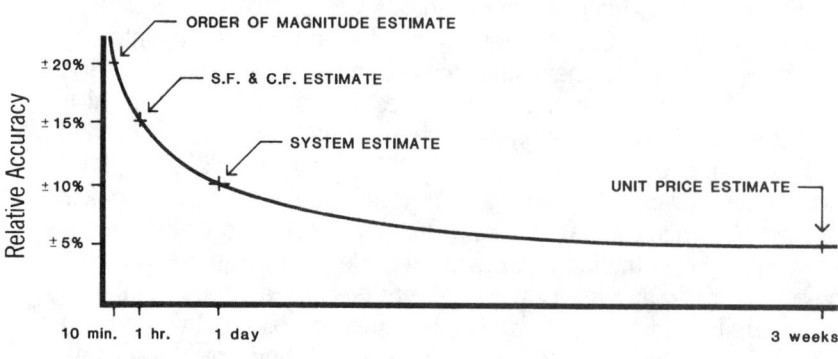

2. **Square Foot and Cubic Foot Estimates:** This type is most often useful when only the proposed size and use of a planned building is known. Very little information is required. Accuracy is plus or minus 15%.
3. **Assemblies Systems Estimate:** An essemblies estimate is best used as a budgetary tool in the planning stages of a project. Accuracy is expected at plus or minus 10%.
4. **Unit Price Estimate:** Working drawings and full specifications are required to complete a unit price estimate. It is the most accurate of the four types but is also the most time consuming. Used primarily for bidding purposes, accuracy is plus or minus 5%.

Order of Magnitude Estimates: The Order of Magnitude estimate can be completed when only minimal information is available. The proposed use and size of the planned structure should be known and may be the only requirement. The "units" can be very general and need not be well defined. For example: "An office building for a small service company in a suburban industrial park will cost about $500,000." This type of statement (or estimate) can be made after a few minutes of thought used to draw upon experience and to make comparisons with similar projects from the past. While this rough figure might be appropriate for a project in one region of the country, an adjustment may be required for a change of location and for cost changes over time (price changes, inflation, etc.).

Square Foot and Cubic Foot Estimates: The use of Square Foot and Cubic Foot estimates is most appropriate prior to the preparation of plans or preliminary drawings, when budgetary parameters are being analyzed and established. Costs may be broken down into different construction components, and then into the relationship of each component to the project as a whole, in terms of costs per square foot. This breakdown enables the designer, planner or estimator to adjust certain components according to the unique requirements of the proposed project.

Historical data for square foot costs of new construction are plentiful. However, the best source of square foot costs is the estimator's own cost records for similar projects, adjusted to the parameters of the project in question. While helpful for preparing preliminary budgets, Square Foot and Cubic Foot estimates can also be useful as checks against other, more detailed estimates. While slightly more time is required than with Order of Magnitude estimates, a greater accuracy (plus or minus 15%) is achieved due to more specific definition of the project.

Assemblies (or Systems) Estimates: Ever increasing design and construction costs make budgeting and cost efficiency increasingly important in the early stages of building projects. Never before has the estimating process had such a crucial role in the initial planning. Unit Price estimating, because of the time and detailed information required, is not suited as a budgetary or planning tool. A faster and more cost effective method is needed for the planning phase of a building project; this is the "Systems," or "Assemblies" estimate.

The Assemblies method is a logical, sequential approach which reflects how a building is constructed. Twelve "Uniformat" divisions organize building construction into major components that can be used in Assemblies estimates. These Uniformat divisions are listed below:

Assemblies Estimating Divisions

Division 01—Foundations
Division 02—Substructures
Division 03—Superstructure
Division 04—Exterior Closure
Division 05—Roofing
Division 06—Interior Construction
Division 07—Conveying
Division 08—Mechanical
Division 09—Electrical
Division 10—General Conditions
Division 11—Special
Division 12—Site Work

Each division is further broken down into individual assemblies. Each individual assembly incorporates several different items into a system that is commonly used in building construction.

In the Assemblies format, a construction component may appear within more than one division. For example, concrete is found in Division 1— Foundations, as well as in Divisions 2, 3 and 12 (see list above). Conversely, each division may incorporate many different areas of construction, and the labor of different trades.

A great advantage of the Assemblies estimate is that the estimator/designer is able to substitute one system for another during design development and can quickly determine the cost differential. The owner can then anticipate accurate budgetary requirements before final details and dimensions are established.

Final design details of the building project are required for a Unit Price estimate. The Assemblies method does not require such details, but the estimators who use it must have a solid background knowledge of construction materials and methods, Building Code requirements, design options, and budgetary restrictions.

The Assemblies estimate should not be used as a substitute for the Unit Price estimate. While the Assemblies approach can be an invaluable tool in the planning stages of a project, it should be supported by Unit Price estimating when greater accuracy is required.

Unit Price Estimates

The Unit Price estimate is the most accurate and detailed of the four estimate types and therefore takes the most time to complete. Detailed working drawings and specifications must be available to the unit price estimator. All decisions regarding the building's materials and methods must have been made before this type of estimate can be completed. There are fewer variables, and the estimate can, therefore, be more accurate. The working drawings and specifications are needed to determine the quantities of materials, equipment, and labor. Current and accurate costs for these items (unit prices) are also necessary.

Because of the detail involved and the need for accuracy, Unit Price estimates require a great deal of time and expense to complete properly. For this reason, Unit Price estimating is often used for construction bidding. It can also be effective for determining certain detailed costs in conceptual budgets or during design development.

Most construction specification manuals and cost reference books, such as Means' *Building Construction Cost Data,* divide all Unit Price estimating information into the sixteen CSI MASTERFORMAT divisions as described and listed in the Foreword of this book.

Before Starting the Estimate

In recent years, drawings and specifications have become massive volumes containing a wealth of information. It is of utmost importance that the estimator read all contract documents thoroughly. They exist to protect all parties involved in the construction process. The contract documents are prepared so that the estimators will be bidding equally and competitively, ensuring that all items in a project are included. The contract documents protect the designer (the architect or engineer) by ensuring that all work is supplied and installed as specified. The owner also benefits from thorough and complete construction documents, being guaranteed a measure of quality control and a complete job. Finally, the contractor benefits because the scope of work is well defined, eliminating the gray areas of what is implied but not stated. "Extras" are more readily avoided. Change orders, if required, are accepted with less argument if the original contract documents are complete, well stated, and most importantly, read by all concerned parties.

During the first review of the specifications, all items to be estimated should be identified and noted. The General Conditions, Supplemental Conditions, and Special Conditions sections of the specifications should be examined carefully. These sections describe the items that have a direct bearing on the proposed project, but may not be part of the actual, physical construction. An office trailer, temporary utilities, and testing are examples of these kinds of items.

While analyzing the drawings and specifications, the estimator should evaluate the different portions of the project to determine which areas warrant the most attention. For example, if a building is to have a steel framework with a glass and aluminum frame skin, then more time should be spent estimating Division 5—Metals, and Division 8—Doors, Windows and Glass, than Division 6—Wood and Plastics.

Figures 0.2 and 0.3 are charts showing the relative percentage of different construction components by MASTERFORMAT Division and Assemblies (UNIFORMAT) Division, respectively. These charts have been developed to represent the average percentages for new construction as a whole. All commonly used building types are included. The estimator should determine, for a given project, the relative proportions of each component, and estimating time should be allocated accordingly. More time and care should be given to estimating those areas which contribute more to the cost of the project.

Perhaps the best way for an estimator to approach a project is to begin with a clear mind and a clear desk. Clutter and confusion can have detrimental effects on the efficiency and accuracy of the estimate.

Figure 0.2   Component Contribution by MASTERFORMAT Division

No.	Division	%	No.	Division	%	No.	Division	%
1.6	Contractor Equip.*	6.6%	5.1	Structural Metals	2.2%	9.2	Lath & Plaster	0.4%
			5.2, 3	Metal Joists & Deck	1.5	9.2	Drywall	4.7
2.2	Earthwork	1.7	5.5	Misc. & Ornamental		9.3, 4	Tile & Terrazzo	1.1
2.3	Caissons & Pilings	0.7		Metals	2.6	9.5	Acoustical Work	0.8
2.5	Roads & Walks	0.8	5	Metals	6.3	9.5, 6	Flooring	1.3
2.6, 7	Site Utilities	0.6				9.9	Painting	2.1
2.8	Site Improvements	0.1	6.1	Rough Carpentry	1.6			
2.9	Landscaping	1.2	6.2	Finish Carpentry	0.1	9	Finishes	10.4
2	Sitework	5.1	6	Wood & Plastics	1.7		Divisions 10-14	8.5
3.1	Formwork	5.8	7.1	Water &		15.1, 2, 3	Plumbing	5.2
3.2	Reinforcing	2.5		Dampproofing	0.7	15.4	Fire Protection	2.6
3.3	Cast in Place Conc.	9.7	7.2	Insulation	0.6	15.5	Heating	7.1
3.4	Precast Concrete	0.3	7.3	Shingles	0.3	15.6, 7	Air Condit. & Vent.	4.4
3.1	Cementitious Decks	0.1	7.4, 5	Roofing & Siding	3.1			
3	Concrete	18.4	7.6	Sheet Metal Work	0.2	15	Mechanical	19.3
4.1	Mortar & Access.	0.3	7	Moisture Protection	4.9	16	Electrical	9.5
4.2	Brick Masonry	2.7	8.1,2,3	Doors & Frames	1.5		Total	100.0%
4.2	Block & Tile Work	6.7	8.5	Windows	2.5			
4.4	Stone	0.4	8.8, 9	Glass & Glazing	1.8	*Percentage for contractor equipment		
4	Masonry	10.1	8	Doors, Windows, Glass	5.8	is spread among divisions and included above for information only.		

Figure 0.3   Cost Distribution by Assemblies (Uniformat) Division

Division No.	Building System	Percent-age	Division No.	Building System	Percent-age
1 & 2	Foundation & Substructure	9.7%	7	Conveying	5.1%
3	Superstructure	17.4	8	Mechanical	19.3
4	Exterior Closure	15.8	9	Electrical	9.5
5	Roofing	3.3	11	Equipment	3.6
6	Interior Construction	12.0	12	Sitework	4.3
			Total weighted index (Div. 1–12)		100.0%

The Quantity Takeoff

Quantities may be taken off by one person if the project is not too large and time allows. For larger projects, the plans are often split into several disciplines (or divisions) and the work assigned to two or more quantity surveyors. In this case, a project leader is assigned to coordinate and assemble the estimate.

When working with the plans during the quantity takeoff, consistency is the most important consideration. If each job is approached in the same manner, a pattern will develop, such as moving from the lower floors to the top, clockwise or counterclockwise. The choice of method is not important, but consistency is. The purpose of being consistent is to avoid duplications as well as omissions and errors. Pre-printed forms provide an excellent means for developing consistent patterns.

An example of a quantity takeoff form from *Means Forms for Building Construction Professionals* is shown in Figure 0.4.

General Rules for the Quantity Takeoff: General rules have been established for improving the speed, ease, and accuracy of the takeoff process. Quantity estimators should adhere to these rules. An accurate quantity takeoff is critical to the accuracy of a cost estimate, since no estimate will be reliable if a mistake is made in the quantity takeoff, no matter how precise the unit price information may be.

General Rule 1

When taking off quantities, follow the guidelines provided by the person who will be applying unit prices to the quantities (the cost estimator). The takeoff should be clear and informative to prevent misinterpretation. Use symbols, sketches, or footnotes to clarify ambiguities in the takeoff. The quantity estimator should think of him- or herself as an assistant to the cost estimator.

General Rule 2

A takeoff list is not just a list of materials, but a list of measurements separated into categories to which unit prices are applied. The quantity sheets should be relatively simple to use. (See Figure 0.4.) The name of the building component is written in the far left-hand column labeled "Description." The number of components called out on the plans is listed next, followed by their dimensions (such as length, width, and depth or height). Quantities of items that are taken off of the component are listed in the subsequent columns. When taking off strip footings, for example, the associated items include structural excavation, concrete, formwork, backfill, and disposal. Appropriate units of measure, such as cubic yards of concrete, square feet of forms, and linear feet of pour strips, are applied to each item.

If more than one building component is listed on the same quantity sheet, or if there are several different sizes of the same component, then quantities are listed in the appropriate columns and totalled at the bottom of the page. In this way the quantity estimator can calculate, for instance, the total number of cubic yards of concrete needed for strip footings for the entire building and write it in one sum at the bottom of the column labeled "Concrete."

Figure 0.4   Quantity Sheet

🔺 Means Forms

QUANTITY SHEET

SHEET NO.

PROJECT

ESTIMATE NO.

LOCATION   ARCHITECT   DATE

TAKE OFF BY   EXTENSIONS BY: CHECKED BY:

DESCRIPTION	NO.	DIMENSIONS		UNIT		UNIT		UNIT		UNIT

General Rule 3

A quantity estimator may begin the takeoff with any building
component and proceed in the order of his or her choice. A good
approach is to follow roughly the order of the actual field construction,

such as from the footings upward to the roof. This provides the quantity estimator with the clearest mental picture of the project. If a project consists of more than one building, each structure should be taken off separately, since unit costs may vary from structure to structure.

General Rule 4

Check the drawings and details carefully for notes such as "NTS" (Not To Scale), changes in the scale as it is used throughout the drawings, drawings reduced to one-half or one-quarter their original size, or discrepancies in the specifications and the plans. Be consistent when listing dimensions.

General Rule 5

Where possible, use the dimensions stated in the drawings instead of measuring by scale, but make a habit of frequently checking printed dimensions with a scale or with mental arithmetic to spot draftsman's errors. Always express dimensions in the same order, such as:

length x width x height (or depth)

General Rule 6

Use a systematic procedure when working with the drawings. For instance, take measurements in a clockwise direction around a floor or roof plan, first recording the measurements of items displayed horizontally on the drawings, and then recording those shown vertically. This method is most useful when taking off two-dimensional areas.

General Rule 7

Whenever possible, the items in a quantity takeoff should be identified by their location on the drawings.

General Rule 8

A quantity estimate is not intended to be used for direct purchasing of materials. In fact, many items on a quantity estimate have no material value. These are called *work items* and are simply areas that require labor, such as fine grading gravel or finishing concrete surfaces. Work items may not appear on the drawings but are nonetheless required to complete the job. The quantity estimator should pay close attention to areas that may contain labor requirements. Items that *do* have material value are called *material items*. Both material items and work items are assembled on the same form for eventual pricing out, or cost estimating.

Any item that has a cost value should be assigned a unit of measure, even if it is only in *lump sum* (LS) form. The term "lump sum" is used for certain work items that cannot be measured or expressed in any other way. "LS" calls the estimator's attention to an item that requires a judged cost allowance.

General Rule 9
Decimals are used in quantity takeoff instead of fractions because they are faster, more precise, and easy to use on a calculator. Drawing dimensions that are given in feet and inches are converted to *decimal feet*, that is, feet and tenths of a foot.

General Rule 10
Quantity takeoff is performed for cost *estimating* purposes. Since estimating is not an exact science, the lists of quantities need not be overly precise. An example of unnecessary precision is calculating excavation quantities to 1/8". However, a reasonable degree of precision is expected. No detailed estimator wants to be accused of ballpark estimating.

In most cases, the use of two decimal places is sufficient for quantity surveying purposes (12' 4-1/2" = 12.38') and easy to enter into a calculator. However, when writing the product of the calculation, decimals are usually meaningless. Develop rules for precision that are consistent with measurement capabilities. Below is an example.

Item	Input	Output
Earthwork	Nearest 0.1 feet	Nearest C.F. or C.Y.
Concrete	Nearest .01 feet	Nearest C.F. or C.Y.
Formwork	Nearest .01 feet	Nearest S.F.
Finishing and Precast	Nearest .01 feet	Nearest S.F.
Lumber	Nearest 0.1 feet	Nearest B.F.
Finishes	Nearest 0.1 feet	Nearest S.F. or S. Y.

The quantity estimator must also learn the standards of each industry. For instance, a lumber dimension of 12' 1-1/2" must be rounded up to 14' due to standard sawmill cutting practices.

Finally, do not convert units until all items in a column are totaled. For instance, keep concrete in cubic feet (C.F.) until all of the quantities listed in the concrete column have been added together. Then convert to cubic yards (C.Y.).

General Rule 11
The quantity estimator should add an allowance for waste to certain quantities. Before the waste allowance is made, the quantities are referred to as *net quantities*. After the allowance for waste is added, the quantities are considered *gross quantities*.

General Rule 12
Ideally, a second quantity takeoff should be performed by a separate individual or team to ensure that no items have been omitted or duplicated. Unfortunately, the personnel to perform a second estimate are usually not available, or the cost to hire additional help is

prohibitive. Typically, the quantity estimator systematically must check his or her own work. In fact, the *dimensions* taken from drawings *should* be checked by the quantity estimator while the *extensions* should be checked by someone *other than* the original quantity estimator.

General Rule 13

One way to avoid omissions and duplications is to mark the drawings as items are taken off. Make colored pencil shadings and check marks directly on the drawings as items are taken off. Most quantity estimators have their own methods of marking drawings. Usually a combination of methods is used, rather than a single method, as one kind of mark may be effective in taking off one particular category, and different marks effective for other categories. The quantity estimator may assume that any item on a drawing that has not been marked has not been taken off yet.

When work is interrupted, for whatever reason, select a natural stopping point and mark it clearly so that when work resumes, no items are missed or duplicated.

Systematic application of these rules will make the quantity estimator's job faster, easier, and more accurate. Refer to the general rules in this chapter whenever necessary.

Pricing the Estimate

When the quantities have been determined, then prices, or unit costs, must be applied in order to determine the total costs. Depending upon the chosen estimating method (and thus the degree of accuracy required) and the level of detail, these unit costs may be direct or bare costs, or may include overhead, profit or contingencies. In Unit Price estimating, the unit costs most commonly used are "bare," or "unburdened." Items such as overhead and profit are usually added to the total direct costs on the bottom line, at the time of the estimate summary.

No matter which source of cost information is used, the system and sequence of pricing should be the same as those used for the quantity takeoff. This consistent approach should continue through both accounting and cost control during construction of the project.

Types of Costs: Unit price estimates for building construction may be organized according to the 16 divisions of the CSI MASTERFORMAT. Within each division, the components or individual construction items are identified, listed, and priced. This kind of definition and detail is necessary to complete an accurate estimate. In addition, each "item" can be broken down further into material, labor, and equipment components.

All costs included in a Unit Price estimate can be divided into two types: direct and indirect. Direct costs are those directly linked to the physical construction of a project, those costs without which the project could not be completed. The material, labor, and equipment costs mentioned above, as well as subcontract costs, are all direct costs. These may also be referred to as "bare," or "unburdened" costs.

Indirect costs are usually added to the estimate at the summary stage and are most often calculated as a percentage of the direct costs. They include such items as sales tax on materials, overhead, profit and contingencies, etc. It is the indirect costs that generally account for the greatest variation in estimating.

Types of Costs in a Construction Estimate

Direct Costs	Indirect Costs
Material	Taxes and Insurance
Labor	Overhead
Equipment	Profit
Subcontractors	Contingencies
Project Overhead	

The Paperwork: At the pricing stage of the estimate, there is typically a large amount of paperwork that must be assembled, analyzed, and tabulated. Generally, the information contained in this paperwork is covered by the following major categories:

- Quantity takeoff sheets for all general contractor items
- Material supplier written quotations
- Material supplier telephone quotations
- Subcontractor written quotations
- Equipment supplier quotations
- Cost Analysis or Consolidated Cost Analysis sheets
- Estimate Summary sheet

A system is needed to efficiently handle this mass of paperwork and to ensure that everything will get transferred (and only once) from the quantity takeoff to the Cost Analysis sheets. Some general rules for this procedure are:

- Write on only one side of any document where possible.
- Code each sheet with a large division number in a consistent place, preferably near one of the upper corners.
- Use Telephone Quotation forms for uniformity in recording prices received from any source, not only telephone quotes.
- Document the source of every quantity and price.
- Keep each type of document in its pile (Quantities, Material, Subcontractors, Equipment) filed in order by division number.

- Keep the entire estimate in one or more compartmented folders.
- When an item is transferred to the Cost Analysis sheet, check it off.
- If gross subcontractor quantities are known, pencil in the resultant unit prices to serve as a guide for future projects.

All subcontract costs should be properly noted and listed separately. These costs contain the subcontractor's markups, and will be treated differently from other direct costs when the estimator calculates the general contractor's overhead, profit, and contingency allowance.

The Estimate Summary: When the pricing of all direct costs is complete, the estimator has two choices: all further price changes and adjustments can be made on the Cost Analysis or Consolidated Estimate sheets, *or* total costs for each subdivision can be transferred to an Estimate Summary sheet so that all further price changes, until bid time, will be done on one sheet.

Unless the estimate has a limited number of items, it is recommended that costs be transferred to an Estimate Summary sheet. This step should be double-checked, since an error of transposition may easily occur. Pre-printed forms can be useful, though a plain columnar form may suffice.

If a company has certain standard listings that are used repeatedly, it would save valuable time to have a custom Estimate Summary sheet printed with the items that need to be listed. Appropriate column headings or categories for any estimate summary form are:

- Material
- Labor
- Equipment
- Subcontractor
- Total

As items are listed in the proper columns, each category is added, and appropriate markups applied to the total dollar values. Generally, the sum of each column has different percentages added near the end of the estimate for indirect costs:

- Sales tax
- Overhead
- Profit
- Contingencies

Division One
General Requirements

Introduction

The "General Requirements" section of the CSI MASTERFORMAT is used as the clearinghouse, for items that do not apply directly to the construction, the cost of which are customarily spread out over the entire project, such as overhead or the cost of a site superintendent.

The term "General Conditions" is sometimes interchanged with "General Requirements." Purists would argue that General Conditions are a "portion of the contract document in which the rights, responsibilities, and relationships of the involved parties to that contract are itemized," whereas General Requirements are cost items described above. For estimating purposes, either term is acceptable. The important thing is to include it in your estimate.

This chapter provides tables and charts for assistance in calculating the general requirements items listed below. Instructions are given for the use of each chart, along with some background information on the circumstances under which each might be used. Following the estimating charts, in this and all succeeding chapters, are an estimator's checklist and lists of estimating precautions.
- Overhead and profit
- Construction time requirements
- Architectural fees
- Architectural fees for repair and/or remodeling projects
- Engineering fees
- Engineering fees for repair and/or remodeling projects
- Construction management
- Construction management fees for repair and/or remodeling projects
- General contractor's overhead and profit
- Installing contractor's overhead and profit
- Main office expense
- Main office expense for repair and/or remodeling projects
- Insurance
- Insurance for repair and/or remodeling projects
- Bonds
- Permits
- Tools
- Unemployment/Social Security taxes
- Scheduling costs
- Surveying costs
- Testing services costs
- Temporary utility installation
- Project clean-up

Estimating Data

The following tables present guidelines for overhead and general conditions items. Please note that these percentages should be used as indicators of what may be expected, but that each project should be evaluated individually.

Table of Contents

Figure 1.1   Architectural Fees

Building Type	Total Project Size in Thousands of Dollars						
	100	250	500	1,000	2,500	5,000	10,000
Factories, garages, warehouses repetitive housing	9.0%	8.0%	7.0%	6.2%	5.6%	5.3%	4.9%
Apartments, banks, schools, libraries, offices, municipal buildings	11.7	10.8	8.5	7.3	6.7	6.4	6.0
Churches, hospitals, homes, laboratories, museums, research	14.0	12.8	11.9	10.9	9.5	8.5	7.8
Memorials, monumental work, decorative furnishings	—	16.0	14.5	13.1	11.3	10.0	9.0

In this figure, typical percentage fees are tabulated by project size, for good professional architectural service. Fees may vary from those listed depending upon the degree of design difficulty and economic conditions in a particular area.

Rates can be interpolated horizontally and vertically. Various portions of the same project requiring different rates should be adjusted proportionately. For alterations, add 50% to the fee for the first $500,000 of project cost and add 25% to the fee for the project cost over $500,000.

Architectural fees tabulated above include Engineering Fees.

Figure 1.2   Architectural Fees for Smaller Projects

Architectural Fees for project sizes of:	
Up to $10,000	15%
to $25,000	13%
to $100,000	10%
to $500,000	8%
to $1,000,000	7%

The listed fees are approximate for smaller projects, such as repair work and/or remodeling existing structures.

Figure 1.3   Engineering Fees as Percentages of a Project

Engineering Fees for:	Minimum	Maximum
Planning consultant/project	0.5%	2.5%
Type of Contract:		
Electrical	4.1%	10.1%
Elevator/Conveying Systems	2.5%	5.0%
Food Service/Kitchen Equipment	8.0%	12.0%
Landscaping & Site Development	2.5%	6.0%
Mechanical, Plumbing & HVAC	4.1%	10.1%
Structural/per Project	1.0%	2.5%

Figure 1.4   Structural Engineering Fees

Structural engineering fees based upon the type and size of a construction project.				
	Total Project Size			
Type of Construction	To $250,000	$250,000 to $500,000	$1,000,000	$5,000,000 & Over
Industrial buildings, factories & warehouses	Technical payroll times 2.0 to 2.5	1.60%	1.25%	1.00%
Hotels, apartments, offices, dormitories, hospitals, public buildings and food stores		2.00%	1.70%	1.20%
Museums, banks, churches and cathedrals		2.00%	1.75%	1.25%
Thin shells, prestressed concrete, earthquake resistive		2.00%	1.75%	1.50%
Parking ramps, auditoriums, stadiums, convention halls, hangars and boiler houses		2.50%	2.00%	1.75%
Special buildings, major alterations, underpinning and future expansion		Add to above 0.5%	Add to above 0.5%	Add to above 0.5%

For complex reinforced concrete or unusually complicated structures, add 20% to 50%.

Figure 1.5   Mechanical and Electrical Engineering Fees

Mechanical and electrical engineering fees based on the size of the subcontract.

Type of Construction	Subcontract Size							
	$25,000	$50,000	$100,000	$225,000	$350,000	$500,000	$750,000	$1,000,000
Simple structures	6.4%	5.7%	4.8%	4.5%	4.4%	4.3%	4.2%	4.1%
Intermediate structures	8.0	7.3	6.5	5.6	5.1	5.0	4.9	4.8
Complex structures	12.0	9.0	9.0	8.0	7.5	7.5	7.0	7.0

For renovations, add 15% to 25% to applicable fee.

These fees are for engineering services only, unless otherwise noted. These figures are included in the Architectural Fees shown in Figures 1.1 and 1.2. If Engineering Fees are to be listed separately, they must be deducted from the Architectural Fees.

Figure 1.6   Engineering Fees for Smaller Projects as a Percentage of Project Costs

Project Type	Minimum	Maximum
Educational planning consultant	4.1%	10.1%
Elevator/Conveying Systems	2.5%	5.0%
Mechanical (Plumbing & HVAC)	4.1%	10.1%
Structural	1.0%	2.5%

The listed fees are approximate for smaller projects, such as repair work and/or the remodeling of existing structures.

Figure 1.7   Construction Management Fees as a Percentage of Project Costs

Project Size	Minimum	Maximum
$1,000,000 job	4.5%	7.5%
$5,000,000 job	2.5%	4.0%

These fees reflect the range of costs charged by construction management for normally incurred construction management services.

Figure 1.8   Construction Management Fees for Smaller Projects

Project Size	Average Fee
For work to $10,000	10%
To $25,000	9%
To $100,000	6%
To $500,000	5%
To $1,000,000	4%

These fees reflect the range of costs charged by construction firms for management of smaller projects, such as the repair and/or remodeling of existing structures.

Figure 1.9 Fees for Progress Schedules

Type of Schedule	Minimum	Maximum
Critical path, as % of architectural fee	2%	4%
Rule of thumb, CPM, per job	.05%	.10%
Cost control, per job	.04%	.15%

These items are usually included in the fees of a construction management firm, if one is used. If an independent firm is contracted to prepare the progress schedules, this table indicates the range of fees one can expect to be charged.

Figure 1.10 General Contractor's Overhead

This table shows a contractor's overhead as a percentage of direct costs in two ways. The figures on the right are for the overhead, with the markup based on both material and labor. The figures on the left are based on the entire overhead applied only to the labor. This figure is used if the owner supplies the materials or if a contract is for labor only.

Items of General Contractor's Indirect Costs	% of Direct Costs	
	As a Markup of Labor Only	As a Markup of Both Material and Labor
Field Supervision	6.0%	2.4%
Main Office Expense (see details below)	9.2	7.7
Tools and Minor Equipment	1.0	0.4
Workers' Compensation & Employers' Liability	13.9	5.6
Field Office, Sheds, Photos, etc.	2.0	0.8
Performance and Payment Bond, 0.5% to 0.9%	0.7	0.7
Unemployment Tax (combined Federal and State)	6.2	2.5
Social Security and Medicare	7.6	3.0
Sales Tax — add if applicable 48/80 x % as markup of total direct costs including both material and labor.		
Subtotal	46.6%	23.1%
*Builder's Risk Insurance ranges from 0.151% to 0.586%.	0.3	0.3
*Public Liability Insurance	1.5	1.5
Grand Total	48.4%	24.9%

*Paid by Owner or Contractor

Figure 1.11   General Contractor's Main Office Expense

A general contractor's main office expense consists of many items. The percentage of main office expense declines with the increased annual volume of the contractor. Typical main office expenses range from 2% to 20%, with the median about 7.2% of total volume. This equals about 7.7% of direct costs. The following are approximate percentages of total overhead for different items usually included in a general contractor's main office overhead. With different accounting procedures, these percentages may vary.

Item	Typical Range	Average
Managers' clerical and estimators' salaries	40% to 55%	48%
Profit sharing, pension and bonus plans	2 to 20	12
Insurance	5 to 8	6
Estimating and project management (not including salaries)	5 to 9	7
Legal, accounting and data processing	0.5 to 5	3
Automobile and light truck expense	2 to 8	5
Depreciation of overhead capital expenditures	2 to 6	4
Maintenance of office equipment	0.1 to 1.5	1
Office rental	3 to 5	4
Utilities, including phone and light	1 to 3	2
Miscellaneous	5 to 15	8
Total		100%

Figure 1.12   General Contractor's Main Office Expense as a Percentage of Annual Volume

This table represents approximate ranges of the cost of maintaining the main office, including salaries, as a percentage of the total dollar volume a contractor expects to bill for in their fiscal year.

Annual Volume	% of Annual Volume
Under $1,000,000	13.6%
Up to $2,500,000	8.0%
Up to $4,000,000	6.8%
Up to $7,000,000	5.6%
Up to $10,000,000	5.1%
Over $10,000,000	3.9%

Figure 1.13 General Contractor's Main Office Expense for Smaller Projects

This table provides average main office expenses (as a percentage of annual volume) for contractors specializing in smaller projects, such as repair and/or remodeling of existing structures.

Annual Volume	% of Annual Volume	
	Minimum	Maximum
To $50,000	20%	30%
To $100,000	17%	22%
To $250,000	16%	19%
To $500,000	14%	16%
To $1,000,000	8%	10%

Figure 1.14 Overhead and Profit as a Percentage of Project Costs

Overhead is defined as costs that are associated with a construction project, but not directly with the actual construction. This table shows percentages that can be used as a rule of thumb for estimating overhead costs.

Overhead as a Percentage of Direct Costs		Overhead and Profit Allowance — Add to Items That Do Not Include Subcontractor's O&P — Average	Allowance to Add to Items That Do Include Subcontractor's O&P		Typical by Size of Project	
Minimum	5%	25%	Minimum	5%	under $100,000	30%
Average	12%		Average	10%	$500,000	25%
Maximum	22%		Maximum	15%	$2,000,000	20%
					over $10,000,000	15%

Figure 1.15 Installing Contractor's Overhead and Profit

		A	B	C	D	E
Abbr.	**Trade**	**Workers' Comp. Ins.**	**Average Fixed Overhead**	**Overhead**	**Profit**	**Total Overhead & Profit**
Skwk	Skilled Workers Average (35 trades)	13.7%	15.7%	12.8%	10%	52.2%
	Helpers Average (5 trades)	15.0		13.0		53.7
	Foremen Average, Inside (50¢ over trade)	13.7		12.8		52.2
	Foremen Average, Outside ($2.00 over trade)	13.7		12.8		52.2
Clab	Common Building Laborers	15.1		11.0		51.8
Asbe	Asbestos Workers	12.4		16.0		54.1
Boil	Boilermakers	8.7		16.0		50.4
Bric	Bricklayers	12.6		11.0		49.3
Brhe	Bricklayer Helpers	12.6		11.0		49.3
Carp	Carpenters	15.1		11.0		51.8
Cefi	Cement Finishers	8.7		11.0		45.4
Elec	Electricians	5.5		16.0		47.2
Elev	Elevator Constructors	7.2		16.0		48.9
Eqhv	Equipment Operators, Crane or Shovel	9.7		14.0		49.4
Eqmd	Equipment Operators, Medium Equipment	9.7		14.0		49.4
Eqlt	Equipment Operators, Light Equipment	9.7		14.0		49.4
Eqol	Equipment Operators, Oilers	9.7		14.0		49.4
Eqmm	Equipment Operators, Master Mechanics	9.7		14.0		49.4
Glaz	Glaziers	11.1		11.0		47.8
Lath	Lathers	9.2		11.0		45.9
Marb	Marble Setters	12.6		11.0		49.3
Mill	Millwrights	9.2		11.0		45.9
Mstz	Mosaic and Terrazzo Workers	7.6		11.0		44.3
Pord	Painters, Ordinary	11.3		11.0		48.0
Psst	Painters, Structural Steel	38.1		11.0		74.8
Pape	Paper Hangers	11.3		11.0		48.0
Pile	Pile Drivers	23.7		16.0		65.4
Plas	Plasterers	12.2		11.0		48.9
Plah	Plasterer Helpers	12.2		11.0		48.9
Plum	Plumbers	6.7		16.0		48.4
Rodm	Rodmen (Reinforcing)	24.4		14.0		64.1
Rofc	Roofers, Composition	27.6		11.0		64.3
Rots	Roofers, Tile & Slate	27.6		11.0		64.3
Rohe	Roofer Helpers (Composition)	27.6		11.0		64.3
Shee	Sheet Metal Workers	9.4		16.0		51.1
Spri	Sprinkler Installers	7.2		16.0		48.9
Stpi	Steamfitters or Pipefitters	6.7		16.0		48.4
Ston	Stone Masons	12.6		11.0		49.3
Sswk	Structural Steel Workers	30.2		14.0		69.9
Tilf	Tile Layers (Floor)	7.6		11.0		44.3
Tilh	Tile Layer Helpers	7.6		11.0		44.3
Trlt	Truck Drivers, Light	12.7		11.0		49.4
Trhv	Truck Drivers, Heavy	12.7		11.0		49.4
Sswl	Welders, Structural Steel	30.2		14.0		69.9
Wrck	*Wrecking	31.4		11.0		68.1

*Not included in Averages.

Listed are the **average** installing contractor's percentage mark-ups applied to base labor rates to arrive at typical billing rates.

Column A: Workers' Compensation rates are the national average of state rates established for each trade in 1990.

Column B: Column B lists average fixed overhead figures for all trades (based on 1990 figures.). Included are federal and state unemployment costs set at 6.2%; Social Security taxes (FICA) set at 7.56%; Builder's Risk Insurance costs set at 0.34%; and public liability costs set at 1.55%. All percentages except those for Social Security taxes vary from state to state as well as from company to company.

Columns C and D: Percentages in Columns C and D are based on the assumption that the installing contractor has annual billing of $500,000 and up. Overhead percentages may increase with smaller annual billing. The overhead percentages for any given contractor may vary greatly and depend on a number of factors, such as the contractor's annual volume, engineering and logistical support costs, and staff requirements. The figures for overhead and profit will also vary depending on the type of job, the job location, and the prevailing economic conditions. All factors should be examined very carefully for each job.

Column E: Column E lists the total of columns A, B, C, and D.

Figure 1.16   Workers' Compensation

The table below tabulates the national averages for Workers' Compensation insurance rates by trade and type of building. The average "Insurance Rate" is multiplied by the "% of Building Cost" for each trade. This produces the "Workers' Compensation Cost" by % of total labor cost, to be added for each trade by building type to determine the weighted average Workers' Compensation rate for the building types analyzed.

Trade	Insurance Rate (% of Labor Cost)		% of Building Cost			Workers' Compensation Cost		
	Range	Average	Office Bldgs.	Schools & Apts.	Mfg.	Office Bldgs.	Schools & Apts.	Mfg.
Excavation, Grading, etc.	3.5% to 26.8%	9.7%	4.8%	4.9%	4.5%	.46%	.47%	.44%
Piles & Foundations	5.0 to 52.0	23.7	7.1	5.2	8.7	1.68	1.23	2.06
Concrete	5.0 to 32.7	13.9	5.0	14.8	3.7	.69	2.06	.51
Masonry	3.6 to 36.3	12.6	6.9	7.5	1.9	.87	.94	.24
Structural Steel	5.0 to 118.4	30.2	10.7	3.9	17.6	3.23	1.18	5.31
Miscellaneous & Ornamental Metals	2.7 to 27.4	10.0	2.8	4.0	3.6	.28	.40	.36
Carpentry & Millwork	5.0 to 40.1	15.1	3.7	4.0	0.5	.56	.60	.08
Metal or Composition Siding	5.0 to 34.2	12.7	2.3	0.3	4.3	.29	.04	.55
Roofing	5.0 to 86.2	27.6	2.3	2.6	3.1	.63	.72	.85
Doors & Hardware	3.5 to 20.9	9.0	0.9	1.4	0.4	.08	.13	.04
Sash & Glazing	3.2 to 22.5	11.1	3.5	4.0	1.0	.39	.44	.11
Lath & Plaster	3.3 to 38.5	12.2	3.3	6.9	0.8	.40	.84	.10
Tile, Marble & Floors	2.6 to 23.3	7.6	2.6	3.0	0.5	.20	.23	.04
Acoustical Ceilings	3.1 to 22.3	9.2	2.4	0.2	0.3	.22	.02	.03
Painting	4.2 to 32.2	11.3	1.5	1.6	1.6	.17	.18	.18
Interior Partitions	5.0 to 40.1	15.1	3.9	4.3	4.4	.59	.65	.66
Miscellaneous Items	2.1 to 98.0	13.7	5.2	3.7	9.7	.71	.51	1.33
Elevators	2.2 to 15.8	7.2	2.1	1.1	2.2	.15	.08	.16
Sprinklers	2.2 to 15.1	7.2	0.5	—	2.0	.04	—	.14
Plumbing	2.5 to 14.2	6.7	4.9	7.2	5.2	.33	.48	.35
Heat., Vent., Air Conditioning	3.2 to 20.4	9.4	13.5	11.0	12.9	1.27	1.03	1.21
Electrical	2.3 to 10.9	5.5	10.1	8.4	11.1	.55	.46	.61
Total	2.1% to 118.4%	—	100.0%	100.0%	100.0%	13.79%	12.69%	15.36%
Overall Weighted Average							13.95%	

Figure 1.17   Workers' Compensation
Rates — Weighted Averages

This table lists the weighted average Workers' Compensation base rate for each state (for 1990), with a factor for comparing each to the national average of 13.7%. The weighted average skilled worker rate for 35 trades is 13.7%. For bidding purposes, apply the full value of Workers' Compensation directly to the total labor costs. For example, if labor is 32%, materials 48%, and overhead and profit 20% of the total cost, carry 32/80 x 13.7% = 5.5% of cost (before overhead and profit) into overhead. Rates vary not only from state to state, but also with the experience rating of the contractor.

State	Weighted Average	Factor	State	Weighted Average	Factor	State	Weighted Average	Factor
Alabama	11.4%	83	Kentucky	13.1%	96	North Dakota	8.0%	58
Alaska	22.6	165	Louisiana	12.6	92	Ohio	8.9	65
Arizona	13.9	101	Maine	18.8	137	Oklahoma	12.3	90
Arkansas	10.0	73	Maryland	15.3	112	Oregon	27.6	201
California	15.6	114	Massachusetts	22.1	161	Pennsylvania	14.7	107
Colorado	21.9	160	Michigan	14.9	109	Rhode Island	18.4	134
Connecticut	23.3	170	Minnesota	24.9	182	South Carolina	10.5	77
Delaware	12.0	88	Mississippi	9.9	72	South Dakota	9.1	66
District of Columbia	19.7	144	Missouri	7.6	55	Tennessee	7.7	56
Florida	22.3	163	Montana	31.8	232	Texas	19.0	139
Georgia	13.9	101	Nebraska	7.8	57	Utah	8.5	62
Hawaii	16.7	122	Nevada	14.7	107	Vermont	9.0	66
Idaho	10.8	79	New Hampshire	18.6	136	Virginia	9.0	66
Illinois	19.1	139	New Jersey	7.5	55	Washington	9.7	71
Indiana	4.9	36	New Mexico	18.3	134	West Virginia	7.8	57
Iowa	11.3	82	New York	9.6	70	Wisconsin	12.6	92
Kansas	8.6	63	North Carolina	6.4	47	Wyoming	5.4	39
Weighted Average for U.S. is 13.9% of payroll = 100								

Figure 1.18   Insurance Rates

Type	Minimum	Maximum
Builder's Risk	.22%	.59%
All-risk Type	.25%	.62%
Contractor's Equipment Floater	.50%	1.50%
Public Liability, Average	—	1.55%

This table represents approximate values relative to total project cost for the most common types of basic insurance coverages.

Figure 1.19   Builder's Risk Insurance Rates

Coverage	Frame Construction (Class 1)		Brick Construction (Class 4)		Fire Resistive (Class 6)	
	Range	Average	Range	Average	Range	Average
Fire Insurance	.300 to .420%	.394%	.132 to .189%	.174%	.062 to .090%	.081%
Extended Coverage	.115 to .150	.144	.080 to .105	.101	.081 to .105	.100
Vandalism	.012 to .016	.015	.008 to .011	.011	.008 to .011	.010
Total Annual Rate	.427 to .586%	.553%	.220 to .305%	.286%	.151 to .206%	.191%

Builder's Risk Insurance is insurance on a building during construction. Premiums are paid by the owner or the contractor. Blasting, collapse, and underground insurance would raise total insurance costs above those listed. A floater policy for materials delivered to the job runs .75% to 1.25% of the material value. Contractor equipment insurance runs .50% to 1.50% of the value of the equipment.

Tabulated are approximate percentages for Builder's Risk Insurance rates in percentages of the policy value for $1,000 deductible. For $25,000 deductible, rates can be reduced 13% to 34%. On contracts over $1,000,000, rates may be lower than those tabulated. Policies are written annually for the total completed value in place. For "all risk" insurance (excluding flood, earthquake, and certain other perils), add .025% to the total rates given.

Figure 1.20   Performance Bond Rates

This table shows the performance bond rate for a job scheduled to be completed in 12 months (in 1990). Add 1% of the premium cost per month for jobs requiring more than 12 months to complete. The rates are "preferred" rates offered to contractors considered by the bonding company to be financially sound and capable of doing the work. The rates quoted are suggested averages. Actual rates vary from contractor to contractor, and from bonding company to bonding company. Contractors should prequalify through a bonding company agency before submitting a bid on a contract that requires a bond.

Contract Amount	Building Construction Class B Projects	Highways & Bridges	
		Class A New Construction	Class A-1 Highway Resurfacing
First $ 100,000 bid	$25.00 per M	$15.00 per M	$ 9.40 per M
Next 400,000 bid	$ 2,500 plus $15.00 per M	$ 1,500 plus $10.00 per M	$ 940 plus $7.20 per M
Next 2,000,000 bid	8,500 plus 10.00 per M	5,500 plus 7.00 per M	3,820 plus 6.00 per M
Next 2,500,000 bid	28,500 plus 7.50 per M	19,500 plus 5.50 per M	15,820 plus 5.00 per M
Next 2,500,000 bid	47,250 plus 7.00 per M	33,250 plus 5.00 per M	28,320 plus 4.50 per M
Over 7,500,000 bid	64,750 plus 6.00 per M	45,750 plus 4.50 per M	39,570 plus 4.00 per M

Figure 1.21   Permit Rates

Project Permits	Minimum	Maximum
Rule of thumb, most cities	.50%	2%

Permit costs vary greatly depending on many factors, such as type of project, proposed occupancy, location, local codes, need for variances or change of zoning, etc. This table provides a "rule of thumb" to use when local conditions cannot be determined.

Figure 1.22 Small Tools Allowance

Small Tools Allowance	Minimum	Maximum
As % of contractor's work	.50%	2%

A variety of small tools must be purchased in the course of almost every project, whether to complete small tasks or to replace tools that have "mysteriously disappeared." This table provides a "rule of thumb" that can be used to assign a value to this often overlooked item.

Figure 1.23 Construction Time Requirements

Table at left is average construction time in months for different types of building projects. Table at right is the construction time in months for different size projects. Design time runs 25% to 40% of construction time.

Type Building	Construction Time	Project Value	Construction Time
Industrial Buildings	12 Months	Under $1,400,000	10 Months
Commercial Buildings	15 Months	Up to $3,800,000	15 Months
Research & Development	18 Months	Up to $19,000,000	21 Months
Institutional Buildings	20 Months	Over $19,000,000	28 Months

In order to estimate the General Requirements of a construction project, it is necessary to have an approximate project duration time. Duration must be determined because many items, such as supervision and temporary facilities, are directly time variable. The average durations presented in this chart will vary depending on such factors as location, complexity, time of year started, local economic conditions, materials required, or the need for the completed project.

Figure 1.24   Survey Data

Survey Data	Crew Makeup	Daily Output	Man-Hours	Unit
Surveying, Conventional, topographical, Minimum	1 Chief of Party 1 Instrument Man 1 Rodman/Chainman	3.30	7.270	Acre
Maximum	1 Chief of Party 1 Instrument Man 2 Rodmen/Chainmen	.60	53.330	Acre
Lot location and lines, for large quantities Minimum	1 Chief of Party 1 Instrument Man 1 Rodman/Chainman	2	12.000	Acre
Average	"	1.25	19.200	Acre
Maximum, for small quantities	1 Chief of Party 1 Instrument Man 2 Rodmen/Chainmen	1	32.000	Acre
Monuments, 3' long	1 Chief of Party 1 Instrument Man 1 Rodman/Chainman	10	2.400	Ea.
Property lines, perimeter, cleared land	"	1.000	.024	L.F.
Crew for building layout, 2 man crew	1 Chief of Party 1 Instrument Man	1	16.000	Day
3 man crew	1 Chief of Party 1 Instrument Man 1 Rodman/Chainman	1	24.000	Day
4 man crew	1 Chief of Party 1 Instrument Man 2 Rodmen/Chainmen	1	32.000	Day

This table provides the information needed to compute the time required to survey a parcel or to perform the layout for a planned structure. To estimate the costs of collecting survey data, multiply the number of units on the plans (number of acres, L.F., etc.) by the amount listed in the man-hours column for the appropriate task. Multiply this number by the local labor rate. The result should be the approximate bare cost (without overhead and profit) for surveying.

Figure 1.25   Testing Services

Type of Building	Minimum	Maximum
Concrete, costing $1,000,000	0.5%	5%
Steel	0.5%	1.0%
Building $10,000,000	0.35%	0.5%

This table lists approximate values for testing services for different types of structures.

Figure 1.26 Final Cleaning

Final Cleaning	Crew Makeup	Daily Output	Man-Hours	Unit
Cleanup of floor area, continuous, per day	2 Building Laborers .25 Truck Driver (light) .25 Light Truck, 1.5 Ton	12	1.500	M.S.F.
Final	"	11.50	1.570	M.S.F.
Cleanup after job completion, allow .30% per total job				

To estimate the cleanup costs for a structure, whether continuous (daily) or the final cleanup, determine the floor area (total), divide by 1,000, and multiply this number by the appropriate number listed in the man-hour column for your task. The result will be the number of man-hours needed for that task. Multiply this by your local labor rate to determine the bare costs (without overhead and profit) for your task.

Checklist

For an estimate to be reliable, all items must be accounted for. A complete estimate can also eliminate the need to include contingencies. The following checklist can be used to help ensure that all items are properly accounted for.

Direct Overhead Costs

Personnel

- ☐ Superintendent
- ☐ Project Manager (if for that project only)
- ☐ Field Engineer (if for that project only)
- ☐ Cost Engineer (if for that project only)
- ☐ Warehouse personnel (if for that project only)
- ☐ Watchman/guard dogs
- ☐ Tool room keeper (if for that job only)
- ☐ Timekeeper (if for that job only)
- ☐ Foreman (working directly for the contractor)

Temporary Facilities

- ☐ Field office expense
  - _____ Set-up and removal
  - _____ Light
  - _____ Heat
  - _____ Water
  - _____ Telephone
  - _____ Supplies
  - _____ Equipment
    - _____ Fax machine
    - _____ Copy machine
    - _____ Blueprint machine
    - _____ Coffee machine
- ☐ Temporary light and power
- ☐ Temporary heat
- ☐ Temporary water
- ☐ Pay telephones
- ☐ Toilet facilities
- ☐ Enclosures
- ☐ Storage trailers
- ☐ Fencing
- ☐ Barricades and signals
- ☐ Construction road
- ☐ Job sign

Indirect Costs

Salaries

- ☐ President
- ☐ Executives
- ☐ Secretaries/Reception
- ☐ Estimators
- ☐ Project Managers
- ☐ Construction Manager
- ☐ Cost Engineers
- ☐ Purchasing Agent
- ☐ Cost/Bookkeeping
- ☐ Engineers
- ☐ Other office personnel
- ☐ Yard personnel
  - _____ Tool Manager
  - _____ Mechanics/Maintenance
  - _____ Drivers
  - _____ Equipment operators

Office

- ☐ Rent/cost of ownership
- ☐ Electricity
- ☐ Gas
- ☐ Water
- ☐ Sewer
- ☐ Telephone
- ☐ Postage
- ☐ Office equipment
- ☐ Furniture/furnishings
- ☐ Office supplies
- ☐ Advertising
- ☐ Literature
- ☐ Club/association dues

Professional Services

- ☐ Legal
- ☐ Accounting
- ☐ Architectural
- ☐ Engineering

Vehicles

- ☐ Cars/trucks
- ☐ Cost of operation
- ☐ Mileage expenses

Insurance
- ☐ Fire
- ☐ Property damage
- ☐ Vehicles
- ☐ Public liability
- ☐ Windstorm
- ☐ Workers' Compensation
- ☐ Unemployment
- ☐ Social Security
- ☐ Flood
- ☐ Theft
- ☐ Elevator

Bonds
- ☐ Bid
- ☐ Payment
- ☐ Performance
- ☐ Surety
- ☐ Lien

Miscellaneous
- ☐ Vehicles
- ☐ Permits
- ☐ Licenses
- ☐ Tools and equipment
- ☐ Photographs
- ☐ Surveying
- ☐ Testing
- ☐ Job signs
- ☐ Pumping
- ☐ Dust control
- ☐ Lifting/hoisting
- ☐ Cleanup (periodical)
- ☐ Final cleanup
- ☐ Damage/repair to adjoining buildings and/or public ways

Tips

- In figuring general requirements, do not include the salaries of the project managers or of any other personnel who are not directly site-related. Their wages are included in the "Main Office" expenses.

- Always allow for cleanup in the estimate. No matter how clean a subcontractor leaves an area, it is almost always necessary to clean the area again.

- In locations where snow is likely, allow for the expense of snowplowing if the project begins, ends, or works through the winter season. Another consideration is melting snow, which will inevitably end up in trenches, pits, or other low areas. Consequently, pumping costs should also be carried.

- Always visit a proposed site. Do not rely on someone else's judgment unless statements in the contract require it.

Check for:

☐ Site Access—Can loaded trucks move into and out of the site easily? Is the site in/near a residential area? Are there height/weight restrictions?

☐ How far away are the utilities that can be hooked into for temporary power?

☐ Site Drainage—Is the area marshy? Will there be water problems when it rains?

☐ Do any utilities need to be relocated?

☐ Will any adjacent structures be affected?

☐ If any of these items apply to the project, the associated costs must be estimated and included in the project bid/estimate.

The cost for installing temporary utilities, especially lighting, may be included in the specifications for the respective trades. Check this out to avoid adding an unnecessary cost to the estimate.

Division Two
Site Work

Introduction

Division 2—Site Work covers a wide variety of site-related items, including demolition, site preparation, earthwork, pilings, pavings, utilities, sewerage and drainage, site improvements, and landscaping.

The scope of work included in demolition can range from the removal of a window or boxes found in a corner, to the complete dismantling and removal of existing structures. The individual estimator must assess the particular project and determine whether the takeoff should be performed by the piece, square foot, cubic foot, or the whole.

Site preparation involves preparing a site to start other work. This includes clearing the land of all trees, shrubs, stumps, etc. Some estimators also include general site grading in this category.

Earthwork can be a nebulous area that includes mass excavation, trench excavation, hauling, backfill, compaction, grading—all tasks that involve working or shaping the earth, including their related materials. On many sites, earthwork consumes a great amount of time and money. It is an area well worth the added time and care it will take to create a complete estimate.

Pilings and caissons are utilized when the ground cannot adequately support the structures designed. These are usually taken off individually by type, length, and material.

Pavings, site improvements, and landscaping all involve finishing the land area not directly covered by the main structure. While these may not be big money items, generally they should not be minimized, as they are the areas of the project that the public will see most and by which the company occupying the building may be judged. Pavings are generally taken off by the square yard of area to be surfaced. Site improvements and landscaping are so individual that their takeoff methods cannot be generalized. Each site must be evaluated individually to determine how to handle the estimate.

Sewerage and drainage are generally taken off by the foot and/or by the unit of whatever system is used. Keep in mind that the local municipality will usually be closely involved in the selection and installation plan for sewerage and drainage, as more than just the immediate site is affected. Good information for this part of the project can usually be found at the local authority's office.

To assist in estimating Division 2—Site Work, the following categories of charts and tables have been included:
- Site Preparation
- Earthwork
- Piles and Caissons
- Paving and Surfacing
- Site Improvements
- Landscaping

Estimating Data

The following tables present effective estimating guidelines for items found in Division 2 — Site Work. Please note that these guidelines can be used as indicators of what may be expected, but that each project must be evaluated individually.

Table of Contents

Figure 2.1   Landscape Systems and Graphics

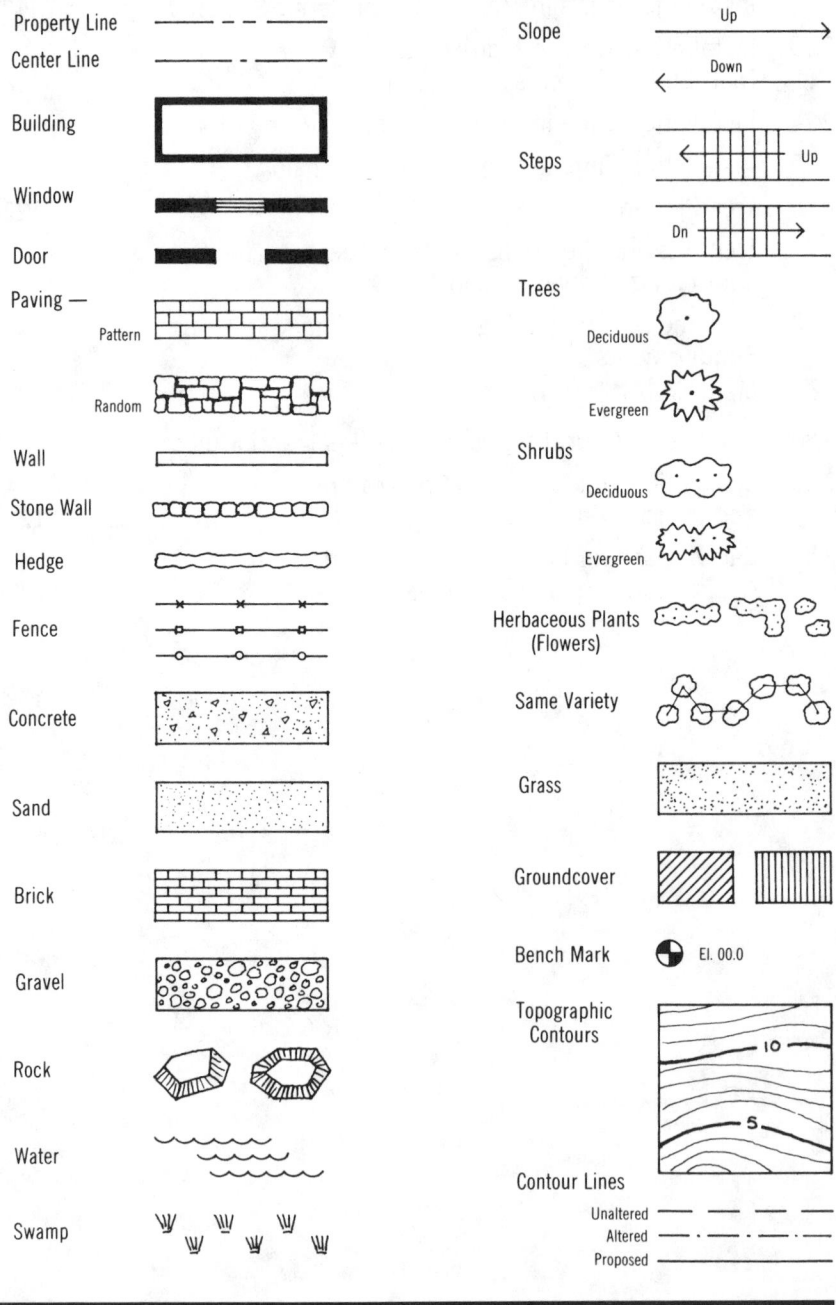

Figure 2.2   Determining Available Workdays

The chart below is a sample format used to determine the number of workdays available per month for exterior (landscape) construction. The example shown might be typical for the mid-Atlantic region of the U.S. The basic steps involved are:
1. Determine total available workdays per month, counting weekdays and Saturdays, if applicable, and deducting holidays.
2. Deduct average time allowed per month for vacations and sick days (based on historical data).
3. Obtain weather data from the local office of the National Weather Service, and subtract anticipated lost days due to poor weather.

| | Available Workdays | | | |
Month	No. of Potential Workdays	Lost Days Due to Weather	Lost Days Due to Vaca./Sick Time	Actual Workdays per Worker
January	22	10	2	10
February	19	10	3	6
March	22	8	3	11
April	21	8	2	11
May	22	6	1	15
June	21	4	2	15
July	21	3	3	15
August	23	3	4	16
September	19	4	3	12
October	22	6	2	14
November	21	8	3	10
December	20	10	4	6
	253	80	32	141

Average number of hours per workday: 8 hrs./day

Figure 2.3   Soil Bearing Capacity in Kips per S.F.

This table can be used to determine foundation footing size. Once the load on a footing has been determined, the footing size is determined by dividing the load by the Allowable Bearing Capacity in this table.

Bearing Material	Typical Allowable Bearing Capacity
Hard sound rock	120 KSF
Medium hard rock	80
Hardpan overlaying rock	24
Compact gravel and boulder–gravel; very compact sandy gravel	20
Soft rock	16
Loose gravel; sandy gravel; compact sand; very compact sand–inorganic silt	12
Hard dry consolidated clay	10
Loose coarse to medium sand; medium compact fine sand	8
Compact sand–clay	6
Loose fine sand; medium compact sand–inorganic silts	4
Firm or stiff clay	3
Loose saturated sand–clay; medium soft clay	2

One kip = 1,000 lbs.

Figure 2.4   Weights and Characteristics of Materials

Approximate Material Characteristics*				
Material	Loose (Lbs./C.Y.)	Bank (Lbs./C.Y.)	Swell (%)	Load Factor
Clay, dry	2,100	2,650	26	0.79
Clay, wet	2,700	3,575	32	0.76
Clay and gravel, dry	2,400	2,800	17	0.85
Clay and gravel, wet	2,600	3,100	17	0.85
Earth, dry	2,215	2,850	29	0.78
Earth, moist	2,410	3,080	28	0.78
Earth, wet	2,750	3,380	23	0.81
Gravel, dry	2,780	3,140	13	0.88
Gravel, wet	3,090	3,620	17	0.85
Sand, dry	2,600	2,920	12	0.89
Sand, wet	3,100	3,520	13	0.88
Sand and gravel, dry	2,900	3,250	12	0.89
Sand and gravel, wet	3,400	3,750	10	0.91

*Exact values will vary with grain size, moisture content, compaction, etc. Test to determine exact values for specific soils.

Typical Soil Volume Conversion Factors				
Soil Type	Initial Soil Condition	Bank	Converted to: Loose	Compacted
Clay	Bank	1.00	1.27	0.90
	Loose	0.79	1.00	0.71
	Compacted	1.11	1.41	1.00
Common earth	Bank	1.00	1.25	0.90
	Loose	0.80	1.00	0.72
	Compacted	1.11	1.39	1.00
Rock (blasted)	Bank	1.00	1.50	1.30
	Loose	0.67	1.00	0.87
	Compacted	0.77	1.15	1.00
Sand	Bank	1.00	1.12	0.95
	Loose	0.89	1.00	0.85
	Compacted	1.05	1.18	1.00

$$\text{Swell (\%)} = \left(\frac{\text{Wt./bank C.Y.}}{\text{Wt./loose C.Y.}} - 1 \right) \times 100$$

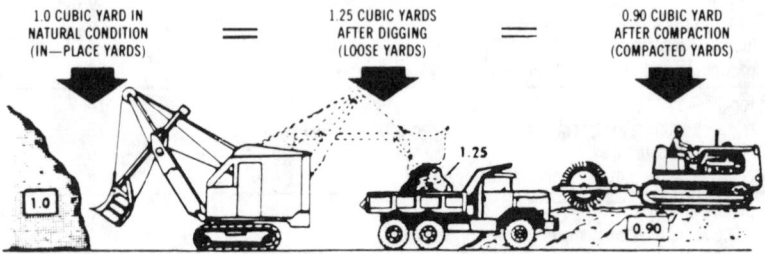

1.0 CUBIC YARD IN NATURAL CONDITION (IN—PLACE YARDS) = 1.25 CUBIC YARDS AFTER DIGGING (LOOSE YARDS) = 0.90 CUBIC YARD AFTER COMPACTION (COMPACTED YARDS)

Figure 2.5 Material Weights

Material	Wt. in Bank per C.Y.	Percent of Swell	Swell Factor	Loose Wt. per C.Y.
Ashes, Hard Coal	700-1000 lbs.	8%	.93	650-930 lbs.
Ashes, Soft Coal with Clinkers	1000-1515 lbs.	8%	.93	930-1410 lbs.
Ashes, Soft Coal, Ordinary	1080-1215 lbs.	8%	.93	1000-1130 lbs.
Bauxite	2700-4325 lbs.	33%	.75	2020-3240 lbs.
Brick				2700 lbs.
Cement, Portland	94 lbs. per bag			
Cement, Portland	2970 lbs. (packed)	20%	.83	2450 lbs.
Coke, Lump, Loose				620-865 lbs.
Coke, Solvay, Egg, Chestnut or Pea				840 lbs.
Coke, Gas, Egg, Chestnut or Pea				785 lbs.
Coke, Gas Furnace				730 lbs.
Concrete	3240-4185 lbs.	40%	.72	2330-3000 lbs.
Concrete Mix, Wet				3500-3750 lbs.
Copper Ore	3800 lbs.	35%	.74	2800 lbs.
Gasoline, 56° Gaume	6.3 lbs. per gallon			
Granite	4500 lbs.	50 to 80%	.67 to .56	1520-3000 lbs.
Iron Ore, Hematite	6500-8700 lbs.		.45	3900 lbs.
Iron Ore, Limonite	6400 lbs.			
Iron Ore, Magnetite	8500 lbs			
Kaolin	2800 lbs.	30%	.77	2160 lbs.
Lead Ore, Galina	12,550 lbs.			
Lime				1400 lbs.
Limestone, Blasted	4200 lbs.	67 to 75%	.60 to .57	2400-2520 lbs.
Limestone, Loose, Crushed				2600-2700 lbs.
Limestone, Marble	4600 lbs.	67 to 75%	.60 to .57	2620-2760 lbs.
Mud, Dry (Close)	2160-2970 lbs.	20%	.83	1790-2460 lbs.
Mud, Wet (Moderately packed)	2970-3510 lbs.	20%	.83	2470-2910 lbs.
Oil, Crude	6.42 lbs. per gallon			
Phosphate Rock	5400 lbs.			
Sand, Dry	3250 lbs.	12%	.89	2900 lbs.
Sand, Wet	3600 lbs.	14%	.88	3200 lbs.
Sandstone	4140 lbs.	40 to 60%	.72 to .63	2610-2980 lbs.
Shale, Riprap	2800 lbs.	33%	.75	2100 lbs.
Slag, Sand	1670 lbs.	12%	.89	1485 lbs.
Slag, Solid	4320-4860 lbs.	33%	.75	2640-3240 lbs.
Slag, Crushed				1900 lbs.
Slag, Furnace, Granulated	1600 lbs.	12%	.89	1430 lbs.

(continued on next page)

Figure 2.5   Material Weights (continued)

Material	Wt. in Bank per C.Y.	Percent of Swell	Swell Factor	Loose Wt. per C.Y.
Slate	4590-4860 lbs.	30%	.77	3530-3740 lbs.
Trap Rock	5075 lbs.	50%	.67	3400 lbs.
Wood and Lumber Weights/per Cord				
Beechwood Chestnut Elm Hemlock	3250 lbs. per cord 2350 lbs. per cord 2350 lbs. per cord 2200 lbs. per cord	Hickory Pine, Norway or White Poplar		4500 lbs. per cord 2000 lbs. per cord 2350 lbs. per cord

Figure 2.6   Quantities for Wellpoint Systems

	Description for 200' System with 8" Header	Quantities
Equipment & Material	Wellpoints 25' long, 2" diameter @ 5' O.C. Header pipe, 8" diameter Discharge pipe, 8" diameter 8" Valves Combination Jetting and Wellpoint pump (standby) Wellpoint pump, 8" diameter Transportation to and from site Fuel 30 days x 60 gal./day Lubricants for 30 days x 16 lbs./day Sand for points	40 Ea. 200 L.F. 100 L.F. 3 Ea. 1 Ea. 1 Ea. 1 day 1800 gal. 480 Lbs. 40 C.Y.
Labor	Technician to supervise installation Labor for installation and removal of system 4 Operators straight time 40 hrs./wk. for 4.33 wks. 4 Operators overtime 2 hrs./wk. for 4.33 wks.	1 week 300 man-hours 693 hrs. 35 hrs.

Figure 2.7 Installation Time in Man-Hours for Dewatering

Description	Man-Hours	Unit
Excavate Drainage Trench, 2' Wide		
2' Deep	.178	C.Y.
3' Deep	.160	C.Y.
Sump Pits, by Hand		
Light Soil	1.130	C.Y.
Heavy Soil	2.290	C.Y.
Pumping 8 Hours, Diaphragm or Centrifugal Pump		
Attended 2 hours per day	3.000	Day
Attended 8 hours per day	12.000	Day
Pumping 24 Hours, Attended 24 Hours, 4 Men at		
6 Hour Shifts, 1 Week Minimum	25.140	Day
Relay Corrugated Metal Pipe, Including		
Excavation, 3' Deep		
12" Diameter	.209	L.F.
18" Diameter	.240	L.F.
Sump Hole Construction, Including Excavation,		
with 12" Gravel Collar		
Corrugated Pipe		
12" Diameter	.343	L.F.
18" Diameter	.480	L.F.
Wood Lining, Up to 4'x4'	.080	SFCA
Wellpoint System, Single Stage, Install and		
Remove, per Length of Header		
Minimum	.750	L.F.
Maximum	2.000	L.F.
Wells, 10' to 20' Deep with Steel Casing		
2' Diameter		
Minimum	.145	V.L.F.
Average	.245	V.L.F.
Maximum	.490	V.L.F.

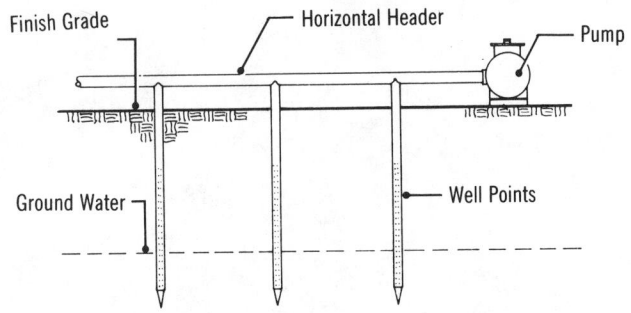

Figure 2.8 Area Clearing Equipment
Selection Table by Size of Area,
Vegetation to be Cleared, and Method

This table suggests equipment requirements for Light, Intermediate, and Heavy Clearing. The productivity of the equipment will depend upon the density and the type of growth.

Light Clearing, Vegetation up to 2 in. (5 cm) Diameter				
	Uprooting Vegetation	Cutting Vegetation At or Above Ground Level	Knocking the Vegetation to the Ground	Incorporation of Vegetation into the Soil
Small areas 10 acres (4.0 hectares)	Bulldozer blade, axes, grub hoes and mattocks	Axes, machetes, brush hooks, grub hoes and mattocks, wheel-mounted circular saws	Bulldozer blade	Moldboard plows, disc plows, disc harrows
Medium areas 100 acres (40 hectares)	Bulldozer blade,	Heavy-duty sickle mowers (up to 1-1/2" (3.7 cm) diameter), tractor-mounted circular saws; suspended rotary mowers	Bulldozer blade, rotary mowers, flail-type rotary cutters, rolling brush cutters	Moldboard plows, disc plows, disc harrows
Large areas 1,000 acres (400 hectares)	Bulldozer blade, root rake, grubber, root plow, anchor chain drawn between two crawler tractors, rails		Rolling brush cutter, flail-type cutter, anchor chain drawn between two crawler tractors, rails	Undercutter with disc, moldboard plows, disk plows, disk harrows

Figure 2.8 Area Clearing Equipment Selection
Table by Size of Area, Vegetation
to be Cleared, and Method (continued)

Intermediate Clearing, Vegetation 2 to 8 in. (5 - 20 cm) Diameter				
	Uprooting Vegetation	Cutting Vegetation at or Above Ground Level	Leveling Vegetation	Tilling Vegetation into Soil
Small areas 10 acres (4.0 hectares)	Bulldozer blade	Axes, crosscut saws, power chain saws, wheel-mounted circular saws	Bulldozer blade	Heavy-duty disc plow; disc harrow
Medium areas 100 acres (40 hectares)	Bulldozer blade	Power chain saws, tractor-mounted circular saws	Bulldozer blade, rolling brush cutter (up to 5 in. (12 cm) diameter), rotary mower (up to 4 in. (10 cm) diameter)	Heavy-duty disc plow, disc harrow
Large areas 1000 acres (400 hectares)	Shearing blade, angling (tilted), bulldozer blade, rakes, anchor chain drawn between two crawler tractors, root plow	Shearing blade (angling or V-type)	Bulldozer blade, flail-type rotary cutter, anchor chain	Bulldozer blade with heavy-duty harrow

Heavy Clearing, Vegetation 8 in. (20 cm) Diameter or Larger			
	Uprooting Vegetation	Cutting Vegetation at or Above Ground Level	Knocking the Vegetation to the Ground
Small areas 10 acres (4.0 hectares)	Bulldozer blade	Axes, crosscut saws, power chain saws	Bulldozer blade
Medium areas 100 acres (40 hectares)	Shearing blade, angling (tilted) knockdown beam, rakes, tree stumper	Shearing blade (angling or V-type), tree shear (up to 26 in. (65 cm) softwood; 14 in. (35 cm) hardwood), shearing blade - power saw combination	Bulldozer blade
Large areas 1,000 acres (400 hectares)	Shearing blade, angling (tilted), knockdown beam, rakes, tree stumper, anchor chain with ball drawn between two crawler tractors	Shearing, blade (angling or V-type) shearing blade-power saw combination	Anchor chain with ball drawn between two crawler tractors

Figure 2.9 Excavating

The selection of equipment used for structural excavation and bulk excavation or for grading is determined by the following factors.

1. Quantity of material
2. Type of material
3. Depth or height of cut
4. Length of haul
5. Condition of haul road
6. Accessibility of site
7. Moisture content and dewatering requirements
8. Availability of excavating and hauling equipment

Some additional costs must be allowed for hand trimming the sides and bottom of concrete pours and other excavation below the general excavation.

When planning excavation and fill, the following should also be considered.

1. Swell factor
2. Compaction factor
3. Moisture content
4. Density requirements

A typical example for scheduling and estimating the cost of excavation of a 15' deep basement on a dry site when the material must be hauled off the site, is outlined below.

Assumptions:

1. Swell factor, 18%
2. No mobilization or demobilization
3. Allowance included for idle time and moving on job
4. No dewatering, sheeting, or bracing
5. No truck spotter or hand trimming

Number of B.C.Y. per truck

$$= 1.5 \text{ C.Y. bucket} \times 8 \text{ passes} = 12 \text{ loose C.Y.}$$

$$= 12 \times \frac{100}{118} = 10.2 \text{ B.C.Y. per truck}$$

Truck haul cycle:

Load truck 8 passes	= 4 minutes
Haul distance 1 mile	= 9 minutes
Dump time	= 2 minutes
Return 1 mile	= 7 minutes
Spot under machine	= 1 minute
	23 minute cycle

Fleet Haul Production per Day in B.C.Y.

4 trucks x $\dfrac{50 \text{ min. hr.}}{23 \text{ min. haul cycle}}$ x 8 hrs. x 10.2 B.C.Y. = 4 x 2.2 x 8 x 10.2 = 718 B.C.Y./day

Note: B.C.Y. = Bank Measure Cubic Yards

Figure 2.10   Excavation Equipment

The table below lists *theoretical* hourly production in C.Y./hr. bank measure for some typical excavation equipment. Figures assume 50 minute hours, 83% job efficiency, 100% operator efficiency, 90° swing and properly sized hauling units, which must be modified for adverse digging and loading conditions. Actual production costs average about 50% of the theoretical values listed here.

Equipment	Soil Type	B.C.Y. Wt.	% Swell	1 C.Y.	1-1/2 C.Y.	2 C.Y.	2-1/2 C.Y.	3 C.Y.	3-1/2 C.Y.	4 C.Y.
Hydraulic Excavator "Backhoe" 15' deep cut	Moist loam, sandy clay	3400 lbs.	40%	85	125	175	220	275	330	380
	Sand and gravel	3100	18	80	120	160	205	260	310	365
	Common earth	2800	30	70	105	150	190	240	280	330
	Clay, hard, dense	3000	33	65	100	130	170	210	255	300
Power Shovel Optimum Cut (Ft.)	Moist loam, sandy clay	3400	40	170 (6.0)	245 (7.0)	295 (7.8)	335 (8.4)	385 (8.8)	435 (9.1)	475 (9.4)
	Sand and gravel	3100	18	165 (6.0)	225 (7.0)	275 (7.8)	325 (8.4)	375 (8.8)	420 (9.1)	460 (9.4)
	Common earth	2800	30	145 (7.8)	200 (9.2)	250 (10.2)	295 (11.2)	335 (12.1)	375 (13.0)	425 (13.8)
	Clay, hard, dense	3000	33	120 (9.0)	175 (10.7)	220 (12.2)	255 (13.3)	300 (14.2)	335 (15.1)	375 (16.0)
Drag line Optimum Cut (Ft.)	Moist loam, sandy clay	3400	40	130 (6.6)	180 (7.4)	220 (8.0)	250 (8.5)	290 (9.0)	325 (9.5)	385 (10.0)
	Sand and gravel	3100	18	130 (6.6)	175 (7.4)	210 (8.0)	245 (8.5)	280 (9.0)	315 (9.5)	375 (10.0)
	Common earth	2800	30	110 (8.0)	160 (9.0)	190 (9.9)	220 (10.5)	250 (11.0)	280 (11.5)	310 (12.0)
	Clay, hard, dense	3000	33	90 (9.3)	130 (10.7)	160 (11.8)	190 (12.3)	225 (12.8)	250 (13.3)	280 (12.0)

Loading Tractors — Wheel Loaders

Equipment	Soil Type	B.C.Y. Wt.	% Swell	3 C.Y.	4 C.Y.	6 C.Y.	8 C.Y.
Loading Tractors	Moist loam, sandy clay	3400	40	260	340	510	690
	Sand and gravel	3100	18	245	320	480	650
	Common earth	2800	30	230	300	460	620
	Clay, hard, dense	3000	33	200	270	415	560
	Rock, well blasted	4000	50	180	245	380	520

Loading Tractors — Track Loaders

2-1/4 C.Y.	3 C.Y.	3-1/2 C.Y.	4 C.Y.
135	180		250
130	170		235
120	155		220
110	145		200
100	130		180

Figure 2.11   Material Excavation

Volume of Excavated Material*	
Depth in Inches and Feet	Cubic Yards per Square Surface Foot
2"	.006
4"	.012
6"	.019
8"	.025
10"	.031
1'	.037
2'	.074
3'	.111
4'	.148
5'	.185
6'	.222
7'	.259
8'	.296
9'	.333
10'	.370

*No swellage factor applied
Example: Excavation Required: 20' x 30' = 600 x .148 = 88.8 C.Y.

Figure 2.12   Hand Excavation

Task	C.Y. per Hour	Hours per C.Y.
Excavating sandy loam	1–2	0.5–1.0
Shoveling loose earth into truck	1/2–1	1.0–2.0
Loosening with pick	1/4–1/2	2.0–4.0
Shoveling from trenches to 6'-0" deep	1/2–1	1.0–2.0
Shoveling from pits to 6'-0" deep	1/2–1	1.0–2.0
Backfilling	1-1/2–2-1/2	0.4–0.7
Spreading loose earth	4–7	0.15–0.25

Note: The lower values should be used for sandy loam and the higher values for heavy soils, such as clay.

Figure 2.13 Compacting Backfill

Compaction of fill in embankments, around structures, in trenches, and under slabs is important to control settlement. Factors affecting compaction are:

1. Soil gradation
2. Moisture content
3. Equipment used
4. Depth of fill per lift
5. Density required

Example: Compact granular fill around a building foundation using a 21" wide x 24" vibratory plate in 8" lifts. Operator moves at 50 FPM working a 50 minute hour to develop 95% Modified Proctor Density with 4 passes

Production Rate:

$$\frac{1.75' \text{ plate width x 50 FPM x 50 min./hr. x .67' lift}}{27 \text{ C.F. per C.Y.}} = 108.5 \text{ C.Y./hr.}$$

Production Rate for 4 Passes:

$$\frac{108.5 \text{ C.Y.}}{4 \text{ Passes}} = 27 \text{ C.Y./hr. x 8 hrs.} = 216 \text{ C.Y./day}$$

Figure 2.14 Trenching Machine Data

Type of Trenching Machine	Trench Depth (in Feet)	Trench Width (in Inches)	Digging Speed (Feet/Hour)
Wheel Type	2–4	16, 18, 20	150/600
		22, 24, 26	90/300
		28, 30	60/180
	4–6	16, 18, 20	40/120
		22, 24, 26	25/90
		28, 30	15/40
Ladder Type	4–6	16, 20, 24	100/300
		22, 26, 30	75/200
		28, 32, 36	40/125
	6–8	16, 20, 24	40/125
		22, 26, 30	30/60
		28, 32, 36	25/50
	8–12	18, 24, 30	30/75
		30, 33, 36	15/40
Chair Boom Type	2	4, 6	100/250
	3	4, 8, 12	50/200
	5	4, 8, 12	50/100
	6	6, 12, 18	30/90
	8	6, 12, 24	30/75

Figure 2.15   Trench Bottom Widths for Various Outside Diameters of Buried Pipe

Proper bedding of buried pipe is important. When figuring trench excavation and bedding material, use this table to estimate quantities. The side slopes will depend upon type of soil and whether or not sheeting is used.

Outside Diameter in Inches	Trench Bottom Width in Feet
24	4.1
30	4.9
36	5.6
42	6.3
48	7.0
60	8.5
72	10.0
84	11.4

Figure 2.16 Trench Shoring—Minimum Requirements

Depth of Trench (Feet)	Kind of Condition of Earth	Uprights Min. Dimension (Inches)	Uprights Max. Spacing (Feet)	Stringers Min. Dimension (Inches)	Stringers Max. Spacing (Feet)	Size and Spacing of Members — Cross Braces,* Width of Trench Up to 3 Feet (Inches)	3 to 6 Feet (Inches)	6 to 9 Feet (Inches)	9 to 12 Feet (Inches)	12 to 15 Feet (Inches)	Maximum Spacing Vertical (Feet)	Horizontal (Feet)
5 to 10	Hard, compact	3 x 4 or 2 x 6	6			2 x 6	4 x 4	4 x 6	6 x 6	6 x 8	4	6
	Likely to crack	3 x 4 or 2 x 6	3	4 x 6	4	2 x 6	4 x 4	4 x 6	6 x 6	6 x 8	4	6
	Soft, sandy, or filled	3 x 4 or 2 x 6	Close sheeting	4 x 6	4	4 x 4	4 x 6	6 x 6	6 x 8	8 x 8	4	6
	Hydrostatic pressure	3 x 4 or 2 x 6	Close sheeting	6 x 8	4	4 x 4	4 x 6	6 x 6	6 x 8	8 x 8	4	6
10 to 15	Hard	3 x 4 or 2 x 6	4	4 x 6	4	4 x 4	4 x 6	6 x 6	6 x 8	8 x 8	4	6
	Likely to crack	3 x 4 or 2 x 6	2	4 x 6	4	4 x 4	4 x 6	6 x 6	6 x 8	8 x 8	4	6
	Soft, sandy, or filled	3 x 4 or 2 x 6	Close sheeting	4 x 6	4	4 x 6	6 x 6	6 x 8	8 x 8	8 x 10	4	6
	Hydrostatic pressure	3 x 6	Close sheeting	8 x 10	4	4 x 6	6 x 6	6 x 8	8 x 8	8 x 10	4	6
15 to 20	All kinds or conditions	3 x 6	Close sheeting	4 x 12	4	4 x 12	6 x 8	8 x 8	8 x 10	10 x 10	4	6
Over 20	All kinds or conditions	3 x 6	Close sheeting	6 x 8	4	4 x 12	8 x 8	8 x 10	10 x 10	10 x 12	4	6

*Trench jacks may be used in lieu of, or in combination with, cross braces.
Shoring is not required in solid rock, hard shale, or hard slag.
Where desirable, steel sheet piling and bracing of equal strength may be substituted for wood.

Figure 2.17 Troughed Conveyor Belts, Carrying Capacities in Tons per Hour at 100 Feet per Minute

Belt Width (inches)	Max Lumps Sized (inches)	Max Lumps Unsized (inches)	Weight, Lbs. per C.F. 30	50	90	100	125	150	160	180	200
14	2	2-1/2	9	15	28	31	39	46	49	56	62
16	2-1/2	3	13	21	38	42	52	63	67	75	83
18	3	4	16	27	48	54	67	81	86	97	107
20	3-1/2	5	20	33	60	67	83	100	107	120	133
24	4-1/2	8	30	50	90	100	125	150	160	180	200
30	7	14	47	79	142	158	197	236	252	284	315
36	9	18	70	117	210	234	292	351	374	421	467
42	11	20	100	167	300	333	417	500	534	600	667
48	14	24	138	230	414	460	575	690	736	828	920
54	15	28	178	297	534	593	741	890	948	1,070	1,190
60	16	30	222	369	664	738	922	1,110	1,180	1,330	1,480

Figure 2.18 Conveyor Belts, Maximum Speeds in Feet per Minute

Type and Condition of Material	Belt Width (in inches) 14	16	18	20	24	30	36	42	48	54	60
	Feet Per Minute										
Unsized coal, gravel, stone, ashes, ore, or similar material	300	300	350	350	400	450	500	550	600	600	600
Sized coal, coke, or other breakable material	250	250	250	300	300	350	350	400	400	400	400
Wet or dry sand	400	400	500	600	600	700	800	800	800	800	800
Crushed coke, crushed slag, or other fine abrasive material	250	250	300	400	400	500	500	500	500	500	500
Large lump ore, rock, slag, or other large abrasive material	—	—	—	—	350	350	400	400	400	400	400

Figure 2.19 Drilling Rock Using Different Size Bits and Types of Drills

Typical Rates						
		Rate of Drilling (Feet/Hour)				
Rock Classification	Hole Size	Jack-hammer	Wagon Drill	Churn Drill	Rotary Drill	Diamond Drill
Soft	1-3/4"	15/20	30/45	—	—	5/8
Medium		10/15	25/35	—	—	3/5
Hard		5/10	15/30	—	—	2/4
Soft	2-3/8"	10/15	30/50	—	—	5/8
Medium		7/10	20/35	—	—	3/5
Hard		4/8	15/30	—	—	2/4
Soft	3"	—	30/50	—	—	4/7
Medium		—	15/30	—	—	3/5
Hard		—	8/20	—	—	2/4
Soft	4"	—	10/25	—	—	3/6
Medium		—	5/15	—	—	2/4
Hard		—	2/8	—	—	1/3
Soft	6"	—	—	4/7	25/50	3/5
Medium		—	—	2/5	10/25	2/4
Hard		—	—	1/2	6/10	1/3

Figure 2.20 Data for Drilling and Blasting Rock

Hole Size (in Inches)	Hole Dimensions (in Feet)	Hole Area (in S.F.)	Rock Volume (per L.F. of Hole, in C.Y.)	Lbs. of Explosive Required (per L.F. of Hole)	Lbs. of Explosive per C.Y. of Rock % of Hole Filled		
					100%	75%	50%
1-1/2"	4 x 4	16	0.59	0.9	1.52	1.14	0.76
	5 x 5	25	0.93	0.9	0.97	0.73	0.48
	6 x 6	36	1.33	0.9	0.68	0.51	0.34
	7 x 7	49	1.81	0.9	0.50	0.38	0.25
2"	5 x 5	25	0.93	1.7	1.83	1.37	0.92
	6 x 6	36	1.33	1.7	1.28	0.96	0.64
	7 x 7	49	1.81	1.7	0.94	0.71	0.47
	8 x 8	64	2.37	1.7	0.72	0.54	0.36
3"	7 x 7	49	1.81	3.9	2.15	1.61	1.08
	8 x 8	64	2.37	3.9	1.65	1.24	0.83
	9 x 9	81	3.00	3.9	1.30	0.97	0.65
	10 x 10	100	3.70	3.9	1.05	0.79	0.53
	11 x 11	121	4.48	3.9	0.87	0.65	0.44
4"	8 x 8	64	2.37	7.5	3.16	2.37	1.58
	10 x 10	100	3.70	7.5	2.03	1.52	1.02
	12 x 12	144	5.30	7.5	1.42	1.06	0.71
	14 x 14	196	7.25	7.5	1.03	0.77	0.52
	16 x 16	256	9.50	7.5	0.79	0.59	0.40
5"	12 x 12	144	5.30	10.9	2.05	1.54	1.02
	14 x 14	196	7.25	10.9	1.50	1.13	0.75
	16 x 16	256	9.50	10.9	1.15	0.86	0.58
	18 x 18	324	12.00	10.9	0.91	0.68	0.46
	20 x 20	400	14.85	10.9	0.73	0.55	0.37
6"	12 x 12	144	5.30	15.6	2.94	2.20	1.47
	14 x 14	196	7.25	15.6	2.05	1.54	1.02
	16 x 16	256	9.50	15.6	1.64	1.23	0.82
	18 x 18	324	12.00	15.6	1.30	0.97	0.65
	20 x 20	400	14.85	15.6	1.05	0.89	0.53
	24 x 24	576	21.35	15.6	0.73	0.55	0.37
9"	20 x 20	400	14.85	35.0	2.36	1.77	1.18
	24 x 24	576	21.35	35.0	1.64	1.23	0.82
	28 x 28	784	29.00	35.0	1.21	0.91	0.61
	30 x 30	900	33.30	35.0	1.05	0.79	0.53
	32 x 32	1,024	37.90	35.0	0.92	0.69	0.46

Figure 2.21   Tunnel Excavation

Bored tunnel excavation is common in rock for diameters from 4 feet for sewer and utilities, to 60 feet for vehicles. Production varies from a few linear feet per day to over 200 linear feet per day. In the smaller diameters, the productivity is limited by the restricted area for mucking or the removal of excavated material.

Most of the tunnels in rock today are excavated by boring machines called moles. Preparation for starting the excavation or setting up the mole is very costly. Shafts must be excavated to the invert of the proposed tunnel and the mole must be lowered into the shaft. If excavating a portal tunnel, that is starting at an open face, the cost is reduced considerably both for mobilization and mucking.

In soft ground and mixed material, special bucket excavators and rotary excavators are used inside a shield. Tunnel liners must follow directly behind the shield to support the earth and prevent cave-ins.

Traditional muck haulage operations are performed by rail with locomotives and muck cars. Sometimes conveyors are more economical and require less ventilation of the tunnel.

Ventilation and air compression are other important cost factors to consider in tunnel excavation. Continuous ventilation ducts are sometimes fabricated at the tunnel site.

Tunnel linings are steel, cast-in-place reinforced concrete, shotcrete, or a combination of these. When required, contact grouting is performed by pumping grout between the lining and the excavation. Intermittent holes are drilled into the lining and separate costs are determined for drilling per hole, grout pump connecting per hole, and grout per cubic foot. Consolidation grouting and roof bolts may also be required where the excavation is unstable or faulting occurs.

Tunnel boring is usually done 24 hours per day. A typical crew for rock boring is:

Tunneling Crew based on three 8 hour shifts	**Surface Crew based on normal 8 hour shift**
1 Shifter	2 Shop Mechanics
1 Walker	1 Electrician
1 Machine Operator for mole	1 Shifter
1 Oiler	2 Laborers
1 Mechanic	1 Operator with 18 ton cherry picker
3 Locomotives with operators	1 Operator with front end loader
5 Miners for rails, vent ducts, and roof bolts	
1 Electrician	
2 Pumps	
2 Laborers for hoisting	
1 Hoist operator for muck removal	
1 Oiler	

Figure 2.22   Piles

Piles are used to transmit foundation loads to strata of adequate bearing capacity and to eliminate settlement from the consolidation of overlying materials. This table lists nine principal pile categories of the three structural materials: wood, steel, and concrete. No exact criteria for the applicability of the various pile types can be given. The selection of types should be based on factors listed in the figures and on comparative costs.

Pile Type	Timber	Steel
Consider for length of	30-60 ft.	40-100 ft
Applicable material specifications	TS-2P3	TS-P67
Maximum stresses	Measured at most critical point, 1200 psi for Southern Pine and Douglas Fir. See U.S.D.A. Wood Handbook No. 72 for stress values of other species.	12,000 psi
Consider for design loads of	10-50 tons	40-120 tons
Disadvantages	Difficult to splice. Vulnerable to damage in hard driving. Vulnerable to decay unless treated, when piles are intermittently submerged.	Vulnerable to corrosion where exposed. BP section may be damaged or deflected by major obstructions.
Advantages	Comparatively low initial cost. Permanently submerged piles are resistant to decay. Easy to handle.	Easy to splice. High capacity. Small displacement. Able to penetrate through light obstructions.
Remarks	Best suited for friction pile in granular material.	Best suited for endbearing on rock. Reduce allowable capacity for corrosive locations.
Typical illustrations	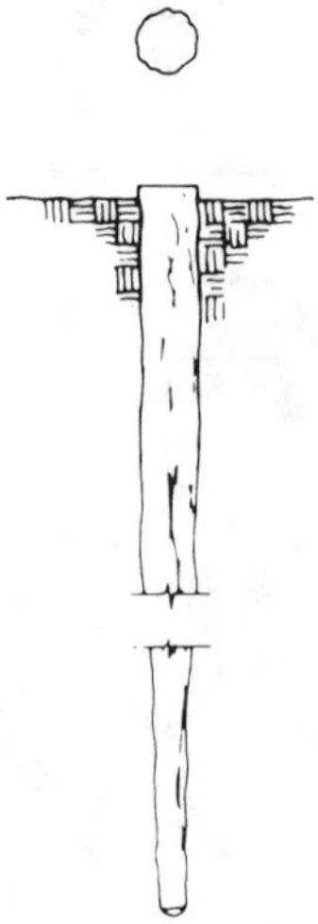	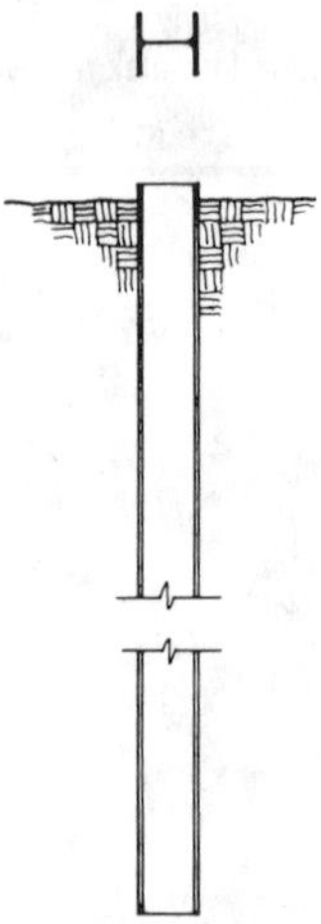

Figure 2.22   Piles (continued)

Pile Type	Precast Concrete (including prestressed)	Cast-in-Place Concrete (thin shell driven with mandrel)
Consider for length of	40-50 ft. for precast. 60-100 ft. for prestressed.	100 ft.
Applicable material specifications	TS-P57	ACI Code 318-for concrete
Maximum stresses	For precast-15% of 28-day strength of concrete, but no more than 700 psi. For prestressed—20% of 28-day strength of concrete, but no more than 1,000 psi in excess of prestress.	25% of 28-day strength of concrete with 1,000 psi maximum, measured at midpoint of length in bearing stratum.
Specifically designed for a wide range of loads		
Disadvantages	Unless prestressed, vulnerable to handling. High initial cost. Considerable displacement. Prestressed difficult to splice.	Difficult to splice after concreting. Redriving not recommended. Thin shell vulnerable during driving. Considerable displacement.
Advantages	High load capacities. Corrosion resistance can be attained. Hard driving possible.	Initial economy. Tapered sections provide higher bearing resistance in granular stratum.
Remarks	Cylinder piles in particular are suited for bending resistance.	Best suited for medium load friction piles in granular materials.
Typical illustrations		

(continued on next page)

Figure 2.22 Piles (continued)

Pile Type	Cast-in-place piles (shells driven without mandrel)	Pressure Injected Footings
Consider for length of	30-80 ft.	10-60 ft.
Applicable material specifications	ACI Code 318	TS-F16
Maximum stresses	25% of 28-day strength of concrete with maximum of 1,000 psi measured at midpoint of length in bearing stratum. 9,000 psi in shell.	25% of 28-day strength of concrete with a minimum of 1,000 psi. 9,000 psi for pipe shell if thickness greater than 1/8".
Consider for design loads of	50-70 tons	60–120 tons
Disadvantages	Hard to splice after concreting. Considerable displacement.	Base of footing cannot be made in clay. When clay layers must be penetrated to reach suitable material, special precautions are required for shafts if in groups.
Advantages	Can be redriven. Shell not easily damaged.	Provides means of placing high capacity footing on bearing stratum without necessity for excavation or dewatering. Required depths can be predicted accurately. High blow energy available for overcoming obstructions. Great uplift resistance if suitably reinforced.
Remarks	Best suited for friction piles of medium length.	Best suited for granular soils where bearing achieved through compaction around base. Minimum spacing 4'-6" on center. For further design requirements see your local building code.
Typical illustrations		

Figure 2.22   Piles (continued)

Pile Type	Concrete Filled Steel Pipe Piles	Composite Piles
Consider for length of	40-120 ft.	60-120 ft.
Applicable material specifications	ASTM A7—for Core ASTM A252—for Pipe ACI Code 318—for Concrete	ACI Code 318—for Concrete ASTM-36—for Structural Section. ASTM A252—for Steel Pipe TS-P2—for Timber.
Maximum stresses	9,000 psi for pipe shell. 25% of 28-day strength of concrete with a maximum of 1,000 psi. 12,000 psi on Steel Cores.	25% of 28-day strength of concrete with 1,000 psi maximum. 9,000 psi for structural and pipe sections. Same as timber piles for wood composite.
Consider for design loads of	80-120 tons without cores. 500-1,500 tons with cores.	30-80 tons
Disadvantages	High initial cost. Displacement for closed end pipe.	Difficult to attain good joint between two materials.
Advantages	Best control during installation. No displacement for open end installation. Open end pipe best against obstructions. High load capacities. Easy to splice.	Considerable length can be provided at comparatively low cost.
Remarks	Provides high bending resistance where unsupported length is loaded laterally.	The weakest of any material used shall govern allowable stresses and capacity.
Typical illustrations		

(continued on next page)

Figure 2.22 Piles (continued)

Pile Type	Concrete Filled Steel Pipe Piles	General Notes
Consider for length of	30-60 ft.	1. Stresses given for steel piles are for noncorrosive locations. For corrosive locations, estimate possible reduction in steel cross section or provide protection from corrosion.
Applicable material specifications	TS-2P69	
Maximum stresses	25% of 28-day strength of concrete with a maximum of 1,000 psi.	
Consider for design loads of	35-70 tons	2. Lengths and loads indicated are for feasibility guidance only. They generally represent current practice.
Disadvantages	More than average dependence on quality workmanship. Not suitable thru peat or similar highly compressible material.	3. Design load capacity should be determined by soil mechanics principles limiting stresses in piles and type and function of structure.
Advantages	Economy. Completely nondisplacement. No driving vibration to endanger adjacent structures. High skin friction. Good contact on rock for end bearing. Convenient for low-headroom underpinning work. Visual inspection of augered material. No splicing required.	
Remarks	Process patented	
Typical illustrations		

Figure 2.23 Pile Type, Classification, Use, and Support Determination

Pile Type		Classification*		Use	Support*		Comments
Material	Form	Displacement	Non-Displacement		End Brng.	Friction	
General	Preformed	√	√	Above ground. Through water.	√	√	Determines some pile types.
Timber — General	Preformed	√	?	Structures with moderate load. Waterfront structures and protection. Trestles and bents. Temporary structures.	?	√	Relatively inexpensive per foot. Good impact absorption. May be jetted in pure sand. Easy to handle and cut off. Use only single length needed (extension is hard and expensive). Prone to hard driving damage. Not driven thru hard stratum or boulders. Limited in size and capacity. Scour and ice action injure pile. without concrete protection. Test piles determine length.
Treated				Above permanent water level. Moist soil (>20%).			Creosote pressure treated. Treat field cuts. 50-yr. life.
Untreated				Below permanent water level. Dry soil (<20%).			Lower material cost. 25-yr. life. Vulnerable to marine borers.
Concrete — General				Building structures of moderate to heavy load. Bridge foundations.			Bracing is not easily attached.
C.I.P.	General						Permanent. Can be treated for sea water. Easy to alter lengths. Damage due to handling eliminated. Easily bonded into pile cap. Specialist contractor.

* √ = Most frequent application
? = Least frequent or questionable application

(continued on next page)

Figure 2.23 Pile Type, Classification, Use, and Support Determination (continued)

| Pile Type | | Classification* | | Use | Support* | | Comments |
Material	Form	Displacement	Non-Displacement		End Brng.	Friction	
Concrete (cont.)							
C.I.P. (cont.)	Uncased		✓	Firm soils. Length, less than 25'	✓	?	No storage space required. Can be made before excavation. Can eliminate vibration and noise. Soft crumbly soils cave in. Inspection dificult. Concrete completed can be damaged by subsequent driving. Bored or extracted casing used.
	Shell left in ground		Open end	Soft soils. All lengths.	✓		Allows inspection before concreting. Easy to cut off or extend. Thin casings may be damaged by handling or soil pressure. Less economical as steel cost increases. Mandrel may be required.
		Closed end	?		?	✓	Same as open end plus below. Increased lateral soil pressures. Thick casings carry part of load. Thin casings support concrete. Mandrel may be required.
Pre-cast (pre-formed)	General			Trestles and bents. Waterfront.			Reinforced for handling. Space required for casting and storage.
	Solid	✓	?		✓	✓	Requires heavy equipment for handling and driving.
	Cylinder		Open end		✓	✓	Durable, can be treated for salt water. Difficult and expensive to cut off, extend or bond into pile cap.
		Plugged end	?		?		Convenient with concrete superstructure.

* ✓ = Most frequent application
  ? = Least frequent or questionable application

Figure 2.23 Pile Type, Classification, Use, and Support Determination (continued)

Pile Type			Classification*		Use	Support*		Comments
	Material	Form	Displace-ment	Non-Dis-placement		End Brg.	Fric-tion	
Steel	General				Large structures of heavy loads. Trestles and bents.			Easy to handle, cut off, extend or bond into pile cap. Available any length by size. Convenient to combine with steel superstructure. Can stand rough handling. Possible damage from corrosion. Requires marine environment protection. Relatively expensive except where bearing stratum can develop large pile capacity. General contractor installs.
	HP	Preformed	Small		Driving through dense layers. When bottom is irregular.	√	?	Seldom used as friction pile. Add 1/16" to thickness for corrosion or 2% copper added. Reinforce bottom for driving into rock or hardpan for stability when muddy soil above. If rock is too hard for substantial penetration with little support above, don't use. Hard driving with no lateral support tends to bend pile. Boulders tend to force piles out of plumb. Do not use in cinder or ash fill May need to batter for lateral resistance in water.
	Pipe	General (preformed)			Good in battered applications.			Considerable structural strength Flexurally strong. Driven with standard hammer rather than mandrel. Normally concrete filled.

* √ = Most frequent application
? = Least frequent or questionable application

(continued on next page)

Figure 2.23   Pile Type, Classification, Use, and Support Determination (continued)

Pile Type		Classification*		Use	Support*		Comments
Material	Form	Displace-ment	Non-Dis-placement		End Brng.	Fric-tion	
Steel (cont.) / Pipe (cont.)	Concrete fill		Open end	Decrease disturbance to adj. structures. Avoid heaving when driving next piles. Driving thru obstructions.	?		Strength and rigidity. Sometimes classified as "composite."
		Closed end		No firm strata at reasonable depth. Desire to compact soil. Water bearing strata punctured or ended in.	?	√	Lower capacity than above if not driven to refusal. May be preferred without concrete fill, if friction support.

* √ = Most frequent application
? = Least frequent or questionable application

Figure 2.24 Installation Time in Man-Hours for Treated Wood Piles

Description	Man-Hours	Unit
Wood Piles Treated 12 lb. Creosote per C.F. 12" Butts 8" Points up to 30' Long	.102	V.L.F.
Boot for Pile Tip	.300	Ea.
Point for Pile Tip	.450	Ea.
Mobilization for 10,000 L.F. Job	.019	

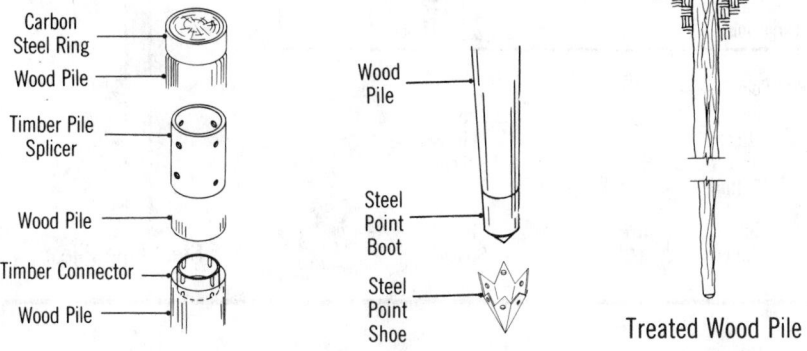

Carbon Steel Ring

Wood Pile

Timber Pile Splicer

Wood Pile

Timber Connector

Wood Pile

Wood Pile

Steel Point Boot

Steel Point Shoe

Treated Wood Pile

Figure 2.25 Installation Time in Man-Hours for Step-Tapered Piles

Description	Man-Hours	Unit
Cast-in-Place Step-Tapered Pile 12" Diameter	.107	V.L.F.
Mobilization		
Small Job	142.000	
Large Job	237.000	

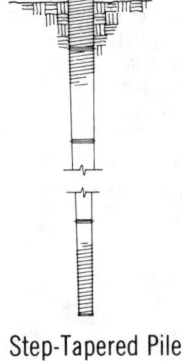

Shell

Steel Ring

Core Step

Shell

Shell

Core

Closure Plate

Step-Tapered Pile

Figure 2.26A Installation Time in Man-Hours for Steel Pipe Piles

Description	Man-Hours	Unit
Pipe Piles 44 lb. per L.F.		
No Concrete	.135	V.L.F.
Concrete Filled	.154	V.L.F.
Splices	2.111	Ea.
Standard Points	1.975	Ea.
Heavy Duty Points	3.960	Ea.
Mobilization		
Small Job	142.000	
Large Job	237.000	

Pipe Point

Rock Insert Cutting

Cutting Shoe

Outside Pipe Splicer

Inside Pipe Splicer

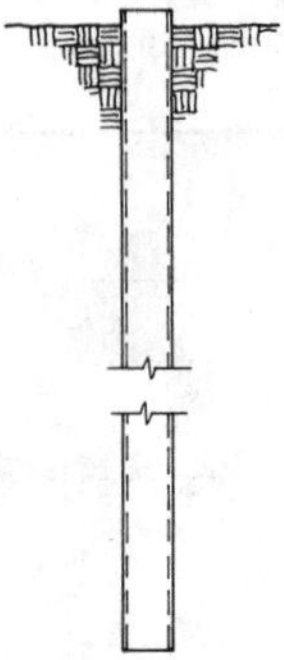

Steel Pipe Pile

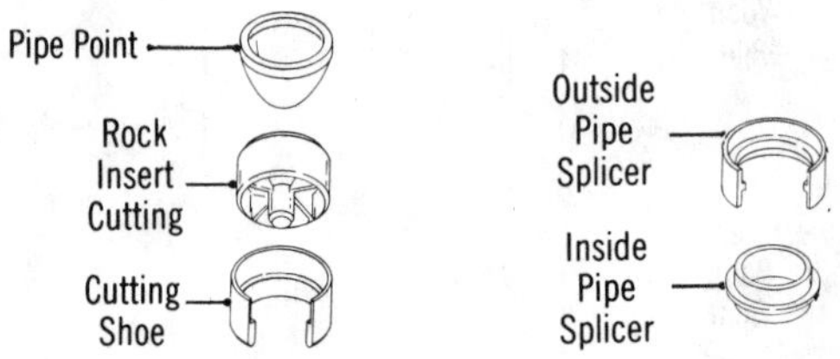

Figure 2.26B Installation Time in Man-Hours for Steel HP Piles

Description	Man-Hours	Unit
H Sections 12″ x 12″, 53 lb. per L.F.	.108	V.L.F.
Splice or Standard Points	2.000	Ea.
Heavy Duty Points	2.286	Ea.
Mobilization		
Small Job	142.000	
Large Job	237.000	

HP Steel Pile

Pipe Slicer

HP Steel Pile

HP Steel Pile

Square End Point

Tapered Point

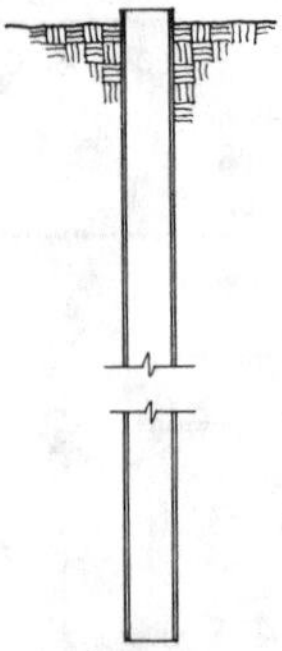

Steel HP Pile

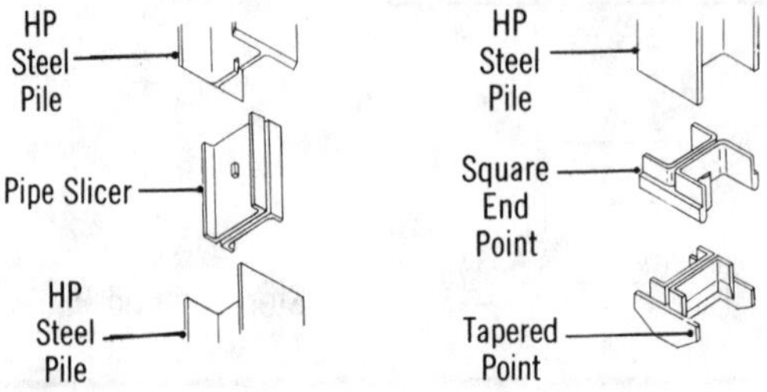

Figure 2.27   Caissons — General

General: Caissons, as covered in this section, are drilled cylindrical foundation shafts which function primarily as short column-like compression members. They transfer superstructure loads through inadequate soils to bedrock or hard stratum. They may be either reinforced or unreinforced and either straight or belled out at the bearing level.

Shaft diameters range in size from 20" to 84" with the most usual sizes beginning at 34". If inspection of bottom is required, the minimum diameter practical is 30". If handwork is required (in addition to mechanical belling, etc.) the minimum diameter is 32". The most frequently used shaft diameter is probably 36" with a 5' or 6' bell diameter. The maximum bell diameter practical is three times the shaft diameter.

Plain concrete is commonly used, poured directly against the excavated face of soil. Permanent casings add to cost and economically should be avoided. Wet or loose strata are undesirable. The associated installation sometimes involves a mudding operation with bentonite clay slurry to keep walls of excavation stable (costs not included here).

Reinforcement is sometimes used, especially for heavy loads. It is required if uplift, bending moment, or lateral loads exist. A small amount of reinforcement is desirable at the top portion of each caisson, even if the above conditions theoretically are not present. This will provide for construction eccentricities and other possibilities. Reinforcement, if present, should extend below the soft strata. Horizontal reinforcement is not required for belled bottoms.

There are three basic types of caisson bearing details:
1. Belled, which are generally recommended to provide reduced bearing pressure on soil. These are not for shallow depths or poor soils. Good soils for belling include most clays, hardpan, soft shale, and decomposed rock.

    Soils requiring handwork include hard shale, limestone, and sandstone.

    Soils not recommended include sand, gravel, silt, and igneous rock. Compact sand and gravel above water table may stand. Water in the bearing strata is undesirable.

2. Straight shafted, which have no bell but the entire length is enlarged to permit safe bearing pressures. They are most economical for light loads on high bearing capacity soil.

3. Socketed (or keyed), which are used for extremely heavy loads. They involve sinking the shaft into rock for combined friction and bearing support action. Reinforcement of shaft is usually necessary. Wide flange cores are frequently used here.

Advantages include:
a. Shafts can pass through soils that piles cannot
b. No soil heaving or displacement during installation
c. No vibration during installation
d. Less noise than pile driving
e. Bearing strata can be visually inspected and tested

Uses include:
a. Situations where unsuitable soil exists to moderate depth
b. Tall structures
c. Heavy structures
d. Underpinning (extensive use)

(continued on next page)

Figure 2.27   Caissons — General (continued)

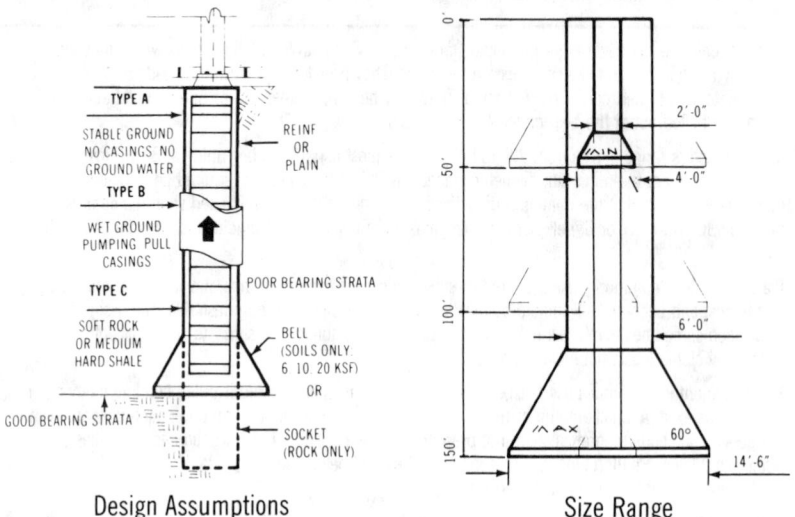

Design Assumptions   Size Range

Figure 2.28   Caissons — Types

The three principal types of caissons are (1) Belled Caissons which, except for shallow depths and poor soil conditions, are generally recommended. They provide more bearing than shaft area. Because of its conical shape, no horizontal reinforcement of the bell is required.
(2) Straight Shaft Caissons are used where relatively light loads are to be supported by a caisson that rests on high value bearing strata. While the shaft is larger in diameter than for belled types this is more than offset by the saving in time and labor. (3) Keyed Caissons are used when extremely heavy loads are to be carried. A keyed or socketed caisson transfers its load into rock by a combination of end-bearing and shear reinforcing of the shaft. The most economical shaft often consists of a steel casing, a steel wide flange core and concrete. Allowable compressive stresses of 0.225 f'c for concrete, 16,000 psi for the wide flange core, and 9,000 psi for the steel casing are commonly used. The usual range of shaft diameter is from 18" to 84". The number of sizes specified for any one project should be limited due to the problems of casing and auger storage. When handwork is to be performed, shaft diameters should not be less than 32". When inspection of borings is required a minimum shaft diameter of 30" is recommended. Concrete caissons are intended to be poured against earth excavation so permanent forms which add to cost should not be used if the excavation is clean and the earth sufficiently impervious to prevent excessive loss of concrete.

Soil Conditions for Belling		
Good	Requires Handwork	Not Recommended
Clay	Hard Shale	Silt
Sandy Clay	Limestone	Sand
Silty Clay	Sandstone	Gravel
Clayey Silt	Weathered Mica	Igneous Rock
Hard-Pan		
Soft Shale		
Decomposed rock		

Figure 2.29 Table of Shaft Diameters, Perimeters, Areas and Volumes; and Shaft or Bell End-Bearing Areas

Shaft Diameter		Shaft Perimeter		Shaft or Bell End Area		Shaft Volume	
in.	cm.	ft.	m	ft.^{2}	m^2	yd^3/ft.	m^3/m
18	45.7	4.71	1.44	1.77	0.164	0.06	0.16
20	50.8	5.24	1.60	2.18	0.203	0.08	0.20
22	55.9	5.76	1.75	2.64	0.245	0.10	0.24
24	61.0	6.28	1.92	3.14	0.292	0.12	0.29
26	66.0	6.81	2.08	3.69	0.343	0.14	0.34
28	71.1	7.33	2.23	4.28	0.397	0.16	0.40
30	76.2	7.85	2.39	4.91	0.456	0.18	0.46
32	81.3	8.38	2.55	5.58	0.519	0.21	0.52
34	86.4	8.90	2.71	6.30	0.586	0.23	0.59
36	91.4	9.42	2.87	7.07	0.657	0.26	0.66
38	96.5	9.95	3.03	7.88	0.732	0.29	0.73
40	102	10.47	3.19	8.73	0.811	0.32	0.81
42	107	11.00	3.35	9.62	0.894	0.34	0.89
44	112	11.52	3.51	10.56	0.981	0.39	0.98
46	117	12.04	3.67	11.54	1.072	0.43	1.07
48	122	12.57	3.83	12.57	1.167	0.46	1.17
50	127	13.09	3.99	13.64	1.267	0.50	1.27
52	132	13.61	4.15	14.75	1.370	0.55	1.37
54	137	14.14	4.31	15.90	1.478	0.59	1.48
56	142	14.66	4.47	17.10	1.589	0.63	1.59
58	147	15.18	4.63	18.35	1.705	0.68	1.70
60	152	15.71	4.79	19.64	1.824	0.73	1.82
62	158	16.23	4.95	20.97	1.948	0.78	1.95
64	163	16.76	5.11	22.34	2.075	0.83	2.08
66	168	17.28	5.27	23.76	2.207	0.88	2.21
68	173	17.80	5.43	25.22	2.343	0.93	2.34
70	178	18.33	5.59	26.72	2.483	0.99	2.48
72	183	18.85	5.74	28.27	2.627	1.05	2.63
74	188	19.37	5.90	29.87	2.775	1.11	2.78
76	193	19.90	6.06	31.50	2.927	1.17	2.93
78	198	20.42	6.22	33.18	3.083	1.23	3.08
80	203	20.94	6.38	34.91	3.243	1.29	3.24
82	208	21.47	6.54	36.67	3.407	1.36	3.41
84	213	21.99	6.70	38.48	3.575	1.42	3.58

Figure 2.30 Bell or Underream Volumes — Dome Type

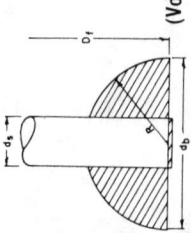

(Volume of shaded portion is volume shown on table.)

Bell Diameter	18" Diameter Shaft		24" Diameter Shaft		30" Diameter Shaft		36" Diameter Shaft		42" Diameter Shaft		48" Diameter Shaft	
	ft.³	m³	ft.³	m³	ft.³	m³	ft.³	m³	ft.³	m³	ft.³	m³
2'-0" 61 cm	2.5	0.07										
2'-6" 76 cm	4.0	0.11	4.0	0.11								
3'-0" 91 cm	6.5	0.18	6.0	0.17	6.0	0.17						
3'-6" 1.07 m	10.0	0.28	9.0	0.25	8.5	0.22	8.5	0.22				
4'-0" 1.22 m	15.0	0.42	14.0	0.40	12.5	0.35	11.5	0.33	13.0	0.37		
4'-6" 1.37 m			20.5	0.58	18.5	0.52	16.0	0.45	15.0	0.42	18.0	0.51
5'-0" 1.52 m			28.5	0.81	26.0	0.74	23.5	0.67	20.5	0.58	21.0	0.59
5'-6" 1.68 m					35.5	1.00	32.5	0.92	29.0	0.82	24.0	0.68
6'-0" 1.83 m					47.0	1.33	43.5	1.23	39.0	1.10	34.5	0.98
6'-6" 1.98 m					61.5	1.74	57.0	1.61	52.0	1.47	46.5	1.32
7'-0" 2.13 m					78.0	2.21	72.5	2.05	67.5	1.91	61.0	1.73
7'-6" 2.29 m					97.5	2.76	92.0	2.60	85.0	2.41	78.0	2.21
8'-0" 2.44 m							114.0	3.23	106.5	3.01	98.5	2.79
8'-6" 2.59 m							138.5	3.92	131.0	3.71	122.0	3.45
9'-0" 2.74 m							167.0	4.73	158.5	4.49	149.0	4.22
9'-6" 2.90 m									189.5	5.36	179.5	5.08
10'-0" 3.05 m									224.5	6.35	213.0	6.03
10'-6" 3.20 m									263.5	7.46	251.5	7.12
11'-0" 3.35 m											293.5	8.31
11'-6" 3.51 m											340.0	9.62
12'-0" 3.66 m											391.0	11.07

Figure 2.31 Bell or Underream Volumes — 45° Type

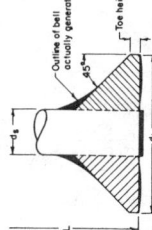

Bell Diameter	18" Diameter Shaft		24" Diameter Shaft		30" Diameter Shaft		36" Diameter Shaft		42" Diameter Shaft		48" Diameter Shaft		52" Diameter Shaft		60" Diameter Shaft	
	ft.³	m³	ft.³	m³	ft.³	m³	ft.³	m³	ft.³	m³	ft.³	m³	ft.³	m³	ft.³	m³
2'-0" 61 cm	0.9	0.02														
2'-6" 76 cm	2.3	0.06	1.1	0.03												
3'-0" 91 cm	4.4	0.13	2.9	0.08	1.3	0.04										
3'-6" 1.07 m	7.3	0.21	5.4	0.15	3.5	0.10	1.6	0.04								
4'-0" 1.22 m	11.1	0.31	8.9	0.25	6.5	0.18	4.1	0.11	1.8	0.05						
4'-6" 1.37 m	15.9	0.45	13.3	0.38	10.5	0.30	7.5	0.21	4.6	0.13	2.1	0.06				
5'-0" 1.52 m			18.8	0.53	15.5	0.44	12.0	0.34	8.5	0.24	5.2	0.15	2.3	0.07		
5'-6" 1.68 m			25.5	0.72	21.8	0.62	17.7	0.50	13.6	0.39	9.6	0.27	5.8	0.16	2.6	0.07
6'-0" 1.83 m			33.5	0.95	29.3	0.83	24.7	0.70	20.0	0.57	15.2	0.43	10.6	0.30	6.4	0.18
6'-6" 1.98 m					38.2	1.08	33.1	0.94	27.7	0.78	22.2	0.63	16.7	0.47	11.6	0.33
7'-0" 2.13 m					48.6	1.38	42.9	1.22	36.9	1.04	30.6	0.87	24.4	0.69	18.3	0.52
7'-6" 2.29 m					60.5	1.71	54.3	1.54	47.6	1.35	40.6	1.15	33.6	0.95	26.6	0.75
8'-0" 2.44 m							67.4	1.91	60.1	1.70	52.3	1.48	44.4	1.26	36.5	1.03
8'-6" 2.59 m							82.2	2.33	74.2	2.10	65.8	1.86	57.0	1.62	48.2	1.36
9'-0" 2.74 m							98.9	2.80	90.3	2.56	81.1	2.30	71.5	2.03	61.8	1.75
9'-6" 2.90 m									180.0	3.07	98.4	2.74	88.0	2.49	77.3	2.19
10'-0" 3.05 m									128.0	3.64	118.0	3.33	106.0	3.02	94.9	2.69
11'-0" 3.35 m											163.0	4.62	150.0	4.25	137.0	3.87
12'-0" 3.66 m											218.0	6.17	203.0	5.75	188.0	5.32
13'-0" 3.96 m													266.0	7.54	249.0	7.05
14'-0" 4.27 m															322.0	9.10
15'-0" 4.57 m															406.0	11.50

Figure 2.32   Bell or Underream Volumes — 30° Type

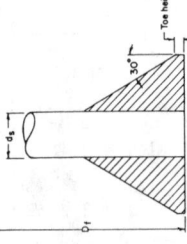

Bell Diameter	18" Diameter Shaft		24" Diameter Shaft		30" Diameter Shaft		36" Diameter Shaft		42" Diameter Shaft		48" Diameter Shaft	
	ft.³	m³	ft.³	m³	ft.³	m³	ft.³	m³	ft.³	m³	ft.³	m³
2'-0" 61 cm	1.0	0.03										
2'-6" 76 cm	2.8	0.08	1.2	0.04								
3'-0" 91 cm	5.7	0.16	3.6	0.10	1.5	0.04						
3'-6" 1.07 m	9.8	0.28	7.1	0.20	4.3	0.12	1.8	0.05				
4'-0" 1.22 m	15.3	0.43	12.0	0.33	8.4	0.24	5.0	0.14	2.1	0.06		
4'-6" 1.37 m	22.4	0.63	18.4	0.52	14.1	0.40	9.8	0.28	5.8	0.16	2.4	0.06
5'-0" 1.52 m			26.6	0.75	21.5	0.61	16.3	0.46	11.1	0.31	6.5	0.18
5'-6" 1.68 m			36.7	1.04	30.8	0.87	24.6	0.70	18.4	0.52	12.5	0.35
6'-0" 1.83 m			48.8	1.38	42.2	1.19	35.1	0.99	27.8	0.79	20.6	0.58
6'-6" 1.98 m					55.8	1.58	47.8	1.35	39.3	1.11	30.8	0.87
7'-0" 2.13 m					71.9	2.01	62.9	1.78	53.3	1.51	43.6	1.23
7'-6" 2.29 m					90.5	2.56	80.5	2.28	70.0	1.98	58.8	1.66
8'-0" 2.44 m							101.0	2.86	89.2	2.52	76.9	2.18
8'-6" 2.59 m							124.5	3.52	111.0	3.14	97.8	2.77
9'-0" 2.74 m							151.0	4.26	137.0	3.87	122.0	3.45
9'-6" 2.90 m									165.0	4.68	149.0	4.22
10'-0" 3.05 m									197.0	5.57	180.0	5.09
10'-6" 3.20 m									233.0	6.60	214.0	6.06
11'-0" 3.35 m											252.0	7.13
12'-0" 3.66 m											340.0	9.61

Toe height

30°

Figure 2.32 Bell or Underream Volumes — 30° Type (continued)

	54" Diameter Shaft		60" Diameter Shaft		66" Diameter Shaft		72" Diameter Shaft		78" Diameter Shaft		84" Diameter Shaft	
5'-0" 1.52 m	2.7	0.08										
5'-6" 1.68 m	7.2	0.20	2.9	0.09								
6'-0" 1.83 m	13.8	0.39	8.0	0.23	3.2	0.09						
6'-6" 1.98 m	22.7	0.64	15.2	0.43	8.7	0.25	3.5	0.10				
7'-0" 2.13 m	34.0	0.96	24.8	0.70	16.6	0.47	9.4	0.27	3.8	0.11		
8'-0" 2.44 m	64.4	1.82	52.0	1.47	40.2	1.14	29.1	0.82	19.2	0.54	10.9	0.31
9'-0" 2.74 m	106.0	3.02	90.0	2.57	75.5	2.14	60.5	1.71	46.4	1.31	33.4	0.95
10'-0" 3.05 m	162.0	4.57	143.0	4.04	124.0	3.50	105.0	2.97	86.6	2.45	69.0	1.95
11'-0" 3.35 m	231.0	6.54	209.0	5.92	187.0	5.28	164.0	4.64	141.0	4.00	119.0	3.36
12'-0" 3.66 m	316.0	8.95	291.0	8.24	265.0	7.50	238.0	6.75	211.0	5.97	185.0	5.24

Figure 2.33   Installation Time in Man-Hours for Caissons

Description	Man-Hours	Unit
Caissons Open Style Machine Drilled to 50' Deep in Stable Ground, No Casings or Ground Water		
36" Diameter	.384	V.L.F.
8' Bell Diameter Add	20.000	Ea.
Open Style Machine Drilled to 50' Deep in Wet Ground Pulled Casing and Pumping		
36" Diameter	.933	V.L.F.
8' Bell Diameter Add	23.333	Ea.
Open Style Machine Drilled to 50' Deep in Soft Rock and Medium Hard Shale		
36" Diameter	5.867	V.L.F.
8' Bell Diameter Add	67.692	Ea.
Mobilization	65.000	
Bottom Inspection	6.667	Ea.

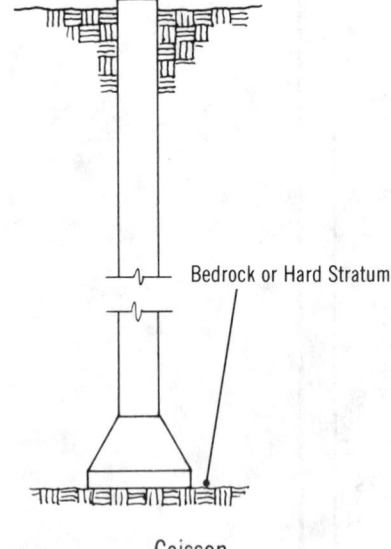

Bedrock or Hard Stratum

Caisson

Figure 2.34   Installation Time in Man-Hours for
Pressure Injected Footings

Description	Man-Hours	Unit
Pressure Injected Footings 12" Diameter Shaft	.400	V.L.F.
Pressure Injected Footing 12" to 18" Diameter to 40'	.278	V.L.F.
Pile Cutoff Concrete Pile with Thin Shell	.211	Ea.
Mobilization		
Small Job	142.000	
Large Job	237.000	

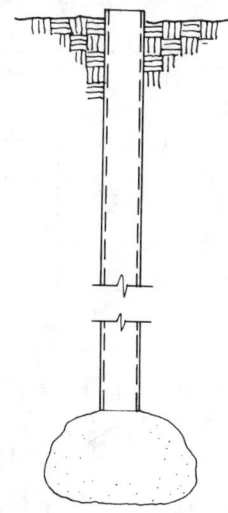

Pressure Injected Footing

Figure 2.35 Installation Time in Man-Hours for Sheet Piling (Sheeting)

Description	Man-Hours	Unit
Wood Sheeting Including Wales, Braces and Spacers		
8' Deep Excavation Pull and Salvage	.121	S.F.
Left in Place	.091	S.F.
12' Deep Pull and Salvage	.148	S.F.
Left in Place	.111	S.F.
16' Deep Pull and Salvage	.167	S.F.
Left in Place	.125	S.F.
20' Deep Pull and Salvage	.190	S.F.
Left in Place	.143	S.F.
Steel Sheet Piling		
15' Deep Excavation Pull and Salvage	.098	S.F.
Left in Place	.065	S.F.
20' Deep Pull and Salvage	.100	S.F.
Left in Place	.067	S.F.
25' Deep Pull and Salvage	.096	S.F.
Left in Place	.064	S.F.
Tieback, Based on Total Length		
Minimum	.553	L.F.
Maximum	1.250	L.F.

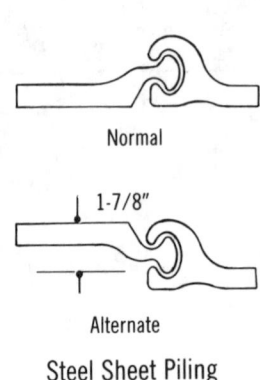

Normal

1-7/8"

Alternate

Steel Sheet Piling Interlocking Connections

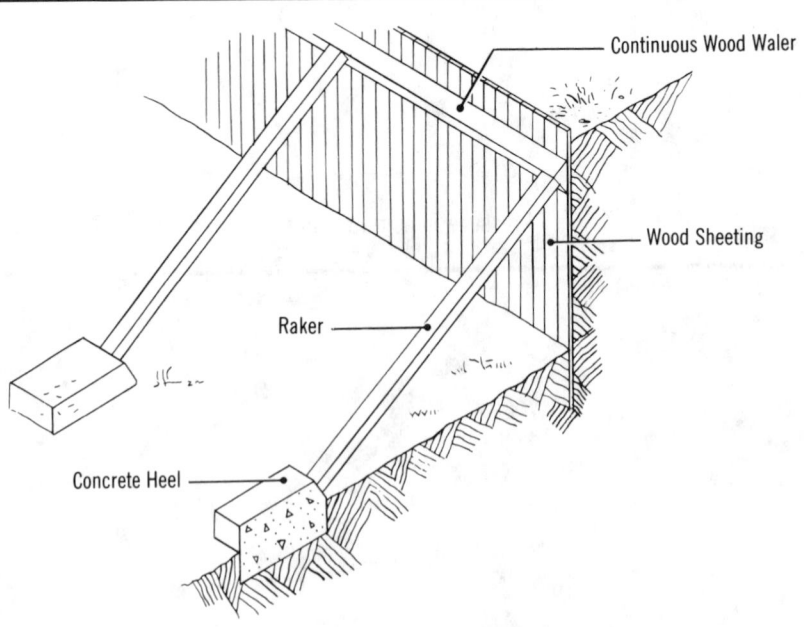

Continuous Wood Waler

Wood Sheeting

Raker

Concrete Heel

Wood Sheet Piling System

Figure 2.35 Installation Time in Man-Hours for
Sheet Piling (Sheeting) (continued)

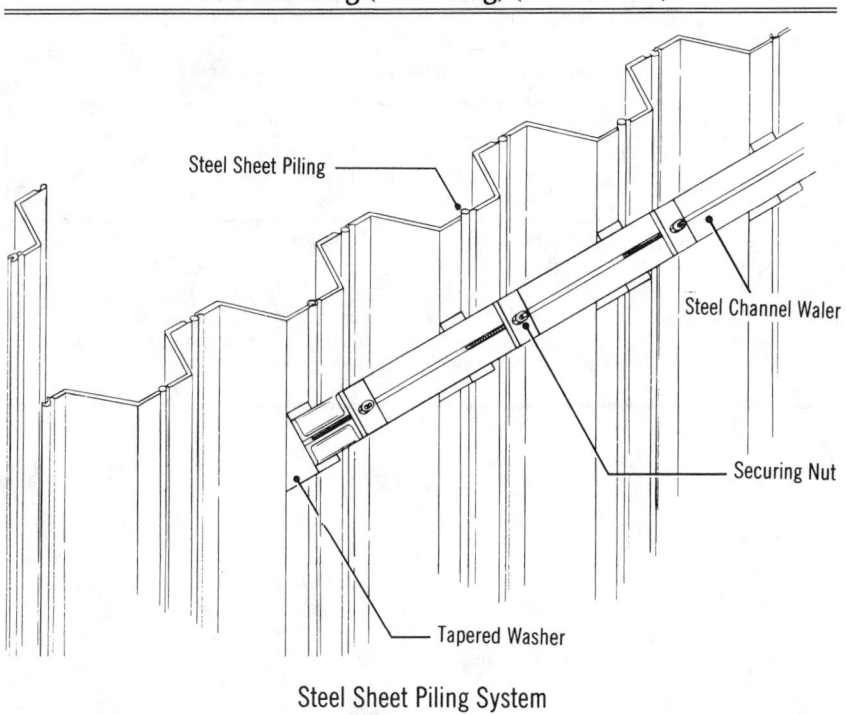

Steel Sheet Piling

Steel Channel Waler

Securing Nut

Tapered Washer

Steel Sheet Piling System

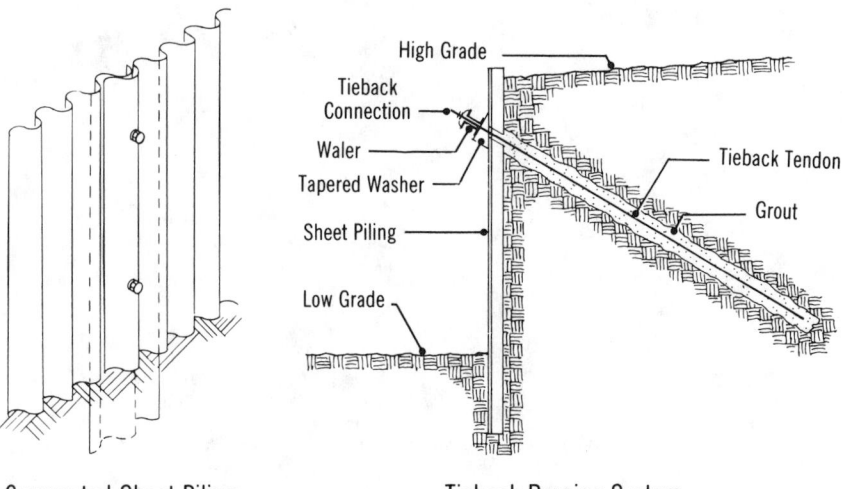

High Grade

Tieback
Connection

Waler

Tapered Washer

Sheet Piling

Low Grade

Tieback Tendon

Grout

Corrugated Sheet Piling   Tieback Bracing System

Figure 2.36 Wood Sheet Piling

Wood sheet piling may be used for depths to 20' where there is no ground water. If moderate ground water is encountered Tongue and Groove sheeting will help to keep it out. When considerable ground water is present, steel sheeting must be used.

For estimating purposes on trench excavation, sizes are as follows:

Depth	Sheeting	Wales	Braces	B.F. per S.F.
To 8'	3 x 12's	6 x 8's, 2 Line	6 x 8's, @ 10'	4.0 @ 8'
8' x 12'	3 x 12's	10x 10's, 2 Line	10 x 10's, @ 9'	5.0 average
12' x 20'	3 x 12's	12x 12's, 3 Line	12 x 12's, @ 8'	7.0 average

Sheeting to be toed in at least 2' depending upon soil conditions. A five-man crew with an air compressor and sheeting driver can drive and brace 440 S.F./day at 8' deep, 360 S.F./day at 12' deep, and 320 S.F./day at 16' deep. For normal soils, piling can be pulled in 1/3 the time to install. Pulling difficulty increases with the time in the ground. Production can be increased by high pressure jetting. Figures below assume 50% of lumber is salvaged and includes pulling costs. Some jurisdictions require an additional equipment operator.

Figure 2.37 Installation Time in Man-Hours for Erosion Control

Description	Man-Hours	Unit
Mulch		
Hand Spread		
Wood Chips, 2" Deep	.004	S.F.
Oat Straw, 1" Deep	.002	S.F.
Excelsior w/Netting	.001	S.F.
Polyethytlene Film	.001	S.F.
Shredded Bark, 3" Deep	.009	S.F.
Pea Stone	.643	C.Y.
Marble Chips	2.400	C.Y.
Polypropylene Fabric	.001	S.F.
Jute Mesh	.001	S.F.
Machine Spread		
Wood Chips, 2" Deep	1.970	MSF
Oat Straw, 1" Deep	.089	MSF
Shredded Bark, 3" Deep	2.960	MSF
Pea Stone	.047	C.Y.
Hydraulic Spraying		
Wood Cellulose	.200	MSF
Rip Rap		
Filter Stone, Machine Placed	.258	C.Y.
1/3 C.Y. Pieces, Crane Set, Grouted	.700	S.Y.
18" Thick, Crane Set, Not Grouted	1.060	S.Y.
Gabion Revetment Mats, Stone Filled,		
12" Deep	.366	S.Y.
Precast Interlocking Concrete Block Pavers	.078	S.F.

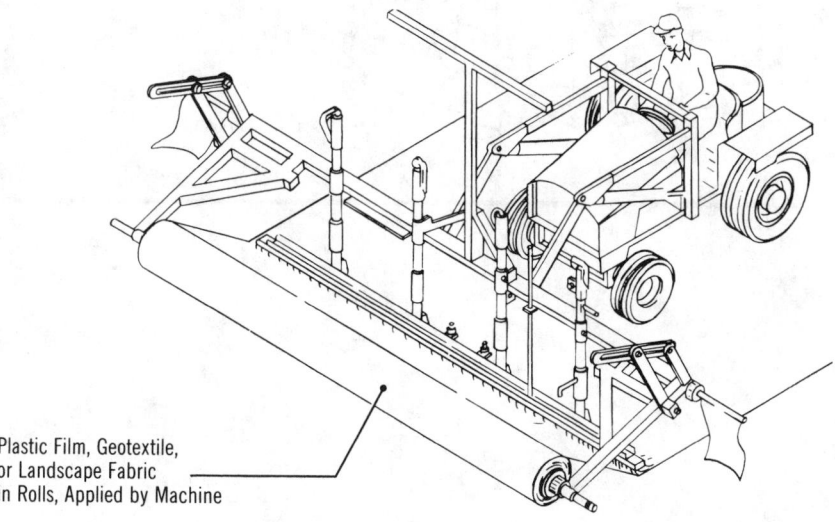

Plastic Film, Geotextile,
or Landscape Fabric
in Rolls, Applied by Machine

Figure 2.38   Installation Time in Man-Hours for Erosion Control: Gabion Retaining Walls

Description	Man-Hours	Unit
Gabion with Stone Fill		
3′ x 6′ x 1′	.280	Ea.
3′ x 6′ x 3′	1.020	Ea.
3′ x 9′ x 1′	.431	Ea.
3′ x 9′ x 3′	1.510	Ea.
3′ x 12′ x 1′	.560	Ea.
3′ x 12′ x 3′	2.240	Ea.
Drainage Stone, 3/4″ Diameter	.092	C.Y.
Backfill Dozer	.010	C.Y.
Compaction, Roller, 12″ Lifts	.014	C.Y.

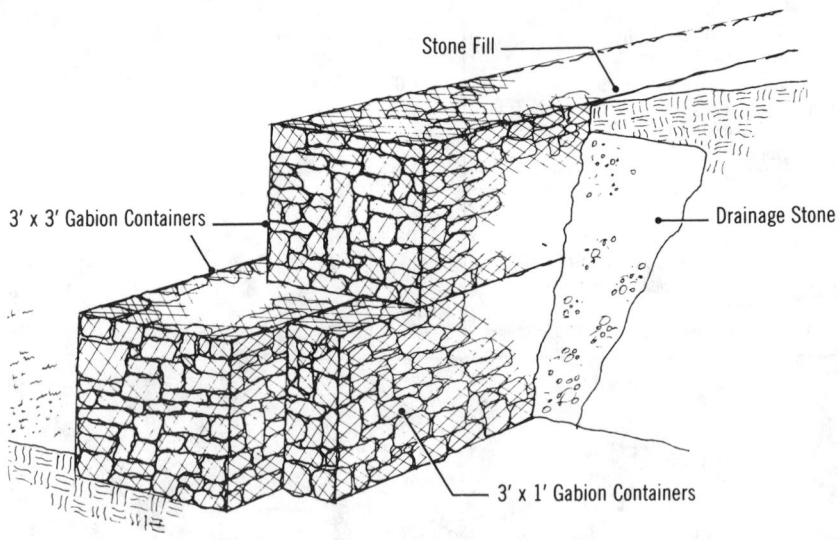

Stone Fill

3′ x 3′ Gabion Containers

Drainage Stone

3′ x 1′ Gabion Containers

Figure 2.39   Installation Time in Man-Hours for Erosion Control: Concrete Retaining Walls

Description	Man-Hours	Unit
Concrete Retaining Wall		
Footing Formwork	.066	sfca
Wall Formwork, Under 8' High	.049	sfca
Under 16' High	.079	sfca
Wall Formwork, Battered to 16' High	.150	sfca
Footing Reinforcing Bars	15.240	ton
Wall Reinforcing Bars	10.670	ton
Footing Concrete, Direct Chute	.400	C.Y.
Pumped	.640	C.Y.
Crane and Bucket	.711	C.Y.
Wall Concrete, Direct Chute	.480	C.Y.
Pumped	.674	C.Y.
Crane and Bucket	.711	C.Y.
Perforated, Clay, 4" Diameter	.060	L.F.
6" Diameter	.076	L.F.
8" Diameter	.083	L.F.
Bituminous, 4" Diameter	.032	L.F.
6" Diameter	.035	L.F.
Porous Concrete, 4" Diameter	.072	L.F.
6" Diameter	.076	L.F.
8" Diameter	.090	L.F.
Drainage Stone, 3/4" Diameter	.092	C.Y.
Backfill, Dozer	.010	C.Y.
Compaction, Vibrating Plate, 12" Lifts	.044	C.Y.

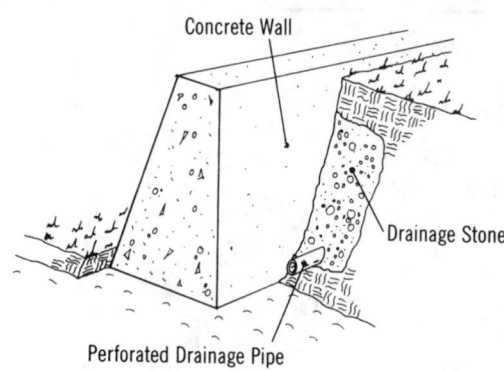

Concrete Wall

Drainage Stone

Perforated Drainage Pipe

Gravity Retaining Wall

Figure 2.40 Installation Time in Man-Hours for Erosion Control: Masonry Retaining Walls

Description	Man-Hours	Unit
Masonry Retaining Wall, 8" Thick, 4' High Continuous Wall Footing	.140	L.F.
Concrete Block Wall Including Reinforcing and Grouting	.610	L.F.
Fill in Trench Crushed Bank Run	.010	L.F.
Perforated Bituminous Underdrain, 4" Diameter	.062	L.F.
Masonry Retaining Wall, 10" Thick, 6' High Continuous Wall Footing	.218	L.F.
Concrete Block Wall Including Reinforcing and Grouting	.957	L.F.
Fill in Trench Crushed Bank Run	.015	L.F.
Perforated Bituminous Underdrain, 4" Diameter	.062	L.F.
Masonry Retaining Wall, 12" Thick, 8' High Continuous Wall Footing	.278	L.F.
Concrete Block Wall Including Reinforcing and Grouting	1.548	L.F.
Fill in Trench Crushed Bank Run	.020	L.F.
Perforated Bituminous Underdrain, 4" Diameter	.062	L.F.
Stone Retaining Wall, 3' Above Grade, Dry Set	2.742	L.F.
Mortar Set	2.400	L.F.
6' Above Grade, Dry Set	4.114	L.F.
Mortar Set	3.600	L.F.

Note: Units are per L.F. of wall.

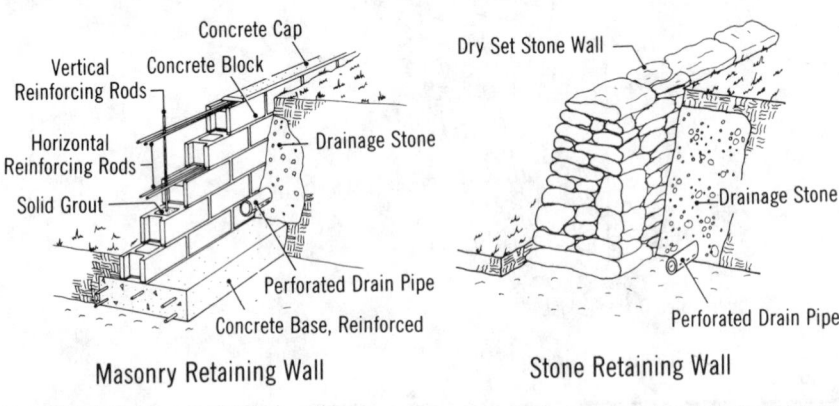

Masonry Retaining Wall

Stone Retaining Wall

Figure 2.41 Area of Road Surface for Various Road Widths

Area in Square Yards

Road Width	Per Lineal Foot	Per 100 Feet	Per Mile
6 Ft.	0.67	66.67	3,520
7	0.78	77.78	4,107
8	0.89	88.89	4,693
9	1.00	100.00	5,280
10	1.11	111.11	5,867
11	1.22	122.22	6,453
12	1.33	133.33	7,040
13	1.44	144.44	7,627
14	1.56	155.56	8,213
15	1.67	166.67	8,800
16	1.78	177.78	9,387
17	1.89	188.89	9,973
18	2.00	200.00	10,560
20	2.22	222.22	11,733
22	2.44	244.44	12,907
24	2.67	266.67	14,080
25	2.78	277.78	14,667
26	2.89	288.89	15,253
28	3.11	311.11	16,427
30	3.33	333.33	17,600
32	3.56	355.56	18,773
34	3.78	377.78	19,947
36	4.00	400.00	21,120
38	4.22	422.22	22,293
40	4.44	444.44	23,467
50	5.56	555.56	29,333
60	6.67	666.67	35,200
70	7.78	777.78	41,067
75	8.33	833.33	44,000
80	8.89	888.89	46,933

(continued on next page)

Figure 2.41   Area of Road Surface
for Various Road Widths (continued)

Area in Square Meters

Road Width	Per Lineal Meter	Per 50 m	Per Kilometer
2 m	2 m^2	100 m^2	2,000 m^2
2.5	2.5	125	2,500
3	3	150	3,000
3.5	3.5	175	3,500
4	4	200	4,000
4.5	4.5	225	4,500
5	5	250	5,000
5.5	5.5	275	5,500
6	6	300	6,000
6.5	6.5	325	6,500
7	7	350	7,000
7.5	7.5	375	7,500
8	8	400	8,000
8.5	8.5	425	8,500
9	9	450	9,000
9.5	9.5	475	9,500
10	10	500	10,000
10.5	10.5	525	10,500
11	11	550	11,000
11.5	11.5	575	11,500
12	12	600	12,000
15	15	750	15,000
20	20	1,000	20,000
25	25	1,250	25,000

Figure 2.42 Cubic Yards of Material Required for Various Widths and Depths per 100 Linear Feet and per Mile

	Width in Feet	1	2	3	4	5	6	7	8	9	10	11	12
Per 100 Linear Feet	1	0.31	0.62	0.93	1.23	1.54	1.85	2.16	2.47	2.78	3.09	3.40	3.70
	2	0.62	1.23	1.85	2.47	3.09	3.70	4.32	4.94	5.56	6.17	6.79	7.41
	3	0.93	1.85	2.78	3.70	4.63	5.56	6.48	7.41	8.33	9.26	10.20	11.10
	4	1.23	2.47	3.70	4.94	6.17	7.41	8.64	9.88	11.10	12.30	13.60	14.80
	5	1.54	3.09	4.63	6.17	7.72	9.26	10.80	12.30	13.90	15.40	17.00	18.50
	6	1.85	3.70	5.56	7.41	9.26	11.10	13.00	14.80	16.70	18.50	20.40	22.20
	7	2.16	4.32	6.48	8.64	10.80	13.00	15.10	17.30	19.40	21.60	23.80	25.90
	8	2.47	4.94	7.41	9.88	12.30	14.80	17.30	19.80	22.20	24.70	27.20	29.60
	9	2.78	5.56	8.33	11.10	13.90	16.70	19.40	22.20	25.00	27.80	30.60	33.30
	10	3.09	6.17	9.26	12.30	15.40	18.50	21.60	24.70	27.80	30.90	34.00	37.00
	20	6.17	12.30	18.50	24.70	30.90	37.00	43.20	49.40	55.60	61.70	67.90	74.10
	30	9.26	18.50	27.80	37.00	46.30	55.60	64.80	74.10	83.30	92.60	102.00	111.00
	40	12.30	24.70	37.00	49.40	61.70	74.10	86.40	98.80	111.00	123.00	136.00	148.00
	50	15.40	30.90	46.30	61.70	77.20	92.60	108.00	123.00	139.00	154.00	170.00	185.00
	60	18.50	37.00	55.60	74.10	92.60	111.00	130.00	148.00	167.00	185.00	204.00	222.00
	70	21.60	43.20	64.80	86.40	108.00	130.00	151.00	173.00	194.00	216.00	238.00	259.00
	80	24.70	49.40	74.10	98.80	123.00	148.00	173.00	198.00	222.00	247.00	272.00	296.00
	90	27.80	55.60	83.30	111.00	139.00	167.00	194.00	222.00	250.00	278.00	306.00	333.00
	100	30.90	61.70	92.60	123.00	154.00	185.00	216.00	247.00	278.00	309.00	340.00	370.00
Per Mile	1	16.30	32.60	48.90	65.20	81.50	97.80	114.00	130.00	147.00	163.00	179.00	196.00
	2	32.60	65.20	97.80	130.00	163.00	196.00	228.00	261.00	293.00	326.00	359.00	391.00
	3	48.90	97.80	147.00	196.00	244.00	293.00	342.00	391.00	440.00	489.00	538.00	587.00
	4	65.20	130.00	196.00	261.00	326.00	391.00	456.00	521.00	587.00	652.00	717.00	782.00
	5	81.50	163.00	244.00	326.00	407.00	489.00	570.00	652.00	733.00	815.00	896.00	978.00
	6	97.80	196.00	293.00	391.00	489.00	587.00	684.00	782.00	880.00	978.00	1,076.00	1,173.00

Depth in Inches

(continued on next page)

Figure 2.42 Cubic Yards of Material Required for Various Widths and Depths per 100 Linear Feet and per Mile (continued)

Width in Feet	Depth in Inches											
	1	2	3	4	5	6	7	8	9	10	11	12
7	114.00	228.00	342.00	456.00	570.00	684.00	799.00	913.00	1,027.00	1,141.00	1,255.00	1,369.00
8	130.00	261.00	391.00	521.00	652.00	782.00	913.00	1,043.00	1,173.00	1,304.00	1,434.00	1,564.00
9	147.00	293.00	440.00	587.00	733.00	880.00	1,027.00	1,173.00	1,320.00	1,467.00	1,613.00	1,760.00
10	163.00	326.00	489.00	652.00	815.00	978.00	1,141.00	1,304.00	1,467.00	1,630.00	1,793.00	1,956.00
20	326.00	652.00	978.00	1,304.00	1,630.00	1,956.00	2,281.00	2,607.00	2,933.00	3,259.00	3,585.00	3,911.00
30	489.00	978.00	1,467.00	1,956.00	2,444.00	2,933.00	3,422.00	3,911.00	4,440.00	4,889.00	5,378.00	5,867.00
40	652.00	1,304.00	1,956.00	2,607.00	3,259.00	3,911.00	4,563.00	5,215.00	5,867.00	6,519.00	7,170.00	7,822.00
50	815.00	1,630.00	2,444.00	3,259.00	4,074.00	4,889.00	5,704.00	6,519.00	7,333.00	8,148.00	8,963.00	9,778.00
60	978.00	1,956.00	2,933.00	3,911.00	4,889.00	5,867.00	6,844.00	7,822.00	8,800.00	9,778.00	10,756.00	11,733.00
70	1,141.00	2,281.00	3,422.00	4,563.00	5,704.00	6,844.00	7,985.00	9,126.00	10,267.00	11,407.00	12,548.00	13,689.00
80	1,304.00	2,607.00	3,911.00	5,215.00	6,519.00	7,822.00	9,126.00	10,430.00	11,733.00	13,037.00	14,341.00	15,644.00
90	1,467.00	2,933.00	4,400.00	5,867.00	7,333.00	8,800.00	10,267.00	11,733.00	13,200.00	14,667.00	16,133.00	17,600.00
100	1,630.00	3,259.00	4,889.00	6,519.00	8,148.00	9,778.00	11,407.00	13,037.00	14,667.00	16,296.00	17,926.00	19,556.00

(Per Mile)

Where: q = Quantity of material, cubic yards
         D = Depth, inches
         W = Width, feet
         L = Length

Formulas used for calculations:

$$100 \text{ L.F.: } q = \left[\frac{D}{36}\right]\left[\frac{W}{3}\right]\left[\frac{100}{3}\right] = 0.3086 \; DW$$

$$\text{Mile: } q = \left[\frac{D}{35}\right]\left[\frac{W}{3}\right]\left[\frac{5,280}{3}\right] = 16.2963 \; DW$$

Figure 2.43   Excavation and Fill Volume Chart

This table is a handy quick reference for estimators when calculating base course gravel or stone for walks, driveways, or parking lots.

Compacted Cubic Yards of Run-of-Bank Gravel = Square Yards x Depth in Inches x .033*

Square Yards	2"	3"	4"	5"	6"	8"	10"
1	.066	.099	.132	.165	.198	.264	.330
2	.132	.198	.264	.330	.396	.528	.660
3	.198	.297	.396	.495	.594	.792	.990
4	.264	.396	.528	.660	.792	1.056	1.320
5	.330	.495	.660	.825	.990	1.320	1.650
6	.396	.594	.792	.990	1.188	1.584	1.980
7	.412	.693	.924	1.155	1.386	1.848	2.310
8	.528	.792	1.056	1.320	1.584	2.112	2.640
9	.594	.891	1.188	1.485	1.782	2.376	2.970
10	.660	.990	1.320	1.650	1.980	2.640	3.300
20	1.32	1.98	2.64	3.30	3.96	5.28	6.60
30	1.98	2.97	3.96	4.95	5.94	7.92	9.90
40	2.64	3.96	5.28	6.60	7.92	10.56	13.20
50	3.30	4.95	6.60	8.25	9.90	13.20	16.50
60	3.96	5.94	7.92	9.90	11.88	15.84	19.80
70	4.62	6.93	9.24	11.55	13.86	18.48	23.10
80	5.28	7.92	10.56	13.20	15.84	21.12	26.40
90	5.94	8.91	11.88	14.85	17.82	23.76	29.70
100	6.60	9.90	13.20	16.50	19.80	26.40	33.00
200	13.20	19.80	26.40	33.00	39.60	52.80	66.00
300	19.80	29.70	39.60	49.50	59.40	79.20	99.00
400	26.40	39.60	52.80	66.00	79.20	105.60	132.00
500	33.00	49.50	66.00	82.50	99.00	132.00	165.00
600	39.60	59.40	79.20	99.00	118.00	158.40	198.00
700	46.20	69.30	92.40	115.50	138.60	184.80	231.00
800	52.80	79.20	105.60	132.00	158.40	211.20	264.00
900	59.40	89.10	118.80	148.50	178.20	237.60	297.00
1000	66.00	99.00	132.00	165.00	198.00	264.00	330.00
2000	132.00	198.00	264.00	330.00	396.00	528.00	660.00
3000	198.00	297.00	396.00	495.00	594.00	792.00	990.00
4000	264.00	396.00	528.00	660.00	792.00	1056.00	1320.00
5000	330.00	495.00	660.00	825.00	990.00	1320.00	1650.00
6000	396.00	594.00	792.00	990.00	1180.00	1584.00	1980.00
7000	462.00	693.00	924.00	1155.00	1386.00	1848.00	2310.00
8000	528.00	792.00	1056.00	1320.00	1584.00	2112.00	2640.00
9000	594.00	891.00	1188.00	1485.00	1782.00	2376.00	2970.00
10,000	660.00	990.00	1320.00	1650.00	1980.00	2640.00	3300.00

*Includes 20% compaction factor

Example: 546 Square Yards x 6 Inches Deep: 500 S.Y.   99.000 C.Y.
                                              40 S.Y.   7.920 C.Y.
                                               6 S.Y.   1.188 C.Y.

For crushed stone, add 10% to figures in table.   108.108, say 108 C.Y.
For loose yards of gravel (truck measure) add 30% to figures in table.

(courtesy of Peckham Industries)

Figure 2.44   Options to be Added for
                  Roadway in Design

Description	Quantity
2" Bituminous concrete mix for sidewalk, incl. fine grade & roll	1.1 S.Y.
5" x 16" Granite vertical curbing	2.0 L.F.
9" Loam for tree lawn	.21 C.Y.
1" Sod, in the East	.7 S.Y.
Paint 4" centerline strip	1.0 L.F.
Lighting 30' aluminum pole, 140' on center, 400 watt mercury vapor	.007 Ea.

Figure 2.45   Typical Parking Lot Plan (50 car)

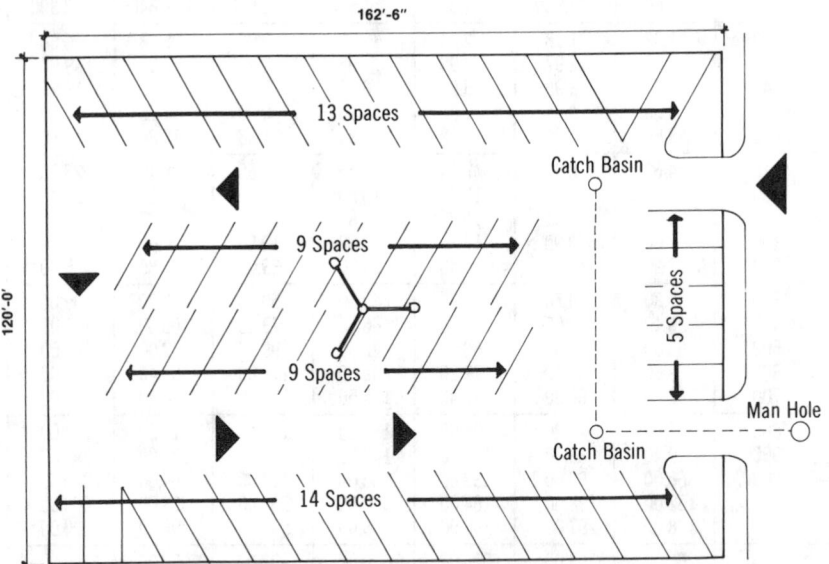

Preliminary Design Data: The space required for parking and maneuvering is between 300 and 400 S.F. per car, depending upon engineering layout and design.

Ninety degree (90°) parking, with a central driveway and two rows of parked cars, will provide the best economy.

Diagonal parking is easier than 90° for the driver and reduces the necessary driveway width, but requires more total space.

Figure 2.46 Parking Lot Layout Data Based on 9' x 19' Parking Stall Size

Φ	P	A	C	W	N	G	D	L	P'	W'
Angle of Stall	Parking Depth	Aisle Width	Curb Length	Width Overall	Net Car Area	Gross Car Area	Distance Last Car	Lost Area	Parking Depth	Width Overall
90°	19'	24'	9	62'	171 S.F.	171 S.F.	9'	0	19'	62'
60°	21'	18'	10.4'	60'	171 S.F.	217 S.F.	7.8'	205 S.F.	18.8'	55.5'
45°	19.8'	13'	12.8'	52.7'	171 S.F.	252 S.F.	6.4'	286 S.F.	16.6'	46.2'

Note: Square foot per car areas do not include the area of the travel lane.

90° Stall Angle: The main reason for use of this stall angle is to achieve the highest car capacity. This may be sound reasoning for employee lots with all day parking, but in most (in & out) lots there is difficulty in entering the stalls and no traffic lane direction. This may outweigh the advantage of high capacity.

60° Stall Angle: This layout is used most often due to the ease of entering and backing out, also the traffic aisle may be smaller.

45° Stall Angle: Requires a small change of direction from the traffic aisle to the stall, so the aisle may be reduced in width.

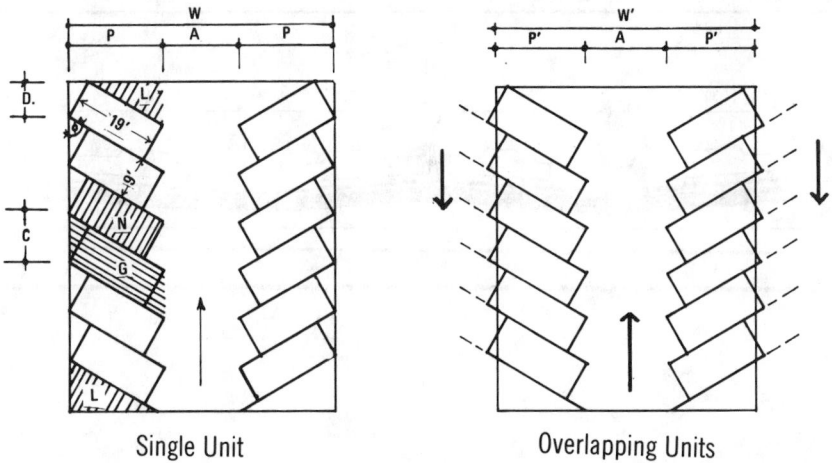

Single Unit   Overlapping Units

Figure 2.47 Options to Be Added for Parking Lot Appurtenances

Description	Unit
8" x 6" Asphalt berm curbing	L.F.
5" x 16" Vertical granite curbing	L.F.
4-1/2" x 12" Sloped granite edging	L.F.
6" x 18" Precast concrete curb	L.F.
24" Wide 6" high concrete curb and gutter	L.F.
4'-0" Wide bituminous concrete walk, 2" thick, 6" gravel base	S.Y.
4'-0" Wide concrete walk 4" thick, 6" gravel base	S.Y.
2-Coat sealcoating, petroleum resistant, under 1000 S.Y.	S.Y.
1/4 to 3/8 C.Y. rip rap slope protection, 18" thick, not grouted	S.Y.
35'-0' High aluminum pole with (3) – 1,000 watt mercury vapor roadway type fixtures	Ea.
6'-0" Long concrete car bumpers, 6" x 10"	Ea.
6'-0" High chain link fence, 1-5/8" top rail, 2" line posts, 10'-0" O.C., 6 gal.	L.F.
3'-0" High corrugated steel guardrail post, 6'-3" O.C.	L.F.
18" Deep perimeter planter bed with shrubs @ 3' O.C.	S.F.
6" Topsoil, including fine grading and seeding	S.Y.

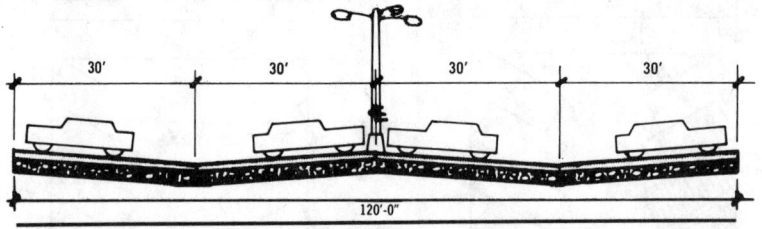

Figure 2.48   Costs per Square Yard of Granular Base and Subbase Courses — per Ton

Where cost per ton is the pay unit—assumed density: 133.3 lb./C.F. (100 lb. S.Y.-in.)

Cost per Ton	Thickness in Inches									
	3	4	5	6	7	8	9	10	11	12
5.00	0.75	1.00	1.25	1.50	1.75	2.00	2.25	2.50	2.75	3.00
6.00	0.90	1.20	1.50	1.80	2.10	2.40	2.70	3.00	3.30	3.60
7.00	1.05	1.40	1.75	2.10	2.45	2.80	3.15	3.50	3.85	4.20
8.00	1.20	1.60	2.00	2.40	2.80	3.20	3.60	4.00	4.40	4.80
9.00	1.35	1.80	2.25	2.70	3.15	3.60	4.05	4.50	4.95	5.40
10.00	1.50	2.00	2.50	3.00	3.50	4.00	4.50	5.00	5.50	6.00
11.00	1.65	2.20	2.75	3.30	3.85	4.40	4.95	5.50	6.05	6.60
12.00	1.80	2.40	3.00	3.60	4.20	4.80	5.40	6.00	6.60	7.20
13.00	1.95	2.60	3.25	3.90	4.55	5.20	5.85	6.50	7.15	7.80
14.00	2.10	2.80	3.50	4.20	4.90	5.60	6.30	7.00	7.70	8.40

Example:
10,000 S.Y. of base, 6" thick at $12 per ton
10,000 S.Y. x $3.60/S.Y. = $36,000

Figure 2.49   Costs per Square Yard of Granular Base and Subbase Courses — per C.Y.

Where cost per cubic yard is the pay unit.

Cost per C.Y.	Thickness in Inches									
	3	4	5	6	7	8	9	10	11	12
5.00	0.42	0.56	0.69	0.83	0.97	1.11	1.25	1.39	1.53	1.67
6.00	0.50	0.67	0.83	1.00	1.17	1.33	1.50	1.67	1.83	2.00
7.00	0.58	0.78	0.97	1.17	1.36	1.56	1.75	1.94	2.14	2.33
8.00	0.67	0.89	1.11	1.33	1.56	1.78	2.00	2.22	2.44	2.67
9.00	0.75	1.00	1.25	1.50	1.75	2.00	2.25	2.50	2.75	3.00
10.00	0.83	1.11	1.39	1.67	1.94	2.22	2.50	2.78	3.06	3.33
11.00	0.92	1.22	1.53	1.83	2.14	2.44	2.75	3.06	3.36	3.67
12.00	1.00	1.33	1.67	2.00	2.33	2.67	3.00	3.33	3.67	4.00
13.00	1.08	1.44	1.81	2.17	2.53	2.89	3.25	3.61	3.97	4.33
14.00	1.17	1.56	1.94	2.33	2.72	3.11	3.50	3.89	4.28	4.67

Example:
10,000 S.Y. of base, 6" thick at $15 per C.Y.
10,000 S.Y. x $2.09/S.Y. = $20,900

Figure 2.50   Costs per Square Yard of Prime Coats, Tack Coats — Gallons

Where cost per gallon is the pay unit.

Cost per Gal.	Application Rate, Gal. per S.Y.					
	0.05	0.10	0.15	0.20	0.25	0.30
0.30	0.02	0.03	0.04	0.06	0.08	0.09
0.40	0.02	0.04	0.06	0.08	0.10	0.12
0.50	0.02	0.05	0.08	0.10	0.12	0.15
0.60	0.03	0.06	0.09	0.12	0.15	0.18
0.70	0.04	0.07	0.10	0.14	0.18	0.21
0.80	0.04	0.08	0.12	0.16	0.20	0.24
0.90	0.04	0.09	0.14	0.18	0.22	0.27
1.00	0.05	0.10	0.15	0.20	0.25	0.30
1.10	0.06	0.11	0.16	0.22	0.28	0.33
1.20	0.06	0.12	0.18	0.24	0.30	0.36
1.30	0.06	0.13	0.20	0.26	0.32	0.39
1.40	0.07	0.14	0.21	0.28	0.35	0.42
1.50	0.08	0.15	0.22	0.30	0.38	0.45
1.60	0.08	0.16	0.24	0.32	0.40	0.48
1.70	0.08	0.17	0.26	0.34	0.42	0.51
1.80	0.09	0.18	0.27	0.36	0.45	0.54
1.90	0.10	0.19	0.28	0.38	0.48	0.57
2.00	0.10	0.20	0.30	0.40	0.50	0.60

Example:
10,000 S.Y. of prime coat, .15 Gal. per S.Y. at $1.00 per Gal.
10,000 S.Y. x $.15/S.Y. = $1,500

Figure 2.51   Costs per Square Yard of Prime
Coats, Tack Coats — Ton

Where cost per ton is the pay unit—assume 1 ton = 241 gal.

Cost per Ton	Application Rate, Gal. per S.Y.					
	0.05	0.10	0.15	0.20	0.25	0.30
70.00	0.01	0.03	0.04	0.06	0.07	0.09
80.00	0.02	0.03	0.05	0.07	0.08	0.10
90.00	0.02	0.04	0.06	0.07	0.09	0.11
100.00	0.02	0.04	0.06	0.08	0.10	0.12
110.00	0.02	0.05	0.07	0.09	0.11	0.14
120.00	0.02	0.05	0.07	0.10	0.12	0.15
130.00	0.03	0.05	0.08	0.11	0.13	0.16
140.00	0.03	0.06	0.09	0.12	0.15	0.17
150.00	0.03	0.06	0.09	0.12	0.16	0.19
160.00	0.03	0.07	0.10	0.13	0.17	0.20
170.00	0.04	0.07	0.11	0.14	0.18	0.21
180.00	0.04	0.07	0.11	0.15	0.19	0.22
190.00	0.04	0.08	0.12	0.16	0.20	0.24
200.00	0.04	0.08	0.12	0.17	0.21	0.25
210.00	0.04	0.09	0.13	0.17	0.22	0.26
240.00	0.05	0.10	0.15	0.20	0.25	0.30
265.00	0.06	0.11	0.16	0.22	0.28	0.33
290.00	0.06	0.12	0.18	0.24	0.30	0.36
315.00	0.06	0.13	0.20	0.26	0.32	0.39
335.00	0.07	0.14	0.21	0.28	0.35	0.42
360.00	0.08	0.15	0.22	0.30	0.38	0.45
385.00	0.08	0.16	0.24	0.32	0.40	0.48
410.00	0.08	0.17	0.26	0.34	0.42	0.51
435.00	0.09	0.18	0.27	0.36	0.45	0.54
460.00	0.10	0.19	0.28	0.38	0.48	0.57
485.00	0.10	0.20	0.30	0.40	0.50	0.60

Example:
10,000 S.Y. of prime coat, .15 Gal. per S.Y. at $140 per ton
10,000 S.Y. x $.09/S.Y. = $900

Figure 2.52   Costs per Square Yard of Asphalt Concrete Pavement Courses

Where cost per ton is the pay unit—assumed density: 145 lbs. per ft.³

Cost per Ton	Thickness of Pavement Course in Inches												
	1/2	1	2	3	4	5	6	7	8	9	10	11	12
15.00	0.41	0.82	1.63	2.45	3.26	4.08	4.89	5.71	6.52	7.34	8.16	8.97	9.79
16.00	0.44	0.87	1.74	2.61	3.48	4.35	5.22	6.09	6.96	7.83	8.70	9.57	10.44
17.00	0.46	0.92	1.85	2.77	3.70	4.62	5.55	6.47	7.40	8.32	9.24	10.17	11.09
18.00	0.49	0.98	1.96	2.94	3.92	4.89	5.87	6.85	7.83	8.81	9.79	10.77	11.74
19.00	0.52	1.03	2.07	3.10	4.13	5.17	6.20	7.23	8.26	9.30	10.33	11.36	12.40
20.00	0.54	1.09	2.18	3.26	4.35	5.44	6.52	7.61	8.70	9.79	10.88	11.96	13.05
21.00	0.57	1.14	2.28	3.43	4.57	5.71	6.85	7.99	9.14	10.28	11.42	12.56	13.70
22.00	0.60	1.20	2.39	3.59	4.79	5.98	7.18	8.37	9.57	10.77	11.96	13.16	14.36
23.00	0.63	1.25	2.50	3.75	5.00	6.25	7.50	8.75	10.00	11.26	12.51	13.76	15.01
24.00	0.65	1.30	2.61	3.92	5.22	6.52	7.83	9.14	10.44	11.74	13.05	14.36	15.66
25.00	0.68	1.36	2.72	4.08	5.44	6.80	8.16	9.52	10.88	12.23	13.59	14.95	16.31
30.00	0.82	1.63	3.26	4.89	6.52	8.16	9.79	11.42	13.05	14.68	16.31	17.94	19.58
35.00	0.95	1.90	3.81	5.71	7.61	9.52	11.42	13.32	15.22	17.13	19.03	20.93	22.84
40.00	1.09	2.18	4.35	6.52	8.70	10.88	13.05	15.22	17.40	15.58	21.75	23.92	26.10
45.00	1.22	2.45	4.89	7.34	9.79	12.23	14.68	17.13	19.58	22.02	24.47	26.92	29.36
50.00	1.36	2.72	5.44	8.16	10.88	13.59	16.31	19.03	21.75	24.47	27.19	29.91	32.62

Example:
10,000 S.Y. of pavement, 4" thick at $30 per ton
10,000 S.Y. x $6.52/S.Y. = $65,200

Figure 2.53 Costs per Square Yard of Cement Treated Bases

Where cost per cubic yard is the pay unit

Cost per C.Y.	Thickness in Inches							
	3	4	5	6	7	8	9	10
0.50	0.04	0.06	0.07	0.08	0.10	0.11	0.12	0.14
1.00	0.08	0.11	0.14	0.17	0.19	0.22	0.25	0.28
2.00	0.17	0.22	0.28	0.33	0.39	0.44	0.50	0.56
3.00	0.25	0.33	0.42	0.50	0.58	0.67	0.75	0.83
4.00	0.33	0.44	0.56	0.67	0.78	0.89	1.00	1.11
5.00	0.42	0.56	0.69	0.83	0.97	1.11	1.25	1.39
10.00	0.83	1.11	1.39	1.67	1.94	2.22	2.50	2.78
11.00	0.92	1.22	1.53	1.83	2.14	2.44	2.75	3.06
12.00	1.00	1.33	1.67	2.00	2.33	2.67	3.00	3.33
13.00	1.08	1.44	1.81	2.17	2.53	2.89	3.25	3.61
14.00	1.17	1.56	1.94	2.33	2.72	3.11	3.50	3.89
15.00	1.25	1.67	2.08	2.50	2.92	3.33	3.75	4.17
20.00	1.67	2.22	2.78	3.33	3.89	4.44	5.00	5.56
25.00	2.08	2.78	3.47	4.17	4.86	5.56	6.25	6.94
30.00	2.50	3.33	4.17	5.00	5.83	6.67	7.50	8.33
35.00	2.92	3.89	4.86	5.83	6.81	7.78	8.75	9.72

Example:
10,000 S.Y. of base, 8" thick at $15 per C.Y.
10,000 S.Y. x $3.33/S.Y. = $33,300

Figure 2.54 Costs per Square Yard of Portland Cement Concrete Pavement Courses

Where cost per cubic yard is the pay unit

Cost per C.Y.	Pavement Thickness in Inches							
	4	6	7	8	9	10	11	12
15.00	1.67	2.50	2.92	3.33	3.75	4.17	4.58	5.00
20.00	2.22	3.33	3.89	4.44	5.00	5.56	6.11	6.67
25.00	2.78	4.17	4.86	5.56	6.25	6.94	7.64	8.33
30.00	3.33	5.00	5.83	6.67	7.50	8.33	9.17	10.00
35.00	3.89	5.83	6.81	7.78	8.75	9.72	10.69	11.67
40.00	4.44	6.67	7.78	8.89	10.00	11.11	12.22	13.33
45.00	5.00	7.50	8.75	10.00	11.25	12.50	13.75	15.00
50.00	5.56	8.33	9.72	11.11	12.50	13.89	15.28	16.67
55.00	6.11	9.17	10.69	12.22	13.75	15.26	16.31	18.33
60.00	6.67	10.00	11.67	13.33	15.00	16.67	18.33	20.00
65.00	7.22	10.83	12.64	14.44	16.25	18.06	19.86	21.67

Example:
10,000 S.Y. of pavement, 8" thick at $50 per C.Y.
10,000 S.Y. x $11.11/S.Y. = $1,111,000

Figure 2.55 Costs per Square Yard for Steel — 1-Inch Pavement Thickness

This table reflects continuously reinforced concrete pavement courses where cost per pound is the pay unit (based on cross-sectional area of steel expressed as a percentage of cross-sectional area of pavement course).

For 1-Inch Pavement Thickness				
Cost per Lb.	Cross-Sectional Area of Steel			
	0.5%	0.6%	0.7%	0.8%
0.05	.09	.11	.13	.15
0.10	.19	.22	.25	.30
0.15	.28	.33	.39	.44
0.20	.37	.44	.51	.59
0.25	.46	.55	.64	.73
0.30	.55	.66	.77	.88
0.35	.64	.77	.90	1.03
0.40	.73	.88	1.03	1.17
0.45	.83	.99	1.16	1.32
0.50	.92	1.10	1.28	1.47
0.55	1.01	1.21	1.41	1.61
0.60	1.10	1.32	1.54	1.76

Figure 2.56 Costs per Square Yard for Steel — 6-Inch Pavement Thickness

This table reflects continuously reinforced concrete pavement courses where cost per pound is the pay unit (based on cross-sectional area of steel expressed as a percentage of cross-sectional area of pavement course).

For 6-Inch Pavement Thickness				
Cost per Lb.	Cross-Sectional Area of Steel			
	0.5%	0.6%	0.7%	0.8%
0.05	.55	.66	.77	.88
0.10	1.10	1.32	1.54	1.76
0.15	1.65	1.98	2.31	2.64
0.20	2.20	2.64	3.08	3.52
0.25	2.75	3.30	3.85	4.40
0.30	3.30	3.96	4.62	5.28
0.35	3.85	4.62	5.39	6.16
0.40	4.40	5.28	6.16	7.04
0.45	4.95	5.94	6.93	7.92
0.50	5.50	6.60	7.70	8.80
0.55	6.05	7.26	8.47	9.68
0.60	6.60	7.92	9.24	10.56

Figure 2.57 Costs per Square Yard for
Steel — 7-Inch Pavement Thickness

This table reflects continuously reinforced concrete pavement courses where cost per pound is the pay unit (based on cross-sectional area of steel expressed as a percentage of cross-sectional area of pavement course).

Cost per Lb.	Cross-Sectional Area of Steel			
	0.5%	0.6%	0.7%	0.8%
0.05	.64	.77	.90	1.03
0.10	1.29	1.54	1.80	2.06
0.15	1.93	2.31	2.70	3.08
0.20	2.57	3.08	3.59	4.11
0.25	3.21	3.85	4.49	5.13
0.30	3.85	4.62	5.39	6.16
0.35	4.49	5.39	6.29	7.19
0.40	5.13	6.16	7.19	8.22
0.45	5.78	6.93	8.09	9.24
0.50	6.42	7.70	8.99	10.27
0.55	7.06	8.47	9.88	11.30
0.60	7.70	9.24	10.78	12.32

For 7-Inch Pavement Thickness

Example:
10,000 S.Y. of pavement, 7" thick, 0.5% steel, $.25/lb.
10,000 S.Y. x $3.21/S.Y. = $32,100

Figure 2.58 Costs per Square Yard for
Steel — 8-Inch Pavement Thickness

This table reflects continuously reinforced concrete pavement courses where cost per pound is the pay unit (based on cross-sectional area of steel expressed as a percentage of cross-sectional area of pavement course).

Cost per Lb.	Cross-Sectional Area of Steel			
	0.5%	0.6%	0.7%	0.8%
0.05	.73	.88	1.03	1.17
0.10	1.47	1.76	2.06	2.35
0.15	2.20	2.64	3.08	3.52
0.20	2.93	3.52	4.11	4.69
0.25	3.67	4.40	5.13	5.87
0.30	4.40	5.28	6.16	7.04
0.35	5.13	6.16	7.19	8.22
0.40	5.87	7.04	8.22	9.39
0.45	6.60	7.92	9.24	10.56
0.50	7.34	8.80	10.27	11.74
0.55	8.07	9.68	11.30	12.91
0.60	8.80	10.56	12.32	14.08

For 8-Inch Pavement Thickness

Figure 2.59 Cost per Square Yard for Sawed Joints — 24-Foot Pavement Width

In this table the pay unit reflects the cost per lineal foot.

For Pavements 24 Ft. Wide Having 1 Longitudinal Joint										
Cost per L.F.	Transverse Joint Spacing, Ft.									
	10	15	20	25	30	35	40	45	50	60
0.15	0.19	0.15	0.12	0.11	0.10	0.09	0.09	0.09	0.08	0.08
0.16	0.20	0.16	0.13	0.12	0.11	0.10	0.10	0.09	0.09	0.08
0.17	0.22	0.17	0.14	0.12	0.11	0.11	0.10	0.10	0.09	0.09
0.18	0.23	0.18	0.15	0.13	0.12	0.11	0.11	0.10	0.10	0.09
0.19	0.24	0.19	0.16	0.14	0.13	0.12	0.11	0.11	0.11	0.10
0.20	0.26	0.20	0.16	0.15	0.14	0.13	0.12	0.12	0.11	0.10
0.21	0.27	0.20	0.17	0.15	0.14	0.13	0.13	0.12	0.12	0.11
0.22	0.28	0.21	0.18	0.16	0.15	0.14	0.13	0.13	0.12	0.12
0.23	0.29	0.22	0.19	0.17	0.16	0.15	0.14	0.13	0.13	0.12
0.24	0.31	0.23	0.20	0.18	0.16	0.15	0.14	0.14	0.13	0.13
0.25	0.32	0.24	0.21	0.18	0.17	0.16	0.15	0.14	0.14	0.13

Example:
10,000 S.Y. pavement, 24' wide, joints at 30', $.22/L.F.
10,000 S.Y. x $.15/S.Y. = $1,500

Figure 2.60 Cost per Square Yard for Sawed Joints — 36-Foot Pavement Width

In this table the pay unit reflects the cost per lineal foot.

For Pavements 36 Ft. Wide Having 2 Longitudinal Joints										
Cost per L.F.	Transverse Joint Spacing, Ft.									
	10	15	20	25	30	35	40	45	50	60
0.15	0.21	0.16	0.14	0.13	0.12	0.11	0.11	0.10	0.10	0.10
0.16	0.22	0.18	0.15	0.14	0.13	0.12	0.12	0.11	0.11	0.10
0.17	0.24	0.19	0.16	0.15	0.14	0.13	0.12	0.12	0.12	0.11
0.18	0.25	0.20	0.17	0.15	0.14	0.14	0.13	0.13	0.12	0.12
0.19	0.27	0.21	0.18	0.16	0.15	0.14	0.14	0.13	0.13	0.12
0.20	0.28	0.22	0.19	0.17	0.16	0.15	0.14	0.14	0.14	0.13
0.21	0.29	0.23	0.20	0.18	0.17	0.16	0.15	0.15	0.14	0.14
0.22	0.31	0.24	0.21	0.19	0.18	0.17	0.16	0.15	0.15	0.14
0.23	0.32	0.25	0.22	0.20	0.18	0.17	0.17	0.16	0.16	0.15
0.24	0.34	0.26	0.23	0.21	0.19	0.18	0.17	0.17	0.16	0.16
0.25	0.35	0.28	0.24	0.22	0.20	0.19	0.18	0.18	0.17	0.16

Example:
10,000 S.Y. pavement, 36' wide, joints at 30', $.22/L.F.
10,000 S.Y. x $.18/S.Y. = $1,800

Figure 2.61   Cost per Square Yard for Sawed Joints — 48-Foot Pavement Width

In this table the pay unit reflects the cost per lineal foot.

Cost per L.F.	For Pavements 48 Ft. Wide Having 3 Longitudinal Joints									
	Transverse Joint Spacing, Ft.									
	10	15	20	25	30	35	40	45	50	60
0.15	0.22	0.17	0.15	0.14	0.13	0.12	0.12	0.11	0.11	0.11
0.16	0.23	0.19	0.16	0.15	0.14	0.13	0.13	0.12	0.12	0.11
0.17	0.25	0.20	0.17	0.16	0.14	0.14	0.13	0.13	0.12	0.12
0.18	0.26	0.21	0.18	0.17	0.16	0.15	0.14	0.14	0.13	0.13
0.19	0.28	0.22	0.19	0.18	0.16	0.16	0.15	0.14	0.14	0.14
0.20	0.29	0.23	0.20	0.18	0.17	0.16	0.16	0.15	0.15	0.14
0.21	0.31	0.24	0.21	0.19	0.18	0.17	0.17	0.16	0.16	0.15
0.22	0.32	0.26	0.22	0.20	0.19	0.18	0.17	0.17	0.16	0.16
0.23	0.34	0.27	0.23	0.21	0.20	0.19	0.18	0.18	0.17	0.16
0.24	0.35	0.28	0.24	0.22	0.21	0.20	0.18	0.18	0.18	0.17
0.25	0.37	0.29	0.25	0.23	0.22	0.20	0.20	0.19	0.19	0.18

Example:
10,000 S.Y. pavement, 48' wide, joints at 30', $.22/L.F.
10,000 S.Y. x $.19/S.Y. = $1,900

Figure 2.62 Installation Time in Man-Hours for Asphalt Pavement

Description	Man-Hours	Unit
Subgrade, Grade and Roll		
Small Area	.024	S.Y.
Large Area	.011	S.Y.
Base Course		
Bank Run Gravel, Spread Compact		
6" Deep	.004	S.Y.
18" Deep	.013	S.Y.
Crushed Stone, Spread Compact		
6" Deep	.016	S.Y.
18" Deep	.029	S.Y.
Asphalt Concrete Base		
4" Thick	.053	S.Y.
8" Thick	.089	S.Y.
Stabilization Fabric, Polypropylene, 6 oz./S.Y.	.002	S.Y.
Asphalt Pavement, Wearing Course		
1-1/2" Thick	.026	S.Y.
3" Thick	.052	S.Y.

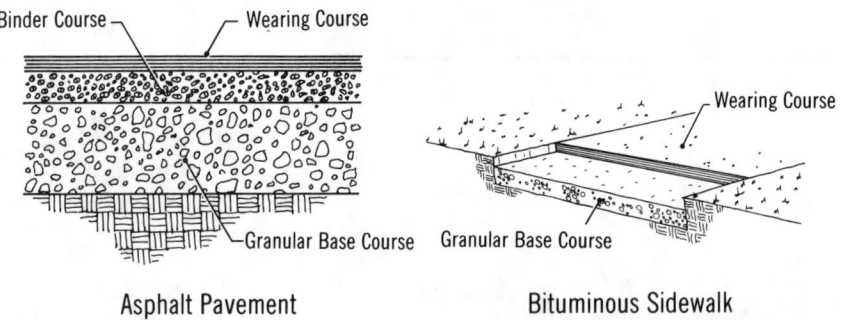

Asphalt Pavement   Bituminous Sidewalk

Figure 2.63   Installation Time in Man-Hours for Curbs

Description	Man-Hours	Unit
Curbs, Bituminous, Plain, 8" Wide, 6" High,		
50 L.F./ton	.032	L.F.
8" Wide, 8" High, 44 L.F./ton	.036	L.F.
Bituminous Berm, 12" Wide, 3" to 6" High,		
35 L.F./ton, before Pavement	.046	L.F.
12" Wide, 1-1/2" to 4" High, 60 L.F./ton,		
Laid with Pavement	.030	L.F.
Concrete, 6" x 18", Cast-in-Place, Straight	.096	L.F.
6" x 18" Radius	.106	L.F.
Precast, 6" x 18", Straight	.160	L.F.
6" x 18" Radius	.172	L.F.
Granite, Split Face, Straight, 5" x 16"	.112	L.F.
6" x 18"	.124	L.F.
Radius Curbing, 6" x 18", Over 10' Radius	.215	L.F.
Corners, 2' Radius	.700	Ea.
Edging, 4-1/2" x 12", Straight	.187	L.F.
Curb Inlets, (Guttermouth) Straight	1.366	Ea.
Monolithic Concrete Curb and Gutter		
Cast-in-Place with 6' High Curb and		
6" Thick Gutter		
24" Wide, .055 C.Y. per L.F.	.128	L.F.
30" Wide, .066 C.Y. per L.F.	.141	L.F.

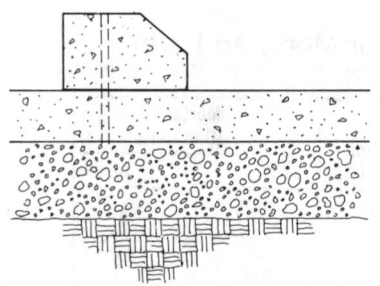

Precast Concrete Parking Bumper

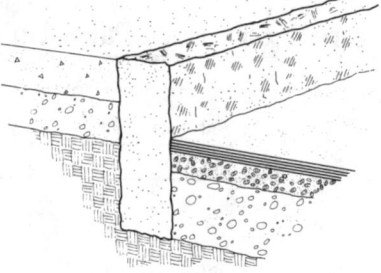

Granite Curb

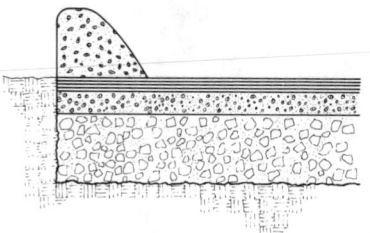

Bituminous Curb

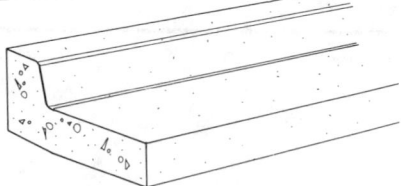

Cast-in-Place Concrete Curb and Gutter

Figure 2.64 Installation Time in Man-Hours for Pavement Recycling

Description	Man-Hours	Unit
Asphalt Pavement Demolition		
Hydraulic Hammer	.035	S.Y.
Ripping Pavement, Load and Sweep	.007	S.Y.
Crush and Screen, Traveling Hammermill		
3" Deep	.004	S.Y.
6" Deep	.007	S.Y.
12" Deep	.012	S.Y.
Pulverizing, Crushing, and Blending		
into Base 4" Pavement		
Over 15,000 S.Y.	.027	S.Y.
5,000 to 15,000 S.Y.	.029	S.Y.
8" Pavement		
Over 15,000 S.Y.	.029	S.Y.
5,000 to 15,000 S.Y.	.032	S.Y.
Remove, Rejuvenate and spread, Mixer-Paver		
Profiling, Load and Sweep		
1" Deep	.002	S.Y.
3" Deep	.006	S.Y.
6" Deep	.011	S.Y.
12" Deep	.019	S.Y.

Figure 2.65 Brick Paver Units for Mortared Paving

Brick Paver Unit Dimensions					Paver Units per S.F.	C.F. of Mortar Joints per 1000 Units	
W	x	L	x	H		3/8" Joint	1/2" Joint
3-5/8		8		2-1/4*	4.3	5.68	—
3-5/8		7-5/8		2-1/4	4.5	5.49	—
3-3/4		8		2-1/4	4.0	—	7.65
3-5/8		7-5/8		1-1/4	4.5	3.05	—
3-3/4		8		1-1/8	4.0	—	3.82

The table does not include provisions for waste. Allow at least 5% for brick and 10% to 25% for mortar waste.

*Running Bond Pattern Only

(courtesy Brick Institute of America)

Figure 2.66 Mortar Material Quantities for 3/8" and 1/2" Setting Beds[a,b]

Mortar Type and Material	C.F. Mortar per 100 S.F.		Material Weight Lb./Ft.³	Material Quantity by Weight/100 S.F. (Lb.)[c]	
	3/8" Joint	1/2" Joint		3/8" Joint	1/2" Joint
Type N (1:1:6)	3.13	4.17			
Portland Cement			15.67	49.05	65.34
Hydrated Lime			6.67	20.88	27.81
Sand			80.00	250.40	333.60
Type S (1:1/2:4-1/2)	3.13	4.17			
Portland Cement			20.89	65.38	87.11
Hydrated Lime			4.44	13.90	18.51
Sand			80.00	250.00	333.60
Type M (1:1/4:3)	3.13	4.17			
Portland Cement			31.33	98.06	130.64
Hydrated Lime			3.33	10.42	1˙˙˙
Sand			80.00	250.40	333.60

[a]The quantities are only for setting bed.
[b]The table does not include provisions for waste. Allow 10% to 25% for waste.
[c]The quantities represent material weights required for 100 S.F., depending on the mortar setting bed thickness. The provisions do not include waste.

(courtesy Brick Institute of America)

Figure 2.67   Brick Quantities for Use in Paving

Estimating Mortarless Paving Units			
Paver Face Dimensions (Actual Inches) W x L		Paver Face Area Sq. In.	Paver Units per S.F.
4	8	32.0	4.5
3-3/4	8	30.0	4.8
3-5/8	7-5/8	27.6	5.2
3-7/8	8-1/4	32.0	4.5
3-7/8	7-3/4	30.0	4.8
3-3/4	7-1/2	28.2	5.1
3-3/4	7-3/4	29.1	5.0
3-5/8	11-5/8	42.1	3.4
3-5/8	8	29.0	5.0
3-5/8	11-3/4	42.6	3.4
3-9.16	8	28.5	5.1
3-1/2	7-3/4	27.1	5.3
3-1/2	7-1/2	26.3	5.5
3-3/8	7-1/2	25.3	5.7
4	4	16.0	9.0
6	6	36.0	4.0
7-5/8	7-5/8	58.1	2.5
7-3/4	7-3/4	60.1	2.4
8	8	64.0	2.3
8	16	128.0	1.1
12	12	144.0	1.0
16	16	256.0	0.6
6	6 Hexagon	31.2	4.6
8	8 Hexagon	55.4	2.6
12	12 Hexagon	124.7	1.2

Note: The above table does not include waste. Allow at least 5% for waste and breakage.

(courtesy Brick Institute of America)

Figure 2.68   Installation Time in Man-Hours for Brick, Stone, and Concrete Paving

Description	Man-Hours	Unit
Brick Paving without Joints (4.5 Brick/S.F.)	.145	S.F.
Grouted, 3/8" Joints (3.9 Brick/S.F.)	.178	S.F.
Sidewalks		
Brick on 4" Sand Bed		
Laid on Edge (7.2/S.F.)	.229	S.F.
Flagging		
Bluestone, Irregular, 1" Thick	.198	S.F.
Snapped Random Rectangular 1" Thick	.174	S.F.
1-1/2" Thick	.188	S.F.
2" Thick	.193	S.F.
Slate		
Natural Cleft, Irregular, 3/4" Thick	.174	S.F.
Random Rectangular, Gauged, 1/2" Thick	.152	S.F.
Random Rectangular, Butt Joint, Gauged,		
1/4" Thick	.107	S.F.
Granite Blocks, 3-1/2" x 3-1/2" x 3-1/2"	.174	S.F.
4" to 12" Long, 3" to 5" Wide, 3" to 5" Thick	.163	S.F.
6" to 15" Long, 3" to 6" Wide, 3" to 5" Thick	.152	S.F.

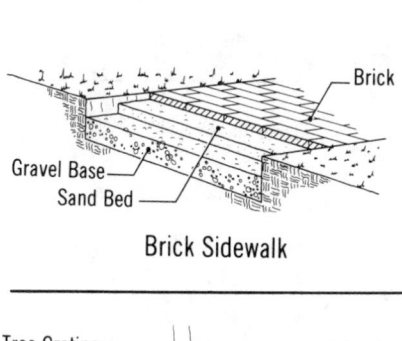

Brick Sidewalk

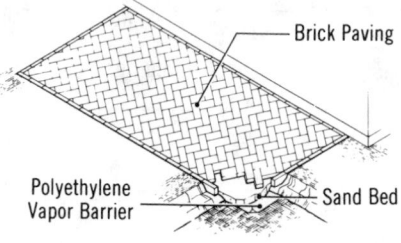

Brick Paving on Sand Bed

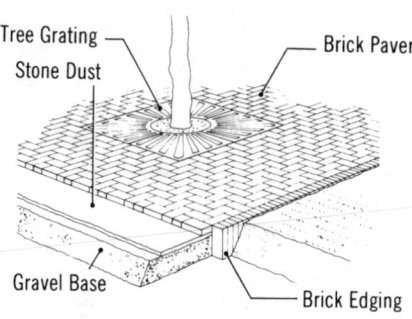

Plaza Brick Paving System

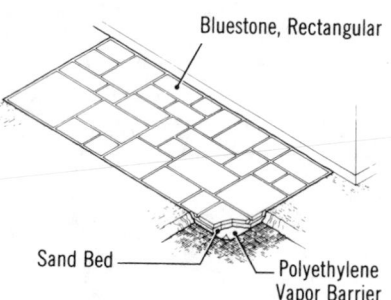

Stone Paving on Sand Bed

(continued on next page)

Figure 2.68 Installation Time in Man-Hours for Brick, Stone, and Concrete Paving (continued)

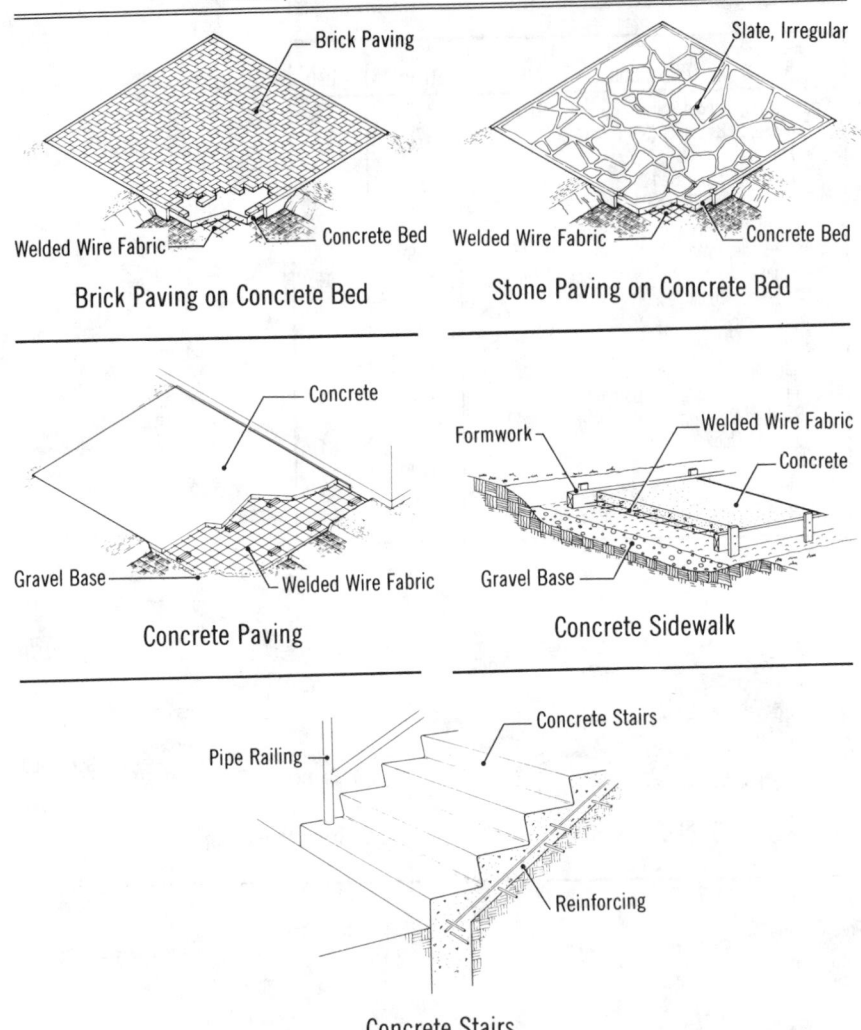

Brick Paving on Concrete Bed

Stone Paving on Concrete Bed

Concrete Paving

Concrete Sidewalk

Concrete Stairs

Figure 2.69 Installation Time in Man-Hours for Open Site Drainage Systems

Description	Man-Hours	Unit
Paving, Asphalt, Ditches	.185	S.Y.
Concrete, Ditches	.360	S.Y.
Filter Stone Rubble	.258	C.Y.
Paving, Asphalt, Aprons	.320	S.Y.
Concrete, Aprons	.620	S.Y.
Drop Structure	8.000	Ea.
Culverts, Reinforced Concrete, 12" Diameter	.162	L.F.
24" Diameter	.183	L.F.
48" Diameter	.280	L.F.
72" Diameter	.431	L.F.
96" Diameter	.560	L.F.
Flared Ends, 12" Diameter	1.080	Ea.
24" Diameter	1.750	Ea.
Corrugated Metal, 12" Diameter	.114	L.F.
24" Diameter	.175	L.F.
48" Diameter	.560	L.F.
72" Diameter	1.240	L.F.
Reinforced Plastic, 12" Diameter	.280	L.F.
Precast Box Culvert, 6' x 3'	.343	L.F.
8' x 8'	.480	L.F.
12' x 8'	.716	L.F.
Aluminum Arch Culvert, 17" x 11"	.150	L.F.
35" x 24"	.300	L.F.
57" x 38"	.800	L.F.
Multi-Plate Arch Steel	.014	lb.
Headwall, Concrete, 30" Diameter Pipe,		
3' Wing Walls	28.750	Ea.
4'-3" Wing Walls	33.500	Ea.
60" diameter Pipe, 5'-6" Wing Walls	63.250	Ea.
8'-0" Wing Walls	76.650	Ea.
Stone, 30" Diameter Pipe, 3' Wing Walls	12.650	Ea.
4'-3" Wing Walls	14.800	Ea.
60" Diameter Pipe, 5'-6" Wing Walls	30.200	Ea.
8'-0" Wing Walls	37.200	Ea.

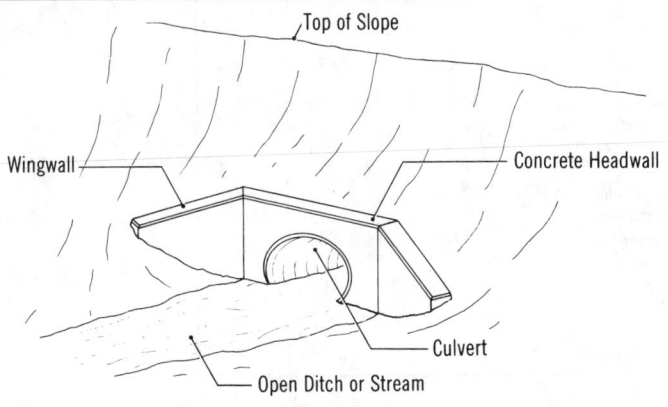

Top of Slope

Wingwall

Concrete Headwall

Culvert

Open Ditch or Stream

Figure 2.70   Installation Time in Man-Hours for Sewage and Drainage Collection Systems

Description	Man-Hours	Unit
Catch Basins or Manholes, Not Including Excavation and Backfill, Frame and Cover		
Brick, 4' I.D., 6' Deep	23.881	Ea.
8' Deep	32.000	Ea.
10' Deep	44.444	Ea.
Concrete Block, 4' I.D., 6' Deep	16.000	Ea.
8' Deep	21.333	Ea.
10' Deep	29.630	Ea.
Precast Concrete, 4' I.D., 6' Deep	3.000	Ea.
8' Deep	4.000	Ea.
10' Deep	6.000	Ea.
Cast-in-place Concrete, 4' I.D., 6' Deep	10.667	Ea.
8' Deep	16.000	Ea.
10' Deep	32.000	Ea.
Frames and Covers, 18" Square, 160 lbs.	2.400	Ea.
270 lbs.	2.791	Ea.
24" Square, 220 lbs.	2.667	Ea.
400 lbs.	3.077	Ea.
26" D Shape, 600 lbs.	3.429	Ea.
Roll Type Curb, 24" Square, 400 lbs.	3.077	Ea.
Light Traffic, 18" Diameter, 100 lbs.	2.526	Ea.
24" Diameter, 300 lbs.	2.759	Ea.
36" Diameter, 900 lbs.	4.138	Ea.
Heavy Traffic, 24" Diameter, 400 lbs.	3.077	Ea.
36" Diameter, 1150 lbs.	8.000	Ea.
Raise Frame and Cover 2", for Resurfacing,		
20" to 26" Frame	3.640	Ea.
30" to 36" Frame	4.440	Ea.
Drainage and Sewage Piping, Not Including Excavation and Backfill		
Concrete, up to 8" Diameter	.140	L.F.
10" to 18" Diameter	.168	L.F.
21" to 24" Diameter	.183	L.F.
30" Diameter	.212	L.F.
36" Diameter	.250	L.F.
Currugated Metal, 8" Diameter	.073	L.F.
12" Diameter	.114	L.F.
18" Diameter	.147	L.F.
24" Diameter	.175	L.F.
30" Diameter	.233	L.F.
36" Diameter	.280	L.F.
Ductile Iron, 6" Diameter	.190	L.F.
8" Diameter	.259	L.F.
12" Diameter	.389	L.F.
16" Diameter	.609	L.F.
Polyvinyl Chloride, 4" Diameter	.064	L.F.
8" Diameter	.072	L.F.
12" Diameter	.088	L.F.
15" Diameter	.147	L.F.
Vitrified Clay, 4" Diameter	.063	L.F.
6" Diameter	.071	L.F.
8" Diameter	.093	L.F.
12" Diameter	.147	L.F.

Figure 2.70 Installation Time in Man-Hours for Sewage and Drainage Collection Systems (continued)

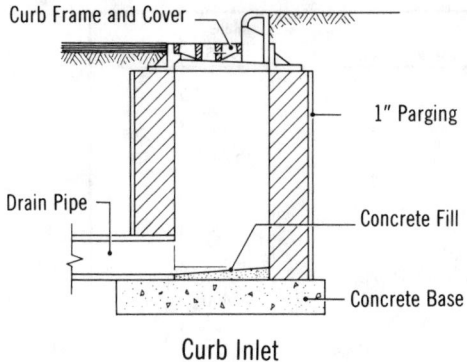

Curb Inlet

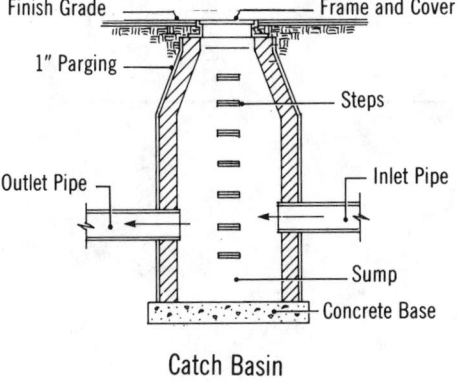

Catch Basin

Figure 2.71 Installation Time in Man-Hours for Trench Drains

Description	Man-Hours	Unit
Trench Forms		
1 Use	.200	sfca
4 Uses	.173	sfca
Reinforcing	15.240	ton
Concrete		
Direct Chute	.320	C.Y.
Pumped	.492	C.Y.
With Crane and Bucket	.533	C.Y.
Trench Cover, Including Angle Frame		
To 18" Wide	.400	L.F.
Cover Frame Only		
For 1" Grating	.178	L.F.
For 2" Grating	.229	L.F.
Geotextile Fabric in Trench		
Ideal Conditions	.007	S.Y.
Adverse Conditions	.010	S.Y.
Drainage Stone		
3/4"	.092	C.Y.
Pea Stone	.092	C.Y.

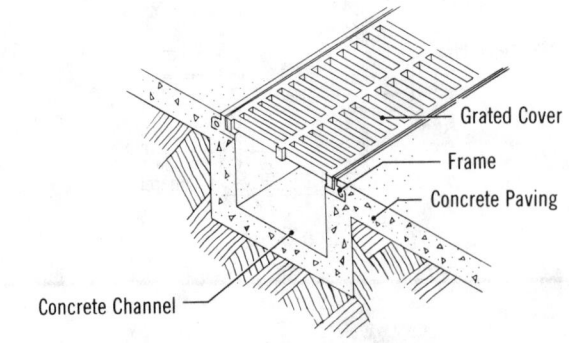

Grated Cover
Frame
Concrete Paving
Concrete Channel

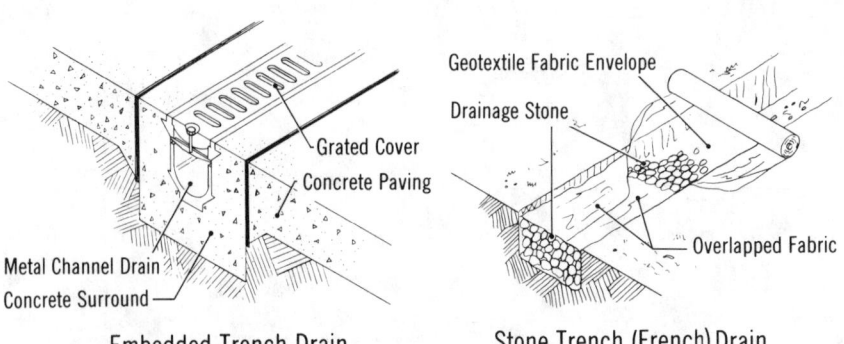

Grated Cover
Concrete Paving
Metal Channel Drain
Concrete Surround

Embedded Trench Drain

Geotextile Fabric Envelope
Drainage Stone
Overlapped Fabric

Stone Trench (French) Drain

Figure 2.72   Installation Time in Man-Hours for Septic Systems

Description	Man-Hours	Unit
Septic Tanks Not Including Excavation or		
Piping, Precast, 1000 Gallon	3.500	Ea.
2000 Gallon	5.600	Ea.
5000 Gallon	16.471	Ea.
Fiberglass 1000 Gallon	4.667	Ea.
1500 Gallon	7.000	Ea.
Excavation, 3/4 C.Y. Backhoe	.110	C.Y.
Distribution Boxes, Concrete, 5 Outlets	1.000	Ea.
12 Outlets	2.000	Ea.
Leaching Field Chambers, 13' x 3'-7" x 1'-4"		
Standard	3.500	Ea.
Heavy Duty, 8' x 4' x 1'-6"	4.000	Ea.
13' x 3'-9" x 1'-6"	4.667	Ea.
20' x 4' x 1'-6"	11.200	Ea.
Leaching Pit, Precast Concrete, 3' Pit	3.500	Ea.
6' Pit	5.957	Ea.
Disposal Field		
Excavation, 4' Trench, 3/4 C.Y. Backhoe	.048	L.F.
Crushed Stone	.160	C.Y.
Piping, Bituminous Underdrain, Perforated,		
4" Diameter	.062	L.F.
6" Diameter	.063	L.F.
Bituminous Fiber Perforated, 4" Diameter	.042	L.F.
6" Diameter	.047	L.F.
Vitrified Clay, Perforated, 4" Diameter	.060	L.F.
6" Diameter	.076	L.F.
PVC, Perforated, 4" Diameter	.056	L.F.
6" Diameter	.060	L.F.

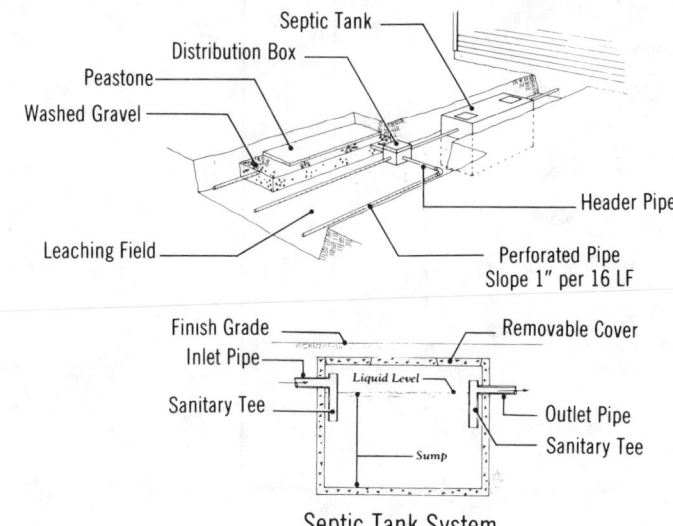

Septic Tank System

Figure 2.73 Installation Time in Man-Hours for Water Distribution Systems

Description	Man-Hours	Unit
Water Distribution Piping, Not Including Excavation and Backfill		
Mains, Ductile Iron, 4" Diameter	.167	L.F.
6" Diameter	.190	L.F.
8" Diameter	.259	L.F.
12" Diameter	.389	L.F.
16" Diameter	.609	L.F.
Polyvinyl Chloride, 4" Diameter	.120	L.F.
6" Diameter	.133	L.F.
8" Diameter	.175	L.F.
12" Diameter	.280	L.F.
Concrete, 10" Diameter	.104	L.F.
12" Diameter	.112	L.F.
16" Diameter	.156	L.F.
Fittings for Mains, Ductile Iron, Bend,		
4" Diameter	.649	Ea.
8" Diameter	1.143	Ea.
16" Diameter	2.000	Ea.
Wye, 4" Diameter	.960	Ea.
8" Diameter	1.714	Ea.
16" Diameter	3.500	Ea.
Increaser, 4" x 6"	2.000	Ea.
6" x 16"	4.670	Ea.
Flange, 4" Diameter	1.600	Ea.
8" Diameter	3.080	Ea.
12" Diameter	4.000	Ea.
Polyvinyl Chloride, Bend, 4" Diameter	.240	Ea.
8" Diameter	.300	Ea.
12" Diameter	.800	Ea.
Wye, 4" Diameter	.267	Ea.
8" Diameter	.343	Ea.
12" Diameter	1.200	Ea.
Concrete, Bend, 12" Diameter	1.170	Ea.
16" Diameter	1.560	Ea.
Wye, 12" Diameter	1.560	Ea.
16" Diameter	2.800	Ea.
Service, Copper, Type K, 3/4" Diameter	.050	L.F.
1" Diameter	.060	L.F.
2" Diameter	.080	L.F.
4" Diameter	.150	L.F.
Polyvinyl Chloride, 1-1/2" Diameter	.080	L.F.
2-1/2" Diameter	.096	L.F.
Fittings for Service, Copper, General,		
3/4" Diameter	.421	Ea.
2" Diameter	.727	Ea.
Wye, 3/4" Diameter	.667	Ea.
2" Diameter	1.140	Ea.
Curb Box, 3/4" Diameter Service	.667	Ea.
2" Diameter Service	1.000	L.F.
Valves for Mains, 4" Diameter	4.000	Ea.
8" Diameter	7.000	Ea.
12" Diameter	9.330	Ea.
Valves for Service, Curb Stop, 3/4" Diameter	.421	Ea.
1" Diameter	.500	Ea.
2" Diameter	.727	Ea.

Figure 2.73 Installation Time in Man-Hours for Water Distribution Systems (continued)

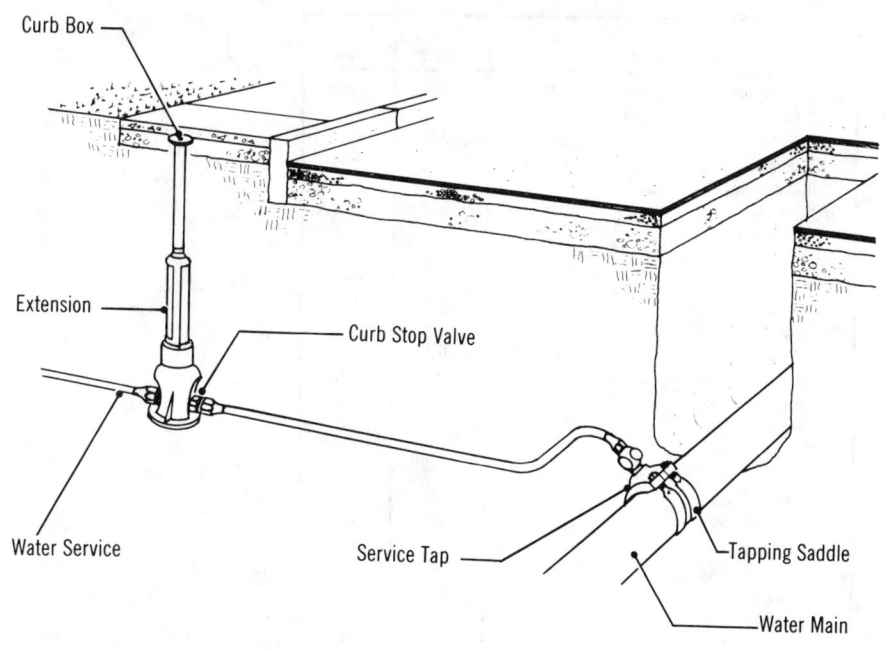

Curb Box

Extension

Curb Stop Valve

Water Service

Service Tap

Tapping Saddle

Water Main

Figure 2.74 Installation Time in Man-Hours for Site Irrigation

Description	Man-Hours	Unit
Sprinkler System, Golf Course, Fully Automatic	.600	9 holes
12' Radius Heads, 15' Spacing		
Minimum	.343	Head
Maximum	.600	Head
30' Radius Heads, Automatic		
Minimum	.857	Head
Maximum	1.040	Head
Sprinkler Heads		
Minimum	.267	Head
Maximum	.320	Head
Trenching, Chain Trencher, 12 H.P.		
4" Wide, 12" Deep	.010	L.F.
6" Wide, 24" Deep	.015	L.F.
Backfill and Compact		
4" Wide, 12" Deep	.010	L.F.
6" Wide, 12" Deep	.030	L.F.
Trenching and Backfilling, Chain Trencher, 40 H.P.		
6" Wide, 12" Deep	.007	L.F.
8" Wide, 36" Deep	.010	L.F.
Compaction		
6" Wide, 12" Deep	.003	L.F.
8" Wide, 36" Deep	.005	L.F.
Vibrating Plow		
8" Deep	.004	L.F.
12" Deep	.006	L.F.
Automatic Valves, Solenoid		
3/4" Diameter	.363	Ea.
2" Diameter	.500	Ea.
Automatic Controllers		
4 Station	1.500	Ea.
12 Station	2.000	Ea.

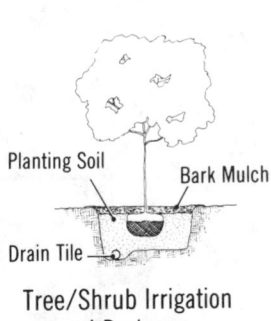

Planting Soil

Bark Mulch

Drain Tile

Tree/Shrub Irrigation and Drainage

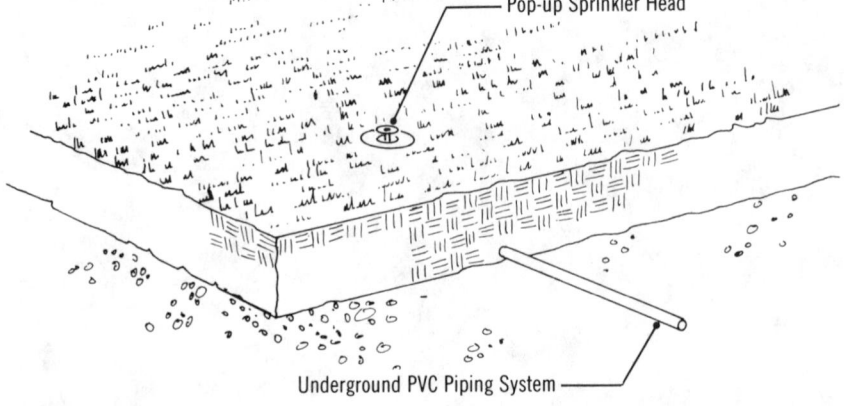

Pop-up Sprinkler Head

Underground PVC Piping System

Figure 2.75   Installation Time in Man-Hours for Fencing

Description	Man-Hours	Unit
Fence, Chain Link, Industrial Plus 3 Strands Barbed Wire, 2" Line Post @ 10' O.C.		
1-5/8" Top Rail, 6' High	.096	L.F.
Corners, Add	.600	Ea.
Braces, Add	.300	Ea.
Gate, Add	.686	Ea.
Residential, 11 Gauge Wire, 1-5/8" Line Post @ 10' O.C. 1-3/8"		
Top Rail, 3' High	.048	L.F.
4' High	.060	L.F.
Gate, Add	.400	Ea.
Tennis Courts, 11 Gauge Wire, 1-3/4" Mesh, 2-1/2" Line Posts, 1-5/8" Top Rail,		
10' High	.155	L.F.
12' High	.185	L.F.
Corner Posts, 3" Diameter, Add	.800	Ea.
Fence, Security, 12' High	.960	L.F.
16' High	1.200	L.F.
Fence, Wood, Cedar Picket, 2 Rail, 3' High	.150	L.F.
Gate, 3'-6", Add	.533	Ea.
Cedar Picket, 3 Rail, 4' High	.160	L.F.
Gate, 3'-6", Add	.585	Ea.
Open Rail Rustic, 2 Rail, 3' High	.150	L.F.
Stockade, 6' High	.150	L.F.
Board, Shadow Box, 1" x 6" Treated Pine, 6' High	.150	L.F.

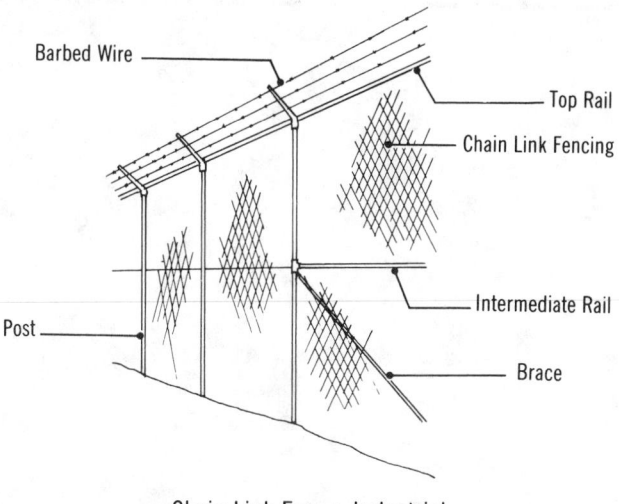

Chain Link Fence, Industrial

Figure 2.76   Tree Planting

This table can be entered either with the number of trees available, to determine the area (Acres or Hectares) required; or with the area, to determine the number of trees that can be planted.

Example:

2000 trees to be planted at 10′ x 10′.
2000 trees/436 trees/Ac = 4.59 Ac

or:

4 Acres to be planted at 10′ x 10′.
436 trees/Ac x 4 Acres = 1744 trees

Tree Spacing, Feet (Meters)	Number Trees per Acre (Hectares)		Tree Spacing, Feet (Meters)	Number Trees per Acre (Hectares)	
2 x 2 (.6 x.6)	10,890	(26,909)	8 x 10 (2.4 x 3.0)	544	(1,344)
3 x 3 (.91 x .91)	4,840	(11,959)	8 x 12 (2.4 x 3.7)	454	(1,122)
4 x 4 (1.2 x 1.2)	2,722	(6,726)	8 x 25 (2.4 x 7.6)	218	(539)
4 x 5 (1.2 x 1.5)	2,178	(5,382)	9 x 9 (2.7 x 2.7)	538	(1,329)
4 x 6 (1.2 x 1.8)	1,815	(4,485)	9 x 12 (2.7 x 3.7)	403	(996)
4 x 8 (1.2 x 2.4)	1,362	(3,365)	10 x 10 (3.0 x 3.0)	436	(1,077)
4 x 10 (1.2 x 3.0)	1,089	(2,691)	10 x 12 (3.0 x 3.7)	363	(897)
5 x 5 (1.5 x 1.5)	1,742	(4,304)	11 x 11 (3.4 x 3.4)	360	(890)
5 x 6 (1.5 x 1.8)	1,452	(3,588)	12 x 12 (3.7 x 3.7)	302	(746)
5 x 8 (1.5 x 2.4)	1,089	(2,691)	12 x 18 (3.7 x 5.5)	202	(499)
5 x 10 (1.5 x 3.0)	871	(2,152)	13 x 13 (4.0 x 4.0)	258	(638)
6 x 6 (1.8 x 1.8)	1,210	(2,990)	14 x 14 (4.3 x 4.3)	222	(549)
6 x 8 (1.8 x 2.4)	908	(2,244)	15 x 15 (4.6 x 4.6)	194	(479)
6 x 10 (1.8 x 3.0)	726	(1,794)	16 x 16 (4.9 x 4.6)	170	(420)
6 x 12 (1.8 x 3.7)	605	(1,495)	18 x 18 (5.5 x 5.5)	134	(331)
7 x 7 (2.1 x 2.1)	889	(2,197)	18 x 20 (5.5 x 6.0)	121	(299)
7 x 10 (2.1 x 3.0)	622	(1,537)	20 x 20 (6.0 x 6.0)	109	(269)
8 x 8 (2.4 x 2.4)	681	(1,683)	25 x 25 (7.6 x 7.6)	70	(173)

Figure 2.77   Tree Pits and Tree Balls — Cubic Feet per Tree for Estimating Excavation and Top Soil

	Depth						Diameters					
		1'	1-1/4'	1-1/2'	1-3/4'	2'	2-1/4'	2-1/2'	2-3/4'	3'	3-1/4'	3-1/2'
Tree Pit	1'	.94	1.47	2.13	2.88	3.77	4.78	5.89	7.13	8.48	9.96	11.5
Ball		.68	1.07	1.54	2.10	2.73	3.42	4.30	5.20	6.19	7.38	8.4
Tree Pit	1-1/4'	1.16	1.85	2.65	3.60	4.71	5.93	7.37	8.90	10.6	12.4	14.4
Ball		.85	1.35	1.93	2.63	3.42	4.29	5.36	6.49	7.7	9.2	10.5
Tree Pit	1-1/2'	1.40	2.22	3.08	4.32	5.65	7.16	8.83	10.7	12.7	15.0	17.3
Ball		1.02	1.62	2.32	3.15	4.10	5.19	6.39	7.8	9.2	10.9	12.7
Tree Pit	1-3/4'	1.65	2.58	3.70	5.04	6.60	8.20	10.3	12.5	14.8	17.4	20.2
Ball		1.20	1.88	3.72	3.68	4.78	6.03	7.5	9.1	10.7	12.7	14.7
Tree Pit	2'	1.87	2.95	4.26	5.76	7.54	9.55	11.8	14.3	17.0	19.9	23.0
Ball		1.38	2.15	3.10	4.20	5.49	6.92	8.5	10.4	12.3	14.5	16.8
Tree Pit	2-1/4'	2.10	3.32	4.78	6.48	8.48	10.7	13.2	16.0	19.1	22.4	26.0
Ball		1.55	2.43	3.48	4.73	6.19	7.5	9.6	11.7	13.8	16.4	18.9
Tree Pit	2-1/2'	2.34	3.69	5.30	7.20	9.42	11.9	14.7	17.8	21.2	24.9	28.9
Ball		1.70	2.70	3.87	5.25	6.89	8.7	10.7	13.0	15.4	18.2	21.0
Tree Pit	2-3/4'	2.57	4.06	5.83	7.92	10.3	13.1	16.2	19.6	23.3	27.4	31.8
Ball		1.87	2.98	4.25	5.77	7.6	9.6	11.8	14.3	17.0	20.0	23.1
Tree Pit	3'	2.81	4.43	6.37	8.64	11.3	14.3	17.7	21.4	25.4	29.9	34.6
Ball		2.05	3.24	4.65	6.30	8.3	10.5	12.9	15.6	18.6	21.8	25.2
Tree Pit	3-1/4'	3.00	4.80	6.90	9.36	12.2	15.5	19.2	23.2	27.5	32.4	37.5
Ball		2.21	3.50	5.03	6.83	8.9	11.4	14.0	16.9	20.1	23.6	27.3
Tree Pit	3-1/2'	3.28	5.17	7.41	10.1	13.2	16.7	20.7	25.0	29.6	34.9	40.4
Ball		2.39	3.77	5.42	7.4	9.6	12.2	15.0	18.2	21.7	25.4	29.4
Tree Pit	3-3/4'	3.50	5.53	7.96	10.8	14.2	17.9	22.2	26.7	31.7	37.1	43.3
Ball		2.56	4.04	3.80	7.9	10.3	13.1	16.1	19.5	23.2	27.2	31.5
Tree Pit	4'	3.74	5.90	8.50	11.5	15.1	19.1	23.7	28.5	33.8	39.9	46.2
Ball		2.73	4.31	6.19	8.4	11.0	14.0	17.2	20.8	24.8	29.0	33.6
Tree Pit	4-1/4'	3.97	6.28	9.01	12.3	16.0	20.3	25.1	30.3	36.0	42.3	49.1
Ball		2.90	4.58	6.58	8.9	11.8	14.8	18.3	22.1	26.3	30.8	35.7
Tree Pit	4-1/2'	4.20	6.64	9.55	13.0	17.0	21.5	26.6	32.0	38.1	44.8	52.0
Ball		3.07	4.85	6.97	9.5	12.4	15.7	19.4	23.4	27.8	32.6	37.8
Tree Pit	4-3/4'	4.43	7.00	10.0	13.7	17.9	22.7	28.0	33.8	40.2	47.3	54.9
Ball		3.24	5.12	7.4	10.0	13.0	16.5	20.4	24.7	29.1	34.2	40.0
Tree Pit	5'	4.68	7.38	10.6	14.4	18.8	23.9	29.8	35.6	42.3	49.9	57.7
Ball		3.41	5.38	7.8	10.5	13.7	17.4	21.5	26.0	30.9	36.1	42.1
Tree Pit	5-1/2'	5.14	8.12	11.7	15.8	20.7	26.3	32.4	39.2	46.7	54.8	63.5
Ball		3.76	5.93	8.5	11.6	15.1	19.1	23.7	28.6	34.0	40.0	46.3
Tree Pit	6'	5.41	8.86	12.7	17.3	22.6	28.7	35.4	42.8	50.9	57.8	69.3
Ball		4.10	6.46	9.3	12.6	16.5	20.9	25.9	31.2	37.1	43.6	50.5

This table lists the cubic feet of excavation necessary for holes for planting trees. The diameters of holes necessary are listed across the top. The depths desired are listed down the second column. For example, if a planting guide calls for a hole 2' in diameter and 2' deep, the pit will have 7.54 C.F. of earth to be excavated and, therefore, you will need 2.05 C.F. of topsoil for the planting.

Figure 2.78 Tree and Shrub Planting Data Including Man-Hour Requirements

Ball Size Diam. x Depth	Soil in Ball	Weight of Ball	Hole Diam. Req'd.	Hole Exca-vation	Amount of Soil Displ.	Topsoil Handled	Time Required in Man-Hours					
							Dig & Lace	Handle Ball	Dig Hole	Plant & Prune	Water & Guy	Total M.H.
Inches	C.F.	Lbs.	Feet	C.F.	C.F.	C.F.						
12 x 12	.7	56	2	4	3	11	.25	.17	.33	.25	.07	1.1
18 x 16	2	160	2.5	8	6	21	.50	.33	.47	.35	.08	1.7
24 x 18	4	320	3	13	9	38	1.00	.67	1.08	.82	.20	3.8
30 x 21	7.5	600	4	27	19.5	76	.82	.71	.79	1.22	.26	3.8
36 x 24	12.5	980	4.5	38	25.5	114	1.08	.95	1.11	1.32	.30	4.76
42 x 27	19	1520	5.5	64	45	185	1.90	1.27	1.87	1.43	.34	6.8
48 x 30	28	2040	6	85	57	254	2.41	1.60	2.06	1.55	.39	8.0
54 x 33	38.5	3060	7	127	88.5	370	2.86	1.90	2.39	1.76	.45	9.4
60 x 36	52	4160	7.5	159	107	474	3.26	2.17	2.73	2.00	.51	10.7
66 x 39	68	5440	8	196	128	596	3.61	2.41	3.07	2.26	.58	11.9
72 x 42	87	7160	9	267	180	785	3.90	2.60	3.71	2.78	.70	13.7

Figure 2.79 Plant Spacing Chart

This chart may be used when plants are to be placed equidistant from each other, staggering their position in each row.

Plant Spacing (Inches)	Row Spacing (Inches)	Plants per C.S.F.	Plants Spacing (Feet)	Row Spacing (Feet)	Plants per M.S.F.
			4	3.46	72
6	5.20	462	5	4.33	46
8	6.93	260	6	5.20	32
10	8.66	166	8	6.93	18
12	10.39	115	10	8.66	12
15	12.99	74	12	10.39	8
18	15.59	51.32	15	12.99	5.13
20	18.19	37.70	20	17.32	2.89
24	20.78	28.87	25	21.65	1.85
30	25.98	18.48	30	25.98	1.28
36	31.18	12.83	40	34.64	0.72

Figure 2.80 Conversion Chart for Loam, Baled Peat, and Mulch

Loam Requirements		Mulch Requirements (Bulk Peat, Crushed Stone, Shredded Bark, Wood Chips)	
Depth	Cubic yards per 10,000 S.F.	Depth	Cubic yards per 1,000 S.F.
2"	70	1"	3-1/2
4"	140	2"	7
6"	210	3"	10-1/2
		4"	14
	Cubic yards per 1,000 S.F.		Cubic yards per 100 S.F.
2"	7	1"	1/3
4"	14	2"	2/3
6"	21	3"	1
		4"	1-1/3
Bales (or Compressed) Peat			
Depth		Bales Required per 100 S.F.	
1"		one 4-cubic foot bale	
2"		two 4-cubic foot bales	
3"		two 6-cubic foot bales	
4"		one 4-cubic foot bales and two 6-cubic foot bales	

Note: Opened and spread, peat from bales will be about 2-1/2 times the volume of the unopened bale.

Figure 2.81 Calculations for Lawn and Soil Material Application — per Acre

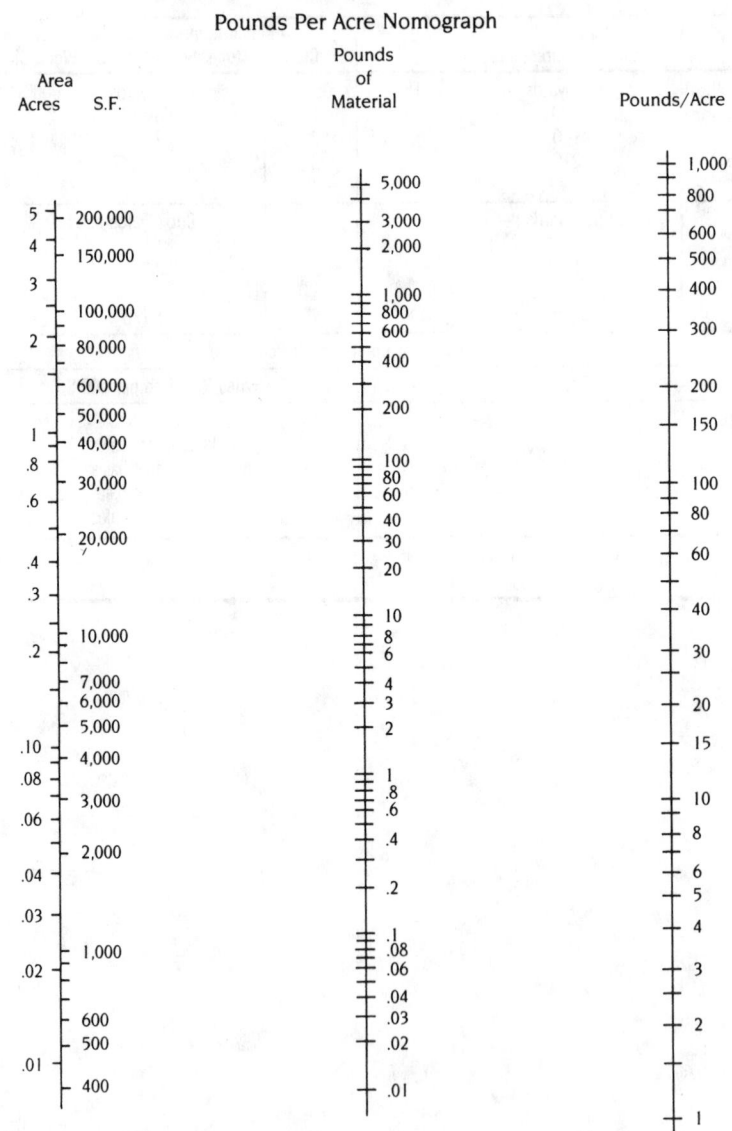

Pounds Per Acre Nomograph

Use by placing a straight edge to known or given points on any 2 scales and reading the unknown on the remaining scale. Example: 80 pounds of material applied at 200 lbs./acre will treat .4 acre.

Prepared By: Harold Davidson, Department of Horticulture, Michigan State University

(courtesy of the National Landscape Association, "Landscape Designer and Estimator's Guide")

Figure 2.82 Calculations for Lawn and Soil Material Application — per 1000 Square Feet

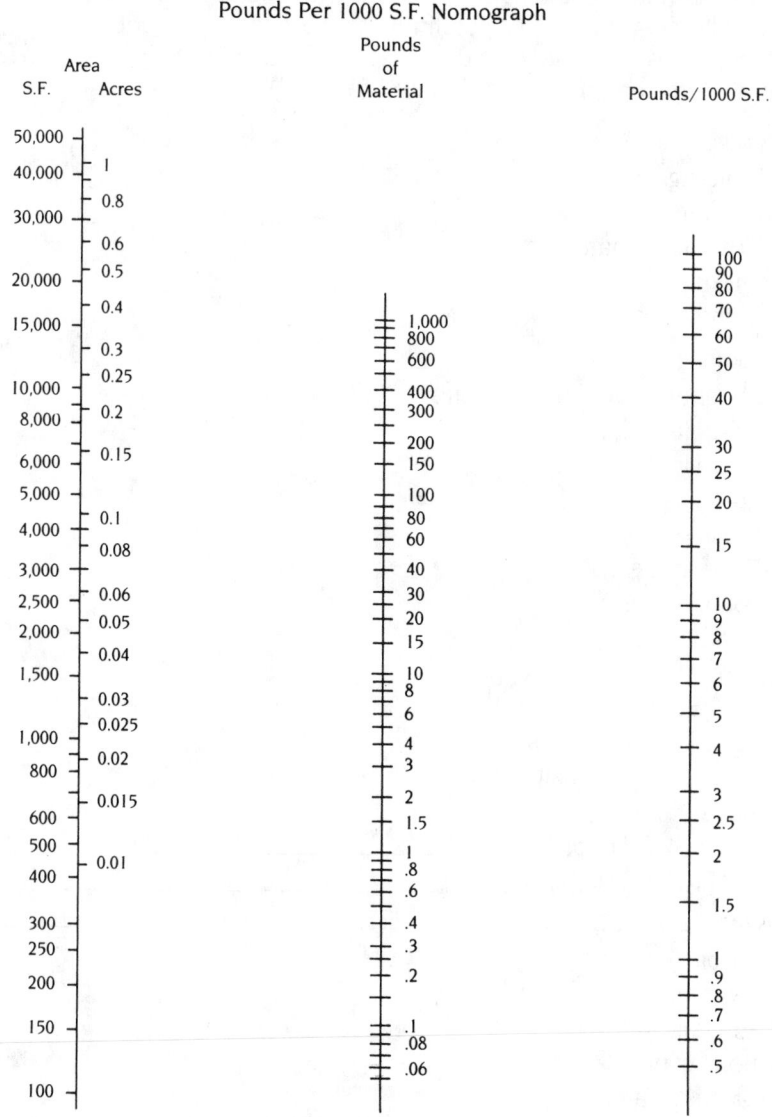

Pounds Per 1000 S.F. Nomograph

Use by placing a straight edge to known or given points on any 2 scales and reading the unknown on the remaining scale.

Example: How many pounds of material will be needed to treat an 8000 S.F. lawn at 5 lbs./1000 S.F.?

Answer: 40 lbs.

Prepared by: Harold Davidson, Department of Horticulture, Michigan State University

(courtesy of the National Landscape Association, "Landscape Designer and Estimator's Guide")

Checklist

For an estimate to be reliable, all items must be accounted for.
A complete estimate can also eliminate the need to include
contingencies. The following checklist can be used to help ensure
that all items are properly accounted for.

Demolition
- ☐ Building
- ☐ Fencing
- ☐ Hazardous materials
- ☐ Interior
- ☐ Remove from site

Site Clearing
- ☐ Materials abandoned on site
  - ____ Hazardous
  - ____ Non-hazardous
- ☐ Boulders
- ☐ Brush
- ☐ Clear and grub
- ☐ Foundations
- ☐ Shrubs
- ☐ Topsoil
  - ____ Stockpile
  - ____ Remove from site
  - ____ Dispose on site
- ☐ Trees

Disconnect Existing Utilities _____

Underpin Existing Buildings/Facilities _____

Blasting _____

Excavation
- ☐ Mass
- ☐ Trench
- ☐ Borrow pit
- ☐ Ground water expected
  - ____ None
  - ____ Pump
  - ____ Wellpoint system
  - ____ Trench away from excavation
- ☐ Sheeting/Shoring

☐ Disposal
  ____ Off site
  ____ On site
  ____ Use for fill
  ____ Other _____

Backfill
☐ Mass
☐ Trench
☐ Berms
☐ Topsoil
☐ Gravel
☐ Stone
☐ Sand
☐ Source

  ____ Borrow pit
  ____ Commercial pit
  ____ On site
  ____ Other _____

Compaction
☐ Hand
☐ Mechanical

Water Control
☐ Dams
☐ Ditching
☐ Pumping
☐ Wellpoint system
☐ Sheet piling

Pestilence Control _____

Piles/Piling _____

Caissons _____

Pressure-Injected Footings _____

Special Footings _____

Drainage
☐ Storm drains
☐ Catch basins
☐ Manholes
☐ Piping systems
☐ Footing drains
☐ French drains
☐ Trenches
☐ Swales

☐   Rip rap
☐   Other _____

Utilities
☐ Gas
☐ Water
☐ Electricity
         _____ permanent
         _____ temporary
☐ Telephone
☐ Other communications _____
☐ Sewer
☐ Steam
☐ Other _____
Driveways _____
Sidewalks _____
Curbs _____

Paving
☐ Roads
☐ Walks
☐ Patios
☐ Other _____
Flagpoles _____
Parking Bumpers _____
Pavement Markings _____
Signage _____
Building Signs _____
Site Lighting _____

Traffic Control
☐ Signals
☐ Lights
☐ Booths
Fences _____
Fountains _____
Planters _____
Playground Equipment _____
Playing Fields
☐ Benches
☐ Sports specialties
☐ Bike racks

Railroad Work _____

Retaining Walls _____

Trash Enclosures _____

Irrigation System _____

Landscaping
- ☐ Lawn
- ☐ Seeding
- ☐ Sodding
- ☐ Trees
- ☐ Bushes
- ☐ Shrubs
- ☐ Plantings
- ☐ Ground cover
- ☐ Mulch
- ☐ Edgings
- ☐ Maintenance agreements
- ☐ Other _____

Special considerations _____

_____

_____

Tips

Pit Excavation

When figuring the slope areas of pit excavation, remember that the slope areas usually form a right triangle when viewed as a section. Since the area of two right triangles equals the area of one rectangle with equal base and height dimensions, the volume of the slope area will equal the depth of the pit (the height) times the cut back distance of the slope (the base) times ½ the perimeter of the pit.

Cut and Fill

If a project has a large quantity of cut or fill, consider inquiring whether the site grade can be raised or lowered to economize the earthwork portion of the estimate.

If the project requires a large quantity of fill, consider creating a borrow pit on the site. This can lower the estimate significantly, as the cost for suitable fill material can be quite high in some areas of the country. The borrow pit can also be helpful as a receptacle for on-site spoil or other excavated materials, in this case saving dump charges.

It is very important to keep in mind while estimating earthwork the location and availability of dump sites for spoil. Unexpectedly large costs can accrue if the only site open to receive spoil is quite a distance away.

Clearing and Grubbing

Often one of the most expensive parts of clearing and grubbing is the disposal of tree stumps. It is a good idea to know how and where this task can be carried out before completing the estimate.

Site Access

One of the most common mistakes made in site work estimating involves site access. Estimators should visit the site not only to review site conditions, but to trace the route that the general or subcontractors' equipment will take to get to the site. If the equipment that is to be used on this project cannot get to the site, the estimate may be completely inaccurate.

Pavings

When estimating paving, keep in mind the overall project schedule. Although it is common to wait until the end of the project before paving, consider what time of year the project is scheduled for completion. In colder climates, many concrete plants close for the season in late November and are thus very busy just before then. The supply or availability may be limited, and the prices may be higher.

Quick Quantity for Asphaltic Concrete Paving

A quick rule of thumb for Asphaltic Concrete is that for each inch of pavement thickness, 1 square yard = 110 lbs.

Sewerage and Drainage

One should not automatically assume that the sewerage and drainage lines will go in at the early stages of the project. It may appear to make sense from a scheduling standpoint, but the inconvenience of having the site divided in half, with open trenches, loose pipe, and restricted access may ultimately cost more in lost time.

Subsurface Investigations

Many companies, eager to get started on their projects, shortchange the site investigation process. For the relatively short time and small amount of money involved, it is not a good idea to skimp on this important item. The untimely discovery of even one subsurface "abnormality" can be a painful lesson. An example is finding that the site was unknowingly used "way back when" as a spoils site for the rest of the industrial park and that there are ten feet of bad soils to excavate from under the stiff clay cap you assumed you would build on. Investigate the site thoroughly!

Miscellaneous Hauling Time and Distance Rule

30 MPH + 44 Feet per Second

Notes

Division Three
Concrete

Introduction

Concrete is one of the most adaptable materials used in the construction industry. It can be custom-mixed to the desired compressive strength or for color applications; placed in almost any type of shape or form using a variety of methods; and it can be finished to resemble masonry, stone formations, and/or a variety of designs. Concrete is durable and virtually maintenance-free. It is also fairly easy to work with and installs relatively quickly. For these and many other reasons, concrete is one of the most widely used materials in construction projects.

Concrete estimating is generally divided into five basic areas:
- Concrete materials
- Formwork
- Reinforcing
- Finishing
- Precast concrete

Concrete quantities are generally taken off in measures of cubic yards, as this is the way the material supply companies charge for it. Formwork is taken off in square feet (generally in square feet of area in contact with the concrete as opposed to the actual area of the forms used). Reinforcing is normally taken off in tons of steel. Finishing is taken off in square feet of concrete surface that will be finished. Precast concrete quantities are taken off in either square feet or by the piece, depending on its use and the type of unit being considered.

For estimating concrete quantities, this division includes charts and tables in the following areas:
- General information
- General formulas
- Proportionate quantities for concrete forms and reinforcing
- Concrete handling
- Material quantities for concrete
- Metric equivalents
- Concrete volume for various applications
- Formwork
- Reinforcing
- Installation times for concrete in man-hours

Estimating Data

The following tables show the expected installation time for various specialty items. Each table lists the specialty item, the typical installation crew mix, the expected average daily output of that crew (the number of units the crew can be expected to install in one eight-hour day), and the man-hours required for the installation of one unit.

Table of Contents

Figure 3.1 Cast-in-Place Formed Concrete

Although the concrete and reinforcing material costs for a grade beam on the ground and exterior spandrel beam on the twentieth floor may be the same, forming, placing, stripping, and rubbing costs are not. This is the reason for performing a detailed breakdown of costs for an estimate.

The cost of formed concrete-in-place is always a function of the area of forms or square feet of contact area required. Per cubic yard of concrete, an 8" foundation wall will require 50% more form material, form labor, stripping, rubbing, and waterproofing than a 12" wall. Still, many contractors continue to estimate formed concrete-in-place by the cubic yard without considering the wall thickness.

Quantities can be compiled and listed in various formats, but EACH of the quantities MUST be computed. It is important that once a format for compiling quantities is established, the same format should be used as a standard operating estimating procedure. This standardization minimizes errors and omissions and presents the various quantities in an orderly manner for pricing, cost control, and future estimating.

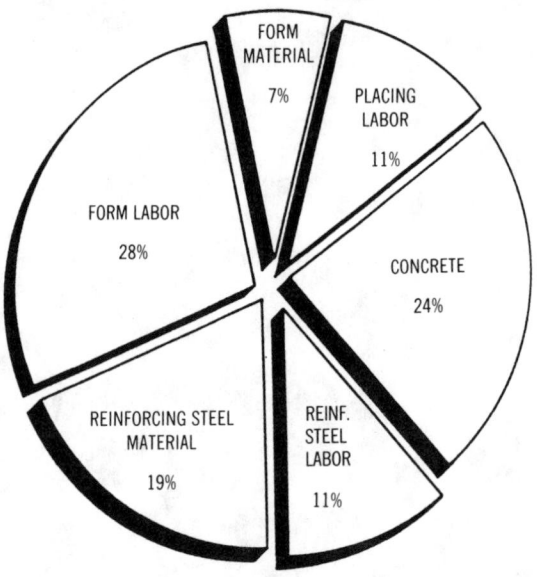

Figure 3.2 Formwork: Ratio of Board Feet to Square Feet of Net Contact Area

Job-built forms for different concrete components have various lumber requirements. This table provides the board feet of lumber required for each square foot of form contact area. In this table the surface of the form is figured at 1 B.F. per S.F., and the studs and bracing are figured in the usual method of calculating board feet.

Formwork	Ratio
Wall or spread footings	2.0 B.F./S.F.
Column and pier footings	2.5 B.F./S.F.
Columns and piers	3.0 B.F./S.F.
Walls – to 12′ 0″ high	2.5 B.F./S.F.
Walls – over 12′ 0″ high	3.5 B.F./S.F.
Wall pilasters	4.0 B.F./S.F.
Beams (without shores)	4.0 B.F./S.F.
Supported slabs (without shores)	2.0 B.F./S.F.
Metal pan centering (without shores)	1.5 B F./S.F.
Wood shores	1.8 B.F./S.F.
Stairs, without soffits	2.5 B.F./S.F.
Stairs, with soffits	6.0 B.F./S.F.

Figure 3.3   Formwork Labor Hours

Man-hours for formwork vary greatly from one component to another. This table is convenient for planning the number of forms to be fabricated or provided, depending on the number of reuses anticipated.

Item	Unit	Fabricate	Erect & Strip	Clean & Move	Total Hours 1 Use	2 Use	3 Use	4 Use
Beam and Girder, interior beams, 12" wide	100 S.F.	6.4	8.3	1.3	16.0	13.3	12.4	12.0
Hung from steel beams		5.8	7.7	1.3	14.8	12.4	11.6	11.2
Beam sides only, 36" high		5.8	7.2	1.3	14.3	11.9	11.1	10.7
Beam bottoms only, 24" wide		6.6	13.0	1.3	20.9	18.1	17.2	16.7
Box out for openings		9.9	10.0	1.1	21.0	16.6	15.1	14.3
Buttress forms, to 8' high		6.0	6.5	1.2	13.7	11.2	10.4	10.0
Centering, steel, 3/4" rib lath			1.0		1.0			
3/8" rib lath or slab form	▼		0.9		0.9			
Chamfer strip or keyway	100 L.F.		1.5		1.5	1.5	1.5	1.5
Columns, fiber tube 8" diameter			20.6		20.6			
12"			21.3		21.3			
16"			22.9		22.9			
20"			23.7		23.7			
24"			24.6		24.6			
30"	▼		25.6		25.6			
Round Steel, 12" diameter			22.0		22.0	22.0	22.0	22.0
16"			25.6		25.6	25.6	25.6	25.6
20"			30.5		30.5	30.5	30.5	30.5
24"	▼		37.7		37.7	37.7	37.7	37.7
Plywood 8" x 8"	100 S.F.	7.0	11.0	1.2	19.2	16.2	15.2	14.7
12" x 12"		6.0	10.5	1.2	17.7	15.2	14.4	14.0
16" x 16"		5.9	10.0	1.2	17.1	14.7	13.8	13.4
24" x 24"		5.8	9.8	1.2	16.8	14.4	13.6	13.2
Steel framed plywood 8" x 8"			10.0	1.0	11.0	11.0	11.0	11.0
12" x 12"			9.3	1.0	10.3	10.3	10.3	10.3
16" x 16"			8.5	1.0	9.5	9.5	9.5	9.5
24" x 24"			7.8	1.0	8.8	8.8	8.8	8.8
Drop head forms, plywood		9.0	12.5	1.5	23.0	19.0	17.7	17.0
Coping forms		8.5	15.0	1.5	25.0	21.3	20.0	19.4
Culvert, box			14.5	4.3	18.8	18.8	18.8	18.8
Curb forms, 6" to 12" high, on grade		5.0	8.5	1.2	14.7	12.7	12.1	11.7
On elevated slabs	▼	6.0	10.8	1.2	18.0	15.5	14.7	14.3
Edge forms to 6" high, on grade	100 L.F.	2.0	3.5	0.6	6.1	5.6	5.4	5.3
7" to 12" high	100 S.F.	2.5	5.0	1.0	8.5	7.8	7.5	7.4
Equipment foundations	"	10.0	18.0	2.0	30.0	25.5	24.0	23.3
Flat slabs, including drops		3.5	6.0	1.2	10.7	9.5	9.0	8.8
Hung from steel		3.0	5.5	1.2	9.7	8.7	8.4	8.2
Closed deck for domes		3.0	5.8	1.2	10.0	9.0	8.7	8.5
Open deck for pans		2.2	5.3	1.0	8.5	7.9	7.7	7.6
Footings, continuous, 12" high		3.5	3.5	1.5	8.5	7.3	6.8	6.6
Spread, 12" high		4.7	4.2	1.6	10.5	8.7	8.0	7.7
Pile caps, square or rectangular		4.5	5.0	1.5	11.0	9.3	8.7	8.4
Grade beams, 24" deep		2.5	5.3	1.2	9.0	8.3	8.0	7.9
Lintel or Sill forms		8.0	17.0	2.0	27.0	23.5	22.3	21.8
Spandrel beams, 12" wide		9.0	11.2	1.3	21.5	17.5	16.2	15.5
Stairs			25.0	4.0	29.0	29.0	29.0	29.0
Trench forms in floor		4.5	14.0	1.5	20.0	18.3	17.7	17.4
Walls, Plywood, at grade, to 8' high		5.0	6.5	1.5	13.0	11.0	9.7	9.5
8' to 16'		7.5	8.0	1.5	17.0	13.8	12.7	12.1
16' to 20'		9.0	10.0	1.5	20.5	16.5	15.2	14.5
Foundation walls, to 8' high		4.5	6.5	1.0	12.0	10.3	9.7	9.4
8' to 16' high		5.5	7.5	1.0	14.0	11.8	11.0	10.6
Retaining wall to 12' high, battered		6.0	8.5	1.5	16.0	13.5	12.7	12.3
Radial walls to 12' high, smooth		8.0	9.5	2.0	19.5	16.0	14.8	14.3
But in 2' chords		7.0	8.0	1.5	16.5	13.5	12.5	12.0
Prefabricated modular, to 8' high		—	4.3	1.0	5.3	5.3	5.3	5.3
Steel, to 8' high		—	6.8	1.2	8.0	8.0	8.0	8.0
8' to 16' high		—	9.1	1.5	10.6	10.3	10.2	10.2
Steel framed plywood to 8' high		—	6.8	1.2	8.0	7.5	7.3	7.2
8' to 16' high	▼	—	9.3	1.2	10.5	9.5	9.2	9.0

Figure 3.4   Typical Metal Pan Erection Procedures

Metal or fiberglass pans are used for joist slab construction. They can be erected either on a solid platform or on centering as described below. Each manufacturer provides information on erection and concrete volumes for its products. The following information (from CECO Corporation) is the type of information provided by pan manufacturers.

1. All beam, girder and column forms are placed and braced by the general contractor.

2. The shores and stringers are erected.

3. The soffit forms are placed on the stringers and the shores are adjusted to support the soffit forms at the elevation set by the beam and girder forms.

4. The wood headers for bridging joists, beam tees and special headers for electrical outlets, etc., are then placed.

5. Lines for joist edges are marked with a chalkline.

6. The end forms are placed first and are nailed to the soffit with barbed nails through the nailholes in the flanges. Note that the soffit form does not require special fillers to accommodate the tapered end form.

7. The 3' intermediates are set and nailed. Overlaps are 1" minimum and 5" maximum.

8. Various joist lengths are accommodated by closing the center gap with 1', 2' or 3' long intermediates. For the special 10" and 15" filler widths, which are furnished in 3' lengths only, joist length variations are accommodated by overlapping.

9. Pans and soffit boards are oiled to facilitate removal.

10. All reinforcing steel and mesh, all hanger wires, plumbing sleeves, anchors, electrical outlet boxes, etc., are placed by the general contractor or his subcontractors.

11. Concrete is placed by the general contractor.

12. Shores and centering are removed when the concrete has gained its proper strength.

13. Steel forms are removed and reconditioned for the next use.

Figure 3.5   Symbols and Abbreviations Used on Reinforcing Steel Drawings

#	Indicates size of deformed bar number
ø	Round, used mainly for plain round bars
(a	Spacing, center to center
⟷	Direction in which bars extend
⟷	Limits of area covered by bars

Pl	Plain bar	Of	Outside face
Bt	Bent	Nf	Near face
Str	Straight	Ff	Far face
Stir	Stirrup	Ef	Each face
Sp	Spiral	Bot	Bottom
Ct	Column tie	Ew	Each way
If	Inside face	T	Top

Figure 3.6   Reinforcing Steel Weights and Measures

Bar Desig-nation No.**	Nominal Weight, Lb./Ft.	U.S. Customary Units Nominal Dimensions*			Nominal Weight kg/m	SI Units Nominal Dimensions*		
		Diameter in.	Cross Sectional Area, in.²	Perimeter in.		Diameter, mm	Cross Sectional Area, cm²	Perimeter mm
3	0.376	0.375	0.11	1.178	0.560	9.52	0.71	29.9
4	0.668	0.500	0.20	1.571	0.994	12.70	1.29	39.9
5	1.043	0.625	0.31	1.963	1.552	15.88	2.00	49.9
6	1.502	0.750	0.44	2.356	2.235	19.05	2.84	59.8
7	2.044	0.875	0.60	2.749	3.042	22.22	3.87	69.8
8	2.670	1.000	0.79	3.142	3.973	25.40	5.10	79.8
9	3.400	1.128	1.00	3.544	5.059	28.65	6.45	90.0
10	4.303	1.270	1.27	3.990	6.403	32.26	8.19	101.4
11	5.313	1.410	1.56	4.430	7.906	35.81	10.06	112.5
14	7.65	1.693	2.25	5.32	11.384	43.00	14.52	135.1
18	13.60	2.257	4.00	7.09	20.238	57.33	25.81	180.1

*The nominal dimensions of a deformed bar are equivalent to those of a plain round bar having the same weight per foot as the deformed bar.
**Bar numbers are based on the number of eighths of an inch included in the nominal diameter of the bars.

Figure 3.7 ASTM A615-81, Metric Rebar Specification

	Grade 300 (300 MPa* = 43,560 psi; + 8.7% vs. Grade 40) Grade 400 (400 MPa* = 58,000 psi; −3.4% vs. Grade 60)			
Bar No.	Diameter mm	Area mm²	Equivalent in.²	Comparison with U.S. Customary Bars
10	11.3	100	0.16	between #3 & #4
15	16.0	200	0.31	#5 (0.31 in²)
20	19.5	300	0.47	#6 (0.44 in²)
25	25.2	500	0.78	#8 (0.79 in²)
30	29.9	700	1.09	#9 (1.00 in²)
35	35.7	1000	1.55	#11 (1.56 in²)
45	43.7	1500	2.33	#14 (2.25 in²)
55	56.4	2500	3.88	#18 (4.00 in²)

*MPa = megapascals
Grade 300 bars are furnished only in sizes 10 through 35

Figure 3.8   Reinforcing Bar Supports

Reinforcing bar supports, their symbols, use, and sizes are shown in the table below.

Symbol	Bar support Illustration	Bar Support Illustration Plastic Capped or dipped	Type of Support	Sizes
SB	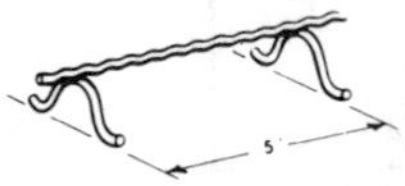	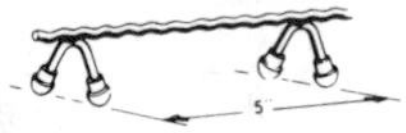 Capped	Slab Bolster	3/4", 1", 1/2", and 2" heights in 5' and 10' lengths
SBU*	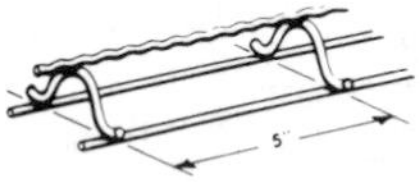		Slab Bolster Upper	Same as SB
BB	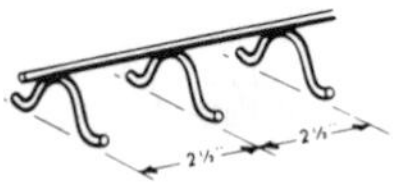	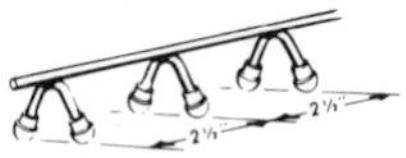 Capped	Beam Bolster	1", 1-1/2", 2", over 2" to 5" heights in increments of 1/4" in lengths of 5'
BBU*	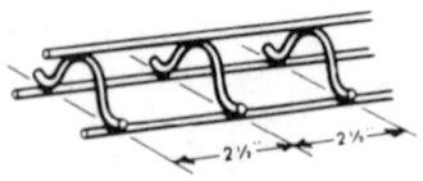		Beam Bolster Upper	Same as BB
BC		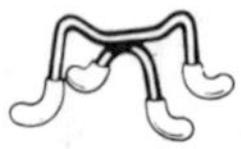 Dipped	Individual Bar Chair	3/4", 1", 1-1/2", and 1-3/4" heights
JC	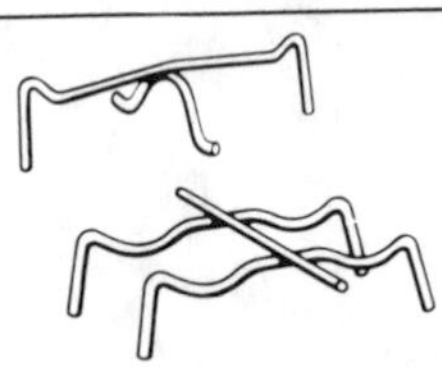	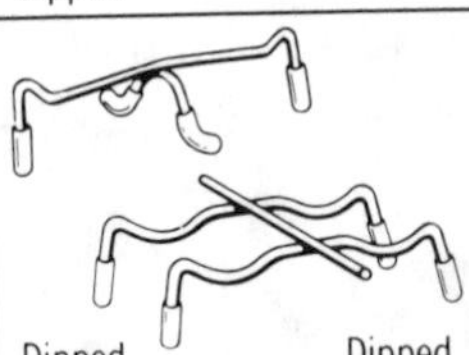 Dipped Dipped	Joist Chair	4", 5", and 6" widths and 3/4", 1", and 1-1/2" heights

*Usually available in Class 3 only, except on special order.
**Usually available in Class 3 only, with upturned or end bearing legs.

Figure 3.8 Reinforcing Bar Supports (continued)

Reinforcing bar supports, their symbols, use, and sizes are shown in the table below.

Symbol	Bar Support Illustration	Bar Support Illustration Plastic Capped or Dipped	Type of Support	Sizes
HC		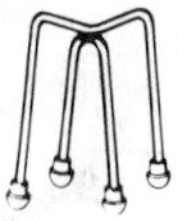 Capped	Individual High Chair	2" to 15" heights in increments of 1/4"
HCM*	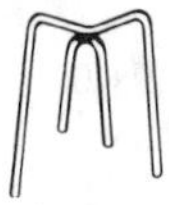		High Chair for Metal Deck	2" to 15" heights in increments of 1/4"
CHC	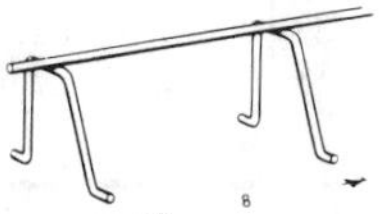	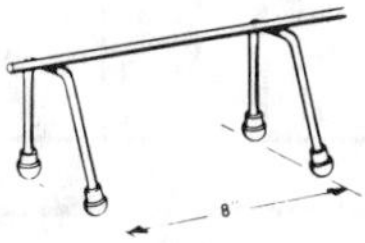 Capped	Continuous High Chair	Same as HC in 5' and 10' lenghts
CHCU*	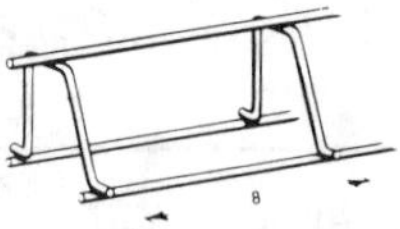		Continuous High Chair Upper	Same as CHC
CHCM*	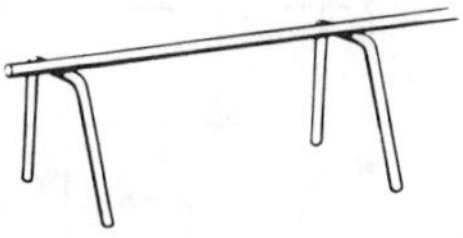		Continuous High Chair for Metal Deck	Up to 5" heights in increments of 1/4"
JCU**	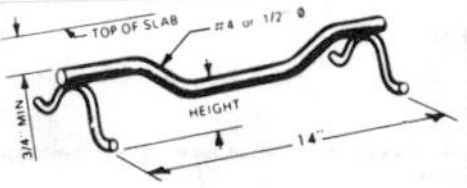	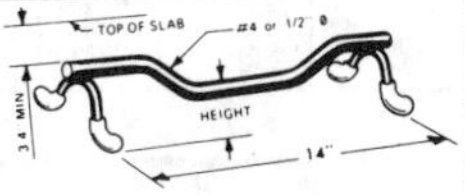 Dipped	Joist Chair Upper	14" span, Heights- 1" through +3-1/2" vary in 1/4" increments

*Usually available in Class 3 only, except on special order.
**Usually available in Class 3 only, with upturned or end bearing legs.

Figure 3.9 Weight of Steel Reinforcing per S.F. in Wall (PSF)

Reinforced Weights: Table below lists the weight per S.F. for reinforcing steel in walls. Weights will be correct for any grade steel reinforcing. For bars in two directions, add weights for each size and spacing.

C/C Spacing in Inches	Bar Size #3 Wt. (PSF)	#4 Wt. (PSF)	#5 Wt. (PSF)	#6 Wt. (PSF)	#7 Wt. (PSF)	#8 Wt. (PSF)	#9 Wt. (PSF)	#10 Wt. (PSF)	#11 Wt. (PSF)
2"	2.26	4.01	6.26	9.01	12.27				
3"	1.50	2.67	4.17	6.01	8.18	10.68	13.60	17.21	21.25
4"	1.13	2.01	3.13	4.51	6.13	8.10	10.20	12.91	15.94
5"	.90	1.60	2.50	3.60	4.91	6.41	8.16	10.33	12.75
6"	.752	1.34	2.09	3.00	4.09	5.34	6.80	8.61	1ſ ͻ3
8"	.564	1.00	1.57	2.25	3.07	4.01	5.10	6.46	7.97
10"	.451	.802	1.25	1.80	2.45	3.20	4.08	5.16	6.38
12"	.376	.668	1.04	1.50	2.04	2.67	3.40	4.30	5.31
18"	.251	.445	.695	1.00	1.32	1.78	2.27	2.86	3.54
24"	.188	.334	.522	.751	1.02	1.34	1.70	2.15	2.66
30"	.150	.267	.417	.600	.817	1.07	1.36	1.72	2.13
36" ·	.125	.223	.348	.501	.681	.890	1.13	1.43	1.77
42"	.107	.191	.298	.429	.584	.763	.97	1.17	1.52
48"	.094	.167	.261	.376	.511	.668	.85	1.08	1.33

Figure 3.10 Minimum Wall Reinforcement Weight (PSF)

The table below lists minimum wall reinforcement weights per S.F. of wall according to ACI 318-83 specifications of 0.12% of gross area for vertical bars, and 0.20% of gross area for horizontal bars.

Loca-tion	Wall Thick-ness	Horizontal Steel Bar Size	Spacing C/C	Sq. In. per L.F.	Total Wt. per S.F.	Vertical Steel Bar Size	Spacing C/C	Sq. In. per L.F.	Total Wt. per S.F.	Horizontal & Vertical Steel Total Wt. per S.F.
Both Faces	10"	#4	18"	.13	.891#	#3	18"	.07	.501#	1.392#
	12"	#4	16"	.15	1.002	#3	15"	.09	.602	1.604
	14"	#4	14"	.17	1.145	#3	13"	.10	.694	1.839
	16"	#4	12"	.20	1.336	#3	11"	.12	.820	2.156
	18"	#5	17"	.22	1.472	#4	18"	.13	.891	2.363
One Face	6"	#3	9"	.15	.501	#3	18"	.07	.251	.752
	8"	#4	12"	.20	.668	#3	11"	.12	.410	1.078
	10"	#5	15"	.25	.834	#4	16"	.15	.501	1.335

Figure 3.11   Common Stock Styles of Welded Wire Fabric

The following graphic and table provide the specifications, sizes, and weights of welded wire fabric used for slab reinforcement.

	New Designation	Old Designation	Steel Area per Foot				Approximate Weight per 100 S.F.	
	Spacing – Cross Sectional Area (in.) – (Sq. In. 100)	Spacing Wire Gauge (in.) – (AS & W)	Longitudinal		Transverse			
			in.	cm	in.	cm	Lbs.	kg
Rolls	6 x 6 – W1.4 x W1.4	6 x 6 – 10 x 10	0.028	0.071	0.028	0.071	21	9.53
	6 x 6 – W2.0 x W2.0	6 x 6 – 8 x 8 (1)	0.040	0.102	0.040	0.102	29	13.15
	6 x 6 – W2.9 x W2.9	6 x 6 – 6 x 6	0.058	0.147	0.053	0.147	42	19.05
	6 x 6 – W4.0 x W4.0	6 x 6 – 4 x 4	0.080	0.203	0.080	0.203	58	26.31
	4 x 4 – W1.4 x W1.4	4 x 4 – 10 x 10	0.042	0.107	0.042	0.107	31	14.06
	4 x 4 – W2.0 x W2.0	4 x 4 – 8 x 8 (1)	0.060	0.152	0.060	0.152	43	19.50
	4 x 4 – W2.9 x W2.9	4 x 4 – 6 x 6	0.087	0.221	0.087	0.221	62	28.12
	4 x 4 – W4.0 x W4.0	4 x 4 – 4 x 4	0.120	0.305	0.120	0.305	85	38.56
Sheets	6 x 6 – W2.9 x W2.9	6 x 6 – 6 x 6	0.058	0.147	0.058	0.147	42	19.05
	6 x 6 – W4.0 x W4.0	6 x 6 – 4 x 4	0.080	0.203	0.080	0.203	58	26.31
	6 x 6 – W5.5 x W5.5	6 x 6 – 2 x 2 (2)	0.110	0.279	0.110	0.279	80	36.29
	6 x 6 – W4.0 x W4.0	4 x 4 – 4 x 4	0.120	0.305	0.120	0.305	85	38.56

Notes:
1. Exact W-number size for 8 gauge is W2.1
2. Exact W-number size for 2 gauge is W5.4

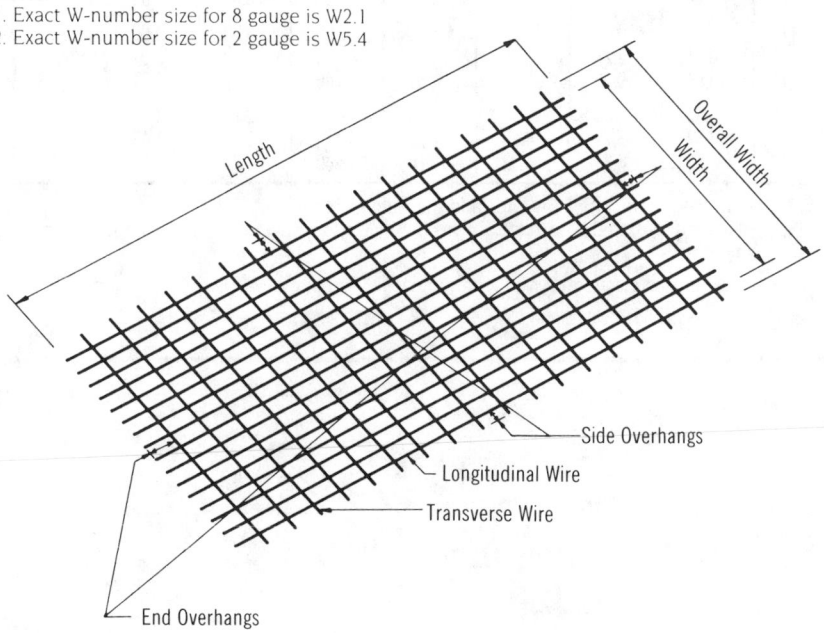

Welded Wire Fabric

Figure 3.12a Estimated Weight of Welded Wire Fabric for All Styles Having Uniform Spacings and Gauges of Transverse Members

Steel Wire Gauge Numbers	Approximate Weights in Lbs. per 100 S.F.—Based on 60" Width C. to C. of Outside Longitudinal Wires							
	Weight of Transverse Members*							
	Spacing							
	2"	3"	4"	6"	8"	10"	12"	16"
0000	256.95	170.95	128.22	85.48	61.11	51.29	42.74	32.05
000	217.31	144.87	108.66	72.44	54.33	43.46	36.22	27.16
00	181.16	120.78	90.58	60.39	45.29	36.23	30.19	22.65
0	155.37	103.58	77.69	51.79	38.84	31.07	25.90	19.42
1	132.43	88.29	66.22	44.14	33.11	26.49	22.07	16.55
2	113.96	75.97	56.98	37.99	28.49	22.79	18.99	14.24
1/4"	103.33	68.89	51.67	34.44	25.83	20.67	17.22	12.92
3	98.21	65.47	49.10	32.74	24.55	19.64	16.37	12.28
4	83.95	55.97	41.97	27.98	20.99	16.79	13.99	10.49
5	70.87	47.24	35.43	23.62	17.72	14.17	11.81	8.86
6	60.96	40.64	30.48	20.32	15.24	12.19	10.16	7.62
7	51.81	34.54	25.90	17.27	12.95	10.36	8.63	6.48
8	43.40	28.93	21.70	14.47	10.85	8.68	7.23	5.43
9	36.37	24.25	18.18	12.12	9.09	7.27	6.06	4.55
10	30.14	20.09	15.07	10.05	7.53	6.03	5.02	3.77
11	24.01	16.01	12.01	8.00	6.00	4.80	4.00	3.00
12	18.41	12.27	9.20	6.14	4.60	3.68	3.07	2.30
13	13.84	9.23	6.92	4.61	3.46	2.77	2.31	1.73
14	10.58	7.06	5.29	3.53	2.65	2.12	1.76	1.32
15	8.57	5.72	4.29	2.86	2.14	1.71	1.43	1.07
16	6.46	4.31	3.23	2.15	1.62	1.29	1.08	.81

Note: To determine the weight of any type, add the weights of the longitudinal and transverse members, adjust the total to the nearest lb., considering 0.5 lb. or over to be 1 lb., and less than 0.5 lb. to be zero.

*Includes weight of 1" projection beyond longitudinal selvage wires.

(courtesy Concrete Reinforcing Steel Institute)

Figure 3.12b Estimated Weight of Welded Wire Fabric for All Styles Having Uniform Spacings and Gauges of Longitudinal Members

Steel Wire Gauge Numbers	Approximate Weights in Lbs. per 100 S. F.—Based on 60" Width C. to C. of Outside Longitudinal Wires						
	Weight of Longitudinal Members						
	Spacing						
	2"	3"	4"	6"	8"	10"	12"
0000000	397.05	268.97	204.93	140.89	108.87	89.66	76.85
000000	352.22	238.60	181.79	124.98	96.58	79.53	68.17
00000	306.47	207.61	158.18	108.75	84.03	69.20	59.31
0000	256.43	173.71	132.35	90.99	70.31	57.90	49.63
000	217.31	147.21	112.16	77.11	59.59	49.07	42.06
00	181.16	122.72	93.50	64.28	49.67	40.91	35.06
0	155.37	105.25	80.19	55.13	42.60	35.08	30.07
1	132.43	89.71	68.35	46.99	36.31	29.90	25.63
2	113.96	77.20	58.82	40.44	31.25	25.73	22.06
1/4"	103.33	70.00	53.33	36.67	28.33	23.33	20.00
3	98.21	66.53	50.69	34.85	26.93	22.18	19.01
4	83.95	56.87	43.33	29.79	23.02	18.96	16.25
5	70.87	48.01	36.58	25.15	19.43	16.00	13.72
6	60.96	41.29	31.46	21.63	16.71	13.76	11.80
7	51.81	35.10	26.74	18.38	14.21	11.70	10.03
8	43.40	29.40	22.40	15.40	11.90	9.80	8.40
9	36.37	24.64	18.77	12.91	9.97	8.21	7.04
10	30.14	20.42	15.56	10.69	8.26	6.81	5.83
11	24.01	16.27	12.39	8.52	6.58	5.42	4.65
12	18.41	12.47	9.50	6.53	5.05	4.16	3.56
13	13.84	9.38	7.15	4.91	3.80	3.13	2.68
14	10.58	7.17	5.46	3.76	2.90	2.39	2.05
15	8.57	5.81	4.43	3.04	2.35	1.94	1.66
16	6.46	4.38	3.33	2.29	1.77	1.46	1.25

Note: To determine the weight of any type, add the weights of the longitudinal and transverse members, adjust the total to the nearest lb., considering 0.5 lb. or over to be 1 lb., and less than 0.5 lb. to be zero.

(courtesy Concrete Reinforcing Steel Institute)

Figure 3.13 Wire Gauge, Designation, and Dimensions in Welded Wire Fabric for Concrete Reinforcing

American Steel and Wire Gauge Number	W & D Size Numbers		Area		Nominal Diameter	
	Smooth	Deformed	in.²	cm²	in.	mm
	W31	D31	0.310	2.000	0.628	15.95
	W30	D30	0.300	1.936	0.618	15.70
	W28	D28	0.280	1.807	0.597	15.16
	W26	D26	0.260	1.679	0.575	14.61
	W24	D24	0.240	1.549	0.553	14.05
	W22	D22	0.220	1.419	0.529	13.44
	W20	D20	0.200	1.290	0.504	12.80
0000000			0.189	1.219	0.490	12.45
	W18	D18	0.180	1.167	0.478	12.14
000000			0.167	1.076	0.4615	11.72
	W16	D16	0.160	1.032	0.451	11.46
00000			0.146	0.942	0.4305	10.94
	W14	D14	0.140	0.903	0.422	10.72
0000			0.122	0.787	0.394	10.01
	W12	D12	0.120	0.774	0.390	9.91
	W11	D11	0.110	0.710	0.374	9.54
	W10.5		0.105	0.678	0.366	9.30
000			0.103	0.665	0.3625	9.21
	W10	D10	0.100	0.645	0.356	9.04
	W9.5		0.095	0.613	0.348	8.84
	W9	D9	0.090	0.581	0.338	8.59
00			0.086	0.556	0.331	8.41
	W8.5		0.085	0.548	0.329	8.36
	W8	D8	0.080	0.516	0.319	8.10
	W7.5		0.075	0.484	0.309	7.85
0			0.074	0.478	0.3065	7.79
	W7	D7	0.070	0.452	0.298	7.57
	W6.5		0.065	0.419	0.288	7.32
1			0.063	0.407	0.283	7.19
	W6	D6	0.060	0.387	0.276	7.01
	W5.5		0.055	0.355	0.264	6.71
2			0.054	0.348	0.2625	6.67
	W5	D5	0.050	0.323	0.252	6.40
3			0.047	0.303	0.244	6.20
	W4.5		0.045	0.290	0.240	6.10
4	W4	D4	0.040	0.258	0.225	5.72
	W3.5		0.035	0.226	0.211	5.36
5			0.034	0.219	0.207	5.26
	W3		0.030	0.194	0.195	4.95
	W2.9		0.029	0.187	0.192	4.88
6	W2.5		0.025	0.161	0.177	4.37
7	W2.1		0.021	0.136	0.162	4.12
8	W2		0.020	0.129	0.159	4.04
9			0.017	0.111	0.148	3.76
	W1.5		0.015	0.097	0.138	3.51
10	W1.4		0.014	0.090	0.135	3.43

Figure 3.14 Types of Welded Wire Fabric Most Commonly Used in Construction

Designation	Size of Square Mesh		Wire Designation	Sectional Area		Weight	
	in.	mm		in.²/Ft.	cm²/m	Lbs./100 Ft.²	kg/m²
6 x 6 – 10/10	6	152.4	W1.4	0.029	.613	21	1.026
6 x 6 – 8/8	6	152.4	W2.1	0.041	.868	30	1.4660
6 x 6 – 6/6	6	152.4	W2.9	0.058	1.227	42	2.053
6 x 6 – 4/4	6	152.4	W4	.080	1.693	58	2.834
4 x 4 – 10/10	4	101.6	W1.4	.043	0.910	31	1.515
4 x 4 – 8/8	4	101.6	W2.1	.062	1.312	44	2.150
4 x 4 – 6/6	4	101.6	W2.9	.087	1.841	61	2.981
4 x 4 – 4/4	4	101.6	W4	.119	2.518	85	4.154

Figure 3.15 Preferred and Acceptable Methods for Transporting Concrete to the Job Site and at the Job Site

Job		Screw Spreaders	Tremies	Drop Chutes	Chutes	Barrows, Buggies	Concrete Pumps	Pneumatic Guns	Belt Conveyors	Elevators	Cranes, Buckets	Mobile Cont. Mixers	Nonagitating Trucks	Truck Mixers	Truck Agitators
Job Site Mix												●		X	
Short Haul													●	X	●
Long Haul														●	X
Horizontally	Short				●	●	●	●		●	●	●	●	●	●
Horizontally	Medium						●	X	X	X					
Horizontally	Long						X		●						
Vertically	Short			●	●		●	●	●		X				
Vertically	Medium						X			X	●				
Vertically	Long						X			●	X				
Footings					●		X				X	●	X	X	X
Foundations					●		X				X	●	X	X	X
Walls				●	X		X	●			X	●	X	X	X
Columns						X	●				●				
Beams						X	●				●				
Underwater			●												
Elevated Slabs						●	●		X	●	●				
Different Shapes							X	●							
Mass Concrete							X		●		X		X	X	X
Slabs on Ground		●										X	●	●	●
Repairs								●			X				

● Preferred Method   X Acceptable Method

Figure 3.16   Data on Handling and Transporting Concrete

On most construction jobs, the method of concrete placement is an important consideration for the estimator and the project manager. This table is a guide for planning time and equipment for concrete placement.

Quantity	Rate
Unloading 6 cubic yard mixer truck, 1 operator	Minimum time 2 minutes, average time 7 minutes
Wheeling, using 4-1/2 cubic feet wheelbarrows, 1 laborer	
up to 100'	Average 1-1/2 cubic yards per hour
up to 200'	Average 1 cubic yard per hour
Wheeling, using 8 cubic feet hand buggies, 1 laborer	
up to 100'	Average 5 cubic yards per hour
up to 200'	Average 3 cubic yards per hour
Transporting, using 28 cubic feet power buggies, 1 laborer	
up to 500'	Average 20 cubic feet per hour
up to 1000'	Average 15 cubic feet per hour
Portable conveyor, 16" belt, 30° elevation, 1 laborer	
100 FPM belt speed	30 C.Y./Hr. max., 15 C.Y./Hr. average
200 FPM belt speed	60 C.Y./Hr. max., 30 C.Y./Hr. average
300 FPM belt speed	90 C.Y./Hr. max., 45 C.Y./Hr. average
400 FPM belt speed	120 C.Y./Hr. max., 60 C.Y./Hr. average
500 FPM belt speed	150 C.Y./Hr. max., 75 C.Y./Hr. average
600 FPM belt speed	180 C.Y./Hr. max., 90 C.Y./Hr. average
Feeder conveyors, up to 600', 5 laborers	
500 FPM belt speed	150 C.Y./Hr. max., 50 C.Y./Hr. average
600 FPM belt speed	180 C.Y./Hr. max., 60 C.Y./Hr. average
Side discharge conveyor, 1 laborer	
fed by portable conveyor	40–60 C.Y./Hr. average
fed by feeder conveyor	80–100 C.Y./Hr. average
fed by crane & bucket	35–60 C.Y./Hr. average
Mobile crane, 1 operator* 2 laborers	
1 cubic yard bucket	40 C.Y./Hr. average
2 cubic yard bucket	60 C.Y./Hr. average
Tower crane 1 operator* 2 laborers	
1 cubic yard bucket	35–40 C.Y./Hr. average
Small line pumping systems, 1 foot vertical = 6 feet horizontally 1 90 degree bend = 40 feet horizontally 1 45 degree bend = 20 feet horizontally 1 30 degree bend = 13 feet horizontally 1 foot rubber hose = 1-1/2 feet of steel tubing 1 operator*	
average output, 1000' horizontally	40–50 C.Y./Hr. Average

*In some regions an oiler or helper may be needed, according to union rules.

(courtesy Concrete Estimating Handbook, Michael F. Kenny, Van Nostrand Reinhold Company).

Figure 3.17 Types of Cement and Concrete and Their Major Uses

Type of Portland Cement	Type of Concrete	Major Use
Type I, Normal*	Standard-type concrete	General construction
Type IA, air-entraining	Standard with air-entraining (more workability, and resistance to freezing and thawing.)	General construction
Portland blast-furnace slag IS	Standard-type concrete	General construction
Portland blast-furnace slag, air-entraining IS-A	Standard-type concrete with better workability and resistance to freezing and thawing	General construction
Type II, Moderate*	Slower setting, lower heat generation, and smaller volume change than Types I and IA; develops strength in 28 days	General construction and situations involving moderate sulfate action
Type IIA, air-entraining	Same as concrete using Type II moderate cement, but with better workability and resistance to freezing and thawing	General construction and situations involving moderate sulfate action
Type III, High-Early-Strength*	Rapid setting, higher heat generation (offsets freezing), some volume change, develops strength in 7 days	Construction requiring rapid development of strength
Type IIIA, air-entraining	Same as concrete using Type III high-early-strength, but with better workability and resistance to freezing and thawing	Construction requiring rapid development of strength
Type IV, Low Heat of Hydration*	Slow setting, low heat generation, small volume change, good strength with age	Massive concrete construction
Type V, Sulfate-Resisting*	High resistance to sulfate attack, fairly low heat generation, high strength with age	Construction exposed to ground water or soil that contains sulfates
Portland-pozzolan P and PIP	A hydraulic concrete	Large hydraulic structures
Portland-pozzolan, air-entraining P-A and IP-A	A hydraulic concrete with air-entraining	Large hydraulic structures

*CSA (Canadian Standards Association) designations for ASTM types I, II, III, IV, and V.

Figure 3.18 Quantities of Cement, Sand, and Stone for One Cubic Yard of Various Concrete Mixes

Use this table to determine the quantities of cement, sand, and stone for small quantities of site-mixed concrete. Cement is listed in sacks or cubic feet, and aggregate in cubic yards.

Concrete (C.Y.)	Mix 1 1:1:1-3/4			Mix 2 1:2:2.25			Mix 3 1:2.25:3			Mix 4 1:3:4		
	Cement (Sacks)	Sand (C.Y.)	Stone (C.Y.)	Cement (Sacks)	Sand (C.Y.)	Stone (C.Y.)	Cement (Sacks)	Sand (C.Y.)	Stone (C.Y.)	Cement (Sacks)	Sand (C.Y.)	Stone (C.Y.)
1	10	.37	.63	7.75	.56	.65	6.25	.52	.70	5.0	.56	.74
2	20	.74	1.26	15.50	1.12	1.30	12.50	1.04	1.40	10.0	1.12	1.48
3	30	1.11	1.89	23.25	1.68	1.95	18.75	1.56	2.10	15.0	1.68	2.22
4	40	1.48	2.52	31.00	2.24	2.60	25.00	2.08	2.80	20.0	2.24	2.96
5	50	1.85	3.15	38.75	2.80	3.25	31.25	2.60	3.50	25.0	2.80	3.70
6	60	2.22	3.78	46.50	3.36	3.90	37.50	3.12	4.20	30.0	3.36	4.44
7	70	2.59	4.41	54.25	3.92	4.55	43.75	3.64	4.90	35.0	3.92	5.18
8	80	2.96	5.04	62.00	4.48	5.20	50.00	4.16	5.60	40.0	4.48	5.92
9	90	3.33	5.67	69.75	5.04	5.85	56.25	4.68	6.30	45.0	5.04	6.66
10	100	3.70	6.30	77.50	5.60	6.50	62.50	5.20	7.00	50.0	5.60	7.40
11	110	4.07	6.93	85.25	6.16	7.15	68.75	5.72	7.70	55.0	6.16	8.14
12	120	4.44	7.56	93.00	6.72	7.80	75.00	6.24	8.40	60.0	6.72	8.88
13	130	4.82	8.20	100.76	7.28	8.46	81.26	6.76	9.10	65.0	7.28	9.62
14	140	5.18	8.82	108.50	7.84	9.10	87.50	7.28	9.80	70.0	7.84	10.36
15	150	5.56	9.46	116.26	8.40	9.76	93.76	7.80	10.50	75.0	8.40	11.10
16	160	5.92	10.08	124.00	8.96	10.40	100.00	8.32	11.20	80.0	8.96	11.84
17	170	6.30	10.72	131.76	9.52	11.06	106.26	8.84	11.90	85.0	9.52	12.58
18	180	6.66	11.34	139.50	10.08	11.70	112.50	9.36	12.60	90.0	10.08	13.32
19	190	7.04	11.98	147.26	10.64	12.36	118.76	9.84	13.30	95.0	10.64	14.06
20	200	7.40	12.60	155.00	11.20	13.00	125.00	10.40	14.00	100.0	11.20	14.80

(courtesy, Estimating Tables for Home Building, Paul I. Thomas, Craftsman Book Company)

Figure 3.18   Quantities of Cement, Sand, and
Stone for One Cubic Yard of
Various Concrete Mixes (continued)

Concrete (C.Y.)	Mix 1 1:1:1-3/4			Mix 2 1:2:2.25			Mix 3 1:2.25:3			Mix 4 1:3:4		
	Cement (Sacks)	Sand (C.Y.)	Stone (C.Y.)	Cement (Sacks)	Sand (C.Y.)	Stone (C.Y.)	Cement (Sacks)	Sand (C.Y.)	Stone (C.Y.)	Cement (Sacks)	Sand (C.Y.)	Stone (C.Y.)
21	210	7.77	13.23	162.75	11.76	13.65	131.25	10.92	14.70	105.0	11.76	15.54
22	220	8.14	13.86	170.05	12.32	14.30	137.50	11.44	15.40	110.0	12.32	16.28
23	230	8.51	14.49	178.25	12.88	14.95	143.75	11.96	16.10	115.0	12.88	17.02
24	240	8.88	15.12	186.00	13.44	15.60	150.00	12.48	16.80	120.0	13.44	17.76
25	250	9.25	15.75	193.75	14.00	16.25	156.25	13.00	17.50	125.0	14.00	18.50
26	260	9.64	16.40	201.52	14.56	16.92	162.52	13.52	18.20	130.0	14.56	19.24
27	270	10.00	17.00	209.26	15.12	17.56	168.76	14.04	18.90	135.0	15.02	20.00
28	280	10.36	17.64	217.00	15.68	18.20	175.00	14.56	19.60	140.0	15.68	20.72
29	290	10.74	18.28	224.76	16.24	18.86	181.26	15.08	20.30	145.0	16.24	21.46

Figure 3.19   Metric Equivalents of Cement Content for Concrete Mixes

94-Pound Bags per Cubic Yard	Kilograms per Cubic Meter	94-Pound Bags per Cubic Yard	Kilograms per Cubic Meter
1.0	55.77	7.0	390.4
1.5	83.65	7.5	418.3
2.0	111.5	8.0	446.2
2.5	139.4	8.5	474.0
3.0	167.3	9.0	501.9
3.5	195.2	9.5	529.8
4.0	223.1	10.0	557.7
4.5	251.0	10.5	585.6
5.0	278.8	11.0	613.5
5.5	306.7	11.5	641.3
6.0	334.6	12.0	669.2
6.5	362.5	12.5	697.1

(a) If you know cement content in pounds per cubic yard, multiply by 0.5933 to obtain kilograms per cubic meter.

(b) If you know cement content in 94-pound bags per cubic yard, multiply by 55.77 to obtain kilograms per cubic meter.

Figure 3.20 Metric Equivalents of Common Concrete Strengths (To convert other psi values to megapascals, multiply by 0.006895)

U.S. Value, psi	SI Value, Megapascals	Non-SI Metric Value, kgf/cm²*
2,000	14	140
2,500	17	175
3,000	21	210
3,500	24	245
4,000	28	280
4,500	31	315
5,000	34	350
6,000	41	420
7,000	48	490
8,000	55	560
9,000	62	630
10,000	69	705

*Kilograms force per square centimeter

Figure 3.21 Concrete, Forms, and Reinforcing Proportionate Quantities

The following tables show both quantities per S.F. of floor area and form and reinforcing quantities per C.Y. Unusual structural requirements would increase these ratios. High strength reinforcing would reduce the steel weights. Figures are for 3,000 psi concrete and 60,000 psi reinforcing unless specified otherwise.

Type of Construction	Live Load	Span	Per S.F. of Floor Area				Per C.Y. of Concrete		
			Concrete	Forms	Reinf.	Pans	Forms	Reinf.	Pans
Flat Plate	50 psf	15 Ft.	.46 C.F.	1.06 S.F.	1.71 lb.		62 S.F.	101 lb.	
		20	.63	1.02	2.4		44	104	
		25	.79	1.02	3.03		35	104	
	100	15	.46	1.04	2.14		61	126	
		20	.71	1.02	2.72		39	104	
		25	.83	1.01	3.47		33	113	
Flat Plate (waffle construction) 20" domes	50	20	.43	1.0	2.1	.84 S.F.	63	135	53 S.F.
		25	.52	1.0	2.9	.89	52	150	46
		30	.64	1.0	3.7	.87	42	155	37
	100	20	.51	1.0	2.3	.84	53	125	45
		25	.64	1.0	3.2	.83	42	135	35
		30	.76	1.0	4.4	.81	36	160	29
Waffle Construction 30" domes	50	25	.69	1.06	1.83	.68	42	72	40
		30	.74	1.06	2.39	.69	39	87	39
		35	.86	1.05	2.71	.69	33	85	39
		40	.78	1.0	4.8	.68	35	165	40
Flat Slab (two way with drop panels)	50	20	.62	1.03	2.34		45	102	
		25	.77	1.03	2.99		36	105	
		30	.95	1.03	4.09		29	116	
	100	20	.64	1.03	2.83		43	119	
		25	.79	1.03	3.88		35	133	
		30	.96	1.03	4.66		29	131	
	200	20	.73	1.03	3.03		38	112	
		25	.86	1.03	4.23		32	133	
		30	1.06	1.03	5.3		26	135	
One Way Joists 20" pans	50	15	.36	1.04	1.4	.93	78	105	70
		20	.42	1.05	1.8	.94	67	120	60
		25	.47	1.05	2.6	.94	60	150	54
	100	15	.38	1.07	1.9	.93	77	140	66
		20	.44	1.08	2.4	.94	67	150	58
		25	.52	1.07	3.5	.94	55	185	49
One Way Joist. 8" x 16" filler blocks	50	15	.34	1.06	1.8	.81 Ea.	84	145	64 Ea.
		20	.40	1.08	2.2	.82	73	145	55
		25	.46	1.07	3.2	.83	63	190	49
	100	15	.39	1.07	1.9	.81	74	130	56
		20	.46	1.09	2.8	.82	64	160	48
		25	.53	1.10	3.6	.83	56	190	42
One Way Beam & Slab	50	15	.42	1.30	1.73		84	111	
		20	.51	1.28	2.61		68	138	
		25	.64	1.25	2.78		53	117	
	100	15	.42	1.30	1.9		84	122	
		20	.54	1.35	2.69		68	154	
		25	.69	1.37	3.93		54	154	
	200	15	.44	1.31	2.24		80	137	
		20	.58	1.40	3.30		65	163	
		25	.69	1.42	4.89		53	183	
Two Way Beam & Slab	100	15	.47	1.20	2.26		69	130	
		20	.63	1.29	3.06		55	131	
		25	.83	1.33	3.79		43	123	
	200	15	.49	1.25	2.70		41	149	
		20	.66	1.32	4.04		54	165	
		25	.88	1.32	6.08		41	187	

Figure 3.21 Concrete, Forms, and Reinforcing Proportionate Quantities (continued)

		4000 psi Concrete and 60,000 psi Reinforcing — Form and Reinforcing Quantities per C.Y.				
Item	Size	Forms		Reinforcing	Minimum	Maximum
Columns (square tied)	10" x 10"	130 S.F.C.A.		#5 to #11	220 lbs.	875 lbs.
	12" x 12"	108		#6 to #14	200	955
	14" x 14"	92		#7 to #14	190	900
	16" x 16"	81		#6 to #14	187	1082
	18" x 18"	72		#6 to #14	170	906
	20" x 20"	65		#7 to #18	150	1080
	22" x 22"	59		#8 to #18	153	902
	24" x 24"	54		#8 to #18	164	884
	26" x 26"	50		#9 to #18	169	994
	28" x 28"	46		#9 to #18	147	864
	30" x 30"	43		#10 to #18	146	983
	32" x 32"	40		#10 to #18	175	866
	34" x 34"	38		#10 to #18	157	772
	36" x 36"	36		#10 to #18	175	852
	38" x 38"	34		#10 to #18	158	765
	40" x 40"	32		#10 to #18	143	692

Item	Size	Forms	Spirals	Reinforcing	Minimum	Maximum
Columns (spirally reinforced)	12" diameter	34.5 L.F. 34.5	190 lbs. 190	#4 to #11 #14 & #18	165 lbs. —	1505 lbs. 1100
	14"	25 25	170 170	#4 to #11 #14 & #18	150 800	970 1000
	16"	19 19	160 160	#4 to #11 #14 & #18	160 605	950 1080
	18"	15 15	150 150	#4 to #11 #14 & #18	160 480	915 1075
	20"	12 12	130 130	#4 to #11 #14 & #18	155 385	865 1020
	22"	10 10	125 125	#4 to #11 #14 & #18	165 320	775 995
	24"	9 9	120 120	#4 to #11 #14 & #18	195 290	800 1150
	26"	7.3 7.3	100 100	#4 to #11 #14 & #18	200 235	729 1035
	28"	6.3 6.3	95 95	#4 to #11 #14 & #18	175 200	700 1075
	30"	5.5 5.5	90 90	#4 to #11 #14 & #18	180 175	670 1015
	32"	4.8 4.8	85 85	#4 to #11 #14 & #18	185 155	615 955
	34"	4.3 4.3	80 80	#4 to #11 #14 & #18	180 170	600 855
	36"	3.8 3.8	75 75	#4 to #11 #14 & #18	165 155	570 865
	40"	3.0 3.0	70 70	#4 to #11 #14 & #18	165 145	500 765

(continued on next page)

Figure 3.21 Concrete, Forms, and Reinforcing Proportionate Quantities (continued)

3000 psi Concrete and 60,000 psi Reinforcing — Form and Reinforcing Quantities per C.Y.						
Item	Type	Loading	Height	C.Y./L.F.	Forms/C.Y.	Reinf./C.Y.
Retaining Walls	Cantilever	Level Backfill	4 Ft.	0.2 C.Y.	49 S.F.	35 lbs.
			8	0.5	42	45
			12	0.8	35	70
			16	1.1	32	85
			20	1.6	28	105
		Highway Surcharge	4	0.3	41	35
			8	0.5	36	55
			12	0.8	33	90
			16	1.2	30	120
			20	1.7	27	155
		Railroad Surcharge	4	0.4	28	45
			8	0.8	25	65
			12	1.3	22	90
			16	1.9	20	100
			20	2.6	18	120
	Gravity, with Vertical Face	Level Backfill	4	0.4	37	None
			7	0.6	27	
			10	1.2	20	
		Sloping Surcharge	4	0.3	31	
			7	0.8	21	
			10	1.6	15	

		Live Load in Kips per Linear Foot						
	Span	Under 1 Kip		2 to 3 Kips		4 to 5 Kips		6 to 7 Kips	
		Forms	Reinf.	Forms	Reinf.	Forms	Reinf.	Forms	Reinf.
Beams	10 Ft.	—	—	90 S.F.	170#	85 S.F.	175#	75 S.F.	185#
	16	130 S.F.	165#	85	180	75	180	65	225
	20	110	170	75	185	62	200	51	200
	26	90	170	65	215	62	215	—	—
	30	85	175	60	200	—	—	—	—

Figure 3.21 Concrete, Forms, and Reinforcing Proportionate Quantities (continued)

Item	Size	Type	Forms per C.Y.	Reinforcing per C.Y.
Spread Footings	Under 1 C.Y.	1,000 psf soil 5,000 10,000	24 S.F. 24 24	44 lb. 42 52
	1 C.Y. to 5 C.Y.	1,000 5,000 10,000	14 14 14	49 50 50
	Over 5 C.Y.	1,000 5,000 10,000	9 9 9	54 52 56
Pile Caps (30 Ton Concrete Piles)	Under 5 C.Y.	shallow medium deep	20 20 20	65 50 40
	5 C.Y. to 10 C.Y.	shallow medium deep	14 15 15	55 45 40
	10 C.Y. to 20 C.Y.	shallow medium deep	11 11 12	60 45 35
	Over 20 C.Y.	shallow medium deep	9 9 10	60 45 40

3000 psi Concrete and 60,000 psi Reinforcing — Form and Reinforcing Quantities per C.Y.							
Item	Size	Pile Spacing	50 T Pile	100 T Pile	50 T Pile	100 T Pile	
Pile Caps (Steel H Piles)	Under 5 C.Y.	24" O.C. 30" 36"	24 S.F. 25 24	24 S.F. 25 24	75 lb. 80 80	90 lb. 100 110	
	5 C.Y. to 10 C.Y.	24" 30" 36"	15 15 15	15 15 15	80 85 75	110 110 90	
	Over 10 C.Y.	24" 30" 36"	13 11 10	13 11 10	85 85 85	90 95 90	

		8" Thick		10" Thick		12" Thick		15" Thick	
	Height	Forms	Reinf.	Forms	Reinf.	Forms	Reinf.	Forms	Reinf.
Basement Walls	7 Ft. 8 9 10	81 S.F.	44 lb. 44 46 57	65 S.F.	45 lb. 45 45 45	54 S.F.	44 lb. 44 44 44	41 S.F.	43 lb. 43 43 43
	12 14 16 18		83 116		50 65 86		52 64 90 106		43 51 65 70

Figure 3.22   Concrete Quantities for Pile Caps — 3'0" O.C.

Load	Number of Piles @ 3'-0" O.C. per Footing Cluster									
Work-ing (K)	2 (C.Y.)	4 (C.Y.)	6 (C.Y.)	8 (C.Y.)	10 (C.Y.)	12 (C.Y.)	14 (C.Y.)	16 (C.Y.)	18 (C.Y.)	20 (C.Y.)
50	(.9)	(1.9)	(3.3)	(4.9)	(5.7)	(7.8)	(9.9)	(11.1)	(14.4)	(16.5)
100	(1.0)	(2.2)	(3.3)	(4.9)	(5.7)	(7.8)	(9.9)	(11.1)	(14.4)	(16.5)
200	(1.0)	(2.2)	(4.0)	(4.9)	(5.7)	(7.8)	(9.9)	(11.1)	(14.4)	(16.5)
400	(1.1)	(2.6)	(5.2)	(6.3)	(7.4)	(8.2)	(13.7)	(11.1)	(14.4)	(16.5)
800		(2.9)	(5.8)	(7.5)	(9.2)	(13.6)	(17.6)	(15.9)	(19.7)	(22.1)
1200			(5.8)	(8.3)	(9.7)	(14.2)	(18.3)	(20.4)	(21.2)	(22.7)
1600				(9.8)	(11.4)	(14.5)	(19.5)	(20.4)	(24.6)	(27.2)
2000				(9.8)	(11.4)	(16.6)	(24.1)	(21.7)	(26.0)	(28.8)
3000						(17.5)		(26.5)	(30.3)	(32.9)
4000								(30.2)	(30.7)	(36.5)

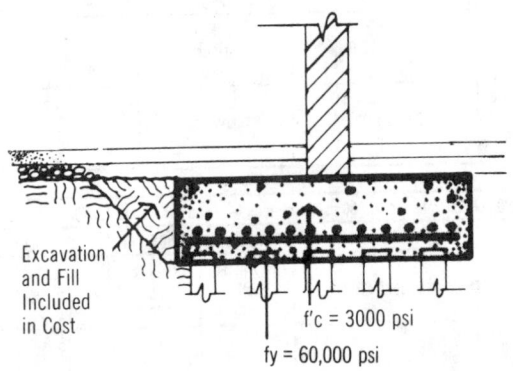

Excavation and Fill Included in Cost

f'c = 3000 psi

fy = 60,000 psi

Section Through Pile Cap

Figure 3.23   Concrete Quantities for Pile Caps — 4'6" O.C.

Load	Number of Piles @ 4'-6" O.C. per Footing					
Working (K)	2 (C.Y.)	3 (C.Y.)	4 (C.Y.)	5 (C.Y.)	6 (C.Y.)	7 (C.Y.)
50	(2.3)	(3.6)	(5.6)	(11.0)	(13.7)	(12.9)
100	(2.3)	(3.6)	(5.6)	(11.0)	(13.7)	(12.9)
200	(2.3)	(3.6)	(5.6)	(11.0)	(13.7)	(12.9)
400	(3.0)	(3.6)	(5.6)	(11.0)	(13.7)	(12.9)
800			(6.2)	(11.5)	(14.0)	(12.9)
1200				(13.0)	(13.7)	(13.4)
1600						(14.0)

Figure 3.24   Concrete Footings/Volume — 6" and 8" Deep

| Ground Floor Area | Cubic Yards of Concrete Needed for Size of Footing Shown | | | | | | | | | |
| | 6" Deep by: | | | | | 8" Deep by: | | | | |
	12"	16"	18"	20"	24"	12"	16"	18"	20"	24"
200	1.11	1.48	1.67	1.85	2.22	1.48	1.97	2.22	2.47	2.96
300	1.37	1.83	2.06	2.28	2.74	1.83	2.44	2.74	3.04	3.65
400	1.57	2.10	2.36	2.62	3.15	2.10	2.80	3.15	3.49	4.19
500	1.76	2.35	2.64	2.93	3.52	2.35	3.13	3.52	3.90	4.68
600	1.92	2.57	2.89	3.20	3.85	2.57	3.43	3.85	4.27	5.13
700	2.07	2.77	3.11	3.45	4.14	2.77	3.69	4.14	4.60	5.52
800	2.22	2.96	3.34	3.70	4.44	2.96	3.96	4.44	4.93	5.92
900	2.35	3.14	3.53	3.91	4.70	3.14	4.19	4.70	5.22	6.26
1000	2.48	3.31	3.73	4.13	4.96	3.31	4.42	4.96	5.51	6.61
1100	2.61	3.48	3.92	4.34	5.22	3.48	4.65	5.22	5.80	6.95
1200	2.72	3.63	4.09	4.53	5.44	3.63	4.85	5.44	6.04	7.25
1300	2.83	3.78	4.25	4.71	5.66	3.78	5.05	5.66	6.29	7.54
1400	2.94	3.93	4.42	4.90	5.88	3.93	5.25	5.88	6.53	7.84
1500	3.03	4.05	4.56	5.05	6.07	4.05	5.41	6.08	6.74	8.09
1600	3.15	4.20	4.73	5.24	6.29	4.20	5.61	6.29	6.99	8.38
1700	3.24	4.32	4.87	5.39	6.48	4.32	5.77	6.48	7.19	8.63
1800	3.33	4.45	5.00	5.54	6.66	4.45	5.94	6.66	7.40	8.87
1900	3.42	4.57	5.14	5.70	6.85	4.57	6.10	6.85	7.60	9.12
2000	3.52	4.69	5.28	5.85	7.03	4.69	6.27	7.03	7.81	9.37
2100	3.59	4.79	5.39	5.98	7.18	4.79	6.40	7.18	7.97	9.56
2200	3.68	4.92	5.53	6.13	7.36	4.92	6.57	7.36	8.18	9.81
2300	3.76	5.01	5.64	6.25	7.51	5.01	6.70	7.51	8.34	10.01
2400	3.85	5.14	5.78	6.41	7.70	5.14	6.86	7.70	8.55	10.25
2500	3.92	5.24	5.89	6.53	7.84	5.24	6.99	7.84	8.71	10.45
2600	4.00	5.34	6.00	6.65	7.99	5.34	7.13	7.99	8.88	10.65
2700	4.07	5.43	6.12	6.78	8.14	5.43	7.26	8.14	9.04	10.85
2800	4.14	5.53	6.23	6.90	8.29	5.53	7.39	8.29	9.21	11.04
2900	4.22	5.63	6.34	7.02	8.44	5.63	7.52	8.44	9.37	11.24
3000	4.31	5.76	6.48	7.18	8.62	5.76	7.69	8.62	9.58	11.49

(courtesy, Estimating Tables for Home Building, Paul I. Thomas, Craftsman Book Company)

Figure 3.25   Concrete Footings/Volume — 10" and 12" Deep

Ground Floor Area	Cubic Yards of Concrete Needed for Size of Footing Shown									
	10" Deep by:					12" Deep by:				
	12"	16"	18"	20"	24"	12"	16"	18"	20"	24"
200	1.85	2.47	2.78	3.08	3.70	2.22	2.96	3.33	3.70	4.44
300	2.28	3.04	3.43	3.80	4.57	2.74	3.65	4.11	4.57	5.48
400	2.62	3.49	3.94	4.37	5.24	3.15	4.19	4.72	5.24	6.29
500	2.93	3.90	4.40	4.88	5.86	3.52	4.68	5.27	5.86	7.03
600	3.20	4.27	4.82	5.35	6.42	3.85	5.13	5.77	6.42	7.70
700	3.45	4.60	5.19	5.76	6.91	4.14	5.52	6.22	6.91	8.29
800	3.70	4.93	5.56	6.17	7.40	4.44	5.92	6.66	7.40	8.88
900	3.91	5.22	5.88	6.53	7.84	4.70	6.26	7.05	7.84	9.40
1000	4.13	5.51	6.20	6.89	8.27	4.96	6.61	7.44	8.27	9.92
1100	4.34	5.80	6.53	7.25	8.70	5.22	6.95	7.83	8.70	10.43
1200	4.53	6.04	6.81	7.56	9.07	5.44	7.25	8.16	9.07	10.89
1300	4.71	6.29	7.08	7.86	9.44	5.66	7.54	8.49	9.44	11.32
1400	4.90	6.53	7.36	8.17	9.81	5.88	7.84	8.82	9.81	11.77
1500	5.05	6.74	7.59	8.43	10.12	6.08	8.09	9.10	10.12	12.14
1600	5.24	6.99	7.87	8.74	10.49	6.29	8.38	9.44	10.49	12.58
1700	5.39	7.19	8.10	9.00	10.80	6.48	8.63	9.71	10.80	12.95
1800	5.54	7.40	8.33	9.25	11.11	6.66	8.87	9.99	11.11	13.32
1900	5.70	7.60	8.57	9.51	11.41	6.85	9.12	10.27	11.41	13.69
2000	5.85	7.81	8.80	9.77	11.72	7.03	9.37	10.55	11.72	14.06
2100	5.98	7.97	8.98	9.97	11.97	7.18	9.56	10.77	11.97	14.36
2200	6.13	8.18	9.21	10.23	12.28	7.36	9.81	11.04	12.28	14.73
2300	6.25	8.34	9.40	10.43	12.53	7.51	10.01	11.27	12.53	15.02
2400	6.41	8.55	9.63	10.69	12.83	7.70	10.25	11.54	12.83	15.39
2500	6.53	8.71	9.82	10.90	13.08	7.84	10.45	11.77	13.08	15.69
2600	6.65	8.88	10.00	11.10	13.33	7.99	10.65	11.99	13.33	15.98
2700	6.78	9.04	10.19	11.31	13.57	8.14	10.85	12.21	13.57	16.28
2800	6.90	9.21	10.37	11.51	13.82	8.29	11.04	12.43	13.82	16.58
2900	7.02	9.37	10.56	11.72	14.07	8.44	11.24	12.65	14.07	16.87
3000	7.18	9.58	10.79	11.98	14.38	8.62	11.49	12.93	14.38	17.24

(courtesy, Estimating Tables for Home Building, Paul I. Thomas, Craftsman Book Company)

Figure 3.26 Cubic Yards of Concrete per Lineal Foot of Wall

To use this table, determine the height of the wall, find the factor in the "Thickness" column, and multiply this factor by the length of the wall.

Example: Calculate the volume of concrete in a wall 40 feet long, 10 inches thick and 8 feet high. Factor x length = .2480 x 40 = 9.9, say 10 C.Y.

Wall Height In Feet	Thickness of Wall in Inches						
	6"	7"	8"	9"	10"	11"	12"
1	.0185	.0217	.0248	.0279	.0310	.0341	.0372
2	.0360	.0434	.0496	.0558	.0620	.0682	.0744
3	.0556	.0651	.0744	.0837	.0930	.1023	.1116
4	.0741	.0868	.0992	.1116	.1240	.1364	.1488
5	.0926	.1085	.1240	.1395	.1550	.1705	.1860
6	.1111	.1302	.1488	.1674	.1860	.2046	.2232
7	.1296	.1519	.1736	.1953	.2170	.2387	.2606
8	.1482	.1736	.1984	.2232	.2480	.2728	.2976
9	.1667	.1953	.2232	.2511	.2790	.3069	.3348
10	.1852	.2170	.2480	.2790	.3100	.3410	.3720
11	.2046	.2387	.2728	.3069	.3410	.3751	.4092
12	.2232	.2604	.2976	.3348	.3720	.4092	.4464

Figure 3.27   Concrete Walls/Volume

(No deduction has been made for openings.) Interpolate for intermediate heights of wall.

| Ground Floor Area | Cubic Yards of Concrete Needed for Size Wall Shown | | | | | | | | | |
| | 8" Wall by Height of: | | | | | 9" Wall by Height of: | | | | |
	1'	2'	4'	6'	8'	1'	2'	4'	6'	8'
200	1.49	2.98	5.95	8.93	11.90	1.67	3.25	6.70	10.04	13.39
300	1.83	3.67	7.34	11.01	14.68	2.06	4.13	8.26	12.39	16.52
400	2.11	4.22	8.43	12.65	16.86	2.37	4.74	9.49	14.23	18.97
500	2.36	4.72	9.42	14.14	18.85	2.65	5.30	10.60	15.90	21.20
600	2.58	5.16	10.32	15.48	20.63	2.90	5.80	11.61	17.41	23.21
700	2.78	5.56	11.11	16.67	22.22	3.13	6.25	12.50	18.75	25.00
800	2.97	5.95	11.90	17.86	23.81	3.35	6.70	13.39	20.09	26.78
900	3.15	6.30	12.60	18.90	25.20	3.54	7.09	14.17	21.26	28.35
1000	3.33	6.65	13.29	19.94	26.59	3.74	7.48	14.95	22.43	29.91
1100	3.50	6.99	13.99	20.98	27.97	3.93	7.87	15.74	23.60	31.47
1200	3.64	7.29	14.58	21.87	29.16	4.10	8.20	16.41	24.61	32.81
1300	3.79	7.59	15.18	22.77	30.36	4.27	8.54	17.07	25.61	34.15
1400	3.95	7.89	15.77	23.66	31.55	4.44	8.87	17.74	26.62	35.49
1500	4.06	8.13	16.27	24.40	32.54	4.56	9.15	18.30	27.45	36.60
1600	4.22	8.43	16.86	25.30	33.73	4.75	9.49	18.97	28.46	37.94
1700	4.34	8.68	17.39	26.04	34.72	4.89	9.77	19.53	29.30	39.06
1800	4.47	8.93	17.86	26.78	35.71	5.02	10.04	20.09	30.13	40.18
1900	4.59	9.18	18.35	27.53	36.70	5.16	10.32	20.65	30.97	41.29
2000	4.71	9.42	18.85	28.27	37.70	5.30	10.60	21.20	31.81	42.41
2100	4.81	9.62	19.24	28.87	38.49	5.42	10.83	21.65	32.48	43.30
2200	4.93	9.87	19.74	29.61	39.48	5.55	11.10	22.21	33.31	44.42
2300	5.03	10.07	20.14	30.21	40.28	5.67	11.33	22.65	33.98	45.31
2400	5.16	10.32	20.63	30.95	41.27	5.81	11.61	23.21	34.82	46.43
2500	5.26	10.52	21.03	31.55	42.06	5.92	11.83	23.66	35.49	47.32
2600	5.35	10.71	21.43	32.14	42.85	6.02	12.05	24.11	36.16	48.21
2700	5.45	10.91	21.82	32.74	43.65	6.14	12.28	24.55	36.83	49.10
2800	5.56	11.11	22.22	33.33	44.44	6.25	12.50	25.00	37.50	50.00
2900	5.66	11.31	22.62	33.93	45.24	6.36	12.72	25.44	38.17	50.89
3000	5.78	11.56	23.11	34.67	46.23	6.50	13.00	26.00	39.00	52.00

(courtesy, Estimating Tables for Home Building, Paul I. Thomas, Craftsman Book Company)

Figure 3.27   Concrete Walls/Volume (continued)

(No deduction has been made for openings.) Interpolate for intermediate heights of wall.

Ground Floor Area	10" Wall by Height of:					12" Wall by Height of:				
	1'	2'	4'	6'	8'	1'	2'	4'	6'	8'
200	1.86	3.72	7.44	11.16	14.88	2.23	4.46	8.93	13.39	17.86
300	2.30	4.59	9.18	13.76	18.35	2.75	5.51	11.01	16.52	22.08
400	2.63	5.27	10.54	15.81	21.08	3.16	6.32	12.65	18.97	25.30
500	2.94	5.89	11.78	17.67	23.56	3.53	7.07	14.14	21.20	28.27
600	3.23	6.45	12.90	19.34	25.79	3.87	7.74	15.48	23.21	30.95
700	3.47	6.94	13.89	20.83	27.78	4.16	8.33	16.67	25.00	33.33
800	3.72	7.44	14.88	22.32	29.76	4.46	8.93	17.86	26.78	35.71
900	3.93	7.87	15.75	23.62	31.50	4.73	9.45	18.90	28.35	37.80
1000	4.15	8.31	16.62	24.92	33.23	4.99	9.97	19.94	29.91	39.88
1100	4.37	8.74	17.48	26.23	34.97	5.25	10.49	20.98	31.47	41.96
1200	4.56	9.11	18.23	27.34	36.46	5.47	10.94	21.87	32.81	43.75
1300	4.74	9.49	18.97	28.46	37.94	5.69	11.38	22.77	34.15	45.53
1400	4.93	9.86	19.72	29.57	39.43	5.92	11.83	23.66	35.49	47.32
1500	5.09	10.17	20.34	30.50	40.67	6.10	12.20	24.40	36.60	48.81
1600	5.27	10.54	21.08	31.62	42.16	6.33	12.65	25.30	37.94	50.59
1700	5.43	10.85	21.70	32.55	43.40	6.51	13.02	26.04	39.06	52.08
1800	5.53	11.16	22.32	33.48	44.64	6.70	13.39	26.78	40.18	53.57
1900	5.70	11.47	22.94	34.41	45.88	6.88	13.76	27.53	41.29	55.06
2000	5.89	11.78	23.56	35.34	47.12	7.07	14.14	28.27	42.41	56.54
2100	6.01	12.03	24.06	36.08	48.11	7.22	14.43	28.87	43.30	57.73
2200	6.17	12.34	24.68	37.01	49.35	7.40	14.81	29.61	44.42	59.22
2300	6.30	12.59	25.17	37.76	50.34	7.55	15.10	30.21	45.31	60.41
2400	6.45	12.90	25.79	38.69	51.58	7.74	15.48	30.95	46.43	61.90
2500	6.57	13.14	26.29	39.43	52.58	7.89	15.77	31.55	47.32	63.09
2600	6.70	13.39	26.78	40.18	53.57	8.04	16.07	32.14	48.21	64.28
2700	6.82	13.64	27.28	40.92	54.56	8.19	16.37	32.74	49.10	65.47
2800	6.95	13.89	27.78	41.66	55.55	8.33	16.67	33.33	50.00	66.66
2900	7.07	14.14	28.27	42.41	56.54	8.48	16.96	33.93	50.89	67.85
3000	7.22	14.45	28.89	43.34	57.78	8.67	17.34	34.67	52.00	69.34

Figure 3.28   Square, Rectangular and Round Column Forms

Multiply the factors by the length of columns to determine the volume of concrete for forms.

Square and Rectangular Column Forms							
Column Size (in Inches)	Volume-C.F./L.F.						
	12"	14"	16"	18"	20"	22"	24"
12	.994	1.161	1.327	1.494	1.661	1.827	1.994
14	1.161	1.355	1.550	1.744	1.938	2.133	2.327
16	1.327	1.550	1.772	1.994	2.216	2.438	2.661
18	1.494	1.744	1.994	2.244	2.494	2.744	2.994
20	1.661	1.938	2.216	2.494	2.772	3.050	3.327
22	1.827	2.133	2.438	2.744	3.050	3.355	3.661
24	1.994	2.327	2.661	2.994	3.327	3.661	3.994

Round Column Forms			
Column Diameter (in Inches)	Volume (C.F./L.F.)	Column Diameter (in Inches)	Volume (C.F./L.F.)
12	0.785	20	2.182
14	1.069	24	3.142
16	1.396	30	4.909
18	1.767	36	7.069

Figure 3.29 Typical Range of Risers for Various Story Heights

This table is to be used as a quick reference for stair design. Use it to determine concrete finish area, form area, and concrete volume.

General Design: Maximum height between landings is 12'; usual stair angle is 20° to 50°, with 30° to 35° the ideal. Usual relation of riser to treads is:

Riser + tread = 17.5

2x (Riser) + tread = 25

Riser x tread = 70 or 75

Maximum riser height is 7" for commercial, 8-1/4" for residential.

Usual riser height is 6-1/2" to 7-1/4".

Minimum tread width is 11" for commercial, 9" for residential.

Story Height	Minimum Risers	Maximum Riser Ht.	Tread Width	Maximum Risers	Mininum Riser Ht.	Tread Width	Average Risers	Average Riser Ht.	Tread Width
7'-6"	12	7.50"	10.00"	14	6.43"	11.07"	13	6.92"	10.58"
8'-0"	13	7.38	10.12	15	6.40	11.10	14	6.86	10.64
8'-6"	14	7.29	10.21	16	6.38	11.12	15	6.80	10.70
9'-0"	15	7.20	10.30	17	6.35	11.15	16	6.75	10.75
9'-6"	16	7.13	10.37	18	6.33	11.17	17	6.71	10.79
10'-0"	16	7.50	10.00	19	6.32	11.18	18	6.67	10.83
10'-6"	17	7.41	10.09	20	6.30	11.20	18	7.00	10.50
11'-0"	18	7.33	10.17	21	6.29	11.21	19	6.95	10.55
11'-6"	19	7.26	10.24	22	6.27	11.23	20	6.90	10.60
12'-0"	20	7.20	10.30	23	6.26	11.24	21	6.86	10.64
12'-6"	20	7.50	10.00	24	6.25	11.25	22	6.82	10.68
13'-0"	21	7.43	10.07	25	6.24	11.26	22	7.09	10.41
13'-6"	22	7.36	10.14	25	6.48	11.02	23	7.04	10.46
14'-0"	23	7.30	10.20	26	6.46	11.04	24	7.00	10.50

Figure 3.30　Floor Selection Guide

This table presents general information to assist in the selection of a concrete floor surface to meet specific use requirements. Service conditions are listed in order of increasing severity, and types of floors in order of increasing durability to permit approximate comparisons. Special considerations (move-in abuse, future use, safety factor, etc.) may justify a highly durable floor even for mild day-to-day conditions.

Service Require- ments	Type of Floor	Relative Cost	Strength of Surface	Relative Abrasion Resistance	Other Characteristics of Floor Surface
Light Traffic	Floor No. 1: Monolothic Concrete Floor. Adequate thickness, reinforcement and strength; properly proportioned mix, 3"–4" slump and minimum bleeding. Good placing and finishing practice with proper tools and proper timing; thorough curing.	100	3500– 4500 psi	100	Susceptible to dusting from (1) laitance caused by use of high slump concrete and/or over-troweling, (2) improper curing, or (3) fracturing of aggregate at the surface under heavy usage, load or impact.
Light Traffic	Floor No. 2: Liquid Hardener Applied to Floor No. 1. Membrane-type compounds should not be used for curing when liquid hardener is to be used. Apply hardener no sooner than 28 days after floor is installed.	105	3500– 4500 psi	100	Liquid hardener temporarily arrests dusting caused by (1) and (2) above. Requires repeated application of hardeners. Does not arrest dusting from (3).
Light and/or Moderate Duty Traffic	Floor No. 3: Dry Shake of Natural Aggregate Applied Over Freshly Floated Surface of Floor No. 1. Well-graded quartz, traprock, emery or granite aggregate is dry mixed with portland cement. Proprietary products contain plasticizing agents and, when desired, colorfast pigments. Heavy applications of shake increases thickness of high-strength surface and service life of floor. Proper timing of finishing procedures and thorough curing are essential.	non- colored: 115–120 colored: 125–135	8000– 12,000 psi	200	Provides "scuff-proofing" for high frequency foot traffic. Built-in colored surface, if desired. Dense surface is more resistant to (1) scaling from cycles of freeze and thaw and use of de-icing salts, and (2) mild corrosive materials. Easy to clean. Aggregate at surface fractures under high frequency, hard wheeled traffic and impact.
Moderate Duty Traffic	Floor No. 4: High Strength Natural Aggregate Topping (two course) Applied Over Fresh or Set Base Slag. 3/4" to 2" thick topping; properly proportioned mix containing 1/4" – 3/8" aggregate, designed for 1" to 2" slump and 8000–10,000 psi 28 day . Aggregates commonly used are quartz, traprock, emery or granite.	over fresh slab: 130–180 over set slab: 140–185	8000– 12,000 psi	200	Same as Floor No. 3 plus ability of surface to withstand heavier loads. Aggregate at surface fractures under high frequency, hard wheeled traffic and impact.

Figure 3.30   Floor Selection Guide (continued)

Service Require-ments	Type of Floor	Relative Cost	Strength of Surface	Relative Abrasion Resistance	Other Characteristics of Floor Surface
Moderate and/or Heavy Duty Traffic	Floor No. 5: Dry Shake of Metallic Aggregate Applied Over Freshly Floated Surface of Floor No. 1 or Floor No. 4. Scientifically graded metallic aggregate, free of rust, oil and non-ferrous particles is combined with plasticizing agents and, where desired, colorfast pigments. Mix dry with Portland cement. Heavy application of shake increases thickness of high-strength iron-armoured surface and service life of floor. Proper timing of finishing procedures and proper curing are essential.	monolithic floor: 130–140 two course floor: 160–170	12,000 psi	800	Ductile metallic surface is non-dusting, dense and resistant to oil and grease. Easy to clean. Provides built-in color and/or durable, slip-resistant finish where desired. Withstands impact and high frequency or heavy load traffic from hard wheeled material handling equipment.
Heavy Duty Traffic and/or Extra Heavy Duty Traffic (Key Area Floors)	Floor No. 6: High Strength Metallic Aggregate Topping Applied Over Fresh or Set Base Slab. Specially formulated all-iron aggregate topping, Anvil Top, requires only addition of water. Apply 1/2"–1" thickness using procedures recommended by field service man.	1/2" over fresh base slab: 380–400 1" over set base slab: 490–550	12,000 psi	800* *Plus greater thickness of iron-armour	Service life under high concentration of heavy industrial traffic is 10 to 15 times greater than high-strength concrete toppings. Withstands heavy impact, heavy abrasion and high point loads. Dense surface is easy to clean.

(Courtesy Modern Plant Operation and Maintenance, Winter 1971/72)

Figure 3.31   Concrete Quantities for Metal Pan Floor Slabs

Concrete Quantities/20" Widths*				
Depth of Metal Pan	Width of Joist	C.F. of Concrete per S.F. for Various Slab Thicknesses		Additional Concrete for Tapered End Forms, C.F. per L.F. of Bearing Wall or Beam (one side only)
		2-1/2"	3"	4-1/2"	
6"	4"	.303	.345	.470	.13
	5"	.319	.361	.486	.12
	6"	.334	.376	.501	.12
8"	4"	.339	.381	.506	.17
	5"	.361	.402	.527	.16
	6"	.380	.422	.547	.16
10"	4"	.377	.419	.544	.21
	5"	.404	.445	.570	.20
	6"	.428	.470	.595	.19
12"	4"	.418	.459	.584	.25
	5"	.449	.491	.616	.24
	6"	.479	.520	.645	.23
14"	5"	.497	.538	.664	.28
	6"	.531	.573	.698	.27
	7"	.562	.604	.729	.26
16"	6"	.585	.627	.752	.31
	7"	.621	.663	.788	.30
	8"	.654	.695	.820	.29
20"	7"	.744	.786	.911	.37
	8"	.785	.826	.951	.36
	9"	.822	.864	.989	.35

*Apply only for areas over flange type forms and joists between them. Bridging joists, special headers, beams, tees, etc., not included.

(Courtesy CECO Corporation.)

Figure 3.32 Concrete Quantities for Metal Pan Floor Slabs (continued)

Concrete Quantities/30" Widths*				
Depth of Metal Pan	Width of Joist	C.F. of Concrete per S.F. for Various Slab Thicknesses		Additional Concrete for Tapered End Forms, C.F. per L.F. of Bearing Wall or Beam (one side only)
		2-1/2"	3"	4-1/2"	
6"	5"	.288	.329	.454	.11
	6"	.299	.341	.466	.10
	7"	.310	.352	.477	.10
8"	5"	.317	.359	.484	.14
	6"	.333	.374	.499	.14
	7"	.347	.389	.514	.14
10"	5"	.348	.390	.515	.18
	6"	.367	.409	.534	.17
	7"	.386	.427	.552	.17
12"	5"	.381	.422	.547	.21
	6"	.404	.445	.570	.21
	7"	.425	.464	.592	.20
14"	5"	.415	.456	.581	.25
	6"	.441	.483	.608	.24
	7"	.467	.508	.633	.24
16"	6"	.481	.522	.647	.28
	7"	.509	.551	.676	.27
	8"	.537	.578	.703	.26
20"	7"	.599	.641	.766	.34
	8"	.633	.675	.800	.33
	9"	.665	.707	.832	.32

*Apply only for areas over flange type forms and joists between them. Bridging joists, special headers, beams, tees, etc., not included.

(Courtesy CECO Corporation.)

Figure 3.32 Concrete Quantities for Metal Pan Floor Slabs Using Adjustable Forms

		Concrete Quantities/20" Widths*			
Depth of Metal Pan	Width of Joist	C.F. of Concrete per S.F. for Various Slab Thicknesses			Additional Concrete for Tapered End Forms, C.F. per L.F. of Bearing Wall or Beam (one side only)
		2"	2-1/2"	3"	
6"	4-1/2"	.279	.321	.362	.12
	5-1/2"	.295	.337	.378	.12
	7-1/2"	.322	.364	.405	.11
8"	4-1/2"	.309	.350	.393	.16
	5-1/2"	.331	.372	.414	.16
	7-1/2"	.367	.409	.451	.15
10"	4-1/2"	.340	.381	.424	.20
	5-1/2"	.367	.408	.450	.19
	7-1/2"	.413	.455	.496	.18
12"	4-1/2"	.371	.412	.455	.24
	5-1/2"	.403	.444	.486	.23
	7-1/2"	.458	.500	.542	.22
14"	4-1/2"	.402	.443	.485	.28
	5-1/2"	.438	.480	.522	.27
	7-1/2"	.504	.545	.587	.26

*Apply only for areas over flange type forms and joists between them. Bridging joists, special headers, beams, tees, etc., not included.

(Courtesy CECO Corporation.)

Figure 3.32   Concrete Quantities for Metal Pan Floor
Slabs Using Adjustable Forms (continued)

Concrete Quantities/30" Widths*					
Depth of Metal Pan	Width of Joist	C.F. of Concrete per S.F. for Various Slab Thicknesses			Additional Concrete for Tapered End Forms, C.F. per L.F. of Bearing Wall or Beam (one side only)
		2-1/2"	3"	3-1/2"	
6"	5-1/2"	.300	.341	.383	.11
	6-1/2"	.311	.353	.395	.11
	7-1/2"	.322	.364	.406	.11
8"	5-1/2"	.326	.368	.408	.15
	6-1/2"	.341	.383	.424	.15
	7-1/2"	.355	.397	.438	.14
10"	5-1/2"	.352	.394	.434	.19
	6-1/2"	.371	.413	.454	.18
	7-1/2"	.389	.430	.472	.18
12"	5-1/2"	.378	.420	.461	.22
	6-1/2"	.401	.443	.484	.22
	7-1/2"	.422	.464	.505	.21
14"	5-1/2"	.404	.446	.486	.26
	6-1/2"	.430	.472	.514	.25
	7-1/2"	.456	.497	.539	.24

*Apply only for areas over flange type forms and joists between them. Bridging joists, special headers, beams, tees, etc., not included.

(*Courtesy* CECO *Corporation.*)

Figure 3.33   Voids Created by Various Size Metal Pan Forms

Depth of Pan Form	C.F. of Void Created per L.F. (for various widths of metal pan forms)				**C.F. per Tapered End	
	30" Width	20" Width	15" Width	10" Width	30" Width	20" Width
8"	1.628	1.072	.794	.516	4.47	2.88
10"	2.023	1.329	.982	.634	5.55	3.57
12"	2.414	1.581	1.165	.748	6.62	4.24
14"	2.801	1.829	1.343	.857	7.67	4.89
16"	3.183	2.072	1.516	.961	8.72	5.54
20"	3.933	2.544	1.850	1.155	10.76	6.78

**Total void created by standard 3'-0" length tapered end form.

(*Courtesy* CECO *Corporation.*)

Figure 3.34   Steel Domes — Two-Way Joist Construction

Two-way joist or waffle slabs are formed with metal or fiberglass domes. These are erected on a flat form in accordance with the manufacturer's specifications. The information in this chart is typical of that found in sheets from manufacturers.

Voids Created with 2'-0" Design Module		
Depth of Dome	**Overall Plan Size 24" x 24"**	
	Plan Size of Void	**C.F. of Void**
8"	19" x 19"	1.41
10"	19" x 19"	1.90
12"	19" x 19"	2.14
14"	19" x 19"	2.44

Voids Created with 3'-0" Design Module		
Standard Sizes	**Overall Plan Size 36" x 36"**	
Depth of Dome	Plan Size of Void	C.F. of Void
8"	30" x 30"	3.85
10"	30" x 30"	4.78
12"	30" x 30"	5.53
14"	30" x 30"	6.54
16"	30" x 30"	7.44
20"	30" x 30"	9.16

Filler Sizes	Overall Plan Size 26" x 36"		Overall Plan Size 26" x 26"	
Depth of Dome	**Plan Size of Void**	**C.F. of Void**	**Plan Size of Void**	**C.F. of Void**
8"	20" x 30"	2.54	20" x 20"	1.65
10"	20" x 30"	3.13	20" x 20"	2.06
12"	20" x 30"	3.63	20" x 20"	2.41
14"	20" x 30"	4.27	20" x 20"	2.87
16"	20" x 30"	4.85	20" x 20"	3.14
20"	20" x 30"	5.90	20" x 20"	3.81

U.S. Patent No. 2,850,785 applies to all sizes
(*Courtesy* CECO *Corporation*)

Figure 3.35 Precast Concrete Wall Panels

This table lists the practical size limits for precast concrete wall panels by thickness. Panels are either solid or insulated with plain, colored or textured finishes. Transportation is an important cost factor. Engineering data is available from fabricators to assist with construction details. Usual minimum job size for economical use of panels is about 5000 S.F.

Thickness	Maximum Size	Thickness	Maximum Size
3"	50 S.F.	6"	300 S.F.
4"	150 S.F.	7"	300 S.F.
5"	200 S.F.	8"	300 S.F.

Figure 3.36 Daily Erection Rate of Precast Concrete Panels

This table lists the daily erection rate of precast concrete panels for a six-man crew with a crane for high-rise and low-rise structures.

Panel Size L x H	Area per Panel (S.F.)	Low-Rise		High-Rise	
		Daily Production Range		Daily Production Range	
		Pieces	Area (S.F.)	Pieces	Area (S.F.)
4' x 4'	16	10	160	9	144
		20	320	18	288
4' x 8'	32	10	320	9	288
		19	608	17	544
8' x 8'	64	9	576	8	512
		18	1152	16	1024
10' x 10'	100	8	800	7	700
		15	1500	14	1400
15' x 10'	150	7	1050	6	900
		12	1800	11	1650
20' x 10'	200	6	1200	5	1000
		8	1600	7	1400
30' x 10'	300	5	1500	4	1200
		7	2100	7	2100

Figure 3.37 Installation Time in Man-Hours for Strip Footings

Description	Man-Hours	Unit
Formwork	.066	S.F.C.A.
Formwork Keyway	.015	L.F.
Reinforcing		
#4 to #7	15.238	ton
#8 to #14	8.889	ton
Dowels		
#4	.128	Ea.
#6	.152	Ea.
Place Concrete,		
Direct Chute	.400	C.Y.
Pumped	.640	C.Y.
Crane and Bucket	.711	C.Y.
Concrete in Place		
36" x 12" Reinforced	1.630	C.Y.

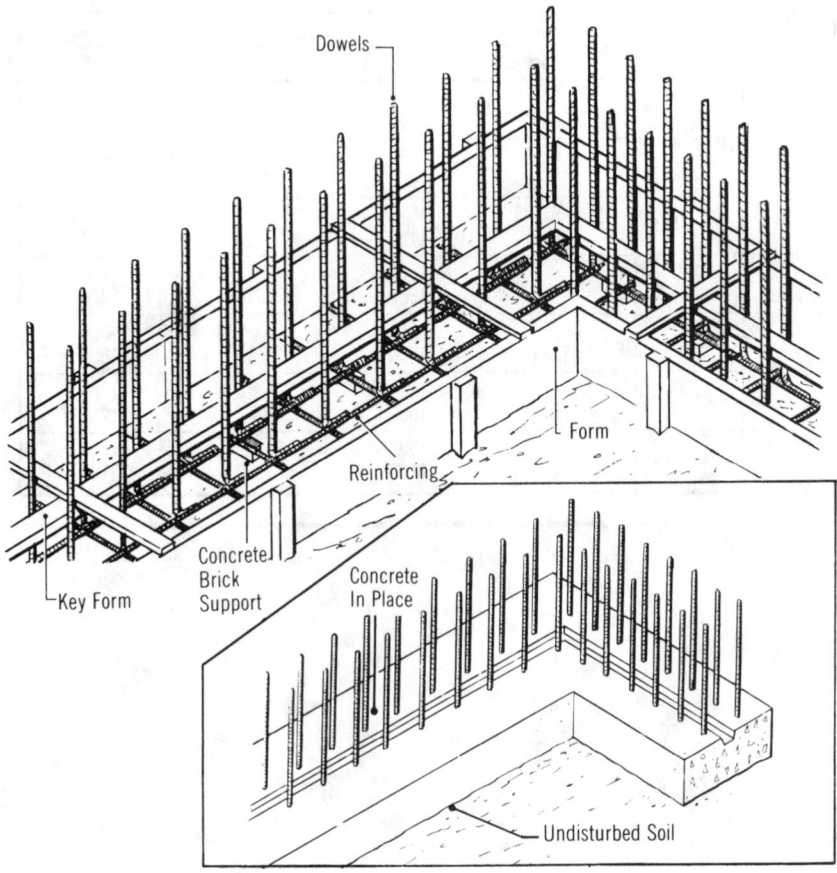

Figure 3.38 Installation Time in Man-Hours for Spread Footings

Description	Man-Hours	Unit
Formwork	.105	S.F.C.A.
Reinforcing		
#4 to #7	15.239	ton
#8 to #14	8.889	ton
Placing Concrete under 1 C.Y.		
Direct Chute	.873	C.Y.
Pumped	1.280	C.Y.
Crane and Bucket	1.422	C.Y.
Over 5 C.Y.		
Direct Chute	.436	C.Y.
Pumped	.610	C.Y.
Crane and Bucket	.640	C.Y.
Anchor bolt or dowel templates	1.000	Ea.

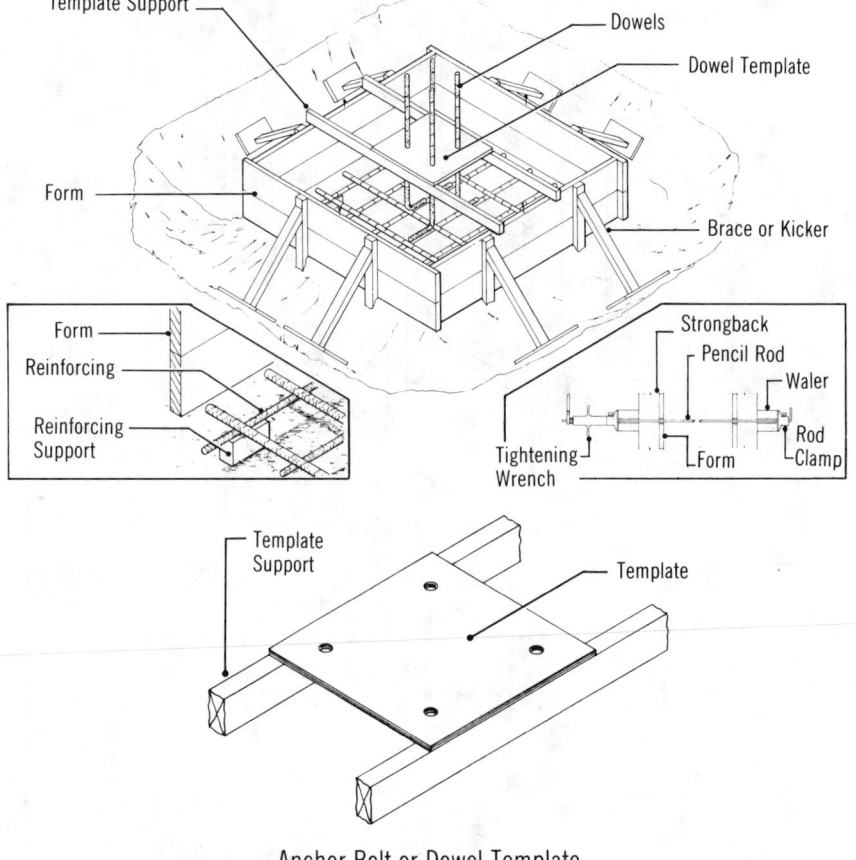

Anchor Bolt or Dowel Template

Figure 3.39 Installation Time in Man-Hours for Stepped Footings

Description	Man-Hours	Unit
Formwork (sides and steps)	.085	S.F.C.A.
Formwork (keyway)	.015	L.F.
Reinforcing #4 to #7	15.238	ton
Reinforcing (dowels, 2' long)	.128	Ea.
Place Concrete,		
Direct Chute	.400	C.Y.
Pumped	.640	C.Y.
Crane and Bucket	.711	C.Y.
Concrete in Place	3.900	C.Y.

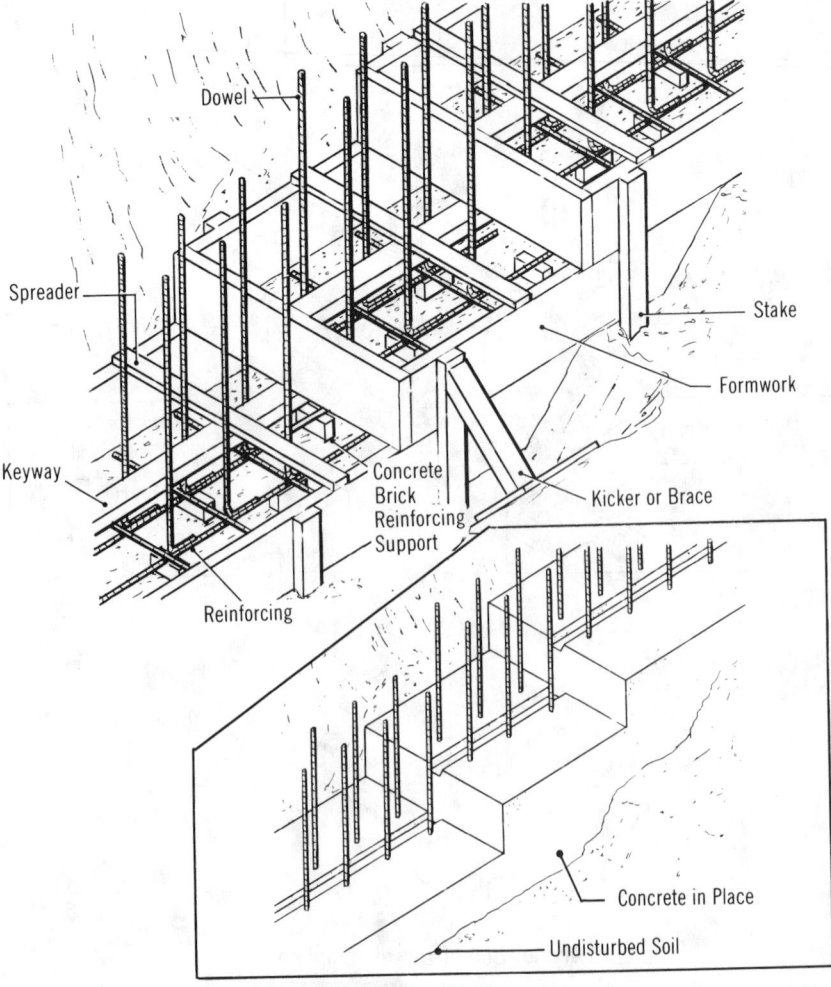

Figure 3.40 Installation Time in Man-Hours for Foundation Walls

Description	Man-Hours	Unit
Forms		
Job-Built Plywood		
1 Use/Month	.130	S.F.C.A.
4 Uses/Month	.095	S.F.C.A.
Modular Prefabricated Plywood		
1 Use/Month	.053	S.F.C.A.
4 Uses/Month	.049	S.F.C.A.
Steel Framed Plywood		
1 Use/Month	.080	S.F.C.A.
4 Uses/Month	.072	S.F.C.A.
Box Out Openings to 10 S.F.	2.000	Ea.
Brick Shelf	.200	S.F.C.A.
Bulkhead	.181	L.F.
Corbel to 12" Wide	.320	L.F.
Pilasters	.178	S.F.C.A.
Waterstop Dumbbell	.055	L.F.
Reinforcing		
#3 to #7	10.667	ton
#8 to #14	8.000	ton
Place Concrete 12" Walls		
Direct Chute	.480	C.Y.
Pumped	.674	C.Y.
With Crane and Bucket	.711	C.Y.
Finish, Break Ties and Patch Voids	.015	S.F.
Burlap Rub with Grout	.018	S.F.
Concrete in Place 12" Thick	5.926	C.Y.

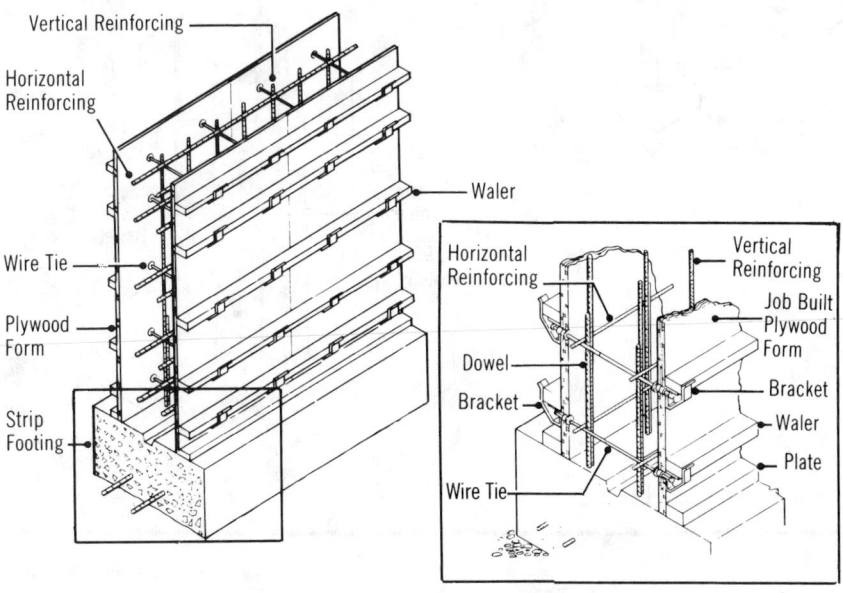

Figure 3.41 Installation Time in Man-Hours for Grade Beams

Description	Man-Hours	Unit
Forms	.083	S.F.C.A.
Brick or Slab Shelf	.200	S.F.C.A.
Bulkhead	.181	L.F.
Reinforcing		
#3 to #7	20.000	ton
#8 to #18	11.852	ton
Place Concrete		
Direct Chute	.320	C.Y.
Pumped	.492	C.Y.
With Crane and Bucket	.533	C.Y.
Finish, Break Ties and Patch Voids	.015	S.F.

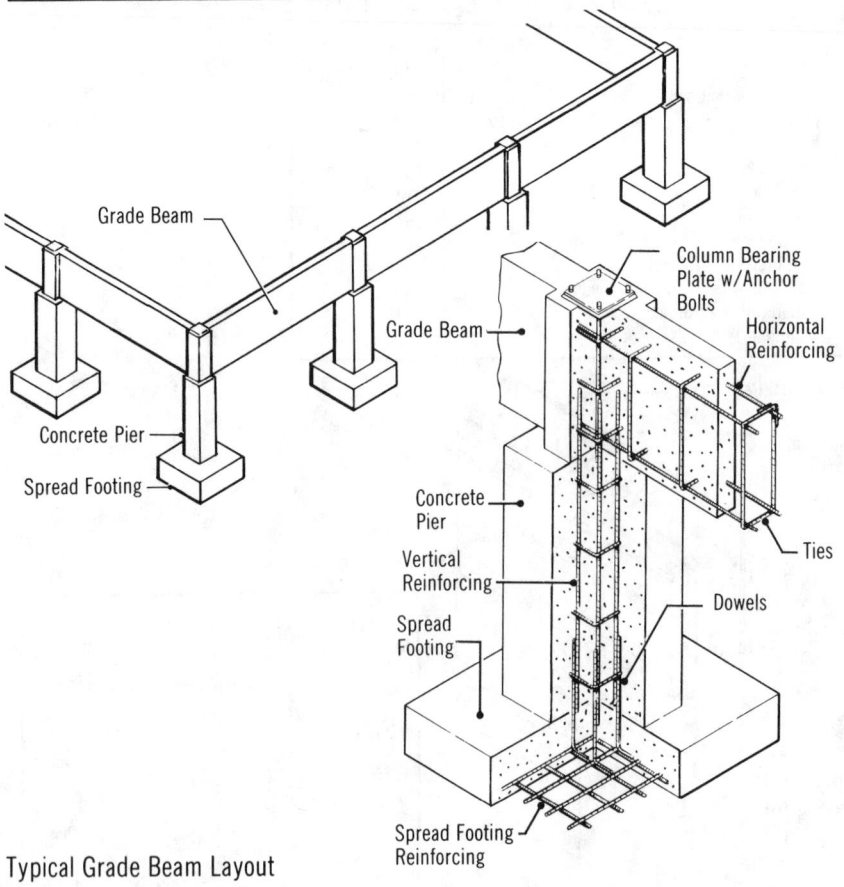

Grade Beam

Column Bearing Plate w/Anchor Bolts

Grade Beam

Horizontal Reinforcing

Concrete Pier

Spread Footing

Grade Beam

Concrete Pier

Ties

Vertical Reinforcing

Dowels

Spread Footing

Spread Footing Reinforcing

Typical Grade Beam Layout

Figure 3.42 Installation Time in Man-Hours for Slabs-on-Grade

Description	Man-Hours	Unit
Slab-on-Grade		
Fine Grade	.010	S.Y.
Gravel Under Floor Slab 6" Deep Compacted	.005	S.F.
Polyethylene Vapor Barrier	.216	Sq.
Reinforcing WWF 6 x 6 (W1.4/W1.4)	.457	C.S.F.
Place and Vibrate Concrete 4" Thick Direct Chute	.436	C.Y.
Expansion Joint Premolded Bituminous Fiber		
1/2" x 6"	.021	L.F.
Edge Forms in Place to 6" High 4 Uses on Grade	.053	L.F.
Curing w/Sprayed Membrane Curing Compound	.168	C.S.F.
Finishing Floor		
Monolithic Screed Finish	.009	S.F.
Steel Trowel Finish	.015	S.F.

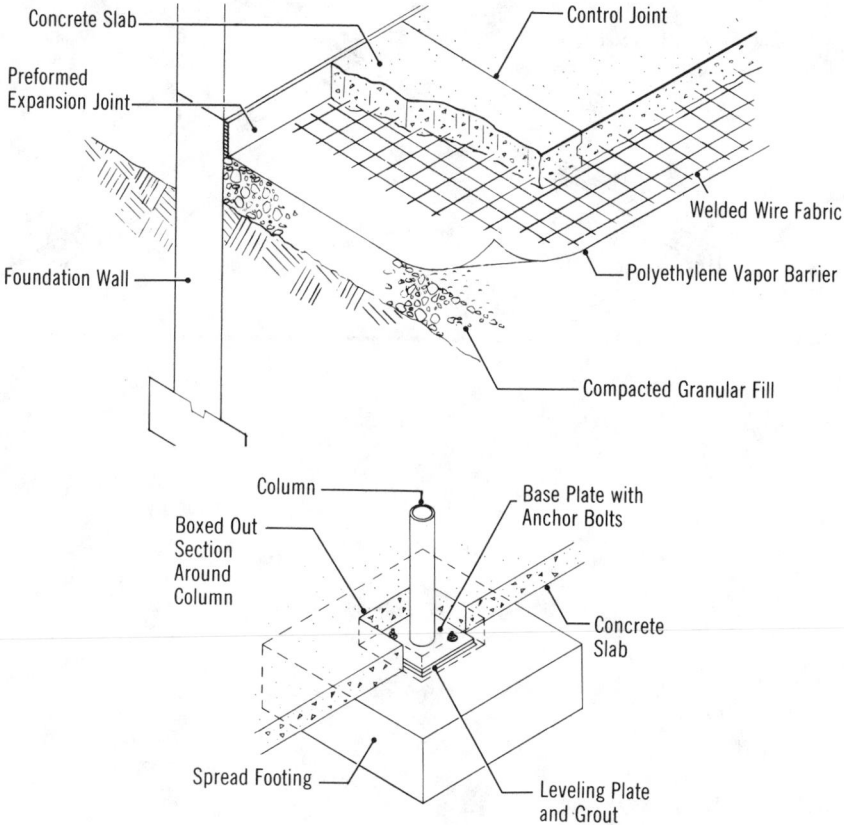

Concrete Slab

Preformed Expansion Joint

Foundation Wall

Control Joint

Welded Wire Fabric

Polyethylene Vapor Barrier

Compacted Granular Fill

Column

Boxed Out Section Around Column

Base Plate with Anchor Bolts

Concrete Slab

Spread Footing

Leveling Plate and Grout

Control Joint Around Column

Figure 3.43 Installation Time in Man-Hours for One-Way Slabs

Description	Man-Hours	Unit
Forms in Place Elevated Slabs		
Column Forms	.134	S.F.C.A.
Floor Slab Hung from Steel Beams	.085	S.F.C.A.
Flat Slab, Shored	.086	S.F.C.A.
Edge Forms	.091	S.F.C.A.
Elevated Slabs Reinforcing in Place	11.034	ton
Columns	13.913	ton
Hoisting Reinforcing	.609	ton
Placing Concrete Pumped	.492	C.Y.
With Crane and Bucket	.582	C.Y.
Steel Trowel Finish	.015	S.F.
Concrete in Place 8" Slab	6.480	C.Y.

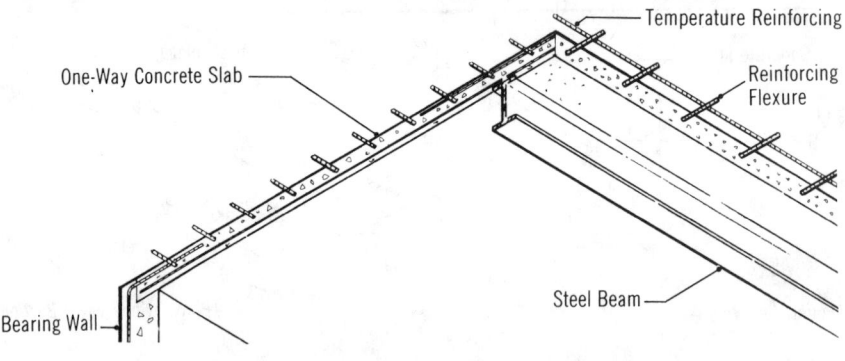

Temperature Reinforcing

One-Way Concrete Slab

Reinforcing Flexure

Steel Beam

Bearing Wall

Figure 3.44 Installation Time in Man-Hours for One-Way Concrete Beams and Slabs

Description	Man-Hours	Unit
Forms in Place		
Square Column	.136	S.F.C.A.
Beam and Girder	.120	S.F.C.A.
Slab	.086	S.F.C.A.
Reinforcing Columns		
#3 to #7	21.333	ton
#8 to #14	13.913	ton
Beams and Girders		
#3 to #7	20.000	ton
#8 to #14	11.852	ton
Elevated Slabs	11.034	ton
Hoisting	.609	ton
Placing Concrete Pumped	.492	C.Y.
With Crane and Bucket	.582	C.Y.
Steel Trowel Finish	.015	S.F.
Concrete in Place Including		
Forms, Reinforcing and Finish	9.880	C.Y.

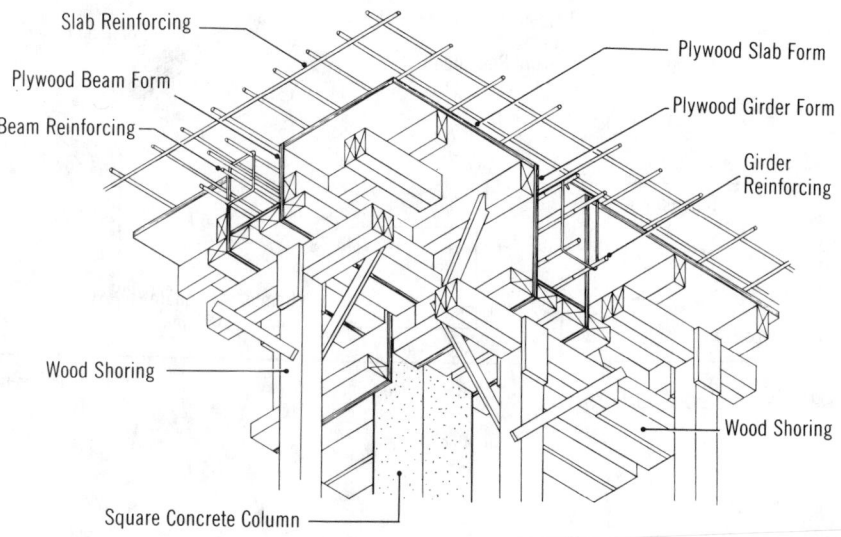

Slab Reinforcing

Plywood Beam Form

Beam Reinforcing

Plywood Slab Form

Plywood Girder Form

Girder Reinforcing

Wood Shoring

Wood Shoring

Square Concrete Column

Figure 3.45   Installation Time in Man-Hours for One-Way Concrete Joist Slabs

Description	Man-Hours	Unit
Forms in Place		
Joists and Pans	.096	S.F.
Beam Bottoms	.166	S.F.C.A.
Beam Sides	.108	S.F.C.A.
Edge Forms	.091	S.F.C.A.
Reinforcing		
Joists and Ribs		
#3 to #11	20.000	ton
Slabs, Bars		
#3 to #7	11.034	ton
WWF 6 x 6 w4/w4	.593	C.S.F.
Placing Concrete		
Pumped	.582	C.Y.
With Crane and Bucket	.674	C.Y.
Steel Trowel Finish	.015	S.F.
Concrete in Place Including		
Forms, Reinforcing and Finish	8.056	C.Y.

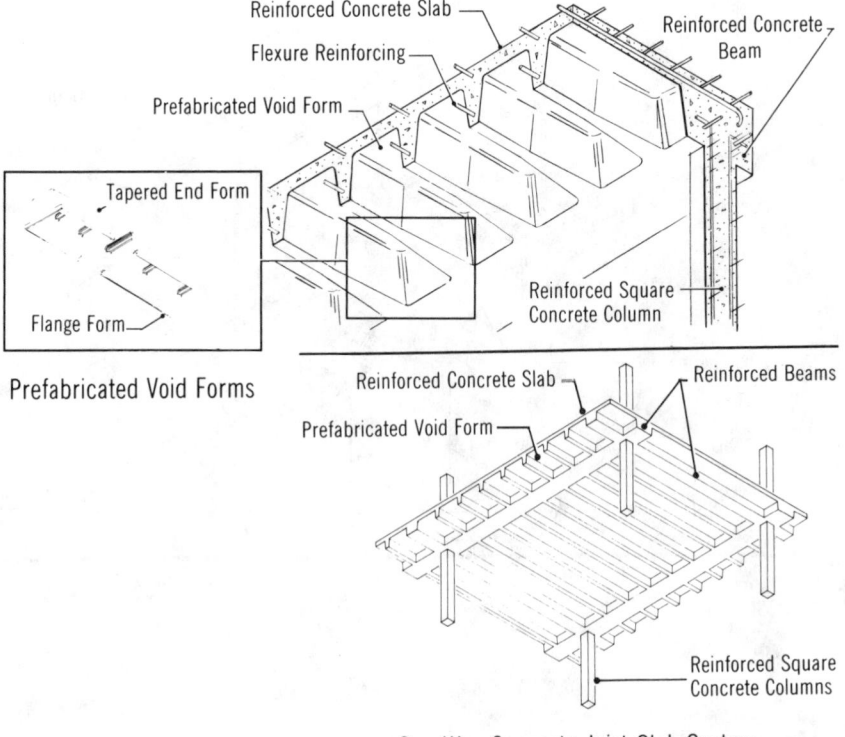

Reinforced Concrete Slab

Flexure Reinforcing

Prefabricated Void Form

Reinforced Concrete Beam

Tapered End Form

Flange Form

Reinforced Square Concrete Column

Prefabricated Void Forms

Reinforced Concrete Slab

Prefabricated Void Form

Reinforced Beams

Reinforced Square Concrete Columns

One-Way Concrete Joist Slab System

Figure 3.46 Installation Time in Man-Hours for Two-Way Concrete Beams and Slabs

Description	Man-Hours	Unit
Forms in Place		
Column Square	.136	S.F.C.A.
Beams	.120	S.F.C.A.
Slab	.086	S.F.C.A.
Reinforcing Columns		
#3 to #7	21.333	ton
#8 to #14	13.913	ton
Beams		
#3 to #7	20.000	ton
#8 to #14	11.852	ton
Elevated Slabs	11.034	ton
Hoisting Reinforcing	.609	ton
Placing Concrete Pumped	.492	C.Y.
With Crane and Bucket	.582	C.Y.
Steel Trowel Finish	.015	S.F.
Concrete in Place Including		
Forms, Reinforcing and Finish	8.630	C.Y.

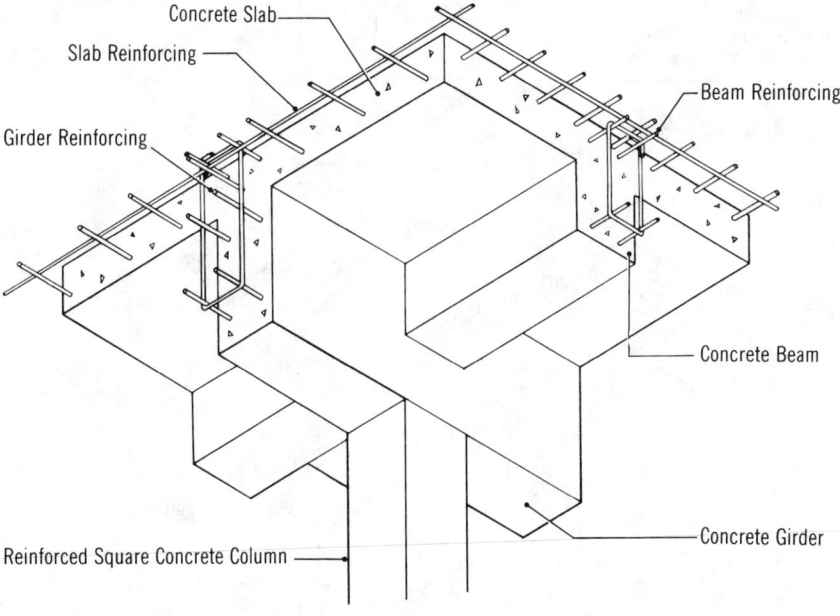

Concrete Slab

Slab Reinforcing

Girder Reinforcing

Beam Reinforcing

Concrete Beam

Concrete Girder

Reinforced Square Concrete Column

Figure 3.47   Installation Time in Man-Hours for Concrete Flat-Plate

Description	Man-Hours	Unit
Forms in Place		
Columns Square	.136	S.F.C.A.
Round Fiber Tube	.221	L.F.
Round Steel	.256	L.F.
Capitols	2.667	Ea.
Flat Plate	.086	S.F.
Edge Forms	.091	S.F.C.A.
Reinforcing Columns		
#3 to #7	21.333	ton
#8 to #14	13.913	ton
Spirals	14.545	ton
Butt Splice, Clamp Sleeve and Wedge	.373	Ea.
Mechanical Full Tension Splice with Filler Metal	.903	Ea.
Elevated Slab	11.034	ton
Hoisting Reinforcing	.609	ton
Place Concrete		
Pumped	.492	C.Y.
With Crane and Bucket	.582	C.Y.
Steel Trowel Finish	.015	S.F.
Concrete in Place Including		
Forms, Reinforcing and Finish	6.467	C.Y.

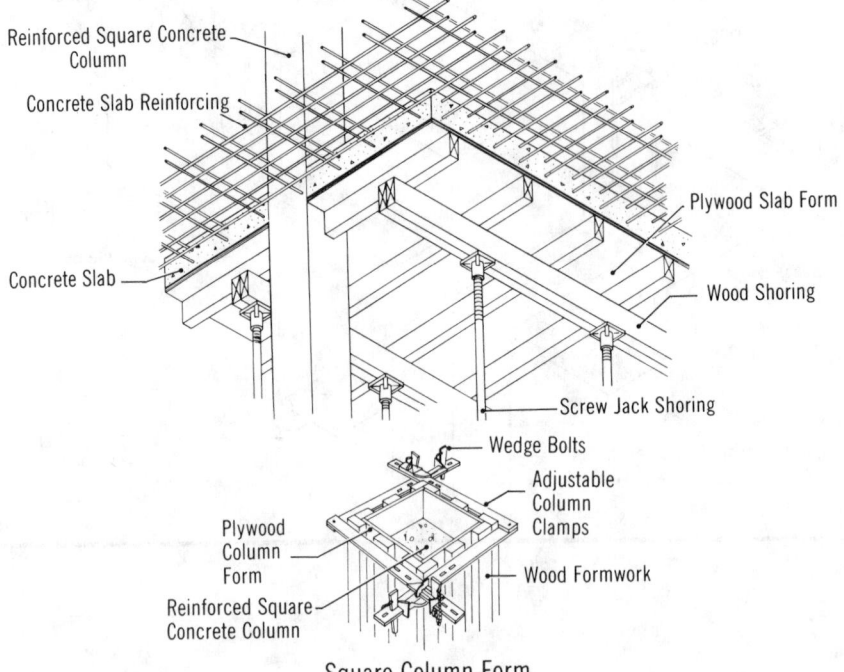

Reinforced Square Concrete Column
Concrete Slab Reinforcing
Plywood Slab Form
Concrete Slab
Wood Shoring
Screw Jack Shoring
Wedge Bolts
Adjustable Column Clamps
Plywood Column Form
Wood Formwork
Reinforced Square Concrete Column

Square Column Form

Figure 3.48 Installation Time in Man-Hours for Concrete Flat Slab with Drop Panels

Description	Man-Hours	Unit
Forms in Place		
Columns Square	.136	S.F.C.A.
Round Fiber Tube	.221	L.F.
Round Steel	.256	L.F.
Capitols	2.667	Ea.
Flat Slab with Drops	.088	S.F.
Edge Forms	.091	S.F.C.A.
Reinforcing Columns		
#3 to #7	21.333	ton
#8 to #14	13.913	ton
Spirals	14.545	ton
Butt Splice Clamp Sleeve and Wedge	.373	Ea.
Mechanical Full Tension Splice with Filler Metal	.903	Ea.
Elevated Slab	11.034	ton
Hoisting Reinforcing	.609	ton
Place Concrete		
Pumped	.492	C.Y.
With Crane and Bucket	.582	C.Y.
Steel Trowel Finish	.015	S.F.
Concrete in Place Including		
Forms, Reinforcing and Finish	5.403	C.Y.

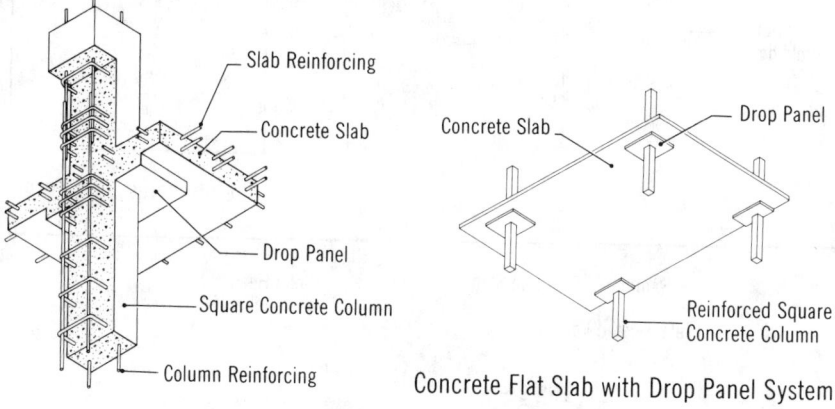

Slab Reinforcing

Concrete Slab

Drop Panel

Square Concrete Column

Column Reinforcing

Concrete Slab

Drop Panel

Reinforced Square Concrete Column

Concrete Flat Slab with Drop Panel System

Figure 3.49 Installation Time in Man-Hours for Concrete Waffle Slabs

Description	Man-Hours	Unit
Forms in Place		
Joists and 19″ Domes	.097	S.F.
Joists and 30″ Domes	.102	S.F.
Edge Forms	.091	S.F.C.A.
Reinforcing		
Joists	20.000	ton
Slabs 6 x 6 w4/w4	.593	C.S.F.
Placing Concrete		
Pumped	.582	C.Y.
With Crane and Bucket	.674	C.Y.
Steel Trowel Finish	.015	S.F.
Concrete in Place Including		
Forms, Reinforcing and Finish	5.950	C.Y.

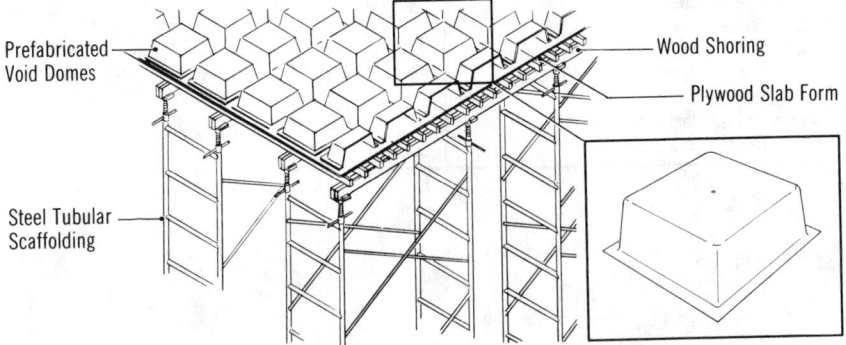

Prefabricated Void Domes

Steel Tubular Scaffolding

Wood Shoring

Plywood Slab Form

Prefabricated Void Form

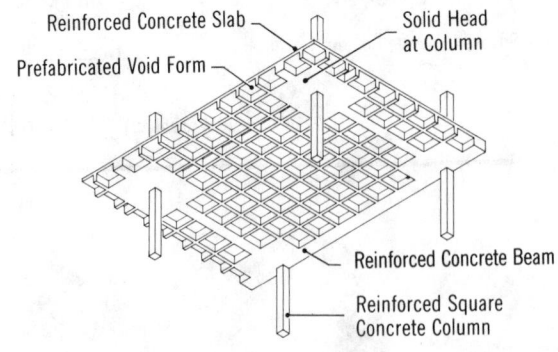

Reinforced Concrete Slab

Prefabricated Void Form

Solid Head at Column

Reinforced Concrete Beam

Reinforced Square Concrete Column

Concrete Waffle Slab System

Figure 3.50   Installation Time in Man-Hours for Precast Concrete Planks

Description	Man-Hours	Unit
Erect and Grout Prestressed Slabs		
6" Deep	1.029	Ea.
8" Deep	1.108	Ea.
10" Deep	1.200	Ea.
12" Deep	1.309	Ea.
Average		
8" Deep	.010	S.F.

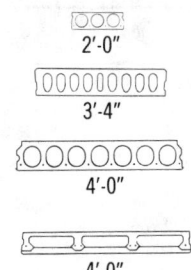

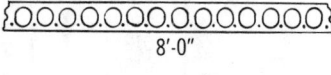

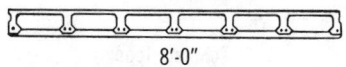

Precast Plank Profiles

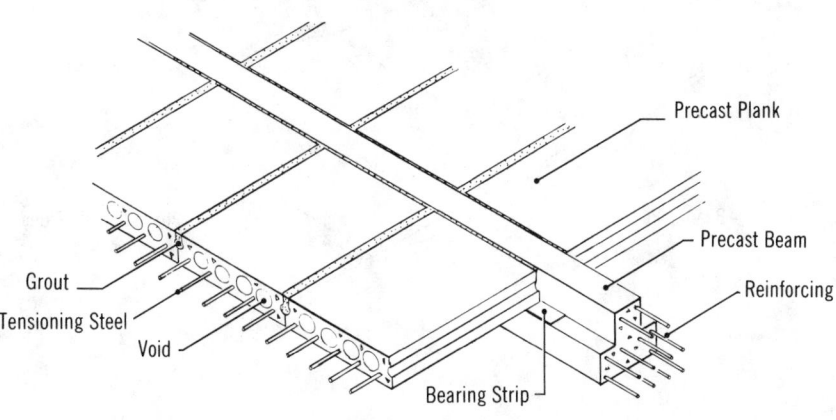

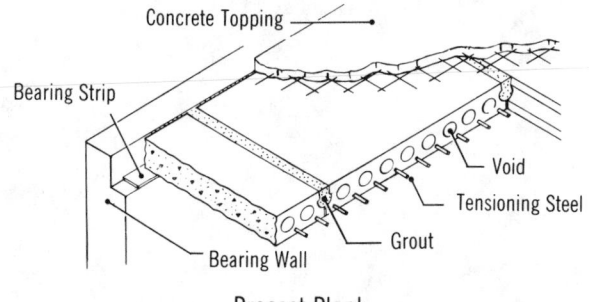

Precast Plank

Figure 3.51 Installation Time in Man-Hours for Precast Single T Beams

Description	Man-Hours	Unit
Single T Erection		
Span		
40'	7.200	Ea.
80'	9.000	Ea.
100'	12.000	Ea.
120'	14.400	Ea.
Welded Wire Fabric 6 x 6-10/10	.457	C.S.F.
Place Concrete		
Pumped	.582	C.Y.
With Crane and Bucket	.674	C.Y.
Broom Finish	.012	S.F.
Steel Trowel Finish	.015	S.F.

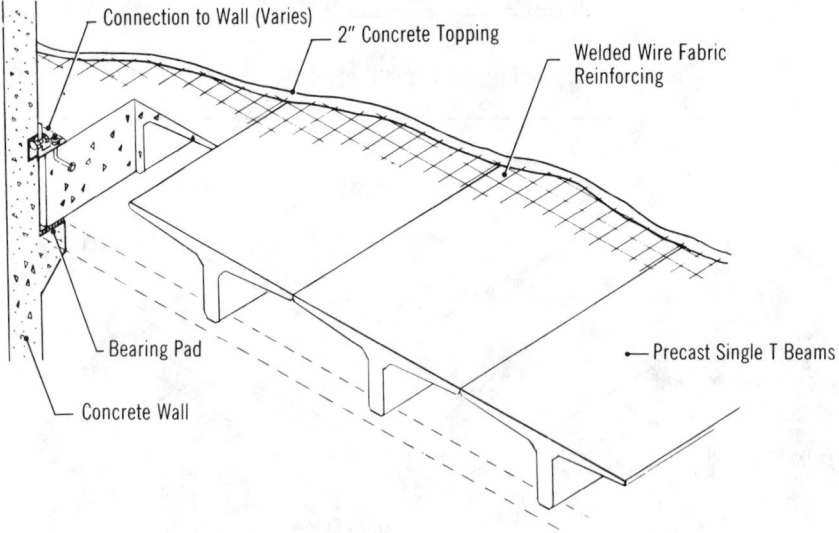

Connection to Wall (Varies)
2" Concrete Topping
Welded Wire Fabric Reinforcing
Bearing Pad
Concrete Wall
Precast Single T Beams

Figure 3.52 Installation Time in Man-Hours for Precast Columns, Beams, and Double T's

Description	Man-Hours	Unit
Columns		
12' High	1.895	Ea.
24' High	2.400	Ea.
Beams		
20' Span		
12" x 20"	2.250	Ea.
18" x 36"	3.000	Ea.
24" x 44"	3.273	Ea.
30' Span		
12" x 36"	3.000	Ea.
18" x 44"	3.600	Ea.
24" x 52"	4.500	Ea.
40' Span		
12" x 52"	3.600	Ea.
18" x 52"	4.500	Ea.
24" x 52"	6.000	Ea.
Double T's		
45' Span		
12" Deep x 8' Wide	3.273	Ea.
18" Deep x 8' Wide	3.600	Ea.
50' Span, 24" Deep x 8' Wide	4.500	Ea.
60' Span, 32" Deep x 10' Wide	5.143	Ea.
Welded Wire Fabric 6 x 6-10/10	.457	C.S.F.
Place Concrete		
Pumped	.582	C.Y.
With Crane and Bucket	.674	C.Y.
Broom Finish	.012	S.F.
Steel Trowel Finish	.015	S.F.

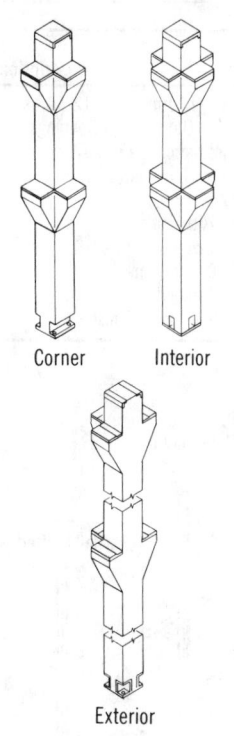

Corner   Interior

Exterior

Precast Concrete Columns

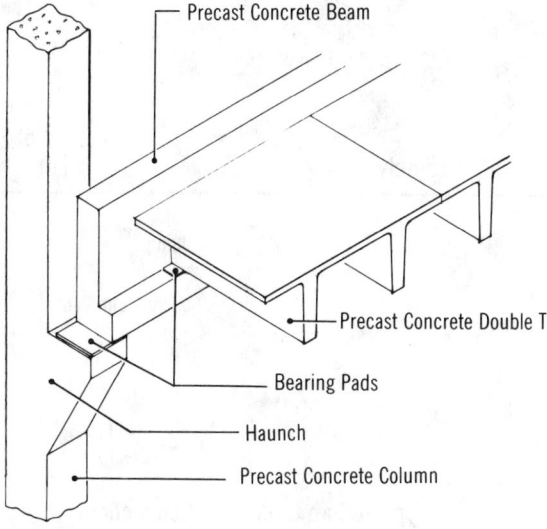

Precast Concrete Beam

Precast Concrete Double T

Bearing Pads

Haunch

Precast Concrete Column

Figure 3.53   Installation Time in Man-Hours for Lift-Slab Construction

Description	Man-Hours	Unit
Edge Forms 7″ to 12″, 4 uses	.074	S.F.C.A.
Box Out for Slab Openings	.120	L.F.
Reinforcing #3 to #7	13.913	ton
Post Tensioning Strand		
100 kip	.021	lb.
200 kip	.019	lb.
Placing Concrete		
Direct Chute	.291	C.Y.
Pumped	.388	C.Y.
Steel Trowel Finish	.015	S.F.

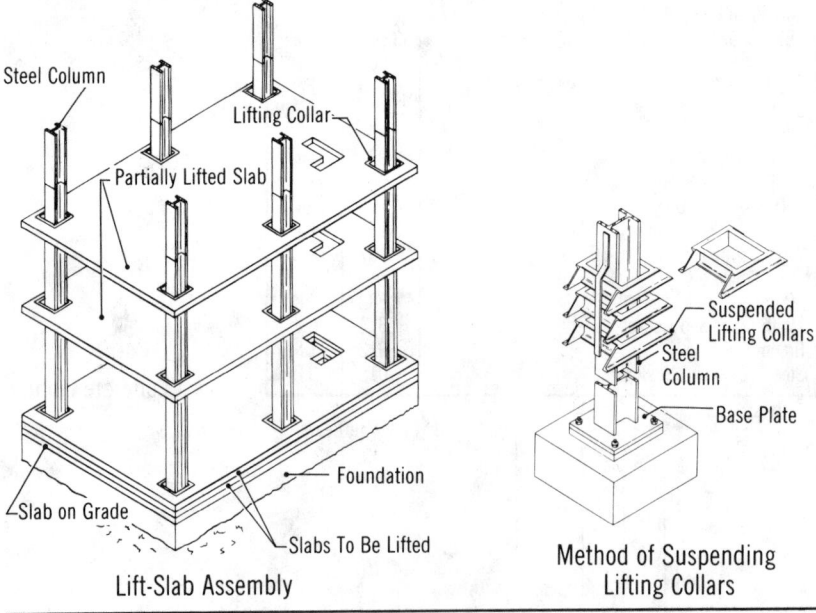

Lift-Slab Assembly

Method of Suspending Lifting Collars

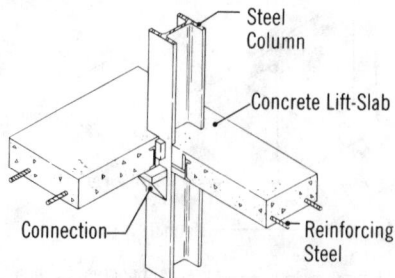

Typical Slab to Column Connection

Figure 3.54 Installation Time in Man-Hours for Prestressed Steel

Man-Hours per Tendon of Prestressed Steel						
Length	100' Beam		75' Beam		100' Slab	
Type Steel	Strand		Bars		Strand	
Diameter	0.5"		3/4"	1-3/8"	0.5"	.06"
Number	12	24	1	1	1	1
Force in Kips	298	595	42	143	25	35
Prep and Placing Cables Stressing Cables Grouting, if Required	6.6 2.4 3.0	10.0 3.0 3.5	0.7 0.6 0.5	2.3 1.3 1.0	0.8 0.4	0.8 0.4
Total Man-hours	12.0	16.5	1.8	4.6	1.2	1.2
Prestressing Steel Wts.	640#	1280#	115#	380#	53#	74#

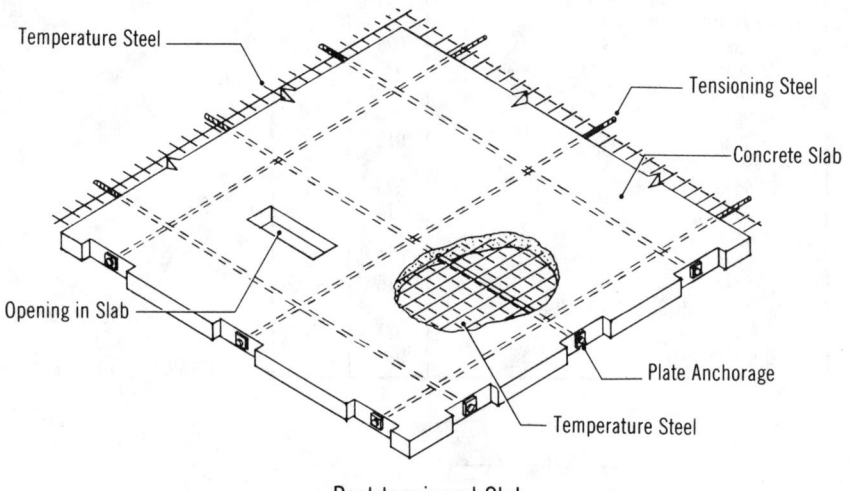

Post-tensioned Slab

Figure 3.55 Installation Time in Man-Hours for Post-Tensioned Slabs and Beams

Description	Man-Hours	Unit
Prestressing Steel Post-Tensioned in Field		
Grouted Strand 100 kip		
50' Span	.053	lb.
100' Span	.038	lb.
200' Span	.024	lb.
300 kip		
50' Span	.024	lb.
100' Span	.020	lb.
200' Span	.018	lb.
Ungrouted Strand 100 kip		
50' Span	.025	lb.
100' Span or 200' Span	.021	lb.
200 kip		
50' Span	.022	lb.
100' Span or 200' Span	.019	lb.
Grouted Bars 42 kip		
50' Span	.025	lb.
75' Span	.020	lb.
143 kip		
50' Span	.020	lb.
75' Span	.015	lb.
Ungrouted Bars 42 kip		
50' Span	.023	lb.
75' Span	.018	lb.
143 kip		
50' Span	.019	lb.
75' Span	.015	lb.
Ungrouted Single Strand, 100' Slab		
25 kip	.027	lb.
35 kip	.022	lb.

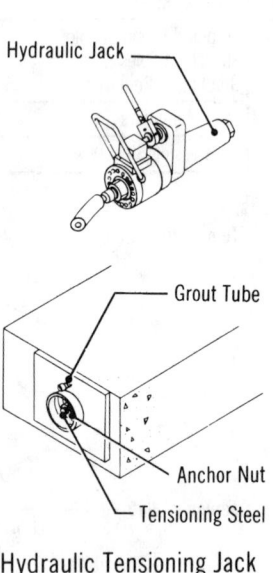

Hydraulic Tensioning Jack

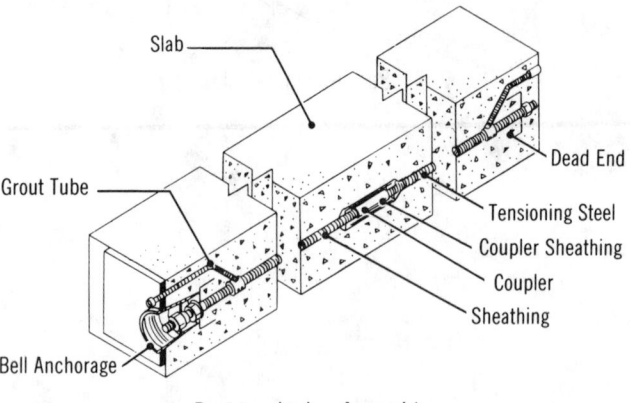

Post-tensioning Assembly

Figure 3.56 Installation Time in Man-Hours for Slipforms

Description	Man-Hours	Unit
Slipforms		
Silos Average	.047	S.F.C.A.
Buildings Average	.057	S.F.C.A.
Reinforcing		
#3 to #7	21.333	ton
#8 to #14	13.913	ton
Posttensioning Grouted Strand		
100 kips	.053	lb.
300 kips	.024	lb.
Placing Concrete		
Minimum	.500	C.Y.
Maximum	1.250	C.Y.

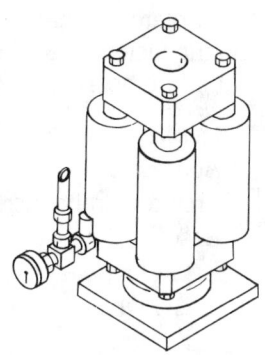

Four-Cylinder Hydraulic Jack

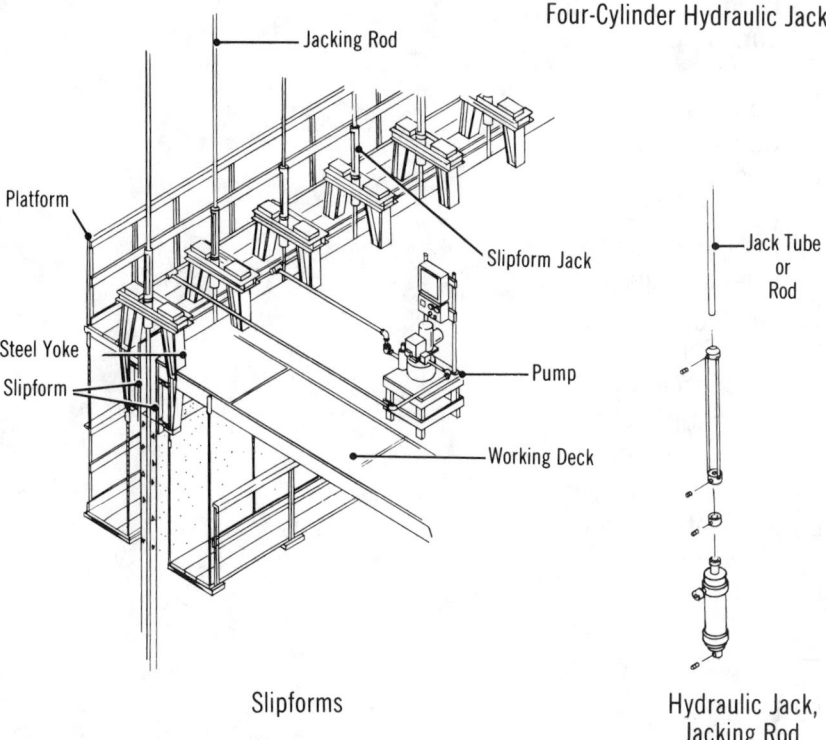

Slipforms

Hydraulic Jack, Jacking Rod

Checklist

For an estimate to be reliable, all items must be accounted for.
A complete estimate can also eliminate the need to include
contingencies. The following checklist can be used to help ensure
that all items are properly accounted for.

Applications
- ☐ Aprons
- ☐ Architectural concrete
- ☐ Bases
  - _____ Flagpole
  - _____ Light
  - _____ Sign
- ☐ Beams
- ☐ Cant strips
- ☐ Columns
- ☐ Copings
- ☐ Curbs
- ☐ Drives
- ☐ Elevated slabs
- ☐ Equipment pads
- ☐ Exterior walls
- ☐ Footings
  - _____ Column
  - _____ Spread
  - _____ Strip (wall)
  - _____ Grade beams
  - _____ Pile caps
- ☐ Foundation mats
- ☐ Foundation walls
- ☐ Girders
- ☐ Gutters
- ☐ Panels
- ☐ Piers
- ☐ Platforms
- ☐ Ramps
- ☐ Retention walls
- ☐ Sills
- ☐ Slab on grade
- ☐ Stair fill

- ☐ Stairs
  - ____ On grade
  - ____ Planters
- ☐ Sump pits
- ☐ Toppings
- ☐ Walks

Concrete

- ☐ Aggregate
- ☐ Air entraining agent
- ☐ Cement
- ☐ Color
- ☐ Compressive strength
- ☐ Enhancer
- ☐ Hardeners
- ☐ Heated
- ☐ Mix design
- ☐ Other admixtures
- ☐ Retarder
- ☐ Sand
- ☐ Testing
- ☐ Water

Formwork

- ☐ Block-outs
- ☐ Bracing
- ☐ Chamfer strips
- ☐ Clamps
- ☐ Clean
- ☐ Erection
- ☐ Expansion joints
- ☐ Inserts
- ☐ Keyways
- ☐ Liners
- ☐ Oil
- ☐ Release agent
- ☐ Removal
- ☐ Repair
- ☐ Scaffolding
- ☐ Shoring
- ☐ Storage
- ☐ Ties

- [ ] Transportation
- [ ] Vapor barrier
- [ ] Beam forms
- [ ] Column forms
- [ ] Curb forms
- [ ] Domes
- [ ] Edge forms
- [ ] Fuel pump forms
- [ ] Gutter forms
- [ ] Pans
- [ ] Ramp forms
- [ ] Slab forms
- [ ] Stair forms
- [ ] Wall forms
- [ ] Joints
  - _____ Expansion
  - _____ Construction

Reinforcing
- [ ] Bars
  - _____ Grade of steel
  - _____ Galvanized
  - _____ Epoxy coating
- [ ] Wire mesh
  - _____ Galvanized
  - _____ Epoxy coating
- [ ] Forming
- [ ] Ties
- [ ] Accessories
- [ ] Cutting
- [ ] Chairs
- [ ] Stirrups

Curing and protection
- [ ] Admixtures
- [ ] Sprays
- [ ] Straw
- [ ] Blankets
- [ ] Heating
- [ ] Canvas
- [ ] Plastic sheeting

Finish
- [ ] Screed

- ☐ Darby
- ☐ Float
- ☐ Broom
- ☐ Trowel

Toppings
- ☐ Abrasive
- ☐ Aggregate
- ☐ Colors
- ☐ Dustproof
- ☐ Epoxy
- ☐ Granolithic
- ☐ Hardeners
- ☐ Integral

Patch _____

Rub _____

Grind _____

Fill _____

Sandblast _____

Caulk _____

Precast Concrete
- ☐ Architectural
- ☐ Beams
  - _____ T's
  - _____ L's
- ☐ Columns
- ☐ Floor slabs
  - _____ Slabs
  - _____ Planks
  - _____ Hollow core
  - _____ Tees
  - _____ Double tees
  - _____ Multiple tees
- ☐ Joists
- ☐ Lift slabs
- ☐ Lintels
- ☐ Lightweight
- ☐ Stairs
- ☐ Wall panels
  - _____ Tilt-ups
  - _____ Precast
  - _____ Architectural

Tips

Concrete Block-outs

When estimating quantities of concrete for floor slabs or walls, do not bother to deduct small areas (two square feet or so) unless there are a large number of these areas, as this can take up more estimating time than the areas are worth. Also remember that you will be adding approximately 3% to the total volume for waste, thereby making these small areas even less significant.

Reinforcing Steel Adds

When estimating the amount of reinforcing steel, either bar or mesh, if no lap specifications are given, add 10% to your quantities for lapping, splices, and waste.

Check All Plans

It cannot be stressed enough that all plans for concrete must be checked. Concrete has been known to show up in all sections of construction drawings in various forms and uses (e.g., equipment pads located *only* on mechanical or electrical drawings, grouting requirements on steel drawings, etc.). Assuming all concrete requirements are indicated on the structural and architectural drawings can be a costly error.

Cold Weather Pouring of Footings

When placing concrete in cold weather, it may not always be necessary to use heating devices to keep concrete warm while it cures. In many cases, insulating blankets and straw are all that are needed. Each situation must be evaluated individually, a worthwhile exercise in view of the considerable expense that can be saved.

Concrete Placement — Direct Chute

When estimating the placement of concrete by direct chute, the forms available generally determine the volume of concrete placed per hour. The more you have available, the more concrete can be placed in the course of a day.

Form Uses

When estimating the number of forms and reuses for a job, remember to have enough forms on hand to keep the forming crew(s) busy while the previously poured concrete sets and starts to cure.

Concrete Placement — Bucket, Pump, or Conveyor

When placing concrete by methods involving a bucket and crane, pumping system, conveyor belt system or other mechanical system, set up enough forms to keep the above systems productive for the entire day. Usually the cost for the use of the equipment for a full day will be charged to you even if you only use it for part of a day.

Sequencing the Pour

During the estimating phase of the project, whatever method or sequence you envision for placing concrete, *document it thoroughly*. The main reasons for complete documentation are that it lets the field know how you arrived at the estimated cost, manpower, and equipment costs; and what methods they should use or try to out-do (in terms of cost effectiveness). Another reason for detailed documentation is that during the documenting process, it may become evident that you cannot place the concrete as you planned or that you can do it differently, more efficiently, or at less cost.

Testing Concrete

When estimating the amount of concrete compression testing that will be necessary for a project, figure on a minimum of one test per pour on smaller pours and a minimum of one test for each fifty yards of concrete placed. Each test should consist of taking a set of two cylinders minimum, with three cylinders the ideal.

Finishing Concrete

Rule of thumb for finishing concrete: Allow 1 man for each 1,500 S.F. of slab.

Drilling Anchor Bolts

Consider drilling anchor bolts into the concrete to save time, manpower, and materials on layout and templates during the pour.

Notes

Division Four
Masonry

Introduction

The word "masonry" may refer to a variety of items, including brick, block, structural tile, glass block, fieldstone, cut stone, and their respective installations. Brick, block, and tile are often called masonry units, and working with them is commonly referred to as "unit masonry." Masonry construction is popular for a variety of reasons. It is fireproof, capable of withstanding abuse, and requires relatively little maintenance. It is also an aesthetically pleasing and a relatively economical form of construction.

The two most common types of masonry units in use are brick and concrete block. Brick is probably the best known type of unit masonry. It is very adaptable, and can be installed in a variety of pleasing patterns, textures, and colors.

The major classifications of brick include face brick, common brick, glazed brick, fire brick, and pavers. Face brick have a better appearance and are used when the bricks are to be in view. Common bricks are used when appearance is not as important, such as for the backup of a face brick veneer wall. Glazed bricks are bricks with a ceramic or other glazing material on one or two of the faces. They are used in a variety of circumstances, such as in restrooms, cafeterias, and kitchens, where durability and cleanliness are important. They may also be used as decorative wainscoting or for an attractive building exterior. Fire bricks are used in areas where high temperatures are encountered. Brick pavers are used as a wearing surface or flooring.

Concrete masonry units—concrete block—are used primarily for walls, partitions, or backup materials for other veneers, such as face brick. The basic concrete block has actual face dimensions of 7-5/8" by 15-5/8", but is referred to as an 8" by 16" unit. Standard block thicknesses include 2, 3, 4, 6, 8, and 12" units. Concrete block also is available in a variety of textures and colors.

Masonry is generally priced by the piece and/or converted into a cost per square foot. The following topics are covered by the estimating charts and tables in the next section:

- Weather factors
- Scaffolding
- Mortar and accessories
- Brick masonry
- Concrete unit masonry
- Glass block

Estimating Data

The following tables present guidelines for overhead and general conditions items. Please note that these percentages should be used as indicators of what may be expected, but that each project should be evaluated individually.

Table of Contents

Figure 4.1   Weather Factors Affecting Masonry Productivity

	Temperature °F											
	40°	45°	50°	55°	60°	65°	70°	75°	80°	85°	90°	95°
100°		.626	.635	.659	.673	.683	.683	.666	.620			
95°	.605	.637	.650	.664	.683	.689	.705	.680	.635	.630		
90°	.615	.635	.657	.671	.697	.717	.726	.727	.690	.640		
85°	.615	.650	.670	.688	.724	.737	.752	.760	.725	.673	.600	
80°	.636	.650	.675	.710	.745	.765	.790	.795	.750	.760	.670	
75°	.636	.665	.690	.727	.765	.790	.835	.860	.780	.679	.610	
70°	.636	.675	.700	.740	.790	.816	.909	.930	.781	.680	.613	
65°	.636	.670	.700	.747	.795	.855	.950	.948	.781	.680	.610	
60°	.619	.670	.708	.747	.786	.820	.947	.973	.779	.673	.613	
55°	.619	.655	.700	.737	.775	.818	.909	.854	.885	.705	.640	.614
50°	.614	.655	.685	.727	.760	.790	.844	.908	.945	.765	.772	.623
45°	.610	.630	.673	.700	.742	.762	.790	.830	.820	.755	.683	.628
40°	.593	.614	.637	.685	.727	.742	.762	.768	.754	.697	.645	.615
35°		.604	.630	.660	.688	.706	.727	.717	.694	.670	.635	
30°			.612	.628	.648	.667	.685	.670	.648	.648	.620	
25°				.610	.630	.631	.620	.620	.625			
	40°	45°	50°	55°	60°	65°	70°	75°	80°	85°	90°	95°

Relative Humidity %

Weather often affects the productivity of masons by its effects on the "workability" of mortar. This table can be used to factor in the effects of weather on productivity. Apply the following steps:

1. Obtain the average temperature and humidity ("local climatological data") for the month or time period needed for the region in question. (This information can be obtained from the U.S. Weather Services in the area, local airports, or possibly the local newspapers.)
2. Enter the table at the top with the average temperature obtained in Step 1.
3. Follow that column down to the average humidity reading obtained in Step 1.
4. Read the productivity factor at that intersection of humidity and temperature.
5. Multiply your average expected productivity for masonry installation by this factor. The result is your productivity adjusted for temperature conditions.

Figure 4.2 Productivity in Scaffolding Assembly

This chart is a summary of the labor involved in assembling scaffolding and staging. The chart shows typical crews, expected daily outputs, and the expected man-hours per unit for various areas of scaffolding erection.

Scaffolding	Crew Makeup	Daily Output	Man-Hours	Unit
SCAFFOLD, Steel tubular, rented, no plank, 1 use per month				
Building exterior 2 stories	3 Carpenters	17.72	1.350	C.S.F.
4 stories	"	17.72	1.350	C.S.F.
6 stories	4 Carpenters	22.60	1.420	C.S.F.
8 stories		20.25	1.580	C.S.F.
10 stories		19.10	1.680	C.S.F.
12 stories	↓	17.70	1.810	C.S.F.
One tier 3' high x 7' long x 5' wide 1 use per month	1 Carpenter	14.90	.537	C.S.F.
2 uses per month		14.90	.537	C.S.F.
4 uses per month		14.90	.537	C.S.F.
8 uses per month		14.90	.537	C.S.F.
5' high x 7' long x 5' wide 1 use per month		20.65	.387	C.S.F.
2 uses per month		20.65	.387	C.S.F.
4 uses per month		20.65	.387	C.S.F.
8 uses per month		20.65	.387	C.S.F.
6'-6" high x 7' long x 5' wide 1 use per month		26.85	.298	C.S.F.
2 uses per month		26.85	.298	C.S.F.
4 uses per month		26.85	.298	C.S.F.
8 uses per month	↓	26.85	.298	C.S.F.
Scaffold steel tubular, suspended slab form supports to 8'-2" high				
1 use per month	4 Carpenters	31	1.030	C.S.F.
2 uses per month		43	.744	C.S.F.
3 uses per month	↓	43	.744	C.S.F.
Steel tubular, suspended slab form supports to 14'-8" high				
1 use per month	4 Carpenters	16	2.000	C.S.F.
2 uses per month		22	1.450	C.S.F.
3 uses per month	↓	22	1.450	C.S.F.
SCAFFOLDING SPECIALTIES				
Sidewalk bridge, heavy duty steel posts & beams, including parapet protection & waterproofing				
8' to 10' wide, 2 posts	3 Carpenters	15	1.600	L.F.
3 posts	"	10	2.400	L.F.
Sidewalk bridge using tubular steel scaffold frames, including planking	3 Carpenters	45	.533	L.F.
Stair unit, interior, for scaffolding, buy				Ea.
Rent per month				Ea.
SWING STAGING for masonry, 5' wide x 7', hand operated Cable type with 150' cables, rent & installation, per week	1 Struc. Steel Foreman 3 Struc. Steel Workers 1 Gas Welding Machine	18	1.780	L.F.
Catwalks, no handrails, 3 joists, 2" x 4"	2 Carpenters	55	.291	L.F.
3 joists, 3" x 6"	"	40	.400	L.F.
Move swing staging	1 Struc. Steel Foreman 3 Struc. Steel Workers 1 Gas Welding Machine	37	.865	L.F.

Figure 4.3   General Mortar Information

General information concerning mortar materials, mixes, and mixing.

Cement
A bag of Portland Cement contains 94 lbs. When packed, Portland Cement weighs 108-1/2 lbs. per cubic foot.
A cubic foot of cement paste requires about 94 lbs. of cement.

Hydrated Lime
One cubic foot weighs about 40 lbs.
One paper package contains about 200 lbs. net.
One 100 lbs. package makes about 2-1/4 cubic feet lime putty.

Prepared Mortar
A pre-mixed mixture of cement, lime and gypsum, in varying proportions, depending on the manufacturer.
One paper package contains 70 lbs. net, and is equal to one cubic foot.

Proportioning of Mortar Materials
1:3 Cement mortar (by volume):
    Cement: 27 cubic feet ÷ 3 = 9 cubic feet = 9 bags
    Sand: 27 cubic feet   = 1 cubic yard
    Mortar yield: = 1 cubic yard
    (The cement fills the voids in the sand and there is little increase in bulk.)

1:2:9 Cement-lime-mortar (by volume):
    Cement: 27 cubic feet ÷ 9 = 3 cubic feet = 3 bags
    Lime: 27 cubic feet ÷ 2 = 13-1/2 cubic feet = 6 bags
    Sand: 27 cubic feet   = 1 cubic yard
    Mortar yield: = 1 cubic yard
    (This is actually a 1:3 cementitious-sand mixture)

1:3 Prepared mortar (by volume):
    Prepared mortar: 27 cubic feet ÷ 3 = 9 cubic feet = 9 bags
    Sand: 27 cubic feet   = 1 cubic yard
    Mortar yield: = 1 cubic yard

Job Mixing
A two-bag mortar mixer, with two bags of cementing agent and 30 number 2 shovels of sand will yield two wheelbarrows of mortar or about seven cubic feet.

A three-bag mixer, with three bags of cementing agent and 45 number 2 shovels of sand will yield three wheelbarrows of mortar, or about 10-1/2 cubic feet.

A two-bag mixer is adequate for jobs using up to 25 masons, and they can be supplied by an experienced mortar maker. Assistance in bringing materials to the machine is necessary.

On larger jobs a three-bag mixer would be required. Mortar would be deposited on a mortar box and then wheeled to the masons, or the box lifted and deposited by fork-lift. Additional labor would be required.

Masonry Estimating Handbook, Michael F. Kenny, Construction Publishing Company, Inc.

Figure 4.4 Brick Mortar Mix Proportions

This chart shows some common mortar types, the mixing proportions and their general uses.

Brick Mortar Mixes*					
Type	Portland Cement	Hydrated Lime	Sand (maximum)**	Strength	Use
M	1	1/4	3-3/4	High	General use where high strength is required, especially good compressive strength; work that is below grade and in contact with earth.
S	1	1/2	4-1/2	High	Okay for general use, especially good where high lateral strength is desired.
N	1	1	6	Medium	General use when masonry is exposed above grade; best to use when high compressive and lateral strengths are not required.
0	1	2	9	Low	Do not use when masonry is exposed to severe weathering; acceptable for non-loadbearing walls of solid units and interior non-loadbearing partitions of hollow units.

*The water used should be of the quality of drinking water. Use as much as is needed to bring the mix to a suitably plastic and workable state.

**The sand should be damp and loose. A general rule for sand content is that it should not be less than 2-1/4 or more than 3 times the sum of the cement and lime volumes.

Figure 4.5 Compressive Strengths of Mortar

This table lists the expected 28 day compressive strengths for some common types of mortar mixes.

Mortar Type	Average Compressive Strength at 28 Days
M	2500 p.s.i.
S	1800 p.s.i.
N	750 p.s.i.
0	350 p.s.i.
K	75 p.s.i.

Masonry and Concrete Construction, Ken Nolan, Craftsman Book Co.

Figure 4.6 Mortar Quantities for Brick Masonry

This table can be used to determine the amounts of mortar needed for brick masonry walls for various joint thicknesses and types of common brick.

Brickwork	Joint Thickness (in inches)	Actual Requirement (in C.F. per 1,000)	Requirement with 15% Waste (in C.Y. per 1,000)
Standard	3/8	8.6	0.4
Standard	1/2	11.7	0.5
Modular	3/8	7.6	0.3
Modular	1/2	10.4	0.4
Roman	3/8	10.7	0.5
Roman	1/2	14.4	0.6
Norman	3/8	11.2	0.5
Norman	1/2	15.1	0.6

Type	Thickness (in inches)	Actual Requirement (in C.F. per S.F.)	Requirement with 15% Waste (in C.Y. per 1,000 S.F.)
Parging or Backplastering	3/8	0.03	1.3
	1/2	0.04	1.7

Note: Quantities are based on 4" thick brickwork. For each additional 4" brickwork, or where masonry units are used in backup, add parging or backplastering.

Modular dimensions are measured from center to center of masonry joints. To fit into the system, the unit or number of units must be the size of the module, less the thickness of the joint. A standard face brick with a 1/2" joint would measure 8-1/2" long, whereas a modular brick with the same joint would measure 8".

Figure 4.7 Mortar Quantities for Concrete Block Masonry

This table can be used to determine the amounts of mortar needed for concrete block masonry walls for joint thicknesses for various types of masonry units.

Type	Joint Thickness (in inches)	Actual Requirement (in C.Y. per 1,000)	Requirement with 15% Waste (in C.Y. per 1,000)
Concrete Block			
Shell Bedding			
All thicknesses	3/8	0.6	0.7
Full Bedding			
12 x 8 x 16, 3-Core	3/8	0.9	1.0
12 x 8 x 16, 2-Core	3/8	0.8	0.9
8 x 8 x 16, 3-Core	3/8	0.7	0.8
8 x 8 x 16, 2-Core	3/8	0.7	0.8
6 x 8 x 16, 3-Core	3/8	0.7	0.8
6 x 8 x 16, 2-Core	3/8	0.7	0.8
4 x 8 x 16, 3-Core	3/8	0.7	0.8
4 x 8 x 16, Solid	3/8	0.8	0.9

Type	Joint Thickness (in inches)	Actual Requirement (in C.F. per 1,000)	Requirement with 15% Waste (in C.Y. per 1,000)
4S Glazed Structural Units			
(nominal 2-1/2" x 8")			
4 SA (2" thick)	1/4	2.6	0.1
4 S (4" thick)	1/4	5.6	0.2
4D Glazed Structural Units			
(nominal 5" x 8")			
4 DCA (2" thick)	1/4	3.3	0.1
4 DC (4" thick)	1/4	7.1	0.3
4 DC 60 (6" thick)	1/4	10.9	0.5
4 DC 80 (8" thick)	1/4	13.7	0.6
6T Glazed Structural Units			
(nominal 5" x 12")			
6 TCA (2" thick)	1/4	4.2	0.2
6 TC (4" thick)	1/4	9.3	0.4
6 TC 60 (6" thick)	1/4	14.2	0.6
6 TC 80 (8" thick)	1/4	19.1	0.8
6P Glazed Structural Units –			
1/4" Joints (nominal 4" x 12")			
6 PCA (2" thick)		4.0	0.2
6 PC (4" thick)		8.6	0.4
6 PC 60 (6" thick)		13.1	0.6
6 PC 80 (8" thick)		17.7	0.8
8W Glazed Structural Units			
(nominal 8" x 16")			
8 WCA (2" thick)	1/4	6.0	0.3
8 WCA (2" thick)	3/8	9.1	0.4
8 WC (4" thick)	1/4	12.9	0.6
8 WC (4" thick)	3/8	19.4	0.8

Note: Gypsum partition tile cement should be used. 900 lbs. are required per cubic yard, plus 1 cubic yard of sand. 1,000 units equal 2,610 square feet.

(continued on next page)

Figure 4.7 Mortar Quantities for Concrete Block Masonry (continued)

This table can be used to determine the amounts of mortar needed for concrete block masonry walls for joint thicknesses for various types of masonry units.

Type	Joint Thickness (in inches)	Actual Requirement (in C.F. per 1,000)	Requirement with 15% Waste (in C.Y. per 1,000)
Spectra-Glaze® Units –			
3/8" Joints (nominal 4" x 16")			
44S (4" thick)		16.0	0.7
64S (6" thick)		24.5	1.0
84S (8" thick)		33.0	1.4
Spectra-Glaze® Units –			
3/8" Joints (nominal 8" x 16")			
2S (2" thick)		9.0	0.4
4S (4" thick)		19.2	0.8
6S (6" thick)		29.5	1.3
8S (8" thick)		39.7	1.7
10S (10" thick)		50.0	2.1
12S (12" thick)		60.2	2.6
Structural Clay Backup and Wall Tile			
5 x 12 (4" thick)	1/2	20.0	0.9
5 x 12 (6" thick)	1/2	30.4	1.3
5 x 12 (8" thick)	1/2	40.5	1.7
8 x 12 (6" thick)	1/2	35.6	1.5
8 x 12 (8" thick)	1/2	47.5	2.0
8 x 12 (12" thick)	1/2	71.2	3.0
Structural Clay Partition Tile			
12 x 12 (4" thick)	1/2	28.4	1.2
12 x 12 (6" thick)	1/2	42.5	1.8
12 x 12 (8" thick)	1/2	56.7	2.4
12 x 12 (10" thick)	1/2	70.9	3.0
12 x 12 (12" thick)	1/2	85.1	3.6
Gypsum Block (all 12" x 30")*			
2" thick	1/4	12.2	0.5
	3/8	18.4	0.8
	1/2	24.6	1.0

Figure 4.7 Mortar Quantities for
Concrete Block Masonry (continued)

Type	Joint Thickness (in inches)	Actual Requirement (in C.F. per 1,000)	Requirement with 15% Waste (in C.Y. per 1,000)
Gypsum Block (all 12" x 30")* cont.			
3" thick	1/4	18.3	0.8
	3/8	27.6	1.2
	1/2	36.9	1.6
4" thick	1/4	24.5	1.0
	3/8	36.8	1.6
	1/2	49.2	2.1
6" thick	1/4	36.7	1.6
	3/8	55.2	2.4
	1/2	73.8	3.1
Glass Block (all 3-7/8" thick)			
5-3/4 x 5-3/4	1/4		
7-3/4 x 7-3/4	1/4		
11-3/4 x 11-3/4	1/4		
Firebrick Thin joints, including waste 300 lbs. fireclay per 1,000.			

*Note: Gypsum partition tile cement should be used. 900 lbs. are required per cubic yard, plus 1 cubic yard of sand. 1,000 units equals 2,610 square feet.

Figure 4.8 Brick Nomenclature

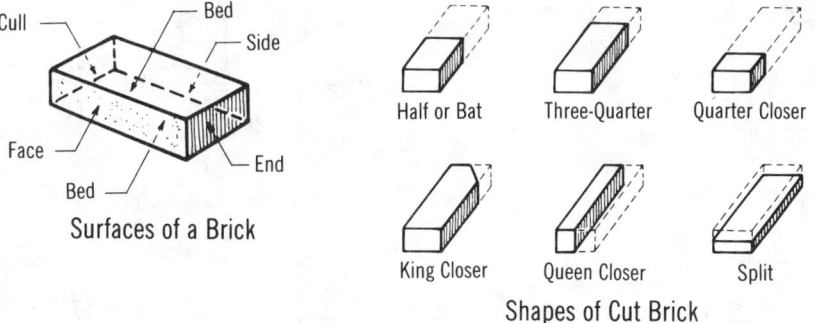

Surfaces of a Brick

Half or Bat   Three-Quarter   Quarter Closer

King Closer   Queen Closer   Split

Shapes of Cut Brick

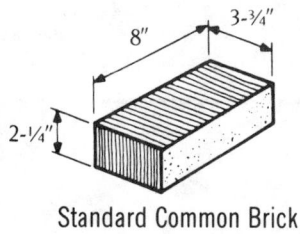

Standard Common Brick

(continued on next page)

Figure 4.8 Brick Nomenclature (continued)

Stretchers
Wythe
Course
Bed Joints
Head Joints

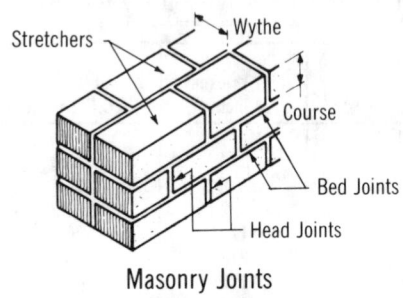

Masonry Joints

Header
Header Course
Stretcher
Stretcher Course
Bull Header
Rowlock Course
Bull Stretcher
Shiner Course
Soldier
Soldier Course
Sailor
Sailor Course
Quoins

Brick Positions and Courses

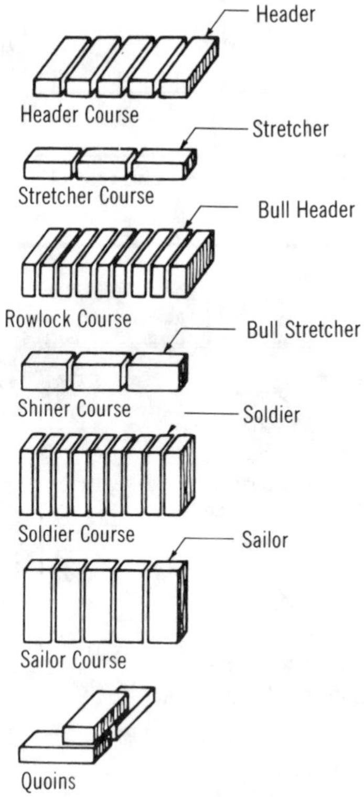

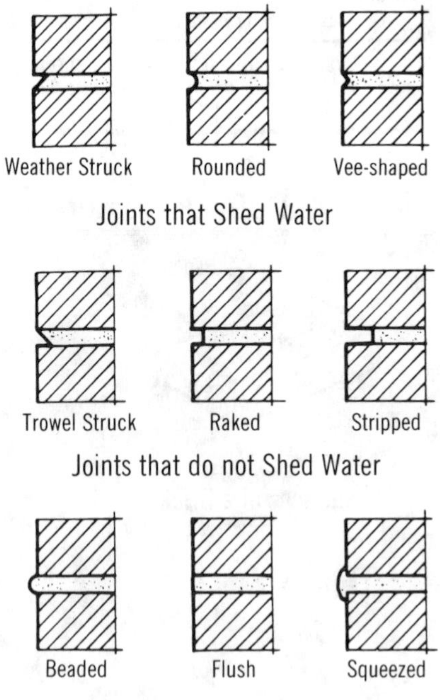

Weather Struck
Rounded
Vee-shaped
Joints that Shed Water
Trowel Struck
Raked
Stripped
Joints that do not Shed Water
Beaded
Flush
Squeezed

Common Mortar Joints

Figure 4.9   Heights of Brick Masonry Walls by Course

Course No.	Standard 2-1/4" Brick (3/8" Joint)	Course No.	Standard 2-1/4" Brick (3/8" joint)
1	2-5/8"	26	5' 8-1/4"
2	5-1/4"	27	5' 10-7/8"
3	7-7/8"	28	6' 1-1/2"
4	10-1/2"	29	6' 4-1/8"
5	1' 1-1/8"	30	6' 6-3/4"
6	1' 1-3/4"	31	6' 9-3/8"
7	1' 6-3/8"	32	7' 0"
8	1' 9"	33	7' 2-5/8"
9	1' 11-5/8"	34	7' 5-1/4"
10	2' 2-1/4"	35	7' 7-7/8"
11	2' 4-7/8"	36	7' 10-1/2"
12	2' 7-1/2"	37	8' 1-1/8"
13	2' 10-1/8"	38	8' 3-3/4"
14	3' 0-3/4"	39	8' 6-3/8"
15	3' 3-3/8"	40	8' 9"
16	3' 6"	41	8' 11-5/8"
17	3' 8-5/8"	42	9' 2-1/4"
18	3' 11-1/4"	43	9' 4-7/8"
19	4' 1-7/8"	44	9' 7-1/2"
20	4' 4-1/2"	45	9' 10-1/8"
21	4' 7-1/8"	46	10' 0-3/4"
22	4' 9-3/4"	47	10' 3-3/8"
23	5' 0-3/8"	48	10' 6"
24	5' 3"	49	10' 8-5/8"
25	5' 5-5/8"	50	10' 11-1/4"

Note: When standard brick is used in conjunction with concrete block, the brick coursing corresponds to height modules, i.e., 3 courses of brick lay up to 8".

Figure 4.10   Sizes of Modular Brick

Unit Designation	Nominal Dimensions (in inches)			Joint Thickness (in inches)	Manufactured Dimensions (in inches)			Modular Coursing (in inches)
	t	h	l		t	h	l	
Standard Modular	4	2-2/3	8	3/8	3-5/8	2-1/4	7-5/8	3C = 8
				1/2	3-1/2	2-1/4	7-1/2	
Engineer	4	3-1/5	8	3/8	3-5/8	2-13/16	7-5/8	5C = 16
				1/2	3-1/2	2-11/16	7-1/2	
Economy 8 or Jumbo Closure	4	4	8	3/8	3-5/8	3-5/8	7-5/8	1C = 4
				1/2	3-1/2	3-1/2	7-1/2	
Double	4	5-1/3	8	3/8	3-5/8	4-15/16	7-5/8	3C = 16
				1/2	3-1/2	4-13/16	7-1/2	
Roman	4	2	12	3/8	3-5/8	1-5/8	11-5/8	2C = 4
				1/2	3-1/2	1-1/2	11-1/2	
Norman	4	2-2/3	12	3/8	3-5/8	2-1/4	11-5/8	3C = 8
				1/2	3-1/2	2-1/4	11-1/2	
Norwegian	4	3-1/5	12	3/8	3-5/8	2-13/16	11-5/8	5C = 16
				1/2	3-1/2	2-11/16	11-1/2	
Economy 12 or Jumbo Utility	4	4	12	3/8	3-5/8	3-5/8	11-5/8	1C = 4
				1/2	3-1/2	3-1/2	11-1/2	
Triple	4	5-1/3	12	3/8	3-5/8	4-15/16	11-5/8	3C = 16
				1/2	3-1/2	4-13/16	11-1/2	
SCR brick	6	2-2/3	12	3/8	5-5/8	2-1/4	11-5/8	3C = 8
				1/2	5-1/2	2-1/4	11-1/2	
6-in. Norwegian	6	3-1/5	12	3/8	5-5/8	2-13/16	11-5/8	5C = 16
				1/2	5-1/2	2-11/16	11-1/2	
6-in. Jumbo	6	4	12	3/8	5-5/8	3-5/8	11-5/8	1C = 4
				1/2	5-1/2	3-1/2	11-1/2	
8-in. Jumbo	8	4	12	3/8	7-5/8	3-5/8	11-5/8	1C = 4
				1/2	7-1/2	3-1/2	11-1/2	

(courtesy Masonry and Concrete Construction, Ken Nolan, Craftsman Book Company)

Figure 4.11 Brick Bonding Patterns

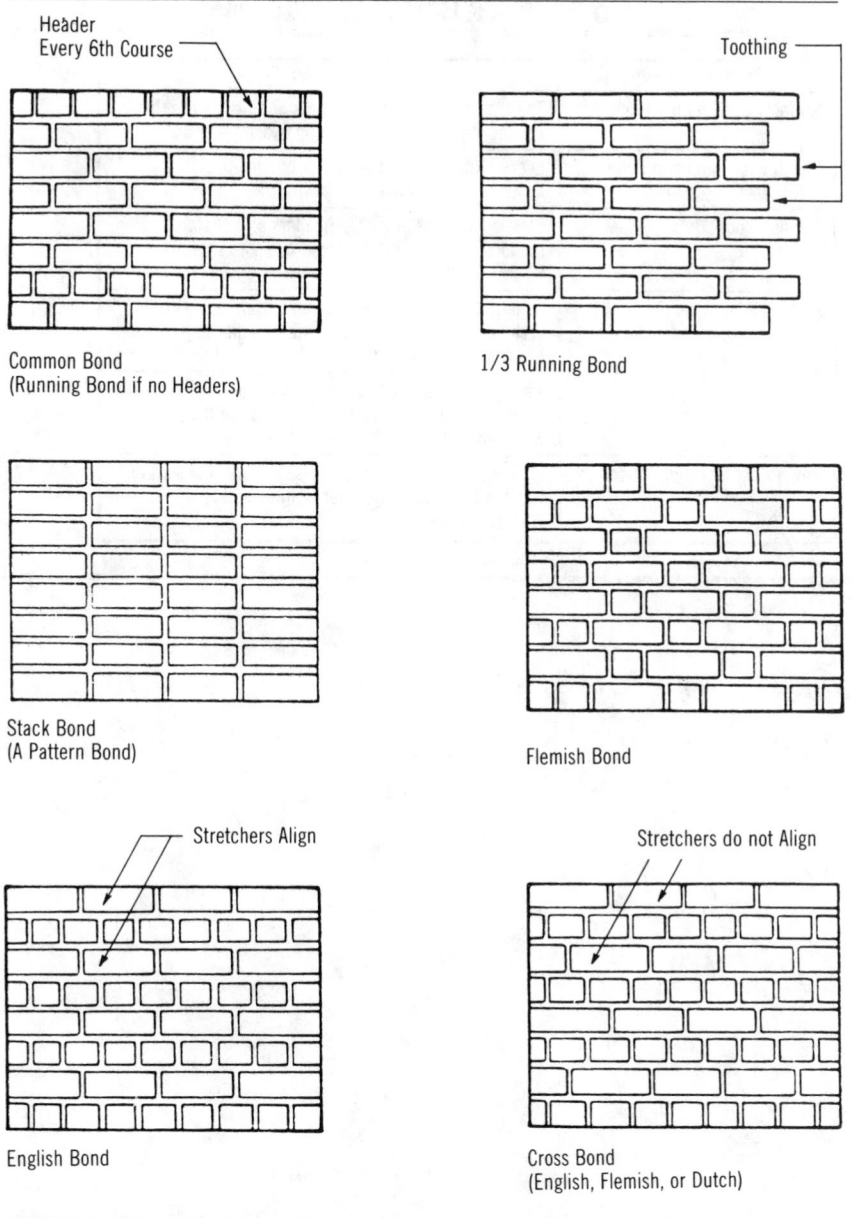

Header
Every 6th Course

Common Bond
(Running Bond if no Headers)

Toothing

1/3 Running Bond

Stack Bond
(A Pattern Bond)

Flemish Bond

Stretchers Align

English Bond

Stretchers do not Align

Cross Bond
(English, Flemish, or Dutch)

Figure 4.12   Waste Allowances for
Various Brick Bonding Patterns

Type of Pattern	% of Waste
Running or Stretcher Bond	The face brick are all stretchers and are tied to the backing by metal or reinforcing. Waste — 5%
Common or American Bond	Every sixth course of stretcher bond is usually a header course. Waste — 4%.
Flemish Bond	Each course has alternate headers and stretchers with the alternate headers centered over the stretcher. Waste — 3 to 5 %.
English Bond	Consists of alternate headers and stretchers with the vertical joints in the header and stretcher aligning or breaking over each other. Waste — 8 to 15%
Stack Bond	Has no overlapping of units since all vertical joints are aligned. Usually this pattern is bonded to the backing with rigid steel ties. Waste — 3%.
English Cross or Dutch Bond	Built up on interlocking crosses. This wall consists of two headers and a stretcher forming a cross. Waste — 8%.

Figure 4.13   Brick Requirements and Waste Factors

This table provides the number of bricks required per S.F., and waste factors based on brick type and bonding pattern.

Type of Brick	Description	Waste		Bricks/S.F.
Standard	Running Bond	net		6.55
		5% waste		6.88
	Common Bond	net		7.64
	(6th course headers)	5% waste		8.02
	English Bond	net		9.83
	(full headers every other course)	5% waste		10.32
	Flemish Bond	net		9.83
	(alternate headers and stretchers			
	each course)	5% waste		10.32
	8" Wall, Running Bond	net		13.10
		5% waste		13.76
	8" Wall, Common Bond	net		13.10
		5% waste		13.76
	12" Wall, Running Bond	net	(Face Brick)	13.10
			(Common Brick)	6.55
		5% waste	(Face Brick)	13.76
	12" Wall, Common Bond	net	(Face Brick)	15.28
			(Common Brick)	4.37
		5% waste	(Face Brick)	16.04
			(Common Brick)	4.59
Concrete	Running Bond	net		6.86
		5% waste		7.20
Modular	Running Bond	net		7.19
		5% waste		7.55
Modular Roman	Running Bond	net		6.00
		5% waste		6.30
Modular Norman	Running Bond	net		4.57
		5% waste		4.80
Fire	(2-1/4" x 8")	net		6.98
		5% waste		7.33
	(2-1/2" x 9")	net		5.67
		5% waste		5.95

Figure 4.14   Brick Quantities per S.F.

Number of bricks per S.F. based on brick type and size of joints.

Brick Type & Size	Size of Joint (in inches)					
	1/4	1/3	3/8	1/2	5/8	3/4
Standard face brick (8" x 2-1/4")	6.98	6.70	6.55	6.16	5.81	5.49
Standard common brick (8" x 2-1/4")	6.98	6.70	6.55	6.16	5.81	5.49
Concrete brick (7-5/8" x 2-1/4")	7.31	7.00	6.86	6.45	6.07	5.73
Modular brick (7-1/2" x 2-1/6")	7.68	7.35	7.19	6.73	6.35	5.98
Modular Roman brick (11-5/8" x 1-5/8")	6.47	6.15	6.00	5.59	5.22	4.90
Modular Norman brick (11-5/8" x 2-1/4")	4.85	4.66	4.57	4.32	4.09	3.88

Note: Above constants are net, i.e., no waste is included. (See Figure 4.13)

Figure 4.15   Adjustments to Brick Quantity Factors

This table provides factors for determining the additional quantities of brick needed when specific bonding patterns are used.

For Other Bonds Standard Size Add to S.F. Quantities				
Bond Type	Description	Factor	Description	Factor
Common	Full header every fifth course	+20%	Header = W x H exposed	+100%
	Full header every sixth course	+16.7%	Rowlock = H x W exposed	+100%
English	Full header every second course	+50%	Rowlock stretcher = L x W exposed	+33.3%
			Soldier = H x L exposed	—
Flemish	Alternate headers every course	+33.3%	Sailor = W x L exposed	-33.3%
	every sixth course	+5.6%		

(See "Brick Quantities per S.F." table above for basic quantities of brick per S.F.).

Figure 4.16   Man-Hours Required for the Installation of Brick Masonry

Description	Man-Hours	Unit
Brick Wall		
Veneer		
4" Thick		
Running Bond		
Standard Brick (6.75/S.F.)	.182	S.F.
Engineer Brick (5.63/S.F.)	.154	S.F.
Economy Brick (4.50/S.F.)	.129	S.F.
Roman Brick (6.00/S.F.)	.160	S.F.
Norman Brick (4.50/S.F.)	.125	S.F.
Norwegian Brick (3.75/S.F.)	.107	S.F.
Utility Brick (3.00/S.F.)	.089	S.F.
Common Bond, Standard Brick (7.88/S.F.)	.216	S.F.
Flemish Bond, Standard Brick (9.00/S.F.)	.267	S.F.
English Bond, Standard Brick (10.13/S.F.)	.286	S.F.
Stack Bond, Standard Brick (6.75/S.F.)	.200	S.F.
6" Thick		
Running Bond		
S.C.R. Brick (4.50/S.F.)	.129	S.F.
Jumbo Brick (3.00/S.F.)	.092	S.F.
Backup		
4" Thick		
Running Bond		
Standard Brick (6.75/S.F.)	.167	S.F.
Solid, Unreinforced		
8" Thick Running Bond (13.50/S.F.)	.296	S.F.
12" Thick, Running Bond (20.25/S.F.)	.421	S.F.
Solid, Rod Reinforced		
8" Thick, Running Bond	.308	S.F.
12" Thick, Running Bond	.444	S.F.
Cavity		
4" Thick		
4" Backup	.242	S.F.
6" Backup	.276	S.F.
Brick Chimney		
16" x 16", Standard Brick w/8" x 8" Flue	.889	V.L.F.
16" x 16", Standard Brick w/8" x 12" Flue	1.000	V.L.F.
20" x 20", Standard Brick w/12" x 12" Flue	1.140	V.L.F.
Brick Column		
8" x 8", Standard Brick 9.0 V.L.F.	.286	V.L.F.
12" x 12", Standard Brick 20.3 V.L.F.	.640	V.L.F.
20" x 20", Standard Brick 56.3 V.L.F.	1.780	V.L.F.
Brick Coping		
Precast, 10" Wide, or Limestone, 4" Wide	.178	L.F.
Precast 14" Wide, or Limestone, 6" Wide	.200	L.F.

(continued on next page)

Figure 4.16 Man-Hours Required for the Installation of Brick Masonry (continued)

Description	Man-Hours	Unit
Brick Fireplace		
30" x 24" Opening, Plain Brickwork	40.000	Ea.
Firebox Only, Fire Brick (110/Ea.)	8.000	Ea.
Brick Prefabricated Wall Panels, 4" Thick		
Minimum	.093	S.F.
Maximum	.144	S.F.
Brick Steps	53.330	M
Window Sill		
Brick on Edge	.200	L.F.
Precast, 6" Wide	.229	L.F.
Needle Brick and Shore, Solid Brick		
8" Thick	6.450	Ea.
12" Thick	8.160	Ea.
Repoint Brick		
Hard Mortar		
Running Bond	.100	S.F.
English Bond	.123	S.F.
Soft Mortar		
Running Bond	.080	S.F.
English Bond	.098	S.F.
Toothing Brick		
Hard Mortar	.267	V.L.F.
Soft Mortar	.200	V.L.F.
Sandblast Brick		
Wet System		
Minimum	.024	S.F.
Maximum	.057	S.F.
Dry System		
Minimum	.013	S.F.
Maximum	.027	S.F.
Sawing Brick, Per Inch of Depth	.027	L.F.
Steam Clean, Face Brick	.033	S.F.
Wash Brick, Smooth	.014	S.F.

Figure 4.17   Brick Masonry Systems

The following table shows typical quantities of brick and mortar, and installation time per S.F. and per thousand bricks for the systems described. The crew used for these examples consists of 3 Bricklayers and 2 Bricklayer-helpers.

Brick in Place			
Item	8" Common Brick Wall 8" x 2-2/3" x 4"	Select Common Face 8" x 2-2/3" x 4"	Red Face Brick 8" x 2-2/3" x 4"
1030 brick delivered Type N mortar Installation	12.5 C.F. 0.556 Days	10.3 C.F. 0.667 Days	10.3 C.F. 0.667 Days
Total per S.F. of wall	13.5 bricks/S.F.	6.75 bricks/S.F.	6.75 bricks/S.F.

This table is for common bond with 3/8" concave joints and includes 3% waste for brick and 25% waste for mortar.

Both the number of bricks per S.F. and the waste factors are based on the brick type and the type of bonding pattern chosen.

Brick Veneer in Place			
Item	8" Common Brick Wall 8" x 2-2/3" x 4"	Select Common Face 8" x 2-2/3" x 4"	Red Face Brick 8" x 2-2/3" x 4"
1030 brick delivered Type N mortar Installation using 3 Bricklayers and 2 Helpers	12.5 C.F. 0.556 Days	10.3 C.F. 0.667 Days	10.3 C.F. 0.667 Days
Total per S.F. of wall	13.5 bricks/S.F.	6.75 bricks/S.F.	6.75 bricks/S.F.

This table is for common bond with 3/8" concave joints and includes 3% waste for brick and 25% waste for mortar.

Reinforced Brick in Walls		
Item	8" to 9" Thick Wall	16" to 17" Thick Wall
1030 brick (select common) Type PL mortar Reinforcing bars delivered Installation using 3 Bricklayers and 2 Helpers	12.5 C.F. 50 lb. 0.571 Days	13.9 C.F. 50 lb. 0.513 Days
Total per S.F. of wall	8" wall, 13.5 brick/S.F.	16" wall, 27.0 brick/S.F.

This table is for common bond with 3/8" concave joints and includes 3% waste. Standard 8" x 2-2/3" x 4" bricks.

(continued on next page)

Figure 4.17   Brick Masonry Systems (continued)

Brick Chimneys				
Quantities	16″ x 16″	20″ x 20″	20″ x 24″	20″ x 32″
Brick	28 brick	37 brick	42 brick	51 brick
Type M mortar	.5 C.F.	.6 C.F.	1.0 C.F.	1.3 C.F.
Flue tile (square)	8″ x 8″	12″ x 12″	2 @ 8″ x 12″	2 @ 12″ x 12″
Install tile & brick using				
1 Bricklayer and				
1 Helper	.055 day	.073 day	.083 day	.10 day

Note: Use 2/3 of the above amounts if the chimney butts against a straight wall. Labor for chimney brick using 1 Bricklayer and 1 Bricklayer Helper is 31 hours per thousand bricks. An 8″ x 12″ flue takes 33 bricks and two 8″ x 8″ flues take 37 bricks.

Figure 4.18   Cleaning Face Brick

A worker can clean 70 S.F. of smooth brick an hour, and 50 S.F. of rough brick per hour. Use one gallon muriatic acid to 20 gallons of water for 1000 S.F. Do not use acid solution until wall is at least seven days old. A mild soap solution may be used after two days.

Figure 4.19   Typical Concrete Block Systems and Nomenclature

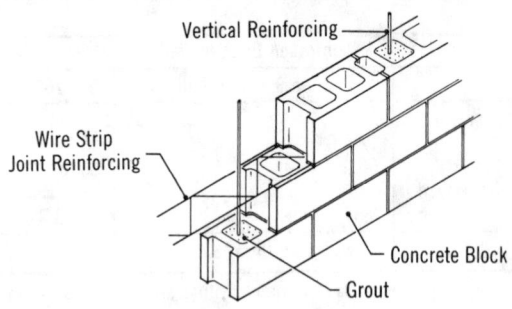

Vertical Reinforcing

Wire Strip
Joint Reinforcing

Concrete Block

Grout

Grouted and Reinforced Block

Figure 4.19   Typical Concrete Block Systems and Nomenclature (continued)

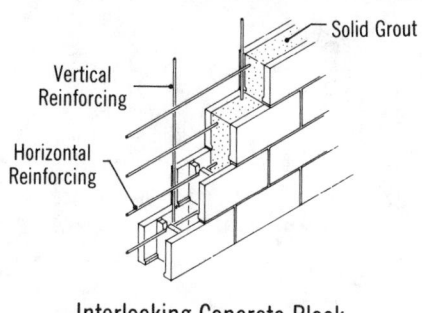

Interlocking Concrete Block

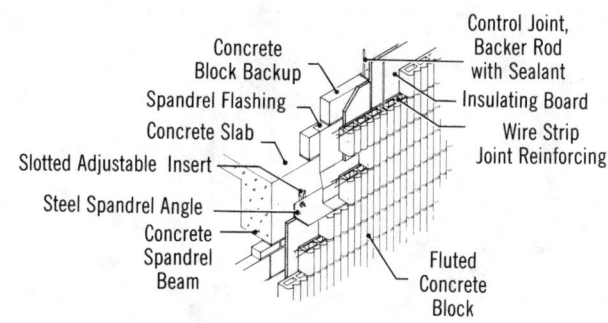

Block Face Cavity Wall System

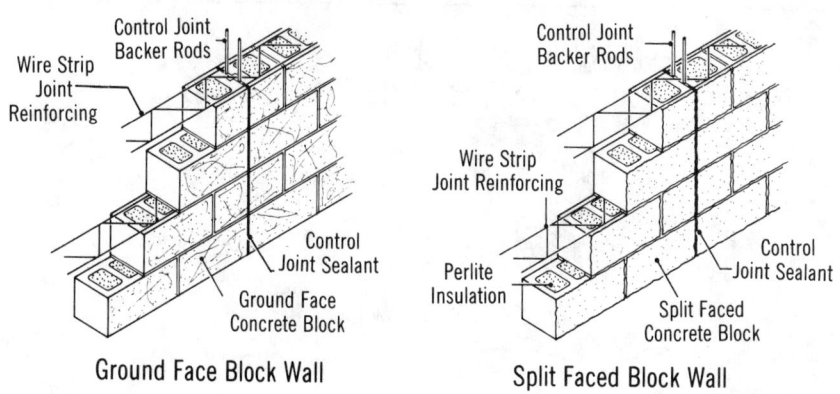

Ground Face Block Wall

Split Faced Block Wall

(continued on next page)

Figure 4.19   Typical Concrete Block Systems
and Nomenclature (continued)

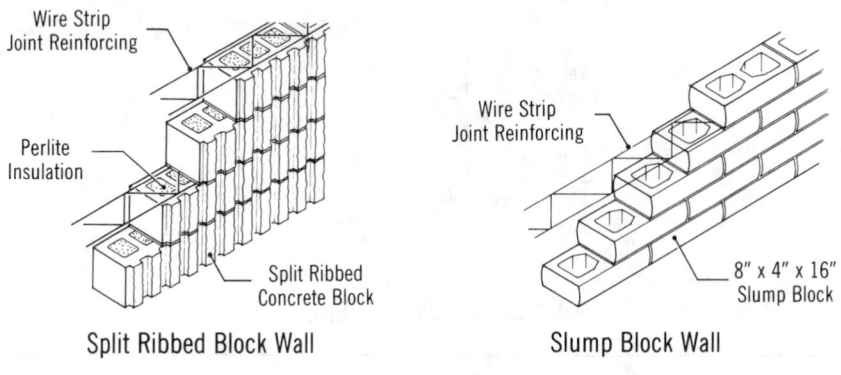

Wire Strip
Joint Reinforcing

Perlite
Insulation

Split Ribbed
Concrete Block

Wire Strip
Joint Reinforcing

8" x 4" x 16"
Slump Block

Split Ribbed Block Wall   **Slump Block Wall**

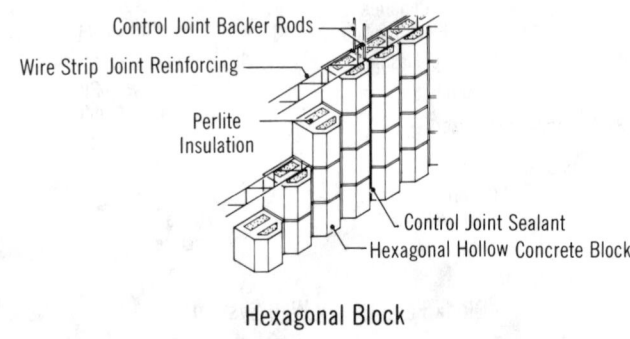

Control Joint Backer Rods

Wire Strip Joint Reinforcing

Perlite
Insulation

Control Joint Sealant
Hexagonal Hollow Concrete Block

Hexagonal Block

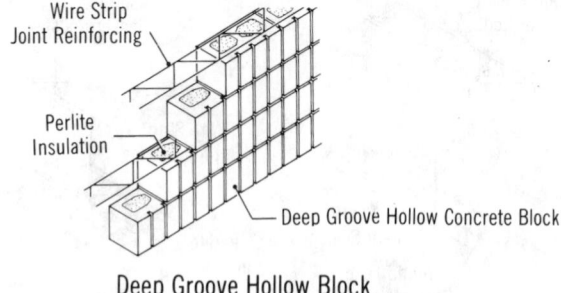

Wire Strip
Joint Reinforcing

Perlite
Insulation

Deep Groove Hollow Concrete Block

Deep Groove Hollow Block

Figure 4.19 Typical Concrete Block Systems and Nomenclature (continued)

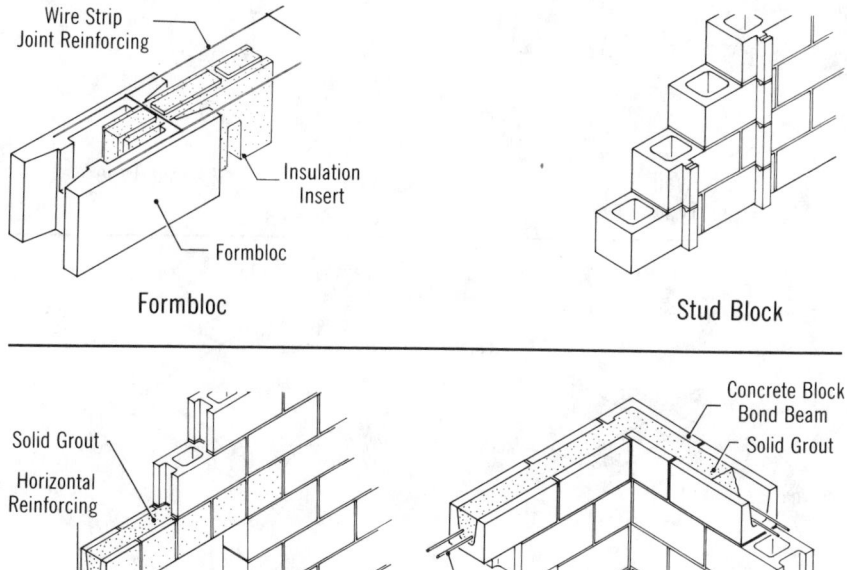

Wire Strip
Joint Reinforcing

Insulation
Insert

Formbloc

Formbloc

Stud Block

Solid Grout

Horizontal
Reinforcing

Concrete Lintel
Block

Lintels

Concrete Block
Bond Beam

Solid Grout

Continuous
Horizontal
Reinforcing

Bond Beam

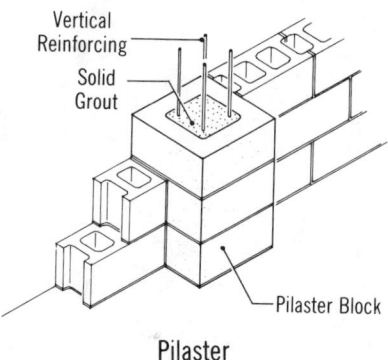

Vertical
Reinforcing

Solid
Grout

Pilaster Block

Pilaster

(continued on next page)

Figure 4.19   Typical Concrete Block Systems and Nomenclature (continued)

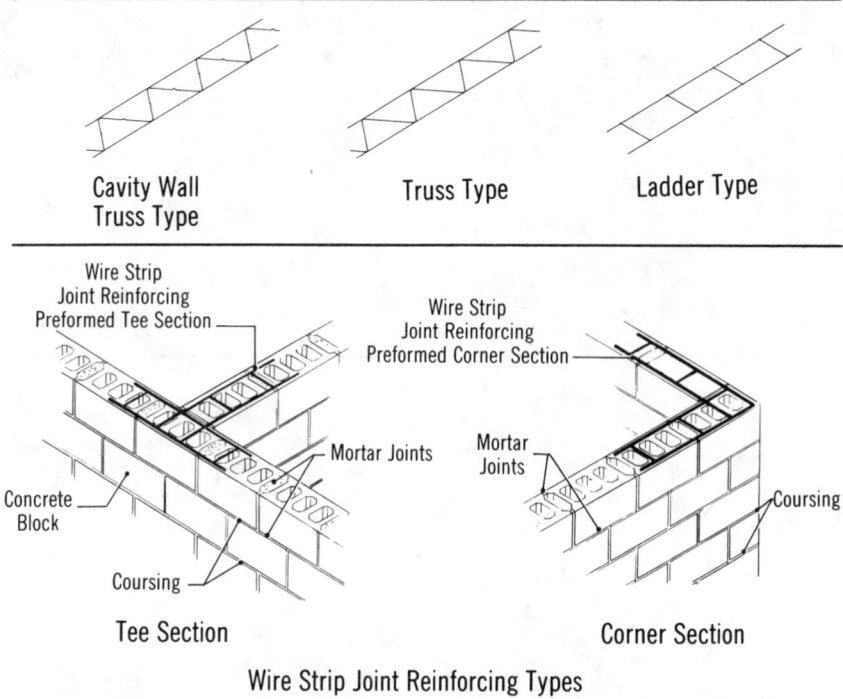

Cavity Wall
Truss Type

Truss Type

Ladder Type

Tee Section

Corner Section

Wire Strip Joint Reinforcing Types

Figure 4.20   Concrete Block Masonry Conversion Factors

Block Type	Description	Waste	Quantity
8" x 16" Concrete Block	All Thicknesses	net	1.125
		5% waste	1.181
4S Glazed Structural Units		net	6.71
		5% waste	7.05
	Horizontal Bullnose	net	1.49*
		5% waste	1.56*
	Vertical Bullnose	net	4.50*
		5% waste	4.73*
4D Glazed Structural Units		net	3.37
		5% waste	3.54
	Horizontal Bullnose	net	1.50*
		5% waste	1.58*
	Vertical Bullnose	net	2.25*
		5% waste	2.36*
6P Glazed Structural Units		net	3.00
		5% waste	3.15
	Horizontal Bullnose	net	1.00*
		5% waste	1.05*
	Vertical Bullnose	net	3.00
		5% waste	3.15
6T Glazed Structural Units		net	2.25
		5% waste	2.36
	Horizontal Bullnose	net	1.00*
		5% waste	1.05*
	Vertical Bullnose	net	2.25*
		5% waste	2.36*
8W Glazed Structural Units		net	1.125
		5% waste	1.181
	Horizontal Bullnose	net	0.75*
		5% waste	0.79*
	Vertical Bullnose	net	1.50*
		5% waste	1.58*
Spectra-Glaze ® Units	8" x 16" (nominal)	net	1.125
		5% waste	1.181
	Horizontal Bullnose	net	0.75*
		5% waste	0.79*
	Vertical Bullnose	net	1.50*
		5% waste	1.58*

**Factors given in units per square feet except those where * occurs, and these are in units per lineal foot.

Quantities of block and waste factors per square foot based upon the type of block.

(courtesy Masonry Estimating Handbook, Michael F. Kenny, Construction Publishing Company, Inc.)

(continued on next page)

Figure 4.20   Concrete Block Masonry
Conversion Factors (continued)

Block Type	Description	Waste	Quantity
Structural Clay Tile	5" x 12"	net	2.09
		5% waste	2.20
	8" x 12"	net	1.36
		5% waste	1.42
	12" x 12"	net	0.92
		5% waste	0.97
Gypsum Block	12" x 30"	net	0.38
		5% waste	0.40
Glass Block	5-3/4" x 5-3/4"	net	4.00
	7-3/4" x 7-3/4"	net	2.25
	11-3/4" x 11-3/4"	net	1.00

**Factors given in units per square feet except those where * occurs, and these are in units per lineal foot.

Quantities of block and waste factors per square foot based upon the type of block.

(courtesy Masonry Estimating Handbook, Michael F. Kenny, Construction Publishing Company, Inc.)

Figure 4.21   Weights of Block Partitions

The approximate weight of various sized block walls per square foot area of wall.

Concrete Blocks Nominal Size	Approximate Weight per S.F.	
	Standard	Lightweight
2" x 8" x 16"	20 PSF	15 PSF
4"	30	20
6"	42	30
8"	55	38
10"	70	47
12"	85	55

Figure 4.22 Height of Concrete Block Walls by Course

Course No.	Concrete Block (3/8" joint)	6T Series G.S.U. (1/4" joint)	8W Series G.S.U. (1/4" joint)
1	0' 8"	5-1/3"	0' 8"
2	1' 4"	10-2/3"	1' 4"
3	2' 0"	1' 4"	2' 0"
4	2' 8"	1' 9-1/3"	2' 8"
5	3' 4"	2' 2-2/3"	3' 4"
6	4' 0"	2' 8"	4' 0"
7	4' 8"	3' 1-1/3"	4' 8"
8	5' 4"	3' 6-2/3"	5' 4"
9	6' 0"	4' 0"	6' 0"
10	6' 8"	4' 5-1/3"	6' 8"
11	7' 4"	4' 10-2/3"	7' 4"
12	8' 0"	5' 4"	8' 0"
13	8' 8"	5' 9-1/3"	8' 8"
14	9' 4"	6' 2-2/3"	9' 4"
15	10' 0"	6' 8"	10' 0"
16	10' 8"	7' 1-1/3"	10' 8"
17	11' 4"	7' 6-2/3"	11' 4"
18	12' 0"	8' 0"	12' 0"
19	12' 8"	8' 5-1/3"	12' 8"
20	13' 4"	8' 10-2/3"	13' 4"
21	14' 0"	9' 4"	14' 0"
22	14' 8"	9' 9-1/3"	14' 8"
23	15' 4"	10' 2-2/3"	15' 4"
24	16' 0"	10' 8"	16' 0"
25	16' 8"	11' 1-1/3"	16' 8"
26	17' 4"	11' 6-2/3"	17' 4"
27	18' 0"	12' 0"	18' 0"
28	18' 8"	12' 5-1/3"	18' 8"
29	19' 4"	12' 10-2/3"	19' 4"
30	20' 0"	13' 4"	20' 0"
31	20' 8"	13' 9-1/3"	20' 8"
32	21' 4"	14' 2-2/3"	21' 4"
33	22' 0"	14' 8"	22' 0"
34	22' 8"	15' 1-1/3"	22' 8"
35	23' 4"	15' 6-2/3"	23' 4"
36	24' 0"	16' 0"	24' 0"
37	24' 8"	16' 5-1/3"	24' 8"
38	25' 4"	16' 10-2/3"	25' 4"
39	26' 0"	17' 4"	26' 0"
40	26' 8"	17' 9-1/3"	26' 8"
41	27' 4"	18' 2-2/3"	27' 4"
42	28' 0"	18' 8"	28' 0"
43	28' 8"	19' 1-1/3"	28' 8"
44	29' 4"	19' 6-2/3"	29' 4"
45	30' 0"	20' 0"	30' 0"
46	30' 8"	20' 5-1/2"	30' 8"
47	31' 4"	20' 10-2/3"	31' 4"
48	32' 0"	21' 4"	32' 0"
49	32' 8"	21' 9-1/3"	32' 8"
50	33' 4"	22' 2-2/3"	33' 4"

Note: When standard brick is used in conjunction with concrete block, the brick coursing corresponds to block height modules, i.e., 3 courses of brick lay up to 8".

(courtesy Masonry Estimating Handbook, Michael F. Kenny, Construction Publishing Company, Inc.)

Figure 4.23   Concrete Block Required for Walls

Ground Floor Area	Number of 8" x 16" Blocks Needed for Height of Wall Shown									
	2'-0"	2'-8"	3'-4"	4'-0"	4'-8"	5'-4"	6'-0"	6'-8"	7'-4"	8'-0"
200	135	180	225	270	315	360	405	450	495	540
300	167	222	278	333	389	444	500	555	611	666
400	191	255	319	383	446	510	574	638	701	765
500	214	285	356	428	499	570	641	713	784	855
600	234	312	390	468	546	624	702	780	858	936
700	252	336	420	504	588	672	756	840	924	1008
800	270	360	450	540	630	720	810	900	990	1080
900	286	381	476	572	667	762	857	953	1048	1143
1000	302	402	503	604	704	804	905	1005	1106	1206
1100	317	423	529	634	740	846	952	1058	1163	1269
1200	331	441	551	662	772	882	992	1103	1213	1323
1300	344	459	574	689	803	918	1033	1148	1262	1377
1400	358	477	596	716	835	954	1073	1193	1312	1431
1500	369	492	615	738	861	984	1107	1230	1353	1476
1600	383	510	638	765	893	1020	1148	1275	1403	1530
1700	394	525	636	788	919	1050	1181	1313	1444	1575
1800	405	540	675	810	945	1080	1215	1350	1485	1620
1900	416	555	694	833	971	1110	1249	1388	1526	1665
2000	428	570	713	855	998	1140	1283	1425	1568	1710
2100	437	582	728	873	1019	1164	1310	1455	1601	1746
2200	448	597	746	896	1045	1194	1343	1493	1642	1791
2300	457	609	761	914	1066	1218	1370	1523	1675	1827
2400	468	624	780	936	1092	1248	1404	1560	1716	1872
2500	477	636	795	954	1113	1272	1431	1590	1749	1908
2600	486	648	810	972	1134	1296	1458	1620	1782	1944
2700	495	660	825	990	1155	1320	1485	1650	1815	1980
2800	504	672	840	1008	1176	1344	1512	1680	1848	2016
2900	513	684	855	1026	1197	1368	1539	1710	1881	2052
3000	524	699	874	1049	1223	1398	1573	1748	1922	2097

This table provides the approximate number of 8" x 16" concrete blocks needed for the foundation walls for structures of various sizes. To use this chart:
1. In the left-most column, find the ground floor area for the structure.
2. Read across the page until you reach the column that matches the height of your foundation wall. The number in that box represents the number of blocks needed for that foundation.

Note: No deductions have been made for openings.

(courtesy How to Estimate Building Losses and Construction Costs, Paul I. Thomas, Prentice Hall)

Figure 4.24 Installation Time in Man-Hours for
Block Walls, Partitions, and Accessories

Description	Man-Hours	Unit
Foundation Walls, Trowel Cut Joints, Parged 1/2" Thick, 1 Side, 8" x 16" Face		
Hollow		
8" Thick	.093	S.F.
12" Thick	.122	S.F.
Solid		
8" Thick	.096	S.F.
12" Thick	.126	S.F.
Backup Walls, Tooled Joint 1 Side, 8" x 16" Face		
4" Thick	.091	S.F.
8" Thick	.100	S.F.
Partition Walls, Tooled Joint 2 Sides 8" x 16" Face		
Hollow		
4" Thick	.093	S.F.
8" Thick	.107	S.F.
12" Thick	.141	S.F.
Solid		
4" Thick	.096	S.F.
8" Thick	.111	S.F.
12" Thick	.148	S.F.
Stud Block Walls, Tooled Joints 2 Sides 8" x 16" Face		
6" Thick and 2", Plain	.098	S.F.
Embossed	.103	S.F.
10" Thick and 2", Plain	.108	S.F.
Embossed	.114	S.F.
6" Thick and 2" Each Side, Plain	.114	S.F.
Acoustical Slotted Block Walls Tooled 2 Sides		
4" Thick	.127	S.F.
8" Thick	.151	S.F.
Glazed Block Walls, Tooled Joint 2 Sides 8" x 16", Glazed 1 Face		
4" Thick	.116	S.F.
8" Thick	.129	S.F.
12" Thick	.171	S.F.
8" x 16", Glazed 2 Faces		
4" Thick	.129	S.F.
8" Thick	.148	S.F.
8" x 16", Corner		
4" Thick	.140	Ea.

(continued on next page)

Figure 4.24   Installation Time in Man-Hours for Block Walls, Partitions, and Accessories (continued)

Description	Man-Hours	Unit
Structural Facing Tile, Tooled 2 Sides		
5" x 12", Glazed 1 Face		
4" Thick	.182	S.F.
8" Thick	.222	S.F.
5" x 12", Glazed 2 Faces		
4" Thick	.205	S.F.
8" Thick	.246	S.F.
8" x 16", Glazed 1 Face		
4" Thick	.116	S.F.
8" Thick	.129	S.F.
8" x 16", Glazed 2 Faces		
4" Thick	.123	S.F.
8" Thick	.137	S.F.
Exterior Walls, Tooled Joint 2 Sides, Insulated		
8" x 16" Face, Regular Weight		
8" Thick	.110	S.F.
12" Thick	.145	S.F.
Lightweight		
8" Thick	.104	S.F.
12" Thick	.137	S.F.
Architectural Block Walls, Tooled Joint 2 Sides		
8" x 16" Face		
4" Thick	.116	S.F.
8" Thick	.138	S.F.
12" Thick	.181	S.F.
Interlocking Block Walls, Fully Grouted		
Vertical Reinforcing		
8" Thick	.131	S.F.
12" Thick	.145	S.F.
16" Thick	.173	S.F.
Bond Beam, Grouted, 2 Horizontal Rebars		
8" x 16" Face, Regular Weight		
8" Thick	.133	L.F.
12" Thick	.192	L.F.
Lightweight		
8" Thick	.131	L.F.
12" Thick	.188	L.F.
Lintels, Grouted, 2 Horizontal Rebars		
8" x 16" Face, 8" Thick	.119	L.F.
16" x 16" Face, 8" Thick	.131	L.F.
Control Joint 4" Wall	.013	L.F.
8" Wall	.020	L.F.
Grouting Bond Beams and Lintels		
8" Deep Pumped, 8" Thick	.018	L.F.
12" Thick	.025	L.F.
Concrete Block Cores Solid		
4" Thick By Hand	.035	S.F.
8" Thick Pumped	.038	S.F.
Cavity Walls 2" Space Pumped	.016	S.F.
6" Space	.034	S.F.

Figure 4.24 Installation Time in Man-Hours for Block Walls, Partitions, and Accessories (continued)

Description	Man-Hours	Unit
Joint Reinforcing		
Wire Strips Regular Truss to 6" Wide	.267	C.L.F.
12" Wide	.400	C.L.F.
Cavity Wall with Drip Section to 6" Wide	.267	C.L.F.
12" Wide	.400	C.L.F.
Lintels Steel Angles Minimum	.008	lb.
Maximum	.016	lb.
Wall Ties	.762	C
Coping For 12" Wall Stock Units, Aluminum	.200	L.F.
Precast Concrete	.188	L.F.
Structural Reinforcing, Placed Horizontal,		
#3 and #4 Bars	.018	lb.
#5 and #6 Bars	.010	lb.
Placed Vertical, #3 and #4 Bars	.023	lb.
#5 and #6 Bars	.012	lb.
Acoustical Slotted Block		
4" Thick	.127	S.F.
6" Thick	.138	S.F.
8" Thick	.151	S.F.
12" Thick	.163	S.F.
Lightweight Block		
4" Thick	.090	S.F.
6" Thick	.095	S.F.
8" Thick	.100	S.F.
10" Thick	.103	S.F.
12" Thick	.130	S.F.
Regular Block		
Hollow		
4" Thick	.093	S.F.
6" Thick	.100	S.F.
8" Thick	.107	S.F.
10" Thick	.111	S.F.
12" Thick	.141	S.F.
Solid		
4" Thick	.095	S.F.
6" Thick	.105	S.F.
8" Thick	.113	S.F.
12" Thick	.150	S.F.
Glazed Concrete Block		
Single Face 8" x 16"		
2" Thick	.111	S.F.
4" Thick	.116	S.F.
6" Thick	.121	S.F.
8" Thick	.129	S.F.
12" Thick	.171	S.F.
Double Face		
4" Thick	.129	S.F.
6" Thick	.138	S.F.
8" Thick	.148	S.F.

(continued on next page)

Figure 4.24   Installation Time in Man-Hours for Block Walls, Partitions, and Accessories (continued)

Description	Man-Hours	Unit
Joint Reinforcing Wire Strips		
4" and 6" Wall	.267	C.L.F.
8" Wall	.320	C.L.F.
10" and 12" Wall	.400	C.L.F.
Steel Bars Horizontal		
#3 and #4	.018	lb.
#5 and #6	.010	lb.
Vertical		
#3 and #4	.023	lb.
#5 and #6	.012	lb.
Grout Cores Solid		
By Hand 6" Thick	.035	S.F.
Pumped 8" Thick	.038	S.F.
10" Thick	.039	S.F.
12" Thick	.040	S.F.

Figure 4.25   Volume of Grout Fill for Concrete Block Walls

Center to Center Spacing Grouted Cores	6" C.M.U. Per S.F. Volume in C.F.		8" C.M.U. Per S.F. Volume in C.F.		12" C.M.U. Per S.F. Volume in C.F.	
	40% Solid	75% Solid	40% Solid	75% Solid	40% Solid	75% Solid
All cores grouted solid	.27	.11	.36	.15	.55	.23
cores grouted 16" O.C.	.14	.06	.18	.08	.28	.12
cores grouted 24" O.C.	.09	.04	.12	.05	.18	.08
cores grouted 32" O.C.	.07	.03	.09	.04	.14	.06
cores grouted 40" O.C.	.05	.02	.07	.03	.11	.05
cores grouted 48" O.C.	.04	.02	.06	.03	.09	.04

Note: Costs are based on high-lift grouting method.

Low-lift grouting is used when the wall is built to a maximum height of 5'. The grout is pumped or poured into the cores of the concrete block. The operation is repeated after each five additional feet of wall height has been completed. High-lift grouting is used when the wall has been built to the full story height. Some of the advantages are: the vertical reinforcing steel can be placed after the wall is completed, and the grout can be supplied by a ready-mix concrete supplier so that it may be pumped in a continuous operation.

Figure 4.26   Quantities for Glass Block Partitions

Quantities needed for the installation of glass block partitions on a per square foot basis

Size	Per 100 S.F.				Per 1000 Block		
	No. of Block	Mortar 1/4" Joint	Asphalt Emulsion	Caulk	Expansion Joint	Panel Anchors	Wall Mesh
6" x 6"	410 ea.	5.0 C.F.	.17 gal.	1.5 gal.	80 L.F.	20 ea.	500 L.F.
8" x 8"	230	3.6	.33	2.8	140	36	670
12" x 12"	102	2.3	.67	6.0	312	80	1000
Approximate quantity per 100 S.F.			.07 gal.	.6 gal.	32 L.F.	9 ea.	51, 68, 102 L.F.

Figure 4.27   Typical Installation Practices for Reinforcing Masonry Walls

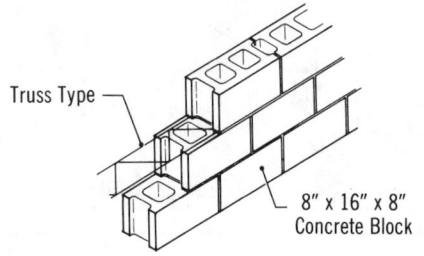

Truss Type

8" x 16" x 8" Concrete Block

Wire Strip Joint Reinforcing

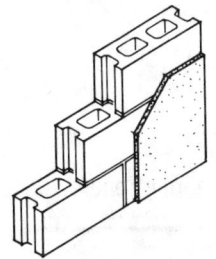

Plaster Direct to Block

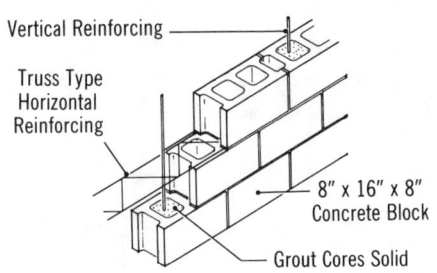

Vertical Reinforcing

Truss Type Horizontal Reinforcing

8" x 16" x 8" Concrete Block

Grout Cores Solid

Reinforced Concrete Block Wall

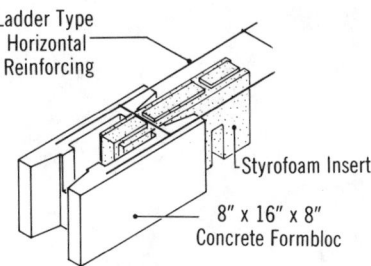

Ladder Type Horizontal Reinforcing

Styrofoam Insert

8" x 16" x 8" Concrete Formbloc

Insulated Concrete Block

(continued on next page)

Figure 4.27   Typical Installation Practices
for Reinforcing Masonry Walls (continued)

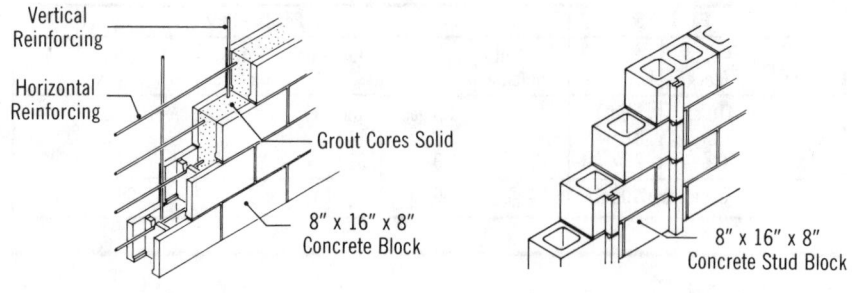

Vertical
Reinforcing

Horizontal
Reinforcing

Grout Cores Solid

8" x 16" x 8"
Concrete Block

8" x 16" x 8"
Concrete Stud Block

Interlocking Concrete Block

Self-furring Concrete Block

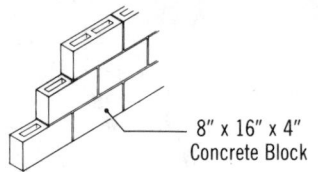

8" x 16" x 4"
Concrete Block

Nonbearing Concrete Block Partition

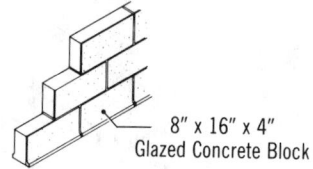

8" x 16" x 4"
Glazed Concrete Block

Glazed Concrete Block

Figure 4.28   Quantities for Joint Reinforcing and Masonry Ties

Number of L.F. of masonry joint reinforcement needed per unit.

Joint Reinforcement			
Number of Courses Between Reinforcement	**L.F. of Reinforcement**		
2	0.67 L.F.	per block	(net)
2	0.74 L.F.		(incl. 10% waste)
2	0.75 L.F.	per S.F. block	(net)
2	0.83 L.F.		(incl. 10% waste)
3	0.44 L.F.	per block	(net)
3	0.48 L.F.		(incl. 10% waste)
3	0.50 L.F.	per S.F. block	(net)
3	0.55 L.F.		(incl. 10% waste)

Masonry joint ties can be estimated on a per S.F. basis as shown in this table.

Masonry Ties		
Ties		**Ties per S.F. of Wall Area**
One tie every 6 courses of brick	1' 0" O.C.	1.5
	2' 0" O.C.	0.75
One tie 2' 0" O.C., each way		0.25

Checklist

For an estimate to be reliable, all items must be accounted for.
A complete estimate can also eliminate the need to include
contingencies. The following checklist can be used to help ensure
that all items are properly accounted for.

Applications _____

Exterior Walls
- ☐ Load bearing
- ☐ Non-load bearing

Interior Walls
- ☐ Load bearing
- ☐ Non-load bearing

Solid Walls _____

Cavity Walls _____

Veneer Walls _____

Flooring _____

Brick
- ☐ Common
- ☐ Face
- ☐ Cement
- ☐ Fire
- ☐ Other _____
- ☐ Color
  - ____ Special color
  - ____ Glazed
  - ____ Special finish
  - ____ Other _____
- ☐ Size
  - ____ Standard
  - ____ Jumbo
  - ____ Norman
  - ____ Roman
  - ____ Engineer
  - ____ Double
  - ____ Other _____
- ☐ Bond pattern
  - ____ Common
  - ____ Running
  - ____ English

____ Flemish
____ Stack
____ Other _____
☐ Reinforcing
☐ Wall ties
☐ Grouting/fill

Concrete Block

☐ Exterior
☐ Interior
☐ Regular
☐ Lightweight
☐ Solid
☐ Hollow core
☐ Finish

____ 1 side
____ 2 side
____ Ribbed
____ Fluted
____ Split face
____ Ground
____ Colored
____ Glazed
____ Other _____

☐ Bond beams
☐ Lintels
☐ Pilasters
☐ Grouting/fill
☐ Reinforcing

____ Bars
____ Wall reinforcing (truss type)

☐ Wall ties
☐ Anchor bolts

Glass Block _____

Structural Tile _____

Terra Cotta _____

Stone

☐ Ashlar
☐ Rubble
☐ Cut Stone

____ Bases
____ Curbs

_____ Facing panels
_____ Flooring
_____ Soffits
_____ Simulated stone
_____ Showers
_____ Stairs
_____ Stair treads
_____ Thresholds
_____ Window sills
_____ Window stools
☐ Other _____

Mortar
☐ Type
    _____ K
    _____ O
    _____ N
    _____ S
    _____ M
    _____ Other _____
☐ Color _____
☐ Admixtures _____

Joints
☐ Concave
☐ Struck
☐ Flush
☐ Raked
☐ Weathered
☐ Stripped
☐ Other _____

Cleaning Masonry
☐ Sandblasting
☐ Steam clean
☐ Acid wash
☐ Power wash
☐ Final clean

Miscellaneous
☐ Anchors
☐ Bolts
☐ Control joints
☐ Copings/flashings
☐ Dowels
☐ Expansion joints

- ☐ Flashings
- ☐ Insulation
- ☐ Inserts
- ☐ Reglets
- ☐ Pointing
- ☐ Waterproofing
- ☐ Weep holes
- ☐ Window sills
- ☐ Window stools
- ☐ Vents

Tips

Masonry Wall Block-outs

When estimating quantities for masonry walls, do not deduct areas less than two square feet in area. They will more than likely use cut whole block and should be figured as such.

Masonry Accessories

Remember to include miscellaneous items in your masonry estimate; they tend to get overlooked. These items include but are not limited to flashing, reinforcing, anchors, wall ties, inserts, bearing plates, lintels, support angles and channels, allowance for joist pockets, waterproofing, cleanup, final cleaning and pointing, steam cleaning, acid or power wash, color for mortar, and control joints. For a more detailed list, see the checklist located at the end of this chapter.

Site Cleanup

If the·plan at the time of the estimate is to erect masonry walls after concrete slabs have been poured, keep in mind that the masonry contractor will be responsible for cleaning the dropped mortar off the concrete slab. Allow for this cleanup.

Bracing

One commonly overlooked item in masonry estimating is an allowance for bracing walls. Until the structural system is tied into the masonry wall systems, the walls can be blown over relatively easily.

Panelization

Plan ahead. Could your next project, which is not scheduled to start until later, be built with pre-assembled panels? If you have the manpower available today, it might be economical to have the contractor pre-build the walls in panelized sections in their yard, then deliver and quickly erect them at the site. This could save quite a bit of time on the project.

Special Brick

When a project calls for special brick such as utility sized or glazed, remember that these (especially glazed) more than likely will be special order. The order and manufacture time can be surprisingly long. Paying a premium or extra charges may be the only way to ensure faster "on-time" delivery.

Split Face Block

Split face block will take longer than common block to set. This is due to the fact that these blocks have a somewhat irregular depth dimension (on account of the splitting process). They do not look "right" if set by lining up the squared corners. Adjustments must be made to have them line up properly.

Bricklaying Productivity

The national average productivity for laying brick ranges from 400 bricks per day (considered "low productivity") to 600 bricks per day (considered "high productivity").

Economy in Bricklaying

The following guidelines may be used to obtain economy in bricklaying:

- Plan to have adequate supervision. Be sure bricklayers are always supplied with materials, so there is no waiting. Place the best bricklayers at corners and openings.
- Use only screened sand for mortar. Otherwise, labor time will be wasted picking out pebbles. Use seamless metal tubs for mortar, as they do not leak or catch trowel. Locate stack and mortar for easy wheeling.
- Have brick delivered for stacking. This makes for faster handling, reduces chipping and breakage, and requires less storage space. Many dealers will deliver select common brick in 2' x 3' x 4' pallets, or face brick packaged. This affords quick handling with a crane or forklift, and easy tonging in units of ten, which reduces waste.
- Use wider bricks for one wythe wall construction. Keep scaffolding away from wall to allow mortar to fall clear and not stain wall.
- On large jobs, develop specialized crews for each type of masonry unit.
- Consider designing for prefabricated panel construction on high-rise projects.
- Avoid excessive corners or openings. Each opening adds about 50% to the labor cost for the area of the opening.
- Bolting stone panels and using window frames as stops reduces labor costs and speeds up erection.

Anti-graffiti Products

If not specified, consider adding as an option the application of one of the various anti-graffiti or vandalism products. These coatings are roller, brush or spray-applied to close the pores of the brick, thereby preventing permanent damage. Remember that graffiti and vandalism are present in all cities and towns, not just the major metropolitan areas.

Notes

Division Five
Metals

Introduction

Steel is among the most versatile materials used in the construction industry. It can be designed, formed, pressed, rolled, cut, bolted and welded, and can be utilized in an almost unlimited variety of ways and situations. Steel for the construction industry is usually shop-fabricated to conform with shop drawings and quality standards. The field erection process usually includes delivery, sorting by sequence (shakeout), hoisting into place, temporarily connecting, plumbing (or truing), and the final connecting (weld, bolt, etc.).

The erecting process and sequence must be thought out thoroughly as this sequence will greatly affect the overall erection time, and ultimately the final cost. The sequence will dictate what type of hoisting equipment can and will be used, the interferences that will be created or avoided, and what other work can commence or continue as the steel is being erected in one area of the project.

The following topics are covered in charts and tables to aid in estimating metals:

- General symbols and nomenclature
- Steel quantity guidelines
- Weights and dimensions of steel shapes and sections
- Weights and depths of open web steel joists
- Steel erecting man-hour tables
- Weights and dimensions of steel bars, plates, and sheets

Estimating Data

The following tables present effective estimating guidelines for items found in Division 5 — Metals. Please note that these guidelines can be used as indicators of what may be expected, and that each project must be evaluated individually.

Table of Contents

Figure 5.1   Standard Welding Symbols

The upper portion of this figure illustrates the standard symbols used to represent the various types of welds specified on construction drawings. The lower portion shows the standard configuration of welding symbols.

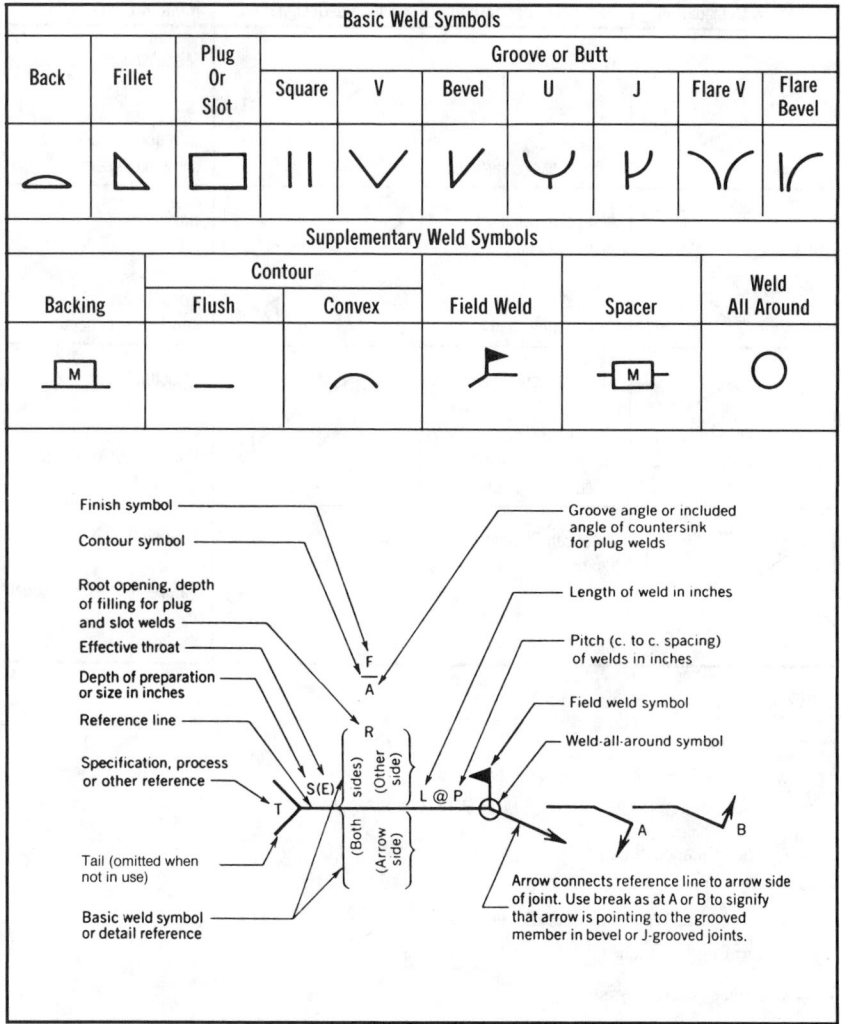

(courtesy of American Institute of Steel Construction, Inc.)

Figure 5.2   Common Steel Sections

The upper portion of this table shows the name, shape, common designation and basic characteristics of commonly used steel sections. The lower portion explains how to read the designations used for the above illustrated common sections.

Shape & Designation	Name & Characteristics	Shape & Designation	Name & Characteristics
W	Wide Flange Parallel flange surfaces	M C	Miscellaneous Channel Infrequently rolled by some producers
S	American Standard Beam (I Beam) Sloped inner flange	L	Angle Equal or unequal legs, constant thickness
M	Miscellaneous Beams Cannot be classified as W, HP or S; infrequently rolled by some producers	T	Structural Tee Cut from W, M or S on center of web
C	American Standard Channel Sloped inner flange	H P	Bearing Pile Parallel flanges and equal flange and web thickness

Common Drawing Designations follow.

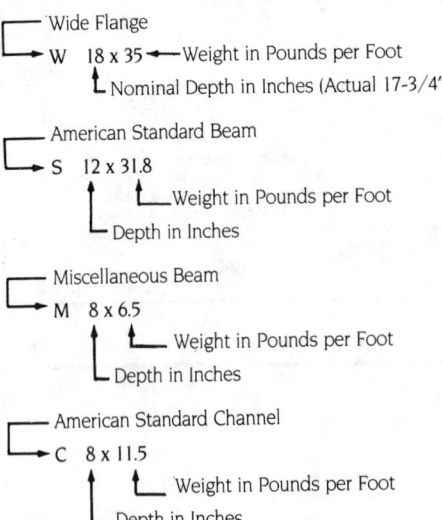

Wide Flange
W   18 x 35 ← Weight in Pounds per Foot
       └ Nominal Depth in Inches (Actual 17-3/4")

American Standard Beam
S   12 x 31.8
              └ Weight in Pounds per Foot
         └ Depth in Inches

Miscellaneous Beam
M   8 x 6.5
              └ Weight in Pounds per Foot
         └ Depth in Inches

American Standard Channel
C   8 x 11.5
              └ Weight in Pounds per Foot
         └ Depth in Inches

Figure 5.2   Common Steel Sections (continued)

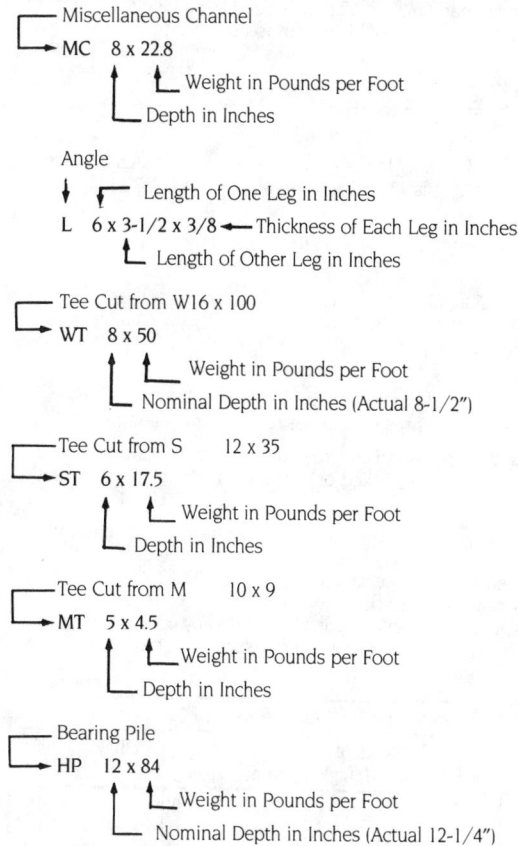

Miscellaneous Channel
MC   8 x 22.8
└ Weight in Pounds per Foot
└ Depth in Inches

Angle
L   6 x 3-1/2 x 3/8 ◄─ Thickness of Each Leg in Inches
┌ Length of One Leg in Inches
└ Length of Other Leg in Inches

Tee Cut from W16 x 100
WT   8 x 50
└ Weight in Pounds per Foot
└ Nominal Depth in Inches (Actual 8-1/2")

Tee Cut from S   12 x 35
ST   6 x 17.5
└ Weight in Pounds per Foot
└ Depth in Inches

Tee Cut from M   10 x 9
MT   5 x 4.5
└ Weight in Pounds per Foot
└ Depth in Inches

Bearing Pile
HP   12 x 84
└ Weight in Pounds per Foot
└ Nominal Depth in Inches (Actual 12-1/4")

Figure 5.3   Common Structural Steel Specifications

ASTM Designation	Yield Stress in KSI	Description
A36	36	Most common carbon steel, all shape groups and plates and bars up to 8"
A529	42	All shape groups, as A36, but plates and bars up to 1/2" thick
A441	40–50	High-strength, low-alloy steel, shapes and plates, but in lesser variety
A572	42–65	High-strength, low-alloy steel, all shapes and plates
A242 & A588	42–50	High-strength, corrosion-resistant; drop in strength as sizes increase
A514	90–100	Quenched and tempered alloy; plates and bars only, with special care so as not to impair the heat treatment

Figure 5.4   Estimating Steel Quantities

One estimate on erection is that a crane can handle 35 to 60 pieces per day. Say the average is 45. With usual sizes of beams, girders, and columns, this would amount to about 20 tons. The type of connection greatly affects the speed of erection. Moment connections for continuous design slow down production and increase erection costs.

Short open web bar joists can be set at the rate of 75 to 80 per day, with 50 per day being the average for setting long span joists.

After main members are calculated, add the following for usual allowances: base plates 2% to 3%; column splices 4% to 5%; and miscellaneous details 4% to 5%, for a total of 10% to 13% in addition to main members.

Ratio of column to beam tonnage varies depending on type of steels used, typical spans, story heights and live loads.

It is more economical to keep the column size constant and to vary the strength of the column by using high strength steels. This also saves floor space. Buildings have recently gone as high as ten stories with 8" high strength columns. For light columns under W8X31 lb. sections, concrete filled steel columns are economical.

High strength steels may be used in columns and beams to save floor space and to meet head room requirements. High strength steels in some sizes sometimes require long lead times.

Round, square and rectangular columns, both plain and concrete filled, are readily available and save floor area, but are higher in cost per pound than rolled columns. For high unbraced columns, tube columns may be less expensive.

Below are average minimum figures for the weights of the structural steel frame for different types of buildings using A36 steel, rolled shapes and simple joints. For economy in domes, rise to span ratio = 0.13. Open web joist framing systems will reduce weights by 10% to 40%. Composite design can reduce figures up to 25% but additional concrete floor slab thickness may be required. Continuous design can reduce the weights up to 20%. There are many building codes with different live load requirements and different structural requirements, such as hurricane and earthquake loadings which can alter the figures.

Structural Steel Weights per S.F. of Floor Area									
Type of Building	No. of Stories	Avg. Spans	L.L. #/S.F.	Lbs. per S.F.	Type of Building	No. of Stories	Avg. Spans	L.L. #/S.F.	Lbs. per S.F.
Steel Frame Mfg.	1	20' x 20' 30' x 30' 40' x 40'	30	6 10 14	Apartments	2–8 9–25	20' x 20'	40	8 14
Parking garage	4	Various	80	8.5	Office	to 10 20 30 over 50	Various	80	10 18 26 35
Domes (Schwedler)	1	200' 300'	30	10 15					

Figure 5.5 Dimensions for Various Steel Shapes

The following tables represent the shapes commonly manufactured and available in the steel industry. The tables also give the physical dimensions for the steel shapes.

		Web	Flange	
Designation	Depth (in.)	Thickness (in.)	Width (in.)	Thickness (in.)
W 36x300	36-3/4	15/16	16-5/8	1-11/16
x280	36-1/2	7/8	16-5/8	1-9/16
x260	36-1/4	13/16	16-1/2	1-7/16
x245	36-1/8	13/16	16-1/2	1-3/8
x230	35-7/8	3/4	16-1/2	1-1/4
W 36x210	36-3/4	13/16	12-1/8	1-3/8
x194	36-1/2	3/4	12-1/8	1-1/4
x182	36-3/8	3/4	12-1/8	1-3/16
x170	36-1/8	11/16	12	1-1/8
x160	36	5/8	12	1
x150	35-7/8	5/8	12	15/16
x135	35-1/2	5/8	12	13/16
W 33x241	34-1/8	13/16	15-7/8	1-3/8
x221	33-7/8	3/4	15-3/4	1-1/4
x201	33-5/8	11/16	15-3/4	1-1/8
W 33x152	33-1/2	5/8	11-5/8	1-1/16
x141	33-1/4	5/8	11-1/2	15/16
x130	33-1/8	9/16	11-1/2	7/8
x118	32-7/8	9/16	11-1/2	3/4
W 30x211	31	3/4	15-1/8	1-5/16
x191	30-5/8	11/16	15	1-3/16
x173	30-1/2	5/8	15	1-1/16
W 30x132	30-1/4	5/8	10-1/2	1
x124	30-1/8	9/16	10-1/2	15/16
x116	30	9/16	10-1/2	7/8
x108	29-7/8	9/16	10-1/2	3/4
x 99	29-5/8	1/2	10-1/2	11/16
W 27x178	27-3/4	3/4	14-1/8	1-3/16
x161	27-5/8	11/16	14	1-1/16
x146	27-3/8	5/8	14	1
W 27x114	27-1/4	9/16	10-1/8	15/16
x102	27-1/8	1/2	10	13/16
x 94	26-7/8	1/2	10	3/4
x 84	26-3/4	7/16	10	5/8
W 24x162	25	11/16	13	1-1/4
x146	24-3/4	5/8	12-7/8	1-1/16
x131	24-1/2	5/8	12-7/8	15/16
x117	24-1/4	9/16	12-3/4	7/8
x104	24	1/2	12-3/4	3/4
W 24x 94	24-1/4	1/2	9-1/8	7/8
x 84	24-1/8	1/2	9	3/4
x 76	23-7/8	7/16	9	11/16
x 68	23-3/4	7/16	9	9/16
W 24x 62	23-3/4	7/16	7	9/16
x 55	23-5/8	3/8	7	1/2

(continued on next page)

Figure 5.5   Dimensions for Various Steel Shapes (continued)

Designation	Depth (in.)	Web Thickness (in.)	Flange Width (in.)	Flange Thickness (in.)
W Shapes Dimensions				
W 21x147	22	3/4	12-1/2	1-1/8
x132	21-7/8	5/8	12-1/2	1-1/16
x122	21-5/8	5/8	12-3/8	15/16
x111	21-1/2	9/16	12-3/8	7/8
x101	21-3/8	1/2	12-1/4	13/16
W 21x 93	21-5/8	9/16	8-3/8	15/16
x 83	21-3/8	1/2	8-3/8	13/16
x 73	21-1/4	7/16	8-1/4	3/4
x 68	21-1/8	7/16	8-1/4	11/16
x 62	21	3/8	8-1/4	5/8
W 21x 57	21	3/8	6-1/2	5/8
x 50	20-7/8	3/8	6-1/2	9/16
x 44	20-5/8	3/8	6-1/2	7/16
W 18x119	19	5/8	11-1/4	1-1/16
x106	18-3/4	9/16	11-1/4	15/16
x 97	18-5/8	9/16	11-1/8	7/8
x 86	18-3/8	1/2	11-1/8	3/4
x 76	18-1/4	7/16	11	11/16
W 18x 71	18-1/2	1/2	7-5/8	13/16
x 65	18-3/8	7/16	7-5/8	3/4
x 60	18-1/4	7/16	7-1/2	11/16
x 55	18-1/8	3/8	7-1/2	5/8
x 50	18	3/8	7-1/2	9/16
W 18x 46	18	3/8	6	5/8
x 40	17-7/8	5/16	6	1/2
x 35	17-3/4	5/16	6	7/16
W 16x100	17	9/16	10-3/8	1
x 89	16-3/4	1/2	10-3/8	7/8
x 77	16-1/2	7/16	10-1/4	3/4
x 67	16-3/8	3/8	10-1/4	11/16
W 16x 57	16-3/8	7/16	7-1/8	11/16
x 50	16-1/4	3/8	7-1/8	5/8
x 45	16-1/8	3/8	7	9/16
x 40	16	5/16	7	1/2
x 36	15-7/8	5/16	7	7/16
W 16x 31	15-7/8	1/4	5-1/2	7/16
x 26	15-3/4	1/4	5-1/2	3/8
W 14x730	22-3/8	3-1/16	17-7/8	4-15/16
x665	21-5/8	2-13/16	17-5/8	4-1/2
x605	20-7/8	2-5/8	17-3/8	4-3/16
x550	20-1/4	2-3/8	17-1/4	3-13/16
x500	19-5/8	2-3/16	17	3-1/2
x455	19	2	16-7/8	3-3/16

Figure 5.5   Dimensions for Various Steel Shapes (continued)

		Web	Flange	
Designation	Depth (in.)	Thickness (in.)	Width (in.)	Thickness (in.)
W 14x426	18-5/8	1-7/8	16-3/4	3-1/16
x398	18-1/4	1-3/4	16-5/8	2-7/8
x370	17-7/8	1-5/8	16-1/2	2-11/16
x342	17-1/2	1-9/16	16-3/8	2-1/2
x311	17-1/8	1-7/16	16-1/4	2-1/4
x283	16-3/4	1-5/16	16-1/8	2-1/16
x257	16-3/8	1-3/16	16	1-7/8
x233	16	1-1/16	15-7/8	1-3/4
x211	15-3/4	1	15-3/4	1-9/16
x193	15-1/2	7/8	15-3/4	1-7/16
x176	15-1/4	13/16	15-5/8	1-5/16
x159	15	3/4	15-5/8	1-3/16
x145	14-3/4	11/16	15-1/2	1-1/16
W 14x132	14-5/8	5/8	14-3/4	1
x120	14-1/2	9/16	14-5/8	15/16
x109	14-3/8	1/2	14-5/8	7/8
x 99	14-1/8	1/2	14-5/8	3/4
x 90	14	7/16	14-1/2	11/16
W 14x 82	14-1/4	1/2	10-1/8	7/8
x 74	14-1/8	7/16	10-1/8	13/16
x 68	14	7/16	10	3/4
x 61	13-7/8	3/8	10	5/8
W 14x 53	13-7/8	3/8	8	11/16
x 48	13-3/4	5/16	8	5/8
x 43	13-5/8	5/16	8	1/2
W 14x 38	14-1/8	5/16	6-3/4	1/2
x 34	14	5/16	6-3/4	7/16
x 30	13-7/8	1/4	6-3/4	3/8
W 14x 26	13-7/8	1/4	5	7/16
x 22	13-3/4	1/4	5	5/16
W 12x336	16-7/8	1-3/4	13-3/8	2-15/16
x305	16-3/8	1-5/8	13-1/4	2-11/16
x279	15-7/8	1-1/2	13-1/8	2-1/2
x252	15-3/8	1-3/8	13	2-1/4
x230	15	1-5/16	12-7/8	2-1/16
x210	14-3/4	1-3/16	12-3/4	1-7/8
x190	14-3/8	1-1/16	12-5/8	1-3/4
x170	14	15/16	12-5/8	1-9/16
x152	13-3/4	7/8	12-1/2	1-3/8
x136	13-3/8	13/16	12-3/8	1-1/4
x120	13-1/8	11/16	12-3/8	1-1/8
x106	12-7/8	5/8	12-1/4	1
x 96	12-3/4	9/16	12-1/8	7/8
x 87	12-1/2	1/2	12-1/8	13/16
x 79	12-3/8	1/2	12-1/8	3/8
x 72	12-1/4	7/16	12	11/16
x 65	12-1/8	3/8	12	5/8

(continued on next page)

Figure 5.5   Dimensions for Various Steel Shapes (continued)

		Web	Flange	
W Shapes Dimensions				
Designation	Depth (in.)	Thickness (in.)	Width (in.)	Thickness (in.)
W 12x 58	12-1/4	3/8	10	5/8
x 53	12	3/8	10	9/16
W 12x 50	12-1/4	3/8	8-1/8	5/8
x 45	12	5/16	8	9/16
x 40	12	5/16	8	1/2
W 12x 35	12-1/2	5/16	6-1/2	1/2
x 30	12-3/8	1/4	6-1/2	7/16
x 26	12-1/4	1/4	6-1/2	3/8
W 12x 22	12-1/4	1/4	4	7/16
x 19	12-1/8	1/4	4	3/8
x 16	12	1/4	4	1/4
x 14	11-7/8	3/16	4	1/4
W 10x112	11-3/8	3/4	10-3/8	1-1/4
x100	11-1/8	11/16	10-3/8	1-1/8
x 88	10-7/8	5/8	10-1/4	1
x 77	10-5/8	1/2	10-1/4	7/8
x 68	10-3/8	1/2	10-1/8	3/4
x 60	10-1/4	7/16	10-1/8	11/16
x 54	10-1/8	3/8	10	5/8
x 49	10	5/16	10	9/16
W 10x 45	10-1/8	3/8	8	5/8
x 39	9-7/8	5/16	8	1/2
x 33	9-3/4	5/16	8	7/16
W 10x 30	10-1/2	5/16	5-3/4	1/2
x 26	10-3/8	1/4	5-3/4	7/16
x 22	10-1/8	1/4	5-3/4	3/8
W 10x 19	10-1/4	1/4	4	3/8
x 17	10-1/8	1/4	4	5/16
x 15	10	1/4	4	1/4
x 12	9-7/8	3/16	4	3/16
W 8x 67	9	9/16	8-1/4	15/16
x 58	8-3/4	1/2	8-1/4	13/16
x 48	8-1/2	3/8	8-1/8	11/16
x 40	8-1/4	3/8	8-1/8	9/16
x 35	8-1/8	5/16	8	1/2
x 31	8	5/16	8	7/16
W 8x 28	8	5/16	6-1/2	7/16
x 24	7-7/8	1/4	6-1/2	3/8
W 8x 21	8-1/4	1/4	5-1/4	3/8
x 18	8-1/8	1/4	5-1/4	5/16
W 8x 15	8-1/8	1/4	4	5/16
x 13	8	1/4	4	1/4
x 10	7-7/8	3/16	4	3/16

Figure 5.5 Dimensions for Various Steel Shapes (continued)

		Web	Flange	
Designation	Depth (in.)	Thickness (in.)	Width (in.)	Thickness (in.)
W Shapes Dimensions				
W 6x 25	6-3/8	5/16	6-1/8	7/16
x 20	6-1/4	1/4	6	3/8
x 15	6	1/4	6	1/4
W 6x 16	6-1/4	1/4	4	3/8
x 12	6	1/4	4	1/4
x 9	5-7/8	3/16	4	3/16
W 5x 19	5-1/8	1/4	5	7/16
x 16	5	1/4	5	3/8
W 4x 13	4-1/8	1/4	4	3/8
M Shapes Dimensions				
M 14x 18	14	3/16	4	1/4
M 12x 11.8	12	3/16	3-1/8	1/4
M 10x 9	10	3/16	2-3/4	3/16
M 8x 6.5	8	1/8	2-1/4	3/16
M 6x 20	6	1/4	6	3/8
x 4.4	6	1/8	1-7/8	3/16
M 5x 18.9	5	5/16	5	7/16
M 4x 13	4	1/4	4	3/8
S Shapes Dimensions				
S 24x121	24-1/2	13/16	8	1-1/16
x106	24-1/2	5/8	7-7/8	1-1/16
x100	24	3/4	7-1/4	7/8
x 90	24	5/8	7-1/8	7/8
x 80	24	1/2	7	7/8
S 20x 96	20-1/4	13/16	7-1/4	15/16
x 86	20-1/4	11/16	7	15/16
x 75	20	5/8	6-3/8	13/16
x 66	20	1/2	6-1/4	13/16
S 18x 70	18	11/16	6-1/4	11/16
x 54.7	18	7/16	6	11/16
S 15x 50	15	9/16	5-5/8	5/8
x 42.9	15	7/16	5-1/2	5/8
S 12x 50	12	11/16	5-1/2	11/16
x 40.8	12	7/16	5-1/4	11/16
x 35	12	7/16	5-1/8	9/16
x 31.8	12	3/8	5	9/16
S 10x 35	10	5/8	5	1/2
x 25.4	10	5/16	4-5/8	1/2
S 8x 23	8	7/16	4-1/8	7/16
x 18.4	8	1/4	4	7/16
S 7x 20	7	7/16	3-7/8	3/8
x 15.3	7	1/4	3-5/8	3/8

(continued on next page)

Figure 5.5   Dimensions for Various Steel Shapes (continued)

| | | Web | Flange | |
Designation	Depth (in.)	Thickness (in.)	Width (in.)	Thickness (in.)
S Shapes Dimensions				
S 6x 17.25	6	7/16	3-5/8	3/8
x 12.5	6	1/4	3-3/8	3/8
S 5x 14.75	5	1/2	3-1/4	5/16
x 10	5	3/16	3	5/16
S 4x 9.5	4	5/16	2-3/4	5/16
x 7.7	4	3/16	2-5/8	5/16
S 3x 7.5	3	3/8	2-1/2	1/4
x 5.7	3	3/16	2-3/8	1/4
H P Shapes Dimensions				
HP 14x117	14-1/4	13/16	14-7/8	13/16
x102	14	11/16	14-3/4	11/16
x 89	13-7/8	5/8	14-3/4	5/8
x 73	13-5/8	1/2	14-5/8	1/2
HP 13x100	13-1/8	3/4	13-1/4	3/4
x 87	13	11/16	13-1/8	11/16
x 73	12-3/4	9/16	13	9/16
x 60	12-1/2	7/16	12-7/8	7/16
HP 12x 84	12-1/4	11/16	12-1/4	11/16
x 74	12-1/8	5/8	12-1/4	5/8
x 63	12	1/2	12-1/8	1/2
x 53	11-3/4	7/16	12	7/16
HP 10x 57	10	9/16	10-1/4	9/16
x 42	9-3/4	7/16	10-1/8	7/16
HP 8x 36	8	7/16	8-1/8	7/16
Channels American Standard Dimensions				
MC18x 58	18	11/16	4-1/4	5/8
x 51.9	18	5/8	4-1/8	5/8
x 45.8	18	1/2	4	5/8
x 42.7	18	7/16	4	5/8
MC13x 50	13	13/16	4-3/8	5/8
x 40	13	9/16	4-1/8	5/8
x 35	13	7/16	4-1/8	5/8
x 31.8	13	3/8	4	5/8
MC12x 50	12	13/16	4-1/8	11/16
x 45	12	11/16	4	11/16
x 40	12	9/16	3-7/8	11/16
x 35	12	7/16	3-3/4	11/16
x 37	12	5/8	3-5/8	5/8
x 32.9	12	1/2	3-1/2	5/8
x 30.9	12	7/16	3-1/2	5/8
x 10.6	12	3/16	1-1/2	5/16

Figure 5.5   Dimensions for Various Steel Shapes (continued)

		Web	Flange	
Designation	Depth (in.)	Thickness (in.)	Width (in.)	Thickness (in.)
Channels American Standard Dimensions				
MC10x 41.1	10	13/16	4-3/8	9/16
x 33.6	10	9/16	4-1/8	9/16
x 28.5	10	7/16	4	9/16
x 28.3	10	1/2	3-1/2	9/16
x 25.3	10	7/16	3-1/2	1/2
x 24.9	10	3/8	3-3/8	9/16
x 21.9	10	5/16	3-1/2	1/2
x 8.4	10	3/16	1-1/2	1/4
x 6.5	10	1/8	1-1/8	3/16
MC 9x 25.4	9	7/15	3-1/2	9/16
x 23.9	9	3/8	3-1/2	9/16
MC 8x 22.8	8	7/16	3-1/2	1/2
x 21.4	8	3/8	3-1/2	1/2
x 20	8	3/8	3	1/2
x 18.7	8	3/8	3	1/2
x 8.5	8	3/16	1-7/8	5/16
MC 7x 22.7	7	1/2	3-5/8	1/2
x 19.1	7	3/8	3-1/2	1/2
x 17.6	7	3/8	3	1/2
MC 6x 18	6	3/8	3-1/2	1/2
x 16.3	6	3/8	3	1/2
x 15.3	6	5/16	3-1/2	3/8
x 15.1	6	5/16	3	1/2
x 12	6	5/16	2-1/2	3/8
C 15x 50	15	11/16	3-3/4	5/8
x 40	15	1/2	3-1/2	5/8
x 33.9	15	3/8	3-3/8	5/8
C 12x 30	12	1/2	3-1/8	1/2
x 25	12	3/8	3	1/2
x 20.7	12	5/16	3	1/2
C 10x 30	10	11/16	3	7/16
x 25	10	1/2	2-7/8	7/16
x 20	10	3/8	2-3/4	7/16
x 15.3	10	1/4	2-5/8	7/16
C 9x 20	9	7/16	2-5/8	7/16
x 15	9	5/16	2-1/2	7/16
x 13.4	9	1/4	2-3/8	7/16
C 8x 18.75	8	1/2	2-1/2	3/8
x 13.75	8	5/16	2-3/8	3/8
x 11.5	8	1/4	2-1/4	3/8
C 7x 14.75	7	7/16	2-1/4	3/8
x 12.25	7	5/16	2-1/4	3/8
x 9.8	7	3/16	2-1/8	3/8
C 6x 13	6	7/16	2-1/8	5/16
x 10.5	6	5/16	2	5/16
x 8.2	6	3/16	1-7/8	5/16

(continued on next page)

Figure 5.5   Dimensions for Various Steel Shapes (continued)

Channels American Standard Dimensions				
		Web	Flange	
Designation	Depth (in.)	Thickness (in.)	Width (in.)	Thickness (in.)
C 5x 9	5	5/16	1-7/8	5/16
x 6.7	5	3/16	1-3/4	5/16
C 4x 7.25	4	5/16	1-3/4	5/16
x 5.4	4	3/16	1-5/8	5/16
C 3x 6	3	3/8	1-5/8	1/4
x 5	3	1/4	1-1/2	1/4
x 4.1	3	3/16	1-3/8	1/4

Angles Equal Legs and Unequal Legs Properties for Designing				
Size and Thickness (in.)	Weight per Foot (lb.)		Size and Thickness (in.)	Weight per Foot (lb.)
L 9 x4 x 5/8	26.3		L 4 x4 x 3/4	18.5
9/16	23.8		5/8	15.7
1/2	21.3		1/2	12.8
L 8 x8 x1-1/8	56.9		7/16	11.3
1	51.0		3/8	9.8
7/8	45.0		5/16	8.2
3/4	38.9		1/4	6.6
5/8	32.7		L 4 x3-1/2x 5/8	14.7
9/16	29.6		1/2	11.9
1/2	26.4		7/16	10.6
L 8 x6 x1	44.2		3/8	9.1
7/8	39.1		5/16	7.7
3/4	33.8		1/4	6.2
5/8	28.5		L 3 x2-1/2x 1/2	8.5
9/16	25.7		7/16	7.6
1/2	23.0		3/8	6.6
7/16	20.2		5/16	5.6
L 8 x4 x1	37.4		1/4	4.5
3/4	28.7		3/16	3.39
9/16	21.9		L 3 x2 x 1/2	7.7
1/2	19.6		7/16	6.8
L 7 x4 x 3/4	26.2		3/8	5.9
5/8	22.1		5/16	5.0
1/2	17.9		1/4	4.1
3/8	13.6		3/16	3.07
L 5 x3-1/2x 3/4	19.8		L 2-1/2x2-1/2x 1/2	7.7
5/8	16.8		3/8	5.9
1/2	13.6		5/16	5.0
7/16	12.0		1/4	4.1
3/8	10.4		3/16	3.07
5/16	8.7		L 2-1/2x2 x 3/8	5.3
1/4	7.0		5/16	4.5
L 5 x3 x 5/8	15.7		1/4	3.62
1/2	12.8		3/16	2.75
7/16	11.3		L 2 x2 x 3/8	4.7
3/8	9.8		5/16	3.92
5/16	8.2		1/4	3.19
1/4	6.6		3/16	2.44
			1/8	1.65

Figure 5.5 Dimensions for Various Steel Shapes (continued)

Structural Tees Cut from W Shapes Dimensions				
		Stem	Flange	
Designation	Depth (in.)	Thickness (in.)	Width (in.)	Thickness (in.)
WT 18 x150	18-3/8	15/16	16-5/8	1-11/16
x140	18-1/4	7/8	16-5/8	1-9/16
x130	18-1/8	13/16	16-1/2	1-7/16
x122.5	18	13-16	16-1/2	1-3/8
x115	18	3/4	16-1/2	1-1/4
x105	18-3/8	13/16	12-1/8	1-3/8
x 97	18-1/4	3/4	12-1/8	1-1/4
x 91	18-1/8	3/4	12-1/8	1-3/16
x 85	18-1/8	11/16	12	1-1/8
x 80	18	5/8	12	1
x 75	17-7/8	5/8	12	15/16
x 67.5	17-3/4	5/8	12	13/16
WT 16.5x120.5	17-1/8	13/16	15-7/8	1-3/8
x110.5	17	3/4	15-3/4	1-1/4
x100.5	16-7/8	11/16	15-3/4	1-1/8
x 76	16-3/4	5/8	11-5/8	1-1/16
x 70.5	16-5/8	5/8	11-1/2	15/16
x 65	16-1/2	9/16	11-1/2	7/8
x 59	16-3/8	9/16	11-1/2	3/4
WT 15 x105.5	15-1/2	3/4	15-1/8	1-5/16
x 95.5	15-3/8	11/16	15	1-3/16
x 86.5	15-1/4	5/8	15	1-1/16
x 66	15-1/8	5/8	10-1/2	1
x 62	15-1/8	9/16	10-1/2	15/16
x 58	15	9/16	10-1/2	7/8
x 54	14-7/8	9/16	10-1/2	3/4
x 49.5	14-7/8	1/2	10-1/2	11/16
WT 13.5x 89	13-7/8	3/4	14-1/8	1-3/16
x 80.5	13-3/4	11/16	14	1-1/16
x 73	13-3/4	5/8	14	1
x 57	13-5/8	9/16	10-1/8	15/16
x 51	13-1/2	1/2	10	13/16
x 47	13-1/2	1/2	10	3/4
x 42	13-3/8	7/16	10	5/8
WT 12 x 81	12-1/2	11/16	13	1-1/4
x 73	12-3/8	5/8	12-7/8	1-1/16
x 65.5	12-1/4	5/8	12-7/8	15/16
x 58.5	12-1/8	9/16	12-3/4	7/8
x 52	12	1/2	12-3/4	3/4
x 47	12-1/8	1/2	9-1/8	7/8
x 42	12	1/2	9	3/4
x 38	12	7/16	9	11/16
x 34	11-7/8	7/16	9	9/16
x 31	11-7/8	7/16	7	9/16
x 27.5	11-3/4	3/8	7	1/2

(continued on next page)

Figure 5.5 Dimensions for Various Steel Shapes (continued)

		Stem	Flange	
Designation	Depth (in.)	Thickness (in.)	Width (in.)	Thickness (in.)
Structural Tees Cut from W Shapes Dimensions				
WT 10.5x 73.5	11	3/4	12-1/2	1-1/8
x 66	10-7/8	5/8	12-1/2	1-1/16
x 61	10-7/8	5/8	12-3/8	15/16
x 55.5	10-3/4	9/16	12-3/8	7/8
x 50.5	10-5/8	1/2	12-1/4	13/16
x 46.5	10-3/4	9/16	8-3/8	15/16
x 41.5	10-3/4	1/2	8-3/8	13/16
x 36.5	10-5/8	7/16	8-1/4	3/4
x 34	10-5/8	7/16	8-1/4	11/16
x 31	10-1/2	3/8	8-1/4	5/8
x 28.5	10-1/2	3/8	6-1/2	5/8
x 25	10-3/8	3/8	6-1/2	9/16
x 22	10-3/8	3/8	6-1/2	7/16
WT 9x 59.5	9-1/2	5/8	11-1/4	1-1/16
x 53	9-3/8	9/16	11-1/4	15/16
x 48.5	9-1/4	9/16	11-1/8	7/8
x 43	9-1/4	1/2	11-1/8	3/4
x 38	9-1/8	7/16	11	11/16
x 35.5	9-1/4	1/2	7-5/8	13/16
x 32.5	9-1/8	7/16	7-5/8	3/4
x 30	9-1/8	7/16	7-1/2	11/16
x 27.5	9	3/8	7-1/2	5/8
x 25	9	3/8	7-1/2	9/16
x 23	9	3/8	6	5/8
x 20	9	5/16	6	1/2
x 17.5	8-7/8	5/16	6	7/16
WT 8x 50	8-1/2	9/16	10-3/8	1
x 44.5	8-3/8	1/2	10-3/8	7/8
x 38.5	8-1/4	7/16	10-1/4	3/4
x 33.5	8-1/8	3/8	10-1/4	.11/16
x 28.5	8-1/4	7/16	7-1/8	11/16
x 25	8-1/8	3/8	7-1/8	5/8
x 22.5	8-1/8	3/8	7	9/16
x 20	8	5/16	7	1/2
x 18	7-7/8	5/16	7	7/16
x 15.5	8	1/4	5-1/2	7/16
x 13	7-7/8	1/4	5-1/2	3/8

Figure 5.5 Dimensions for Various Steel Shapes (continued)

Designation	Depth (in.)	Stem Thickness (in.)	Flange Width (in.)	Flange Thickness (in.)
WT 7x365	11-1/4	3-1/16	17-7/8	4-15/16
x332.5	10-7/8	2-13/16	17-5/8	4-1/2
x302.5	10-1/2	2-5/8	17-3/8	4-3/16
x275	10-1/8	2-3/8	17-1/4	3-13/16
x250	9-3/4	2-3/16	17	3-1/2
x227.5	9-1/2	2	16-7/8	3-3/16
x213	9-3/8	1-7/8	16-3/4	3-1/16
x199	9-1/8	1-3/4	16-5/8	2-7/8
x185	9	1-5/8	16-1/2	2-11/16
x171	8-3/4	1-9/16	16-3/8	2-1/2
x155.5	8-1/2	1-7/16	16-1/4	2-1/4
x141.5	8-3/8	1-5/16	16-1/8	2-1/16
x128.5	8-1/4	1-3/16	16	1-7/8
x116.5	8	1-1/16	15-7/8	1-3/4
x105.5	7-7/8	1	15-3/4	1-9/16
x 96.5	7-3/4	7/8	15-3/4	1-7/16
x 88	7-5/8	13/16	15-5/8	1-5/16
x 79.5	7-1/2	3/4	15-5/8	1-3/16
x 72.5	7-3/8	11/16	15-1/2	1-1/16
x 66	7-3/8	5/8	14-3/4	1
x 60	7-1/4	9/16	14-5/8	15/16
x 54.5	7-1/8	1/2	14-5/8	7/8
x 49.5	7-1/8	1/2	14-5/8	3/4
x 45	7	7/16	14-1/2	11/16
x 41	7-1/8	1/2	10-1/8	7/8
x 37	7-1/8	7/16	10-1/8	13/16
x 34	7	7/16	10	3/4
x 30.5	7	3/8	10	5/8
x 26.5	7	3/8	8	11/16
x 24	6-7/8	5/16	8	5/8
x 21.5	6-7/8	5/16	8	1/2
x 19	7	5/16	6-3/4	1/2
x 17	7	5/16	6-3/4	7/16
x 15	6-7/8	1/4	6-3/4	3/8
x 13	7	1/4	5	7/16
x 11	6-7/8	1/4	5	5/16

(continued on next page)

Figure 5.5 Dimensions for Various Steel Shapes (continued)

Structural Tees Cut from W Shapes Dimensions				
		Stem	Flange	
Designation	Depth (in.)	Thickness (in.)	Width (in.)	Thickness (in.)
WT 6 x168	8-3/8	1-3/4	13-3/8	2-15/16
x152.5	8-1/8	1-5/8	13-1/4	2-11/16
x139.5	7-7/8	1-1/2	13-1/8	2-1/2
x126	7-3/4	1-3/8	13	2-1/4
x115	7-1/2	1-5/16	12-7/8	2-1/16
x105	7-3/8	1-3/16	12-3/4	1-7/8
x 95	7-1/4	1-1/16	12-5/8	1-3/4
x 85	7	15/16	12-5/8	1-9/16
x 76	6-7/8	7/8	12-1/2	1-3/8
x 68	6-3/4	13/16	12-3/8	1-1/4
x 60	6-1/2	11/16	12-3/8	1-1/8
x 53	6-1/2	5/8	12-1/4	1
x 48	6-3/8	9/16	12-1/8	7/8
x 43.5	6-1/4	1/2	12-1/8	13/16
x 39.5	6-1/4	1/2	12-1/8	3/4
x 36	6-1/8	7/16	12	11/16
x 32.5	6	3/8	12	5/8
x 29	6-1/8	3/8	10	5/8
x 26.5	6	3/8	10	9/16
x 25	6-1/8	3/8	8-1/8	5/8
x 22.5	6	5/16	8	9/16
x 20	6	5/16	8	1/2
x 17.5	6-1/4	5/16	6-1/2	1/2
x 15	6-1/8	1/4	6-1/2	7/16
x 13	6-1/8	1/4	6-1/2	3/8
x 11	6-1/8	1/4	4	7/16
x 9.5	6-1/8	1/4	4	3/8
x 8	6	1/4	4	1/4
x 7	6	3/16	4	1/4

Figure 5.5 Dimensions for Various Steel Shapes (continued)

Structural Tees Cut from W Shapes Dimensions		Stem		Flange	
Designation	Depth (in.)	Thickness (in.)	Width (in.)	Thickness (in.)
WT 5 x 56	5-5/8	3/4	10-3/8	1-1/4
x 50	5-1/2	11/16	10-3/8	1-1/8
x 44	5-3/8	5/8	10-1/4	1
x 38.5	5-1/4	1/2	10-1/4	7/8
x 34	5-1/4	1/2	10-1/8	3/4
x 30	5-1/8	7/16	10-1/8	11/16
x 27	5	3/8	10	5/8
x 24.5	5	5/16	10	9/16
y 22.5	5	3/8	8	5/8
x 19.5	5	5/16	8	1/2
x 16.5	4-7/8	5/16	8	7/16
x 15	5-1/4	5/16	5-3/4	1/2
x 13	5-1/8	1/4	5-3/4	7/16
x 11	5-1/8	1/4	5-3/4	3/8
x 9.5	5-1/8	1/4	4	3/8
x 8.5	5	1/4	4	5/16
x 7.5	5	1/4	4	1/4
6	4-7/8	3/16	4	3/16
WT 4 x 33.5	4-1/2	9/16	8-1/4	15/16
x 29	4-3/8	1/2	8-1/4	13/16
x 24	4-1/4	3/8	8-1/8	11/16
x 20	4-1/8	3/8	8-1/8	9/16
x 17.5	4	5/16	8	1/2
x 15.5	4	5/16	8	7/16
x 14	4	5/16	6-1/2	7/16
x 12	4	1/4	6-1/2	3/8
x 10.5	4-1/8	1/4	5-1/4	3/8
x 9	4-1/8	1/4	5-1/4	5/16
x 7.5	4	1/4	4	5/16
x 6.5	4	1/4	4	1/4
x 5	4	3/16	4	3/16
WT 3 x 12.5	3-1/4	5/16	6-1/8	7/16
x 10	3-1/8	1/4	6	3/8
x 7.5	3	1/4	6	1/4
x 8	3-1/8	1/4	4	3/8
x 6	3	1/4	4	1/4
x 4.5	3	3/16	4	3/16
WT 2.5x 9.5	2-5/8	1/4	5	7/16
x 8	2-1/2	1/4	5	3/8
WT 2 x 6.5	2-1/8	1/4	4	3/8

(continued on next page)

Figure 5.5   Dimensions for Various Steel Shapes (continued)

Structural Tees Cut from M Shapes Dimensions				
		Stem	Flange	
Designation	Depth (in.)	Thickness (in.)	Width (in.)	Thickness (in.)
MT 7 x 9	7	3/16	4	1/4
MT 6 x 5.9	6	3/16	3-1/8	1/4
MT 5 x 4.5	5	3/16	2-3/4	3/16
MT 4 x 3.25	4	1/8	2-1/4	3/16
MT 3 x 10	3	1/4	6	3/8
x 2.2	3	1/8	1-7/8	3/16
MT 2.5x 9.45	2-1/2	5/16	5	7/16
MT 2 x 6.5	2	1/4	4	3/8
Structural Tees Cut from S Shapes Dimensions				
ST 12 x 60.5	12-1/4	13/16	8	1-1/16
x 53	12-1/4	5/8	7-7/8	1-1/16
x 50	12	3/4	7-1/4	7/8
x 45	12	5/8	7-1/8	7/8
x 40	12	1/2	7	7/8
ST 10 x 48	10-1/8	13/16	7-1/4	15/16
x 43	10-1/8	11/16	7	15/16
x 37.5	10	5/8	6-3/8	13/16
x 33	10	1/2	6-1/4	13/16
ST 9 x 35	9	11/16	6-1/4	11/16
x 27.35	9	7/16	6	11/16
ST 7.5x 25	7-1/2	9/16	5-5/8	5/8
x 21.45	7-1/2	7/16	5-1/2	5/8
ST 6 x 25	6	11/16	5-1/2	11/16
x 20.4	6	7/16	5-1/4	11/16
x 17.5	6	7/16	5-1/8	9/16
x 15.9	6	3/8	5	9/16
ST 5 x 17.5	5	5/8	5	1/2
x 12.7	5	5/16	4-5/8	1/2
ST 4 x 11.5	4	7/16	4-1/8	7/16
x 9.2	4	1/4	4	7/16
ST 3.5x 10	3-1/2	7/16	3-7/8	3/8
x 7.65	3-1/2	1/4	3-5/8	3/8
ST 3 x 8.625	3	7/16	3-5/8	3/8
x 6.25	3	1/4	3-3/8	3/8
ST 2.5x 7.375	2-1/2	1/2	3-1/4	5/16
x 5	2-1/2	3/16	3	5/16
ST 2 x 4.75	2	5/16	2-3/4	5/16
x 3.85	2	3/16	2-5/8	5/16
ST 1.5x 3.75	1-1/2	3/8	2-1/2	1/4
x 2.85	1-1/2	3/16	2-3/8	1/4

Figure 5.6 External Dimensions for Various Steel Shapes

This figure relates the surface areas and box-out areas for various structural steel shapes. This information is especially helpful when estimating fireproofing or boxing out a column with drywall or other materials.

Surface Areas and Box Areas—W Shapes—Square Feet per Foot of Length				
Designation	Case A	Case B	Case C	Case D
W 36x300	9.99	11.40	7.51	8.90
x280	9.95	11.30	7.47	8.85
x260	9.90	11.30	7.42	8.80
x245	9.87	11.20	7.39	8.77
x230	9.84	11.20	7.36	8.73
x210	8.91	9.93	7.13	8.15
x194	8.88	9.89	7.09	8.10
x182	8.85	9.85	7.06	8.07
x170	8.82	9.82	7.03	8.03
x160	8.79	9.79	7.00	8.00
x150	8.76	9.76	6.97	7.97
x135	8.71	9.70	6.92	7.92
W 33x241	9.42	10.70	7.02	8.34
x221	9.38	10.70	6.97	8.29
x201	9.33	10.60	6.93	8.24
x152	8.27	9.23	6.55	7.51
x141	8.23	9.19	6.51	7.47
x130	8.20	9.15	6.47	7.43
x118	8.15	9.11	6.43	7.39
W 30x211	8.71	9.97	6.42	7.67
x191	8.66	9.92	6.37	7.62
x173	8.62	9.87	6.32	7.57
x132	7.49	8.37	5.93	6.81
x124	7.47	8.34	5.90	6.78
x116	7.44	8.31	5.88	6.75
x108	7.41	8.28	5.84	6.72
x 99	7.37	8.25	5.81	6.68
W 27x178	7.95	9.12	5.81	6.98
x161	7.91	9.08	5.77	6.94
x146	7.87	9.03	5.73	6.89
x114	6.88	7.72	5.39	6.23
x102	6.85	7.68	5.35	6.18
x 94	6.82	7.65	5.32	6.15
x 84	6.78	7.61	5.28	6.11
W 24x162	7.22	8.30	5.25	6.33
x146	7.17	8.24	5.20	6.27
x131	7.12	8.19	5.15	6.22
x117	7.08	8.15	5.11	6.18
x104	7.04	8.11	5.07	6.14
x 94	6.16	6.92	4.81	5.56
x 84	6.12	6.87	4.77	5.52
x 76	6.09	6.84	4.74	5.49
x 68	6.06	6.80	4.70	5.45
x 62	5.57	6.16	4.54	5.13
x 55	5.54	6.13	4.51	5.10

Case A: Shape perimeter, minus one flange surface.
Case B: Shape perimeter.
Case C: Box perimeter, equal to one flange surface plus twice the depth.
Case D: Box perimeter, equal to two flange surfaces plus twice the depth.

(continued on next page)

Figure 5.6   External Dimensions for Various Steel Shapes (continued)

Surface Areas and Box Areas—W Shapes—Square Feet per Foot of Length				
Designation	Case A	Case B	Case C	Case D
W 21x147	6.61	7.66	4.72	5.76
x132	6.57	7.61	4.68	5.71
x122	6.54	7.57	4.65	5.68
x111	6.51	7.54	4.61	5.64
x101	6.48	7.50	4.58	5.61
x 93	5.54	6.24	4.31	5.01
x 83	5.50	6.20	4.27	4.96
x 73	5.47	6.16	4.23	4.92
x 68	5.45	6.14	4.21	4.90
x 62	5.42	6.11	4.19	4.87
x 57	5.01	5.56	4.06	4.60
x 50	4.97	5.51	4.02	4.56
x 44	4.94	5.48	3.99	4.53
W 18x119	5.81	6.75	4.10	5.04
x106	5.77	6.70	4.06	4.99
x 97	5.74	6.67	4.03	4.96
x 86	5.70	6.62	3.99	4.91
x 76	5.67	6.59	3.95	4.87
x 71	4.85	5.48	3.71	4.35
x 65	4.82	5.46	3.69	4.32
x 60	4.80	5.43	3.67	4.30
x 55	4.78	5.41	3.65	4.27
x 50	4.76	5.38	3.62	4.25
x 46	4.41	4.91	3.52	4.02
x 40	4.38	4.88	3.48	3.99
x 35	4.34	4.84	3.45	3.95
W 16x100	5.28	6.15	3.70	4.57
x 89	5.24	6.10	3.66	4.52
x 77	5.19	6.05	3.61	4.47
x 67	5.16	6.01	3.57	4.43
x 57	4.39	4.98	3.33	3.93
x 50	4.36	4.95	3.30	3.89
x 45	4.33	4.92	3.27	3.86
x 40	4.31	4.89	3.25	3.83
x 36	4.28	4.87	3.23	3.81
x 31	3.92	4.39	3.11	3.57
x 26	3.89	4.35	3.07	3.53

Case A: Shape perimeter, minus one flange surface.
Case B: Shape perimeter.
Case C: Box perimeter, equal to one flange surface plus twice the depth.
Case D: Box perimeter, equal to two flange surfaces plus twice the depth.

Figure 5.6 External Dimensions for Various Steel Shapes (continued)

| Designation | Surface Areas and Box Areas—W Shapes—Square Feet per Foot of Length | | | |
	Case A	Case B	Case C	Case D
W 14x730	7.61	9.10	5.23	6.72
x665	7.46	8.93	5.08	6.55
x605	7.32	8.77	4.94	6.39
x550	7.19	8.62	4.81	6.24
x500	7.07	8.49	4.68	6.10
x455	6.96	8.36	4.57	5.98
x426	6.89	8.28	4.50	5.89
x398	6.81	8.20	4.43	5.81
x370	6.74	8.12	4.36	5.73
x342	6.67	8.03	4.29	5.65
x311	6.59	7.94	4.21	5.56
x283	6.52	7.86	4.13	5.48
x257	6.45	7.78	4.06	5.40
x233	6.38	7.71	4.00	5.32
x211	6.32	7.64	3.94	5.25
x193	6.27	7.58	3.89	5.20
x176	6.22	7.53	3.84	5.15
x159	6.18	7.47	3.79	5.09
x145	6.14	7.43	3.76	5.05
x132	5.93	7.16	3.67	4.90
x120	5.90	7.12	3.64	4.86
x109	5.86	7.08	3.60	4.82
x 99	5.83	7.05	3.57	4.79
x 90	5.81	7.02	3.55	4.76
x 82	4.75	5.59	3.23	4.07
x 74	4.72	5.56	3.20	4.04
x 68	4.69	5.53	3.18	4.01
x 61	4.67	5.50	3.15	3.98
x 53	4.19	4.86	2.99	3.66
x 48	4.16	4.83	2.97	3.64
x 43	4.14	4.80	2.94	3.61
x 38	3.93	4.50	2.91	3.48
x 34	3.91	4.47	2.89	3.45
x 30	3.89	4.45	2.87	3.43
x 26	3.47	3.89	2.74	3.16
x 22	3.44	3.86	2.71	3.12

Case A: Shape perimeter, minus one flange surface.
Case B: Shape perimeter.
Case C: Box perimeter, equal to one flange surface plus twice the depth.
Case D: Box perimeter, equal to two flange surfaces plus twice the depth.

(continued on next page)

Figure 5.6   External Dimensions for Various Steel Shapes (continued)

Surface Areas and Box Areas—W Shapes—Square Feet per Foot of Length				
	Case A	Case B	Case C	Case D
Designation				
W 12x336	5.77	6.88	3.92	5.03
x305	5.67	6.77	3.82	4.93
x279	5.59	6.68	3.74	4.83
x252	5.50	6.58	3.65	4.74
x230	5.43	6.51	3.58	4.66
x210	5.37	6.43	3.52	4.58
x190	5.30	6.36	3.45	4.51
x170	5.23	6.28	3.39	4.43
x152	5.17	6.21	3.33	4.37
x136	5.12	6.15	3.27	4.30
x120	5.06	6.09	3.21	4.24
x106	5.02	6.03	3.17	4.19
x 96	4.98	5.99	3.13	4.15
x 87	4.95	5.96	3.10	4.11
x 79	4.92	5.93	3.07	4.08
x 72	4.89	5.90	3.05	4.05
x 65	4.87	5.87	3.02	4.02
x 58	4.39	5.22	2.87	3.70
x 53	4.37	5.20	2.84	3.68
x 50	3.90	4.58	2.71	3.38
x 45	3.88	4.55	2.68	3.35
x 40	3.86	4.52	2.66	3.32
x 35	3.63	4.18	2.63	3.18
x 30	3.60	4.14	2.60	3.14
x 26	3.58	4.12	2.58	3.12
x 22	2.97	3.31	2.39	2.72
x 19	2.95	3.28	2.36	2.69
x 16	2.92	3.25	2.33	2.66
x 14	2.90	3.23	2.32	2.65

Case A: Shape perimeter, minus one flange surface.
Case B: Shape perimeter.
Case C: Box perimeter, equal to one flange surface plus twice the depth.
Case D: Box perimeter, equal to two flange surfaces plus twice the depth.

Figure 5.6 External Dimensions for Various Steel Shapes (continued)

Surface Areas and Box Areas—W Shapes—Square Feet per Foot of Length				
	Case A	Case B	Case C	Case D
Designation				
W 10x112	4.30	5.17	2.76	3.63
x100	4.25	5.11	2.71	3.57
x 88	4.20	5.06	2.66	3.52
x 77	4.15	5.00	2.62	3.47
x 68	4.12	4.96	2.58	3.42
x 60	4.08	4.92	2.54	3.38
x 54	4.06	4.89	2.52	3.35
x 49	4.04	4.87	2.50	3.33
x 45	3.56	4.23	2.35	3.02
x 39	3.53	4.19	2.32	2.98
x 33	3.49	4.16	2.29	2.95
x 30	3.10	3.59	2.23	2.71
x 26	3.08	3.56	2.20	2.68
x 22	3.05	3.53	2.17	2.65
x 19	2.63	2.96	2.04	2.38
x 17	2.60	2.94	2.02	2.35
x 15	2.58	2.92	2.00	2.33
x 12	2.56	2.89	1.98	2.31
W 8x 67	3.42	4.11	2.19	2.88
x 58	3.37	4.06	2.14	2.83
x 48	3.32	4.00	2.09	2.77
x 40	3.28	3.95	2.05	2.72
x 35	3.25	3.92	2.02	2.69
x 31	3.23	3.89	2.00	2.67
x 28	2.87	3.42	1.89	2.43
x 24	2.85	3.39	1.86	2.40
x 21	2.61	3.05	1.82	2.26
x 18	2.59	3.03	1.79	2.23
x 15	2.27	2.61	1.69	2.02
x 13	2.25	2.58	1.67	2.00
x 10	2.23	2.56	1.64	1.97
W 6x 25	2.49	3.00	1.57	2.08
x 20	2.46	2.96	1.54	2.04
x 15	2.42	2.92	1.50	2.00
x 16	1.98	2.31	1.38	1.72
x 12	1.93	2.26	1.34	1.67
x 9	1.90	2.23	1.31	1.64
W 5x 19	2.04	2.45	1.28	1.70
x 16	2.01	2.43	1.25	1.67
W 4x 13	1.63	1.96	1.03	1.37

Case A: Shape perimeter, minus one flange surface.
Case B: Shape perimeter.
Case C: Box perimeter, equal to one flange surface plus twice the depth.
Case D: Box perimeter, equal to two flange surfaces plus twice the depth.

Figure 5.7 Dimensions and Weights for Rectangular and Square Structural Steel Tubing

Structural Tubing Rectangular Dimensions		
Nominal Size (in.)	Wall Thickness (in.)	Weight per Foot (Lbs.)
20 x 12	1/2	103.30
	3/8	78.52
	5/16	65.87
20 x 8	1/2	89.68
	3/8	68.31
	5/16	57.36
20 x 4	1/2	76.07
	3/8	58.10
	5/16	48.86
18 x 6	1/2	76.07
	3/8	58.10
	5/16	48.86
16 x 12	1/2	89.68
	3/8	68.31
	5/16	57.36
16 x 8	1/2	76.07
	3/8	58.10
	5/16	48.86
16 x 4	1/2	62.46
	3/8	47.90
	5/16	40.35
14 x 10	1/2	76.07
	3/8	58.10
	5/16	48.86
14 x 6	1/2	62.46
	3/8	47.90
	5/16	40.35
	1/4	32.63
14 x 4	1/2	55.66
	3/8	42.79
	5/16	36.10
	1/4	29.23
12 x 8	5/8	76.33
	1/2	62.46
	3/8	47.90
	5/16	40.35
	1/4	32.63
12 x 6	1/2	55.66
	3/8	42.79
	5/16	36.10
	1/4	29.23
	3/16	22.18
12 x 4	1/2	48.85
	3/8	37.69
	5/16	31.84
	1/4	25.82
	3/16	19.63
12 x 2	1/4	22.42
	3/16	17.08

Figure 5.7 Dimensions and Weights for Rectangular and Square Structural Steel Tubing (continued)

Structural Tubing Rectangular Dimensions		
Nominal Size (in.)	Wall Thickness (in.)	Weight per Foot (Lbs.)
10 x 6	1/2	48.85
	3/8	37.69
	5/16	31.84
	1/4	25.82
	3/16	19.63
10 x 4	1/2	42.05
	3/8	32.58
	5/16	27.59
	1/4	22.42
	3/16	17.08
10 x 2	3/8	27.48
	5/16	23.34
	1/4	19.02
	3/16	14.53
8 x 6	1/2	42.05
	3/8	32.58
	5/16	27.59
	1/4	22.42
	3/16	17.08
8 x 4	1/2	35.24
	3/8	27.48
	5/16	23.34
	1/4	19.02
	3/16	14.53
8 x 3	3/8	24.93
	5/16	21.21
	1/4	17.32
	3/16	13.25
8 x 2	3/8	22.37
	5/16	19.08
	1/4	15.62
	3/16	11.97
7 x 5	1/2	35.24
	3/8	27.48
	5/16	23.34
	1/4	19.02
	3/16	14.53
7 x 4	3/8	24.93
	5/16	21.21
	1/4	17.32
	3/16	13.25
7 x 3	3/8	22.37
	5/16	19.08
	1/4	15.62
	3/16	11.97
6 x 4	1/2	28.43
	3/8	22.37
	5/16	19.08
	1/4	15.62
	3/16	11.97

(continued on next page)

Figure 5.7 Dimensions and Weights for Rectangular and Square Structural Steel Tubing (continued)

Structural Tubing Rectangular Dimensions		
Nominal Size	Wall Thickness	Weight per Foot
(in.)	(in.)	(Lbs.)
6 x 3	3/8 5/16 1/4 3/16	19.82 16.96 13.91 10.70
6 x 2	3/8 5/16 1/4 3/16	17.27 14.83 12.21 9.42
5 x 4	3/8 5/16 1/4 3/16	19.82 16.96 13.91 10.70
5 x 3	1/2 3/8 5/16 1/4 3/16	21.63 17.27 14.83 12.21 9.42
5 x 2	5/16 1/4 3/16	12.70 10.51 8.15
4 x 3	5/16 1/4 3/16	12.70 10.51 8.15
4 x 2	5/16 1/4 3/16	10.58 8.81 6.87
3 x 2	1/4 3/16	7.11 5.59

Figure 5.7 Dimensions and Weights for Rectangular and Square Structural Steel Tubing (continued)

Structural Tubing Square Dimensions		
Nominal Size (in.)	Wall Thickness (in.)	Weight per Foot (Lbs.)
16 x 16	1/2 3/8 5/16	103.30 78.52 65.87
14 x 14	1/2 3/8 5/16	89.68 68.31 57.36
12 x 12	1/2 3/8 5/16 1/4	76.07 58.10 48.86 39.43
10 x 10	5/8 1/2 3/8 5/16 1/4	76.33 62.46 47.90 40.35 32.63
8 x 8	5/8 1/2 3/8 5/16 1/4 3/16	59.32 48.85 37.69 31.84 25.82 19.63
7 x 7	1/2 3/8 5/16 1/4 3/16	42.05 32.58 27.59 22.42 17.08
6 x 6	1/2 3/8 5/16 1/4 3/16	35.24 27.48 23.34 19.02 14.53
5 x 5	1/2 3/8 5/16 1/4 3/16	28.43 22.37 19.08 15.62 11.97
4 x 4	1/2 3/8 5/16 1/4 3/16	21.63 17.27 14.83 12.21 9.42
3.5 x 3.5	5/16 1/4 3/16	12.70 10.51 8.15
3 x 3	5/16 1/4 3/16	10.58 8.81 6.87
2.5 x 2.5	1/4 3/16	7.11 5.59
2 x 2	1/4 3/16	5.41 4.32

Figure 5.8   Steel Floor Grating and Treads

This table lists the material weight per S.F. for welded steel grating.

Bearing Bar Size (in.)	Steel Bearing Bars, 1-3/16" O.C. Cross Bars 4" O.C. Weight	Steel Bearing Bars, 15/16" O.C. Cross Bars 4" O.C. Weight	Steel Bearing Bars, 1-3/16" O.C. Cross Bars 2" O.C. Weight	Steel Bearing Bars, 15/16" O.C. Cross Bars 2" O.C. Weight
3/4 x 1/8"	4.1#	5.0#	4.8#	5.7#
1 x 1/8	5.2	6.4	5.9	7.1
1-1/4 x 1/8	6.3	7.9	7.0	8.6
1-1/4 x 3/16	9.1	11.5	9.8	12.2
1-1/2 x 1/8	7.4	9.3	8.1	10.0
1-3/4 x 3/16	12.5	15.8	13.2	16.5
2 x 3/16	14.1	18.0	14.8	18.7
2-1/4 x 3/16	15.7	20.0	16.4	20.7

Figure 5.9   Aluminum Floor Grating

This table lists the material weight per S.F. for aluminum grating, alloy 6063.

Bearing Bar Size in Inches	Bearing Bars 1-3/16" O.C.		For Close Mesh Add to 4" O.C. Chart	
	Cross Bars 4" O.C. Weight	Cross Bars 2" O.C. Weight	Total Weight	Add to Cost
1 x 1/8	2.0#	2.1#	3.1#	75%
1-1/4 x 1/8	2.4	2.5	3.8	75%
1-1/4 x 3/16	3.3	3.5	5.1	60%
1-1/2 x 1/8	2.9	3.0	4.6	75%
1-3/4 x 3/16	4.6	4.8	7.2	60%
2 x 3/16	5.3	5.5	8.1	60%
2-1/4 x 3/16	5.9	6.1	9.1	60%

Figure 5.10 Coating Structural Steel

This figure shows the expected average coverage of coating structural steel with red oxide rust inhibitive paint or an aluminum paint. On field-welded jobs, shop coat is necessarily omitted. All painting must be done in the field and usually consists of two coats. Table below shows paint coverage and daily production for field painting.

Type Construction	Surface Area per Ton	Coat	One Gallon Covers		In 8 Hrs. Man Covers		Average per Ton Spray	
			Brush	Spray	Brush	Spray	Gallons	Man-Hours
Light Structural	300 S.F. to 500 S.F.	1st	500 S.F.	455 S.F.	640 S.F.	2000 S.F.	0.9 gals.	1.6 M.H.
		2nd	450	410	800	2400	1.0	1.3
		3rd	450	410	960	3200	1.0	1.0
Medium	150 S.F. to 300 S.F.	All	400	365	1600	3200	0.6	0.6
Heavy Structural	50 S.F. to 100 S.F.	1st	400	365	1920	4000	0.2	0.2
		2nd	400	365	2000	4000	0.2	0.2
		3rd	400	365	2000	4000	0.2	0.2
Weighted Average	225 S.F.	All	400	365	1350	3000	0.6	0.6

Figure 5.11 Standard Joist Details

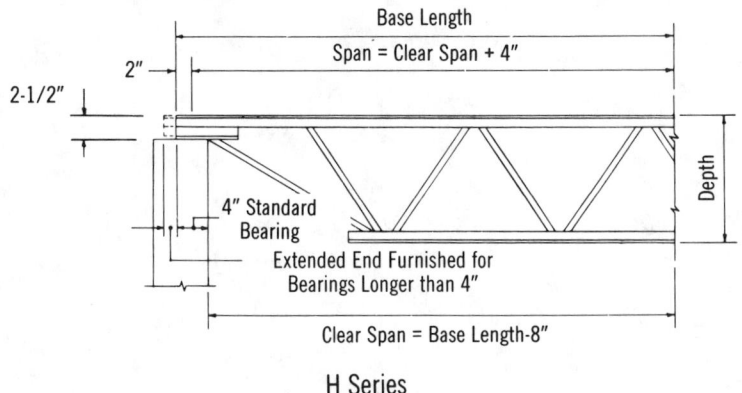

H Series

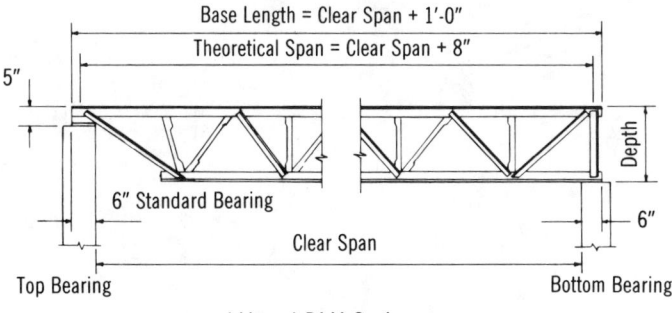

LH and DLH Series

(continued on next page)

Figure 5.11   Standard Joist Details (continued)

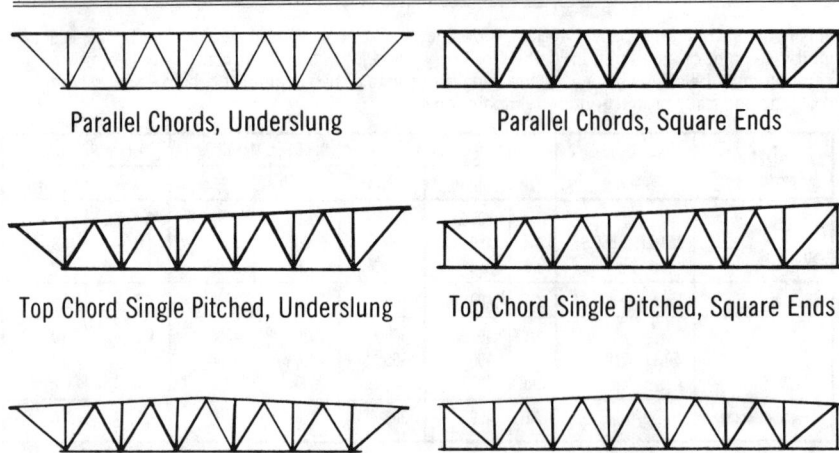

Parallel Chords, Underslung   Parallel Chords, Square Ends

Top Chord Single Pitched, Underslung   Top Chord Single Pitched, Square Ends

Top Chord Double Pitched, Underslung   Top Chord Double Pitched, Square Ends

Figure 5.12 Weights of Open Web Steel Joists (K Series)

The approximate weights per linear foot as shown in the following table do not include any accessories or connection plates.

Joist Designation	8K1	10K1	12K1	12K3	12K5	14K1	14K3	14K4	14K6	16K2	16K3
Depth (in.)	8	10	12	12	12	14	14	14	14	16	16
Approx. Wt. (Lbs./Ft.)	5.1	5.0	5.0	5.7	7.1	5.2	6.0	6.7	7.7	5.5	6.3
Joist Designation	16K4	16K5	16K6	16K7	16K9	18K3	18K4	18K5	18K6	18K7	18K9
Depth (in.)	16	16	16	16	16	18	18	18	18	18	18
Approx. Wt. (Lbs./Ft.)	7.0	7.5	8.1	8.6	10.0	6.6	7.2	7.7	8.5	9.0	10.2
Joist Designation	18K10	20K3	20K4	20K5	20K6	20K7	20K9	20K10	22K4	22K5	22K6
Depth (in.)	18	20	20	20	20	20	20	20	22	22	22
Approx. Wt. (Lbs./Ft.)	11.7	6.7	7.6	8.2	8.9	9.3	10.8	12.2	8.0	8.8	9.2
Joist Designation	22K7	22K9	22K10	22K11	24K4	24K5	24K6	24K7	24K8	24K9	24K10
Depth (in.)	22	22	22	22	24	24	24	24	24	24	24
Approx. Wt. (Lbs./Ft.)	9.7	11.3	12.6	13.8	8.4	9.3	9.7	10.1	11.5	12.0	13.1
Joist Designation	24K12	26K5	26K6	26K7	26K8	26K9	26K10	26K12	28K6	28K7	28K8
Depth (in.)	24	26	26	26	26	26	26	26	28	28	28
Approx. Wt. (Lbs./Ft.)	16.0	9.8	10.6	10.9	12.1	12.2	13.8	16.6	11.4	11.8	12.7
Joist Designation	28K9	28K10	28K12	30K7	30K8	30K9	30K10	30K11	30K12		
Depth (in.)	28	28	28	30	30	30	30	30	30		
Approx. Wt. (Lbs./Ft.)	13.0	14.3	17.1	12.3	13.2	13.4	15.0	16.4	17.6		

Figure 5.13   Weights of Open Web Steel Joists (LH Series)

The approximate weights per linear foot as shown in the following table do not include any accessories or connection plates.

Joist Designation	18LH02	18LH03	18LH04	18LH05	18LH06	18LH07	18LH08	18LH09
Depth (in.)	18	18	18	18	18	18	18	18
Approx. Wt. (Lbs./Ft.)	10	11	12	15	15	17	19	21

Joist Designation	20LH02	20LH03	20LH04	20LH05	20LH06	20LH07	20LH08	20LH09	20LH10
Depth (in.)	20	20	20	20	20	20	20	20	20
Approx. Wt. (Lbs./Ft.)	10	11	12	14	15	17	19	21	23

Joist Designation	24LH03	24LH04	24LH05	24LH06	24LH07	24LH08	24LH09	24LH10	24LH11
Depth (in.)	24	24	24	24	24	24	24	24	24
Approx. Wt. (Lbs./Ft.)	11	12	13	16	17	18	21	23	25

Joist Designation	28LH05	28LH06	28LH07	28LH08	28LH09	28LH10	28LH11	28LH12	28LH13
Depth (in.)	28	28	28	28	28	28	28	28	28
Approx. Wt. (Lbs./Ft.)	13	16	17	18	21	23	25	27	30

Joist Designation	32LH06	32LH07	32LH08	32LH09	32LH10	32LH11	32LH12	32LH13	32LH14	32LH15
Depth (in.)	32	32	32	32	32	32	32	32	32	32
Approx. Wt. (Lbs./Ft.)	14	16	17	21	21	24	27	30	33	35

Joist Designation	36LH07	36LH08	36LH09	36LH10	36LH11	36LH12	36LH13	36LH14	36LH15
Depth (in.)	36	36	36	36	36	36	36	36	36
Approx. Wt. (Lbs./Ft.)	16	18	21	21	23	25	30	36	36

Joist Designation	40LH08	40LH09	40LH10	40LH11	40LH12	40LH13	40LH14	40LH15	40LH16
Depth (in.)	40	40	40	40	40	40	40	40	40
Approx. Wt. (Lbs./Ft.)	16	21	21	22	25	30	35	36	42

Joist Designation	44LH09	44LH10	44LH11	44LH12	44LH13	44LH14	44LH15	44LH16	44LH17
Depth (in.)	44	44	44	44	44	44	44	44	44
Approx. Wt. (Lbs./Ft.)	19	21	22	25	30	31	36	42	47

Joist Designation	48LH10	48LH11	48LH12	48LH13	48LH14	48LH15	48LH16	48LH17
Depth (in.)	48	48	48	48	48	48	48	48
Approx. Wt. (Lbs./Ft.)	21	22	25	29	32	36	42	47

Figure 5.14   Weights of Open Web Steel Joists (DLH Series)

The approximate weights per linear foot as shown in the following table do not include any accessories or connection plates.

Joist Designation	52DLH10	52DLH11	52DLH12	52DLH13	52DLH14	52DLH15	52DLH16	52DLH17
Depth (in.)	52	52	52	52	52	52	52	52
Approx. Wt. (Lbs./Ft.)	25	26	29	34	39	42	45	52
Joist Designation	56DLH11	56DLH12	56DLH13	56DLH14	56DLH15	56DLH16	56DLH17	
Depth (in.)	56	56	56	56	56	56	56	
Approx. Wt. (Lbs./Ft.)	26	30	34	39	42	46	51	
Joist Designation	60DLH12	60DLH13	60DLH14	60DLH15	60DLH16	60DLH17	60DLH18	
Depth (in.)	60	60	60	60	60	60	60	
Approx. Wt. (Lbs./Ft.)	29	35	40	43	46	52	59	
Joist Designation	64DLH12	64DLH13	64DLH14	64DLH15	64DLH16	64DLH17	64DLH18	
Depth (in.)	64	64	64	64	64	64	64	
Approx. Wt. (Lbs./Ft.)	31	34	40	43	46	52	59	
Joist Designation	68DLH13	68DLH14	68DLH15	68DLH16	68DLH17	68DLH18	68DLH19	
Depth (in.)	68	68	68	68	68	68	68	
Approx. Wt. (Lbs./Ft.)	37	40	44	49	55	61	67	
Joist Designation	72DLH14	72DLH15	72DLH16	72DLH17	72DLH18	72DLH19		
Depth (in.)	72	72	72	72	72	72		
Approx. Wt. (Lbs./Ft.)	41	44	50	56	59	70		

Figure 5.15   Weights of Square and Round Bars

This table lists the weights per linear foot of various sized square and round bars.

Size in Inches	Weight in Lbs. per Foot	
	Square	Round
0		
1/16	0.013	0.010
1/8	0.053	0.042
3/16	0.120	0.094
1/4	0.213	0.167
5/16	0.332	0.261
3/8	0.479	0.376
7/16	0.651	0.512
1/2	0.851	0.668
9/16	1.077	0.846
5/8	1.329	1.044
11/16	1.608	1.263
3/4	1.914	1.503
13/16	2.246	1.764
7/8	2.605	2.046
15/16	2.991	2.349
1	3.403	2.673
1/16	3.841	3.017
1/8	4.307	3.382
3/16	4.798	3.769
1/4	5.317	4.176
5/16	5.862	4.604
3/8	6.433	5.053
7/16	7.032	5.523
1/2	7.656	6.013
9/16	8.308	6.525
5/8	8.985	7.057
11/16	9.690	7.610
3/4	10.421	8.185
13/16	11.179	8.780
7/8	11.963	9.396
15/16	12.774	10.032
2	13.611	10.690
1/16	14.475	11.369
1/8	15.366	12.068
3/16	16.283	12.788
1/4	17.227	13.530
5/16	18.197	14.292
3/8	19.194	15.075
7/16	20.217	15.879
1/2	21.267	16.703
9/16	22.344	17.549
5/8	23.447	18.415
11/16	24.577	19.303
3/4	25.734	20.211
13/16	26.917	21.140
7/8	28.126	22.090
15/16	29.362	23.061

Figure 5.15 Weights of Square and Round Bars (continued)

Size in Inches	Weight in Lbs. per Foot	
	Square	Round
3	30.63	24.05
1/16	31.91	25.07
1/8	33.23	26.10
3/16	34.57	27.15
1/4	35.94	28.23
5/16	37.34	29.32
3/8	38.76	30.44
7/16	40.21	31.58
1/2	41.68	32.74
9/16	43.19	33.92
5/8	44.71	35.12
11/16	46.27	36.34
3/4	47.85	37.58
13/16	49.46	38.85
7/8	51.09	40.13
15/16	52.76	41.43
4	54.44	42.76
1/16	56.16	44.11
1/8	57.90	45.47
3/16	59.67	46.86
1/4	61.46	48.27
5/16	63.28	49.70
3/8	65.13	51.15
7/16	67.01	52.63
1/2	68.91	54.12
9/16	70.83	55.63
5/8	72.79	57.17
11/16	74.77	58.72
3/4	76.78	60.30
13/16	78.81	61.90
7/8	80.87	63.51
15/16	82.96	65.15
5	85.07	66.81
1/16	87.21	68.49
1/8	89.38	70.20
3/16	91.57	71.92
1/4	93.79	73.66
5/16	96.04	75.43
3/8	98.31	77.21
7/16	100.61	79.02
1/2	102.93	80.84
9/16	105.29	82.69
5/8	107.67	84.56
11/16	110.07	86.45
3/4	112.50	88.36
13/16	114.96	90.29
7/8	117.45	92.24
15/16	119.96	94.22

(continued on next page)

Figure 5.15   Weights of Square and Round Bars (continued)

| Size in Inches | Weight in Lbs. per Foot | |
	Square	Round
6	122.50	96.21
1/16	125.07	98.23
1/8	127.66	100.26
3/16	130.28	102.32
1/4	132.92	104.40
5/16	135.59	106.49
3/8	138.29	108.61
7/16	141.02	110.75
1/2	143.77	112.91
9/16	146.55	115.10
5/8	149.35	117.30
11/16	152.18	119.52
3/4	155.04	121.77
13/16	157.92	124.03
7/8	160.83	126.32
15/16	163.77	128.63
7	166.74	130.95
1/16	169.73	133.30
1/8	172.74	135.67
3/16	175.79	138.06
1/4	178.86	140.48
5/16	181.96	142.91
3/8	185.08	145.36
7/16	188.23	147.84
1/2	191.41	150.33
9/16	194.61	152.85
5/8	197.84	155.38
11/16	201.10	157.94
3/4	204.38	160.52
13/16	207.69	163.12
7/8	211.03	165.74
15/16	214.39	168.38
8	217.78	171.04
1/16	221.19	173.73
1/8	224.64	176.43
3/16	228.11	179.15
1/4	231.60	181.90
5/16	235.12	184.67
3/8	238.67	187.45
7/16	242.25	190.26
1/2	245.85	193.09
9/16	249.48	195.94
5/8	253.13	198.81
11/16	256.82	201.70
3/4	260.53	204.62
13/16	264.26	207.55
7/8	268.02	210.50
15/16	271.81	213.48

Figure 5.15 Weights of Square and Round Bars (continued)

Size in Inches	Weight in Lbs. per Foot	
	Square	Round
9	275.63	216.48
1/16	279.47	219.49
1/8	283.33	222.53
3/16	287.23	225.59
1/4	291.15	228.67
5/16	295.10	231.77
3/8	299.07	234.89
7/16	303.07	238.03
1/2	307.10	241.20
9/16	311.15	244.38
5/8	315.24	247.59
11/16	319.34	250.81
3/4	323.48	254.06
13/16	327.64	257.33
7/8	331.82	260.61
15/16	336.04	263.92
10	340.28	267.25
1/16	344.54	270.60
1/8	348.84	273.98
3/16	353.16	277.37
1/4	357.50	280.78
5/16	361.88	284.22
3/8	366.28	287.67
7/16	370.70	291.15
1/2	375.16	294.65
9/16	379.64	298.17
5/8	384.14	301.70
11/16	388.67	305.26
3/4	393.23	308.84
13/16	397.82	312.45
7/8	402.43	316.07
15/16	407.07	319.71
11	411.74	323.38
1/16	416.43	327.06
1/8	421.15	330.77
3/16	425.89	334.49
1/4	430.66	338.24
5/16	435.46	342.01
3/8	440.29	345.80
7/16	445.14	349.61
1/2	450.02	353.44
9/16	454.92	357.30
5/8	459.85	361.17
11/16	464.81	365.06
3/4	469.80	368.98
13/16	474.81	372.91
7/8	479.84	376.87
15/16	484.91	380.85
12	490.00	384.85

Figure 5.16   Weights of Rectangular Sections

This table lists weights per linear foot of rectangular steel sections.

Width in Inches	Weight of Rectangular Sections in Pounds per Linear Foot													
	Thickness in Inches													
	3/16	1/4	5/16	3/8	7/16	1/2	9/16	5/8	11/16	3/4	13/16	7/8	15/16	1
1/4	0.16	0.21	0.27	0.32	0.37	0.43	0.48	0.53	0.58	0.64	0.69	0.74	0.80	0.85
1/2	0.32	0.43	0.53	0.64	0.74	0.85	0.96	1.06	1.17	1.28	1.38	1.49	1.60	1.70
3/4	0.48	0.64	0.80	0.96	1.12	1.28	1.44	1.60	1.75	1.91	2.07	2.23	2.39	2.55
1	0.64	0.85	1.06	1.28	1.49	1.70	1.91	2.13	2.34	2.55	2.76	2.98	3.19	3.40
1-1/4	0.80	1.06	1.33	1.60	1.86	2.13	2.39	2.66	2.92	3.19	3.46	3.72	3.99	4.25
1-1/2	0.96	1.28	1.60	1.91	2.23	2.55	2.87	3.19	3.51	3.83	4.15	4.47	4.79	5.10
1-3/4	1.12	1.49	1.86	2.23	2.61	2.98	3.35	3.72	4.09	4.47	4.84	5.21	5.58	5.95
2	1.28	1.70	2.13	2.55	2.98	3.40	3.83	4.25	4.68	5.10	5.53	5.95	6.38	6.81
2-1/4	1.44	1.91	2.39	2.87	3.35	3.83	4.31	4.79	5.26	5.74	6.22	6.70	7.18	7.66
2-1/2	1.60	2.13	2.66	3.19	3.72	4.25	4.79	5.32	5.85	6.38	6.91	7.44	7.98	8.51
2-3/4	1.75	2.34	2.92	3.51	4.09	4.68	5.26	5.85	6.43	7.02	7.60	8.19	8.77	9.36
3	1.91	2.55	3.19	3.83	4.47	5.10	5.74	6.38	7.02	7.66	8.29	8.93	9.57	10.2
3-1/4	2.07	2.76	3.46	4.15	4.84	5.53	6.22	6.91	7.60	8.29	8.99	9.68	10.4	11.1
3-1/2	2.23	2.98	3.72	4.47	5.21	5.95	6.70	7.44	8.19	8.93	9.68	10.4	11.2	11.9
3-3/4	2.39	3.19	3.99	4.79	5.58	6.38	7.18	7.98	8.77	9.57	10.4	11.2	12.0	12.8
4	2.55	3.40	4.25	5.10	5.95	6.81	7.66	8.51	9.36	10.2	11.1	11.9	12.8	13.6
4-1/4	2.71	3.62	4.52	5.42	6.33	7.23	8.13	9.04	9.94	10.8	11.8	12.7	13.6	14.5
4-1/2	2.87	3.83	4.79	5.74	6.70	7.66	8.61	9.57	10.5	11.5	12.4	13.4	14.4	15.3
4-3/4	3.03	4.04	5.05	6.06	7.07	8.08	9.09	10.1	11.1	12.1	13.1	14.1	15.2	16.2
5	3.19	4.25	5.32	6.38	7.44	8.51	9.57	10.6	11.7	12.8	13.8	14.9	16.0	17.0
5-1/4	3.35	4.47	5.58	6.70	7.82	8.93	10.0	11.2	12.3	13.4	14.5	15.6	16.7	17.9
5-1/2	3.51	4.68	5.85	7.02	8.19	9.36	10.5	11.7	12.9	14.0	15.2	16.4	17.5	18.7
5-3/4	3.67	4.89	6.11	7.34	8.56	9.78	11.0	12.2	13.5	14.7	15.9	17.1	18.3	19.6
6	3.83	5.10	6.38	7.66	8.93	10.2	11.5	12.8	14.0	15.3	16.6	17.9	19.1	20.4
6-1/4	3.99	5.32	6.65	7.98	9.30	10.6	12.0	13.3	14.6	16.0	17.3	18.6	19.9	21.3
6-1/2	4.15	5.53	6.91	8.29	9.68	11.1	12.4	13.8	15.2	16.6	18.0	19.4	20.7	22.1
6-3/4	4.31	5.74	7.18	8.61	10.0	11.5	12.9	14.4	15.8	17.2	18.7	20.1	21.5	23.0
7	4.47	5.95	7.44	8.93	10.4	11.9	13.4	14.9	16.4	17.9	19.4	20.8	22.3	23.8
7-1/4	4.63	6.17	7.71	9.25	10.8	12.3	13.9	15.4	17.0	18.5	20.0	21.6	23.1	24.7
7-1/2	4.79	6.38	7.98	9.57	11.2	12.8	14.4	16.0	17.5	19.1	20.7	22.3	23.9	25.5
7-3/4	4.94	6.59	8.24	9.89	11.5	13.2	14.8	16.5	18.1	19.8	21.4	23.1	24.7	26.4
8	5.10	6.81	8.51	10.2	11.9	13.6	15.3	17.0	18.7	20.4	22.1	23.8	25.5	27.2
8-1/2	5.42	7.23	9.04	10.8	12.7	14.5	16.3	18.1	19.9	21.7	23.5	25.3	27.1	28.9
9	5.74	7.66	9.57	11.5	13.4	15.3	17.2	19.1	21.1	23.0	24.9	26.8	28.7	30.6
9-1/2	6.06	8.08	10.1	12.1	14.1	16.2	18.2	20.2	22.2	24.2	26.3	28.3	30.3	32.3
10	6.38	8.51	10.6	12.8	14.9	17.0	19.1	21.3	23.4	25.5	27.6	29.8	31.9	34.0
10-1/2	6.70	8.93	11.2	13.4	15.6	17.9	20.1	22.3	24.6	26.8	29.0	31.3	33.5	35.7
11	7.02	9.36	11.7	14.0	16.4	18.7	21.1	23.4	25.7	28.1	30.4	32.8	35.1	37.4
11-1/2	7.34	9.78	12.2	14.7	17.1	19.6	22.0	24.5	26.9	29.3	31.8	34.2	36.7	39.1
12	7.66	10.2	12.8	15.3	17.9	20.4	23.0	25.5	28.1	30.6	33.2	35.7	38.3	40.8

Figure 5.17   Weights of Flat Rolled Steel

Thick-ness (in inches)	Weight			Thick-ness (in inches)	Weight		
	Lbs./S.F.	Lbs./ Sq. Meter	Kgs./ Sq. Meter		Lbs./S.F.	Lbs./ Sq. Meter	Kgs./ Sq. Meter
.01	.408	4.392	1.992	.51	20.808	223.975	101.593
.02	.816	8.783	3.984	.52	21.216	228.366	103.585
.03	1.224	13.175	5.976	.53	21.624	232.758	105.577
.04	1.632	17.567	7.968	.54	22.032	237.150	107.569
.05	2.040	21.958	9.960	.55	22.440	241.541	109.561
.06	2.448	26.350	11.952	.56	22.848	245.933	111.553
.07	2.856	30.742	13.944	.57	23.256	250.324	113.545
.08	3.264	35.133	15.936	.58	23.664	254.716	115.537
.09	3.672	39.525	17.928	.59	24.072	259.108	117.529
.10	4.080	43.917	19.920	.60	24.480	263.499	119.521
.11	4.488	48.308	21.912	.61	24.888	267.891	121.513
.12	4.896	52.700	23.904	.62	25.296	272.283	123.505
.13	5.304	57.092	25.896	.63	25.704	276.674	125.497
.14	5.712	61.483	27.888	.64	26.112	281.066	127.489
.15	6.120	65.875	29.880	.65	26.520	285.458	129.481
.16	6.528	70.267	31.872	.66	26.928	289.849	131.474
.17	6.936	74.658	33.864	.67	27.336	294.241	133.466
.18	7.344	79.050	35.856	.68	27.744	298.633	135.458
.19	7.752	83.441	37.848	.69	28.152	303.024	137.450
.20	8.160	87.833	39.840	.70	28.560	307.416	139.442
.21	8.568	92.225	41.832	.71	28.968	311.808	141.434
.22	8.976	96.616	43.825	.72	29.376	316.199	143.426
.23	9.384	101.008	45.817	.73	29.784	320.591	145.418
.24	9.792	105.400	47.809	.74	30.192	324.983	147.410
.25	10.200	109.791	49.801	.75	30.600	329.374	149.402
.26	10.608	114.183	51.793	.76	31.008	333.766	151.394
.27	11.016	118.575	53.785	.77	31.416	338.158	153.386
.28	11.424	122.966	55.777	.78	31.824	342.549	155.378
.29	11.832	127.358	57.769	.79	32.232	346.941	157.370
.30	12.240	131.750	59.761	.80	32.640	351.333	159.362
.31	12.648	136.141	61.753	.81	33.048	355.724	161.354
.32	13.056	140.533	63.745	.82	33.456	360.116	163.346
.33	13.464	144.925	65.737	.83	33.864	364.508	165.338
.34	13.872	149.316	67.729	.84	34.272	368.899	167.330
.35	14.280	153.708	69.721	.85	34.680	373.291	169.322
.36	14.688	158.100	71.713	.86	35.088	377.683	171.314
.37	15.096	162.491	73.705	.87	35.496	382.074	173.306
.38	15.504	166.883	75.697	.88	35.904	386.466	175.298
.39	15.912	171.275	77.689	.89	36.312	390.858	177.290
.40	16.320	175.666	79.681	.90	36.720	395.249	179.282
.41	16.728	180.058	81.673	.91	37.128	399.641	181.274
.42	17.136	184.450	83.665	.92	37.536	404.033	183.266
.43	17.544	188.841	85.657	.93	37.944	408.424	185.258
.44	17.952	193.233	87.649	.94	38.352	412.816	187.250
.45	18.360	197.625	89.641	.95	38.760	417.207	189.242
.46	18.768	202.016	91.633	.96	39.168	421.599	191.234
.47	19.176	206.408	93.625	.97	39.576	425.991	193.226
.48	19.584	210.800	95.617	.98	39.984	430.382	195.218
.49	19.992	215.191	97.609	.99	40.392	434.774	197.210
.50	20.400	219.583	99.601	1.00	40.800	439.166	199.202

Figure 5.18   Weights of Flat Rolled Steel—Metric

Thickness (in mm)	Thickness (in inches)	Weight Lbs./S.F.	Weight Lbs./Sq. Meter	Weight Kgs./Sq. Meter
1	.03937	1.606	17.290	7.843
2	.07874	3.213	34.580	15.685
3	.11811	4.819	51.870	23.528
4	.15748	6.425	69.160	31.370
5	.19685	8.031	86.450	39.213
6	.23622	9.638	103.740	47.056
7	.27559	11.244	121.030	54.898
8	.31496	12.850	138.320	62.741
9	.35433	14.457	155.610	70.583
10	.39370	16.063	172.900	78.426
11	.43307	17.669	190.190	86.269
12	.47244	19.276	207.479	94.111
13	.51181	20.882	224.769	101.954
14	.55118	22.488	242.059	109.796
15	.59055	24.094	259.349	117.639
16	.62992	25.701	276.639	125.481
17	.66929	27.307	293.929	133.324
18	.70866	28.913	311.219	141.167
19	.74803	30.520	328.509	149.009
20	.78740	32.126	345.799	156.852
21	.82677	33.732	363.089	164.694
22	.86614	35.339	380.379	172.537
23	.90551	36.945	397.669	180.380
24	.94488	38.551	414.959	188.222
25	.98425	40.157	432.249	196.065
25.4	1.0	40.800	439.166	199.202

Figure 5.19   Weights of Rolled Floor Plates

This table lists weights for floor plates. Note that various mills offer various raised patterns and surface projections in a variety of widths. A maximum width of 96" and a maximum thickness of 1" are available.

Rolled Floor Plates: Standard Thicknesses and Weights					
Gauge	Theoretical Weight per S.F. (in Lbs.)	Nominal Thickness (in inches)	Theoretical Weight per S.F. (in Lbs.)	Nominal Thickness (in inches)	Theoretical Weight per S.F. (in Lbs.)
18	2.40	1/8"	6.16	1/2"	21.47
16	3.00	3/16"	8.71	9/16"	24.02
14	3.75	1/4"	11.26	5/8"	26.58
13	4.50	5/16"	13.81	3/4"	31.68
12	5.25	3/8"	16.37	7/8"	36.78
		7/16"	18.92	1"	41.89

Thickness of rolled floor plates is measured near the edge and does not take the raised pattern into account.

Figure 5.20   Dimensions and Weights of Sheet Steel

Gauge No.	Approximate Thickness				Weight		
	Inches (in fractions)	Inches (in decimal parts)		Millimeters			
	Wrought Iron	Wrought Iron	Steel	Steel	per S.F. in Ounces	per S.F. in Lbs.	per Square Meter in Kg.
0000000	1/2″	0.5	0.4782	12.146	320	20	97.65
000000	15/32″	.46875	.4484	11.389	300	18.75	91.55
00000	7/16″	.4375	.4185	10.630	280	17.50	85.44
0000	13/32″	.40625	.3886	9.870	260	16.25	79.33
000	3/8″	.375	.3587	9.111	240	15	73.24
00	11/32″	.34375	.3288	8.352	220	13.75	67.13
0	5/16″	.3125	.2989	7.592	200	12.50	61.03
1	9/32″	.28125	.2690	6.833	180	11.25	54.93
2	17/64″	.265625	.2541	6.454	170	10.625	51.88
3	1/4″	.25	.2391	6.073	160	10	48.82
4	15/64″	.234375	.2242	5.695	150	9.375	45.77
5	7/32″	.21875	.2092	5.314	140	8.75	42.72
6	13/64″	.203125	.1943	4.935	130	8.125	39.67
7	3/16″	.1875	.1793	4.554	120	7.5	36.32
8	11/64″	.171875	.1644	4.176	110	6.875	33.57
9	5/32″	.15625	.1495	3.797	100	6.25	30.52
10	9/64″	.140625	.1345	3.416	90	5.625	27.46
11	1/8″	.125	.1196	3.038	80	5	24.41
12	7/64″	.109375	.1046	2.657	70	4.375	21.36
13	3/32″	.09375	.0897	2.278	60	3.75	18.31
14	5/64″	.078125	.0747	1.897	50	3.125	15.26
15	9/128″	.0703125	.0673	1.709	45	2.8125	13.73
16	1/16″	.0625	.0598	1.519	40	2.5	12.21
17	9/160″	.05625	.0538	1.367	36	2.25	10.99
18	1/20″	.05	.0478	1.214	32	2	9.765
19	7/160″	.04375	.0418	1.062	28	1.75	8.544
20	3/80″	.0375	.0359	.912	24	1.50	7.324
21	11/320″	.034375	.0329	.836	22	1.375	6.713
22	1/32″	.03125	.0299	.759	20	1.25	6.103
23	9/320″	.028125	.0269	.683	18	1.125	5.49
24	1/40″	.025	.0239	.607	16	1	4.882
25	7/320″	.021875	.0209	.531	14	0.875	4.272
26	3/160″	.01875	.0179	.455	12	.75	3.662
27	11/640″	.0171875	.0164	.417	11	.6875	3.357
28	1/64″	.015625	.0149	.378	10	.625	3.052

Figure 5.21   Installation Time in Man-Hours for Various Structural Steel Building Components

The following tables show the expected average installation times for various structural steel shapes. Table A presents installation times for column shapes; Table B for beams; Table C for light framing and bolts; and Table D for structural steel for various project types.

Table A		
Description	**Man-Hours**	**Unit**
Columns		
Steel Concrete Filled		
3-1/2" Diameter	.933	Ea.
6-5/8" Diameter	1.120	Ea.
Steel Pipe		
3" Diameter	.933	Ea.
8" Diameter	1.120	Ea.
12" Diameter	1.244	Ea.
Structural Tubing		
4" x 4"	.966	Ea.
8" x 8"	1.120	Ea.
12" x 8"	1.167	Ea.
Wide Flange 2 Tier		
W8 x 31	.052	L.F.
W8 x 67	.057	L.F.
W10 x 45	.054	L.F.
W10 x 112	.058	L.F.
W12 x 50	.054	L.F.
W12 x 190	.061	L.F.
W14 x 74	.057	L.F.
W14 x 176	.061	L.F.

Table B				
Description	**Man-Hours**	**Unit**	**Man-Hours**	**Unit**
Beams Wide Flange				
W6 x 9	.949	Ea.	.093	L.F.
W10 x 22	1.037	Ea.	.085	L.F.
W12 x 26	1.037	Ea.	.064	L.F.
W14 x 34	1.333	Ea.	.069	L.F.
W16 x 31	1.333	Ea.	.062	L.F.
W18 x 50	2.162	Ea.	.088	L.F.
W21 x 62	2.222	Ea.	.077	L.F.
W24 x 76	2.353	Ea.	.072	L.F.
W27 x 94	2.581	Ea.	.067	L.F.
W30 x 108	2.857	Ea.	.067	L.F.
W33 x 130	3.200	Ea.	.071	L.F.
W36 x 300	3.810	Ea.	.077	L.F.

Figure 5.21   Installation Time in Man-Hours for Various Structural Steel Building Components (continued)

Table C		
Description	**Man-Hours**	**Unit**
Light Framing		
Angles 4" and Larger	.011	lbs.
Less than 4"	.018	lbs.
Channels 8" and Larger	.009	lbs.
Less than 8"	.016	lbs.
Cross Bracing Angles	.009	lbs.
Rods	.034	lbs.
Hanging Lintels	.028	lbs.
High-Strength Bolts in Place		
3/4" Bolts	.097	Ea.
7/8" Bolts	.100	Ea.

Table D				
Description	**Man-Hours**	**Unit**	**Man-Hours**	**Unit**
Apartments, Nursing Homes, etc.				
1-2 Stories	4.211	Piece	7.767	ton
3-6 Stories	4.444	Piece	7.921	ton
7-15 Stories	4.923	Piece	9.014	ton
Over 15 Stories	5.333	Piece	9.209	ton
Offices, Hospitals, etc.				
1-2 Stories	4.211	Piece	7.767	ton
3-6 Stories	4.741	Piece	8.889	ton
7-15 Stories	4.923	Piece	9.014	ton
Over 15 Stories	5.120	Piece	9.209	ton
Industrial Buildings				
1 Story	3.478	Piece	6.202	ton

Figure 5.22 Installation Time in Man-Hours for the Erection of Steel Superstructure Systems

Description	Man-Hours	Unit
Steel Columns Concrete Filled		
4" Diameter	.072	L.F.
5" Diameter	.055	L.F.
6-5/8" Diameter	.047	L.F.
Steel Pipe		
6" Diameter	6.000	ton
12" Diameter	2.000	ton
Structural Tubing		
6" x 6"	6.000	ton
10" x 10"	2.000	ton
Wide Flange		
W8 x 31	3.355	ton
W10 x 45	2.412	ton
W12 x 50	2.171	ton
W14 x 74	1.538	ton
Beams WF Average	4.000	ton
Steel Joists H Series Horizontal Bridging		
To 30' Span	6.667	ton
30' to 50' Span	6.353	ton
(Includes One Row of Bolted Cross Bridging for Spans Over 40' Where Required)		
LH Series Bolted Cross Bridging		
Spans to 96'	6.154	ton
DLH Series Bolted Cross Bridging		
Spans to 144' Shipped in 2 Pieces	6.154	ton
Joist Girders	6.154	ton
Trusses, Factory Fabricated, with Chords	7.273	ton
Metal Decking Open Type		
1-1/2" Deep		
22 Gauge	.007	S.F.
18 and 20 Gauge	.008	S.F.
3" Deep		
20 and 22 Gauge	.009	S.F.
18 Gauge	.010	S.F.
16 Gauge	.011	S.F.
4-1/2" Deep		
20 Gauge	.012	S.F.
18 Gauge	.013	S.F.
16 Gauge	.014	S.F.
7-1/2" Deep		
18 Gauge	.019	S.F.
16 Gauge	.020	S.F.

Figure 5.22 Installation Time in Man-Hours
for the Erection of Steel
Superstructure Systems (continued)

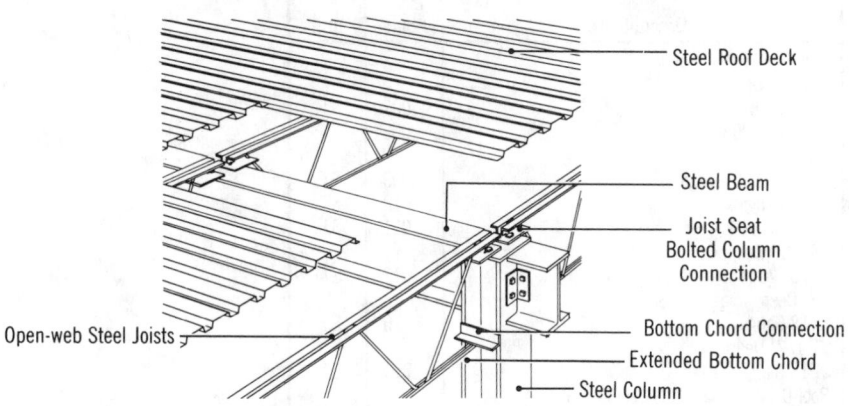

Steel Roof Deck

Steel Beam

Joist Seat
Bolted Column
Connection

Open-web Steel Joists

Bottom Chord Connection

Extended Bottom Chord

Steel Column

Figure 5.23   Installation Time in Man-Hours
for Steel Roof Decks

Description	Man-Hours	Unit
Roof Deck Open Type		
1-1/2" Deep		
22 Gauge	.007	S.F.
18 and 20 Gauge	.008	S.F.
3" Deep		
22 and 20 Gauge	.009	S.F.
18 Gauge	.010	S.F.
16 Gauge	.011	S.F.
4-1/2" Deep		
20 Gauge	.012	S.F.
6" Deep		
18 Gauge	.016	S.F.
7-1/2" Deep		
18 Gauge	.019	S.F.
Roof Deck Cellular Units Galv.		
3" Deep 20-20 Gauge	.023	S.F.
4-1/2" Deep 20-18 Gauge	.029	S.F.

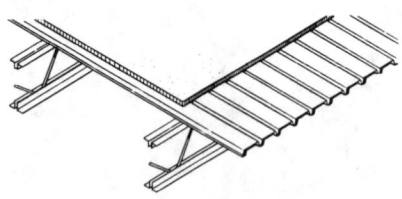

Roof Deck System with Insulation

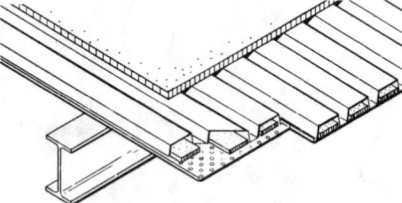

Acoustic Deck System

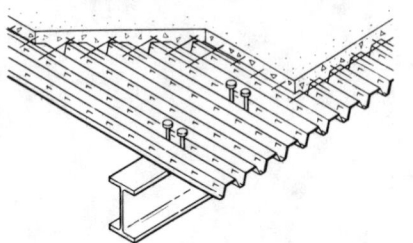

Composite Beam, Deck and Slab

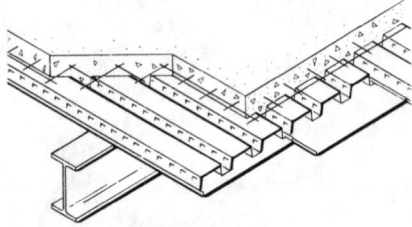

Cellular Deck System

Figure 5.24 Installation Time in Man-Hours for Fireproofing Structural Steel

Description	Man-Hours	Unit
Fireproofing - 10" Column Encasements		
Perlite Plaster	.273	V.L.F.
1" Perlite on 3/8" Gypsum Lath	.345	V.L.F.
Sprayed Fiber	.131	V.L.F.
Concrete 1-1/2" Thick	.716	V.L.F.
Gypsum Board 1/2" Fire Resistant,		
1 Layer	.364	V.L.F.
2 Layer	.428	V.L.F.
3 Layer	.530	V.L.F.
Fireproofing – 16" x 7" Beam Encasements		
Perlite Plaster on Metal Lath	.453	L.F.
Gypsum Plaster on Metal Lath	.408	L.F.
Sprayed Fiber	.079	L.F.
Concrete 1-1/2" Thick	.554	L.F.
Gypsum Board 5/8" Fire Resistant	.488	L.F.

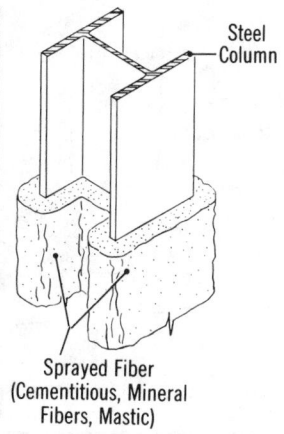

Sprayed Fiber on Columns

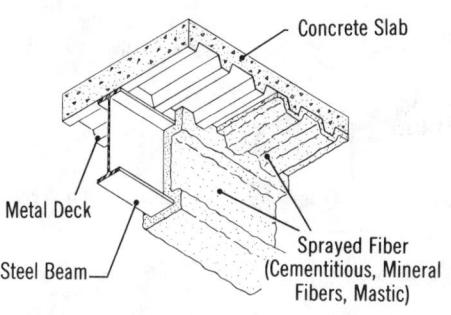

Sprayed Fiber on Beams and Girders

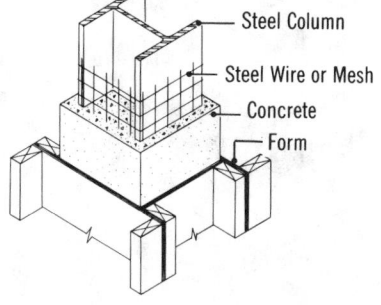

Concrete Encasement on Columns

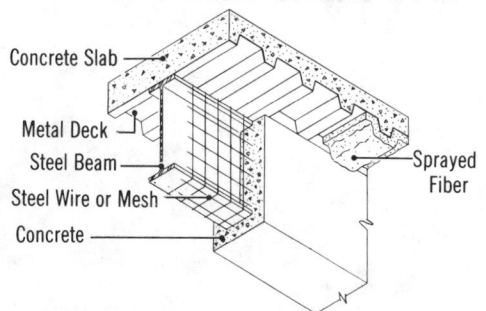

Concrete Encasement on Beams and Girders

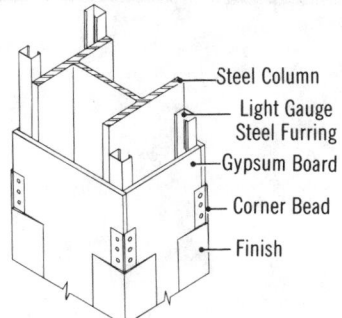

Gypsum Board on Columns

(continued on next page)

Figure 5.24 Installation Time in Man-Hours for Fireproofing Structural Steel (continued)

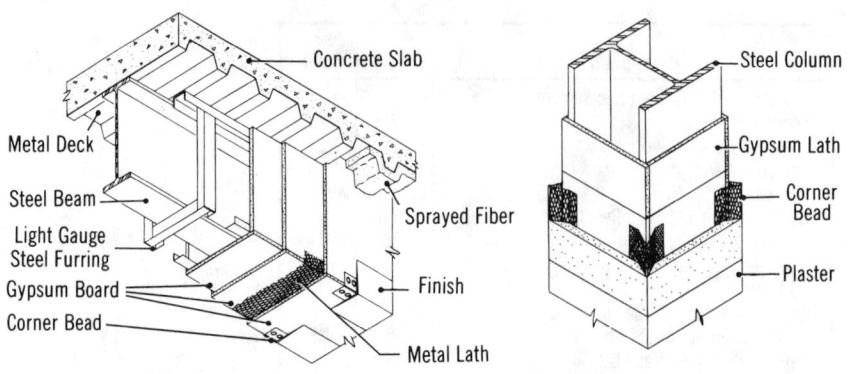

Gypsum Board on Beams and Girders   Plaster on Gypsum Lath — Columns

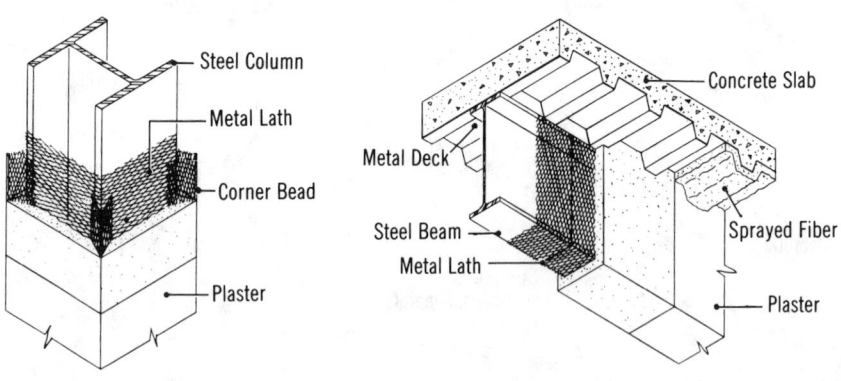

Plaster on Metal Lath — Columns   Plaster on Metal Lath — Beams and Girders

Figure 5.25   Installation Time in Man-Hours for Stair Systems

Description	Man-Hours	Unit
Concrete		
Stairs C.I.P.	.600	L.F. tread
Landings C.I.P.	.253	S.F.
Steel Custom Stair 3'-6" Wide	.914	riser
Steel Pan		
Stair Shop Fabricated 3'-6" Wide	.376	riser
Landing Shop Fabricated	.125	S.F.
Concrete Fill Pans	.110	S.F.
Spiral Stair		
Aluminum 5'-0" Diameter	.711	riser
Cast Iron 4'-0" Diameter	.711	riser
Steel Industrial 6'-0" Diameter	.800	riser
Included Ladder (Ships' Stair) 3'-0" Wide	.320	V.L.F.
Wood Box Stair Prefabricated 3'-6" Wide		
4' High	4.000	flight
8' High	5.333	flight
Open Stair Prefabricated 8" High	5.333	flight
Curved Stair 3'-3" Wide		
Open 1 Side 10' High	22.857	flight
Open 2 Sides 10' High	32.000	flight
Railings 1-1/2" Pipe		
Aluminum	.164	L.F.
Steel	.160	L.F.
Wall-Mounted	.125	L.F.
Rails Ornamental		
Bronze or Stainless	.611	L.F.
Aluminum	.767	L.F.
Wrought Iron	.834	L.F.
Composite Metal, Wood or Glass	1.467	L.F.

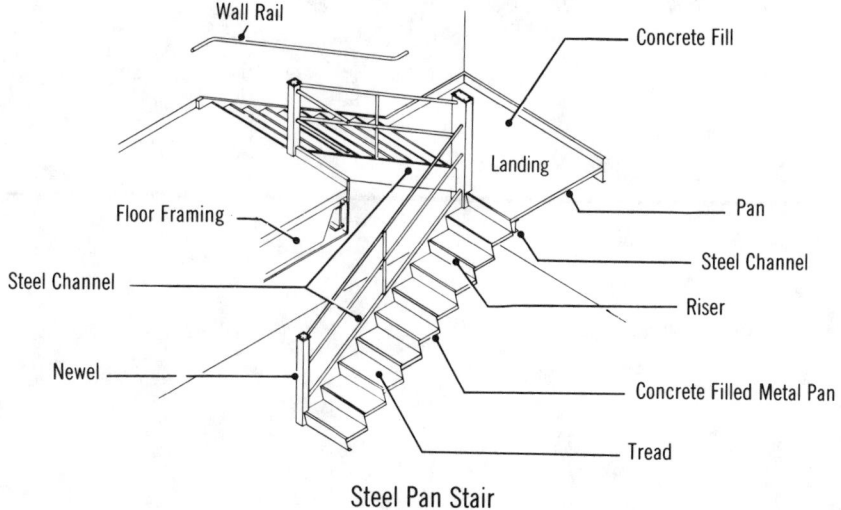

Steel Pan Stair

Figure 5.26 Installation Time in Man-Hours for Platforms and Walkways

Description	Man-Hours	Unit
Light Framing, Steel		
L's 4" and Larger	.011	lb.
Less than 4"	.018	lb.
C's 8" and Larger	.009	lb.
Less than 8"	.016	lb.
Aluminum Shapes	.053	lb.
Grating, Aluminum		
1-1/4" Deep	.032	S.F.
2-1/4" Deep	.037	S.F.
Fiberglass	.064	S.F.
Steel	.070	S.F.
Steel Pipe Railing w/Toe Plate	.160	L.F.

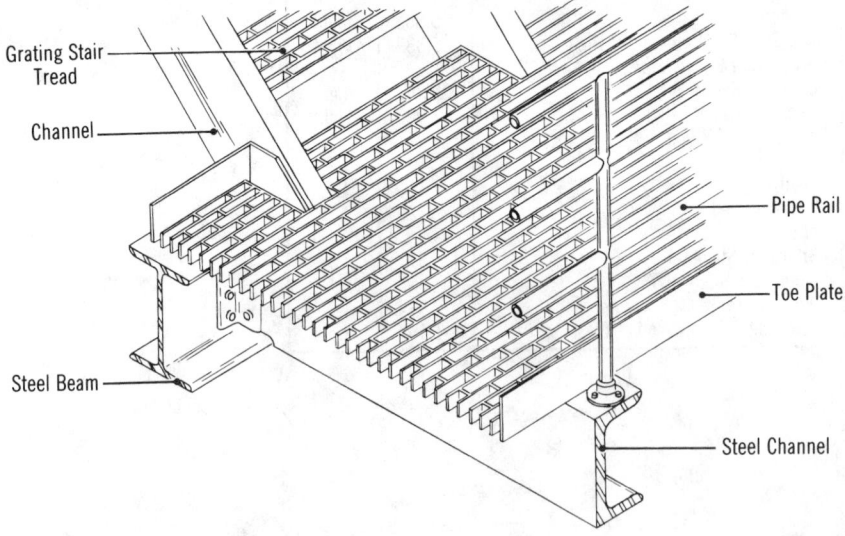

Grating Stair Tread

Channel

Steel Beam

Pipe Rail

Toe Plate

Steel Channel

Figure 5.27   Installation Time in Man-Hours for Pre-Engineered Buildings

Description	Man-Hours	Unit
Pre-Engineered Building Shell		
Above Foundation Average	.044	S.F. floor
Eave Overhang 4' Wide with Soffit	.224	L.F.
End Wall Overhang 4' Wide with Soffit	.112	L.F.
Door 3' x 7'	3.200	opng
Door Framing Only 10' x 10'	2.667	opng
Sash		
3' x 3'	1.714	opng
4' x 3'	1.846	opng
6' x 4'	2.000	opng
Gutter	.050	L.F.
Skylight, Fiberglass Panel to 30 S.F.	2.400	Ea.
Roof Vents, Circular 20" Diameter	2.667	Ea.
Continuous 12" Wide 10' Long	4.000	Ea.

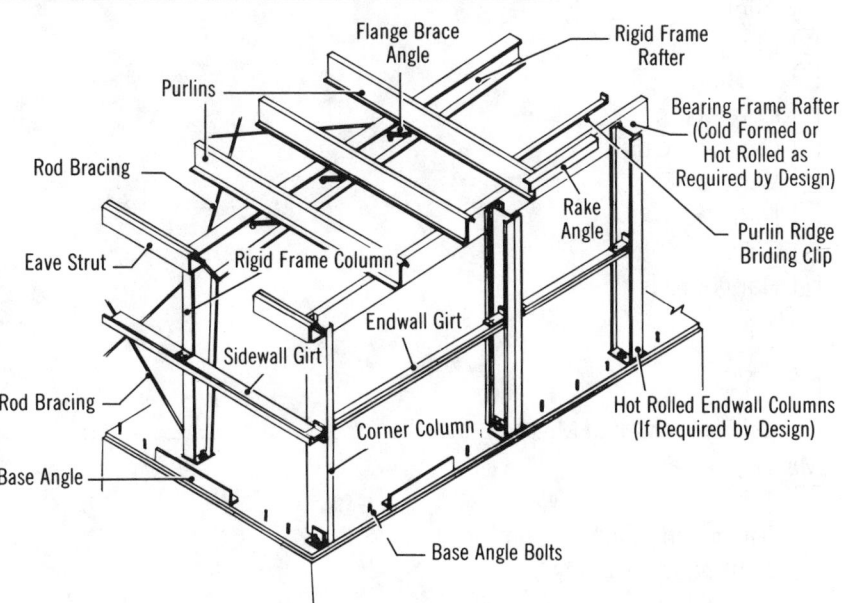

Checklist

For an estimate to be reliable, all items must be accounted for.
A complete estimate can also eliminate the need to include
contingencies. The following checklist can be used to help ensure
that all items are properly accounted for.

Anchor Bolts _____

Base Plates _____

Levelling Plates _____

Structural Steel
- ☐ Columns
- ☐ Beams
- ☐ Girders
- ☐ Joists
- ☐ Angles
- ☐ Clip angles
- ☐ Channels
- ☐ Connectors
- ☐ Cross bracing
- ☐ Wind bracing
- ☐ Bridging
- ☐ Girts
- ☐ Hanger rods

Decking
- ☐ Roof
- ☐ Floor

Galvanizing/Galvanized Materials _____

Miscellaneous Framing
- ☐ Roof frames
- ☐ Equipment support framing
- ☐ Light gauge framing
- ☐ Miscellaneous _____

Fireproofing
- ☐ Columns
- ☐ Beams
- ☐ Decking

Stairs _____

Fastening
- ☐ Welding
- ☐ Bolts

☐ Rivets
☐ Studs
☐ Timber connectors
☐ Machine screws

Painting
☐ Shop
☐ Field
☐ Finish

Seismic Considerations _____

Miscellaneous Metals
☐ Bumper posts
☐ Corner guards
☐ Crane rails
☐ Curb angles
☐ Decorative iron and steel
☐ Expansion joints and covers
☐ Floor gratings
☐ Ladders

Lintels _____

Miscellaneous Frames _____

Pipe Supports _____

Railings
☐ Handrails
☐ Balcony rails
☐ Stair
☐ Wall

Stair Treads _____

Trench Covers _____

Walkways _____

Window Guards _____

Wire Products _____

Tips

Plates and Connections

When estimating the total tonnage of structural steel, add 10% to the total weight to allow for plates, connections, and waste, as a rule of thumb.

Shop-Applied Finish Paint

When the specifications call for the finish coat of paint to be applied prior to installation, allow considerable time for touching up the paint. Every time you lift, move, weld, bolt or alter the position of a piece of steel, you will need to touch up the finish.

Joist Spacings

It is possible to lower overall costs for steel joist-type construction by having greater spacing between joists. The economies of needing fewer joists can well offset the costs of the larger or heavier joists that may be required to replace them.

Joist Bridging

It generally costs less to install joist systems that can utilize horizontal bridging (as opposed to cross or other types of bridging).

Wood &
Plastics

Introduction

Wood, unlike most processed building materials, is an organic material that can be used in its natural state. Factors that influence its strength are density, natural defects (knots, grain, etc.), and moisture content. Wood can be easily shaped or cut to size on the site, or it may be prefabricated in the shop. Structurally, wood may be used for joists, posts, columns, beams, and trusses. It may also be laminated into structural shapes such as columns, beams, rigid frames, arches, vaults, and folded-plate configurations. Various framing systems may be used with fiberboard, particle board, plywood, or wood decking (plain or laminated) to form composite panels. Wood may be exposed and finished for interior or exterior use. It may be purchased pressure-treated to provide resistance to moisture, rot, or exposure.

Structural lumber is stress-graded for bending, tension parallel to the grain, horizontal shear, compression perpendicular to the grain, compression parallel to the grain, and modulus of elasticity.

Structural lumber is usually dressed on four sides and, therefore, reduced in its normal size by approximately 1/2" in each direction. For example, the actual dimension of a 6"x10" beam dressed is 5-1/2" x 9-1/2".

Wood structures are fastened by nails, pins, dowels, screws, bolts, and adhesives, or by fabricated metal connectors, tailored to perform specific connection functions.

This division includes tables for the following areas:

- Framing systems
- Board foot measure of jobs
- Stud requirements
- Interior/exterior framing
- Roof framing systems
- Decking requirements
- Installation times
- Nail quantities
- Millwork

Estimating Data

The following tables present effective estimating guidelines for items found in Division 6 — Wood and Plastics. Please note that these guidelines can be used as indicators of what may be expected, and that each project must be evaluated individually.

Table of Contents

Figure 6.1   Wood Framing Nomenclature

These graphics illustrate the common terms used for the various components of wood framing. Whether for residential (shown here) or commercial, the terminology is the same.

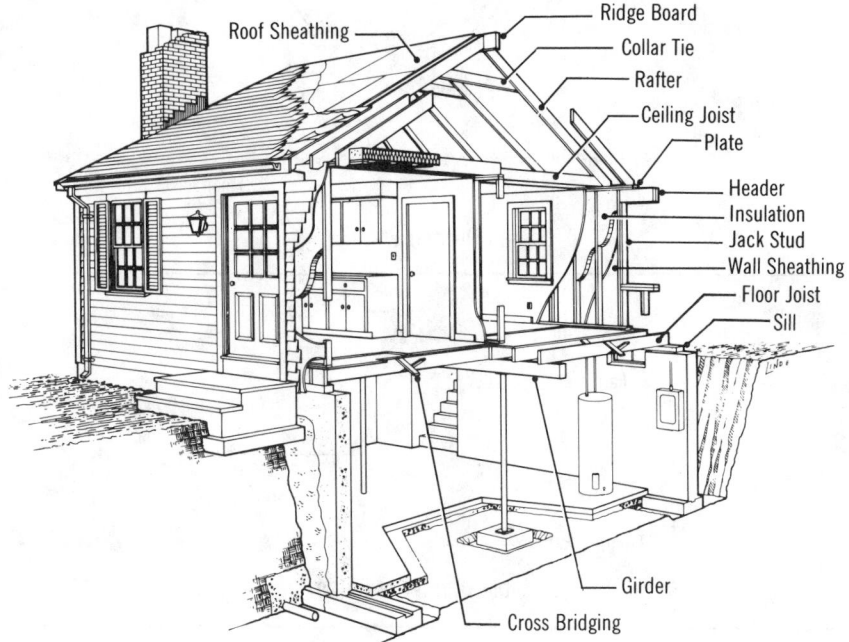

Wood Framing System

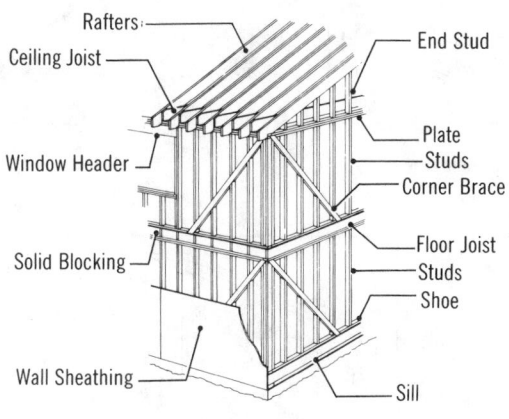

Platform Framing

(continued on next page)

Figure 6.1 Wood Framing Nomenclature (continued)

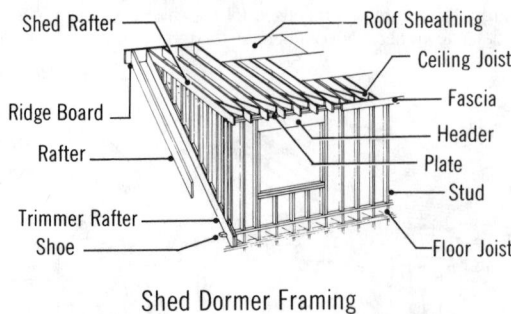

Shed Rafter — Roof Sheathing
Ceiling Joist
Ridge Board — Fascia
Rafter — Header
Plate
Stud
Trimmer Rafter —
Shoe — Floor Joist

Shed Dormer Framing

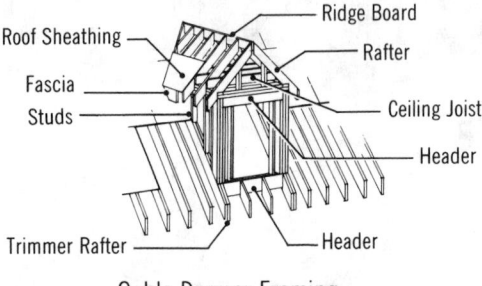

Roof Sheathing — Ridge Board
Rafter
Fascia — Ceiling Joist
Studs — Header
Trimmer Rafter — Header

Gable Dormer Framing

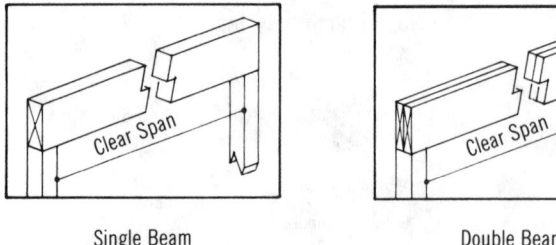

Clear Span   Clear Span

Single Beam   Double Beam

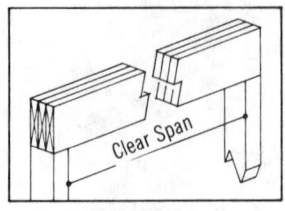

Clear Span

Triple Beam

Wood Beams and Columns

Figure 6.1   Wood Framing Nomenclature (continued)

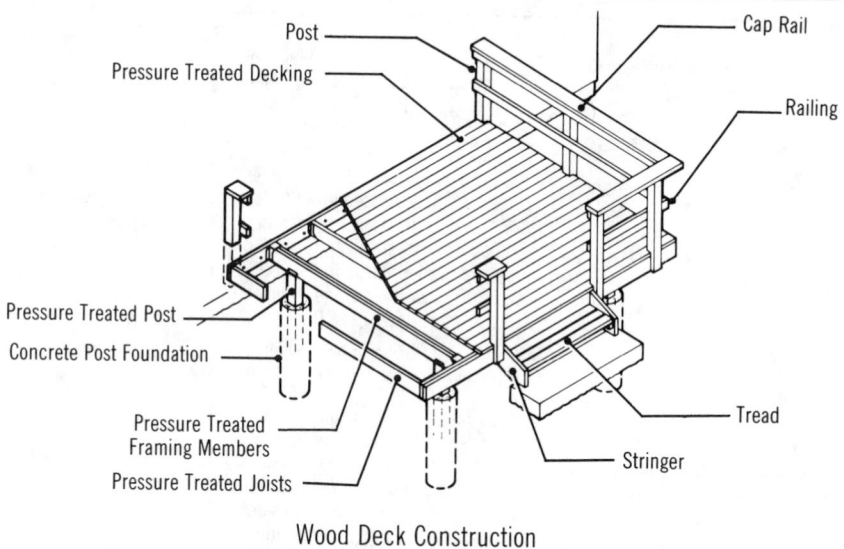

Wood Deck Construction

Figure 6.2   Lumber Industry Abbreviations

The following is a list of abbreviations used in connection with lumber in the construction industry. The abbreviations are commonly used in the form shown, but variations can and do occur.

AD	Air-dried
ADF	After deducting freight
ALS	American Lumber Standards
AV or AVG	Average
Bd	Board
Bd. ft.	Board foot or feet
Bdl	Bundle
Bev	Beveled
B/L	Bill of lading
BM	Board Measure
Btr	Better
B&B or B& Btr	B and better
B&S	Beams and stringers
CB1S	Center bead one side
CB2S	Center bead two sides
CF	Cost and freight
CG2E	Center groove two edges
CIF	Cost, insurance, and freight
CIFE	Cost, insurance, freight, and exchange
Clg	Ceiling
Clr	Clear
CM	Center matched

(continued on next page)

Figure 6.2   Lumber Industry Abbreviations (continued)

Com	Common
CS	Caulking seam
Csg	Casing
Cu. Ft.	Cubic foot or feet
CV1S	Center Vee one side
CV2S	Center Vee two sides
D&H	Dressed and headed
D&M	Dressed and matched
DB. Clg.	Double-beaded ceiling (E&CB2S)
DB. Part	Double-beaded partition (E&CB2S)
DET	Double end trimmed
Dim	Dimension
Dkg	Decking
D/S or D/Sdg	Dropped siding
EB1S	Edge bead one side
EB2S	Edge bead two sides
E&CB1S	Edge and center bead one side
E&CB2S	Edge and center bead two sides
E&CV1S	Edge and center Vee one side
E&CV2S	Edge and center Vee two sides
EE	Eased edges
EG	Edge (vertical) grain
EM	End matched
EV1S	Edge Vee one side
EV2S	Edge Vee two sides
Fac	Factory
FAS	Free alongside (named vessel)
FBM	Foot or feet board measure
FG	Flat (slash) grain
Flg	Flooring
FOB	Free on board (named point)
FOHC	Free of heart center or centers
FOK	Free of knots
Frt	Freight
Ft	Foot or feet
GM	Grade marked
G/R or G/Rfg	Grooved roofing
HB	Hollow back
H&M	Hit-and-miss
H or M	Hit-or-miss
Hrt	Heart
Hrt CC	Heart cubical content
Hrt FA	Heart facial area
Hrt G	Heart girth
IN	Inch or inches
J&P	Joists and planks
KD	Kiln-dried
LCL	Less than carload
LFT or Lin. Ft.	Linear foot or feet
Lgr	Longer
Lgth	Length
Lin	Linear
Lng	Lining
M	Thousand
MBM	Thousand (feet) board measure

Figure 6.2 Lumber Industry Abbreviations (continued)

MC	Moisture content
Merch	Merchantable
Mldg	Moulding
No	Number
N1E	Nosed one edge
N2E	Nosed two edges
Og	Ogee
Ord	Order
Par	Paragraph
Part	Partition
Pat	Pattern
Pc	Piece
Pcs	Pieces
PE	Plain end
PO	Purchase order
P&T	Post and timbers
Reg	Regular
Res	Resawed or resawn
Rfg	Roofing
Rgh	Rough
R/L	Random lengths
R/W	Random widths
R/W&L	Random widths and lengths
Sdg	Siding
Sel	Select
S&E	Side and Edge (surfaced on)
SE Sdg	Square edge siding
SE & S	Square edge and sound
S/L or S/LAP	Shiplap
SL&C	Shipper's load and count
SM. or Std. M	Standard Matched
Spec	Specifications
Std	Standard
Stpg	Stepping
Str. or Struc	Structural
S1E	Surfaced one edge
S1S	Surfaced one side
S1S1E	Surfaced one side and one edge
S1S2E	Surfaced one side and two edges
S2E	Surfaced two edges
S2S	Surfaced two sides
S2S1E	Surfaced two sides and one edge
S2S&CM	Surfaced two sides and center matched
S2S&SM	Surfaced two sides and standard matched
S4S	Surfaced four sides
S4S&CS	Surfaced four sides and caulking seam
T&G	Tongued and grooved
VG	Vertical grain
Wdr	Wider
Wt	Weight

(excerpted from How to Estimate Building Losses and Construction Costs, Paul I. Thomas, Prentice Hall)

Figure 6.3 Number of Board Feet in Various Sized Pieces of Lumber

This table shows how many board feet are in various pieces of lumber for the lengths indicated.

Size (in inches)	Length of Piece							
	8'	10'	12'	14'	16'	18'	20'	22'
2 x 3	4	5	6	7	8	9	10	11
2 x 4	5-1/3	6-2/3	8	9-1/3	10-2/3	12	13-1/3	14-2/3
2 x 6	8	10	12	14	16	18	20	22
2 x 8	10-2/3	13-1/3	16	18-2/3	21-1/3	24	26-2/3	29-1/3
2 x 12	16	20	24	28	32	36	40	44
3 x 4	8	10	12	14	16	18	20	22
3 x 6	12	15	18	21	24	27	30	33
3 x 8	16	20	24	28	32	36	40	44
3 x 10	20	25	30	35	40	45	50	55
3 x 12	24	30	36	42	48	54	60	66
4 x 4	10-2/3	13-1/3	16	18-2/3	21-1/3	24	26-2/3	29-1/3
4 x 6	16	20	24	28	32	36	40	44
4 x 8	21-1/3	26-2/3	32	37-1/3	42-2/3	48	53-1/3	58-2/3
6 x 6	24	30	36	42	48	54	60	66
6 x 8	32	40	48	56	64	72	80	88
6 x 10	40	50	60	70	80	90	100	110
8 x 8	42-2/3	53-1/3	64	74-2/3	85-1/3	96	106-2/3	117-1/3
8 x 10	53-1/3	66-2/3	80	93-1/3	106-2/3	120	133-1/3	146-2/3
8 x 12	64	80	96	112	128	144	160	176
10 x 10	66-2/3	83-1/3	100	116-2/3	133-1/3	150	166-2/3	183-1/3
10 x 12	80	100	120	140	160	180	200	220
12 x 14	112	140	168	196	224	252	280	308

Figure 6.4 Board Foot Multipliers

This table is used for computing the number of board feet in any length of dimension lumber. Find the size of lumber to be used, and read across for the multiplier. For example, to find out how many board feet of lumber are in an 8' length of 2 x 4 lumber, look for the multiplier for 2 x 4's (0.667) and multiply it by the length of lumber (8'). The result is 0.667 x 8, or 5.334 board feet.

Nominal Size (in inches)	Multiply length by	Nominal Size (in inches)	Multiply length by
2 x 2	0.333	4 x 4	1.333
2 x 3	0.500	4 x 6	2.000
2 x 4	0.667	4 x 8	2.667
2 x 6	1.000	4 x 10	3.333
2 x 8	1.333	4 x 12	4.000
2 x 10	1.667		
2 x 12	2.000	6 x 6	3.000
		6 x 8	4.000
3 x 3	0.750	6 x 10	5.000
3 x 4	1.000	6 x 12	6.000
3 x 6	1.500		
3 x 8	2.000	8 x 8	5.333
3 x 10	2.500	8 x 10	6.667
3 x 12	3.000	8 x 12	8.000

Figure 6.5   Factors for the Board Foot
Measure of Floor or Ceiling Joists

This table allows you to compute the amount of lumber (in board feet) for the floor or ceiling
joists of a known area, based on the size of lumber and the spacing. To use this chart, simply
multiply the area of the room or building in question by the factor across from the appropriate
board size and spacing.

Ceiling Joists			
Joist Size	Inches On Center	Board Feet per Square Foot of Ceiling Area	Nails Lbs. per MBM
2" x 4"	12"	.78	17
	16"	.59	19
	20"	.48	19
2" x 6"	12"	1.15	11
	16"	.88	13
	20"	.72	13
	24"	.63	13
2" x 8"	12"	1.53	9
	16"	1.17	9
	20"	.96	9
	24"	.84	9
2" x 10"	12"	1.94	7
	16"	1.47	7
	20"	1.21	7
	24"	1.04	7
3" x 8"	12"	2.32	6
	16"	1.76	6
	20"	1.44	6
	24"	1.25	6

Figure 6.6   Floor and Ceiling Joist Calculation Tables

This table provides the amount of lumber (in board feet) for floor or ceiling joists based on the size of lumber and the spacing.

Floor Area	2" x 4" On Center			2" x 6" On Center			2" x 8" On Center			2" x 10" On Center			2" x 12" On Center		
	16"	20"	24"	16"	20"	24"	16"	20"	24"	16"	20"	24"	16"	20"	24"
200	110	88	73	170	136	114	220	176	147	270	216	180	330	264	220
300	165	132	110	255	204	170	330	264	220	405	324	270	495	396	330
400	220	176	147	340	272	227	440	352	293	540	432	360	660	528	440
500	275	220	183	425	340	283	550	440	367	695	540	450	825	660	550
600	330	264	220	510	408	340	660	528	440	810	648	540	990	792	660
700	385	308	257	595	476	397	770	616	514	945	756	630	1155	924	770
800	440	352	293	680	544	454	880	704	587	1080	864	720	1320	1056	880
900	495	396	330	765	612	510	990	792	660	1215	972	810	1485	1188	990
1000	550	440	367	850	680	567	1100	880	734	1350	1080	900	1650	1320	1100
1100	605	484	404	935	748	624	1210	968	807	1485	1188	990	1815	1415	1210
1200	660	528	440	1020	816	680	1320	1056	880	1620	1296	1080	1980	1584	1320
1300	715	572	477	1105	884	737	1430	1144	954	1755	1404	1170	2145	1716	1430
1400	770	616	514	1190	952	794	1540	1232	1027	1890	1512	1260	2310	1848	1540
1500	825	660	550	1275	1020	850	1650	1320	1100	2025	1620	1350	2475	1980	1650
1600	880	704	587	1360	1088	907	1760	1408	1177	2160	1728	1440	2640	2112	1760
1700	935	748	624	1445	1156	964	1870	1496	1247	2295	1836	1530	2805	2244	1870
1800	990	792	660	1530	1224	1021	1980	1584	1320	2430	1944	1620	2970	2376	1980
1900	1045	836	697	1615	1292	1077	2090	1672	1394	2565	2052	1710	3135	2508	2090
2000	1100	880	734	1700	1360	1134	2200	1760	1467	2700	2160	1800	3300	2640	2200
2100	1155	924	770	1785	1428	1190	2310	1848	1540	2835	2268	1890	3465	2772	2310
2200	1210	968	807	1870	1496	1247	2420	1936	1614	2970	2376	1980	3630	2904	2420
2300	1265	1012	844	1955	1564	1304	2530	2024	1688	3105	2484	2070	3795	3036	2530
2400	1320	1056	880	2040	1632	1360	2640	2112	1760	3240	2592	2160	3960	3168	2640
2500	1375	1100	917	2125	1700	1417	2750	2200	1834	3375	2700	2250	4125	3300	2750
2600	1430	1144	953	2210	1768	1474	2860	2288	1908	3510	2808	2340	4290	3432	2860
2700	1485	1188	990	2295	1836	1530	2970	2376	1980	3645	2916	2430	4455	3564	2970
2800	1540	1232	1027	2380	1904	1587	3080	2464	2054	3780	3024	2520	4620	3696	3080
2900	1595	1276	1064	2465	1972	1644	3190	2552	2128	3915	3132	2610	4785	3828	3190
3000	1650	1320	1100	2550	2040	1700	3300	2640	2200	4050	3240	2700	4950	3960	3300

(courtesy, Estimating Tables for Home Building, Paul I. Thomas, Craftsman Book Company)

Figure 6.7   Pieces of Studding Required for Walls, Floors, or Ceilings

This table lists the number of pieces of studding or furring needed for framing walls, floors, ceilings, partitions, etc. based on the length of the item being framed and its required spacing. Note that no provisions have been made for the doubling of studs at corners, window openings, door openings, etc. These must be added as called for.

Length of Wall, Floor or Ceiling (in feet)	On Center Spacing			
	12"	16"	20"	24"
8	9	7	6	5
9	10	8	6	6
10	11	9	7	6
11	12	9	8	7
12	13	10	8	7
13	14	11	9	8
14	15	12	9	8
15	16	12	10	9
16	17	13	11	9
17	18	14	11	10
18	19	15	12	10
19	20	15	12	11
20	21	16	13	11
21	22	17	14	12
22	23	18	14	12
23	24	18	15	13
24	25	19	15	13
25	26	20	16	14
26	27	21	17	14
27	28	21	17	15
28	29	22	18	15
29	30	23	18	16
30	31	24	19	16
32	33	25	20	17
34	35	27	22	18
36	37	28	23	19

(excerpted from How to Estimate Building Losses and Construction Costs, Paul I. Thomas, Prentice Hall)

Figure 6.8 Multipliers for Computing Required Studding or Furring

This table allows you to compute the number of pieces of studding or furring required for framing walls, floors, roofs, partitions, etc.

Spacing (in inches)	Factor by Which to Multiply the Width of Room, Length of Wall or Roof	
12	1.00	
16	.75	
18	.67	Add one for
20	.60	end in each
24	.50	case

Examples:	Answers:
A room 16' wide requires studs 16" O.C.	16' x .75 + 1 = 13 joists
A wall 30' long requires studs 18" O.C.	30' x .67 + 1 = 21 studs
A gable roof 40' long requires studs 24" O.C.	40' x .50 + 1 = 21 rafters

How to Estimate Building Losses and Construction Costs, Paul I. Thomas, Prentice Hall

Figure 6.9   Partition Framing

This table can be used to compute the board feet of lumber required for each square foot of wall area to be framed, based on the lumber size and spacing design.

Stud Size	Inches on Center	Studs Including Sole and Cap Plates		Horizontal Bracing in All Partitions		Horizontal Bracing in Bearing Partitions Only	
		Board Feet per Square Foot of Partition Area	Lbs. Nails per MBM of Stud Framing	Board Feet per Square Foot of Partition Area	Lbs. Nails per MBM of Bracing	Board Feet per Square Foot of Partition Area	Lbs. Nails per MBM of Bracing
2" x 3"	12"	.91	25	.04	145	.01	145
	16"	.83	25	.04	111	.01	111
	20"	.78	25	.04	90	.01	90
	24"	.76	25	.04	79	.01	79
2" x 4"	12"	1.22	19	.05	108	.02	108
	16"	1.12	19	.05	87	.02	87
	20"	1.05	19	.05	72	.02	72
	24"	1.02	19	.05	64	.02	64
2" x 6"	16"	1.38	19			.04	59
	20"	1.29	16			.04	48
	24"	1.22	16			.04	43
2" x 4" Staggered	8"	1.69	22				
3" x 4"	16"	1.35	17				
2" x 4" 2" Way	16"	1.08	19				

Figure 6.10   Exterior Wall Stud Framing

This table allows you to compute the board feet of lumber required for each square foot of exterior wall to be framed based on the lumber size and spacing design.

Stud Size	Inches On Center	Studs Including Corner Bracing		Horizontal Bracing Midway Between Plates	
		Board Feet per Square Foot of Ext. Wall Area	Lbs. of Nails per MBM of Stud Framing	Board Feet per Square Foot of Ext. Wall Area	Lbs. of Nails per MBM of Bracing
2" x 3"	16"	.78	30	.03	117
	20"	.74	30	.03	97
	24"	.71	30	.03	85
2" x 4"	16"	1.05	22	.04	87
	20"	.98	22	.04	72
	24"	.94	22	.04	64
2" x 6"	16"	1.51	15	.06	59
	20"	1.44	15	.06	48
	24"	1.38	15	.06	43

Figure 6.11   Furring Quantities

This table provides multiplication factors for converting square feet of wall area requiring furring into board feet of lumber. These figures are based on lumber size and spacing.

	Board Feet per Square Feet of Wall Area				Lbs. Nails per MBM of Furring
	Spacing Center to Center				
Size	12"	16"	20"	24"	
1" x 2"	.18	.14	.11	.10	55
1" x 3"	.28	.21	.17	.14	37

Figure 6.12   Floor Framing

This table can be used to compute the board feet required for each square foot of floor area, based on the size of lumber to be used and the spacing design.

Joist Size	Inches On Center	Floor Joists			Block Over Main Bearing	
		Board Feet per Square Foot of Floor Area	Nails Lbs. per MBM		Board Feet per Square Foot of Floor Area	Nails Lbs. per MBM Blocking
2" x 6"	12"	1.28	10		.16	133
	16"	1.02	10		.03	95
	20"	.88	10		.03	77
	24"	.78	10		.03	57
2" x 8"	12"	1.71	8		.04	100
	16"	1.36	8		.04	72
	20"	1.17	8		.04	57
	24"	1.03	8		.05	43
2" x 10"	12"	2.14	6		.05	79
	16"	1.71	6		.05	57
	20"	1.48	6		.06	46
	24"	1.30	6		.06	34
2" x 12"	12"	2.56	5		.06	66
	16"	2.05	5		.06	47
	20"	1.77	5		.07	39
	24"	1.56	5		.07	29
3" x 8"	12"	2.56	5		.04	39
	16"	2.05	5		.05	57
	20"	1.77	5		.06	45
	24"	1.56	5		.06	33
3" x 10"	12"	3.20	4		.05	72
	16"	2.56	4		.07	46
	20"	2.21	4		.07	36
	24"	1.95	4		.08	26

Figure 6.13 Lumber (in Board Feet) Required for Framing a One-Story Structure

This table shows the total number of board feet required to frame a typical one-story structure based on ground floor area, spacing, and building style. The figures include the shoe and plate.

Ground Floor Area	2" x 4" Exterior Wall Studs				2" x 4" Partition Studs		Total FBM Wall & Partition Studs, 16" O.C.	
	Spaced 16" O.C.		Spaced 24" O.C.		Spacing On Center			
	Gables Included	No Gables	Gables Included	No Gables	16"	24"	Gables Included	No Gables
200	419	402	313	300	132	88	551	534
300	521	496	389	370	198	132	719	694
400	603	570	450	425	264	176	867	834
500	678	637	506	475	330	220	1008	967
600	747	697	558	520	396	264	1143	1093
700	809	750	604	560	462	308	1271	1212
800	871	804	650	600	528	352	1400	1332
900	925	850	691	635	594	396	1520	1444
1000	981	898	732	670	660	440	1641	1558
1100	1037	945	774	705	726	484	1763	1671
1200	1085	985	810	735	792	528	1877	1777
1300	1134	1025	847	765	858	572	1992	1883
1400	1182	1065	883	795	924	616	2106	1990
1500	1225	1100	914	820	990	660	2215	2090
1600	1273	1140	950	850	1056	704	2329	2196
1700	1314	1173	981	875	1122	748	2436	2295
1800	1356	1206	1013	900	1188	792	2544	2394
1900	1400	1240	1044	925	1254	836	2654	2494
2000	1440	1273	1075	950	1320	880	2760	2593
2100	1475	1300	1100	970	1386	924	2861	2686
2200	1516	1333	1133	995	1452	968	2968	2785
2300	1550	1360	1158	1015	1518	1012	3068	2878
2400	1594	1394	1190	1040	1584	1056	3178	2978
2500	1628	1420	1216	1060	1650	1100	3278	3070
2600	1664	1447	1242	1080	1716	1144	3380	3163
2700	1700	1474	1269	1100	1782	1188	3482	3256
2800	1733	1500	1295	1120	1848	1232	3581	3348
2900	1768	1528	1320	1140	1914	1276	3682	3442
3000	1811	1561	1353	1165	1980	1320	3791	3541

(courtesy, Estimating Tables for Home Building, Paul I. Thomas, Craftsman Book Company)

Figure 6.14 Lumber (in Board Feet) Required for Framing a One-and-a-Half-Story Structure

This table shows the total number of board feet required to frame a typical one-and-a-half-story structure, based on ground floor area, spacing, and building style. The figures include the shoe and plate.

Ground Floor Area	2" x 4" Exterior Wall Studs				2" x 4" Partition Studs		Total FBM Wall & Partition Studs, 16" O.C.	
	Spaced 16" O.C.		Spaced 24" O.C.		Spacing On Center			
	Gables Included	No Gables	Gables Included	No Gables	16"	24"	Gables Included	No Gables
200	713	696	541	528	234	154	947	930
300	883	858	670	651	351	231	1234	1209
400	1019	986	773	748	468	308	1487	1454
500	1143	1102	867	836	585	385	1728	1687
600	1256	1206	953	915	702	462	1958	1908
700	1360	1300	1053	986	819	539	2178	2119
800	1459	1392	1106	1056	936	616	2395	2328
900	1550	1475	1174	1118	1053	693	2603	2528
1000	1638	1555	1241	1179	1170	770	2808	2725
1100	1728	1636	1310	1241	1287	847	3015	2923
1200	1806	1706	1369	1294	1404	924	3210	3110
1300	1884	1775	1428	1346	1521	1000	3405	3296
1400	1961	1844	1488	1400	1638	1078	3600	3482
1500	2027	1902	1537	1443	1755	1155	3782	3657
1600	2105	1972	1596	1496	1872	1232	3977	3844
1700	2171	2030	1646	1540	1990	1310	4161	4020
1800	2238	2088	1697	1584	2106	1386	4344	4194
1900	2305	2146	1744	1628	2223	1463	4528	4369
2000	2371	2204	1797	1672	2340	1540	4711	4544
2100	2425	2250	1838	1707	2457	1617	4882	4707
2200	2492	2309	1890	1751	2574	1694	5066	4883
2300	2545	2355	1930	1786	2691	1771	5236	5046
2400	2613	2413	1980	1830	2808	1848	5421	5221
2500	2668	2460	2022	1866	2925	1925	5593	5385
2600	2723	2506	2062	1900	3042	2000	5765	5548
2700	2777	2552	2105	1936	3159	2080	5936	5711
2800	2831	2598	2146	1971	3276	2156	6107	5874
2900	2885	2645	2186	2006	3393	2233	6278	6038
3000	2953	2703	2238	2050	3510	2310	6463	6213

(courtesy, Estimating Tables for Home Building, Paul I. Thomas, Craftsman Book Company)

Figure 6.15 Lumber (in Board Feet) Required for Framing a Two-Story Structure

This table shows the total number of board feet required to frame a typical two-story structure, based on ground floor area, spacing, and building style. The figures include the shoe and plate.

Ground Floor Area	2" x 4" Exterior Wall Studs				2" x 4" Partition Studs		Total FBM Wall & Partition Studs, 16" O.C.	
	Spaced 16" O.C.		Spaced 24" O.C.		Spacing On Center			
	Gables Included	No Gables	Gables Included	No Gables	16"	24"	Gables Included	No Gables
200	821	804	613	600	264	176	1085	1068
300	1017	992	759	740	396	264	1413	1388
400	1172	1140	875	850	528	352	1700	1668
500	1315	1274	981	950	660	440	1975	1934
600	1444	1394	1078	1040	792	528	2236	2186
700	1559	1500	1164	1120	924	616	2483	2424
800	1675	1608	1250	1200	1056	704	2731	2664
900	1775	1700	1326	1270	1188	792	2963	2888
1000	1879	1796	1402	1340	1320	880	3200	3116
1100	1982	1890	1479	1410	1452	968	3434	3342
1200	2070	1970	1545	1470	1584	1056	3654	3554
1300	2159	2050	1612	1530	1716	1144	3875	3766
1400	2247	2130	1678	1590	1848	1232	4095	3980
1500	2325	2200	1734	1640	1980	1320	4305	4180
1600	2413	2280	1800	1700	2112	1408	4525	4392
1700	2487	2346	1856	1750	2244	1496	4731	4590
1800	2562	2412	1913	1800	2376	1584	4938	4788
1900	2639	2480	1969	1850	2508	1672	5147	4988
2000	2713	2546	2025	1900	2640	1760	5353	5186
2100	2775	2600	2071	1940	2772	1848	5547	5372
2200	2849	2666	2128	1990	2904	1936	5753	5570
2300	2910	2720	2173	2030	3036	2024	5946	5756
2400	2988	2788	2230	2080	3168	2112	6156	5956
2500	3048	2840	2276	2121	3300	2200	6348	6140
2600	3111	2894	2322	2160	3432	2288	6543	6326
2700	3173	2948	2369	2200	3564	2376	6737	6512
2800	3233	3000	2415	2240	3696	2464	6929	6696
2900	3296	3056	2460	2280	3828	2552	7124	6884
3000	3372	3122	2518	2330	3960	2640	7332	7082

(courtesy, Estimating Tables for Home Building, Paul I. Thomas, Craftsman Book Company)

Figure 6.16 Flat Roof Framing

This table can be used to compute the amount of lumber in board feet required to frame a flat roof, based on the size lumber required.

Flat Roof Framing			
Joist Size	Inches On Center	Board Feet per Square Foot of Ceiling Area	Nails Lbs. per MBM
2" x 6"	12"	1.17	10
	16"	.91	10
	20"	.76	10
	24"	.65	10
2" x 8"	12"	1.56	8
	16"	1.21	8
	20"	1.01	8
	24"	.86	8
2" x 10"	12"	1.96	6
	16"	1.51	6
	20"	1.27	6
	24"	1.08	6
2" x 12"	12"	2.35	5
	16"	1.82	5
	20"	1.52	5
	24"	1.30	5
3" x 8"	12"	2.35	5
	16"	1.82	5
	20"	1.52	5
	24"	1.30	5
3" x 10"	12"	2.94	4
	16"	2.27	4
	20"	1.90	4
	24"	1.62	4

MBM; MFBM = Thousand Feet Board Measure

Figure 6.17 Conversion Factors for Wood Joists and Rafters

This table is used to compute actual lengths or board feet of lumber for roofs of various inclines. To compute actual rafter quantities for incline roofs, multiply the quantities figured for a flat roof by the factors as shown in the table. Please note that this table DOES NOT INCLUDE QUANTITIES FOR CANTILEVERED OVERHANGS.

Roof Slope	Approximate Angle	Factor	Roof Slope	Approximate Angle	Factor
Flat	0°	1.000	12 in 12	45.0°	1.414
1 in 12	4.8°	1.003	13 in 12	47.3°	1.474
2 in 12	9.5°	1.014	14 in 12	49.4°	1.537
3 in 12	14.0°	1.031	15 in 12	51.3°	1.601
4 in 12	18.4°	1.054	16 in 12	53.1°	1.667
5 in 12	22.6°	1.083	17 in 12	54.8°	1.734
6 in 12	26.6°	1.118	18 in 12	56.3°	1.803
7 in 12	30.3°	1.158	19 in 12	57.7°	1.873
8 in 12	33.7°	1.202	20 in 12	59.0°	1.943
9 in 12	36.9°	1.250	21 in 12	60.3°	2.015
10 in 12	39.8°	1.302	22 in 12	61.4°	2.088
11 in 12	42.5°	1.357	23 in 12	62.4°	2.162

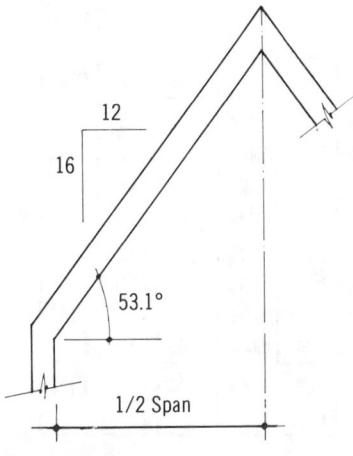

Inclined to Horizontal Sections

Figure 6.18   Allowance Factors for Roof Overhangs

Horizontal Span	Roof Overhang Measured Horizontally							
	0'-6"	1'-0"	1'-6"	2'-0"	2'-6"	3'-0"	3'-6"	4'-0"
6'	1.083	1.167	1.250	1.333	1.417	1.500	1.583	1.667
7'	1.071	1.143	1.214	1.286	1.357	1.429	1.500	1.571
8'	1.063	1.125	1.188	1.250	1.313	1.375	1.438	1.500
9'	1.056	1.111	1.167	1.222	1.278	1.333	1.389	1.444
10'	1.050	1.100	1.150	1.200	1.250	1.300	1.350	1.400
11'	1.045	1.091	1.136	1.182	1.227	1.273	1.318	1.364
12'	1.042	1.083	1.125	1.167	1.208	1.250	1.292	1.333
13'	1.038	1.077	1.115	1.154	1.192	1.231	1.269	1.308
14'	1.036	1.071	1.107	1.143	1.179	1.214	1.250	1.286
15'	1.033	1.067	1.100	1.133	1.167	1.200	1.233	1.267
16'	1.031	1.063	1.094	1.125	1.156	1.188	1.219	1.250
17'	1.029	1.059	1.088	1.118	1.147	1.176	1.206	1.235
18'	1.028	1.056	1.083	1.111	1.139	1.167	1.194	1.222
19'	1.026	1.053	1.079	1.105	1.132	1.158	1.184	1.211
20'	1.025	1.050	1.075	1.100	1.125	1.150	1.175	1.200
21'	1.024	1.048	1.071	1.095	1.119	1.143	1.167	1.190
22'	1.023	1.045	1.068	1.091	1.114	1.136	1.159	1.182
23'	1.022	1.043	1.065	1.087	1.109	1.130	1.152	1.174
24'	1.021	1.042	1.063	1.083	1.104	1.125	1.146	1.167
25'	1.020	1.040	1.060	1.080	1.100	1.120	1.140	1.160
26'	1.019	1.038	1.058	1.077	1.096	1.115	1.135	1.154
27'	1.019	1.037	1.056	1.074	1.093	1.111	1.130	1.148
28'	1.018	1.036	1.054	1.071	1.089	1.107	1.125	1.143
29'	1.017	1.034	1.052	1.069	1.086	1.103	1.121	1.138
30'	1.017	1.033	1.050	1.067	1.083	1.100	1.117	1.133
32'	1.016	1.031	1.047	1.063	1.078	1.094	1.109	1.125

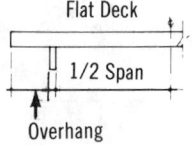

Flat Deck

1/2 Span

Overhang

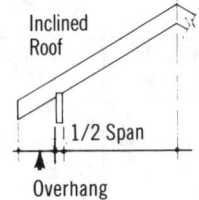

Inclined Roof

1/2 Span

Overhang

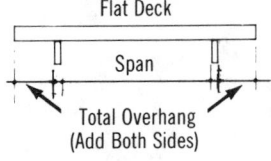

Flat Deck

Span

Total Overhang (Add Both Sides)

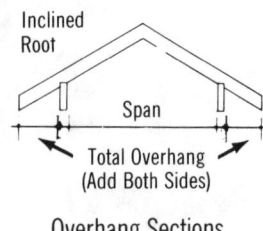

Inclined Roof

Span

Total Overhang (Add Both Sides)

Overhang Sections

Figure 6.19 Roofing Factors When Rate of Rise is Known

No. Inches Rise per Foot of Run	*Pitch	Rafter Length in Inches per Foot of Run	To Obtain Rafter Length, Multiply Run (in feet) By:
3.0 3.5	1/8	12.37 ÷ 12 =	1.030 1.045
4.0 4.5	1/6	12.65	1.060 1.075
5.0 5.5	5/24	13.00	1.090 1.105
6.0 6.5	1/4	13.42	1.120 1.140
7.0 7.5	7/24	13.89	1.160 1.185
8.0 8.5	1/3	14.42	1.201 1.230
9.0 9.5	3/8	15.00	1.250 1.280
10.0 10.5	5/12	15.62	1.301 1.335
11.0 11.5	11/24	16.28	1.360 1.390
12.0	1/2	16.97	1.420

*Pitch is determined by dividing the inches rise-per-foot-of-run by 24. (Most roofs of dwellings are constructed with a pitch of 1/8, 1/6, 1/4, 1/3 or 1/2.)

(excerpted from How to Estimate Building Losses and Construction Costs, Paul I. Thomas, Prentice Hall)

Figure 6.20 Pitched Roof Framing

This table is used to compute the board feet of lumber required per square foot of roof area, based on the required spacing and the lumber size to be used.

Rafters Including Collar Ties, Hip and Valley Rafters, Ridge Poles

Spacing Center to Center

Rafter Size	12" Board Feet per Square Foot of Roof Area	12" Nails Lbs. per MBM	16" Board Feet per Square Foot of Roof Area	16" Nails Lbs. per MBM	20" Board Feet per Square Foot of Roof Area	20" Nails Lbs. per MBM	24" Board Feet per Square Foot of Roof Area	24" Nails Lbs. per MBM
2" x 4"	.89	17	.71	17	.59	17	.53	17
2" x 6"	1.29	12	1.02	12	.85	12	.75	12
2" x 8"	1.71	9	1.34	9	1.12	9	.98	9
2" x 10"	2.12	7	1.66	7	1.38	7	1.21	7
2" x 12"	2.52	6	1.97	6	1.64	6	1.43	6
3" x 8"	2.52	6	1.97	6	1.64	6	1.43	6
3" x 10"	3.13	5	2.45	5	2.02	5	1.78	5

Figure 6.21 Board Feet Required for Roof Framing — Truss Construction

This table shows the number of board feet of lumber required for roof framing using "W" type truss construction. This table is based on a truss spacing of 24" on center, the roof pitch, and the area being covered. Braces and struts are included.

Ground Floor Area	2" x 4" Rafter and Chord						2" x 6" Rafter ... 2" x 4" Chord					
	Roof Pitch						Roof Pitch					
	1/8	1/6	5/24	1/4	1/3	1/2	1/8	1/6	5/24	1/4	1/3	1/2
200	210	212	214	218	224	240	246	248	252	256	264	288
300	315	318	321	327	336	360	369	372	378	384	396	432
400	420	424	428	436	448	480	492	496	504	512	528	576
500	525	530	535	545	560	600	615	620	630	640	660	720
600	630	636	642	654	672	720	738	744	756	768	792	864
700	735	742	749	763	784	840	861	868	882	896	924	1008
800	840	848	856	872	896	960	984	992	1008	1024	1056	1152
900	945	954	963	981	1008	1080	1107	1116	1134	1152	1188	1296
1000	1050	1060	1070	1090	1120	1200	1230	1240	1260	1280	1320	1440
1100	1155	1166	1177	1199	1232	1320	1353	1364	1386	1408	1452	1584
1200	1260	1272	1284	1308	1344	1440	1476	1488	1512	1536	1584	1728
1300	1365	1378	1391	1417	1456	1560	1599	1612	1638	1664	1716	1872
1400	1470	1484	1498	1526	1568	1680	1722	1736	1764	1792	1848	2016
1500	1575	1590	1605	1635	1680	1800	1845	1860	1890	1920	1980	2160
1600	1680	1696	1712	1744	1792	1920	1968	1984	2016	2048	2112	2304
1700	1785	1802	1819	1853	1904	2040	2091	2108	2142	2176	2244	2448
1800	1890	1908	1926	1962	2016	2160	2214	2232	2268	2304	2376	2592
1900	1995	2014	2033	2071	2128	2280	2337	2356	2394	2432	2508	2736
2000	2100	2120	2140	2180	2240	2400	2460	2480	2520	2560	2640	2880
2100	2205	2226	2247	2289	2352	2520	2583	2604	2646	2688	2772	3024
2200	2310	2332	2354	2398	2464	2640	2706	2728	2772	2816	2904	3168
2300	2415	2438	2461	2507	2576	2760	2829	2852	2898	2944	3036	3312
2400	2520	2544	2568	2616	2688	2880	2952	2976	3024	3072	3168	3456
2500	2625	2650	2675	2725	2800	3000	3075	3100	3150	3200	3300	3600
2600	2730	2756	2782	2834	2912	3120	3198	3224	3276	3328	3432	3744
2700	2835	2862	2889	2943	3024	3240	3321	3348	3402	3456	3564	3888
2800	2940	2968	2996	3052	3136	3360	3444	3472	3528	3584	3696	4032
2900	3045	3074	3103	3161	3248	3480	3567	3596	3654	3712	3828	4176
3000	3150	3180	3210	3270	3360	3600	3690	3720	3780	3840	3960	4320

(courtesy, Estimating Tables for Home Building, Paul I. Thomas, Craftsman Book Company)

Figure 6.22 Board Feet Required for Roof Framing — 2 x 4 Construction

This table shows the number of board feet of lumber required for roof framing using 2" x 4" rafters, and including the ridge pole. The table is based on the ground floor area being covered, spacing, and roof pitch.

Ground Floor Area	16" Center to Center						24" Center to Center					
	Roof Pitch						Roof Pitch					
	1/8	1/6	5/24	1/4	1/3	1/2	1/8	1/6	5/24	1/4	1/3	1/2
200	114	116	120	124	132	156	78	80	82	86	92	108
300	171	174	180	186	198	234	117	120	123	129	138	162
400	228	232	240	248	264	312	156	160	164	172	184	216
500	285	290	300	310	330	390	195	200	205	215	230	270
600	342	348	360	372	396	468	234	240	246	258	276	324
700	399	406	420	434	462	546	273	280	287	301	322	378
800	456	464	480	496	528	624	312	320	328	344	368	432
900	513	522	540	558	594	702	351	360	369	387	414	486
1000	570	580	600	620	660	780	390	400	410	430	460	540
1100	627	638	660	682	726	858	429	440	451	473	506	594
1200	684	696	720	744	792	936	468	480	492	516	552	648
1300	741	754	780	806	858	1014	507	520	533	559	598	702
1400	798	812	840	868	924	1092	546	560	574	602	644	756
1500	855	870	900	930	990	1170	585	600	615	645	690	810
1600	912	928	960	992	1056	1248	624	640	656	688	736	864
1700	969	986	1020	1054	1122	1326	663	680	697	731	782	918
1800	1026	1044	1080	1116	1188	1404	702	720	738	774	828	972
1900	1083	1102	1140	1178	1254	1482	741	760	779	817	874	1026
2000	1140	1160	1200	1240	1320	1560	780	800	820	860	920	1080
2100	1197	1218	1260	1302	1386	1638	819	840	861	903	966	1134
2200	1254	1276	1320	1363	1452	1716	858	880	902	946	1012	1188
2300	1311	1334	1380	1426	1518	1794	897	920	943	989	1058	1242
2400	1368	1392	1440	1488	1584	1872	936	960	984	1032	1104	1296
2500	1425	1450	1500	1550	1650	1950	975	1000	1025	1075	1150	1350
2600	1482	1508	1560	1612	1716	2028	1014	1040	1066	1118	1196	1404
2700	1539	1566	1620	1674	1782	2106	1053	1080	1107	1161	1242	1458
2800	1596	1624	1680	1736	1848	2184	1092	1120	1148	1204	1288	1512
2900	1653	1682	1740	1798	1914	2262	1131	1160	1189	1247	1334	1566
3000	1710	1740	1800	1860	1980	2340	1170	1200	1230	1290	1380	1620

(courtesy, Estimating Tables for Home Building, Paul I. Thomas, Craftsman Book Company)

Figure 6.23   Board Feet Required for Roof Framing — 2 x 6 Construction

This table shows the number of board feet of lumber required for roof framing using 2" x 6" rafters, and including the ridge pole. The table is based on the ground floor area being covered, spacing, and roof pitch.

Ground Floor Area	16" Center to Center						24" Center to Center					
	Roof Pitch						Roof Pitch					
	1/8	1/6	5/24	1/4	1/3	1/2	1/8	1/6	5/24	1/4	1/3	1/2
200	164	170	174	180	192	228	114	116	120	124	132	156
300	246	255	261	270	288	342	171	174	180	186	198	234
400	328	340	348	360	384	456	228	232	240	248	264	312
500	410	425	435	450	480	570	285	290	300	310	330	390
600	492	510	522	540	576	684	342	348	360	372	396	468
700	574	595	609	630	672	798	399	406	420	434	462	546
800	656	680	696	720	768	912	456	464	480	496	528	624
900	739	765	783	810	864	1026	513	522	540	558	594	702
1000	820	850	870	900	960	1140	570	580	600	620	660	780
1100	902	935	957	990	1056	1254	627	638	660	682	726	858
1200	984	1020	1044	1080	1152	1368	684	696	720	744	792	936
1300	1066	1105	1131	1170	1248	1482	741	754	780	806	858	1014
1400	1148	1190	1218	1260	1344	1596	798	812	840	868	924	1092
1500	1230	1274	1305	1350	1440	1710	855	870	900	930	990	1170
1600	1312	1360	1392	1440	1536	1824	912	928	960	992	1056	1248
1700	1394	1445	1479	1530	1632	1938	969	986	1020	1054	1122	1326
1800	1476	1530	1566	1620	1728	2052	1026	1044	1080	1116	1188	1404
1900	1558	1615	1653	1710	1824	2166	1083	1102	1140	1178	1254	1482
2000	1640	1700	1740	1800	1920	2280	1140	1160	1200	1240	1320	1560
2100	1722	1785	1827	1890	2016	2394	1197	1218	1260	1302	1386	1638
2200	1804	1870	1914	1980	2112	2508	1254	1276	1320	1364	1452	1716
2300	1886	1955	2001	2070	2208	2622	1311	1334	1380	1426	1518	1794
2400	1968	2040	2088	2160	2304	2736	1368	1392	1440	1488	1584	1872
2500	2050	2125	2175	2250	2400	2850	1425	1450	1500	1550	1650	1950
2600	2132	2210	2262	2340	2496	2964	1482	1508	1560	1612	1716	2028
2700	2214	2295	2349	2430	2592	3078	1539	1566	1620	1674	1782	2106
2800	2296	2380	2436	2520	2688	3192	1596	1624	1680	1736	1848	2184
2900	2378	2465	2523	2610	2784	3306	1653	1682	1740	1798	1914	2262
3000	2460	2550	2610	2700	2880	3420	1710	1740	1800	1860	1980	2340

(courtesy, Estimating Tables for Home Building, Paul I. Thomas, Craftsman Book Company)

Figure 6.24   Material Required for On-the-Job Cut Bridging — Lineal Feet

Based on the size of the lumber used for joists, the total amount of lumber (in board feet) can be obtained for the various spacing of the joists using this table.

Size Joist in Inches	Spacing in Inches	Lineal Feet per Set (2)	Lineal Feet per Foot-of-Row
2 x 6	16	2.57	1.92
2 x 8	16	2.70	2.02
2 x 10	16	2.87	2.15
2 x 12	16	3.06	2.30
2 x 8	20	3.31	2.00
2 x 10	20	3.45	2.07
2 x 12	20	3.61	2.17
2 x 8	24	3.94	2.00
2 x 10	24	4.05	1.97
2 x 12	24	4.19	2.10

Note: Add to the total lineal feet developed from the table at least 10% cutting waste.

Example: A room 20 feet wide has two rows of bridging. The 2" x 12" joists are 16" on center.

2 x 20 L.F. x 2.30 ft. per foot-of-row =   92.0 L.F.
Add 10% waste   9.2
_____

Total L.F. = 101.2
Round out to 101   L.F.

(Per set method would be, 30 sets x 3.06 = 91.8 to which must be added 10% waste.)

(excerpted from How to Estimate Building Losses and Construction Costs, Paul I. Thomas, Prentice Hall)

Figure 6.25   Board Feet Required for On-the-Job Cut Bridging

Based on the size of the lumber used for joists, the total lengths of lumber for various sized bridging can be obtained from this chart.

		Cross Bridging—Board Feet per Square Foot of Floors, Ceiling or Flat Roof Area Nails — Pounds Per MBM of Bridging					
		1" x 3"		1" x 4"		2" x 3"	
Joist Size	Spacing	B.F.	Nails	B.F.	Nails	B.F.	Nails
2" x 8"	12"	.04	147	.05	112	.08	77
	16"	.04	120	.05	91	.08	61
	20"	.04	102	.05	77	.08	52
	24"	.04	83	.05	63	.08	42
2" x 10"	12"	.04	136	.05	103	.08	71
	16"	.04	114	.05	87	.08	58
	20"	.04	98	.05	74	.08	50
	24"	.04	80	.05	61	.08	41
2" x 12"	12"	.04	127	.05	96	.08	67
	16"	.04	108	.05	82	.08	55
	20"	.04	94	.05	71	.08	48
	24"	.04	78	.05	59	.08	39
3" x 8"	12"	.04	160	.05	122	.08	84
	16"	.04	127	.05	96	.08	66
	20"	.04	107	.05	81	.08	54
	24"	.04	86	.05	65	.08	44
3" x 10"	12"	.04	146	.05	111	.08	77
	16"	.04	120	.05	91	.08	62
	20"	.04	102	.05	78	.08	52
	24"	.04	83	.05	63	.08	42

Figure 6.26 Board Feet of Sheathing and Subflooring

This table is used to compute the board feet of sheathing or subflooring required per square foot of roof, ceiling or flooring.

Type	Size	Board Feet per Square Foot of Area	Diagonal			
			Lbs. Nails per MBM Lumber			
			Joist, Stud or Rafter Spacing			
			12"	16"	20"	24"
Surface 4 Sides (S4S)	1" x 4"	1.22	58	46	39	32
	1" x 6"	1.18	39	31	25	21
	1" x 8"	1.18	30	23	19	16
	1" x 10"	1.17	35	27	23	19
Tongue and Groove (T&G)	1" x 4"	1.36	65	51	43	36
	1" x 6"	1.26	42	33	27	23
	1" x 8"	1.22	31	24	20	17
	1" x 10"	1.20	36	28	24	19
Shiplap	1" x 4"	1.41	67	53	45	37
	1" x 6"	1.29	43	33	28	23
	1" x 8"	1.24	31	24	20	17
	1" x 10"	1.21	36	28	24	19

Figure 6.27 Roof Decking for All Buildings with a Roof Pitch of 1/8

Ground Floor Area	Roof Surface Area	Shiplap 1" By		Tongue & Groove Boards 1" By			Square Edge Boards 1" By		Plywood Insulation & Particleboard
		6" & 8"	10"	4"	6" & 8"	10"	4"	6" & 8"	
S.F.	S.F.	B.F.	B.F.	B.F.	B.F.	B.F.	B.F.	B.F.	S.F.
200	206	252	244	274	252	244	246	236	226
300	309	378	366	411	378	366	369	354	339
400	412	504	488	548	504	488	492	472	452
500	515	630	610	685	630	610	615	590	565
600	618	756	632	822	756	732	738	708	678
700	721	882	854	959	882	854	861	826	791
800	824	1008	976	1096	1008	976	984	944	904
900	927	1134	1098	1233	1134	1098	1107	1062	1017
1000	1030	1260	1220	1370	1260	1220	1230	1180	1130
1100	1133	1386	1342	1507	1386	1342	1353	1298	1243
1200	1236	1512	1464	1644	1512	1464	1476	1416	1356
1300	1339	1638	1586	1781	1638	1586	1600	1534	1469
1400	1442	1764	1708	1918	1764	1708	1722	1652	1582
1500	1545	1890	1830	2055	1890	1830	1845	1770	1695
1600	1648	2016	1952	2192	2016	1952	1968	1888	1808
1700	1751	2142	2074	2329	2142	2074	2091	2006	1921
1800	1854	2268	2196	2466	2268	2196	2214	2124	2034
1900	1957	2394	2318	2603	2394	2318	2337	2242	2147
2000	2060	2520	2440	2740	2520	2440	2460	2360	2260
2100	2163	2646	2562	2877	2646	2562	2583	2478	2373
2200	2266	2772	2684	3014	2772	2684	2706	2596	2486
2300	2369	2898	2806	3151	2898	2806	2829	2714	2600
2400	2472	3024	2928	3288	3024	2928	2953	2832	2712
2500	2575	3150	3050	3425	3150	3050	3075	2950	2825
2600	2678	3276	3172	3562	3276	3172	3198	3068	2938
2700	2781	3402	3294	3699	3402	3294	3321	3186	3051
2800	2884	3528	3416	3836	3528	3416	3440	3304	3164
2900	2987	3654	3538	3973	3654	3538	3567	3422	3277
3000	3090	3780	3660	4110	3780	3660	3690	3540	3390

All figures include milling and cutting waste.

(courtesy, Estimating Tables for Home Building, Paul I. Thomas, Craftsman Book Company)

Figure 6.28   Roof Decking for All Buildings
with a Roof Pitch of 1/6

Ground Floor Area	Roof Surface Area	Shiplap 1" By		Tongue & Groove Boards 1" By			Square Edge Boards 1" By		Plywood Insulation & Particleboard
		6" & 8"	10"	4"	6" & 8"	10"	4"	6" & 8"	
S.F.	S.F.	B.F.	B.F.	B.F.	B.F.	B.F.	B.F.	B.F.	S.F.
200	212	258	250	282	258	250	252	244	234
300	318	387	375	423	387	375	378	366	351
400	424	516	500	564	516	500	504	488	468
500	530	645	625	705	645	625	630	610	585
600	636	774	750	846	774	750	756	732	702
700	742	900	875	987	900	875	882	854	819
800	848	1030	1000	1128	1030	1000	1008	976	936
900	954	1160	1125	1269	1160	1125	1134	1098	1053
1000	1060	1290	1250	1410	1290	1250	1260	1220	1170
1100	1166	1420	1375	1551	1420	1375	1386	1342	1287
1200	1272	1551	1500	1692	1550	1500	1512	1464	1404
1300	1378	1680	1625	1833	1680	1625	1638	1586	1521
1400	1484	1810	1750	1974	1810	1750	1764	1708	1638
1500	1590	1935	1875	2115	1935	1875	1890	1830	1755
1600	1696	2064	2000	2256	2064	2000	2016	1952	1872
1700	1802	2195	2125	2397	2195	2125	2142	2074	1989
1800	1908	2322	2250	2538	2322	2250	2268	2196	2106
1900	2014	2450	2375	2679	2450	2375	2394	2318	2223
2000	2120	2580	2500	2820	2580	2500	2520	2440	2340
2100	2226	2710	2625	2961	2710	2625	2646	2562	2457
2200	2332	2840	2750	3102	2840	2750	2772	2684	2574
2300	2438	2970	2875	3243	2970	2875	2898	2806	2691
2400	2544	3100	3000	3384	3100	3000	3024	2928	2808
2500	2650	3225	3125	3525	3225	3125	3150	3050	2925
2600	2756	3354	3250	3666	3354	3250	3276	3172	3042
2700	2862	3483	3375	3807	3483	3375	3402	3294	3159
2800	2968	3610	3500	3948	3610	3500	3528	3416	3276
2900	3074	3740	3625	4089	3740	3625	3654	3528	3393
3000	3180	3870	3750	4230	3870	3750	3780	3660	3510

All figures include milling and cutting waste.

(courtesy, Estimating Tables for Home Building, Paul I. Thomas, Craftsman Book Company)

Figure 6.29 Roof Decking for All Buildings with a Roof Pitch of 5/24

Ground Floor Area	Roof Surface Area	Shiplap 1" By		Tongue & Groove Boards 1" By			Square Edge Boards 1" By		Plywood Insulation & Particleboard
		6" & 8"	10"	4"	6" & 8"	10"	4"	6" & 8"	
S.F.	S.F.	B.F.	B.F.	B.F.	B.F.	B.F.	B.F.	B.F.	S.F.
200	218	266	258	290	266	258	260	250	240
300	327	400	387	435	400	387	390	375	360
400	426	532	516	580	532	516	520	500	480
500	546	665	645	725	665	645	650	625	600
600	654	800	774	870	800	774	780	750	720
700	763	930	900	1015	930	900	910	875	840
800	872	1064	1030	1160	1064	1030	1040	1000	960
900	981	1200	1160	1305	1200	1160	1170	1125	1080
1000	1090	1330	1290	1450	1330	1290	1300	1250	1200
1100	1199	1463	1420	1595	1463	1420	1430	1375	1320
1200	1308	1600	1550	1740	1600	1550	1560	1500	1440
1300	1417	1730	1680	1885	1730	1680	1690	1625	1560
1400	1526	1860	1810	2030	1860	1810	1820	1750	1680
1500	1635	2000	1935	2175	2000	1935	1950	1875	1800
1600	1744	2130	2064	2320	2130	2064	2080	2000	1920
1700	1853	2260	2195	2465	2260	2195	2210	2125	2040
1800	1962	2400	2322	2610	2400	2322	2340	2250	2160
1900	2071	2530	2450	2755	2530	2450	2470	2375	2280
2000	2180	2660	2580	2900	2660	2580	2600	2500	2400
2100	2289	2800	2710	3045	2800	2710	2730	2625	2520
2200	2398	2930	2840	3190	2930	2840	2860	2750	2640
2300	2507	3060	2970	3335	3060	2970	2990	2875	2760
2400	2616	3200	3100	3480	3200	3100	3120	3000	2880
2500	2725	3325	3225	3625	3325	3225	3250	3125	3000
2600	2834	3460	3354	3770	3460	3354	3380	3250	3120
2700	2943	3590	3483	3915	3590	3483	3510	3375	3240
2800	3052	3724	3610	4060	3724	3610	3640	3500	3360
2900	3161	3860	3740	4205	3860	3740	3770	3625	3480
3000	3270	3990	3870	4350	3990	3870	3900	3750	3600

All figures include milling and cutting waste.

(courtesy, Estimating Tables for Home Building, Paul I. Thomas, Craftsman Book Company)

Figure 6.30 Roof Decking for All Buildings with a Roof Pitch of 1/4

Ground Floor Area	Roof Surface Area	Shiplap 1" By		Tongue & Groove Boards 1" By			Square Edge Boards 1" By		Plywood Insulation & Particleboard
		6" & 8"	10"	4"	6" & 8"	10"	4"	6" & 8"	
S.F.	S.F.	B.F.	B.F.	B.F.	B.F.	B.F.	B.F.	B.F.	S.F.
200	224	274	264	298	274	264	266	258	246
300	336	411	396	447	411	396	400	387	369
400	448	548	528	600	548	528	532	516	492
500	560	685	660	745	685	660	665	645	615
600	672	822	792	894	822	792	800	774	738
700	784	959	924	1043	959	924	930	900	861
800	896	1096	1056	1192	1096	1054	1064	1030	984
900	1008	1233	1188	1340	1233	1188	1200	1160	1107
1000	1120	1370	1320	1490	1370	1320	1330	1290	1230
1100	1232	1507	1452	1639	1507	1452	1463	1420	1353
1200	1344	1644	1584	1788	1644	1584	1600	1550	1476
1300	1456	1781	1716	1937	1781	1716	1730	1680	1600
1400	1568	1918	1848	2086	1918	1848	1860	1810	1722
1500	1680	2055	1980	2235	2055	1980	2000	1935	1845
1600	1792	2192	2112	2384	2192	2112	2130	2064	1968
1700	1904	2329	2244	2533	2329	2244	2260	2195	2091
1800	2016	2466	2376	2682	2466	2376	2400	2322	2214
1900	2128	2603	2508	2830	2603	2508	2530	2450	2337
2000	2240	2740	2640	2980	2740	2640	2660	2580	2460
2100	2352	2877	2772	3130	2877	2772	2800	2710	2583
2200	2464	3014	2904	3278	3014	2904	2930	2840	2706
2300	2576	3151	3036	3427	3151	3036	3060	2970	2829
2400	2688	3288	3168	3576	3288	3168	3200	3100	2952
2500	2800	3425	3300	3725	3425	3300	3325	3225	3075
2600	2912	3562	3432	3874	3562	3432	3460	3354	3198
2700	3024	3700	3564	4023	3700	3564	3590	3483	3321
2800	3136	3836	3696	4172	3836	3696	3724	3610	3440
2900	3248	3973	3828	4321	3973	3828	3860	3740	3567
3000	3360	4110	3960	4470	4110	3960	3990	3870	3690

All figures include milling and cutting waste.

(courtesy, Estimating Tables for Home Building, Paul I. Thomas, Craftsman Book Company)

Figure 6.31 Roof Decking for All Buildings with a Roof Pitch of 1/3

Ground Floor Area	Roof Surface Area	Shiplap 1" By		Tongue & Groove Boards 1" By			Square Edge Boards 1" By		Plywood Insulation & Particleboard
		6" & 8"	10"	4"	6" & 8"	10"	4"	6" & 8"	
S.F.	S.F.	B.F.	B.F.	B.F.	B.F.	B.F.	B.F.	B.F.	S.F.
200	240	292	284	320	292	284	286	276	264
300	360	438	426	480	438	426	429	414	396
400	480	584	569	640	584	568	572	552	528
500	600	730	710	800	730	710	715	690	660
600	720	876	852	960	876	852	858	828	792
700	840	1022	994	1120	1022	994	1000	966	924
800	960	1170	1136	1280	1170	1136	1144	1104	1056
900	1080	1314	1278	1440	1314	1278	1287	1242	1188
1000	1200	1460	1420	1600	1460	1420	1430	1380	1320
1100	1320	1610	1562	1760	1610	1562	1573	1518	1452
1200	1440	1752	1704	1920	1752	1704	1716	1656	1584
1300	1560	1900	1846	2080	1900	1846	1859	1794	1716
1400	1680	2044	1990	2240	2044	1990	2000	1932	1848
1500	1800	2190	2130	2400	2190	2130	2145	2070	1980
1600	1920	2336	2272	2560	2336	2272	2288	2208	2112
1700	2040	2480	2414	2720	2480	2414	2430	2346	2244
1800	2160	2630	2556	2880	2630	2556	2574	2484	2376
1900	2280	2774	2700	3040	2774	2700	2717	2622	2508
2000	2400	2920	2840	3200	2920	2840	2860	2760	2640
2100	2520	3066	2982	3360	3066	2982	3000	2898	2772
2200	2740	3212	3124	3520	3212	3124	3146	3036	2904
2300	2760	3358	3266	3680	3358	3266	3290	3174	3036
2400	2880	3504	3410	3840	3504	3410	3430	3312	3168
2500	3000	3650	3550	4000	3650	3550	3575	3450	3300
2600	3120	3800	3692	4160	3800	3692	3720	3588	3432
2700	3240	3942	3834	4320	3942	3834	3860	3726	3564
2800	3360	4090	3976	4480	4090	3976	4000	3864	3696
2900	3480	4234	4118	4640	4234	4118	4150	4000	3828
3000	3600	4380	4260	4800	4380	4260	4290	4140	3960

All figures include milling and cutting waste.

(courtesy, Estimating Tables for Home Building, Paul I. Thomas, Craftsman Book Company)

Figure 6.32 Roof Decking for All Buildings
with a Roof Pitch of 1/2

Ground Floor Area	Roof Surface Area	Shiplap 1" By		Tongue & Groove Boards 1" By			Square Edge Boards 1" By		Plywood Insulation & Particleboard
		6" & 8"	10"	4"	6" & 8"	10"	4"	6" & 8"	
S.F.	S.F.	B.F.	B.F.	B.F.	B.F.	B.F.	B.F.	B.F.	S.F.
200	284	346	336	378	346	336	338	326	312
300	426	520	504	567	520	504	507	489	468
400	568	690	672	756	690	672	676	652	624
500	710	865	840	945	865	840	845	815	780
600	852	1040	1008	1134	1040	1008	1014	978	936
700	994	1210	1176	1323	1210	1176	1183	1141	1092
800	1136	1385	1344	1512	1385	1344	1352	1304	1248
900	1278	1560	1512	1700	1560	1512	1521	1467	1404
1000	1420	1730	1680	1890	1730	1680	1690	1630	1560
1100	1562	1900	1848	2080	1900	1848	1859	1793	1716
1200	1704	2076	2016	2268	2076	2016	2028	1956	1872
1300	1846	2250	2184	2457	2250	2184	2200	2119	2028
1400	1990	2422	2352	2646	2422	2352	2370	2282	2184
1500	2130	2595	2520	2835	2595	2520	2535	2445	2340
1600	2272	2770	2688	3024	2770	2688	2705	2608	2500
1700	2414	2940	2856	3213	2940	2856	2875	2771	2652
1800	2556	3115	3024	3400	3115	3024	3040	2934	2810
1900	2700	3290	3192	3590	3290	3192	3210	3097	2964
2000	2840	3460	3360	3780	3460	3360	3380	3260	3120
2100	2982	3633	3528	3970	3633	3528	3550	3423	3276
2200	3124	3810	3696	4158	3810	3696	3720	3586	3432
2300	3266	3980	3864	4350	3980	3864	3890	3749	3590
2400	3410	4150	4032	4536	4150	4032	4056	3912	3744
2500	3550	4325	4200	4725	4325	4200	4225	4075	3900
2600	3692	4500	4368	4914	4500	4368	4395	4238	4056
2700	3834	4670	4536	5100	4670	4536	4565	4400	4212
2800	3976	4845	4704	5292	4845	4704	4730	4564	4368
2900	4118	5020	4872	5480	5020	4872	4900	4727	4524
3000	4260	5190	5040	5670	5190	5040	5070	4890	4680

All figures include milling and cutting waste.

(courtesy, Estimating Tables for Home Building, Paul I. Thomas, Craftsman Book Company)

Figure 6.33 Wood Siding Factors

This table shows the factor by which area to be covered is multiplied to determine exact amount of surface material needed.

Item	Nominal Size	Width Overall	Face	Area Factor
Shiplap	1" x 6"	5-1/2"	5-1/8"	1.17
	1 x 8	7-1/4	6-7/8	1.16
	1 x 10	9-1/4	8-7/8	1.13
	1 x 12	11-1/4	10-7/8	1.10
Tongue and Grooved	1 x 4	3-3/8	3-1/8	1.28
	1 x 6	5-3/8	5-1/8	1.17
	1 x 8	7-1/8	6-7/8	1.16
	1 x 10	9-1/8	8-7/8	1.13
	1 x 12	11-1/8	10-7/8	1.10
S4S	1 x 4	3-1/2	3-1/2	1.14
	1 x 6	5-1/2	5-1/2	1.09
	1 x 8	7-1/4	7-1/4	1.10
	1 x 10	9-1/4	9-1/4	1.08
	1 x 12	11-1/4	11-1/4	1.07
Solid Paneling	1 x 6	5-7/16	5-7/16	1.19
	1 x 8	7-1/8	6-3/4	1.19
	1 x 10	9-1/8	8-3/4	1.14
	1 x 12	11-1/8	10-3/4	1.12
Bevel Siding*	1 x 4	3-1/2	3-1/2	1.60
	1 x 6	5-1/2	5-1/2	1.33
	1 x 8	7-1/4	7-1/4	1.28
	1 x 10	9-1/4	9-1/4	1.21
	1 x 12	11-1/4	11-1/4	1.17

Note: This area factor is strictly so-called milling waste. The cutting and fitting waste must be added.

*1" lap

(from Western Wood Products Association)

Figure 6.34   Milling and Cutting Waste Factors for Wood Siding

This table lists the milling waste, the cutting waste, and the amount of nails required for various types of wood siding. Amounts are per 1,000 FBM of wood siding.

Type of Siding	Nominal Size in Inches	Lap in Inches 1" Lap	Pounds Nails per 1,000 FBM	Percentage of Waste
Bevel Siding	1 x 4	1	25-6d common	63
	1 x 6	1	25-6d common	35
	1 x 8	1-1/4	20-8d common	35
	1 x 10	1-1/2	20-8d common	30
Rustic and Drop Siding	1 x 4	Matched	40-8d common	33
	1 x 6	Matched	30-8d common	25
	1 x 8	Matched	25-8d common	20
Vertical Siding	1 x 6	Matched	25-8d finish	20
	1 x 8	Matched	20-8d finish	18
	1 x 10	Matched	20-8d finish	15
Batten Siding*	1 x 8	Rough	25	5
	1 x 10	Rough	20	5
	1 x 12	Rough	20	5
	1 x 8	Dressed	25	13
	1 x 10	Dressed	20	11
	1 x 12	Dressed	20	10
Plywood Siding	1/4	Sheets		5-10
	3/8	Sheets	15 per MSF	5-10
	5/8	Sheets		5-10

*For 1" x 10" boards allow 1,334 lineal feet 1" x 2" joint strips for each 1,000 FBM of batten siding. Add 12 pounds 8d common nails.

(excerpted from How to Estimate Building Losses and Construction Costs, Paul I. Thomas, Prentice Hall)

Figure 6.35 Installation Time in Man-Hours for Joists and Decking

Description	Man-Hours	Unit
Joist Framing		
2" x 6"	.013	L.F.
2" x 8"	.015	L.F.
2" x 10"	.018	L.F.
2" x 12"	.018	L.F.
2" x 14"	.021	L.F.
3" x 8"	.017	L.F.
3" x 12"	.027	L.F.
4" x 8"	.026	L.F.
4" x 12"	.036	L.F.
Bridging Wood or Steel, Joists		
16" On Center	.062	Pr.
24" On Center	.057	Pr.
Sub Floor Plywood CDX		
1/2" Thick	.011	S.F.
5/8" Thick	.012	S.F.
3/4" Thick	.013	S.F.
Boards Diagonal		
1" x 8"	.019	S.F.
1" x 10"	.018	S.F.
Wood Fiber T & G		
2' x 8' Planks		
1" Thick	.016	S.F.
1-3/8" Thick	.018	S.F.

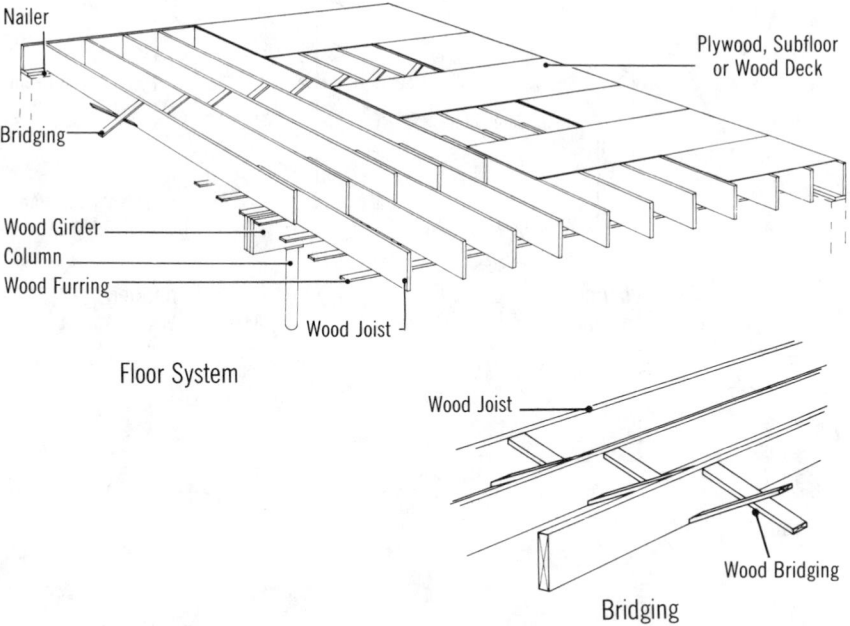

Floor System

Bridging

Figure 6.36   Installation Time in Man-Hours for Columns, Beams, Girders and Decking

Description	Man-Hours	Unit
Columns		
6" x 6"	.074	L.F.
8" x 8"	.122	L.F.
10" x 10"	.178	L.F.
12" x 12"	.240	L.F.
Beams and Girders		
6" x 10"	.073	L.F.
8" x 16"	.142	L.F.
12" x 12"	.240	L.F.
10" x 16"	.213	L.F.
Wood Deck		
3" Nominal	.050	S.F.
4" Nominal	.064	S.F.
Laminated Wood Deck		
3" Nominal	.038	S.F.
4" Nominal	.049	S.F.

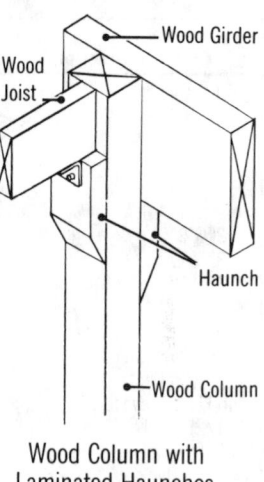

Wood Column with
Laminated Haunches

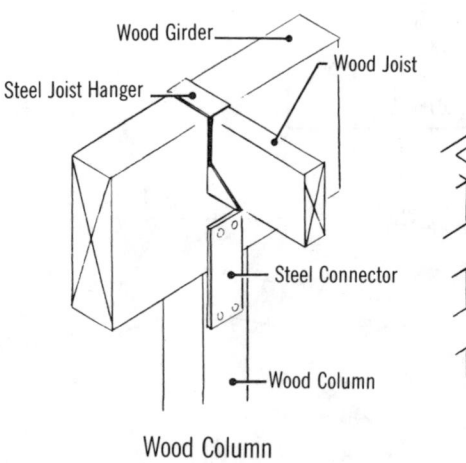

Wood Column
Girder and Joist

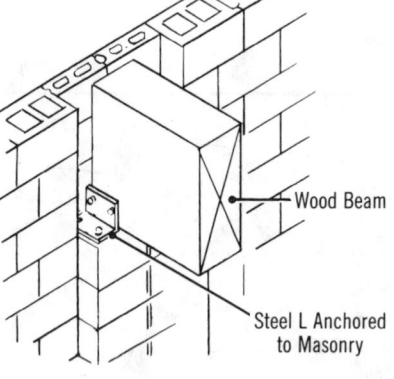

Wood Girder Supported
by Masonry Wall

Figure 6.36 Installation Time in Man-Hours for Columns, Beams, Girders and Decking (continued)

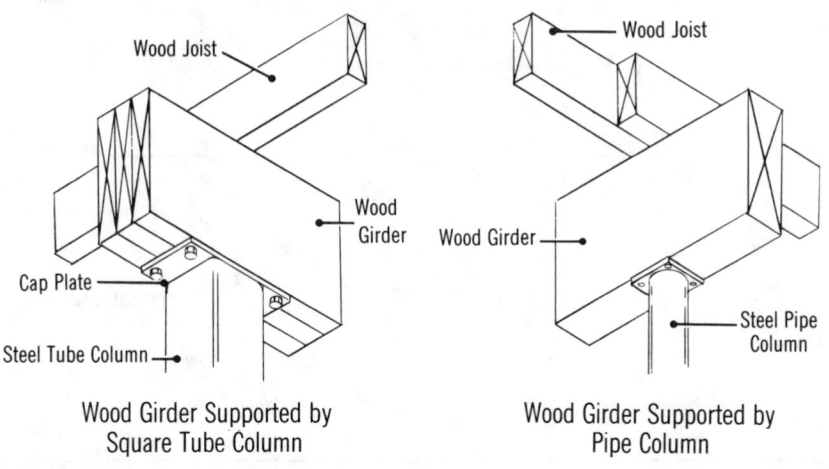

Wood Girder Supported by
Square Tube Column

Wood Girder Supported by
Pipe Column

Figure 6.37 Installation Time in Man-Hours for Laminated Construction

Description	Man-Hours	Unit
Laminated Framing Roof Beams		
20' Span		
8' On Center	.016	S.F. Floor
16' On Center	.013	S.F. Floor
40' Span		
8' On Center	.013	S.F. Floor
16' On Center	.010	S.F. Floor
60' Span		
8' On Center	.017	S.F. Floor
16' On Center	.013	S.F. Floor
Columns	.020	MBF
Beams	.011	MBF
Wood Deck		
3" Nominal	.050	S.F.
4" Nominal	.064	S.F.
Laminated Wood Deck		
3" Nominal	.038	S.F.
4" Nominal	.049	S.F.

(continued on next page)

Figure 6.37   Installation Time in Man-Hours for Laminated Construction (continued)

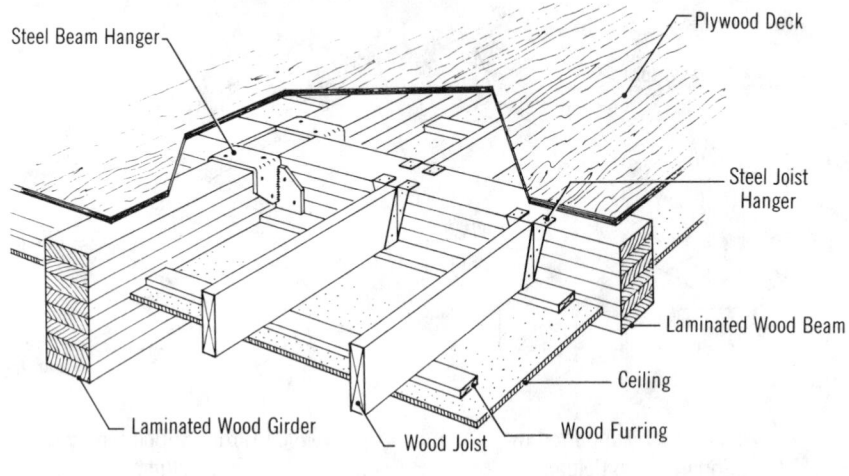

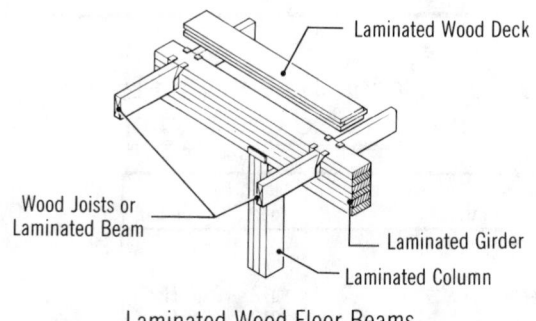

Laminated Wood Floor Beams

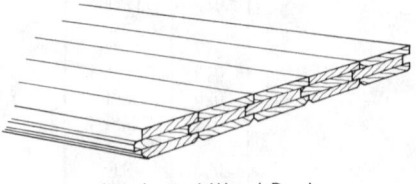

Laminated Wood Deck

Figure 6.38 Installation Time in Man-Hours for
Pre-Fabricated Wood Roof Trusses

Description	Man-Hours	Unit
Wood Roof Trusses		
4/12 Pitch		
20' Span	.645	Ea.
28' Span	.755	Ea.
36' Span	.870	Ea.
8/12 Pitch		
20' Span	.702	Ea.
28' Span	.816	Ea.
36' Span	.976	Ea.
Fink or King Post 30' to 60' Span	.013	S.F. Floor
Plywood Roof Sheathing		
1/2" Thick	.011	S.F.
5/8" Thick	.012	S.F.
3/4" Thick	.013	S.F.
Stressed Skin, Plywood Roof Panels 4' x 8'		
4-1/4" Deep	.019	S.F. Roof
6-1/2" Deep	.023	S.F. Roof
Wood Roof Decks		
3" Thick Nominal	.050	S.F.
4" Thick Nominal	.064	S.F.
Laminated Wood Roof Deck		
3" Thick Nominal	.038	S.F.
4" Thick Nominal	.049	S.F.

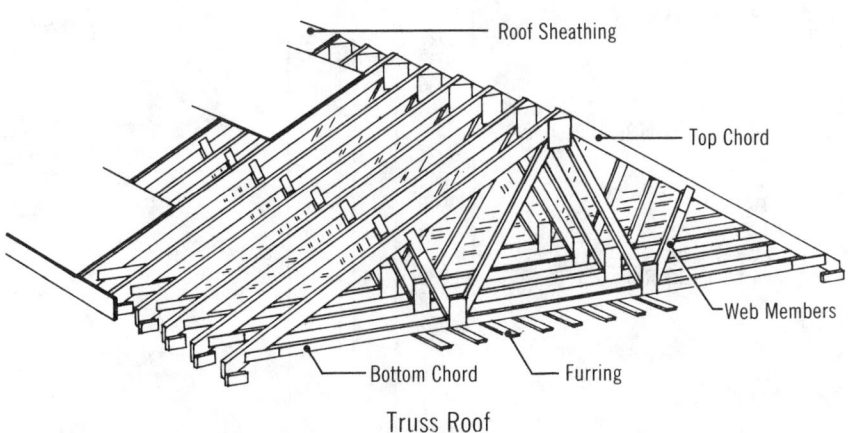

Truss Roof

Figure 6.39 Installation Time in Man-Hours for Rafter Systems

Description	Man-Hours	Unit
Rafters, Common, to 4 in 12 Pitch, 2" x 6"	.016	L.F.
2" x 8"	.017	L.F.
2" x 10"	.025	L.F.
2" x 12"	.028	L.F.
On Steep Roofs, 2" x 6"	.020	L.F.
2" x 8"	.021	L.F.
2" x 10"	.032	L.F.
2" x 12"	.035	L.F.
On Dormers or Complex Roofs, 2" x 6"	.027	L.F.
2" x 8"	.030	L.F.
2" x 10"	.038	L.F.
2" x 12"	.041	L.F.
Hip and Valley, to 4 in 12 Pitch, 2" x 6"	.021	L.F.
2" x 8"	.022	L.F.
2" x 10"	.028	L.F.
2" x 12"	.030	L.F.
Hip and Valley, on Steep Roofs, 2" x 6"	.027	L.F.
2" x 8"	.029	L.F.
2" x 10"	.036	L.F.
2" x 12"	.039	L.F.
Hip and Valley, on Dormers/Complex Roofs,		
2" x 6"	.031	L.F.
2" x 8"	.034	L.F.
2" x 10"	.042	L.F.
2" x 12"	.045	L.F.
Hip and Valley Jacks to 4 in 12 Pitch,		
2" x 6"	.027	L.F.
2" x 8"	.033	L.F.
2" x 10"	.036	L.F.
2" x 12"	.043	L.F.
Hip and Valley Jacks on Steep Roofs,		
2" x 6"	.034	L.F.
2" x 8"	.042	L.F.
2" x 10"	.046	L.F.
2" x 12"	.054	L.F.
Hip and Valley Jacks on Dormers/Complex Roofs, 2" x 6"	.039	L.F.
2" x 8"	.048	L.F.
2" x 10"	.052	L.F.
2" x 12"	.063	L.F.
Collar Ties, 1" x 4"	.020	L.F.
Ridge Board, 1" x 6"	.027	L.F.
1" x 8"	.029	L.F.
1" x 10"	.032	L.F.
2" x 6"	.032	L.F.
2" x 8"	.036	L.F.
2" x 10"	.040	L.F.
Sub-Fascia, 2" x 8"	.071	L.F.
2" x 10"	.089	L.F.

Figure 6.39   Installation Time In
Man-Hours for Rafter Systems (continued)

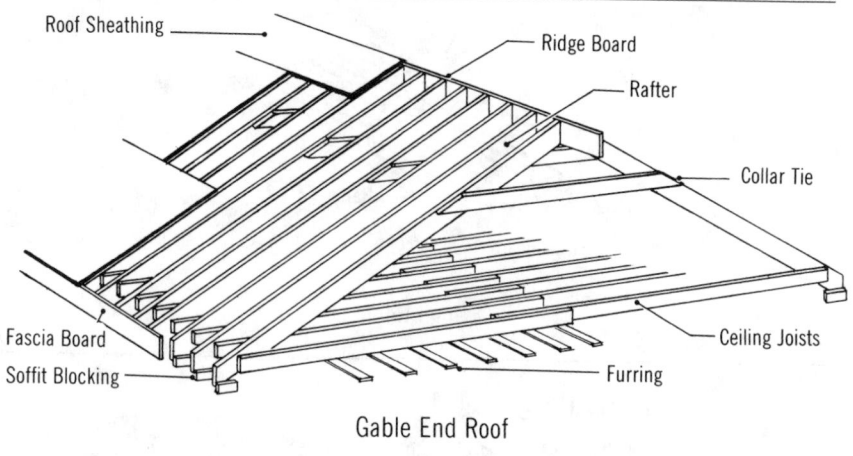

Gable End Roof

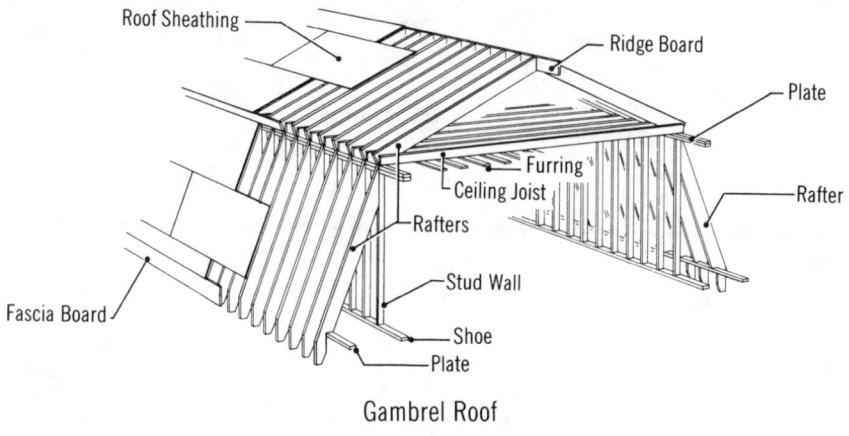

Gambrel Roof

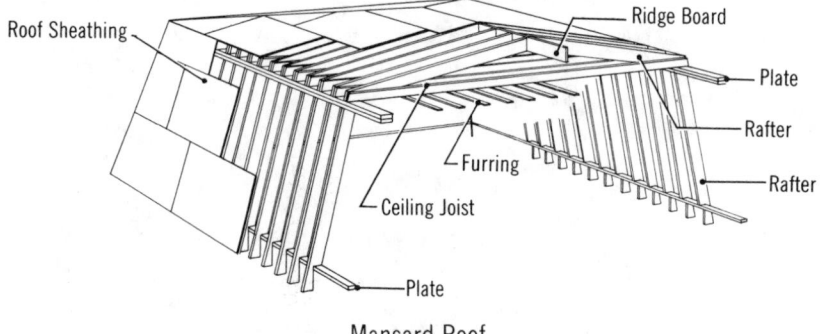

Mansard Roof

(continued on next page)

Figure 6.39 Installation Time In
Man-Hours for Rafter Systems (continued)

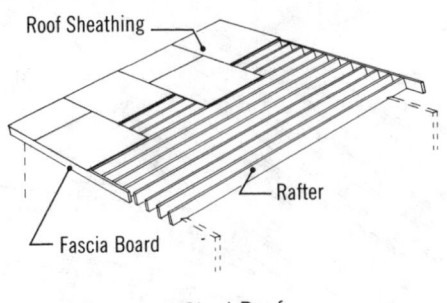

Shed Roof

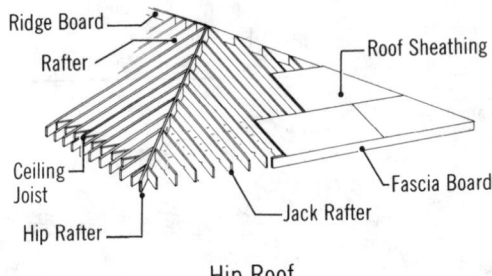

Hip Roof

Figure 6.40 Installation Time in Man-Hours for Wood Siding

Description	Man-Hours	Unit
Siding		
Clapboard		
Cedar Beveled		
1/2" x 6"	.032	S.F.
1/2" x 8"	.029	S.F.
3/4" x 10"	.027	S.F.
Redwood, Beveled		
1/2" x 4"	.040	S.F.
1/2" x 8"	.032	S.F.
3/4" x 10"	.027	S.F.
Board		
Redwood		
Tongue and Groove		
1" x 4"	.053	S.F.
1" x 8"	.043	S.F.
Channel		
1" x 10"	.028	S.F.
Cedar		
Butted		
1" x 4"	.033	S.F.
Channel		
1" x 8"	.032	S.F.
Board and Batten		
1" x 12"	.030	S.F.
Pine		
Butted		
1" x 8"	.029	S.F.
Sheets		
Hardboard, Lapped	.021	S.F.
Plywood		
MDO		
3/8" Thick	.021	S.F.
1/2" Thick	.022	S.F.
3/4" Thick	.024	S.F.
Texture 1-11		
5/8" Thick	.023	S.F.
Sheathing		
Plywood on Walls		
3/8" Thick	.013	S.F.
3/4" Thick	.016	S.F.
Wood Fiber	.013	S.F.
Trim Exterior, Up to 1" x 6"	.040	L.F.
Fascia		
1" x 6"	.032	L.F.
1" x 8"	.035	L.F.

(continued on next page)

Figure 6.40   Installation Time In Man-Hours for Wood Siding (continued)

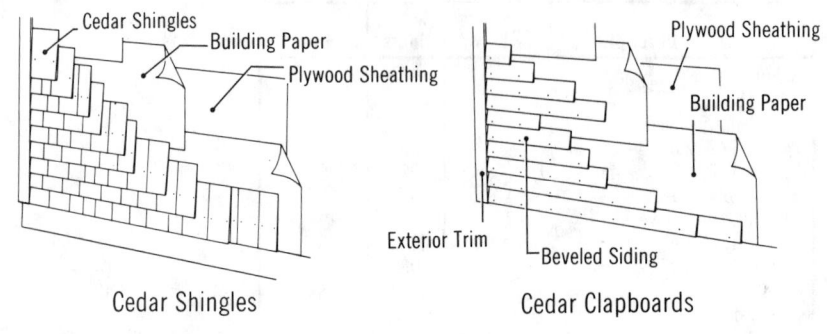

Cedar Shingles   Cedar Clapboards

Figure 6.41   Nails—Dimensions

This table describes the sizes, lengths, gauge number, and approximate number of nails one can expect per pound, based on the type of nail and use.

Penny Nail System					
			Approximate Number to Pound		
Size	**Length in Inches**	**Gauge Number**	**Common**	**Finishing**	**Casing**
2d	1	15	850		
3d	1-1/4	14	550	640	
4d	1-1/2	12-1/2	350	456	
5d	1-3/4	12-1/2	230	328	
6d	2	11-1/2	180	273	228
7d	2-1/4	11	140	170	178
8d	2-1/2	10-1/4	100	151	133
9d	2-3/4	9-1/2	80	125	100
10d	3	9	65	107	96
12d	3-1/4	9	50		60
16d	3-1/2	8	40		50
20d	4	6	31		
30d	4-1/2	5	22		
40d	5	4	18		
50d	5-1/2	3	14		
60d	6	2	12		

(excerpted from How to Estimate Building Losses and Construction Costs, Paul I. Thomas, Prentice Hall)

Figure 6.42   Size and Quantity of Nails for a Job

This table shows the size and approximate quantity of nails needed for various portions of a wood project.

Where a mix of nails is shown, judgment must be exercised as to the different sizes of lumber.

Kind of Framing and Size of Lumber	Size of Nails Used	Lbs. per 1,000 FBM
Sills and plates	10d, 16d & 20d	8
Wall and partition stud	10d & 16d	10
Joists and rafters		
2" x 6"	16d, some 20d	9
2" x 8"	16d, some 20d	8
2" x 10"	16d, some 20d	7
2" x 12"	16d, some 20d	6
Average for total house framing	8d, 10d, 16d, 20d	15
Wood cross bridging, 1" x 3"	8d	1 lb. per 12 sets
Furring (100 S.F.) on masonry	8d	1
Furring (100 S.F.) on studding	8d	1/2
Roof trusses	10d, 20d and 40d	10

Figure 6.43   Rough Hardware Allowances

Average Material Cost Allowances for Rough Hardware as a Percentage of Carpentry Material Costs	
Minimum	0.5%
Maximum	1.5%

Figure 6.44 Installation Time in Man-Hours for Millwork

Description	Man-Hours	Unit
Beams, Decorative 4" x 8"	.100	L.F.
Cabinets, Kitchen Base, 24" x 24" x 35" High	.717	Ea.
Kitchen Wall, 12" x 24" x 30" High	.788	Ea.
Casework Frames		
Base Cabinets, 36" High, Two Bay, 36" Wide	3.636	Ea.
Book Cases, 7' High, Two Bay, 36" Wide	5.000	Ea.
Coat Racks, 7' High, Two Bay, 48" Wide	2.909	Ea.
Wall Mounted Cabinets, 30" High, Two Bay,		
36" Wide	3.721	Ea.
Wardrobe, 7' High, 48" Wide	4.706	Ea.
Cabinet Doors		
Glass Panel, Hardwood Frame, 18" Wide,		
30' High	.276	Ea.
Hardwood, Raised Panel, 18" Wide, 30" High	.571	Ea.
Plastic Laminate, 18" Wide, 30" High	.364	Ea.
Counter Tops, Stock, Plastic Laminate,		
1-1/4" Thick	.286	L.F.
Moldings, Base, 9/16" x 3-1/2"	.033	L.F.
Casing, Apron, 5/8" x 3-1/2"	.036	L.F.
Band, 11/16" x 1-3/4"	.032	L.F.
Casings, 11/16" x 3-1/2"	.037	L.F.
Ceilings, Bed, 9/16" x 2"	.033	L.F.
Cornice, 9/16" x 2-1/4"	.027	L.F.
Cove Scotia, 11/16" x 2-3/4"	.031	L.F.
Crown, 11/16" x 4-5/8"	.036	L.F.
Exterior Cornice Boards, 1" x 6"	.040	L.F.
Corner Board, 1" x 6"	.040	L.F.
Fascia, 1" x 6"	.032	L.F.
Moldings	.032	L.F.
Verge Board, 1" x 6"	.040	L.F.
Trim, Astragal, 1-5/16" x 2-3/16"	.033	L.F.
Chair Rail, 5/8" x 3-1/2"	.033	L.F.
Handrail, 1-1/2" x 1-3/4"	.100	L.F.
Door Moldings	.471	set
Door Trim Including Headers, Stops,		
and Casing 4-1/2" Wide	1.509	Opng.
Stool Caps, 1-1/16" x 3-1/4"	.053	L.F.
Threshold, 3' Long, 5/8" x 3-5/8"	.250	Ea.
Window Trim Sets Including Casings, Header,		
Stops, Stool and Apron	.800	Opng.
Paneling, 1/4" Thick	.040	S.F.
3/4" Thick, Stock	.050	S.F.
Architectural Grade	.071	S.F.
Stairs, Prefabricated		
Box 3" Wide, 8' High	5.333	Flight
Open Stairs, 8' High	5.333	Flight
3 Piece Wood Railings and Baluster, 8" High	1.333	Ea.
Spiral Oak, 4'-6" Diameter Including Rail	10.667	Flight

Figure 6.44   Installation Time In Man Hours
for Millwork (continued)

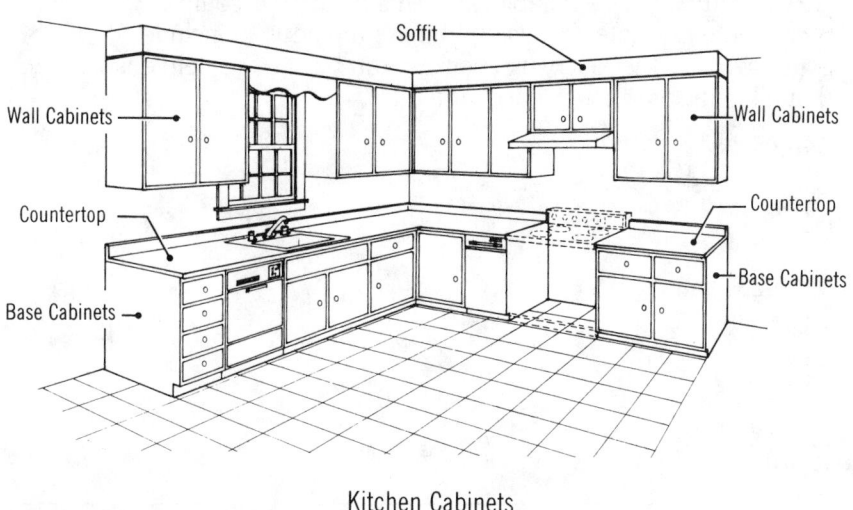

Kitchen Cabinets

Checklist

For an estimate to be reliable, all items must be accounted for.
A complete estimate can also eliminate the need to include
contingencies. The following checklist can be used to help ensure
that all items are properly accounted for.

Structural
☐ Columns
☐ Beams
   ____ Box
   ____ Laminated
☐ Joists
☐ Posts and girts
☐ Purlins
☐ Rafters
☐ Trusses

Bracing _____

Bridging _____

Cabinets _____

Casings _____

Ceiling Beams, Decorative _____

Chair Rails _____

Countertops _____

Door Bucks _____

Finish Carpentry
☐ Casings
☐ Chair rails
☐ Closets
☐ Cornices
☐ Countertops
☐ Cupolas
☐ Mouldings and trim
☐ Paneling
☐ Railings
☐ Shelving
☐ Stairs
☐ Thresholds
☐ Wainscotting

Floor Planking _____

Nailer Plates _____

Siding
☐ Sheathing
    ____ Hardboard
    ____ Particleboard
    ____ Plywood
☐ Cedar
☐ Redwood
☐ Fir
☐ White Pine

Roof Carpentry
☐ Cants
☐ Decking
☐ Sheathing

Sills _____

Soffits _____

Subflooring _____

Underlayment _____

Wall Framing
☐ Studs
☐ Plates
☐ Sills

Tips

Items Listed

Note that on construction documents not all rough carpentry items are listed or noted. Protective treads, inserts, and rails are examples of necessary but not listed items that the contractor must provide.

Item Search

The search for carpentry items must include roofing sections, wall sections, and all detail drawings.

Temporary Construction

Temporary construction should be included in Division 1, though it may sometimes be listed in Division 6. Double check to assure yourself that you are not double-pricing these items.

Treated Lumber

Treated lumber should always be used when the lumber will come in direct contact with concrete, masonry, or the earth.

Bridging

Even when bridging is not shown on joist drawings, always include it, as it helps to distribute concentrated loads to the adjacent joists.

Blocking

One of the most overlooked areas of wood blocking is at roof edges. Almost always, a built-up roof or membrane roof system will require some sort of blocking and/or cant strip system.

Millwork

When budgeting millwork, a rule of thumb is that the total cost of millwork items will be two to three times the cost of the materials required. Do not overlook adding protection to all millwork, especially if the work is by others.

Pricing

Do not rely on yesterday's material quotes. Due to current and predicted shortages of lumber and the resultant fluctuations in the market, caution is warranted.

Estimating Help

Short of time? Many lumber yards retain competent estimators who will provide material lists/estimates from your plans for little or no cost. This is also a good method for checking your own takeoff.

Division Seven
Thermal &
Moisture
Protection

Introduction

This division covers all types of materials used to seal the outside of the building against moisture, thermal and air penetration, plus the associated insulation and accessories. (Note, however, that doors and windows are covered in Division 8.) Included in this chapter are dampproofing, waterproofing, insulation and roofing.

Dampproofing usually consists of one or two layers of coatings applied to a foundation wall up to the finished grade elevation. Dampproofing is used to inhibit the migration of moisture from the outside of the structure to the inside.

Waterproofing usually consists of membrane sheeting used to prevent or stop the flow of water.

Insulation is primarily used to reduce the unwanted effects of heat transfer through the exterior enclosure of a building. The type, form, and material used for insulation depend on the location within the structure, and the size of the space available.

Roofing can be subdivided into various types depending on its application. Shingles are used mostly on residences. Commercial and/or industrial roofs usually consist of one or a combination of built-up (layered tar and asphalt felts, for example), single ply membrane, metal sheets, corrugated metal panels, copper sheets, copper panels, and corrugated fiberglass panels.

Also included in Division 7 is sheet metal work associated with roofing: flashings, trim, gutters and downspouts, gravel stops, and the like.

Roof accessories must also be considered as part of a complete estimate. They include hatches, skylights, and vents.

The items addressed in this chapter include:

- Insulation and dampproofing
- Sidings
- Roofing
- Flashings
- Roof specialties
- Skylights

Estimating Data

The following tables present effective estimating guidelines for items found in Division 7 — Thermal and Moisture Control. Please note that these guidelines can be used as indicators of what may be expected, and that each project must be evaluated individually.

Table of Contents

Figure 7.1 Common Insulating Materials Used in Construction

This table lists the common types of materials used in construction, their forms and their major uses.

Physical Form	Type of Material	Major Uses
Powders	Diatomaceous earth; sawdust; silica aerogel	Filler
Loose fibrous materials	Cork granules; glass wool; mineral wool; shredded bark; vermiculite; perlite; mica	Flat areas such as air spaces above ceilings, adjacent to roof, to reduce conduction and convection
Batt or blanket insulation[a]	Glass wool; mineral wool; wood fibers, etc., enclosed by paper, cloth, wire mesh, or aluminum	Air spaces, particularly in vertical walls and flat surfaces, to reduce conduction and convection
Board, sheet, and integrant insulation[a]	Cork; fiber; paper pulp; cellular glass; glass fiber	Sheathing for walls, to increase strength as well as reduce heat loss; rigid insulation on roofs; perimeter insulation along edges of slab floors
Reflective insulation	Aluminum foil, often combined in layers with one or more adjoining air spaces or combined with sheets of paper	Used principally for all types of refrigerated or controlled environmental spaces
Special types of block and brick insulation and refractories	Insulation block made of cork, expanded glass, 85% magnesia, or vermiculite; insulating refractory block or brick made of diatomaceous earth or kaolin (clay); heavy refractories made of fire clay, magnesite, or silica	Special controlled temperature and high-temperature insulation problems; for example, refrigerator rooms, pipes, ducts, boilers, fire chambers of boilers, and chimneys
Foam-type insulation	Rigid boards; 2-component on-job application with special applicators; polystyrene and polyurethane	Interior-applied insulation for walls above or below grade; roof insulation; air space and perimeter insulation

[a]These are available with vapor barrier as part of insulation.

(courtesy Construction Materials, Caleb Hornbostel, John Wiley & Sons)

Figure 7.2   Thermal and Moisture Protection

This table lists the common types of thermal and moisture protection by category, their most common forms, and an explanation of how and why they are used in buildings. The Section Number and Broadscope refer to the CSI Masterformat Classification.

CSI Section Number	CSI Section Title	Broadscope Explanation

07100 -110	**Waterproofing**
	 Sheet Membrane Waterproofing
	    Bituminous Sheet Membrane
	      Waterproofing
	    Elastomeric/Plastomeric Sheet
	      Membrane Waterproofing
	    Modified Bituminous Sheet Membrane
	      Waterproofing
		Impervious membranes or coatings applied to walls, slabs, decks, or other surfaces subject to continuous or intermittent hydrostatic head or water immersion, and boards or coatings required for protection of waterproofing.

-120
 -130
 -140
 -145
	Fluid Applied Waterproofing
	 Bentonite Waterproofing
	 Metal Oxide Waterproofing
	 Cementitious Waterproofing
		Related Work:
		    Dampproofing: Section 07150.
		    Membrane Roofing: Section 07500.
		    Traffic Topping: Section 07570.

07150
 -160
 -175
 -180
	Dampproofing
	 Bituminous Dampproofing
	 Water Repellent Coatings
	 Cementitious Dampproofing
		Materials installed to provide resistance to moisture penetration through surfaces subject to high humidity, dampness, or direct water contact, but not subject to hydrostatic pressures.
		
		 Related Work:
		    Waterproofing: Section 07100.
		    Separate vapor retarders:
		      Section 07190.
		    Vapor retarders integral with
		      insulation: Section 07200.

07190
 -192
 -196
	Vapor and Air Retarders
	 Vapor Retarders
	 Air Retarders
		Bituminous, laminated, or plastic vapor or air retarders applied separately in wall, roof, or flow assemblies to provide resistance to moisture or air penetration.
		
		 Related Work:
		    Vapor retarders integral with
		      insulation: Section 07200.
		    Vapor resistant primers:
		      Section 09900.
		
		 Note: Vapor retarders under concrete slabs on grade are often specified in Section 03300. Vapor retarders under roof deck insulation are usually specified in Section 07500.

Figure 7.2 Thermal and Moisture Protection (continued)

CSI Section Number	CSI Section Title	Broadscope Explanation
07200 -210	**Insulation** Building Insulation Batt and Blanket Insulation Board Insulation Foamed-in-Place Insulation Loose Fill Insulation Sprayed Insulation	Thermal insulation including organic or inorganic insulation applied to walls, roofs, decks, perimeter of foundations, and under concrete slabs on grade. Vapor retarders integral with insulation are also included.
-220	Roof and Deck Insulation Asphaltic Perlite Concrete Deck Board Insulation	*Related Work:* Vapor and Air Retarders: Section 07190.
-240	Exterior Insulation and Finish Systems	Fireproofing: Section 07250. Acoustical insulation: Division 9. Mechanical insulation: Section 15250. Note: Insulation integral to masonry is often specified in Section 04200. Roof deck insulation is usually specified in Section 07500. Insulating sheathing is usually specified in Section 06100. Vapor retarders integral with insulation may be specified here.
07250 -255 -260 -265	**Fireproofing** Cementitious Fireproofing Thermal Barriers for Plastics Mineral Fiber Fireproofing Mineral Fiberboard Fireproofing Mineral Fiber Sprayed-on Fireproofing	Special coatings, mineral fiber, and cementitious coverings to provide fire resistance to structural members. Also includes firestopping.
-270 -275 -280	Firestopping Intumescent Mastic Fireproofing Magnesium Oxychloride Fireproofing	*Related Work:* Fire retardant treated lumber: Section 06300. Plaster fireproofing: Section 09200. Gypsum board fireproofing: Section 09250. Fire-resistant paints: Section 09800. Note: Penetration sealants are usually specified in this section. Firestopping and fire-safing for mechanical and electrical penetrations of fire-rated assemblies are usually specified in Divisions 15 and 16.
07300 -310	**Shingles and Roofing Tiles** Shingles Asphalt Shingles Fiberglass Shingles Metal Shingles Mineral Fiber-Cement Shingles Porcelain Enamel Shingles Slate Shingles Wood Shingles and Shakes	Lapped roofing shingles, shakes and roofing tiles, including underlayment and fastening products and methods. *Related Work:* Flashing and Sheet Metal: Section 07600. Roofing Specialties and Accessories: Section 07700.
-320	Roofing Tiles Clay Roofing Tiles Concrete Roofing Tiles Metal Roofing Tiles Mineral Fiber-Cement Roofing Tiles Plastic Roofing Tiles	

(continued on next page)

Figure 7.2   Thermal and Moisture Protection (continued)

CSI Section Number	CSI Section Title	Broadscope Explanation
07400 -410	**Preformed Roofing and Cladding/Siding** Preformed Roof and Wall Panels Preformed Roof Panels Preformed Wall Panels	Manufactured or prefabricated products of metal, wood, plywood, plastic, mineral fiber-cement, or composite materials forming the following types of wall, roof, or fascia surfaces: Roof and wall panels: Self-supporting and designed to support superimposed loads. Composite building panels: Self-supporting panels composed of various finishes and backup materials. Cladding/siding: Applied to sheathing, decking, or other supporting devices.
-420	Composite Building Panels Aggregate Coated Glass Fiber Reinforced Concrete Porcelain Enameled Tile Faced	
-460	Cladding/Siding Aluminum Composition Mineral Fiber-Cement Plastic Plywood Wood	*Related Work:* Sheet metal roofing: Section 07600. Louvers and Vents: Section 10200. Note: Cladding is a Canadian term. Wood, plywood composition, and plastic sidings are often specified in Section 06200.
07500 -510	**Membrane Roofing** Built-Up Bituminous Roofing Asphalt Coal Tar	Roofing membranes including surfacing materials, composition or elastomeric flashing, walk boards, and other items integral with roofing.
-515	Cold-Applied Bituminous Roofing Cold-Applied Mastic Roof Membrane Glass Fiber Reinforced Ashpalt Emulsion	*Related Work:* Waterproofing: Section 07100. Vapor retarders: Section 07190. Roof deck insulation: Section 07200. Flashing and Sheet Metal: Section 07600.
-520 -530	Prepared Roll Roofing Elastomeric/Plastomeric Sheet Roofing Fully Adhered Loose Laid/Ballasted Mechanically Attached	Roofing Specialties and Accessories: Section 07700.
-535	Modified Bitumen Sheet Roofing Modified Bitumen Composite Self-Adhering Modified Bitumen Reinforced Composite	Note: Roof deck insulation and vapor retarders are usually specified in this section.
-540 -550 -560	Fluid Applied Roofing Protected Membrane Roofing Roof Maintenance and Repairs Roof Moisture Survey Roofing Re-saturants	
07570	**Traffic Topping**	Surface-applied waterproof, elastomeric, or composition type membrane exposed to weather and suitable for normal or light duty traffic (foot or vehicle), but not intended for heavy industrial use. *Related Work:* Waterproofing: Section 07100. Special Flooring: Section 09700. Floor Treatment: Section 09780.

Figure 7.2 Thermal and Moisture Protection (continued)

CSI Section Number	CSI Section Title	Broadscope Explanation
07800 -810 -820 -840	**Skylights** Plastic Skylights 　Domed 　Pyramid 　Vaulted Metal-Framed Skylights 　Lean-to 　Motorized (Operable) 　Ridge 　Skydomes 　Vaulted Glass Block Skylights	Preformed plastic skylights and metal-framed skylight assemblies or structures with plastic or glass glazing. *Related Work:* 　Sloped glazing: Section 08900. 　Translucent wall and skylight system: Section 08900. 　Roof windows: Section 08650. 　Greenhouses: Section 13120. Note: Glazing for skylights may be specified in Section 08800.
07900 -910 -920	**Joint Sealers** Joint Fillers and Gaskets 　Compression Seals Sealants and Caulking	Elastomeric and nonelastomeric sealants, caulking compounds, compression seals, joint fillers, and related accessories. *Related Work:* 　Paving joint sealants: Section 02500. 　Glazing sealants: Section 08800. Note: Acoustical sealants are usually specified in Division 9. Penetration sealants are usually specified in Section 07270, Division 15, or Division 16.

Figure 7.3   Resistance Values of Construction Materials

This table lists the R Values for commonly used materials.

Material	R Value
Silver	0.00034
Copper	0.00039
Steel	0.0031
Granite	0.051
Slate shingles	0.096
Concrete per 1" (25.4 mm)	0.08
4" (101.6 mm) concrete block	0.71
8" (203.2 mm) concrete block	1.11
12" (304.8 mm) concrete block	1.28
Common brick per 1" (25.4 mm)	0.20
Face brick per 1" (25.4 mm)	0.11
Plate glass	0.18
Window glass	0.89
1/2" (12.7 mm) plasterboard	0.45
3/8" (11.03 mm) plywood	0.59
3/8" (11.03 mm) insulation board	2.06
1" (25.4 mm) polyurethane foam	6.25
2" (50.8 mm) polyurethane foam	14.29
Insulating glass, one air space	1.61
Insulating glass, two air spaces	2.13
Air space 3-1/2" (88.9 mm)	0.91
Air space 3/4" (19.05 mm)	0.92
Fiberglass, paper-faced	11.0
Built-up roofing, 4-ply slag	0.33
Asphalt strip shingles	0.44
White pine V-joint T & G	0.94
2" (50.8 mm) wood decking	2.03
3" (76.2 mm) wood decking	3.28
1" (25.4 mm) fiberboard	2.78
2" (50.8 mm) fiberboard	5.26
3" (76.2 mm) fiberboard	8.33
8" (203.2 mm) lapped beveled wood siding	0.81
10" (254.0 mm) lapped beveled wood siding	1.05
1" (25.4 mm) fiberglass perimeter insulation	4.30
Wood shingles	0.87

The resistance factor (R) is the reciprocal of conductivity (K), conductance (C), or overall heat transfer coefficient values (U):

$$R = \frac{1}{K} \text{ or } \frac{1}{C} \text{ or } \frac{1}{U}$$

(courtesy Construction Materials, Caleb Hornbostel, John Wiley & Sons)

Figure 7.4   Loose Fill Insulation

Square feet covered by a 40 lb. bag of mineral wool or glass fiber. (Includes area occupied by studding or joists.) Divide the factors into the area to determine the number of bags required.

Fill Depth	Fill Density		
	6 Lbs./C.F.	8 Lbs./C.F.	10 Lbs./C.F.
1"	85.0	63.8	51.0
2"	42.5	31.9	25.5
3"	28.4	21.3	17.0
3-1/2"	24.3	18.3	14.5
3-5/8"	23.5	17.6	14.1
4"	21.2	16.0	12.7
6"	14.2	10.6	8.5
10"	8.5	6.4	5.1

(courtesy, Estimating Tables for Home Building, Paul I. Thomas, Craftsman Book Company)

Figure 7.5   Insulation Required for the Exterior
                 Walls of a Typical One-Story Building

This table shows the number of 40 lb. bags of loose fill (mineral wool or glass fiber) insulation
required for exterior walls for various building sizes as described. Quantities listed are to the
nearest full bag. (Note: Gabled ends not included.)

Ground Floor Area S.F.	Ext. Wall Area S.F.	Density 6 Lbs./C.F. Depth of Fill						Density 8 Lbs./C.F. Depth of Fill						Density 10 Lbs./C.F. Depth of Fill					
		1"	2"	3"	3-½"	3-⅝"	4"	1"	2"	3"	3-½"	3-⅝"	4"	1"	2"	3"	3-½"	3-⅝"	4"
200	540	6	13	19	22	23	26	9	17	25	30	31	34	11	21	32	37	39	43
300	666	8	16	24	27	28	31	10	21	31	37	38	42	13	26	39	46	47	52
400	765	9	18	27	32	33	36	12	24	36	42	43	48	15	30	45	53	54	60
500	855	10	20	30	35	37	40	13	27	40	47	49	54	17	34	50	59	61	67
600	936	11	22	33	39	40	44	15	29	44	51	53	59	18	37	55	64	67	73
700	1008	12	24	36	42	43	48	16	32	47	55	57	63	20	40	59	69	72	79
800	1080	13	25	38	44	46	51	17	34	51	59	61	68	21	42	63	74	77	85
900	1143	14	27	40	47	49	54	18	36	54	63	65	72	23	45	67	78	82	90
1000	1206	14	28	43	50	52	57	19	38	57	66	69	76	24	47	71	83	86	95
1100	1269	15	30	45	52	54	60	20	40	60	70	72	79	25	50	75	87	90	99
1200	1323	16	31	47	54	57	62	21	42	62	73	75	83	26	52	78	91	94	104
1300	1377	16	32	49	57	59	65	22	43	65	76	78	86	27	54	81	95	98	108
1400	1431	17	34	50	59	61	67	22	45	67	79	81	90	28	56	84	98	102	112
1500	1476	17	35	52	61	63	70	23	46	69	81	84	93	29	58	87	101	105	116
1600	1530	18	36	54	63	65	72	24	48	72	84	87	96	30	60	90	105	109	120
1700	1575	19	37	56	65	67	74	25	49	74	86	89	99	31	62	93	108	112	123
1800	1620	19	38	57	67	69	76	26	51	76	89	92	102	32	64	95	111	115	127
1900	1665	20	39	59	69	71	78	26	52	78	91	95	105	33	65	98	114	118	131
2000	1710	20	40	60	70	73	80	27	54	80	94	97	107	34	67	100	117	122	134
2100	1746	21	41	62	72	74	82	27	55	82	96	99	109	34	68	103	120	124	137
2200	1791	21	42	63	74	77	84	28	56	84	98	102	113	35	70	105	123	128	141
2300	1827	22	43	64	75	78	86	29	57	86	100	104	115	36	72	107	125	130	143
2400	1872	22	44	66	77	80	88	29	59	88	103	106	117	37	73	110	129	133	147
2500	1908	23	45	67	79	82	90	30	60	90	105	109	120	38	75	112	131	136	150
2600	1944	23	46	69	80	83	92	31	61	91	107	111	122	38	76	114	133	138	153
2700	1980	23	47	70	82	85	93	31	62	93	109	113	124	39	78	116	136	141	155
2800	2016	24	47	71	83	86	95	32	63	95	111	115	126	40	79	118	138	143	158
2900	2052	24	48	72	84	88	97	32	65	96	113	117	129	40	81	121	141	146	161
3000	2097	25	49	74	86	90	99	33	66	98	115	119	132	41	82	123	144	149	165

(courtesy, *Estimating Tables for Home Building*, Paul I. Thomas, Craftsman Book Company)

Figure 7.6 Insulation Required for the Exterior Walls of a Typical One and One Half-Story Building

This table shows the number of 40 lb. bags of loose fill (mineral wool or glass fiber) insulation required for exterior walls for various building sizes as described. Quantities listed are to the nearest full bag. (Note: Gabled ends not included.)

Ground Floor Area S.F.	Ext. Wall Area S.F.	Density 6 Lbs./C.F. Depth of Fill						Density 8 Lbs./C.F. Depth of Fill						Density 10 Lbs./C.F. Depth of Fill					
		1"	2"	3"	3-½"	3-⅝"	4"	1"	2"	3"	3-½"	3-⅝"	4"	1"	2"	3"	3-½"	3-⅝"	4"
200	900	11	21	32	37	38	42	14	28	42	49	51	56	18	36	54	63	65	72
300	1110	12	26	39	46	47	52	17	35	51	60	62	68	22	44	66	77	80	88
400	1275	15	30	45	53	54	60	20	40	60	70	72	80	25	50	75	88	90	100
500	1425	17	34	50	59	61	67	22	45	66	77	80	88	28	56	84	98	102	112
600	1560	18	37	55	64	67	73	24	49	72	84	87	96	31	62	93	109	112	124
700	1680	20	40	59	69	72	79	26	53	78	92	94	104	33	66	99	116	120	132
800	1800	21	42	64	74	77	85	28	57	84	98	102	112	35	70	105	123	127	140
900	1905	22	45	67	78	81	90	30	60	90	105	109	120	37	74	111	130	134	148
1000	2010	24	47	71	83	86	95	32	63	96	112	116	128	39	78	117	137	141	156
1100	2115	25	50	75	87	90	100	33	66	99	116	120	132	41	82	123	144	149	164
1200	2205	26	52	78	91	94	104	35	69	105	123	127	140	43	86	129	151	156	172
1300	2295	27	54	81	95	98	108	36	72	108	126	131	144	45	90	135	158	163	180
1400	2385	28	56	84	98	102	112	37	75	111	130	134	148	47	94	141	165	170	188
1500	2460	29	58	87	101	105	116	39	77	117	137	141	156	48	96	144	168	174	192
1600	2550	30	60	90	105	109	120	40	80	120	140	145	160	50	100	150	175	181	200
1700	2625	31	62	93	108	112	124	41	82	123	144	149	164	51	102	153	179	185	204
1800	2700	32	64	95	111	115	127	42	85	126	147	153	169	53	106	159	186	192	212
1900	2775	33	65	98	114	118	131	44	87	132	154	160	176	54	108	162	189	196	216
2000	2850	34	67	101	117	122	134	45	89	135	158	163	180	56	112	168	196	203	224
2100	2910	34	69	103	120	124	137	46	91	138	161	167	184	57	114	171	200	207	228
2200	2985	35	70	105	123	127	140	47	93	141	165	170	188	59	118	177	207	214	236
2300	3045	36	72	108	125	130	143	48	95	144	168	174	192	60	120	180	210	218	240
2400	3120	37	73	110	129	133	147	49	98	147	172	178	196	61	122	183	214	221	244
2500	3180	37	75	112	131	136	150	50	100	150	175	181	200	62	124	186	217	225	248
2600	3240	38	76	114	133	138	153	51	102	153	179	185	204	64	128	192	224	232	256
2700	3300	39	78	117	136	141	155	52	104	156	182	189	208	65	130	195	228	236	260
2800	3360	40	79	119	138	143	158	53	105	159	186	192	212	66	132	198	231	239	264
2900	3420	40	81	121	141	146	161	54	107	162	189	196	216	67	134	201	235	243	268
3000	3495	41	82	123	144	149	165	55	110	165	193	199	220	69	138	207	242	250	276

(courtesy, Estimating Tables for Home Building, Paul I. Thomas, Craftsman Book Company)

Figure 7.7 Insulation Required for the Exterior Walls of a Typical Two-Story Building

This table shows the number of 40 lb. bags of loose fill (mineral wool or glass fiber) insulation required for exterior walls for various building sizes as described. Quantities listed are to the nearest full bag. (Note: Gabled ends not included.)

Ground Floor Area S.F.	Ext. Wall Area S.F.	Density 6 Lbs./C.F. Depth of Fill						Density 8 Lbs./C.F. Depth of Fill						Density 10 Lbs./C.F. Depth of Fill					
		1"	2"	3"	3-½"	3-⅝"	4"	1"	2"	3"	3-½"	3-⅝"	4"	1"	2"	3"	3-½"	3-⅝"	4"
200	1080	13	25	38	45	46	51	17	34	51	59	61	68	21	42	63	74	78	85
300	1332	16	31	47	55	57	63	21	42	63	73	76	84	26	52	78	92	95	105
400	1530	18	36	54	63	65	72	24	48	72	84	87	96	30	60	90	105	109	120
500	1710	20	40	60	70	73	80	27	53	80	93	97	107	33	66	99	117	121	133
600	1872	22	44	66	77	80	88	29	59	88	103	106	117	37	74	111	128	133	147
700	2016	24	47	71	83	86	95	32	63	95	111	115	127	40	80	120	138	143	158
800	2160	25	51	76	89	92	102	34	68	101	119	123	135	42	84	126	148	153	169
900	2286	27	54	81	94	98	108	36	72	108	125	130	143	45	90	135	157	163	178
1000	2412	28	57	85	99	103	114	38	76	114	133	137	151	47	95	141	166	172	189
1100	2538	30	60	90	105	108	120	40	80	120	140	144	160	50	100	150	175	181	200
1200	2646	31	62	93	109	113	124	42	83	124	145	150	166	52	104	155	182	188	207
1300	2754	32	65	97	113	118	130	43	86	130	151	157	173	54	108	162	189	196	217
1400	2862	34	67	101	118	122	135	45	90	135	157	163	180	56	112	168	197	204	225
1500	2952	35	69	104	122	126	139	46	93	139	162	168	185	58	116	173	203	210	231
1600	3060	36	72	108	126	131	144	48	96	144	168	174	192	60	120	180	210	218	240
1700	3150	37	74	111	130	134	148	49	99	148	173	179	197	62	124	185	216	223	247
1800	3240	38	76	114	133	138	152	51	101	152	177	184	203	63	126	190	222	230	253
1900	3330	39	78	118	137	142	157	52	105	157	183	189	209	65	130	196	229	237	261
2000	3420	40	80	121	141	146	161	54	107	161	188	194	214	67	134	201	235	243	268
2100	3492	41	82	123	144	149	164	55	110	164	192	199	219	69	138	206	240	248	274
2200	3582	42	84	126	147	153	168	56	112	168	196	203	224	70	140	211	246	254	281
2300	3654	43	86	129	151	156	172	57	115	172	201	208	229	72	144	215	251	260	287
2400	3744	44	88	132	154	160	176	59	117	176	205	213	235	73	146	220	257	266	293
2500	3916	45	90	135	157	163	180	60	120	180	209	217	239	75	150	225	262	271	299
2600	3888	46	91	137	160	166	183	61	122	183	213	221	244	76	152	228	267	276	305
2700	3960	47	93	140	163	169	186	62	124	186	217	225	248	78	156	232	272	282	311
2800	4032	47	95	142	166	172	190	63	126	189	221	229	253	79	158	237	277	287	316
2900	4104	48	97	145	169	175	193	64	129	193	225	233	258	81	162	242	282	292	322
3000	4194	49	99	148	173	179	197	66	131	197	230	238	263	82	164	247	288	298	329

(courtesy, Estimating Tables for Home Building, Paul I. Thomas, Craftsman Book Company)

Figure 7.8 Batt Insulation Quantities

To determine the number of insulation batts needed, multiply the factor listed for the size batt to be used by the area (square feet) to be insulated. Note: S.F. area includes that occupied by studding or joists.

Batt Sizes	No. of Batts per S.F.
15" x 24"	.38
15" x 48"	.19
19" x 24"	.30
19" x 48"	.15
23" x 24"	.25
23" x 48"	.125

(courtesy, Estimating Tables for Home Building, Paul I. Thomas, Craftsman Book Company)

Figure 7.9   Insulation Batts Required for Sidewalls

This table shows the number of insulating batts required for various building sizes as described. No deductions for openings have been included.

Ground Floor Area S.F.	1-Story						1-1/2-Story						2-Story					
	Batts 15″ x		Batts 19″ x		Batts 23″ x		Batts 15″ x		Batts 19″ x		Batts 23″ x		Batts 15″ x		Batts 19″ x		Batts 23″ x	
	24″	48″	24″	48″	24″	48″	24″	48″	24″	48″	24″	48″	24″	48″	24″	48″	24″	48″
200	205	103	162	81	135	68	342	171	270	135	225	113	410	205	324	162	270	135
300	253	127	200	100	166	83	422	211	332	166	276	138	506	253	400	200	333	166
400	290	145	230	115	191	96	484	242	382	191	319	161	580	290	460	230	383	192
500	325	162	256	128	214	107	542	271	428	214	356	178	650	325	512	256	428	214
600	356	178	280	140	234	117	592	296	468	234	390	195	712	356	562	281	468	234
700	383	192	302	151	252	126	638	319	504	252	420	210	766	383	604	302	504	252
800	410	205	324	162	270	135	648	342	540	270	450	225	820	410	648	324	540	270
900	434	217	343	171	286	143	724	362	572	286	476	238	868	434	686	343	572	286
1000	458	229	362	181	302	151	764	382	604	302	502	251	916	458	724	362	603	302
1100	482	241	380	190	317	159	804	402	634	317	529	265	964	482	762	381	635	318
1200	502	251	396	198	331	166	838	419	662	331	551	275	1006	503	794	397	662	331
1300	524	262	413	207	344	172	872	436	688	344	574	287	1046	523	826	413	688	344
1400	544	272	429	215	358	179	906	453	716	358	596	298	1088	544	858	429	715	358
1500	560	280	442	221	369	185	934	467	738	369	615	308	1122	561	886	443	738	369
1600	580	290	460	230	383	192	970	485	764	382	638	319	1162	581	918	459	765	383
1700	598	299	472	236	394	197	998	499	788	398	656	328	1198	599	944	472	788	394
1800	615	308	486	243	405	203	1026	513	810	405	675	338	1230	615	972	486	810	405
1900	632	316	500	250	416	208	1054	527	832	416	694	347	1266	633	1000	500	832	416
2000	650	325	513	257	428	214	1084	542	854	427	713	356	1300	650	1026	513	855	427
2100	664	332	524	262	437	218	1106	553	874	437	728	364	1326	663	1048	524	873	436
2200	680	340	537	268	448	224	1134	567	896	448	746	373	1360	680	1074	537	896	448
2300	694	347	548	274	457	228	1156	578	914	457	761	380	1388	694	1096	548	914	457
2400	711	356	562	281	468	234	1186	593	936	468	780	390	1422	711	1123	562	936	468
2500	726	363	572	286	477	238	1208	604	954	477	795	398	1450	725	1144	572	954	477
2600	738	369	584	292	486	243	1232	616	972	486	810	405	1476	738	1166	583	972	486
2700	752	376	594	297	495	248	1254	627	990	495	825	413	1504	752	1188	594	990	495
2800	766	383	604	302	504	252	1276	638	1008	504	840	420	1532	766	1210	605	1008	504
2900	780	390	616	308	513	256	1300	650	1026	513	855	428	1560	780	1232	616	1026	513
3000	797	398	630	315	524	262	1328	664	1048	524	874	437	1594	797	1258	629	1048	524

(courtesy, Estimating Tables for Home Building, Paul I. Thomas, Craftsman Book Company)

Figure 7.10 Insulation Batts Required for Ceilings

This table shows the number of insulating batts required for various building sizes as described.

Ground Floor Area S.F.	Batts 15" x		Batts 19" x		Batts 23" x	
	24"	48"	24"	48"	24"	48"
200	76	38	60	30	50	25
300	114	57	90	45	75	38
400	152	76	120	60	100	50
500	190	95	150	75	125	63
600	228	114	180	90	150	75
700	266	133	210	105	175	86
800	304	152	240	120	200	100
900	342	171	270	135	225	113
1000	380	190	300	150	250	125
1100	418	209	330	165	275	137
1200	456	228	360	180	300	150
1300	494	247	390	195	325	163
1400	532	266	420	210	350	175
1500	570	285	450	225	375	188
1600	608	304	480	240	400	200
1700	646	323	510	255	425	212
1800	684	342	540	270	450	225
1900	722	361	570	285	475	238
2000	760	380	600	300	500	250
2100	798	399	630	315	525	263
2200	836	418	660	330	550	275
2300	874	437	690	345	575	288
2400	912	456	720	360	600	300
2500	950	475	750	375	625	313
2600	988	494	780	390	650	325
2700	1026	513	810	405	675	338
2800	1064	532	840	420	700	350
2900	1102	551	870	435	725	363
3000	1140	570	900	450	750	375

(courtesy, Estimating Tables for Home Building, Paul I. Thomas, Craftsman Book Company)

Figure 7.11 Characteristics of Roof Insulation Materials

This table lists the various types of roofing insulation (across the top) and the desirable characteristics (down the left side), and compares how each type is rated for those characteristics.

Characteristics	Type of Insulating Board								
	Polyiso-cyanurate Foam	Polyure-thane Foam	Extruded Poly-styrene	Molded Poly-styrene	Cellular Glass	Mineral Fiber	Phenolic Foam	Wood Fiber	Glass Fiber
Impact resistant	G	G	G	F	G	E	G	E	F
Moisture resistant	E	G	E	G	E	G	E		
Fire resistant	E				E	E	E		E
Compatible with bitumens	E	G	F	F	E	E	E	E	G
Durable	E	E	E	E	E	G	F	E	E
Stable "k" value			E	E	E	E		E	E
Dimensionally stable	E	E	E	E	E	E	E	E	G
High thermal resistance	E	E	E	G	F	F	E	F	G
Available tapered slabs	Y	Y	Y	Y	Y	Y	Y	Y	Y
"R" value per in. thickness	7.20	6.25	4.76	3.85–4.35	2.86	2.78	8.30	1.75–2.00	4.00
Thicknesses available	1"–3"	1"–4"	1"–3½"	½"–24"	1½"–4"	¾"–3"	1"–4"	1"–3"	¾"–2½"
Density (lb./ft.³)	2.0	1.5	1.8–3.5	1.0–2.0	8.5	16–17	1.5	22–27	49
Remarks	Prone to "thermal drift"	Prone to "thermal drift" Should be overlaid with a thin layer of wood fiber, glass fiber or perlite board, with staggered joints.	Somewhat sensitive to hot bitumen & adhesive vapors Should be overlaid with a thin layer of wood fiber, glass fiber or perlite board, with staggered joints.	Somewhat sensitive to hot bitumen & adhesive vapors Should be overlaid with a thin layer of wood fiber, glass fiber or perlite board, with staggered joints.			Prone to "thermal drift" Relatively new & untested.	Expands with moisture — holds moisture.	Prone to damage from moisture infiltra-tion.

E = Excellent   G = Good   F = Fair
(*excerpted from ASHRAE 1985 Fundamentals Manual*)

Figure 7.12 Installation Time in Man-Hours for Roof Deck Insulation

Description	Man-Hours	Unit
Roof Deck Insulation		
Fiberboard to 2" Thick	.010	S.F.
Fiberglass	.008	S.F.
Fiberglass and Urethane Composite		
1-11/16" Thick	.008	S.F.
2" and 2-5/8" Thick	.010	S.F.
Foamglass		
1-1/2" and 2" Thick	.010	S.F.
3" and 4" Thick	.011	S.F.
Tapered	.013	S.F.
Perlite to 1-1/2" Thick	.010	S.F.
2" Thick	.011	S.F.
Perlite Urethane Composite		
To 1-3/4" Thick	.008	S.F.
2" Thick	.010	S.F.
2-1/2" and 3" Thick	.011	S.F.
Phenolic Foam to 1-3/4" Thick	.008	S.F.
2" to 3" Thick	.010	S.F.
Polystyrene		
1" Thick	.005	S.F.
2" Thick	.006	S.F.
3" Thick	.008	S.F.
Urethane Felt Both Sides		
1" and 1-1/2" Thick	.008	S.F.
2" to 3" Thick	.010	S.F.
Urethane and Gypsum Board Composite		
1-5/8" Thick	.008	S.F.
2" to 3" Thick	.010	S.F.
Sprayed Polystyrene or Urethane		
1" Thick	.031	S.F.
2" Thick	.051	S.F.
Lightweight Cellular Fill		
Portland Cement and Foaming Agent	1.120	C.Y.
Vermiculite or Perlite	1.120	C.Y.
Ready Mix		
2" Thick	.006	S.F.
3" Thick	.007	S.F.

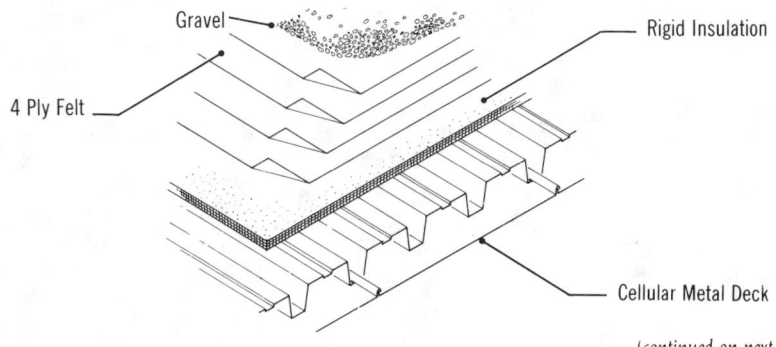

Gravel — Rigid Insulation — 4 Ply Felt — Cellular Metal Deck

(continued on next page)

Figure 7.12   Installation Time in Man-Hours for Roof Deck Insulation (continued)

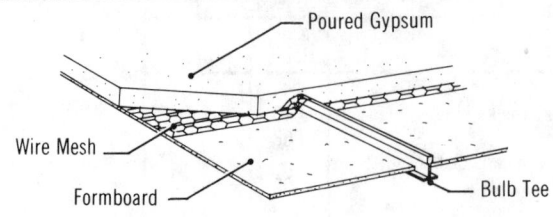

Poured-in-Place Gypsum Concrete and Formboard

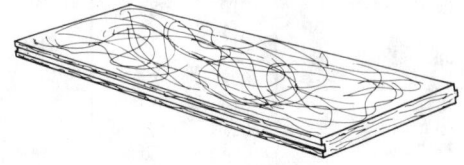

Tongue and Groove Fiberboard Insulation

Figure 7.13 Factors for Converting Horizontal Roof Dimensions to Inclined

This table is useful when a pitched roof is called for, and only the pitch and the horizontal dimension of a roof are known. Multiply the horizontal dimension by the conversion factor shown for the appropriate roof pitch.

Roof Slope	Approximate Angle	Factor
Flat	0	1.000
1 in 12	4.8	1.003
2 in 12	9.5	1.014
3 in 12	14.0	1.031
4 in 12	18.4	1.054
5 in 12	22.6	1.083
6 in 12	26.6	1.118
7 in 12	30.3	1.158
8 in 12	33.7	1.202
9 in 12	36.9	1.250
10 in 12	39.8	1.302
11 in 12	42.5	1.357
12 in 12	45.0	1.414
13 in 12	47.3	1.474
14 in 12	49.4	1.537
15 in 12	51.3	1.601
16 in 12	53.1	1.667
17 in 12	54.8	1.734
18 in 12	56.3	1.803
19 in 12	57.7	1.873
20 in 12	59.0	1.943
21 in 12	60.3	2.015
22 in 12	61.4	2.088
23 in 12	62.4	2.162

Figure 7.14   Roofing Quantities

The following tables show the quantities of roofing materials required for various pitched roofs for the building sizes described.

Roofing for a Roof Pitch of 1/8 (Figures include cutting and fitting waste.)										
Ground Floor Area	Roof Surface Area	Asphalt Strip Shingles	Wood Shingles 18" Long		Wood Shingles 24" Long		Fiberglass Shingles	Clay Roofing Tile	Number of Rolls of Saturated Felt Roofing Paper	
			4" Expo.	5-1/2" Expo.	6" Expo.	7-1/2" Expo.				
S.F.	S.F.	S.F.	S.F.	S.F.	S.F.	S.F.	S.F.	S.F.	15 Lb.	30 Lb.
200	206	226	278	226	284	226	226	216	.57	1.14
300	309	339	417	339	426	339	339	324	.85	1.70
400	412	452	556	452	568	452	452	432	1.13	2.27
500	515	565	695	565	710	565	565	540	1.42	2.84
600	618	678	834	678	852	678	678	648	1.70	3.40
700	721	791	973	791	994	791	791	756	1.98	3.97
800	824	904	1112	904	1136	904	904	864	2.27	4.54
900	927	1010	1251	1010	1278	1010	1010	972	2.55	5.10
1000	1030	1130	1390	1130	1420	1130	1130	1080	2.83	5.66
1100	1133	1243	1529	1243	1562	1243	1243	1188	3.11	6.22
1200	1236	1356	1668	1356	1704	1356	1356	1296	3.40	6.80
1300	1339	1469	1810	1469	1846	1469	1469	1404	3.68	7.36
1400	1442	1582	1946	1582	1990	1582	1582	1512	3.96	7.92
1500	1545	1695	2085	1695	2130	1695	1695	1620	4.25	8.50
1600	1648	1808	2224	1808	2272	1808	1808	1728	4.53	9.06
1700	1751	1921	2363	1921	2414	1921	1921	1836	4.81	9.62
1800	1854	2034	2500	2034	2556	2034	2034	1944	5.09	10.18
1900	1957	2147	2640	2147	2700	2147	2147	2052	5.38	10.76
2000	2060	2260	2780	2260	2840	2260	2260	2160	5.66	11.32
2100	2163	2378	2920	2378	2982	2378	2378	2268	5.94	11.88
2200	2266	2486	3058	2486	3124	2486	2486	2376	6.23	12.46
2300	2369	2600	3200	2600	3266	2600	2600	2484	6.51	13.02
2400	2472	2712	3336	2712	3410	2712	2712	2592	6.80	13.58
2500	2575	2825	3475	2825	3550	2825	2825	2700	7.08	14.16
2600	2678	2938	3615	2938	3692	2938	2938	2808	7.36	14.72
2700	2781	3051	3753	3051	3834	3051	3051	2916	7.64	15.28
2800	2884	3164	3892	3164	3976	3164	3164	3024	7.92	15.84
2900	2987	3277	4030	3277	4118	3277	3277	3132	8.21	16.42
3000	3090	3390	4170	3390	4260	3390	3390	3240	8.49	16.98

(courtesy, Estimating Tables for Home Building, Paul I. Thomas, Craftsman Book Company)

Figure 7.14   Roofing Quantities (continued)

Ground Floor Area	Roof Surface Area	Asphalt Strip Shingles	Wood Shingles 18" Long		Wood Shingles 24" Long		Fiberglass Shingles	Clay Roofing Tile	Number of Rolls of Saturated Felt Roofing Paper	
			4" Expo.	5-1/2" Expo.	6" Expo.	7-1/2" Expo.			15 Lb.	30 Lb.
S.F.	S.F.	S.F.	S.F.	S.F.	S.F.	S.F.	S.F.	S.F.		
200	212	234	286	234	292	234	234	224	.59	1.17
300	318	351	429	351	438	351	351	336	.88	1.76
400	424	468	472	468	584	468	468	448	1.17	2.34
500	530	585	715	585	730	585	585	560	1.46	2.93
600	636	702	858	702	876	702	702	672	1.76	3.51
700	742	819	1000	819	1022	819	819	784	2.05	4.10
800	848	936	1144	936	1170	936	936	896	2.34	4.68
900	954	1053	1287	1050	1314	1050	1050	1008	2.63	5.27
1000	1060	1170	1430	1170	1460	1170	1170	1120	2.93	5.85
1100	1166	1287	1573	1287	1610	1287	1287	1232	3.22	6.44
1200	1272	1404	1716	1404	1752	1404	1404	1344	3.51	7.02
1300	1378	1521	1859	1521	1900	1521	1521	1456	3.80	7.61
1400	1484	1638	2000	1638	2044	1638	1638	1568	4.10	8.19
1500	1590	1755	2145	1755	2190	1755	1755	1680	4.39	8.78
1600	1696	1872	2288	1872	2336	1872	1872	1792	4.68	9.36
1700	1802	1989	2430	1989	2480	1989	1989	1904	4.97	9.95
1800	1908	2106	2574	2106	2630	2106	2106	2016	5.27	10.53
1900	2014	2223	2717	2223	2774	2223	2223	2128	5.56	11.12
2000	2120	2340	2860	2340	2920	2340	2340	2240	5.85	11.70
2100	2226	2457	3000	2457	3066	2457	2457	2352	6.14	12.29
2200	2332	2574	3146	2574	3212	2574	2574	2464	6.44	12.87
2300	2438	2691	3290	2691	3358	2691	2691	2576	6.73	13.46
2400	2544	2808	3430	2808	3504	2808	2808	2688	7.02	14.04
2500	2650	2925	3575	2925	3650	2925	2925	2800	7.31	14.63
2600	2756	3042	3720	3042	3800	3042	3042	2912	7.61	15.21
2700	2862	3159	3860	3159	3942	3159	3159	3024	7.90	15.80
2800	2968	3276	4000	3276	4090	3276	3276	3136	8.19	16.38
2900	3074	3393	4150	3393	4234	3393	3393	3248	8.48	16.96
3000	3180	3510	4290	3510	4380	3510	3510	3360	8.78	17.55

Roofing for a Roof Pitch of 1/6 (Figures include cutting and fitting waste.)

(continued on next page)

Figure 7.14   Roofing Quantities (continued)

Roofing for a Roof Pitch of 5/24 (Figures include cutting and fitting waste.)										
Ground Floor Area	Roof Surface Area	Asphalt Strip Shingles	Wood Shingles 18" Long		Wood Shingles 24" Long		Fiberglass Shingles	Clay Roofing Tile	Number of Rolls of Saturated Felt Roofing Paper	
			4" Expo.	5-1/2" Expo.	6" Expo.	7-1/2" Expo.				
S.F.	S.F.	S.F.	S.F.	S.F.	S.F.	S.F.	S.F.	S.F.	15 Lb.	30 Lb.
200	218	240	294	240	300	240	240	228	.60	1.20
300	327	360	441	360	450	360	360	342	.90	1.80
400	426	480	588	480	600	480	480	456	1.20	2.40
500	545	600	735	600	750	600	600	570	1.50	3.00
600	654	720	882	720	900	720	720	684	1.80	3.60
700	763	840	1029	840	1050	840	840	798	2.10	4.20
800	872	960	1176	960	1200	960	960	912	2.40	4.80
900	981	1080	1323	1080	1350	1080	1080	1026	2.70	5.40
1000	1090	1200	1470	1200	1500	1200	1200	1140	3.00	6.00
1100	1199	1320	1617	1320	1650	1320	1320	1254	3.30	6.60
1200	1308	1440	1764	1440	1800	1440	1440	1368	3.60	7.20
1300	1417	1560	1911	1560	1950	1560	1560	1482	3.90	7.80
1400	1526	1680	2058	1680	2100	1680	1680	1596	4.20	8.40
1500	1635	1800	2205	1800	2250	1800	1800	1710	4.50	9.00
1600	1744	1920	2352	1920	2400	1920	1920	1824	4.80	9.60
1700	1853	2040	2500	2040	2550	2040	2040	1938	5.10	10.20
1800	1962	2160	2646	2160	2700	2160	2160	2052	5.40	10.80
1900	2071	2280	2793	2280	2850	2280	2280	2166	5.70	11.40
2000	2180	2400	2940	2400	3000	2400	2400	2280	6.00	12.00
2100	2289	2520	3087	2520	3150	2520	2520	2394	6.30	12.60
2200	2398	2640	3234	2640	3300	2640	2640	2508	6.60	13.20
2300	2507	2760	3381	2760	3450	2760	2760	2622	6.90	13.80
2400	2616	2880	3528	2880	3600	2880	2880	2736	7.20	14.40
2500	2725	3000	3675	3000	3750	3000	3000	2850	7.50	15.00
2600	2834	3120	3822	3120	3900	3120	3120	2964	7.80	15.60
2700	2943	3240	3969	3240	4050	3240	3240	3078	8.10	16.20
2800	3052	3360	4116	3360	4200	3360	3360	3192	8.40	16.80
2900	3161	3480	4263	3480	4350	3480	3480	3306	8.70	17.40
3000	3270	3600	4410	3600	4500	3600	3600	3420	9.00	18.00

(courtesy, Estimating Tables for Home Building, Paul I. Thomas, Craftsman Book Company)

Figure 7.14 Roofing Quantities (continued)

Roofing for a Roof Pitch of 1/4 (Figures include cutting and fitting waste.)										
Ground Floor Area	Roof Surface Area	Asphalt Strip Shingles	Wood Shingles 18" Long		Wood Shingles 24" Long		Fiberglass Shingles	Clay Roofing Tile	Number of Rolls of Saturated Felt Roofing Paper	
			4" Expo.	5-1/2" Expo.	6" Expo.	7-1/2" Expo.			15 Lb.	30 Lb.
S.F.	S.F.	S.F.	S.F.	S.F.	S.F.	S.F.	S.F.	S.F.		
200	224	246	300	246	310	246	246	236	.62	1.24
300	336	369	450	369	465	369	369	354	.93	1.86
400	448	492	600	492	620	492	492	472	1.24	2.48
500	560	615	750	615	775	615	615	590	1.55	3.10
600	672	738	900	738	930	738	738	708	1.86	3.72
700	784	861	1050	861	1085	861	861	826	2.17	4.34
800	896	984	1200	984	1240	984	984	944	2.48	4.96
900	1008	1107	1350	1107	1395	1107	1107	1062	2.79	5.58
1000	1120	1230	1500	1230	1550	1230	1230	1180	3.10	6.20
1100	1232	1353	1650	1353	1705	1353	1353	1298	3.41	6.82
1200	1344	1476	1800	1476	1860	1476	1476	1416	3.72	7.44
1300	1456	1600	1950	1600	2015	1600	1600	1534	4.03	8.06
1400	1568	1722	2100	1722	2170	1722	1722	1652	4.34	8.68
1500	1680	1845	2250	1845	2325	1845	1845	1770	4.65	9.30
1600	1793	1968	2400	1968	2480	1968	1968	1888	4.96	9.92
1700	1904	2091	2550	2091	2635	2091	2091	2006	5.27	10.54
1800	2016	2214	2700	2214	2790	2214	2214	2124	5.58	11.16
1900	2128	2337	2850	2337	2945	2337	2337	2242	5.89	11.78
2000	2240	2460	3000	2460	3100	2460	2460	2360	6.20	12.40
2100	2352	2583	3150	2583	3255	2583	2583	2478	6.51	13.02
2200	2464	2706	3300	2706	3410	2706	2706	2596	6.82	13.64
2300	2576	2829	3450	2829	3565	2829	2829	2714	7.13	14.26
2400	2688	2952	3600	2952	3720	2952	2952	2832	7.44	14.88
2500	2800	3075	3750	3075	3875	3075	3075	2950	7.75	15.50
2600	2912	3198	3900	3198	4030	3198	3198	3068	8.06	16.12
2700	3024	3321	4050	3321	4185	3321	3321	3186	8.37	16.74
2800	3136	3440	4200	3440	4340	3440	3440	3304	8.68	17.36
2900	3248	3567	4350	3567	4495	3567	3567	3422	9.00	18.00
3000	3360	3690	4500	3690	4650	3690	3690	3540	9.30	18.60

(continued on next page)

Figure 7.14   Roofing Quantities (continued)

			Wood Shingles 18" Long		Wood Shingles 24" Long				Number of Rolls of	
Ground Floor Area	Roof Surface Area	Asphalt Strip Shingles	4" Expo.	5-1/2" Expo.	6" Expo.	7-1/2" Expo.	Fiberglass Shingles	Clay Roofing Tile	Saturated Felt Roofing Paper	
S.F.	S.F.	S.F.	S.F.	S.F.	S.F.	S.F.	S.F.	S.F.	15 Lb.	30 Lb.
200	240	264	324	264	332	264	264	252	.66	1.32
300	360	396	486	396	498	396	396	378	.99	1.98
400	480	528	648	528	664	528	528	504	1.32	2.64
500	600	660	810	660	830	660	660	630	1.65	3.30
600	720	792	972	792	1000	792	792	756	1.98	3.96
700	840	924	1134	924	1162	924	924	882	2.31	4.62
800	960	1056	1296	1056	1328	1056	1056	1008	2.64	5.28
900	1080	1188	1458	1188	1494	1188	1188	1134	2.97	5.94
1000	1200	1320	1620	1320	1660	1320	1320	1260	3.30	6.60
1100	1320	1452	1782	1452	1826	1452	1452	1386	3.63	7.26
1200·	1440	1584	1944	1584	1992	1584	1584	1512	3.96	7.92
1300	1560	1716	2106	1716	2158	1716	1716	1638	4.29	8.58
1400	1680	1848	2268	1848	2324	1848	1848	1764	4.62	9.24
1500	1800	1980	2430	1980	2490	1980	1980	1890	4.95	9.90
1600	1920	2112	2592	2112	2656	2112	2112	2016	5.28	10.56
1700	2040	2244	2754	2244	2822	2244	2244	2142	5.61	11.22
1800	2160	2376	2916	2376	2990	2376	2376	2268	5.94	11.88
1900	2280	2508	3078	2508	3154	2508	2508	2394	6.27	12.54
2000	2400	2640	3240	2640	3320	2640	2640	2520	6.60	13.20
2100	2520	2772	3400	2772	3486	2772	2772	2646	6.93	13.86
2200	2640	2904	3564	2904	3652	2904	2904	2772	7.26	14.52
2300	2760	3036	3726	3036	3820	3036	3036	2898	7.59	15.18
2400	2880	3168	3888	3168	3984	3168	3168	3024	7.92	15.80
2500	3000	3300	4050	3300	4150	3300	3300	3150	8.25	16.50
2600	3120	3432	4212	3432	4316	3432	3432	3276	8.58	17.16
2700	3240	3564	4374	3564	4482	3564	3564	3402	8.91	17.82
2800	3360	3696	4536	3696	4648	3696	3696	3528	9.24	18.48
2900	3480	3828	4700	3828	4814	3828	3828	3654	9.57	19.14
3000	3600	3960	4860	3960	4980	3960	3960	3780	9.90	19.80

Table heading: Roofing for a Roof Pitch of 1/3 (Figures include cutting and fitting waste.)

(courtesy, Estimating Tables for Home Building, Paul I. Thomas, Craftsman Book Company)

Figure 7.14 Roofing Quantities (continued)

Roofing for a Roof Pitch of 1/2 (Figures include cutting and fitting waste.)										
Ground Floor Area	Roof Surface Area	Asphalt Strip Shingles	Wood Shingles 18" Long		Wood Shingles 24" Long		Fiberglass Shingles	Clay Roofing Tile	Number of Rolls of Saturated Felt Roofing Paper	
			4" Expo.	5-1/2" Expo.	6" Expo.	7-1/2" Expo.				
S.F.	S.F.	S.F.	S.F.	S.F.	S.F.	S.F.	S.F.	S.F.	15 Lb.	30 Lb.
200	284	312	384	312	392	312	312	298	.78	1.55
300	426	468	576	468	588	468	468	447	1.16	2.32
400	568	624	768	624	784	624	624	600	1.55	3.10
500	710	780	960	780	980	780	780	745	1.94	3.88
600	852	936	1152	936	1176	936	936	894	2.33	4.65
700	994	1092	1344	1092	1372	1092	1092	1043	2.71	5.43
800	1136	1248	1536	1248	1568	1248	1248	1192	3.10	6.20
900	1278	1404	1728	1404	1764	1404	1404	1340	3.49	6.98
1000	1420	1560	1920	1560	1960	1560	1560	1490	3.88	7.75
1100	1562	1716	2112	1716	2156	1716	1716	1639	4.26	8.53
1200	1704	1872	2304	1872	2352	1872	1872	1788	4.65	9.30
1300	1846	2028	2496	2028	2548	2028	2028	1937	5.04	10.08
1400	1990	2184	2688	2184	2744	2184	2184	2086	5.43	10.85
1500	2130	2340	2880	2340	2940	2340	2340	2235	5.81	11.63
1600	2272	2500	3072	2500	3136	2500	2500	2384	6.20	12.40
1700	2414	2652	3264	2652	3332	2652	2652	2533	6.59	13.18
1800	2556	2810	3456	2810	3528	2810	2810	2682	6.98	13.95
1900	2700	2964	3648	2964	3724	2964	2964	2830	7.36	14.93
2000	2840	3120	3840	3120	3920	3120	3120	2980	7.75	15.50
2100	2982	3276	4032	3276	4116	3276	3276	3130	8.14	16.28
2200	3124	3432	4224	3432	4312	3432	3432	3278	8.53	17.05
2300	3266	3590	4416	3590	4508	3590	3590	3427	8.91	17.83
2400	3410	3744	4608	3744	4704	3744	3744	3576	9.30	18.60
2500	3550	3900	4800	3900	4900	3900	3900	3725	9.69	19.38
2600	3692	4056	4992	4056	5096	4056	4056	3874	10.08	20.15
2700	3834	4212	5184	4212	5292	4212	4212	4023	10.46	20.93
2800	3976	4368	5376	4638	5488	4368	4368	4172	10.85	21.70
2900	4118	4524	5568	4524	5684	4524	4524	4321	11.24	22.48
3000	4260	4680	5760	4680	5880	4680	4680	4470	11.63	23.25

Figure 7.15   Waste Values for Roof Styles

This table lists the approximate values for waste that should be added to the net area for materials of various roof types.

Roof Shape	Plain	Cut-up
Gable	10	15
Hip	15	20
Gambrel	10	20
Gothic	10	15
Mansard (sides)	10	15
Porches	10	—

Figure 7.16   Roofing Materials and Nomenclature

These graphics illustrate various types of roofing systems.

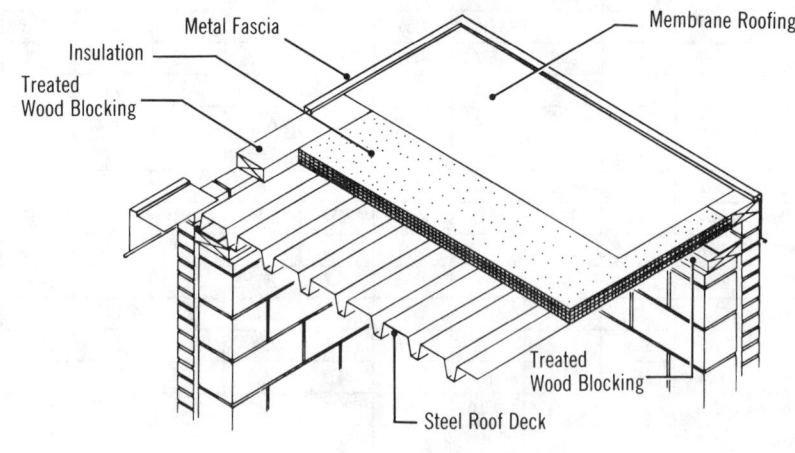

Single Ply Roofing

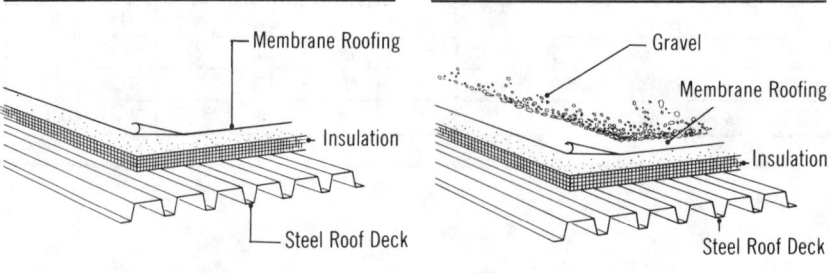

Membrane Roofing-Adhered

Membrane Roofing-Ballasted

Figure 7.16 Roofing Materials and Nomenclature (continued)

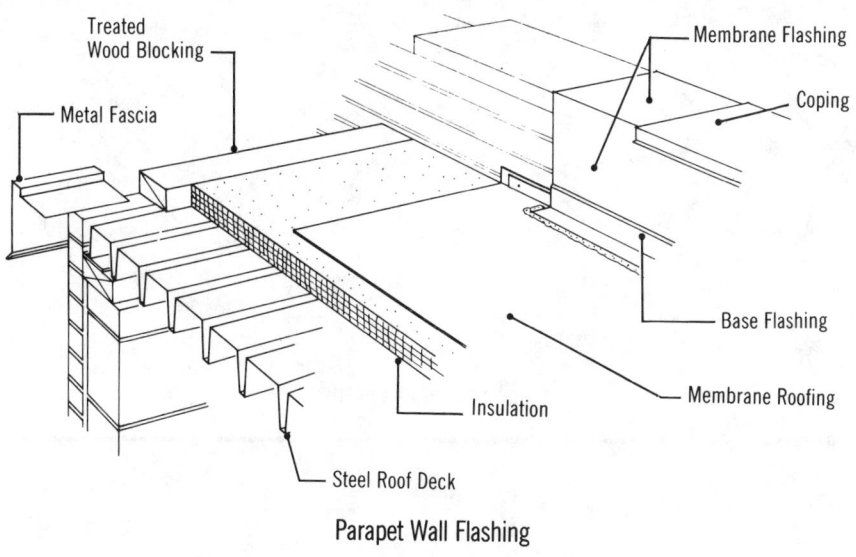

Parapet Wall Flashing

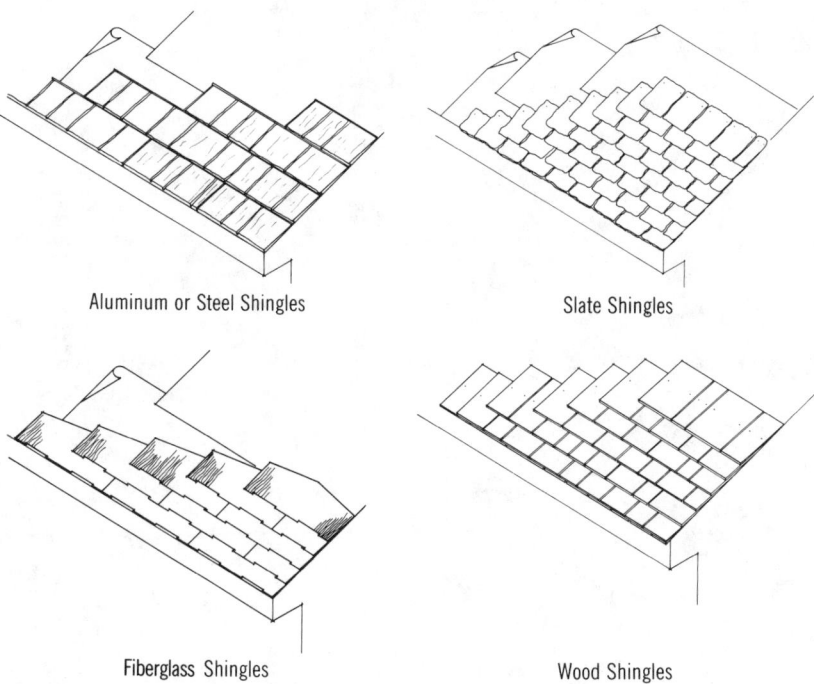

Aluminum or Steel Shingles

Slate Shingles

Fiberglass Shingles

Wood Shingles

Water Shed Roof Coverings

(continued on next page)

Figure 7.16 Roofing Materials and Nomenclature (continued)

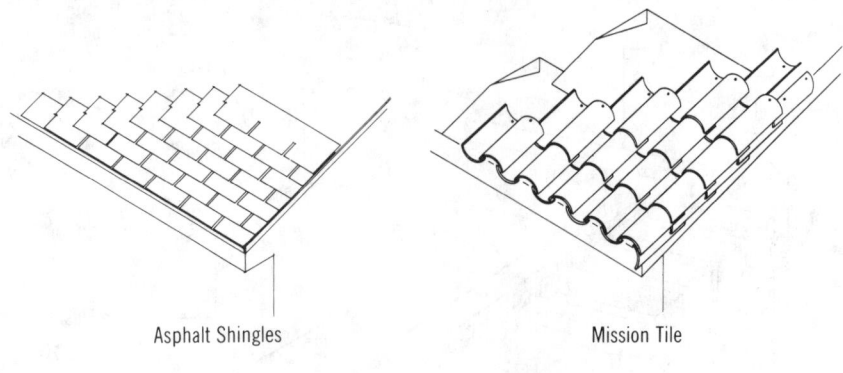

Asphalt Shingles   Mission Tile

Water Shed Roof Coverings

Figure 7.17 Commercial Roofing Systems

This.table lists and compares the more common types of commercial roofing systems. This table is not intended to be a comprehensive listing of available systems but an overview of the most common types.

ROOF SYSTEM TYPE	Low initial cost	Many re-entrant corners & penetrations	High traffic roofs or areas	Resistant to oils, solvents	Mid-rise or high-rise use (over 5 stories)	Ease of maintenance	Compatible with light steel decks	Long record of serviceability	REMARKS
Built-up roof, Asphalt	3	5	5	3	1	5	1	5	Traditional materials.
Built-up roof, Coal Tar Pitch	2	4	5	3	1	5	1	5	Traditional materials.
Modified Bitumen (APP modifier)	2	5	4	4	3	4	2	3	Torch-applied materials.
Modified Bitumen (SBS modifier)	2	5	4	4	3	5	2	3	Mop-applied materials.
Single-ply, ballasted*	4	2	2	2	1	2	1	3	Simple & inexpensive.
Single-ply, adhered*	2	4	2	2	4	3	4	3	High growth in this segment.
Single-ply, mechanically attached*	3	3	2	2	4	3	4	3	Rapid development in this segment.
Non-vulcanized elastomers, mechanically attached	2	2	2	4	4	3	4	3	Relatively small market segment.
Preformed Metal	2	2	2	4	2	2	3	5	Durable materials.
Spray-applied Polyurethane	2	5	1	2	3	1	2	1	Requires training and careful quality control.

*Characteristics listed are for EPDM single-ply sheet; other elastomers have similar characteristcs.

A rating of: 1 = poor; not recommended
2 = fair
3 = average
4 = better than average
5 = excellent; recommended

Figure 7.18   Roofing Material Quantities — Asphalt Rolls and Sheets

This table shows some of the characteristics of asphalt roofing materials.

Product	Approximate Shipping Weight		Squares per Package	Length	Width	Selvage	Ex- posure	ASTM Fire & Wind Ratings
	per Roll	per Square						
Mineral Surface Roll	75# to 90#	75# to 90#	1	36' to 38'	36"	0" to 4"	32" to 34"	C
Mineral Surface Roll Double Coverage	55# to 70#	110# to 140#	1/2	36'	36"	19"	17"	C
Smooth Surface Roll	50# to 86#	40# to 65#	1 to 2	36' to 72'	36"	N/A	34"	None
Saturated Felt Underlayment (non-perforated)	35# to 60#	11# to 30#	2 to 4	72' to 144'	36"	N/A	17" to 34"	*

*May be component in a complete fire-rated system. Check with manufacturer.

(courtesy Asphalt Roofing Manufacturers Association)

Figure 7.19   Roofing Material Quantities — Built-Up Systems

This table shows typical quantities required to apply 100 S.F. (1 square) of built-up asphalt roofing.

Number of Plies	Rolls of 15 Lb. Saturated Felt (432 S.F.)	Pounds of Asphalt	Pounds of Slag	Pounds of Gravel
3	3/4	150	300	400
4	1	180	300	400
5	1-1/4	230	300	400
3 (1 dry ply)	3/4	120	300	400
4 (2 dry plies)	1	120	300	400
5 (2 dry plies)	1-1/4	150	300	400

(For smooth-top finish without slag or gravel, deduct 20 lbs. of asphalt per 100 S.F. of roofing. When 30 lb. saturated felt is used as a first ply, add 1/4 of a roll of 15 lb. felt.)

(courtesy How to Estimate Building Losses and Construction Costs, Paul I. Thomas, Prentice-Hall)

Figure 7.20 Single Ply Roofing Guide

This table lists the common generic classification of single ply roofing materials, compatible substrates, methods of attachment to the roof deck, and methods for sealing the material.

	Generic Materials (Classification)	Compatible Substrates						Attachment Method				Sealing Method				
		Slip-sheet required	Concrete	Existing asphalt membrane	Insulation board	Plywood	Spray urethane foam	Adhesive	Fully adhered	Loosely laid/ballast	Partially-adhered	Adhesive	Hot air gun	Self sealing	Solvent	Torch heating
Thermo Setting	EPDM (Ethylene propylene diene monomer)	X	X	X	X	X	X	X	X	X	X	X		X	X	
	Neoprene (Synthetic rubber)	X	X		X	X		X	X	X		X				
	PIB (Polyisobutylene)	X	X	X	X	X	X	X		X		X	X			X
Thermo Plastic	CSPE (Chlorosulfenated polyethylene)	X	X		X	X	X	X	X	X	X	X	X			
	CPE (Chlorinated polyethylene)	X	X		X	X			X	X	X	X				
	PVC (Polyvinyl chloride)	X	X		X	X	X			X	X	X			X	
Composites	Glass reinforced EPDM/neoprene	X	X		X	X	X			X		X				
	Modified bitumen/polyester	X		X	X	X			X			X	X			X
	Modified bitumen/polyethylene & aluminum	X	X		X	X		X	X			X				X
	Modified bitumen/polyethylene sheet	X	X		X	X				X			X			X
	Modified CPE				X	X			X			X				
	Non woven glass reinforced PVC							X	X	X		X				
	Nylon reinforced PVC		X		X	X				X			X		X	
	Nylon reinforced/butylorneoprene	X							X				X		X	
	Polyester reinforced CPE	X	X	X	X	X	X			X	X	X	X			X
	Polyester reinforced PVC	X	X		X	X	X			X	X	X	X		X	
	Rubber asphalt/plastic sheet	X	X	X	X	X			X					X		

Figure 7.21   Single Ply Membrane Table

This table lists some of the manufacturers of single ply roofing membrane and the characteristics of each type. This is not meant to be a comprehensive listing, but rather a representative listing of the more common types of single ply roofing materials.

Manufacturer	Brand Name	Product Type*	Oldest U.S. Installation (yrs.)	Color****	Thickness (mils)	Weight (Lbs./S.F.)	Roll width range	Roll length range	Reinforcement	Tensile strength (psi) (ASTM D472 & D751)	Elongation (%) ASTM D472 & D751	Low temp. flexibility (%F) (ASTM D746)	Dimensional stability (%)	Contaminants to avoid***	Incompatible with (substrate)**
Carlisle SynTec Systems	Sure-Seal	EPDM	26	w/b	45	.28	54"/50'	50/200'	n	1640	500	-75	<2%	ABC	14₈
Cooley Roofing Systems	Cool Top	CPE	11	w	40	.28	72"	108'	y	135	29	-40	1%	DF	
Duro-Last Roofing	Duro-Last	Reinf. CP	9	w/t	35	.26	†	†	y	†	27	-40	1%	D	
Firestone Bldg. Products	Rubber Gard	EPDM	8	b	45	.28	60"/50'	100/200'	n	1305	300	-49	<2%	DE	
Goodyear Tire & Rubber	Versigard	EPDM	23	w/b	45	.28	54"/50'	to 150'	n	1300	350	-75	<2%	E	37
Gen. Corp. Polymer Prod.	Gen Flex	EPDM	7	w/b	45	.26	42"/50'	100'	n	1500	425	-67	<2%	DE	
Kelly Energy Systems	Premium Whaleskin	EPDM	22	w/b	45	.28	10'/20'	50/100'	n	1400	300	-75	<1%	E	
Kelly Energy Systems	Kelly CPA	CPA	8	w	50	.32	72"	75'	y	250	25	-40	<5%	GHEB	
Manville Roofing Systems	SPM Roof Systems	EPDM	9	b	45	.26	to 50'	150'	n	1305	300	-49	<2%	BCG	358
Republic Powdered Metals	Geoflex	PIB	10	w/b	100	.57	42"	60'	y	550	410	-40	<1%	BCG	
Sarnafil	Sarnafil	PVC	12	c	48	.33	78"	65'	y	230	20	-40	<<1%	BGH	1345₈
Seal-Dry	System 7000	CPA	5	w	40	.25	to 20'	100'	y	400	35	-30	<<1%	†	8
Seaman	Fiber Tite	EIP	20	†	33	.25	58"/20'	to 100'	y	8500	30	-30	<1%	†	
J.P. Stevens & Co.	Hi-Tuff	CSPE	12	w	45	.29	64"	80'	y	1400	400	-40	<1%	CD	
Trocal Roofing Systems	Trocal Membrane	PVC	15	w	60	.49	†	†	y	2400	250	-40	<1.5%	(1)	34

Figure 7.21 Single Ply Membrane Table (continued)

		Carlisle SynTec Systems	Cooley Roofing Systems	Duro-Last Roofing	Firestone Bldg. Products	Goodyear Tire & Rubber	Gen. Corp. Polymer Prod.	Kelly Energy Systems	Kelly Energy Systems	Manville Roofing Systems	Republic Powdered Metals	Sarnafil	Seal-Dry	Seaman	J.P. Stevens & Co.	Trocal Roofing Systems
Warranty & Remarks†		5, 10, 15 yr. warranty avail.; UL Class A & B; FM I-60 & I-90	10 yr. warr.; UL Class A; FM I-60 & I-90	†pretab. up to 2500 S.F. sections; ASTM D882:7200 psi; 15, 20 yr. warr.; UL Class A; FM I-90	5, 10, 15, 20 yr. warr.; UL Class A; FM I-60, I-90	5, 10, 15, 20 yr. warr.; UL Class A; FM I-90	5, 10, 15 yr. warr.; UL Class A; FM I-90	6, 11, 15, 21 yr. mat'l.; 6, 11 yr. lab/mat.; UL Class A; FM Class 1	15 yr. mat'l.; 5, 10 yr. labor; UL Class A & B; FM I-90	5, 10, 15 yr. warr.; UL Class A & B; FM I-60, I-90	10 yr. warr.; UL Class A; FM I-90	10 yr. warr.; UL Class A; FM I-90	†contact factory; 15 yr. warr.; UL (A),FM I-60, I-90	†contact factory; 5, 10 yr. warr.; UL(A); FM I-90	5, 10, 15 yr. warr.; UL Class A; FM I-90	†varies with membrane type; 10 yr. warr.; (1) contact manufacturer, UL Class A; FM I-90
Lap Joining Method	Factory dielectrics											•	•			
	Pre-fab, self-seal							•								
	Solvent weld							•								•
	Hot air weld		•	•				•		•		•	•	•	•	•
	Splice tape	•			•		•	•								
	Adhesive/Sealant	•			•	•	•	•		•						
Installation Method	Protected membrane	•	•		•	•	•	•		•	•	•		•	•	•
	Fully adhered	•			•	•	•	•		•	•			•		
	Mechanically attached	•	•	•	•	•	•	•	•	•	•	•	•	•	•	•
	Loose laid/ballasted	•	•	•	•	•	•			•	•	•		•	•	•

*EPDM: ethylene propylene diene monomer; CPE: chlorinated polyethylene; CSPE: chlorosulfonated polyethylene; CP (CPA): Copolymer alloy; PIB: polysobutylene; PVC: polyvinyl chloride; EIP: ethylene interpolymer alloy

**Substrates & insulation: 1) perlite board; 2) urethane; 3) phenolic; 4) wood fiber; 5) fiberglass; 6) polystyrene; 7) isocyanurate; 8) cellular glass; 9) existing asphaltic roofing

***Contaminants: A) acids; B) oils; C) solvents; D) hydrocarbons; E) petroleum-based substances; F) oxidizers; G) coal tar; H) asphalt

****Colors: b: black; w: white; t: tan; g: gray; c: colors

Note: This chart is not intended to endorse any product or manufacturer, but merely to illustrate the range of materials available and their various properties. It is not in any way a complete listing of products available, or intended to replace thorough research by the specifier of available systems and materials. Due to the rate of product development and improvement in the single-ply market, entries may be superseded.

Figure 7.22   Roofing Material Quantities — Shingles

This table describes various types of roofing shingles and some of their characteristics and expected coverages.

Product	Configuration	Approxi-mate Shipping Weight per Square	Shingles per Square	Bundles per Square	Width	Length	Ex-posure	ASTM* Fire & Wind Ratings
Self-Sealing Random-Tab Strip Shingle 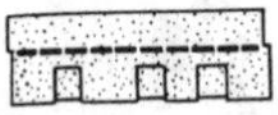 Multi-Thickness	Various Edge, Surface Texture & Application Treatments	240# to 360#	64 to 90	3, 4 or 5	11-1/2" to 14"	36" to 40"	4" to 6"	A or C - Many Wind Resistant
Self-Sealing Random-Tab Strip Shingle  Single Thickness	Various Edge, Surface Texture & Application Treatments	240# to 300#	65 to 80	3 or 4	12" to 13-1/4"	36" to 40"	4" to 5-5/8"	A or C - Many Wind Resistant
Self-Sealing Square-Tab Strip Shingle  Three-Tab	3 Tab or 4 Tab	200# to 300#	65 to 80	3 or 4	12" to 13-1/4"	36" to 40"	5" to 5-5/8"	A or C - All Wind Resistant
Self-Sealing Square-Tab Strip Shingle 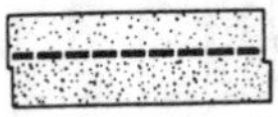 No Cut-Out	Various Edge and Surface Texture Treatments	200# to 300#	65 to 81	3 or 4	12" to 13-1/4"	36" to 40"	5" to 5 5/8"	A or C - All Wind Resistant
Individual Interlocking Shingle  Basic Design	Several Design Variations	180# to 250#	72 to 120	3 or 4	18" to 22-1/4"	20" to 22-1/2"	—	A or C - Many Wind Resistant

Other types available from some manufacturers in certain areas of the country.
Consult your Regional Asphalt Roofing Manufacturers Association manufacturer.
*American Society for Testing and Materials
(courtesy Asphalt Roofing Manufacturers Association)

Figure 7.23 Roofing Material Quantities — Red Cedar Shingles

This table shows expected coverages and materials required for the installation of various sizes of red cedar shingles.

Covering Capacities and Approximate Nail Requirements of Certigrade Red Cedar Shingles

Shingle Exposure in Inches	No. 1 Grade Sixteen Inch Shingles — Nail Size: 3d, 1-1/4" Long				No. 1 Grade Eighteen Inch Shingles — Nail Size: 3d, 1-1/4" Long				No. 1 Grade Twenty-Four Inch Shingles — Nail Size: 4d, 1-1/2" Long			
	Four-Bundle Square		One Bundle		Four-Bundle Square		One Bundle		Four-Bundle Square		One Bundle	
	Coverage in S.F.	Pounds Nails	Coverage in S.F.	Pounds Nails	Coverage in S.F.	Pounds Nails	Coverage in S.F.	Pounds Nails	Coverage in S.F.	Pounds Nails	Coverage in S.F.	Pounds Nails
3-1/2	70	2-7/8	17-1/2	3/4								
4	80	2-1/2	20	5/8	72-1/2	2-1/2	18	5/8				
4-1/2	90	2-1/4	22-1/2	5/8	81-1/2	2-1/4	20	5/8				
5	100*	2	25	1/2	90-1/2	2	22-1/2	1/2				
5-1/2	110	1-3/4	27-1/2	1/2	100*	1-3/4	25	1/2				
6	120	1-2/3	30	3/8	109	1-2/3	27	3/8	80	2-1/3	20	5/8
6-1/2	130	1-1/2	32-1/2	3/8	118	1-1/2	29-1/2	3/8	86-1/2	2-1/8	21-1/2	1/2
7	140	1-2/5	35	1/3	127	1-2/5	31-1/2	1/3	93	2	23	1/2
7-1/2	†150	1-1/3	37-1/2	1/3	136	1-1/3	34	1/3	100*	1-7/8	25	1/2
8	160		40	1/3	145-1/2	1-1/4	36	1/3	106-1/2	1-3/4	26-1/2	1/2
8-1/2	170		42-1/2		†154-1/2	1-1/4	38-1/2	1/4	113	1-2/3	28	1/2
9	180		45		163-1/2		40-1/2		120	1-1/2	30	3/8
9-1/2	190		47-1/2		172-1/2		43		126-1/2	1-1/2	31-1/2	3/8
10	200		50		181-1/2		45		133	1-1/2	33	3/8
10-1/2	210		52-1/2		191		47-1/2		140	1-1/3	35	1/3
11	220		55		200		50		146-1/2	1-1/4	36-1/2	1/3
11-1/2	230		57-1/2		209		52		†153	1-1/4	38	1/3
12	‡240		60		218		54-1/2		160		40	
12-1/2					227		56-1/2		166-1/2		41-1/2	
13					236		59		173		43	
13-1/2					245-1/2		61		180		45	
14					‡254-1/2		63-1/2		186-1/2		46-1/2	
14-1/2									193		48	
15									200		50	
15-1/2									206-1/2		51-1/2	
16									‡213		53	

* Maximum exposure recommended for roofs.

† Maximum exposure recommended for single-coursing on side walls.

‡ Maximum exposure recommended for double-coursing on side walls. Figures in italics are inserted for convenience in estimating quantity of shingles needed for wide exposures in double-coursing, with butt-nailing. In double-coursing, with any exposure chosen, the figures indicate the amount of shingles for the outer courses. Order an equivalent number of shingles for concealed courses. Approximately 1-1/2 lbs. 5d small-headed nails required per square (100 S.F. wall area) to apply outer course of 16-inch shingles at 12-inch weather exposure. Plus 1/2 lb. 3d nails for under course shingles. Figure slightly fewer nails for 18-inch shingles at 14-inch exposure.

(from Certigrade Handbook of Red Cedar Shingles, published by Red Cedar Shingle Bureau (Red Cedar Shingle and Handsplit Shake Bureau), Seattle, Washington, 1957.)

(courtesy of American Institute of Steel Construction, Inc.)

Figure 7.24   Roofing Material Quantities — Standard 3/16″ Thick Slate

This table shows the materials needed for installing various sizes of 3/16″-thick roof slates.

Size of Slate (In.)	Slates per Square	Exposure with 3″ Lap	Nails per Square Lbs.	Nails per Square Ozs.	Size of Slate (In.)	Slates per Square	Exposure with 3″ Lap	Nails per Square Lbs.	Nails per Square Ozs.
26 x 14	89	11-1/2″	1	0	16 x 14	160	6-1/2″	1	13
					16 x 12	184	6-1/2″	2	2
24 x 16	86	10-1/2″	1	0	16 x 11	201	6-1/2″	2	5
24 x 14	98	10-1/2″	1	2	16 x 10	222	6-1/2″	2	8
24 x 13	106	10-1/2″	1	3	16 x 9	246	6-1/2″	2	13
24 x 12	114	10-1/2″	1	5	16 x 8	277	6-1/2″	3	2
24 x 11	125	10-1/2″	1	7					
					14 x 12	218	5-1/2″	2	8
22 x 14	108	9-1/2″	1	4	14 x 11	238	5-1/2″	2	11
22 x 13	117	9-1/2″	1	5	14 x 10	261	5-1/2″	3	3
22 x 12	126	9-1/2″	1	7	14 x 9	291	5-1/2″	3	5
22 x 11	138	9-1/2″	1	9	14 x 8	327	5-1/2″	3	12
22 x 10	152	9-1/2″	1	12	14 x 7	374	5-1/2″	4	4
20 x 14	121	8-1/2″	1	6	12 x 10	320	4-1/2″	3	10
20 x 13	132	8-1/2″	1	8	12 x 9	355	4-1/2″	4	1
20 x 12	141	8-1/2″	1	10	12 x 8	400	4-1/2″	4	9
20 x 11	154	8-1/2″	1	12	12 x 7	457	4-1/2″	5	3
20 x 10	170	8-1/2″	1	15	12 x 6	533	4-1/2″	6	1
20 x 9	189	8-1/2″	2	3					
					11 x 8	450	4″	5	2
18 x 14	137	7-1/2″	1	9	11 x 7	515	4″	5	14
18 x 13	148	7-1/2″	1	11					
18 x 12	160	7-1/2″	1	13	10 x 8	515	3-1/2″	5	14
18 x 11	175	7-1/2″	2	0	10 x 7	588	3-1/2″	7	4
18 x 10	192	7-1/2″	2	3	10 x 6	686	3-1/2″	7	13
18 x 9	213	7-1/2″	2	7					

Figure 7.25 Installation Time in Man-Hours for Built-Up Roofing

Description	Man-Hours	Unit
Built-Up Roofing		
Asphalt Flood Coat with Gravel or Slag		
Fiberglass Base Sheet		
3 Plies Felt Mopped	2.545	Sq.
On Nailable Decks	2.667	Sq.
4 Plies Felt Mopped	2.800	Sq.
On Nailable Decks	2.947	Sq.
Coated Glass Fiber Base Sheet		
3 Plies Felt Mopped	2.545	Sq.
On Nailable Decks	2.667	Sq.
Organic Base Sheet and 3 Plies Felt	2.333	Sq.
On Nailable Decks	2.435	Sq.
Coal Tar Pitch with Gravel or Slag		
Coated Glass Fiber Base Sheet and		
2 Plies Glass Fiber Felt	2.947	Sq.
On Nailable Decks	3.111	Sq.
Asphalt Mineral Surface Roll Roofing	2.074	Sq.
Walkway		
Asphalt Impregnated	.020	S.F.
Patio Blocks 2" Thick	.070	S.F.
Expansion Joints Covers	.048	L.F.

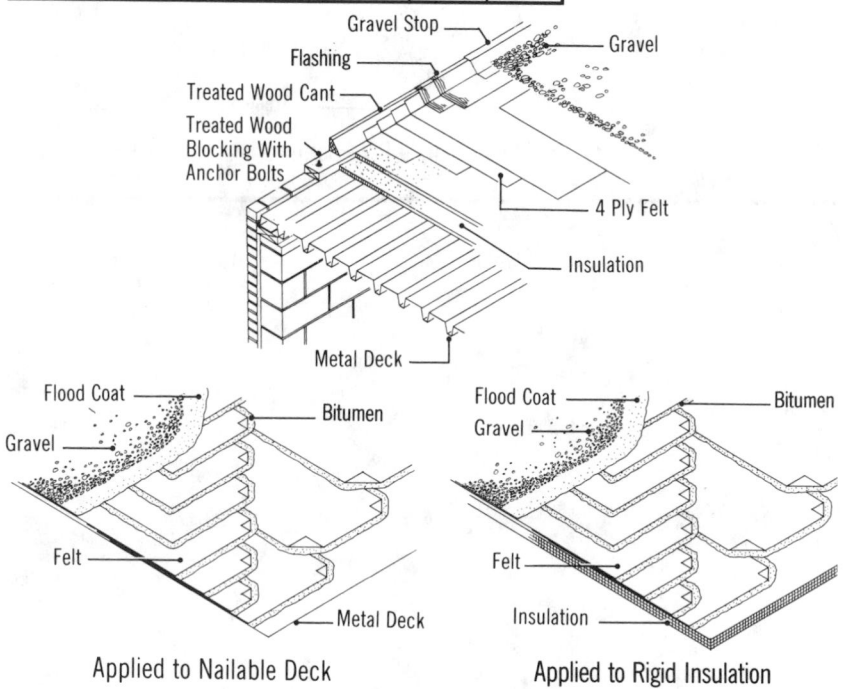

Gravel Stop

Flashing

Treated Wood Cant

Treated Wood Blocking With Anchor Bolts

Gravel

4 Ply Felt

Insulation

Metal Deck

Flood Coat

Bitumen

Gravel

Felt

Metal Deck

Flood Coat

Bitumen

Gravel

Felt

Insulation

Applied to Nailable Deck

Applied to Rigid Insulation

Built-Up Roof

Figure 7.26 Installation Time in Man-Hours for Single Ply Roofing

Description	Man-Hours	Unit
Single Ply Membrane, General		
Loose-Laid and Ballasted	.006	S.F.
Partially Adhered, All Types	.008	S.F.
Fully Adhered, All Types	.011	S.F.
Modified Bitumen		
Loose-Laid and Ballasted	.013	S.F.
Partially Adhered Torch Welding	.016	S.F.
Fully Adhered Torch Welding or Hot		
Asphalt Attachment	.020	S.F.
Uncured Neoprene for Flashing	.013	S.F.
Separator Sheet	.010	S.F.

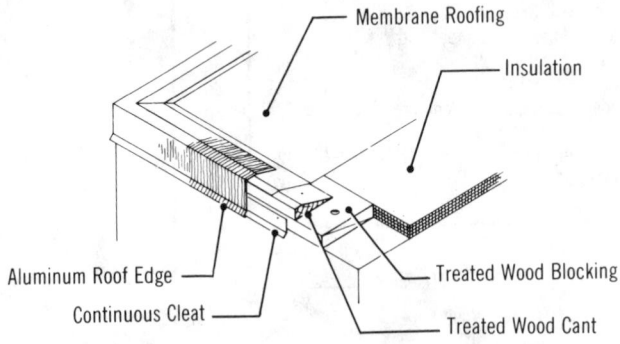

Roof Edge, Single Ply Roofing

Figure 7.27 Installation Time in Man-Hours for Shingle and Tile Roofing

Description	Man-Hours	Unit
Shingles		
Aluminum	3.200	Sq.
Ridge Cap or Valley	.047	L.F.
Shakes	3.478	Sq.
Fiberglass		
325 lb. per Sq.	2.000	Sq.
165 lb. per Sq.	2.286	Sq.
Hip and Ridge	8.000	C.L.F.
Asphalt Standard Strip		
Class A 210 to 235 lb. per Sq.	1.455	Sq.
Class C 235 to 240 lb. per Sq.	1.600	Sq.
Standard Laminated		
Class A 240 to 260 lb. per Sq.	1.778	Sq.
Class C 260 to 300 lb. per Sq.	2.000	Sq.
Premium Laminated		
Class A 260 to 300 lb. per Sq.	2.286	Sq.
Class C 300 to 385 lb. per Sq.	2.667	Sq.
Hip and Ridge Roll	.020	L.F.
Slate Including Felt Underlay	4.571	Sq.
Steel	3.636	Sq.
Wood		
5" Exposure	3.325	Sq.
5-1/2" Exposure	3.034	Sq.
Panelized 8' Strips 7" Exposure	2.667	Sq.
Laps Rakes or Valleys	.040	L.F.
Tiles		
Aluminum		
Mission	3.200	Sq.
Spanish	2.667	Sq.
Clay 8-1/4" x 11"	4.848	Sq.
Spanish	4.444	Sq.
Mission	6.957	Sq.
French	5.926	Sq.
Norman	8.000	Sq.
Williamsburg	5.926	Sq.
Concrete 13" x 16-1/2"	5.926	Sq.
Steel	3.200	Sq.

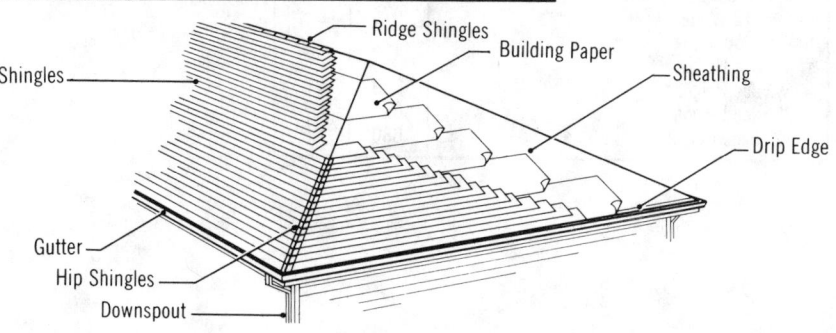

Figure 7.28 Installation Time in Man-Hours for Flashing, Expansion Joints, and Gravel Stops

Description	Man-Hours	Unit
Flashing Aluminum Mill Finish	.055	S.F.
Fabric-Backed or Mastic-Coated, 2 Sides	.024	S.F.
Copper Sheets Under 6000 lbs.		
16 oz. Sheets	.070	S.F.
20 oz. Sheets	.073	S.F.
24 oz. Sheets	.076	S.F.
32 oz. Sheets	.080	S.F.
Paperbacked, Fabric-Backed or		
Mastic-Backed, 2 Sides	.024	S.F.
Lead-Coated Copper, Paperbacked		
Fabric-Backed or Mastic-Backed	.024	S.F.
Lead 2.5 lbs. per S.F.	.059	S.F.
Polyvinyl Chloride or Butyl Rubber	.028	S.F.
Copper-Clad Stainless Steel Under 500 lbs.		
.015" Thick	.070	S.F.
.018" Thick	.080	S.F.
Stainless Steel Sheets or		
Terne Coated Stainless Steel	.052	S.F.
Paperbacked 2 Sides	.024	S.F.
Zinc and Copper Alloy	.052	S.F.
Expansion Joint, Butyl, 1/16" Thick, 29" Wide		
Metal Flanges	.048	L.F.
Neoprene, Double-seal Type with		
Thick Center, 4-1/2" Wide	.064	L.F.
Polyethylene Bellows with		
Galvanized Flanges	.080	L.F.
Roof Expansion, Joint with		
Extruded Aluminum Cover, 2"	.070	L.F.
Roof Expansion Joint, Plastic Curbs,		
Foam Center	.080	L.F.
Transitions, Regular, Minimum	.800	Ea.
Maximum	2.000	Ea.
Large, Minimum	.889	Ea.
Maximum	2.667	Ea.
Roof to Wall Expansion Joint with Extruded		
Aluminum Cover	.070	L.F.
Wall Expansion Joint, Closed Cell Foam on		
PVC Cover, 9" Wide	.064	L.F.
12" Wide	.070	L.F.
Gravel Stop		
4" Face Height	.055	L.F.
6" Face Height	.059	L.F.
8" Face Height	.064	L.F.
12" Face Height	.080	L.F.

Figure 7.28 Installation Time in Man-Hours for Flashing, Expansion Joints and Gravel Stops (continued)

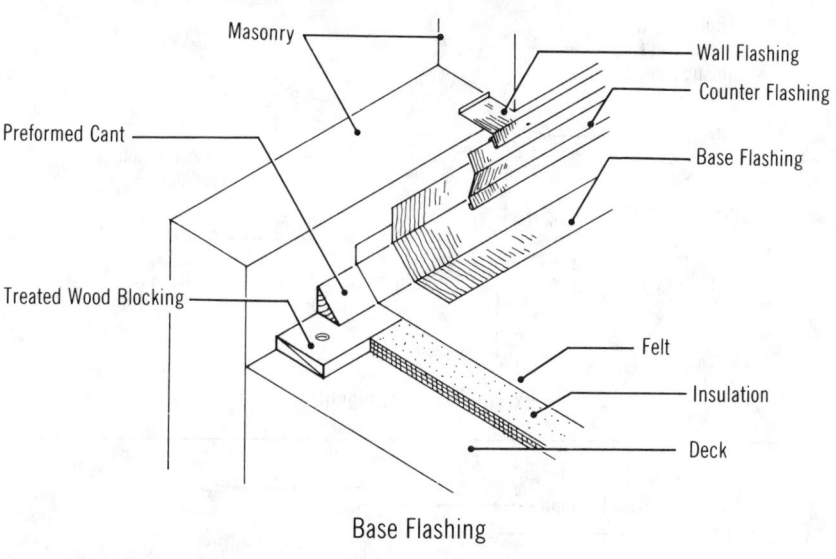

Base Flashing

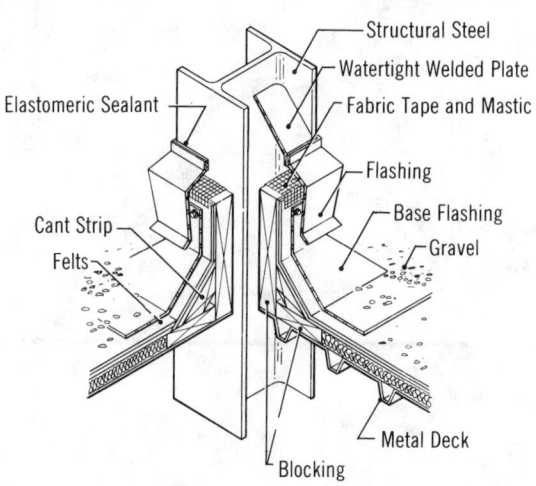

Structural Steel Flashing

(continued on next page)

Figure 7.28   Installation Time in Man-Hours for Flashing, Expansion Joints and Gravel Stops (continued)

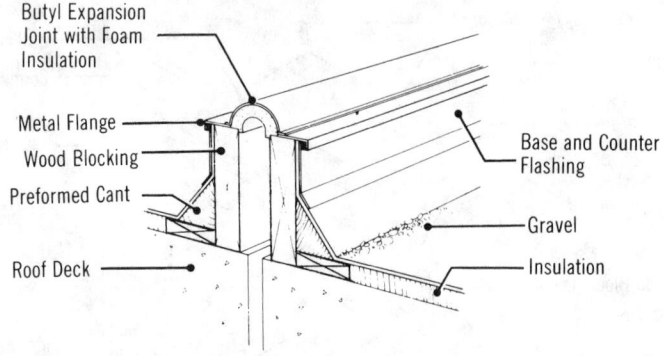

Butyl Expansion Joint with Foam Insulation

Metal Flange

Wood Blocking

Preformed Cant

Roof Deck

Base and Counter Flashing

Gravel

Insulation

Expansion Joint Flashing

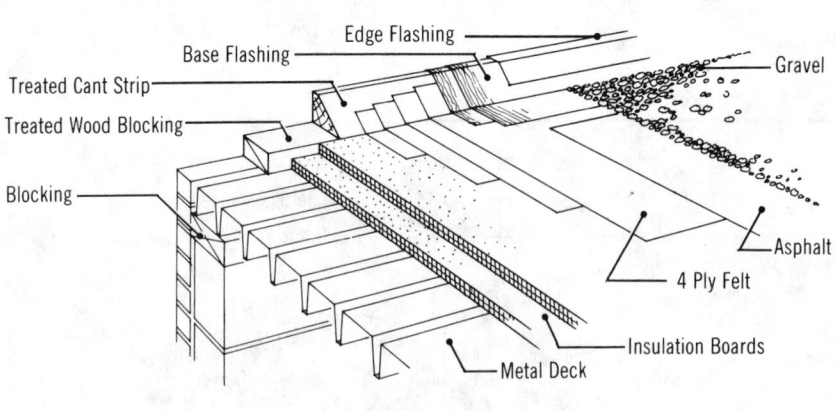

Edge Flashing

Base Flashing

Treated Cant Strip

Treated Wood Blocking

Blocking

Gravel

Asphalt

4 Ply Felt

Insulation Boards

Metal Deck

Roof Edge Flashing

Figure 7.29 Installation Time in Man-Hours for Roof Hatches, Smoke Hatches, and Skylights

Description	Man-Hours	Unit
Roof Hatches with Curb		
2'-6" x 3'-0"	3.200	Ea.
2'-6" x 4'-6"	3.556	Ea.
2'-6" x 8'-0"	4.848	Ea.
Smoke Hatches		
4'-0" x 4'-0"	2.462	Ea.
4'-0" x 8'-0"	4.000	Ea.
Plastic Roof Domes Flush or Curb Mounted		
Under 10 S.F.		
Single	.200	S.F.
Double	.246	S.F.
10 S.F. to 20 S.F.		
Single	.081	S.F.
Double	.102	S.F.
20 S.F. to 30 S.F.		
Single	.069	S.F.
Double	.081	S.F.
30 S.F. to 65 S.F.		
Single	.052	S.F.
Double	.069	S.F.
Ventiliation, Insulated Plexiglass Dome		
Curb Mounted		
30" x 32"	2.667	Ea.
44" x 45"	3.200	Ea.
Skyroofs		
Translucent Panels 2-3/4" Thick	.081	S.F. Horiz.
Continuous Vaulted to 8' Wide		
Single Glazed	.200	S.F. Horiz.
Double Glazed	.221	S.F. Horiz.
To 20' Wide Single Glazed	.183	S.F. Horiz.
Over 20' Wide Single Glazed	.160	S.F. Horiz.
Pyramid Type to 30' Clear Opening	.194	S.F. Horiz.
Grid Type 4' x 10' Modules	.200	S.F. Horiz.
Ridge Units Continuous to 8' Wide		
Double	.246	S.F. Horiz.
Single	.160	S.F. Horiz.

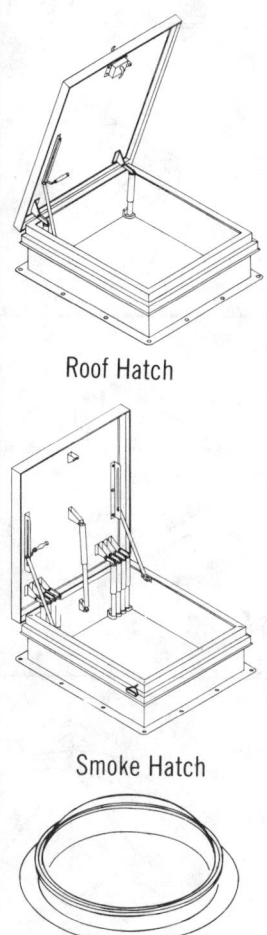

Roof Hatch

Smoke Hatch

Circular Dome Skylight

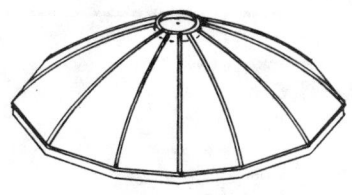

Domed Skylight

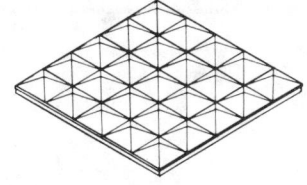

Pyramid Skylights in Grid Form

Skyroofs

(continued on next page)

Figure 7.29   Installation Time in Man-Hours
                  for Roof Hatches, Smoke Hatches,
                  and Skylights (continued)

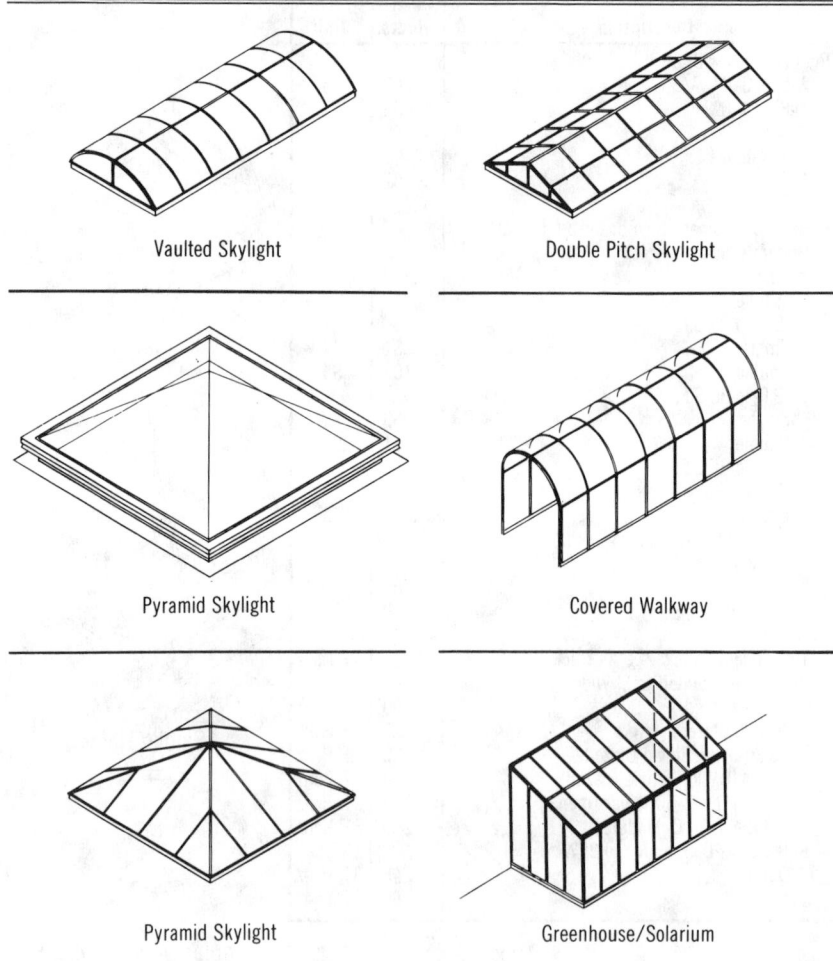

Vaulted Skylight   Double Pitch Skylight

Pyramid Skylight   Covered Walkway

Pyramid Skylight   Greenhouse/Solarium

Skyroofs

Figure 7.30 Installation Time in Man-Hours for Gutters and Downspouts

Description	Man-Hours	Unit
Gutters		
Aluminum	.067	L.F.
Copper Stock Units		
4" Wide	.067	L.F.
6" Wide	.070	L.F.
Steel Galvanized or Stainless	.067	L.F.
Vinyl	.073	L.F.
Wood	.080	L.F.
Downspouts		
Aluminum		
2" x 3"	.042	L.F.
3" Diameter	.042	L.F.
4" Diameter	.057	L.F.
Copper		
2" or 3" Diameter	.042	L.F.
4" Diameter	.055	L.F.
5" Diameter	.062	L.F.
2" x 3"	.042	L.F.
3" x 4"	.055	L.F.
Steel Galvanized		
2" or 3" Diameter	.042	L.F.
4" Diameter	.055	L.F.
5" Diameter	.062	L.F.
6" Diameter	.076	L.F.
2" x 3"	.042	L.F.
3" x 4"	.055	L.F.
Epoxy Painted		
2" x 3"	.042	L.F.
3" x 4"	.055	L.F.
Steel Pipe		
4" Diameter	.400	L.F.
6" Diameter	.444	L.F.
Stainless Steel		
2" x 3" or 3" Diameter	.042	L.F.
3" x 4" or 4" Diameter	.055	L.F.
4" x 5" or 5" Diameter	.059	L.F.
Vinyl		
2" x 3"	.038	L.F.
2-1/2" Diameter	.036	L.F.

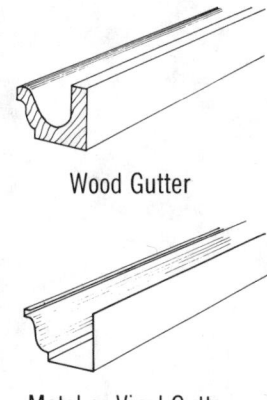

Wood Gutter

Metal or Vinyl Gutter

(continued on next page)

Figure 7.30   Installation Time in Man-Hours
for Gutters and Downspouts (continued)

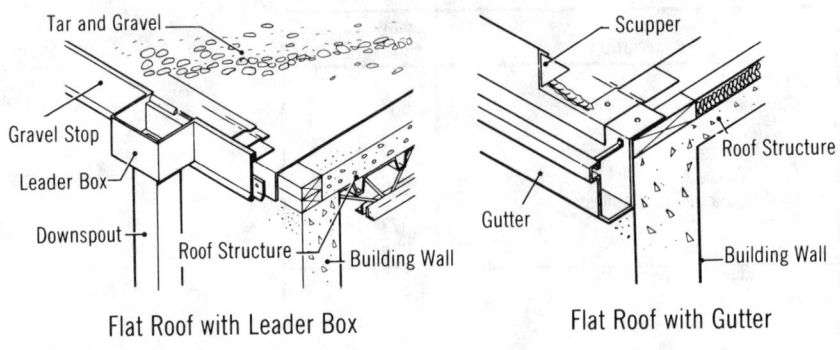

Flat Roof with Leader Box   Flat Roof with Gutter

Figure 7.31 Wood Siding Quantities

This table allows you to compute the amount of wood siding (in board feet required per square foot of wall area to receive siding) based on the style and dimensions of the siding.

Type of Siding	Size	Exposure	Board Feet per S.F. of Wall Area	Lbs. Nails per MBM of Siding		
				Stud Spacing		
				16"	20"	24"
Plain Bevel Siding	1/2" x 4"	2-1/2"	1.60	17	13	11
		2-3/4"	1.45			
	1/2" x 6"	4-1/2"	1.33	11	9	7
		4-3/4"	1.26			
		5"	1.20			
	1/2" x 8"	6-1/2"	1.23	8	7	6
		7"	1.14			
Plain Bevel Bungalow Siding	5/8" x 8"	6-1/2"	1.23	14	11	9
		7"	1.14			
	5/8" x 10"	8-1/2"	1.18	16	13	11
		9"	1.11			
	3/4" x 8"	6-1/2"	1.23	14	11	9
		7"	1.14			
	3/4" x 10"	8-1/2"	1.18	16	13	11
		9"	1.11			
	3/4" x 12"	10-1/2"	1.14	14	11	9
		11"	1.09			
Drop or Rustic Siding	3/4" x 4"	3-1/4"	1.23	27	22	18
	3/4" x 6"	5-1/6"	1.19	18	15	12
		5-3/16"	1.17			

Figure 7.32   Installation Time in Man-Hours for Metal Siding

Description	Man-Hours	Unit
Aluminum Siding		
On Steel Frame	.041	S.F.
On Wood Frame	.030	S.F.
Closure Strips and Flashing	.040	L.F.
Steel Siding, on Steel Frame	.040	S.F.
Metal Siding Panels, Insulated with Liner		
Field Assembled	.164	S.F.
Factory Assembled	.084	S.F.
Metal Fascia, No Furring or Framing Incl.		
Long Panels	.055	S.F.
Short Panels	.069	S.F.
Mansard Roofing,		
With Battens, Custom		
Multi-Sloped Surface	.145	S.F.
Projected Surface	.106	S.F.
Vertical Surface	.069	S.F.
Stock, All Surfaces	.069	S.F.
Framing, to 5' High	.069	L.F.
Soffits, Metal Panel	.064	S.F.
Furring		
Metal, 3/4" Channel		
16" On Center	.030	S.F.
24" On Center	.022	S.F.
Wood Strips		
On Masonry	.016	L.F.
On Concrete	.030	L.F.
Channel Framing, Overhead 24" On Center	.017	S.F.
Stud Framing		
Metal, Overhead,		
16" On Center	.026	S.F.
24" On Center	.017	S.F.
Wood, Overhead,		
2" x 4"	.026	L.F.
2" x 8"	.070	L.F.
Sheathing		
Drywall	.022	S.F.
Plywood	.011	S.F.
Channel Girts, Steel		
Less than 8"	.016	lb.
Greater than 8"	.009	lb.
Slotted Channel Framing System		
Minimum	.006	lb.
Maximum	.010	lb.

Figure 7.32 Installation Time in Man-Hours
for Metal Siding (continued)

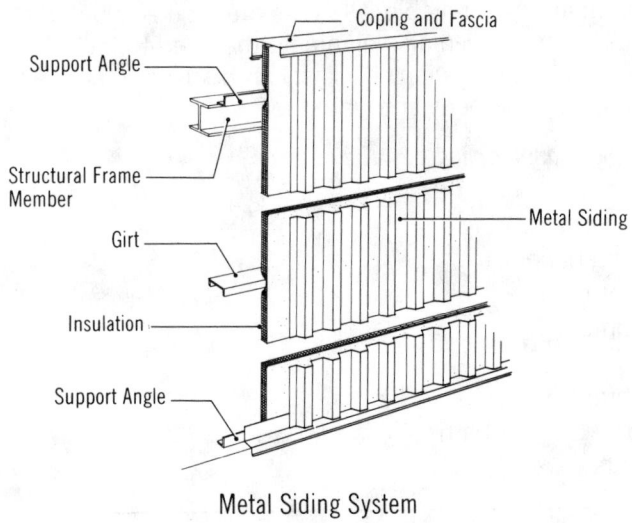

Metal Siding System

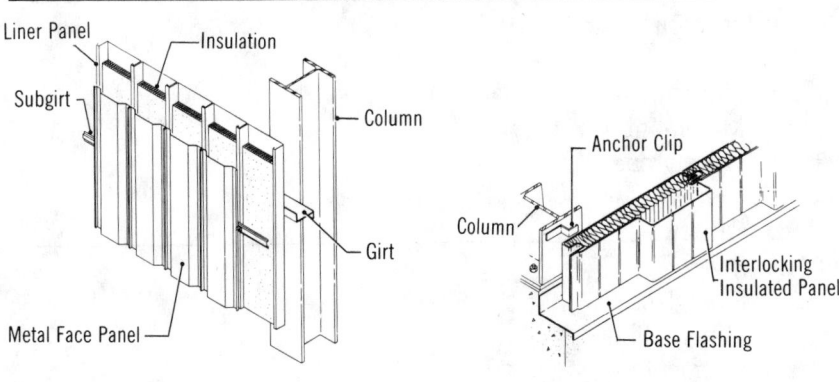

Field Assembled Insulated Metal Wall   Factory Assembled Insulated Metal Wall

Checklist

For an estimate to be reliable, all items must be accounted for. A complete estimate can also eliminate the need to include contingencies. The following checklist can be used to help ensure that all items are properly accounted for.

Waterproofing/Dampproofing
- ☐ Bentonite
- ☐ Bituminous coating
- ☐ Building paper
- ☐ Cementitious
- ☐ Elastomeric
- ☐ Liquid
- ☐ Membrane
- ☐ Preformed vapor barrier
- ☐ Sealants
- ☐ Other _____

Caulking _____

Building Insulation
- ☐ Rigid
  - _____ Fiberglass
  - _____ Mineral fiber
  - _____ Polystyrene
  - _____ Urethane
  - _____ Other _____
- ☐ Non-Rigid
  - _____ Fiberglass
  - _____ Mineral fiber
  - _____ Vermiculite
  - _____ Perlite
  - _____ Other _____
- ☐ Form Board
  - _____ Acoustical
  - _____ Fiberglass
  - _____ Gypsum
  - _____ Mineral fiber
  - _____ Wood fiber
  - _____ Other _____

Masonry Insulation
- ☐ Poured

☐ Rigid
☐ Core
☐ Foam

Roof Deck Insulation
☐ Fiberboard
☐ Fiberglass
☐ Foamglass
☐ Polystyrene
☐ Urethane
☐ Other _____

Spray-On Insulation _____

Special Insulation _____

Shingles for Siding or Roof
☐ Aluminum
☐ Asphalt
☐ Clay tile
☐ Concrete
☐ Slate
☐ Wood

Siding _____

Fiberglass Panels
☐ Roofing
☐ Siding

Metal Panels _____

Steel Panels
☐ Roofing
☐ Siding
☐ Insulated sandwich

Vinyl Siding _____

Membrane Roofing _____

Built-Up Roofing _____

Single Ply Roofing _____

Roll Roofing _____

Metal Roofing
☐ Lead
☐ Copper
☐ Stainless steel
☐ Zinc

Fluid Applied Roofing _____

Roof Accessories
- ☐ Downspouts
- ☐ Gutters
- ☐ Expansion joints
- ☐ Fascia
- ☐ Flashing
- ☐ Gravel stop
- ☐ Louvers
- ☐ Reglet
- ☐ Roof drains
- ☐ Soffit
- ☐ Hatches
- _____ Roof
- _____ Smoke vents
- ☐ Skyroof
- ☐ Ventilators
- ☐ Other _____

Tips

Roof Walkways

When estimating the installation of a membrane roof, be sure to include the cost for walkway pavers to create a path to roof-mounted mechanical equipment. You will more than likely be required to install these in order to obtain a warranty.

Single Ply Roofing Sources

In many locations, manufacturers make their materials available through one distributor/contractor franchise. If only one type of roofing is specified, you may not have much choice as far as acquiring costs or quotes.

Estimating the Specified System

Even though it may be less expensive to use a "suitable substitute," the system that is specified is the one that should be estimated. If you include as part of your estimate a non-specified system and get the job, you may be required to furnish the originally specified, more expensive roof system.

Sheet Metal Work

Do not assume that all sheet metal items are "off-the-shelf" items. Gutters and downspouts as well as termite shields, gravel stops, expansion joints, and reglets may need to be shop-fabricated. Checking into information catalogs could save quite a tidy sum in extras.

Notes

Division Eight

Doors, Windows & Glass

Introduction

Doors, windows, glass, and glazing can represent a considerable portion of a total construction contract. The more "upscale" a building or project, the more emphasis will be placed on the visible effects of the building, including doors, windows, and glass treatments. There is a wide range of door and window materials and hardware. Increasing the quality of any of these components can add a considerable amount to an estimate.

Most architectural plans and specifications will generally include a door, window, and hardware schedule which will describe the choices of the owner/architect and take most of the "guesswork" out of the estimating process. If no such schedule is available at the time the estimate is prepared, it becomes even more important to document the grade of each component used for the bid.

To assist in estimating this division, we have provided tables and charts for the following areas:

- Metal doors/frames
- Wood/plastic doors
- Special doors
- Entrances/store fronts
- Windows — metal
- Windows — wood/plastic
- Hardware/operators/seals
- Glass/glazing
- Curtain walls

Estimating Data

The following tables present effective estimating guidelines for items found in Division 8 — Doors, Windows and Glass. Please note that these guidelines can be used as indicators of what may be expected, and that each project must be evaluated individually.

Table of Contents

Figure 8.1   Standard Door Nomenclature

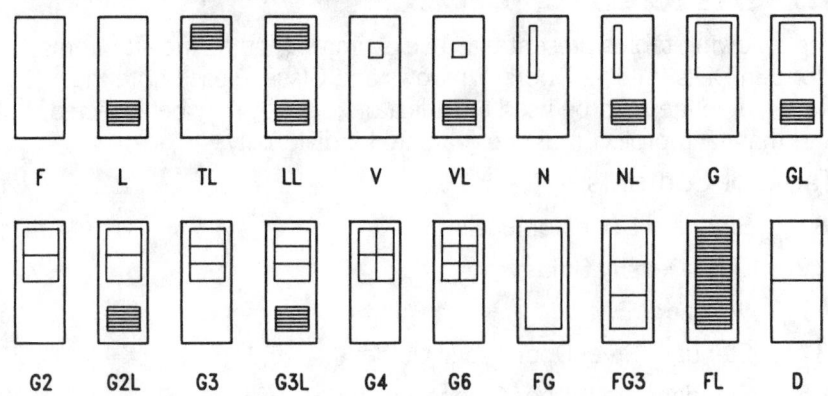

NOMENCLATURE LETTER SYMBOLS

F = Flush

L = Louvered (bottom)

TL = Louvered (top)

LL = Louvered (top and bottom)

V = Vision Lite

VL = Vision Lite and Louvered

N = Narrow Lite

NL = Narrow Lite and Louvered

G = Half Glass (options G2, G3, G4, and G6)

GL = Half Glass and Louvered (options G2L and G3L)

FG = Full Glass (option FG3)

FLI = Full Louver (Inserted)

D = Dutch Door

(courtesy Steel Door Institute)

Figure 8.2   Door Density Guide

This table can be used to estimate the quantity of doors needed when no interior plan is available, such as in a conceptual estimating situation. To use this table, first pick out from the left-most column the type of structure to be estimated. Moving across that major row, from the second column, "Stories," pick out the number of stories in the building. Read across that line to the "Doors" column. These figures are typical ratios of square feet of floor area to each door.

Example: To estimate the number of interior doors for a one-story office building of 7,000 square feet, find "Offices" in the left column. Then move across to the number "1" in the "Stories" column. In the "Doors" column, "200–500 S.F./Door" is listed. If the building is average (no high density of small offices; no high density of large conference/meeting rooms), one would choose a density figure in the middle, say 350 S.F. To find the total number of doors to be expected in the office, divide 7,000 S.F. by 350 S.F. = 20 doors.

Building Type	Stories	S.F./ Door	Building Type	Stories	S.F./ Door
Apartments	1	90	Nursing Home	1	80
	2	80		2–4	80
	3	90	Offices	1	200–500
	5	90		2	200–500
	6-15	80		3–5	200–500
Bakery	1	500		6–10	200–500
	2	500		11–20	200–500
Bank	1	200	Parking Ramp (Open)	2–8	600
	2–4	150	Parking Garage	2–8	600
Bottling Plant	1	500	Pre-Engineered Store	1	600
Bowling Alley	1	500	Office	1	150
Bus Terminal	1	150	Shop	1	150
Cannery	1	1000	Radio & TV Broadcasting	1	250
			& TV Transmitter	1	400
Car Wash	1	180	Self Service Restaurant	1	150
Dairy Plant	1	300	Cafe & Drive-In Restaurant	1	180
Department Store	1	600	Restaurant with seating	1	250
	2–5	600	Supper Club	1	250
			Bar or Lounge	1	240
Dormitory	2	90	Retail Store or Shop	1	600
	3–5	90	Service Station Masonry	1	150
	6–15	90	Metal panel	1	150
Funeral Home	1	150	Frame	1	150
	2	140	Shopping Center (strip)	1	300
Garage Sales & Service	1	300	(group)	1	400
Hotel	3–8	90		2	400
	9–15	90	Small Food Store	1	300
Laundromat	1	250	Store/Apt. above Masonry	2	100
Medical Clinic	1	60	Frame	2	100
	2–4	60	Frame	3	100
Motel	1	70	Supermarkets	1	400
	2–3	70	Truck Terminal	1	0
Movie Theater 200–600 seats	1	180	Warehouse	1	0
601–1400 seats		200			
1401–2200 seats		250			

Figure 8.3   Weight of Doors

Door Thickness	Weight of Doors in Pounds per Square Foot				
	White Pine	Oak	Hollow Core	Solid Core	Hollow Metal
1-3/8″	3 psf	6 psf	1-1/2 psf	3-1/2 – 4 psf	6-1/2 psf
1-3/4″	3-1/2	7	2	4-1/2 – 5-1/4	6-1/2
2-1/4″	4-1/2	9	—	5-1/2 – 6-3/4	6-1/2

Figure 8.4   Standard Steel Door Usage Guide

Building Types	Grade I Standard Duty	Grade ii Heavy Duty (1-3/4" only)	Grade III Extra Heavy Duty (1-3/4" only)	1-3/4"	1-3/4" or 1-3/8"	Flush (F)	Half Glass (G)	Vision Lite (V)	Full Glass (FG)	Narrow Lite (N)	General Remarks
Apartment											
Main Entrance			•	•			•		•	•	Two way vision
Unit Entrance	•	•		•	•	•					Check for fire door rqmts.
Bedroom	•				•	•					
Bathroom	•				•	•					
Closet	•				•	•					
Stairwell		•	•	•				•			Fire door – two way vision
Dormitory											
Main Entrance			•	•			•		•	•	Two way vision
Unit Entrance	•	•		•		•					Check for fire door rqmts.
Bedroom	•				•	•					
Bathroom	•				•	•					
Closet	•				•	•					
Stairwell		•	•	•				•			Fire door – two way vision
Hotel – Motel											
Unit Entrance	•	•		•		•					Check for fire door rqmts.
Bathroom	•				•	•					
Closet	•				•	•					
Stairwell		•	•	•				•			Fire door – two way vision
Storage & Utility	•	•		•		•					Optional type selection

(Courtesy Steel Door Institute)

(continued on next page)

Figure 8.4   Standard Steel Door Usage Guide (continued)

Building Types	Grade I Standard Duty	Grade II Heavy Duty (1-3/4" only)	Grade III Extra Heavy Duty (1-3/4" only)	1-3/4"	1-3/4" or 1-3/8"	Flush (F)	Half Glass (G)	Vision Lite (V)	Full Glass (FG)	Narrow Lite (N)	General Remarks
Hospital—Nursing Home											
Main Entrance		•		•			•		•	•	Two way vision
Patient Room		•		•		•					
Stairwell		•	•	•				•			Fire door – two way vision
Opertg. & Exam.		•	•	•	•						Optional type selection
Bathroom	•				•	•					
Closet	•				•	•					
Recreation		•		•		•		•			Optional design selection
Kitchen		•	•	•				•			Two way vision
Industrial											
Entrance & Exit		•		•			•		•	•	Optional design selection
Office	•	•		•		•					Optional design selection
Production		•		•		•					
Toilet		•	•	•		•					Design option – louver door
Tool		•		•		•					Design option – Dutch door
Trucking		•			•		•				
Monorail		•		•		•	•				Optional design selection

Figure 8.4 Standard Steel Door Usage Guide (continued)

Building Types	Grade I Standard Duty	Grade II Heavy Duty (1-3/4" only)	Grade III Extra Heavy Duty (1-3/4" only)	1-3/4"	1-3/4" or 1-3/8"	Flush (F)	Half Glass (G)	Vision Lite (V)	Full Glass (FG)	Narrow Lite (N)	General Remarks
	Standard Steel Door Grades			**Door Thickness**		**Door Design Nomenclature**					
Office											
Entrance			•	•			•		•	•	Two way vision
Individual Office	•			•	•	•	•				Optional thickness & design
Closet	•				•	•					
Toilet		•	•	•		•					Design option-louver door
Stairwell		•	•	•				•			Fire door – two way vision
Equipment		•	•	•		•					Optional type selection
Boiler		•	•	•		•					Fire Door – flush design
School											
Entrance & Exit			•	•			•		•	•	Two way vision
Classroom		•		•			•			•	Two way vision
Toilet		•	•	•		•					Design option – louver door
Gymnasium		•	•	•		•		•			Optional design
Cafeteria		•	•	•			•				Two way vision
Stairwell		•	•	•				•			Fire door – two way vision
Closet	•				•	•					

(courtesy Steel Door Institute)

Figure 8.5 Standard Steel Door Grades and Types

Grade	Model	Construction	Minimum Thickness Gauge Number Panels— Face Sheet	Minimum Thickness Gauge Number Stiles— Rails
I—Standard Duty (1-3/4" and 1-3/8")	1	Full Flush (Hollow Steel)	20	—
	2	Full Flush (Composite)	20	—
	3	Seamless (Hollow Steel)	20	—
	4	Seamless (Composite)	20	—
II—Heavy Duty (1-3/4" only)	1	Full Flush (Hollow Steel)	18	—
	2	Full Flush (Composite)	18	—
	3	Seamless (Hollow Steel)	18	—
	4	Seamless (Composite)	18	—
III—Extra Heavy Duty (1-3/4" only)	1	Full Flush (Hollow Steel)	16	—
	2	Full Flush (Composite)	16	—
	3	Seamless (Hollow Steel)	16	—
	4	Seamless (Composite)	16	—
	5	Flush Panel (Stile and Rail)	18	16

(Courtesy Steel Door Institute)

Figure 8.6 Fire Door Classifications

This table lists the various fire door ratings by label, time, and temperature rating, plus allowable glass area for a fire-rated door.

Classification	Time Rating (as shown) on label)		Temperature Rise (as shown) on label)	Maximum Glass Area
3 Hour fire doors (A) are for use in openings in walls separating buildings or dividing a single building into fire areas.	3 Hr. 3 Hr 3 Hr. 3 Hr.	(A) (A) (A) (A)	30 Min. 250°F Max 30 Min. 450°F Max 30 Min. 650°F Max *	None
1-1/2 Hour fire doors (B) and (D) are for use in openings in 2 Hour enclosures of vertical communication through buildings (stairs, elevators, etc.) or in exterior walls which are subject to severe fire exposure from outside of the building. 1 Hour fire doors (B) are for use in openings in 1 Hour enclosures of vertical communication through buildings (stairs, elevators, etc.)	1-1/2 Hr. 1-1/2 Hr. 1-1/2 Hr. 1-1/2 Hr. 1 Hr. 1-1/2 Hr. 1-1/2 Hr. 1-1/2 Hr. 1-1/2 Hr.	(B) (B) (B) (B) (B) (D) (D) (D) (D)	30 Min. 250°F Max 30 Min. 450°F Max 30 Min. 650°F Max * 30 Min. 250°F Max 30 Min. 250°F Max 30 Min. 450°F Max 30 Min. 650°F Max *	100 square inches per door None
3/4 Hour fire doors (C) and (E) are for use in openings in corridor and room partitions or in exterior walls which are subject to moderate fire exposure from outside of the building.	3/4 Hr. 3/4 Hr.	(C) (E)	** **	1296 square 720 square inches per light
1/2 Hour fire doors and 1/3 Hour fire doors are for use where smoke control is a primary consideration and are for the protection of openings in partitions between a habitable room and a corridor when the wall has a fire-resistance rating of not more than one hour.	1/2 Hr. 1/3 Hr.		** **	No Limit

* The labels do not record any temperature rise limits. This means that the temperature rise on the unexposed face of the door at the end of 30 minutes of test is in excess of 650°F.

** Temperature rise is not recorded.

Figure 8.7 Installation Time in Man-Hours for Exterior Doors and Entry Systems

Description	Man-Hours	Unit
Glass Entrance Door, Including Frame and Hardware		
Balanced, Including Glass		
3' x 7'		
Economy	17.780	Ea.
Premium	22.860	Ea.
Hinged, Aluminum		
3' x 7'	8.000	Ea.
3' x 7', 3' Transom	8.890	Ea.
6' x 7'	12.310	Ea.
6' x 10', 3' Transom	14.550	Ea.
Stainless Steel, Including Glass		
3' x 7'		
Minimum	10.000	Ea.
Average	11.430	Ea.
Maximum	13.330	Ea.
Tempered Glass		
3' x 7'	8.000	
6' x 7'	11.430	Ea.
Hinged, Automatic, Aluminum		
6' x 7'	22.860	Ea.
Revolving, Aluminum, 7'-0"		
Diameter x 7' High		
Minimum	42.667	Ea.
Average	53.333	Ea.
Maximum	71.111	Ea.
Stainless Steel, 7'-0" Diameter x 7' High	106.667	Ea.
Bronze, 7'-0" Diameter x 7' High	213.333	Ea.
Glass Storefront System, Including Frame and Hardware		
Hinged, Aluminum, Including Glass, 400 S.F.		
w/3' x 7' Door		
Commercial Grade	.107	S.F.
Institutional Grade	.123	S.F.
Monumental Grade	.139	S.F.
w/6' x 7' Door		
Commercial Grade	.119	S.F.
Institutional Grade	.139	S.F.
Monumental Grade	.160	S.F.
Sliding, Automatic, 12' x 7'-6" w/5' x 7' door	22.860	Ea.
Mall Front, Manual, Aluminum		
15' x 9'	12.310	Ea.
24' x 9'	22.860	Ea.
48' x 9' w/Fixed Panels	17.780	Ea.

Figure 8.7 Installation Time in Man-Hours for Exterior Doors and Entry Systems (continued)

Description	Man-Hours	Unit
Tempered All-Glass w/Glass Mullions,		
up to 10' High	.185	S.F.
up to 20' High, Minimum	.218	S.F.
Average	.240	S.F.
Maximum	.300	S.F.
Entrance Frames, Aluminum, 3' x 7'	2.290	Ea.
3' x 7', 3' Transom	2.460	Ea.
6' x 7'	2.670	Ea.
6' x 7', 3' Transom	2.910	Ea.
Glass, Tempered, 1/4" Thick	.133	S.F.
1/2" Thick	.291	S.F.
3/4" Thick	.457	S.F.
Insulating, 1" Thick	.213	S.F.
Overhead Commercial Doors		
Frames Not Included		
Stock Sectional Heavy Duty Wood		
1-3/4" Thick		
8' x 8'	8.000	Ea.
10' x 10'	8.889	Ea.
12' x 12'	10.667	Ea.
Fiberglass and Aluminum Heavy Duty Sectional		
12' x 12'	10.667	Ea.
20' x 20', Chain Hoist	32.000	Ea.
Steel 24' Gauge Sectional Manual		
8' x 8' High	8.000	Ea.
10' x 10' High	8.889	Ea.
12' x 12' High	10.667	Ea.
20' x 14' High, Chain Hoist	22.857	Ea.
For Electric Trolley Operator to 14' x 14'	4.000	Ea.
Over 14' x 14'	8.000	Ea.

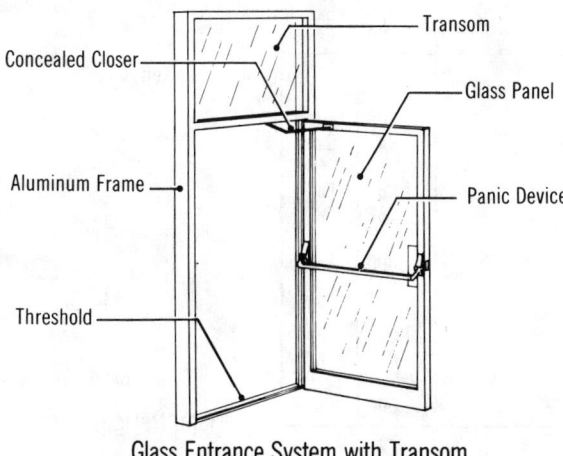

Glass Entrance System with Transom

Figure 8.8 Installation Time in Man-Hours for Interior Doors and Frames

Description	Man-Hours	Unit
Architectural, Flush, Interior, Hollow Core, Veneer Face		
Up to 3'-0" x 7' x 0"	1.020	Ea.
4'-0" x 7'-0"	1.080	Ea.
High Pressured Plastic Laminate Face		
Up to 2'-6" x 6'-8"	1.000	Ea.
3'-0" x 7'-0"	1.153	Ea.
4'-0" x 7'-0"	1.234	Ea.
Particle Core, Veneer Face		
2'-6" x 6'-8"	1.067	Ea.
3'-0" x 6'-8"	1.143	Ea.
3'-0" x 7'-0"	1.231	Ea.
4'-0" x 7'-0"	1.333	Ea.
M.D.O. on Hardboard Face		
3'-0" x 7'-0"	1.333	Ea.
4'-0" x 7'-0"	1.600	Ea.
High Pressure Plastic Laminate Face		
3'-0" x 7'-0"	1.455	Ea.
4'-0" x 7'-0"	2.000	Ea.
Flush, Exterior, Solid Core, Veneer Face		
2'-6" x 7'-0"	1.067	Ea.
3'-0" x 7'-0"	1.143	Ea.
Decorator, Hand Carved		
Solid Wood		
Up to 3'-0" x 7'-0"	1.143	Ea.
3'-6" x 7'-0"	1.231	Ea.
Fire Door, Flush, Mineral Core		
B Label, 1 Hour, Veneer Face		
2'-6" x 6'-8"	1.143	Ea.
3'-0" x 7'-0"	1.333	Ea.
4'-0" x 7'-0"	1.333	Ea.
High Pressure Plastic Laminate Face		
3'-0" x 7'-0"	1.455	Ea.
4'-0" x 7'-0"	1.600	Ea.
Residential, Interior		
Hollow Core or Panel		
Up to 2'-8" x 6'-8"	.889	Ea.
3'-0" x 6'-8"	.941	Ea.
Bi-Folding Closet		
3'-0" x 6'-8"	1.231	Ea.
5'-0" x 6'-8"	1.455	Ea.
Interior Prehung, Hollow Core or Panel		
Up to 2'-8" x 6'-8"	.800	Ea.
3'-0" x 6'-8"	.842	Ea.
Exterior, Entrance, Solid Core or Panel		
Up to 2'-8" x 6'-8"	1.000	Ea.
3'-0" x 6'-8"	1.067	Ea.
Exterior Prehung, Entrance		
Up to 3'-0" x 7'-0"	1.000	Ea.

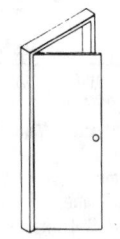

Left Hand Reverse

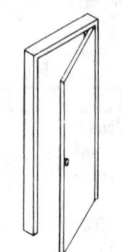

Right Hand Reverse

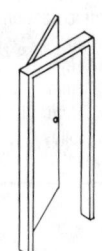

Left Hand

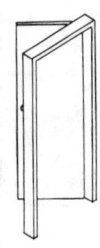

Right Hand

Hand Designations

Figure 8.8 Installation Time in Man-Hours for Interior Doors and Frames (continued)

Description	Man-Hours	Unit
Hollow Metal Doors Flush		
Full Panel, Commercial		
20 Gauge		
2'-0" x 6'-8"	.800	Ea.
2'-6" x 6'-8"	.888	Ea.
3'-0" x 6'-8" or 3'-0" x 7'-0"	.941	Ea.
4'-0" x 7'-0"	1.066	Ea.
18 Gauge		
2'-6" x 6'-8" or 2'-6" x 7'-0"	.941	Ea.
3'-0" x 6'-8" or 3'-0" x 7'-0"	1.000	Ea.
4'-0" x 7'-0"	1.066	Ea.
Residential		
24 Gauge		
2'-8" x 6'-8"	1.000	Ea.
3'-0" x 7'-0"	1.066	Ea.
Bifolding		
3'-0" x 6'- 8"	1.000	Ea.
5'-0" x 6'-8"	1.143	Ea.
Steel Frames		
18 Gauge		
3'-0" Wide	1.000	Ea.
6'-0" Wide	1.142	Ea.
16 Gauge		
4'-0" Wide	1.066	Ea.
8'-0" Wide	1.333	Ea.
Transom Lite Frames		
Fixed Add	.103	S.F.
Movable Add	.123	S.F.

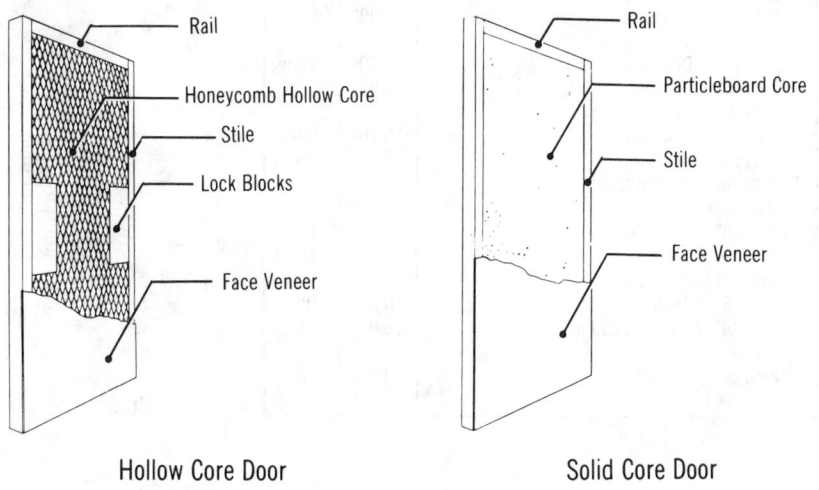

Hollow Core Door   Solid Core Door

Figure 8.9 Installation Time in Man-Hours for Special Doors

Description	Man-Hours	Unit
Access Panels Metal		
12" x 12"	.400	Ea.
12" x 24"	.533	Ea.
18" x 18"	.666	Ea.
24" x 36"	1.000	Ea.
36" x 48"	1.333	Ea.
Cold Strorage Doors		
Single Galvanized Steel Horizontal Sliding		
5' x 7'		
Manual Operation	8.000	Ea.
Power Operation	8.421	Ea.
9' x 10'		
Manual Operation	9.411	Ea.
Power Operation	10.000	Ea.
Hinged Lightweight 3' x 6'-6"		
2" Thick	8.000	Ea.
4" Thick	8.421	Ea.
6" Thick	11.428	Ea.
Bi-parting Electric Operated 6' x8'	20.000	Opng.
For Door Buck Framing and Door Protection Add	6.400	Opng.
Galvanized Batten Door		
4' x 7'	8.000	Opng.
6' x 8'	8.888	Opng.
Fire Door 6' x 8"		
Single Slide	20.000	Opng.
Double Bi-parting	22.857	Opng.
Darkroom Doors Revolving		
2 Way 30" Diameter	4.571	Opng.
3 Way 43" Diameter	5.333	Opng.
4 Way 68" Diameter	8.000	Opng.
Hinged Safety		
2 Way 30" Diameter	5.000	Opng.
3 Way 43" Diameter	5.517	Opng.
Pop Out Safety		
2 Way 30" Diameter	5.161	Opng.
3 Way 43" Diameter	5.714	Opng.
Pass Window Painted Steel		
24" x 36"	10.000	Ea.
48" x 48"	13.333	Ea.
72" x 40"	20.000	Ea.
Vault Front Door and Frame		
32" x 78", 1 Hour, 750 lbs.	10.670	Opng.
40" x 78", 2 Hours, 1130 lbs.	16.000	Opng.
Day Gate		
32" Wide	10.670	Opng.
42" Wide	11.430	Opng.

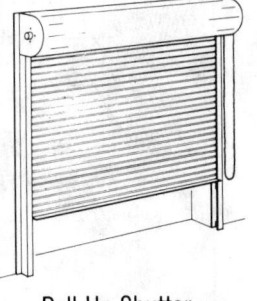

Roll Up Shutter

Figure 8.9   Installation Time in Man-Hours for
Special Doors (continued)

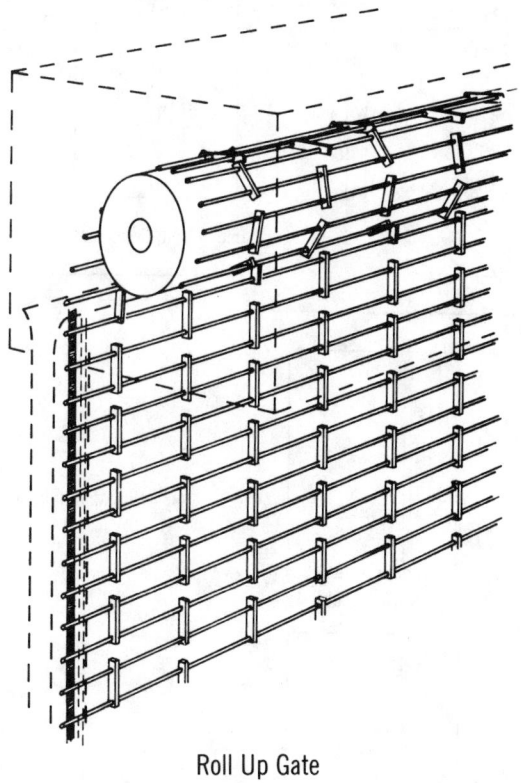

Roll Up Gate

Figure 8.10   Hinge Requirements

All closer equipped doors should have ball bearing hinges. Lead lined or extremely heavy doors require special strength hinges. Usually 1-1/2 pair of hinges are used per door up to 7'-6" high openings. The table below shows typical hinge requirements.

Use Frequency	Type Hinge Required	Type of Opening	Type of Structure
High	Heavy Weight Ball Bearing	Entrances Toilet Rooms	Banks, Office Buildings, Schools, Stores and Theaters Office Buildings and Schools
Average	Standard Weight Ball Bearing	Entrances Corridors Toilet Rooms	Dwellings Office Buildings and Schools Stores
Low	Plain Bearing	Interior	Dwellings

Figure 8.11   Hardware Nomenclature

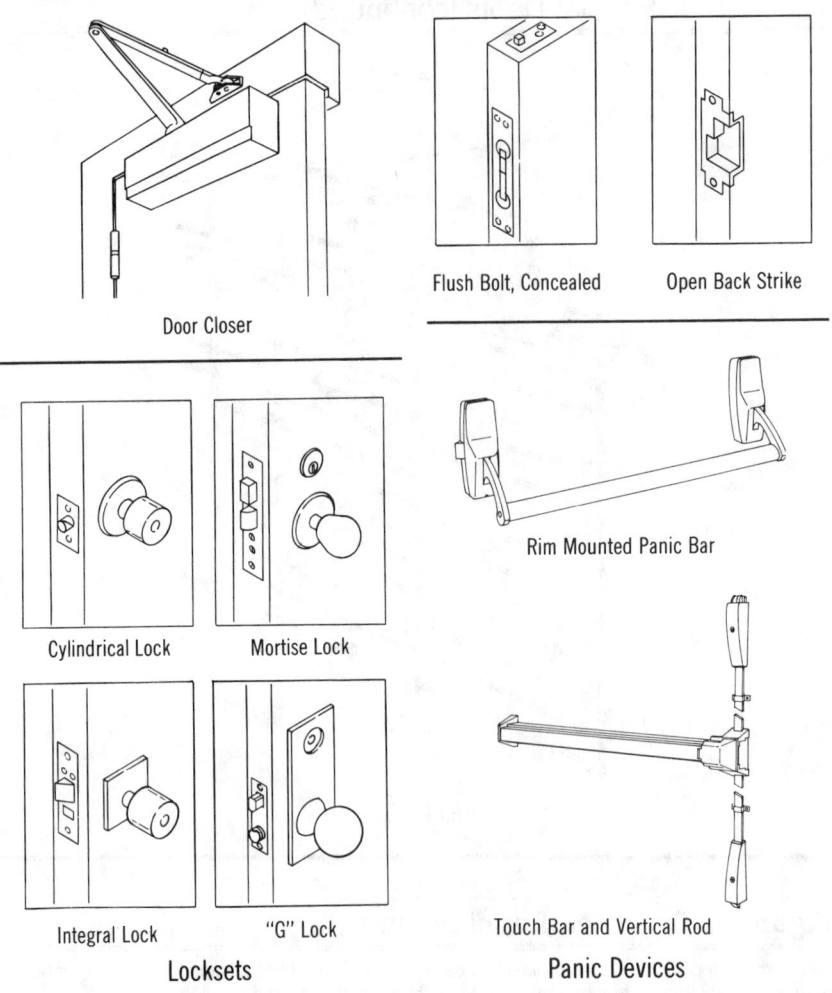

Door Closer

Flush Bolt, Concealed   Open Back Strike

Cylindrical Lock   Mortise Lock

Rim Mounted Panic Bar

Integral Lock   "G" Lock

Touch Bar and Vertical Rod

Locksets

Panic Devices

Figure 8.12   Finish Hardware Allowances

Average Allowances for Finish Hardware as a Percentage of Total Job Costs	
Minimum	0.75%
Maximum	3.50%

Average distribution of total finish hardware costs for a typical building is 85% material, 15% labor.

Figure 8.13   Installation Time in Man-Hours for Finish Hardware

Description	Man-Hours	Unit
Astragals, 1/8" x 3"	.089	L.F.
Spring Hinged Security Seal with Cam	.107	L.F.
Two Piece Overlapping	.133	L.F.
Automatic Openers, Swing Doors, Single	20.000	Ea.
Single Operating, Pair	32.000	Pair
Sliding Doors, 3' Wide Including Track and Hanger, Single	26.667	Opng.
Bi-parting	32.000	Opng.
Handicap Opener Button, Operating	2.000	Ea.
Bolts, Flush, Standard, Concealed	1.143	Ea.
Bumper Plates, 1-1/2" x 3/4" U Channel	.200	L.F.
Door Closer, Adjustable Backcheck, 3 Way Mount	1.333	Ea.
Doorstops	.333	Ea.
Kickplate	533	Ea.
Lockset, Non-Keyed	.667	Ea.
Keyed	.800	Ea.
Dead Locks, Heavy Duty	.889	Ea.
Entrance Locks, Deadbolt	1.000	Ea.
Mortise Lockset, Non Keyed	.889	Ea.
Keyed	1.000	Ea.
Panic Device for Rim Locks		
Single Door, Exit Only	1.333	Ea.
Outside Key and Pull	1.600	Ea.
Bar and Vertical Rod, Exit Only	1.600	Ea.
Outside Key and Pull	2.000	Ea.
Push Pull Plate	.667	Ea.
Weatherstripping, Window, Double Hung	1.111	Opng.
Doors, Wood Frame, 3' x 7'	1.053	Opng.
6' x 7'	1.143	Opng.
Metal Frame, 3' x 7'	2.667	Opng.
6' x 7'	3.200	Opng.

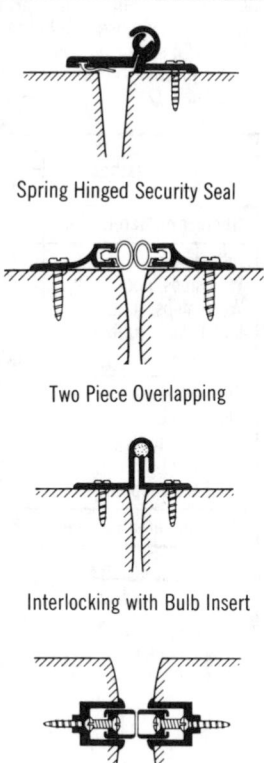

Spring Hinged Security Seal

Two Piece Overlapping

Interlocking with Bulb Insert

Two Piece Adjustable

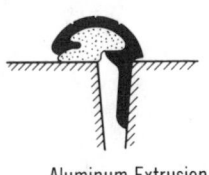

Aluminum Extrusion
with Sponge Insert

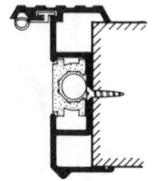

Spring Loaded Locking Bolt

Spring Bronze Strip

Astragals

Figure 8.14   Glazing Labor

Glass sizes are estimated by the "united inch" (height + width). The table below shows the number of lights glazed in an eight hour period by the crew size indicated, for glass up to 1/4" thick. Square or nearly square lights are more economical on a S.F. basis. Long slender lights will have a high S.F. installation cost. For insulated glass reduce production by 33%. For 1/2" plate glass reduce production by 50%. Production time for glazing with two glazers per day averages; 1/4" plate glass 120 S.F.; 1/2" plate glass 55 S.F.; 1/2" insulated glass 95 S.F.; insulated glass 75 S.F.

Glazing Method	United Inches per Light							
	40"	60"	80"	100"	135"	165"	200"	240"
Number of Men in Crew	1	1	1	1	2	3	3	4
Industrial sash, putty	60	45	24	15	18	—	—	—
With stops, putty bed	50	36	21	12	16	8	4	3
Wood stops, rubber	40	27	15	9	11	6	3	2
Metal stops, rubber	30	24	14	9	9	6	3	2
Structural glass	10	7	4	3	—	—	—	—
Corrugated gloass	12	9	7	4	4	4	3	—
Store fronts	16	15	13	11	7	6	4	4
Skylights, puttyglass	60	36	21	12	16	—	—	—
Thiokol set	15	15	11	9	9	6	3	2
Vinyl set, snap on	18	18	13	12	12	7	5	4
Maximum area per light	2.8 S.F.	6.3 S.F.	11.1 S.F.	17.4 S.F.	31.6 S.F.	47 S.F.	69 S.F.	100 S.F.

Figure 8.15 Installation Time in Man-Hours for Exterior Windows/Sash

Description	Man-Hours	Unit
Custom Aluminum Sash, Glazing Not Included	.080	S.F.
Stock Aluminum Windows, Frame and Glazing		
Including Casement, 3'-1" x 3'-2",		
Wood Opening	1.600	Ea.
Masonry Opening	2.667	Ea.
Combination Storm and Screen,		
2'-4" x 3'-2" Wood Opening	.533	Ea.
Masonry Opening	.941	Ea.
3'-4" x 5'-6" Wood Opening	.615	Ea.
Masonry Opening	1.143	Ea.
Projected, with Screen,		
3'-1" x 3'-2" Wood Opening	1.600	Ea.
Masonry Opening	2.667	Ea.
4'-5' x 5'-3" Wood Opening	2.000	Ea.
Masonry Opening	4.000	Ea.
Single Hung, 2'x3' Wood Opening	1.600	Ea.
Masonry Opening	2.667	Ea.
3'-4" x 5'-0" Wood Opening	1.778	Ea.
Masonry Opening	3.200	Ea.
Sliding, 3' x 2" Wood Opening	1.600	Ea.
Masonry Opening	2.667	Ea.
5'x3" Wood Opening	1.778	Ea.
Masonry Opening	3.200	Ea.
8'x4' Wood Opening	2.667	Ea.
Masonry Opening	5.333	Ea.
Jalousies, 2'-3" x 4'-0" Opening	1.600	Ea.
3'-1' x 5'-3" Opening	1.600	Ea.
Custom Steel Sash Units, Glazing and		
Trim Not Included	.080	S.F.
Stock Steel Windows, Frame, Trim and		
Glass Included		
Double Hung, 2'-8" x 4'-6" Opening	1.333	Ea.
Commercial Projected, 3'-9" x 5'-5" Opening	1.600	Ea.
6'-9" x 4'-1" Opening	2.286	Ea.
Intermediate Projected, 2'-9" x 4'-1" Opening	1.333	Ea.
Custom Wood Sash, Including Double/Triple		
Glazing but Not Trim		
5' x 4' Opening	3.200	Ea.
7' x 4'-6" Opening	3.721	Ea.
8'-6" x 5' Opening	4.571	Ea.
Stock Wood Windows, Frame,		
Trim and Glass Included		
Awning Type, 2'-10" x 1'-9" Opening	1.333	Ea.
4'-4" x 6'-0" Opening	2.286	Ea.

(continued on next page)

Figure 8.15   Installation Time in Man-Hours for Exterior Windows/Sash (continued)

Description	Man-Hours	Unit
Bow Bay, Fixed Lights, 8' x 5' Opening	5.333	Ea.
9'-9" x 5'-4" Opening	8.000	Ea.
End Units Vent., 8' x 5' Opening	6.400	Ea.
Casement, 1'-10" x 3'x2" Opening	1.455	Ea.
4'-2" x 4'-2" Opening, 2 Leaf	1.778	Ea.
5'-11" x 5'-2" Opening, 3 Leaf	2.286	Ea.
9'-11" x 6'-3" Opening, 5 Leaf	3.200	Ea.
Double Hung, 2'-8" x 4'-6" Opening	2.353	Ea.
3'-0" x 5'-6" Opening	2.667	Ea.
Picture Window, 4'-6" x 4'-6" Opening	3.200	Ea.
Roof Window, Complete,		
2'-9" x 3'-8" Opening	5.000	Ea.
Sliding, 3'-4" x 2'-7" Opening	2.462	Ea.
5'-4" x 6'-0" Opening	2.667	Ea.

Combination Storm
and Screen Window

Single Hung Window - Aluminum

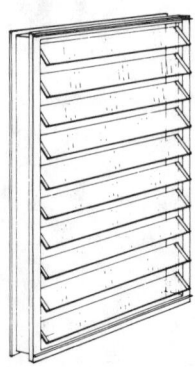

Jalousie Window - Aluminum

Projected Window - Aluminum

Figure 8.15 Installation Time in Man-Hours for
Exterior Windows/Sash (continued)

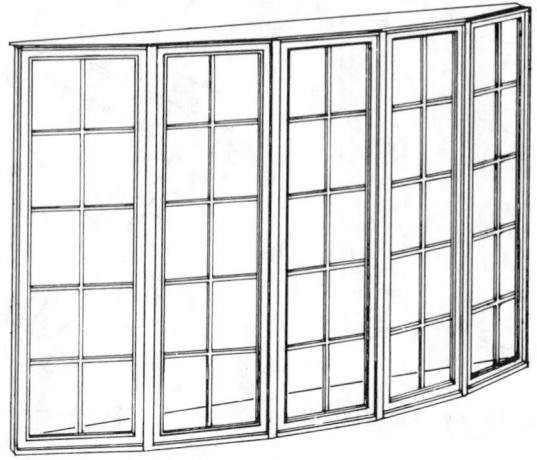

Casement Bow Window - Wood

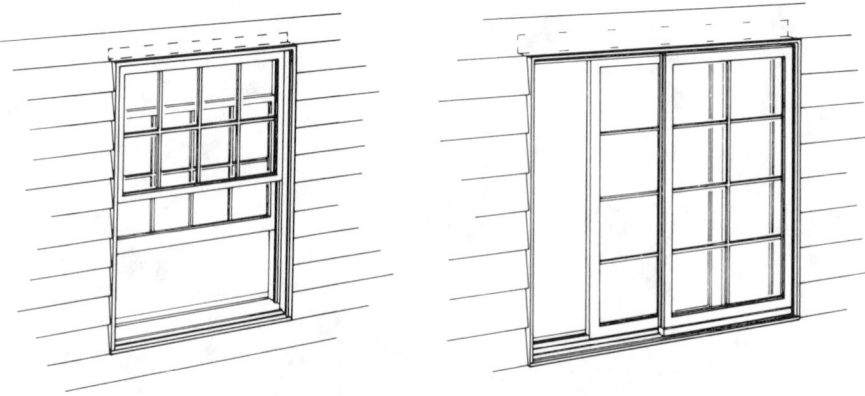

Double Hung Window - Wood

Sliding Window - Wood

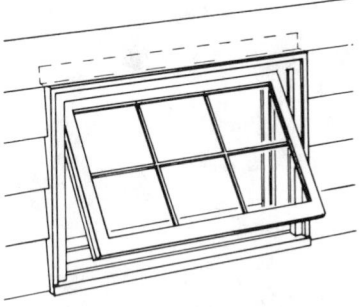

Awning Window - Wood

(continued on next page)

Figure 8.15 Installation Time in Man-Hours for Exterior Windows/Sash (continued)

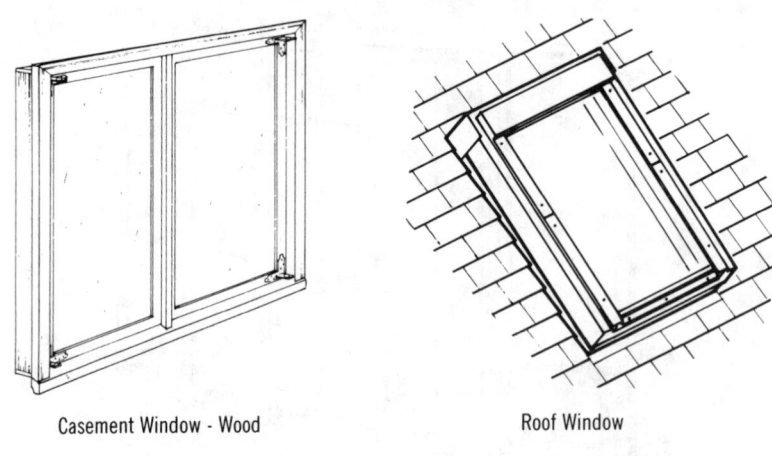

Casement Window - Wood   Roof Window

Checklist

For an estimate to be reliable, all items must be accounted for. A complete estimate can also eliminate the need to include contingencies. The following checklist can be used to help ensure that all items are properly accounted for.

Doors
- ☐ Accordian
- ☐ Acoustical
- ☐ Aluminum
- ☐ Bi-parting
- ☐ Blast
- ☐ Cold storage
- ☐ Fire
  - _____ Rated
  - _____ Rolling industrial
- ☐ Folding
- ☐ Garage
- ☐ Glass
  - _____ Sliding
  - _____ Swing
- ☐ Grilles
  - _____ Rolling
- ☐ Hollow metal
- ☐ Overhead
- ☐ Revolving
- ☐ Rolling
- ☐ Service
- ☐ Sliding
- ☐ Tin-clad
- ☐ Vault
- ☐ Wood

Door Frames for above _____

Entrances _____

Store Fronts _____

Windows
- ☐ Casement
- ☐ Projected
- ☐ Single hung
- ☐ Double hung

- [ ] Jalousie
- [ ] Picture
- [ ] Roof
- [ ] Screens
- [ ] Security
- [ ] Sliding

Window Wall Systems

- [ ] Curtain wall system

Hardware _____

Hardware Allowance _____

Automatic Openers _____

Glass

- [ ] Acoustical
- [ ] Faceted
- [ ] Insulated
- [ ] Laminated
- [ ] Obscure
- [ ] Plexiglass
- [ ] Reflective
- [ ] Sheet
- [ ] Spandrel
- [ ] Stained
- [ ] Window
- [ ] Wire
- [ ] Glazing
- [ ] Allowance for broken glass replacement
- [ ] Allowance for temporary glass

Tips

Interior Door Ratings

For walls to be considered fire-rated, any doors in those walls must be fire-rated. Most plans do not spell out which interior partitions are to be considered fire-rated. In commercial applications, a rule of thumb is that all partitions that have drywall (or masonry) from the floor to the above structure and few, if any, penetrations should be considered fire-rated.

Door Schedule

If the drawings do not include a door schedule, it may be worth the estimator's time to develop one, especially if the project is large or complicated. On the schedule should be the opening number, door type, size material, glass or louver requirements, and remarks.

The door schedule should include a frame schedule listing the frame material, type, and jamb, head and sill details.

Hardware requirements should be listed on the door schedule as well. Keep in mind the fact that hardware can, in some instances, be more costly than the door itself.

Handicap Access

While the drawings may not show it, local codes may require special hardware and opening systems to allow a structure to be accessible to the handicapped. Contact the local authorities for their codes and requirements.

Special Doors

Special attention should be given to any oversized or unusual type of doors. These items should not be priced on a prorated basis, as they are generally special order items. The costs of special doors can skyrocket, especially if they involve exotic woods, special finishes, or special attention (which usually means higher labor costs). Note also that special doors may require a considerable amount of lead time for ordering and shipping.

Window Schedule

As with all doors to be included in the project, all windows should be listed on the drawings in a window schedule. If none is included, it may be well worth the time to create one. The schedule should contain the opening number, window type, window size, glass type, frame material and details, and required accessories and hardware.

Building Hardware

As a rule of thumb, building hardware for an average quality building can be expected to run in the neighborhood of 2% of the entire building cost.

Notes

Division Nine
Finishes

Introduction

From an estimating standpoint, Division 9 — Finishes, does not represent a heavy cost area of the overall project cost. However, from the construction standpoint, it is one of the most important. Upon taking possession of the building, the owner cannot see the piles you had to drive in winter conditions while working on unstable ground. He/she cannot see the good quality of work that went into the structure of the building. The most visible components are those in the finish work: the paint job in the main office area, the carpeting in the executive area, the vinyl flooring in the production area. Therefore, when estimating the finishes for any structure, this is one category that should not be allowed to be "cheapened" to get the job or get back on budget.

Many buildings are built on a "spec" basis, that is, built before they have tenants committed; on pure speculation that someone will lease or purchase the structure, or at least part of it. In this case, the party financing the project may take one of two paths while completing the building. They might seek to attract a client by finishing the building in a handsome, attractive but generic manner, or they may leave the building unfinished, perhaps finishing only the lobby or common areas, if the building warrants it. Either way, finishes are a very important part of the overall project, even if they represent a relatively small portion of the whole picture.

This chapter contains many charts and tables for use in estimating Division 9 — Finishes. Included here are appropriate tables to estimate materials for finishing walls, ceilings, floors, doors, windows, and trim work, as well as installation time information. The areas that are covered include:

- Drywall and plaster construction
- Tile work
- Terrazzo
- Acoustical treatment
- Flooring
- Painting and wallcoverings

Estimating Data

The following tables present effective estimating guidelines for items found in Division 9 — Finishes. Please note that these guidelines can be used as indicators of what may be expected, and that each project must be evaluated individually.

Table of Contents

Figure 9.1 Partition Density Guide

This table allows you to estimate the average quantities of partitions found in various types of buildings when no interior plans are available, such as in a conceptual estimating situation. To use this chart, pick out the type of structure to be estimated from the left-most column. Moving across, choose from the second column the number of stories of the building. Read across to the "Partition Density" column. This figure represents the number of square feet of floor area for every linear foot of partition.

Example: The project to be estimated is a three-story office building of 10,000 square feet per floor (total of 30,000 square feet). To estimate the approximate quantity of interior partitions in the entire building, first look in column 1 for "Office." Reading across to the "Stories" column, find the "3-5 Story" line. The "Partition Density" column for that line indicates 20 S.F./L.F. To find the number of partitions, divide the building area by the density factor: 30,000 S.F./20 = 1,500 L.F. of partition. Note that the right-most column gives a breakdown of average partition types. For our office building, the average mix is 30% concrete block (generally found at stairwells and service area partitions) and 70% drywall. Our drywall total would then be 1,050 L.F. To determine the total square footage of partitions, multiply this linear footage by the specified (or assumed) partition height. Note that the remainder of the partitions (450 L.F.) must be added to the masonry portion of the estimate, or at least accounted for in the estimate.

Building Type	Stories	Partition Density	Description of Partition
Apartments	1 story	9 S.F./L.F.	Plaster, wood doors & trim
	2 story	8 S.F./L.F.	Drywall, wood studs, wood doors & trim
	3 story	9 S.F./L.F.	Plaster, wood studs, and wood doors & trim
	5 story	9 S.F./L.F.	Plaster, wood studs, wood doors & trim
	6–15 story	8 S.F./L.F.	Drywall, wood studs, wood doors & trim
Bakery	1 story	50 S.F./L.F.	Conc. block, paint, door & drywall, wood studs
	2 story	50 S.F./L.F.	Conc. block, paint, door & drywall, wood studs
Bank	1 story	20 S.F./L.F.	Plaster, wood studs, wood doors & trim
	2-4 story	15 S.F./L.F.	Plaster, wood studs, wood doors & trim
Bottling Plant	1 story	50 S.F./L.F.	Conc. block, drywall, wood studs, wood trim
Bowling Alley	1 story	50 S.F./L.F.	Conc. block, wood & metal doors, wood trim
Bus Terminal	1 story	15 S.F./L.F.	Conc. block, ceramic tile, wood trim
Cannery	1 story	100 S.F./L.F.	Drywall on metal studs
Car Wash	1 story	18 S.F./L.F.	Concrete block, painted & hollow metal door
Dairy Plant	1 story	30 S.F./L.F.	Concrete block, glazed tile, insulated cooler doors
Department Store	1 story	60 S.F./L.F.	Drywall, wood studs, wood doors & trim
	2–5 story	60 S.F./L.F.	30% concrete block, 70% drywall, wood studs
Dormitory	2 story	9 S.F./L.F.	Plaster, concrete block, wood doors & trim
	3–5 story	9 S.F./L.F.	Plaster, concrete block, wood doors & trim
	6–15 story	9 S.F./L.F.	Plaster, concrete block, wood doors & trim
Funeral Home	1 story	15 S.F./L.F.	Plaster on concrete block & wood studs, paneling
	2 story	14 S.F./L.F.	Plaster, wood studs, paneling & wood doors
Garage Sales & Service	1 story	30 S.F./L.F.	50% conc. block, 50% drywall, wood studs
Hotel	3–8 story	9 S.F./L.F.	Plaster, conc. block, wood doors & trim
	9–15 story	9 S.F./L.F.	Plaster, conc. block, wood doors & trim
Laundromat	1 story	25 S.F./L.F.	Drywall, wood studs, wood doors & trim
Medical Clinic	1 story	6 S.F./L.F.	Drywall, wood studs, wood doors & trim
	2–4 story	6 S.F./L.F.	Drywall, wood studs, wood doors & trim
Motel	1 story	7 S.F./L.F.	Drywall, wood studs, wood doors & trim
	2–3 story	7 S.F./L.F.	Concrete block, drywall on wood studs, wood paneling

(continued on next page)

Figure 9.1 Partition Density Guide (continued)

Building Type	Stories	Partition Density	Description of Partition
Movie 200–600 seats Theater 601–1400 seats 1401–2200 seats	1 story	18 S.F./L.F. 20 S.F./L.F. 25 S.F./L.F.	Concrete block, wood, metal, vinyl trim Concrete block, wood, metal, vinyl trim Concrete block, wood, metal, vinyl trim
Nursing Home	1 story 2–4 story	8 S.F./L.F. 8 S.F./L.F.	Drywall, wood studs, wood doors & trim Drywall, wood studs, wood doors & trim
Office	1 story 2 story 3–5 story 6–10 story 11–20 story	20 S.F./L.F. 20 S.F./L.F. 20 S.F./L.F. 20 S.F./L.F. 20 S.F./L.F.	30% concrete block, 70% drywall on wood studs 30% concrete block, 70% drywall on wood studs 30% concrete block, 70% movable partitions 30% concrete block, 70% movable partitions 30% concrete block, 70% movable partitions
Parking Ramp (Open) Parking Garage	2–8 story 2–8 story	60 S.F./L.F. 60 S.F./L.F.	Stair and elevator enclosures only Stair and elevator enclosures only

Pre-Engineered   Store
                              Office
                              Shop
	1 story
	1 story
	1 story
		60 S.F./L.F.
		15 S.F./L.F.
		15 S.F./L.F.
			Drywall on wood studs, wood doors & trim
			Concrete block, movable wood partitions
			Movable wood partitions

Radio & TV Broadcasting & TV Transmitter	1 story 1 story	25 S.F./L.F. 40 S.F./L.F.	Concrete block, metal and wood doors Concrete block, metal and wood doors
Self Service Restaurant Cafe & Drive-in Restaurant Restaurant with seating Supper Club Bar or Lounge	1 story 1 story 1 story 1 story 1 story	15 S.F./L.F. 18 S.F./L.F. 25 S.F./L.F. 25 S.F./L.F. 24 S.F./L.F.	Concrete block, wood and aluminum trim Drywall, wood studs, ceramic & plastic trim Concrete block, paneling, wood studs & trim Concrete block, paneling, wood studs & trim Plaster or gypsum lath, wood studs
Retail Store or Shop	1 story	60 S.F./L.F.	Drywall wood studs, wood doors & trim

Service Station   Masonry
                       Metal panel
                       Frame
	1 story
	1 story
	1 story
		15 S.F./L.F.
		15 S.F./L.F.
		15 S.F./L.F.
			Concrete block, paint, door & drywall, wood studs
			Concrete block paint door & drywall, wood studs
			Drywall, wood studs, wood doors & trim

Shopping Center (strip) (group)	1 story 1 story 2 story	30 S.F./L.F. 40 S.F./L.F. 40 S.F./L.F.	Drywall, wood studs, wood doors & trim 50% concrete block, 50% drywall, wood studs 50% concrete block, 50% drywall, wood studs
Small Food Store	1 story	30 S.F./L.F.	Concrete block drywall, wood studs, wood trim
Store/Apt. Masonry above Frame Frame	2 story 2 story 3 story	10 S.F./L.F. 10 S.F./L.F. 10 S.F./L.F.	Plaster, wood studs, wood doors & trim Plaster wood studs, wood doors & trim Plaster, wood studs and wood doors & trim
Supermarkets	1 story	40 S.F./L.F.	Concrete block, paint, drywall & porcelain panel
Truck Terminal	1 story	0	
Warehouse	1 story	0	

Figure 9.2 Surface Area of Rooms

This table provides the total surface area in square feet of room walls and ceilings. Select the room dimensions from the top row and left column. The square at which the appropriate column and row intersect is the total square footage for that room. This chart is especially helpful in determining quantities for painting or drywalling. Note that no deductions have been made for openings (windows or doors).

The last two sections of this table show square footages for single surface areas, such as a single wall or a floor area.

Rooms with 7' Ceilings—Total Area – 4 Walls and Ceiling (in square feet)

	3'	4'	5'	6'	7'	8'	9'	10'	11'	12'	13'	14'	15'	16'	17'	18'	19'	20'	21'	22'
3'	93	110	127	144	161	178	195	212	229	246	263	280	297	314	331	348	365	382	399	416
4'	110	128	146	164	182	200	218	236	254	272	290	308	326	344	362	380	398	416	434	452
5'	127	146	165	184	203	222	241	260	279	298	317	336	355	374	393	412	431	450	469	488
6'	144	164	184	204	224	244	264	284	304	324	344	364	384	404	424	444	464	484	504	524
7'	161	182	203	224	245	266	287	308	329	350	371	392	413	434	455	476	497	518	539	560
8'	178	200	222	244	266	288	310	332	354	376	398	420	442	464	486	508	530	552	574	596
9'	195	218	241	264	287	310	333	356	379	402	425	448	471	494	517	540	563	586	609	632
10'	212	236	260	284	308	332	356	380	404	428	452	476	500	524	548	572	596	620	644	668
11'	229	254	279	304	329	354	379	404	429	454	479	504	529	554	579	604	629	654	679	704
12'	246	272	298	324	350	376	402	428	454	480	506	532	558	584	610	636	662	688	714	740
13'	263	290	317	344	371	398	425	452	479	506	533	560	587	614	641	668	695	722	749	776
14'	280	308	336	364	392	420	448	476	504	532	560	588	616	644	672	700	728	756	784	812
15'	297	326	355	384	413	442	471	500	529	558	587	616	645	674	703	732	761	790	819	848
16'	314	344	374	404	434	464	494	524	554	584	614	644	674	704	734	764	794	824	854	884
17'	331	362	393	424	455	486	517	548	579	610	641	672	703	734	765	796	827	858	889	920
18'	348	380	412	444	476	508	540	572	604	636	668	700	732	764	796	828	860	892	924	956
19'	365	398	431	464	497	530	563	596	629	662	695	728	761	794	827	860	893	926	959	992
20'	382	416	450	484	518	552	586	620	654	688	722	756	790	824	858	892	926	960	994	1028
21'	399	434	469	504	539	574	609	644	679	714	749	784	819	854	889	924	959	994	1029	1064
22'	416	452	488	524	560	596	632	668	704	740	776	812	848	884	920	956	992	1028	1064	1100
23'	433	470	507	544	581	618	655	692	729	766	803	840	877	914	951	988	1025	1062	1099	1136
24'	450	488	526	564	602	640	678	716	754	792	830	868	906	944	982	1020	1058	1096	1134	1172

(continued on next page)

Figure 9.2 Surface Area of Rooms (continued)

Rooms with 7'6" Ceilings—Total Area – 4 Walls and Ceiling (in square feet)

	3'	4'	5'	6'	7'	8'	9'	10'	11'	12'	13'	14'	15'	16'	17'	18'	19'	20'	21'	22'
3'	99	117	135	153	171	189	207	225	243	261	279	297	315	333	351	369	387	405	423	441
4'	117	136	155	174	193	212	231	250	269	288	307	326	345	364	383	402	421	440	459	478
5'	135	155	175	195	215	235	255	275	295	315	335	355	375	395	415	435	455	475	495	515
6'	153	174	195	216	237	258	279	300	321	342	363	384	405	426	447	468	489	510	531	552
7'	171	193	215	237	259	281	303	325	347	369	391	413	435	457	479	501	523	545	567	589
8'	189	212	235	258	281	304	327	350	373	396	419	442	465	488	511	534	557	580	603	626
9'	207	231	255	279	303	327	351	375	399	423	447	471	495	519	543	567	591	615	639	663
10'	225	250	275	300	325	350	375	400	425	450	475	500	525	550	575	600	625	650	675	700
11'	243	269	295	321	347	373	399	425	451	477	503	529	555	581	607	633	659	685	711	737
12'	261	288	315	342	369	396	423	450	477	504	531	558	585	612	639	666	693	720	747	774
13'	279	307	335	363	391	419	447	475	503	531	559	587	615	643	671	699	727	755	783	811
14'	297	326	355	384	413	442	471	500	529	558	587	616	645	674	703	732	761	790	819	848
15'	315	345	375	405	435	465	495	525	555	585	615	645	675	705	735	765	795	825	855	885
16'	333	364	395	426	457	488	519	550	581	612	643	674	705	736	767	798	829	860	891	922
17'	351	383	415	447	479	511	543	575	607	639	671	703	735	767	799	831	863	895	927	959
18'	369	402	435	468	501	534	567	600	633	666	699	732	765	798	831	864	897	930	963	996
19'	387	421	455	489	523	557	591	625	659	693	727	761	795	829	863	897	931	965	999	1033
20'	405	440	475	510	545	580	615	650	685	720	755	790	825	860	895	930	965	1000	1035	1070
21'	423	459	495	531	567	603	639	675	711	747	783	819	855	891	927	963	999	1035	1071	1107
22'	441	478	515	552	589	626	663	700	737	774	811	848	885	922	959	996	1033	1070	1107	1144
23'	459	497	535	573	611	649	687	725	763	801	839	877	915	953	991	1029	1067	1105	1143	1181
24'	477	516	555	594	633	672	711	750	789	828	867	906	945	984	1023	1062	1101	1140	1179	1218

Figure 9.2   Surface Area of Rooms (continued)

Rooms with 8' Ceilings—Total Area – 4 Walls and Ceiling (in square feet)

	3'	4'	5'	6'	7'	8'	9'	10'	11'	12'	13'	14'	15'	16'	17'	18'	19'	20'	21'	22'
3'	105	124	143	162	181	200	219	238	257	276	295	314	333	352	371	390	409	428	447	466
4'	124	144	164	184	204	224	244	264	284	304	324	344	364	384	404	424	444	464	484	504
5'	143	164	185	206	227	248	269	290	311	332	353	374	395	416	437	458	479	500	521	542
6'	162	184	206	228	250	272	294	316	338	360	382	404	426	448	470	492	514	536	558	580
7'	181	204	227	250	273	296	319	342	365	388	411	434	457	480	503	526	549	572	595	618
8'	200	224	248	272	296	320	344	368	392	416	440	464	488	512	536	560	584	608	632	656
9'	219	244	269	294	319	344	369	394	419	444	469	494	519	544	569	594	619	644	669	694
10'	238	264	290	316	342	368	394	420	446	472	498	524	550	576	602	628	654	680	706	732
11'	257	284	311	338	365	392	419	446	473	500	527	554	581	608	635	662	689	716	743	770
12'	276	304	332	360	388	416	444	472	500	528	556	584	612	640	668	696	724	752	780	808
13'	295	324	353	382	411	440	469	498	527	556	585	614	643	672	701	730	759	788	817	846
14'	314	344	374	404	434	464	494	524	554	584	614	644	674	704	734	764	794	824	854	884
15'	333	364	395	426	457	488	519	550	581	612	643	674	705	736	767	798	829	860	891	922
16'	352	384	416	448	480	512	544	576	608	640	672	704	736	768	800	832	864	896	928	960
17'	371	404	437	470	503	536	569	602	635	668	701	734	767	800	833	866	899	932	965	998
18'	390	424	458	492	526	560	594	628	662	696	730	764	798	832	866	900	934	968	1002	1036
19'	409	444	479	514	549	584	619	654	689	724	759	794	829	864	899	934	969	1004	1039	1074
20'	428	464	500	536	572	608	644	680	716	752	788	824	860	896	932	968	1004	1040	1076	1112
21'	447	484	521	558	595	632	669	706	743	780	817	854	891	928	965	1002	1039	1076	1113	1150
22'	466	504	542	580	618	656	694	732	770	808	846	884	922	960	998	1036	1074	1112	1150	1188
23'	485	524	563	602	641	680	719	758	797	836	875	914	953	992	1031	1070	1109	1148	1187	1226
24'	504	544	584	624	664	704	744	784	824	864	904	944	984	1024	1064	1104	1144	1184	1224	1264

(continued on next page)

Figure 9.2   Surface Area of Rooms (continued)

Rooms with 8'6" Ceilings—Total Area – 4 Walls and Ceiling (in square feet)

	3'	4'	5'	6'	7'	8'	9'	10'	11'	12'	13'	14'	15'	16'	17'	18'	19'	20'	21'	22'
3'	111	131	151	171	191	211	231	251	271	291	311	331	351	371	391	411	431	451	471	491
4'	131	152	173	194	215	236	257	278	299	320	341	362	383	404	425	446	467	488	509	530
5'	151	173	195	217	239	261	283	305	327	349	371	393	415	437	459	481	503	525	547	569
6'	171	194	217	240	263	286	309	332	355	378	401	424	447	470	493	516	539	562	585	608
7'	191	215	239	263	287	311	335	359	383	407	431	455	479	503	527	551	575	599	623	647
8'	211	236	261	286	311	336	361	386	411	436	461	486	511	536	561	586	611	636	661	686
9'	231	257	283	309	335	361	387	413	439	465	491	517	543	569	595	621	647	673	699	725
10'	251	278	305	332	359	386	413	440	467	494	521	548	575	602	629	656	683	710	737	764
11'	271	299	327	355	383	411	439	467	495	523	551	579	607	635	663	691	719	747	775	803
12'	291	320	349	378	407	436	465	494	523	552	581	610	639	668	697	726	755	784	813	842
13'	311	341	371	401	431	461	491	521	551	581	611	641	671	701	731	761	791	821	851	881
14'	331	362	393	424	455	486	517	548	579	610	641	672	703	734	765	796	827	858	889	920
15'	351	383	415	447	479	511	543	575	607	639	671	703	735	767	799	831	863	895	927	959
16'	371	404	437	470	503	536	569	602	635	668	701	734	767	800	833	866	899	932	965	998
17'	391	425	459	493	527	561	595	629	663	697	731	765	799	833	867	901	935	969	1003	1037
18'	411	446	481	516	551	586	621	656	691	726	761	796	831	866	901	936	971	1006	1041	1076
19'	431	467	503	539	575	611	647	683	719	755	791	827	863	899	935	971	1007	1043	1079	1115
20'	451	488	525	562	599	636	673	710	747	784	821	858	895	932	969	1006	1043	1080	1117	1154
21'	471	509	547	585	623	661	699	737	775	813	851	889	927	965	1003	1041	1079	1117	1155	1193
22'	491	530	569	608	647	686	725	764	803	842	881	920	959	998	1037	1076	1115	1154	1193	1232
23'	511	551	591	631	671	711	751	791	831	871	911	951	991	1031	1071	1111	1151	1191	1231	1271
24'	531	572	613	654	695	736	777	818	859	900	941	982	1023	1064	1105	1146	1187	1228	1269	1310

Figure 9.2 Surface Area of Rooms (continued)

Rooms with 9' Ceilings—Total Area – 4 Walls and Ceiling (in square feet)

	3'	4'	5'	6'	7'	8'	9'	10'	11'	12'	13'	14'	15'	16'	17'	18'	19'	20'	21'	22'
3'	117	138	159	180	201	222	243	264	285	306	327	348	369	390	411	432	453	474	495	516
4'	138	160	182	204	226	248	270	292	314	336	358	380	402	424	446	468	490	512	534	556
5'	159	182	205	228	251	274	297	320	343	366	389	412	435	458	481	504	527	550	573	596
6'	180	204	228	252	276	300	324	348	372	396	420	444	468	492	516	540	564	588	612	636
7'	201	226	251	276	301	326	351	376	401	426	451	476	501	526	551	576	601	626	651	676
8'	222	248	274	300	326	352	378	404	430	456	482	508	534	560	586	612	638	664	690	716
9'	243	270	297	324	351	378	405	432	459	486	513	540	567	594	621	648	675	702	729	756
10'	264	292	320	348	376	404	432	460	488	516	544	572	600	628	656	684	712	740	768	796
11'	285	314	343	372	401	430	459	488	517	546	575	604	633	662	691	720	749	778	807	836
12'	306	336	366	396	426	456	486	516	546	576	606	636	666	696	726	756	786	816	846	876
13'	327	358	389	420	451	482	513	544	575	606	637	668	699	730	761	792	823	854	885	916
14'	348	380	412	444	476	508	540	572	604	636	668	700	732	764	796	828	860	892	924	956
15'	369	402	435	468	501	534	567	600	633	666	699	732	765	798	831	864	897	930	963	996
16'	390	424	458	492	526	560	594	628	662	696	730	764	798	832	866	900	934	968	1002	1036
17'	411	446	481	516	551	586	621	656	691	726	761	796	831	866	901	936	971	1006	1041	1076
18'	432	468	504	540	576	612	648	684	720	756	792	828	864	900	936	972	1008	1044	1080	1116
19'	453	490	527	564	601	638	675	712	749	786	823	860	897	934	971	1008	1045	1082	1119	1156
20'	474	512	550	588	626	664	702	740	778	816	854	892	930	968	1006	1044	1082	1120	1158	1196
21'	495	534	573	612	651	690	729	768	807	846	885	924	963	1002	1041	1080	1119	1158	1197	1236
22'	516	556	596	636	676	716	756	796	836	876	916	956	996	1036	1076	1116	1156	1196	1236	1276
23'	537	578	619	660	701	742	783	824	865	906	947	988	1029	1070	1111	1152	1193	1234	1275	1316
24'	558	600	642	684	726	768	810	852	894	936	978	1020	1062	1104	1146	1188	1230	1272	1314	1356

(continued on next page)

Figure 9.2 Surface Area of Rooms (continued)

Rooms with 9'6" Ceilings—Total Area - 4 Walls and Ceiling (in square feet)

	3'	4'	5'	6'	7'	8'	9'	10'	11'	12'	13'	14'	15'	16'	17'	18'	19'	20'	21'	22'
3'	123	145	167	189	211	233	255	277	299	321	343	365	387	409	431	453	475	497	519	541
4'	145	168	191	214	237	260	283	306	329	352	375	398	421	444	467	490	513	536	559	582
5'	167	191	215	239	263	287	311	335	359	383	407	431	455	479	503	527	551	575	599	623
6'	189	214	239	264	289	314	339	364	389	414	439	464	489	514	539	564	589	614	639	664
7'	211	237	263	289	315	341	367	393	419	445	471	497	523	549	575	601	627	653	679	705
8'	233	260	287	314	341	368	395	422	449	476	503	530	557	584	611	638	665	692	719	746
9'	255	283	311	339	367	395	423	451	479	507	535	563	591	619	647	675	703	731	759	787
10'	277	306	335	364	393	422	451	480	509	538	567	596	625	654	683	712	741	770	799	828
11'	299	329	359	389	419	449	479	509	539	569	599	629	659	689	719	749	779	809	839	869
12'	321	352	383	414	445	476	507	538	569	600	631	662	693	724	755	786	817	848	879	910
13'	343	375	407	439	471	503	535	567	599	631	663	695	727	759	791	823	855	887	919	951
14'	365	398	431	464	497	530	563	596	629	662	695	728	761	794	827	860	893	926	959	992
15'	387	421	455	489	523	557	591	625	659	693	727	761	795	829	863	897	931	965	999	1033
16'	409	444	479	514	549	584	619	654	689	724	759	794	829	864	899	934	969	1004	1039	1074
17'	431	467	503	539	575	611	647	683	719	755	791	827	863	899	935	971	1007	1043	1079	1115
18'	453	490	527	564	601	638	675	712	749	786	823	860	897	934	971	1008	1045	1082	1119	1156
19'	475	513	551	589	627	665	703	741	779	817	855	893	931	969	1007	1045	1083	1121	1159	1197
20'	497	536	575	614	653	692	731	770	809	848	887	926	965	1004	1043	1082	1121	1160	1199	1238
21'	519	559	599	639	679	719	759	799	839	879	919	959	999	1039	1079	1119	1159	1199	1239	1279
22'	541	582	623	664	705	746	787	828	869	910	951	992	1033	1074	1115	1156	1197	1238	1279	1320
23'	563	605	647	689	731	773	815	857	899	941	983	1025	1067	1109	1151	1193	1235	1277	1319	1361
24'	585	628	671	714	757	800	843	886	929	972	1015	1058	1101	1144	1187	1230	1273	1316	1359	1402

Figure 9.2   Surface Area of Rooms (continued)

Rooms with 10' Ceilings — Total Area – 4 Walls and Ceiling (in square feet)

	3'	4'	5'	6'	7'	8'	9'	10'	11'	12'	13'	14'	15'	16'	17'	18'	19'	20'	21'	22'
3'	129	152	175	198	221	244	267	290	313	336	359	382	405	428	451	474	497	520	543	566
4'	152	176	200	224	248	272	296	320	344	368	392	416	440	464	488	512	536	560	584	608
5'	175	200	225	250	275	300	325	350	375	400	425	450	475	500	525	550	575	600	625	650
6'	198	224	250	276	302	328	354	380	406	432	458	484	510	536	562	588	614	640	666	692
7'	221	248	275	302	329	356	383	410	437	464	491	518	545	572	599	626	653	680	707	734
8'	244	272	300	328	356	384	412	440	468	496	524	552	580	608	636	664	692	720	748	776
9'	267	296	325	354	383	412	441	470	499	528	557	586	615	644	673	702	731	760	789	818
10'	290	320	350	380	410	440	470	500	530	560	590	620	650	680	710	740	770	800	830	860
11'	313	344	375	406	437	468	499	530	561	592	623	654	685	716	747	778	809	840	871	902
12'	336	368	400	432	464	496	528	560	592	624	656	688	720	752	784	816	848	880	912	944
13'	359	392	425	458	491	524	557	590	623	656	689	722	755	788	821	854	887	920	953	986
14'	382	416	450	484	518	552	586	620	654	688	722	756	790	824	858	892	926	960	994	1028
15'	405	440	475	510	545	580	615	650	685	720	755	790	825	860	895	930	965	1000	1035	1070
16'	428	464	500	536	572	608	644	680	716	752	788	824	860	896	932	968	1004	1040	1076	1112
17'	451	488	525	562	599	636	673	710	747	784	821	858	895	932	969	1006	1043	1080	1117	1154
18'	474	512	550	588	626	664	702	740	778	816	854	892	930	968	1006	1044	1082	1120	1158	1196
19'	497	536	575	614	653	692	731	770	809	848	887	926	965	1004	1043	1082	1121	1160	1199	1238
20'	520	560	600	640	680	720	760	800	840	880	920	960	1000	1040	1080	1120	1160	1200	1240	1280
21'	543	584	625	666	707	748	789	830	871	912	953	994	1035	1076	1117	1158	1199	1240	1281	1322
22'	566	608	650	692	734	776	818	860	902	944	986	1028	1070	1112	1154	1196	1238	1280	1322	1364
23'	589	632	675	718	761	804	847	890	933	976	1019	1062	1105	1148	1191	1234	1277	1320	1363	1406
24'	612	656	700	744	788	832	876	920	964	1008	1052	1096	1140	1184	1228	1272	1316	1360	1404	1448

(continued on next page)

Figure 9.2   Surface Area of Rooms (continued)

Rooms with 10'6" Ceilings—Total Area - 4 Walls and Ceiling (in square feet)

	3'	4'	5'	6'	7'	8'	9'	10'	11'	12'	13'	14'	15'	16'	17'	18'	19'	20'	21'	22'
3'	135	159	183	207	231	255	279	303	327	351	375	399	423	447	471	495	519	543	567	591
4'	159	184	209	234	259	284	309	334	359	384	409	434	459	484	509	534	559	584	609	634
5'	183	209	235	261	287	313	339	365	391	417	443	469	495	521	547	573	599	625	651	677
6'	207	234	261	288	315	342	369	396	423	450	477	504	531	558	585	612	639	666	693	720
7'	231	259	287	315	343	371	399	427	455	483	511	539	567	595	623	651	679	707	735	763
8'	255	284	313	342	371	400	429	458	487	516	545	574	603	632	661	690	719	748	777	806
9'	279	309	339	369	399	429	459	489	519	549	579	609	639	669	699	729	759	789	819	849
10'	303	334	365	396	427	458	489	520	551	582	613	644	675	706	737	768	799	830	861	892
11'	327	359	391	423	455	487	519	551	583	615	647	679	711	743	775	807	839	871	903	935
12'	351	384	417	450	483	516	549	582	615	648	681	714	747	780	813	846	879	912	945	978
13'	375	409	443	477	511	545	579	613	647	681	715	749	783	817	851	885	919	953	987	1021
14'	399	434	469	504	539	574	609	644	679	714	749	784	819	854	889	924	959	994	1029	1064
15'	423	459	495	531	567	603	639	675	711	747	783	819	855	891	927	963	999	1035	1071	1107
16'	447	484	521	558	595	632	669	706	743	780	817	854	891	928	965	1002	1039	1076	1113	1150
17'	471	509	547	585	623	661	699	737	775	813	851	889	927	965	1003	1041	1079	1117	1155	1193
18'	495	534	573	612	651	690	729	768	807	846	885	924	963	1002	1041	1080	1119	1158	1197	1236
19'	519	559	599	639	679	719	759	799	839	879	919	959	999	1039	1079	1119	1159	1199	1239	1279
20'	543	584	625	666	707	748	789	830	871	912	953	994	1035	1076	1117	1158	1199	1240	1281	1322
21'	567	609	651	693	735	777	819	861	903	945	987	1029	1071	1113	1155	1197	1239	1281	1323	1365
22'	591	634	677	720	763	806	849	892	935	978	1021	1064	1107	1150	1193	1236	1279	1322	1365	1408
23'	615	659	703	747	791	835	879	923	967	1011	1055	1099	1143	1187	1231	1275	1319	1363	1407	1451
24'	639	684	729	774	819	864	909	954	999	1044	1089	1134	1179	1224	1269	1314	1359	1404	1449	1494

Figure 9.2 Surface Area of Rooms (continued)

Rooms with 11' Ceilings—Total Area – 4 Walls and Ceiling (in square feet)

	3'	4'	5'	6'	7'	8'	9'	10'	11'	12'	13'	14'	15'	16'	17'	18'	19'	20'	21'	22'
3'	141	166	191	216	241	266	291	316	341	366	391	416	441	466	491	516	541	566	591	616
4'	166	192	218	244	270	296	322	348	374	400	426	452	478	504	530	556	582	608	634	660
5'	191	218	245	272	299	326	353	380	407	434	461	488	515	542	569	596	623	650	677	704
6'	216	244	272	300	328	356	384	412	440	468	496	524	552	580	608	636	664	692	720	748
7'	241	270	299	328	357	386	415	444	473	502	531	560	589	618	647	676	705	734	763	792
8'	266	296	326	356	386	416	446	476	506	536	566	596	626	656	686	716	746	776	806	836
9'	291	322	353	384	415	446	477	508	539	570	601	632	663	694	725	756	787	818	849	880
10'	316	348	380	412	444	476	508	540	572	604	636	668	700	732	764	796	828	860	892	924
11'	341	374	407	440	473	506	539	572	605	638	671	704	737	770	803	836	869	902	935	968
12'	366	400	434	468	502	536	570	604	638	672	706	740	774	808	842	876	910	944	978	1012
13'	391	426	461	496	531	566	601	636	671	706	741	776	811	846	881	916	951	986	1021	1056
14'	416	452	488	524	560	596	632	668	704	740	776	812	848	884	920	956	992	1028	1064	1100
15'	441	478	515	552	589	626	663	700	737	774	811	848	885	922	959	996	1033	1070	1107	1144
16'	466	504	542	580	618	656	694	732	770	808	846	884	922	960	998	1036	1074	1112	1150	1188
17'	491	530	569	608	647	686	725	764	803	842	881	920	959	998	1037	1076	1115	1154	1193	1232
18'	516	556	596	636	676	716	756	796	836	876	916	956	996	1036	1076	1116	1156	1196	1236	1276
19'	541	582	623	664	705	746	787	828	869	910	951	992	1033	1074	1115	1156	1197	1238	1279	1320
20'	566	608	650	692	734	776	818	860	902	944	986	1028	1070	1112	1154	1196	1238	1280	1322	1364
21'	591	634	677	720	763	806	849	892	935	978	1021	1064	1107	1150	1193	1236	1279	1322	1365	1408
22'	616	660	704	748	792	836	880	924	968	1012	1056	1100	1144	1188	1232	1276	1320	1364	1408	1452
23'	641	686	731	776	821	866	911	956	1001	1046	1091	1136	1181	1226	1271	1316	1361	1406	1451	1496
24'	666	712	758	804	850	896	942	988	1034	1080	1126	1172	1218	1264	1310	1356	1402	1448	1494	1540

(continued on next page)

Figure 9.2   Surface Area of Rooms (continued)

Rooms with 12' Ceilings—Total Area – 4 Walls and Ceiling (in square feet)

	3'	4'	5'	6'	7'	8'	9'	10'	11'	12'	13'	14'	15'	16'	17'	18'	19'	20'	21'	22'
3'	153	180	207	234	261	288	315	342	369	396	423	450	477	504	531	558	585	612	639	666
4'	180	208	236	264	292	320	348	376	404	432	460	488	516	544	572	600	628	656	684	712
5'	207	236	265	294	323	352	381	410	439	468	497	526	555	584	613	642	671	700	729	758
6'	234	264	294	324	354	384	414	444	474	504	534	564	594	624	654	684	714	744	774	804
7'	261	292	323	354	385	416	447	478	509	540	571	602	633	664	695	726	757	788	819	850
8'	288	320	352	384	416	448	480	512	544	576	608	640	672	704	736	768	800	832	864	896
9'	315	348	381	414	447	480	513	546	579	612	645	678	711	744	777	810	843	876	909	942
10'	342	376	410	444	478	512	546	580	614	648	682	716	750	784	818	852	886	920	954	988
11'	369	404	439	474	509	544	579	614	649	684	719	754	789	824	859	894	929	964	999	1034
12'	396	432	468	504	540	576	612	648	684	720	756	792	828	864	900	936	972	1008	1044	1080
13'	423	460	497	534	571	608	645	682	719	756	793	830	867	904	941	978	1015	1052	1089	1126
14'	450	488	526	564	602	640	678	716	754	792	830	868	906	944	982	1020	1058	1096	1134	1172
15'	477	516	555	594	633	672	711	750	789	828	867	906	945	984	1023	1062	1101	1140	1179	1218
16'	504	544	584	624	664	704	744	784	824	864	904	944	984	1024	1064	1104	1144	1184	1224	1264
17'	531	572	613	654	695	736	777	818	859	900	941	982	1023	1064	1105	1146	1187	1228	1269	1310
18'	558	600	642	684	726	768	810	852	894	936	978	1020	1062	1104	1146	1188	1230	1272	1314	1356
19'	585	628	671	714	757	800	843	886	929	972	1015	1058	1101	1144	1187	1230	1273	1316	1359	1402
20'	612	656	700	744	788	832	876	920	964	1008	1052	1096	1140	1184	1228	1272	1316	1360	1404	1448
21'	639	684	729	774	819	864	909	954	999	1044	1089	1134	1179	1224	1269	1314	1359	1404	1449	1494
22'	666	712	758	804	850	896	942	988	1034	1080	1126	1172	1218	1264	1310	1356	1402	1448	1494	1540
23'	693	740	787	834	881	928	975	1022	1069	1116	1163	1210	1257	1304	1351	1398	1445	1492	1539	1586
24'	720	768	816	864	912	960	1008	1056	1104	1152	1200	1248	1296	1344	1392	1440	1488	1536	1584	1632

Figure 9.2 Surface Area of Rooms (continued)

	3'	4'	5'	6'	7'	8'	9'	10'	11'	12'	13'	14'	15'	16'	17'	18'	19'	20'	21'	22'
3'	9	12	15	18	21	24	27	30	33	36	39	42	45	48	51	54	57	60	63	66
4'	12	16	20	24	28	32	36	40	44	48	52	56	60	64	68	72	76	80	84	88
5'	15	20	25	30	35	40	45	50	55	60	65	70	75	80	85	90	95	100	105	110
6'	18	24	30	36	42	48	54	60	66	72	78	84	90	96	102	108	114	120	126	132
7'	21	28	35	42	49	56	63	70	77	84	91	98	105	112	119	126	133	140	147	154
8'	24	32	40	48	56	64	72	80	88	96	104	112	120	128	136	144	152	160	168	176
9'	27	36	45	54	63	72	81	90	99	108	117	126	135	144	153	162	171	180	189	198
10'	30	40	50	60	70	80	90	100	110	120	130	140	150	160	170	180	190	200	210	220
11'	33	44	55	66	77	88	99	110	121	132	143	154	165	176	187	198	209	220	231	242
12'	36	48	60	72	84	96	108	120	132	144	156	168	180	192	204	216	228	240	252	264
13'	39	52	65	78	91	104	117	130	143	156	169	182	195	208	221	234	247	260	273	286
14'	42	56	70	84	98	112	126	140	154	168	182	196	210	224	238	252	266	280	294	308
15'	45	60	75	90	105	120	135	150	165	180	195	210	225	240	255	270	285	300	315	330
16'	48	64	80	96	112	128	144	160	176	192	208	224	240	256	272	288	304	320	336	352
17'	51	68	85	102	119	136	153	170	187	204	221	238	255	272	289	306	323	340	357	374
18'	54	72	90	108	126	144	162	180	198	216	234	252	270	288	306	324	342	360	378	396
19'	57	76	95	114	133	152	171	190	209	228	247	266	285	304	323	342	361	380	399	418
20'	60	80	100	120	140	160	180	200	220	240	260	280	300	320	340	360	380	400	420	440
21'	63	84	105	126	147	168	189	210	231	252	273	294	315	336	357	378	399	420	441	462
22'	66	88	110	132	154	176	198	220	242	264	286	308	330	352	374	396	418	440	462	484
23'	69	92	115	138	161	184	207	230	253	276	299	322	345	368	391	414	437	460	483	506
24'	72	96	120	144	168	192	216	240	264	288	312	336	360	384	408	432	456	480	504	528

Square Footage for Single Floor, Ceiling or Wall Area

(continued on next page)

Figure 9.2 Surface Area of Rooms (continued)

Square Footage for Single Floor, Ceiling or Wall Area

	23'	24'	25'	26'	27'	28'	29'	30'	31'	32'	33'	34'	35'	36'	37'	38'	39'	40'	41'	42'
25'	575	600	625	650	675	700	725	750	775	800	825	850	875	900	925	950	975	1000	1025	1050
26'	598	624	650	676	702	728	754	780	806	832	858	884	910	936	962	988	1014	1040	1066	1092
27'	621	648	675	702	729	756	783	810	837	864	891	918	945	972	999	1026	1053	1080	1107	1134
28'	644	672	700	728	756	784	812	840	868	896	924	952	980	1008	1036	1064	1092	1120	1148	1176
29'	667	696	725	754	783	812	841	870	899	928	957	986	1015	1044	1073	1102	1131	1160	1189	1218
30'	690	720	750	780	810	840	870	900	930	960	990	1020	1050	1080	1110	1140	1170	1200	1230	1260
31'	713	744	775	806	837	868	899	930	961	992	1023	1054	1085	1116	1147	1178	1209	1240	1271	1302
32'	736	768	800	832	864	896	928	960	992	1024	1056	1088	1120	1152	1184	1216	1248	1280	1312	1344
33'	759	792	825	858	891	924	957	990	1023	1056	1089	1122	1155	1188	1221	1254	1287	1320	1353	1386
34'	782	816	850	884	918	952	986	1020	1054	1088	1122	1156	1190	1224	1258	1292	1326	1360	1394	1428
35'	805	840	875	910	945	980	1015	1050	1085	1120	1155	1190	1225	1260	1295	1330	1365	1400	1435	1470
36'	828	864	900	936	972	1008	1044	1080	1116	1152	1188	1224	1260	1296	1332	1368	1404	1440	1476	1512
37'	851	888	925	962	999	1036	1073	1110	1147	1184	1221	1258	1295	1332	1369	1406	1443	1480	1517	1554
38'	874	912	950	988	1026	1064	1102	1140	1178	1216	1254	1292	1330	1368	1406	1444	1482	1520	1558	1596
39'	897	936	975	1014	1053	1092	1131	1170	1209	1248	1287	1326	1365	1404	1443	1482	1521	1560	1599	1638
40'	920	960	1000	1040	1080	1120	1160	1200	1240	1280	1320	1360	1400	1440	1480	1520	1560	1600	1640	1680
41'	941	984	1025	1066	1107	1148	1189	1230	1271	1312	1353	1394	1435	1476	1517	1558	1599	1640	1681	1722
42'	966	1008	1050	1092	1134	1176	1218	1260	1302	1344	1386	1428	1470	1512	1554	1596	1638	1680	1722	1764
43'	989	1032	1075	1118	1161	1204	1247	1290	1333	1376	1419	1462	1505	1548	1591	1634	1677	1720	1763	1806
44'	1012	1056	1100	1144	1188	1232	1276	1320	1364	1408	1452	1496	1540	1584	1628	1672	1716	1760	1804	1848
45'	1035	1080	1125	1170	1215	1260	1305	1350	1395	1440	1485	1530	1575	1620	1665	1710	1755	1800	1845	1890
46'	1058	1104	1150	1196	1242	1288	1334	1380	1426	1472	1518	1564	1610	1656	1702	1748	1794	1840	1886	1932

Figure 9.3 Stud, Drywall, and Plaster Wall Systems

These figures show the components of common studding drywall, and plaster partition systems.

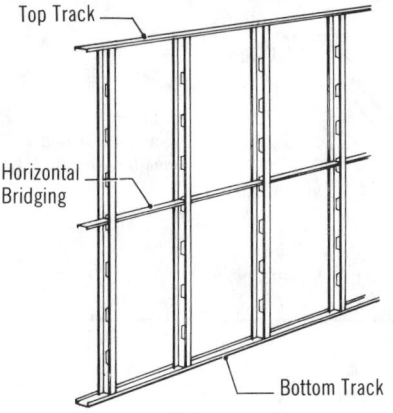

Top Track

Horizontal Bridging

Bottom Track

Load Bearing Steel Studs

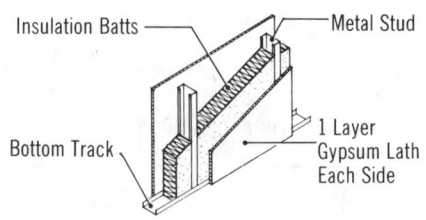

Insulation Batts

Metal Stud

Bottom Track

1 Layer Gypsum Lath Each Side

Fiberglass Batt Insulation

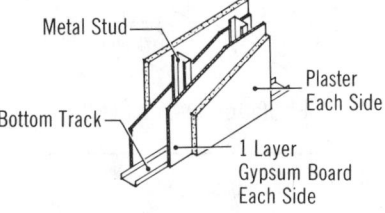

Metal Stud

Plaster Each Side

Bottom Track

1 Layer Gypsum Board Each Side

Plaster on Gypsum Lath

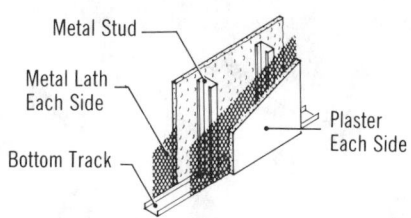

Metal Stud

Metal Lath Each Side

Bottom Track

Plaster Each Side

Plaster on Metal Lath

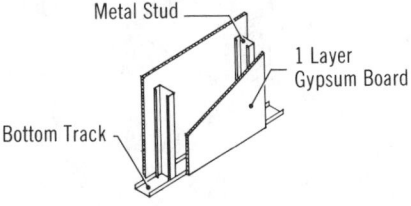

Metal Stud

1 Layer Gypsum Board

Bottom Track

Gypsum Plasterboard

(continued on next page)

Figure 9.3 Stud, Drywall, and Plaster Wall Systems (continued)

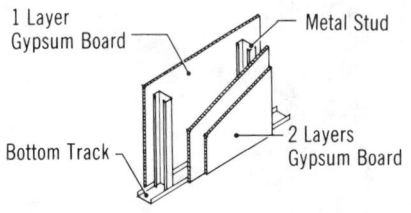

Gypsum Plasterboard, 2 Layers

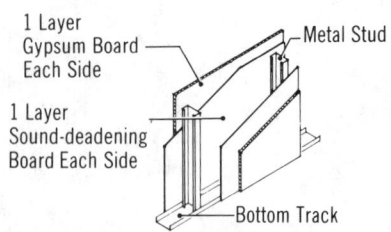

Sound-deadening Board

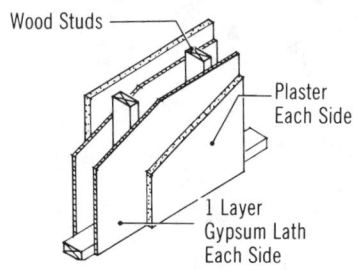

Plaster on Gypsum Lath

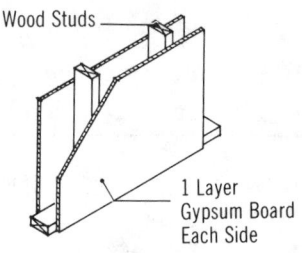

Gypsum Plasterboard

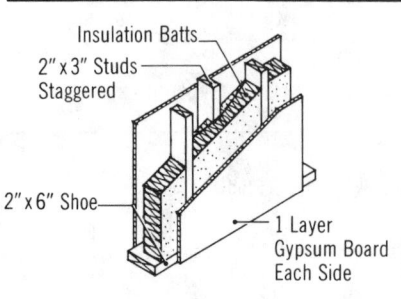

Staggered Stud Wall

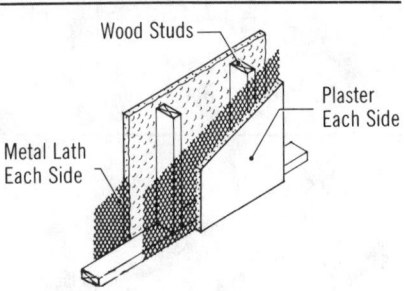

Plaster on Metal Lath

Figure 9.3   Stud, Drywall, and Plaster Wall Systems (continued)

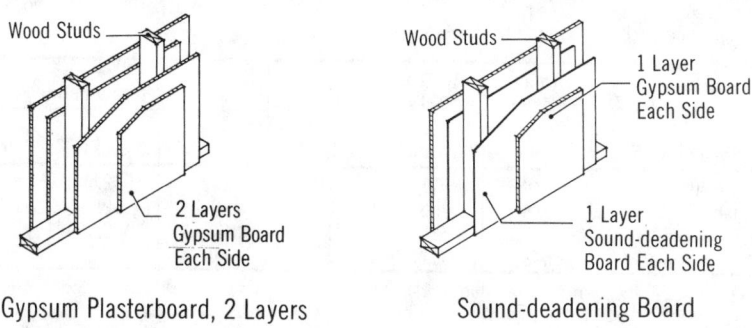

| Gypsum Plasterboard, 2 Layers | Sound-deadening Board |

Figure 9.4   Furring Quantities

This table provides multiplication factors for converting square feet of wall area requiring furring into board feet of lumber. These figures are based on lumber size and spacing.

| | Board Feet per Square Feet of Wall Area | | | | |
| | Spacing Center to Center | | | | Lbs. Nails per MBM |
Size	12"	16"	20"	24"	of Furring
1" x 2"	.18	.14	.11	.10	55
1" x 3"	.28	.21	.17	.14	37

Figure 9.5   Quantities for Perlite Gypsum Plaster

This table shows the approximate number of sacks of prepared perlite gypsum plaster required to cover 100 square yards of surface area. The coverage depends on the type of lathing or base used and the required thickness of plaster.

Type of Base	Total Plaster Thickness (inches)	Number of Sacks
Wood lath	5/8	22 (80 lbs. per sack)
Metal lath	5/8	35 (80 lbs. per sack)
Gypsum lath	1/2	]20 (80 lbs. per sack)
Masonry walls	5/8	24 (67 lbs. per sack)

(courtesy How to Estimate Building Losses and Construction Costs, Paul I. Thomas, Prentice-Hall, Inc.)

Figure 9.6   Quantities for Job-Mixed Plaster

This table presents approximate quantities required to prepare enough job-mixed sanded or perlite plaster to cover 100 square yards of surface area. The coverage depends on the type of base material used and the required thickness of plaster.

Type of Base	Total Plaster Thickness Inches	100 Lb. Sacks of Neat Gypsum	C.Y. Sand 1:2-1/2 Mix	Aggregate Perlite		
				C.F.	4 C.F. Sacks	Mix
Wood lath	5/8	11	1.1	27.5	12.5	1:2-1/2
Metal lath	5/8	20	2.0	50.0	12.5	1:2-1/2
Gypsum lath	1/2	10	1.0	25	6.5	1:2-1/2
Masonry walls	5/8	12	1.2	30	7.5	1:3

ASA permit 250 lbs. damp loose sand or 2-1/2 C.F. of vermiculite or perlite, provided this proportioning is used for both scratch and brown coats on three-coat work.

The number of 4 C.F. sacks of perlite are shown to nearest 1/2 sack.

(Courtesy How to Estimate Building Losses and Construcion Costs, Paul I. Thomas, Prentice-Hall, Inc.)

Figure 9.7   Quantities for White Coat Finish Plaster

This table shows the approximate quantities of materials required to cover 100 square yards of surface area with a 1/16" coat of white coat finish.

Type of Finish	100 Lb. Sacks Neat Gypsum	100 Lb. Sacks Keen's Cement	50 Lb. Sacks Hydrated Lime
Lime putty	2	—	8
Keene's cement	—	4	4

Based on ASA mix of 1 (100-lb.) sack neat gypsum gauging plaster to 4 (50-lb.) sacks hydrated lime, and 1 (100-lb.) sack of Keene's cement to 1 (50-lb.) sack hydrated lime for medium hard finish.

(courtesy How to Estimate Building Losses and Construction Costs, Paul I. Thomas, Prentice-Hall, Inc.)

Figure 9.8   Quantities for Portland Cement Plaster

This table shows the approximate quantities of materials required to cover 100 square yards of surface area with Portland cement plaster. The coverage depends on the base material used and the required thickness of plaster.

Plaster Thickness (inches)	Cubic Yards on Masonry Base	Cubic Yards on Wire Lath Base
1/4	0.75	0.90
1/2	1.50	1.80
5/8	1.37	1.64
3/4	2.25	2.70
1	3.00	3.60

(courtesy How to Estimate Building Losses and Construction Costs, Paul I. Thomas, Prentice Hall, Inc.)

Figure 9.9 Wallboard Quantities

This table lists factors that can be applied to predetermined area figures in order to quantify actual material requirements. Included are waste factors for the intended use of the materials. The table also lists the amount of nails required for the wallboard, based on stud spacing.

Includes Fiber Board, Gypsum Board, and Plywood, Used as Underflooring, Sheathing, Plaster Base, or as Drywall Finish					
Factors		Nails			
Used for Underflooring, Sheathing, and Plaster Base	Used for Exposed Drywall Finish	Pounds of Nails per 1000 Sq. Ft. of Wallboard			
		Joist, Stud or Rafter Spacing			
		12"	16"	20"	24"
1.05	1.10	7	6	5	4

Figure 9.10   Gypsum Panel Coverage

This table shows the expected coverage from the size panels indicated across the top of the table. Inversely, if the area to be covered with gypsum board panels is known (in square feet), you can quickly determine the required number of sheets of drywall. Note that no provision has been made for openings.

No. of Panels	Sizes and Numbers of Panels and S.F. of Area Covered							
	4' x 7'	4' x 8'	4' x 9'	4' x 10'	4' x 11'	4' x 12'	4' x 13'	4' x 14'
10	280	320	360	400	440	480	520	560
11	308	352	396	440	484	528	572	616
12	336	384	432	480	528	576	624	672
13	364	416	468	520	572	624	676	728
14	392	448	504	560	616	672	728	784
15	420	480	540	600	660	720	780	840
16	448	512	576	640	704	768	832	896
17	476	544	612	680	748	816	884	952
18	504	576	648	720	792	864	936	1008
19	532	608	684	760	836	912	988	1064
20	560	640	720	800	880	960	1040	1120
21	588	672	756	840	924	1008	1092	1176
22	616	704	792	880	968	1056	1144	1232
23	644	736	828	920	1012	1104	1196	1288
24	672	768	864	960	1056	1152	1248	1344
25	700	800	900	1000	1100	1200	1300	1400
26	728	832	936	1040	1144	1248	1352	1456
27	756	864	972	1080	1188	1296	1404	1512
28	784	896	1008	1120	1232	1344	1456	1568
29	812	928	1044	1160	1276	1392	1508	1624
30	840	960	1080	1200	1320	1440	1560	1680
31	868	992	1116	1240	1364	1488	1612	1736

Figure 9.11 Drywall Accessories

This table provides the approximate quantities of nails, joint compound and tape needed for the indicated amounts of drywall.

With this Amount of Sheetrock Gypsum Panel	Type GWB-54 Nails Required*	**USE** this Amount of Powder-Type Compound	**OR** this Amount of USG Ready-To-Use Compound-All Purpose	**AND** this Amount of Perf-A-Tape Reinforcing Tape
100 S.F.	.6 lbs.	6 lbs.	1 Gal.	37 Ft.
200 S.F.	1.1 lbs.	12 lbs.	2 Gals.	74 Ft.
300 S.F.	1.6 lbs.	18 lbs.	2 Gals.	111 Ft.
400 S.F.	2.1 lbs.	24 lbs.	3 Gals.	148 Ft.
500 S.F.	2.7 lbs.	30 lbs.	3 Gals.	185 Ft.
600 S.F.	3.2 lbs.	36 lbs.	4 Gals.	222 Ft.
700 S.F.	3.7 lbs.	42 lbs.	5 Gals.	259 Ft.
800 S.F.	4.2 lbs.	48 lbs.	5 Gals.	296 Ft.
900 S.F.	4.8 lbs.	54 lbs.	6 Gals.	333 Ft.
1000 S.F.	5.3 lbs.	60 lbs.	6 Gals.	370 Ft.

*Spaced 7" on ceiling; 8" on wall.

Figure 9.12   Installation Time in Man-Hours for Wood Stud Partition Systems

Description	Man-Hours	Unit
Wood Partitions Studs with Single Bottom Plate and Double Top Plate		
2" x 3" or 2" x 4" Studs		
12" On Center	.020	S.F.
16" On Center	.016	S.F.
24" On Center	.013	S.F.
2" x 6" Studs		
12" On Center	.023	S.F.
16" On Center	.018	S.F.
24" On Center	.014	S.F.
Plates		
2" x 3"	.019	L.F.
2" x 4"	.020	L.F.
2" x 6"	.021	L.F.
Studs		
2" x 3"	.013	L.F.
2" x 4"	.012	L.F.
2" x 6"	.016	L.F.
Blocking	.032	L.F.
Grounds 1" x 2"		
For Casework	.024	L.F.
For Plaster	.018	L.F.
Insulation Fiberglass Batts	.005	S.F.
Metal Lath Diamond Expanded		
2.5 lb. per S.Y.	.094	S.Y.
3.4 lb. per S.Y.	.100	S.Y.
Gypsum Lath		
3/8" Thick	.094	S.Y.
1/2" Thick	.100	S.Y.
Gypsum Plaster		
2 Coats	.381	S.Y.
3 Coats	.460	S.Y.
Perlite or Vermiculite Plaster		
2 Coats	.435	S.Y.
3 Coats	.541	S.Y.
Wood Fiber Plaster		
2 Coats	.556	S.Y.
3 Coats	.702	S.Y.

Figure 9.12 Installation Time in Man-Hours
for Wood Stud Partition Systems (continued)

Description	Man-Hours	Unit
Drywall Gypsum Plasterboard Including Taping		
3/8" Thick	.015	S.F.
1/2" or 5/8" Thick	.017	S.F.
For Thin Coat Plaster Instead of Taping Add	.013	S.F.
Prefinished Vinyl Faced Drywall	.015	S.F.
Sound-deadening Board	.009	S.F.
Walls in Place		
2" x 4" Studs with 5/8"		
Gypsum Drywall Both Sides Taped	.053	S.F.
2" x 4" Studs with 2 Layers Gypsum Drywall		
Both Sides Taped	.078	S.F.

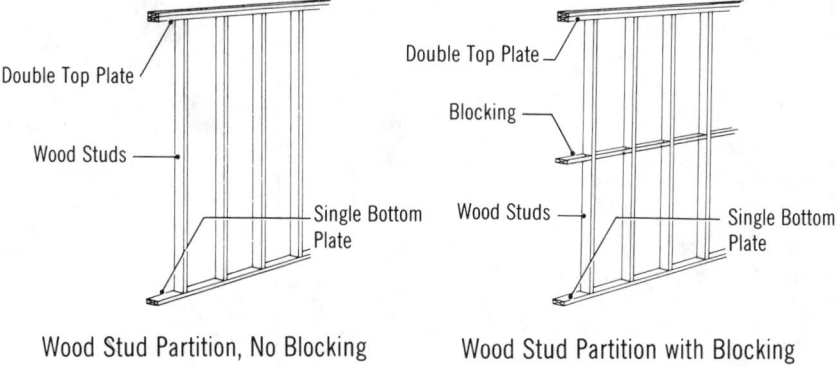

Wood Stud Partition, No Blocking   Wood Stud Partition with Blocking

Figure 9.13   Installation Time in Man-Hours for Steel Stud Partition Systems

Description	Man-Hours	Unit
Non-Loadbearing Stud, Galv., 25 Gauge,		
1-5/8" Wide,		
16" On Center	.019	S.F.
24" On Center	.016	S.F.
2-1/2" Wide,		
16" On Center	.020	S.F.
24" On Center	.016	S.F.
20 Gauge, 1-5/8" Wide,		
16" On Center	.018	S.F.
24" On Center	.016	S.F.
2-1/2" Wide,		
16" On Center	.019	S.F.
24" On Center	.016	S.F.
25 Gauge or 20 Gauge, 3-5/8" or 4" Wide		
16" On Center	.020	S.F.
24" On Center	.017	S.F.
6" Wide,		
16" On Center	.022	S.F.
24" On Center	.018	S.F.
Loadbearing Stud, Galv. or Painted, 18 Gauge,		
2-1/2" Wide,		
16" On Center	.019	S.F.
24" On Center	.016	S.F.
3-5/8" or 4" Wide,		
16" On Center	.020	S.F.
24" On Center	.017	S.F.
6" Wide,		
16" On Center	.022	S.F.
24" On Center	.018	S.F.
Loadbearing Stud, Galv. or Painted,		
16 Gauge, 2-1/2" Wide,		
16" On Center	.020	S.F.
24" On Center	.017	S.F.
3-5/8" or 4" Wide,		
16" On Center	.021	S.F.
24" On Center	.018	S.F.
6" Wide,		
16" On Center	.024	S.F.
24" On Center	.019	S.F.
Insulation Fiberglass Batts	.005	S.F.

Figure 9.13   Installation Time in Man-Hours for Steel Stud Partition Systems (continued)

Description	Man-Hours	Unit
Metal Lath Diamond Expanded		
2.5 lb. Screwed to Studs	.100	S.Y.
3.4 lb. Screwed to Studs or Wired to Framing	.107	S.Y.
Rib Lath Wired to Steel		
2.75 lb.	.107	S.Y.
3.40 lb.	.114	S.Y.
4.00 lb.	.123	S.Y.
Gypsum Lath		
3/8" Thick	.094	S.Y.
1/2" Thick	.100	S.Y.
Gypsum Plaster		
2 Coats	.381	S.Y.
3 Coats	.460	S.Y.
Perlite or Vermiculite Plaster		
2 Coats	.435	S.Y.
3 Coats	.541	S.Y.
Wood Fiber Plaster		
2 Coats	.556	S.Y.
3 Coats	.702	S.Y.
Drywall Gypsum Plasterboard Including Taping		
3/8" Thick	.015	S.F.
1/2" or 5/8" Thick	.017	S.F.
For Thincoat Plaster Instead of Taping, Add	.013	S.F.
Prefinished Vinyl Faced Drywall	.015	S.F.
Sound-deadening Board	.009	S.F.
Walls in Place		
3-5/8" Studs, NLB, 25 Gauge,		
16" On Center with 5/8" Gypsum Drywall		
Both Sides Taped	.047	S.F.
24" On Center	.044	S.F.
3-5/8" Studs, NLB, 25 Gauge,		
16" On Center with 2 Layers Gypsum Drywall		
Both Sides Taped	.065	S.F.
24" On Center	.060	S.F.

Figure 9.14 Installation Time In Man Hours for Ceiling Systems

Description	Man-Hours	Unit
Ceiling Tile Stapled, Cemented or Installed on Suspension System, 12" x 12" or 12" x 24" Not Including Furring		
Mineral Fiber, Plastic Coated	.020	S.F.
Fire Rated, 3/4" Thick, Plain Faced	.020	S.F.
Plastic Coated Face	.021	S.F.
Aluminum Faced, 5/8" Thick, Plain	.021	S.F.
Metal Pan Units, 24 ga. Steel, Not Incl. Pads, Painted, 12" x 12"	.023	S.F.
12" x 36" or 12" x 24", 7% Open Area	.024	S.F.
Aluminum, 12" x 12"	.023	S.F.
12" x 24"	.022	S.F.
Stainless Steel, 12" x 24", 26 ga., Solid	.023	S.F.
5.2% Open Area	.024	S.F.
Suspended Acoustic Ceiling Boards Not Including Suspension System		
Fiberglass Boards, Film Faced, 2' x 2' or 2' x 4', 5/8" Thick	.012	S.F.
3/4" Thick	.016	S.F.
3" Thick, Thermal, R11	.018	S.F.
Glass Cloth Faced Fiberglass, 3/4" Thick	.016	S.F.
1" Thick	.016	S.F.
1-1/2" Thick, Nubby Face	.017	S.F.
Mineral Fiber Boards, 5/8" Thick, Aluminum Faced, 24" x 24"	.013	S.F.
24" x 48"	.012	S.F.
Plastic Coated Face	.020	S.F.
Mineral Fiber, 2 Hour Rating, 5/8" Thick	.012	S.F.
Mirror Faced Panels, 15/16" Thick	.016	S.F.
Air Distributing Ceilings, 5/8" Thick, F.R.D. Water Felted Board	.020	S.F.
Eggcrate, Acrylic, 1/2" x 1/2" x 1/2" Cubes	.016	S.F.
Polystyrene Eggcrate	.016	S.F.
Luminous Panels, Prismatic	.020	S.F.
Perforated Aluminum Sheets, .024" Thick, Corrugated, Painted	.016	S.F.
Mineral Fiber, 24" x 24" or 48", reveal edge, Painted, 5/8" Thick	.013	S.F.
3/4" Thick	.014	S.F.
Wood Fiber in Cementitious Binder, 2' x 2' or 4', Painted, 1" Thick	.013	S.F.
2" Thick	.015	S.F.
2-1/2" Thick	.016	S.F.
3" Thick	.018	S.F.
Access Panels, Metal, 12" x 12"	.400	Ea.
24" x 24"	.800	Ea.

Figure 9.14 Installation Time in Man-Hours for Ceiling Systems (continued)

Description	Man-Hours	Unit
Suspension Systems		
Class A Suspension System, T Bar, 2' x 4' Grid	.010	S.F.
2' x 2' Grid	.012	S.F.
Concealed Z Bar Suspension System, 12"		
Module	.015	S.F.
1-1/2" Carrier Channels, 4' O.C., Add	.017	S.F.
Carrier Channels for Ceiling With Recessed		
Lighting Fixtures, Add	.017	S.F.
Hanging Wire, 12 Ga.	.002	S.F.
Suspended Ceilings, Complete Including		
Standard Suspension System but Not Including		
1-1/2" Carrier Channels		
Air Distributing Ceilings, Including Barriers,		
2' x 2' Board	.027	S.F.
12" x 12" Tile	.033	S.F.
Ceiling Board System, 2' x 4', Plain Faced,		
Supermarkets	.016	S.F.
Offices	.021	S.F.
Wood Fiber, Cementitious Binder, T Bar Susp.		
2' x 2' x 1" Board	.023	S.F.
2' x 4' x 1" Board	.021	S.F.
Luminous Panels, Flat or Ribbed	.031	S.F.
Metal Pan with Acoustic Pad	.039	S.F.
Tile, Z Bar Suspension, 5/8" Mineral Fiber Tile	.034	S.F.
3/4" Mineral Fiber Tile	.035	S.F.
Reveal Tile With Drop, 2' x 2' Grid with Colored		
Suspension System	.023	S.F.
Gypsum Plaster Ceilings 2 Coats, No Lath		
Included	.435	S.Y.
2 Coats on and Incl. 3/8" Gypsum Lath on Steel	.578	S.Y.
3 Coats, No Lath Included	.513	S.Y.
3 Coats on and Including Painted Metal Lath	.627	S.Y.
Metal Lath 2.5 lb. Diamond Painted, on Wood		
Framing	.107	S.Y.
On Ceilings, 3.4 lb. Diamond Painted, on		
Wood Framing	.114	S.Y.
3.4 lb. Diamond Painted, Wired to Steel Framing	.133	S.Y.
Suspended Ceiling System, Incl. 3.4 lb.		
Diamond Lath	.533	S.Y.
Drywall Ceilings Gypsum Drywall, Fire Rated		
Finished		
Screwed to Grid, Channel of Joists	.021	S.F.
Over 8' High, 1/2" Thick	.022	S.F.
5/8" Thick	.023	S.F.
Grid Suspension System, Direct Hung		
1-1/2" C.R.C., With 7/8" Hi Hat Furring		
Channel, 16" O.C.	.027	S.F.
24" O.C.	.018	S.F.
3-5/8" Channel, 25 ga., with Track,		
16" O.C.	.027	S.F.
24" O.C.	.018	S.F.

(continued on next page)

Figure 9.14 Installation Time in Man-Hours for Ceiling Systems (continued)

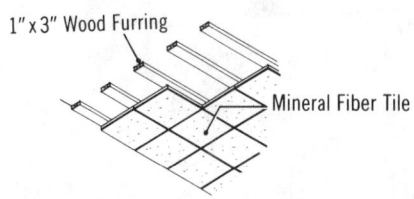

Acoustical Mineral Fiber Tile
on 1" x 3" Wood Furring

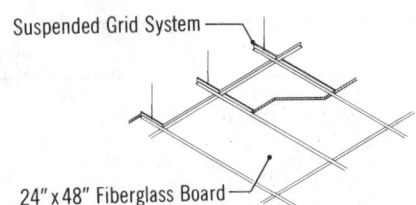

Fiberglass Board
on Suspended Grid System

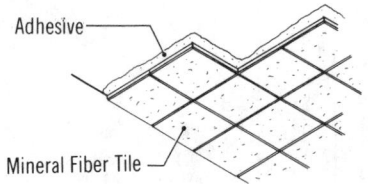

Mineral Fiber Tile Applied with Adhesive

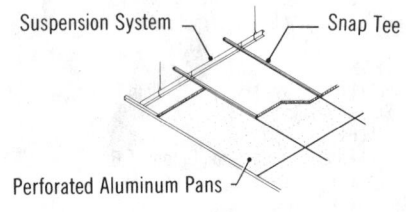

Acoustical Perforated Metal Pans

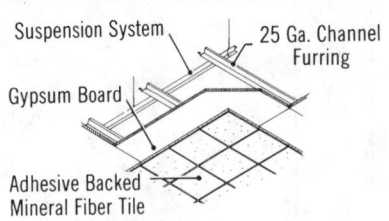

Acoustical Mineral Fiber Tile
on Gypsum Board

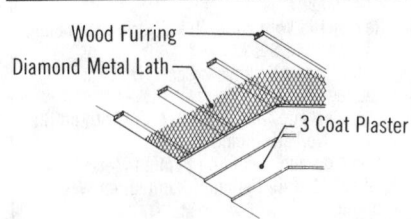

Plaster on Metal Lath
and Wood Furring

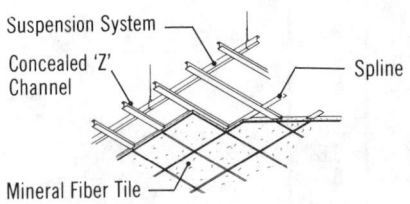

Mineral Fiber Tile
on Concealed 'Z' Channel

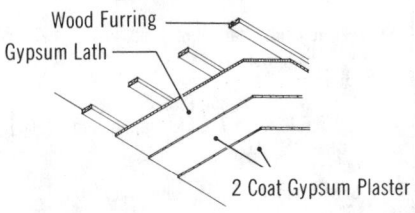

Plaster on Wood Furring

Figure 9.14 Installation Time in Man-Hours for Ceiling Systems (continued)

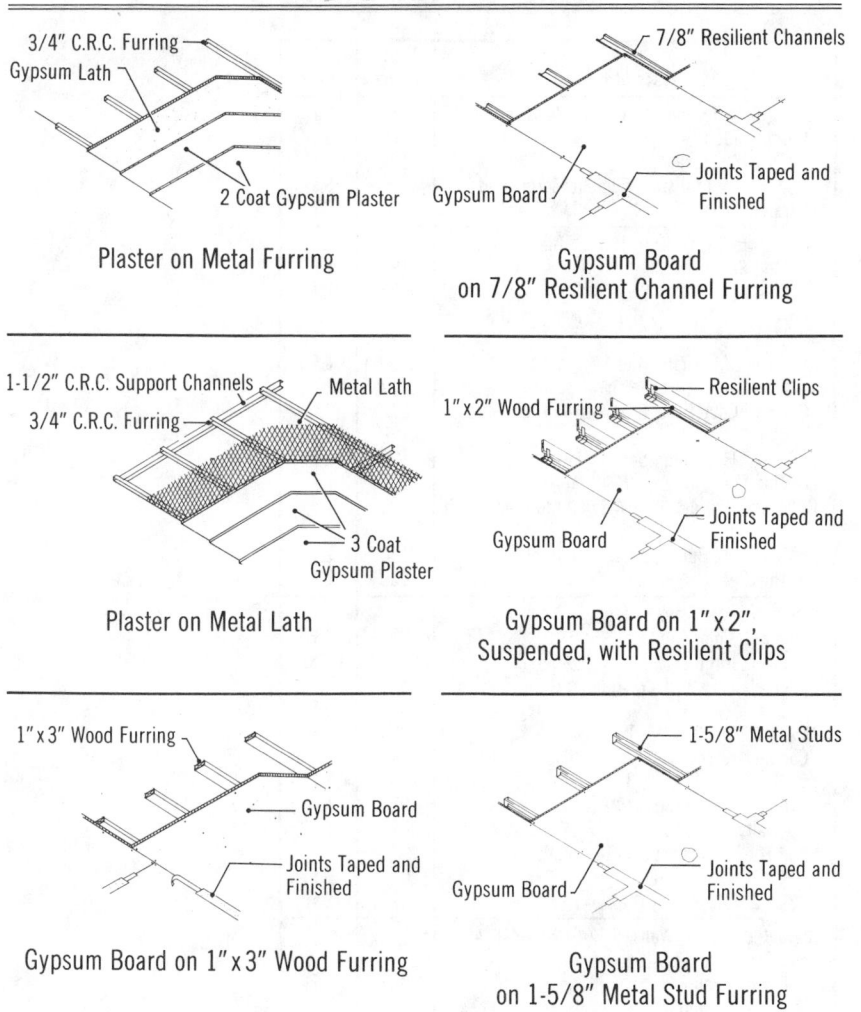

Plaster on Metal Furring

Gypsum Board
on 7/8" Resilient Channel Furring

Plaster on Metal Lath

Gypsum Board on 1"x 2",
Suspended, with Resilient Clips

Gypsum Board on 1"x 3" Wood Furring

Gypsum Board
on 1-5/8" Metal Stud Furring

Figure 9.15   Installation Time in Man-Hours for Floor and Wall Tile Systems

Description	Man-Hours	Unit
Flooring Cast Ceramic 4" x 8" x 3/4" Pressed	.160	S.F.
Hand Molded	.168	S.F.
8" x 3/4" Hexagonal	.188	S.F.
Heavy Duty Industrial Cement Mortar Bed	.200	S.F.
Ceramic Pavers 8" x 4"	.168	S.F.
Ceramic Tile Base, Using 1" x 1" Tiles, 4" High,		
Mud Set	.195	L.F.
Thin Set	.125	L.F.
Cove Base, 4-1/4" x 4-1/4" High, Mud Set	.176	L.F.
Thin Set	.125	L.F.
6" x 4-1/4" High, Mud Set	.160	L.F.
Thin Set	.117	L.F.
Sanitary Cove Base, 6" x 4-1/4" High, Mud Set	.172	L.F.
Thin Set	.129	L.F.
6" x 6" High, Mud Set	.190	L.F.
Thin Set	.137	L.F.
Bullnose Trim, 4-1/4" x 4-1/4", Mud Set	.195	L.F.
Thin Set	.125	L.F.
6" x 4-1/4" Bullnose Trim, Mud Set	.190	L.F.
Thin Set	.129	L.F.
Floors, Natural Clay, Random or Uniform,		
Thin Set, Color Group 1	.087	S.F.
Color Group 2	.087	S.F.
Porcelain Type, 1 Color, Color Group 2,		
1" x 1"	.087	S.F.
2" x 2" or 2" x 1", Thin Set	.084	S.F.
Conductive Tile, 1" Squares, Black	.147	S.F.
4" x 8" or 4" x 4", 3/8" Thick	.133	S.F.
Trim, Bullnose, Etc.	.080	L.F.
Specialty Tile, 3" x 6" x 1/2", Decorator Finish	.087	S.F.
Add For Epoxy Grout, 1/16" Joint,		
1" x 1" Tile	.020	S.F.
2" x 2" Tile	.020	S.F.
Pregrouted Sheets, Walls, 4-1/4", 6" x 4-1/4",		
and 8-1/2" x 4-1/4", S.F. Sheets,		
Silicone Grout	.067	S.F.
Floors, Unglazed, 2 S.F. Sheets		
Urethane Adhesive	.089	S.F.
Walls, Interior, Thin Set, 4-1/4" x 4-1/4" Tile	.084	S.F.
6" x 4-1/4" Tile	.084	S.F.
8-1/2" x 4-1/4" Tile	.084	S.F.
6" x 6" Tile	.080	S.F.

Figure 9.15 Installation Time in Man-Hours
for Floor and Wall Tile Systems (continued)

Description	Man-Hours	Unit
Decorated Wall Tile, 4-1/4"x 4-1/4"		
Minimum	.018	Ea.
Maximum	.028	Ea.
Exterior Walls, Frostproof, Mud Set,		
4-1/4" x 4-1/4"	.157	S.F.
1-3/8" x 1-3/8"	.172	S.F.
Crystalline Glazed, 4-1/4" x 4-1/4", Mud		
Set, Plain	.160	S.F.
4-1/4" x 4-1/4", Scored Tile	.160	S.F.
1-3/8" Squares	.172	S.F.
For Epoxy Grout, 1/16" Joints, 4-1/4" Tile,		
Add	.020	S.F.
For Tile Set in Dry Mortar, Add	.009	S.F.
For Tile Set in Portland Cement Mortar, Add	.055	S.F.
Regrout Tile 4-1/2" x 4-1/2", or Larger, Wall	.080	S.F.
Floor	.073	S.F.
Ceramic Tile Panels Insulated, Over 1000		
Square Feet,		
1-1/2" Thick	.073	S.F.
2-1/2" Thick	.073	S.F.
Glass Mosaics 3/4" Tile on 12" Sheets,		
Color Group 1 and 2 Minimum	.195	S.F.
Maximum (Latex Set)	.219	S.F.
Color Group 3	.219	S.F.
Color Group 4	.219	S.F.
Color Group 5	.219	S.F.
Color Group 6	.219	S.F.
Color Group 7	.219	S.F.
Color Group 8, Gold, Silvers and Specialties	.250	S.F.
Marble Thin Gauge Tile, 12" x 6", 9/32", White		
Carara	.250	S.F.
Filled Travertine	.250	S.F.
Synthetic Tiles, 12" x 12" x 5/8",		
Thin Set, Floors	.250	S.F.
On Walls	.291	S.F.
Metal Tile Cove Base, Standard Colors,		
4-1/4" Square	.053	L.F.
4-1/8" x 8-1/2"	.040	L.F.
Walls, Aluminum, 4-1/4" Square, Thin		
Set, Plain	.100	S.F.
Epoxy Enameled	.107	S.F.
Leather on Aluminum, Colors	.123	S.F.
Stainless Steel	.107	S.F.
Suede on Aluminum	.123	S.F.
Plastic Tile Walls, 4-1/4" x 4-1/4",		
.050" Thick	.064	S.F.
.110" Thick	.067	S.F.

(continued on next page)

Figure 9.15   Installation Time in Man-Hours
for Floor and Wall Tile Systems (continued)

Description	Man-Hours	Unit
Quarry Tile Base, Cove or Sanitary, 2" or 5" High, Mud Set		
1/2" Thick	.145	L.F.
Bullnose Trim, Red, Mud Set, 6" x 6" x 1/2" Thick	.133	L.F.
4" x 4" x 1/2" Thick	.145	L.F.
4" x 8" x 1/2" Thick, Using 8" as Edge	.123	L.F.
Floors, Mud Set, 1000 S.F. Lots, Red,		
4" x 4" x 1/2" Thick	.133	S.F.
6" x 6" x 1/2" Thick	.114	S.F.
4" x 8" x 1/2" Thick	.123	S.F.
Brown Tile, Imported, 6" x 6" x 7/8"	.133	S.F.
9" x 9" x 1-1/4"	.145	S.F.
For Thin Set Mortar Application, Deduct	.023	S.F.
Stair Tread and Riser, 6" x 6" x 3/4", Plain	.320	S.F.
Abrasive	.340	S.F.
Wainscot, 6" x 6" x 1/2", Thin Set, Red	.152	S.F.
Colors Other Than Green	.152	S.F.
Window Sill, 6" Wide, 3/4" Thick	.178	L.F.
Corners	.200	Ea.
Terra Cotta Tile on Walls, Dry Set, 1/2" Thick		
Square, Hexagonal or Lattice Shapes, Unglazed	.059	S.F.
Glazed, Plain Colors	.062	S.F.
Intense Colors	.064	S.F.

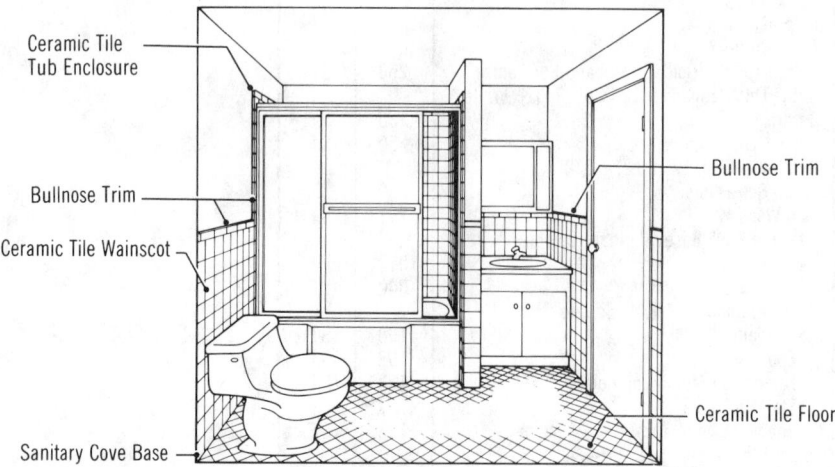

Ceramic Tile Tub Enclosure

Bullnose Trim

Bullnose Trim

Ceramic Tile Wainscot

Ceramic Tile Floor

Sanitary Cove Base

Figure 9.16 Installation Time in Man-Hours for Terrazzo

Description	Man-Hours	Unit
Terrazzo Cast in Place Cove Base, 6" High	.348	L.F.
Curb, 6" High and 6" Wide	.533	L.F.
Floor, Bonded to Concrete, 1-3/4" Thick,		
Gray Cement	.123	S.F.
White Cement	.128	S.F.
Not Bonded, 3" Total Thickness,		
Gray Cement	.160	S.F.
White Cement	.168	S.F.
Bonded Conductive Floor for Hospitals	.145	S.F.
Epoxy Terrazzo, 1/4" Thick, Minimum	.094	S.F.
Average	.123	S.F.
Maximum	.160	S.F.
Monolithic Terrazzo, 5/8" Thick, Including		
3-1/2" Base Slab, 10' Panels, Mesh and Felt	.070	S.F.
Stairs, Cast in Place, Pan Filled Treads	.291	L.F.
Treads and Risers	.800	L.F.
Stair Landings, Add to Floor Prices	.258	S.F.
Stair Stringers and Fascia	.291	S.F.
For Abrasive Metal Nosings on Stairs, Add	.056	L.F.
For Abrasive Surface Finish, Add	.027	S.F.
For Flush Abrasive Strips, Add	.026	L.F.
Wainscot, Bonded, 1-1/2" Thick	.400	S.F.
Epoxy Terrazzo, 1/4" Thick	.229	S.F.
Terrazzo Precast Base, 6" High, Straight	.067	L.F.
Cove	.080	L.F.
8" High Base, Straight	.073	L.F.
Cove	.089	L.F.
Curbs, 4" x 4" High	.145	L.F.
8" x 8" High	.178	L.F.
Floor Tiles, Non-slip, 1" Thick, 12" x 12"	.267	S.F.
1-1/4" Thick, 12" x 12"	.267	S.F.
16" x 16"	.291	S.F.
1-1/2" Thick, 16" x 16"	.320	S.F.
Floor Tiles, 12" x 12", 3/16" Thick	.062	S.F.
Stair Treads, 1-1/2" Thick, Non-slip, Diamond Pat.	.168	L.F.
Line Pattern	.178	L.F.
2" Thick Treads, Straight	.178	L.F.
Curved	.188	L.F.
Stair Risers, 1" Thick, to 6" High,		
Straight Sections	.100	L.F.
Cove	.107	L.F.
Curved, 1" Thick, to 6" High, Vertical	.119	L.F.
Cove	.123	L.F.
Stair Tread and Riser, Single Piece, Straight,		
Minimum	.246	L.F.
Maximum	.267	L.F.
Curved Tread and Riser, Minimum	.267	L.F.
Maximum	.291	L.F.
Stair Stringers, Notched, 1" Thick	.152	L.F.
2" Thick	.267	L.F.
Stair Landings, Structural, Non-slip,		
1-1/2" Thick	.229	S.F.
3" Thick	.168	S.F.
Wainscot, 12" x 12" x 1" Tiles	.229	S.F.
16" x 16" x 1-1/2" Tiles	.267	S.F.

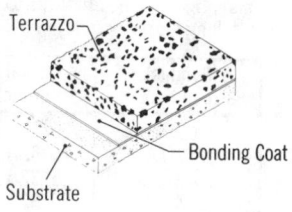

Terrazzo

Bonding Coat

Substrate

Figure 9.17   Wood Flooring Quantities

This table lists the amounts to be added to the measured floor areas for the materials required for installing strip wood flooring. Milling and cutting waste values are included in the figures.

Nominal Size in Inches	Actual Size in Inches	Milling + 5% Cutting Waste	Multiply Area by	Floor Needed per 1,000 FBM
1 x 2	25/32 x 1-1/2	55%	1.55	1,500
1 x 2-1/2	25/32 x 2	42	1.42	1,420
1 x 3	25/32 x 2-1/4	38	1.38	1,380
1 x 4	25/32 x 3-1/4	29	1.29	1,290
1 x 2	3/8 x 1-1/2	38	1.38	1,380
1 x 2-1/2	3/8 x 2	30	1.30	1,300
	1/2 x 1-1/2	38	1.38	1,380
	1/2 x 2	30	1.30	1,300

Figure 9.18 Installation Time in Man-Hours for Wood Flooring

Description	Man-Hours	Unit
Wood Floors Fir, Vertical Grain, 1" x 4", Not Including Finish	.031	S.F.
Gym Floor, in Mastic, Over 2 Ply Felt, #2 and Better 25/32" Thick Maple, Including Finish	.080	S.F.
33/32" Thick Maple, Including Finish	.082	S.F.
For 1/2" Corkboard Underlayment, Add	.011	S.F.
Maple Flooring, Over Sleepers, #2 and Better Including Finish, 25/32" Thick	.094	S.F.
33/32" Thick	.096	S.F.
For 3/4" Subfloor, Add	.023	S.F.
With Two 1/2" Subfloors, 25/32" Thick	.116	S.F.
Maple, Including Finish, #2 and better, 25/32" Thick, on Rubber Sleepers, with Two 1/2" Subfloors	.105	S.F.
With Steel Spline, Double Connection to Channels	.110	S.F.
Portable Hardwood, Prefinished Panels	.096	S.F.
Insulated with Polystyrene, Add	.048	S.F.
Running Tracks, Sitka Spruce Surface	.129	S.F.
3/4" Plywood Surface	.080	S.F.
Maple, Strip, Not Including Finish	.047	S.F.
Oak Strip, White or Red, Not Including Finish	.047	S.F.
Parquetry, Standard, 5/16" Thick, Not Including Finish, Minimum	.050	S.F.
13/16" Thick, Select Grade, Minimum	.050	S.F.
Maximum	.080	S.F.
Custom Parquetry, Including Finish, Minimum	.080	S.F.
Maximum	.160	S.F.
Prefinished White Oak, Prime Grade, 2-1/4" Wide	.047	S.F.
3-1/4" Wide	.043	S.F.
Ranch Plank	.055	S.F.
Hardwood Blocks, 9" x 9", 25/32" Thick	.050	S.F.

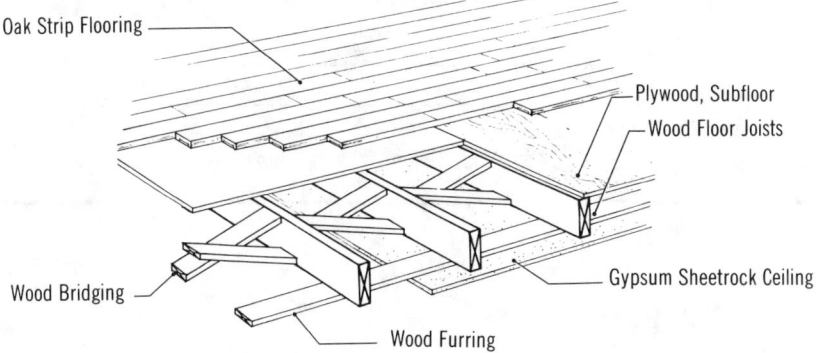

Oak Strip Flooring

Plywood, Subfloor

Wood Floor Joists

Wood Bridging

Gypsum Sheetrock Ceiling

Wood Furring

(continued on next page)

Figure 9.18 Installation Time in Man-Hours for Wood Flooring (continued)

Description	Man-Hours	Unit
Acrylic Wood Parquet Blocks,		
12″ x 12″ x 5/16″, Irradiated, Set in Epoxy	.050	S.F.
Yellow Pine, 3/4″ x 3-1/8″, T & G, C and		
Better, Not Including Finish	.040	S.F.
Refinish Old Floors, Minimum	.020	S.F.
Maximum	.062	S.F.
Sanding and Finishing, Fill, Shellac, Wax	.027	S.F.
Wood Block Flooring End Grain Flooring,		
Creosoted, 2″ Thick	.027	S.F.
Natural Finish, 1″ Thick	.029	S.F.
1-1/2″ Thick	.031	S.F.
2″ Thick	.033	S.F.
Wood Composition Gym Floors		
2-1/4″ x 6-7/8″ x 3/8″, on 2″ Grout Setting		
Bed	.107	S.F.
Thin Set, on Concrete	.064	S.F.
Sanding and Finishing, Add	.040	S.F.

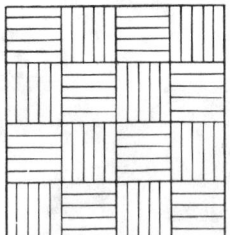
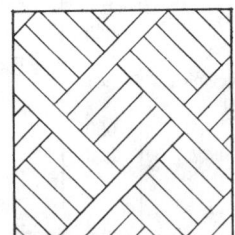
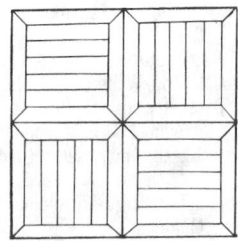

Random Patterns of Parquet Flooring

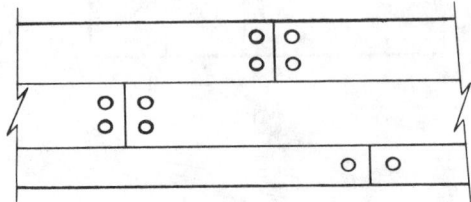

Strip Plank Flooring with Plugs

Figure 9.19 Installation Time in Man-Hours for Carpeting

Description	Man-Hours	Unit
Carpet Commercial Grades, Cemented		
Acrylic, 26 oz., Light to Medium Traffic	.216	S.F.
28 oz., Medium Traffic	.229	S.F.
35 oz., Medium to Heavy Traffic	.242	S.F.
Nylon, Non Anti-Static, 15 oz., Light Traffic	.229	S.F.
Nylon, With Anti-Static, 17 oz., Light to		
Medium Traffic	.211	S.F.
20 oz., Medium Traffic	.216	S.F.
22 oz., Medium Traffic	.216	S.F.
24 oz., Medium to Heavy Traffic	.229	S.F.
26 oz., Medium to Heavy Traffic	.229	S.F.
28 oz., Heavy Traffic	.229	S.F.
32 oz., Heavy Traffic	.242	S.F.
42 oz., Heavy Traffic	.258	S.F.
Needle Bonded, 20 oz., No Padding	.143	S.F.
Polypropylene, 15 oz., Light Traffic	.143	S.F.
22 oz., Medium Traffic	.182	S.F.
24 oz., Medium to Heavy Traffic	.182	S.F.
26 oz., Medium to Heavy Traffic	.182	S.F.
28 oz., Heavy Traffic	.205	S.F.
32 oz., Heavy Traffic	.205	S.F.
42 oz., Heavy Traffic	.216	S.F.
Scrim Installed, Nylon Sponge Back Carpet		
20 oz.	.242	S.Y.
60 oz.	.267	S.Y.
Tile, Foam-Backed, Needle Punch	.014	S.F.
Tufted Loop or Shag	.014	S.Y.
Wool, 30 oz., Medium Traffic	.229	S.Y.
Wool, 36 oz., Medium to Heavy Traffic	.242	S.Y.
Sponge Back, Wool, 36 oz., Medium to		
Heavy Traffic	.143	S.Y.
42 oz., Heavy Traffic	.229	S.Y.
Padding, Sponge Rubber Cushion, Minimum	.108	S.Y.
Maximum	.123	S.Y.
Felt, 32 oz. to 56 oz., Minimum	.108	S.Y.
Maximum	.123	S.Y.
Bonded Urethane, 3/8" Thick, Minimum	.094	S.Y.
Maximum	.107	S.Y.
Prime Urethane, 1/4" Thick, Minimum	.094	S.Y.
Maximum	.107	S.Y.

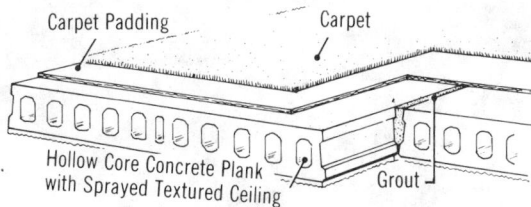

Carpet Padding   Carpet

Hollow Core Concrete Plank
with Sprayed Textured Ceiling   Grout

Carpet on Hollow Core Concrete Plank

Figure 9.20   Approximate Waste Percentages
to be Added for Resilient Tile

This table lists approximate waste figures to be added when estimating resilient tiling. These figures are based on expected averages for the size of the areas to receive the tile.

Area	Waste
Up to 75 S.F.	10–12%
75–150 S.F.	7–10%
150–300 S.F.	6–7%
300–1,000 S.F.	4–6%
1,000–5,000 S.F.	3–4%
5,000 S.F. and Up	2–3%

Figure 9.21   Installation Time in Man-Hours for Resilient Flooring

Description	Man-Hours	Unit
Resilient Asphalt Tile, on Concrete, 1/8" Thick		
Color Group B	.015	S.F.
Color Group C and D	.015	S.F.
For Less Than 500 S.F., Add	.002	S.F.
For Over 5000 S.F., Deduct	.007	S.F.
Base, Cove, Rubber or Vinyl, .080" Thick		
Standard Colors, 2-1/2" High	.026	L.F.
4" High	.027	L.F.
6" High	.027	L.F.
1/8" Thick, Standard Colors,		
2-1/2" High	.026	L.F.
4" High	.027	L.F.
6" High	.027	L.F.
Corners, 2-1/2" High	.026	Ea.
4" High	.027	Ea.
6" High	.027	Ea.
Conductive Flooring, Rubber Tile,		
1/8" Thick	.025	S.F.
Homogeneous Vinyl Tile, 1/8" Thick	.025	S.F.
Cork Tile, Standard Finish, 1/8" Thick	.025	S.F.
3/16" Thick	.025	S.F.
5/16" Thick	.025	S.F.
1/2" Thick	.025	S.F.
Urethane Finish, 1/8" Thick	.025	S.F.
3/16" Thick	.025	S.F.
5/16" Thick	.025	S.F.
1/2" Thick	.025	S.F.
Polyethylene, in Rolls, No Base Incl.,		
Landscape Surfaces	.029	S.F.
Nylon Action Surface, 1/8" Thick	.029	S.F.
1/4" Thick	.029	S.F.
3/8" Thick	.029	S.F.
Golf Tee Surface with Foam Back	.033	S.F.
Practice Putting, Knitted Nylon Surface	.033	S.F.
Polyurethane, Thermoset, Prefabricated in Place, Indoor		
3/8" Thick for Basketball, Gyms, etc.	.080	S.F.

(continued on next page)

Figure 9.21   Installation Time in Man-Hours for Resilient Flooring (continued)

Description	Man-Hours	Unit
Stair Treads and Risers		
Rubber, Molded Tread, 12" Wide, 5/16"		
Thick, Black	.070	L.F.
Colors	.070	L.F.
1/4" Thick, Black	.070	L.F.
Colors	.070	L.F.
Grit Strip Safety Tread, Colors 5/16" Thick	.070	L.F.
3/16" Thick	.067	L.F.
Landings, Smooth Sheet Rubber,		
1/8" Thick	.029	S.F.
3/16" Thick	.030	S.F.
Nosings, 1-1/2" Deep, 3" Wide, Residential	.057	L.F.
Commercial	.057	L.F.
Risers, 7" High, 1/8" Thick, Flat	.046	L.F.
Coved	.046	L.F.
Vinyl, Molded Tread, 12" Wide, Colors,		
1/8" Thick	.070	L.F.
1/4" Thick	.070	L.F.
Landing Material, 1/8" Thick	.040	S.F.
Riser, 7" High, 1/8" Thick, Coved	.046	L.F.
Threshold, 5-1/2" Wide	.080	L.F.
Tread and Riser Combined, 1/8" Thick	.100	L.F.

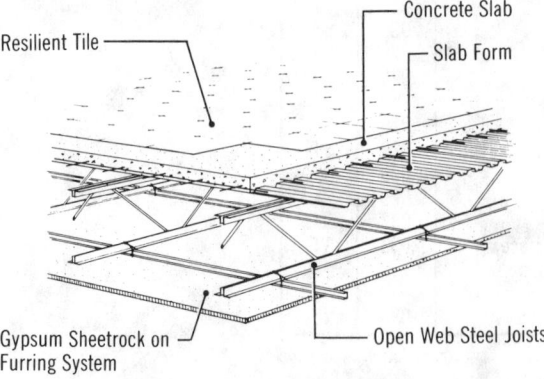

Resilient Tile

Concrete Slab

Slab Form

Gypsum Sheetrock on
Furring System

Open Web Steel Joists

Figure 9.22 Installation Time in Man-Hours for Composition Flooring

Description	Man-Hours	Unit
Composition Flooring, Acrylic, 1/4" Thick	.092	S.F.
3/8" Thick	.107	S.F.
Cupric Oxychloride, on Bond Coat, Minimum	.100	S.F.
Maximum	.114	S.F.
Epoxy, with Colored Quartz Chips,		
Broadcast, Minimum	.071	S.F.
Maximum	.098	S.F.
Trowelled, Minimum	.086	S.F.
Maximum	.100	S.F.
Heavy Duty Epoxy Topping, 1/4" Thick		
500 to 1,000 S.F.	.114	S.F.
1,000 to 2,000 S.F.	.107	S.F.
Over 10,000 S.F.	.100	S.F.
Epoxy Terrazzo, 1/4" Thick, Chemical		
Resistant, Minimum	.128	S.F.
Maximum	.171	S.F.
Conductive, Minimum	.135	S.F.
Maximum	.178	S.F.
Mastic, Hot Laid, 2 Coat, 1-1/2" Thick,		
Standard, Minimum	.070	S.F.
Maximum	.092	S.F.
Acidproof, Minimum	.079	S.F.
Maximum	.137	S.F.
Neoprene, Trowelled on, 1/4" Thick,		
Minimum	.088	S.F.
Maximum	.112	S.F.
Polyacrylate Terrazzo, 1/4" Thick, Minimum	.065	S.F.
Maximum	.100	S.F.
3/8" Thick, Minimum	.077	S.F.
Maximum	.100	S.F.
Conductive Terrazzo, 1/4" Thick, Minimum	.107	S.F.
Maximum	.157	S.F.
3/8" Thick, Minimum	.132	S.F.
Maximum	.188	S.F.
Granite, Conductive, 1/4" Thick, Minimum	.069	S.F.
Maximum	.114	S.F.
3/8" Thick, Minimum	.107	S.F.
Maximum	.126	S.F.
Polyester, with Colored Quartz Chips,		
1/16" Thick, Minimum	.045	S.F.
Maximum	.086	S.F.
1/8" Thick, Minimum	.059	S.F.
Maximum	.071	S.F.
Polyester, Heavy Duty, Compared to		
Epoxy, Add	.019	S.F.
Polyurethane, with Suspended Vinyl Chips,		
Minimum	.045	S.F.
Maximum	.056	S.F.

Figure 9.23 Installation Time in Man-Hours for Stone Flooring

Description	Man-Hours	Unit
Bluestone Flooring, Natural Cleft, Smooth or Thermal Finish 1" Thick	.267	S.F.
1-1/2" Thick	.276	S.F.
Granite Flooring 3/4" to 1-1/2" Thick	.276	S.F.
Granite Pavers 4" x 4" x 4" Blocks	.369	S.F.
Marble Flooring Tiles, 3/8" Thick, Thinset	.267	S.F.
Mortar Bed	.369	S.F.
Marble Travertine; 1-1/4" Thick, Mortar Bed	.308	S.F.
Slate 1/2" Thick 6" x 6" x 1/2", Thinset	.267	S.F.
Mortar Bed	.320	S.F.
24" x 24" x 1/2" Mortar Bed	.267	S.F.

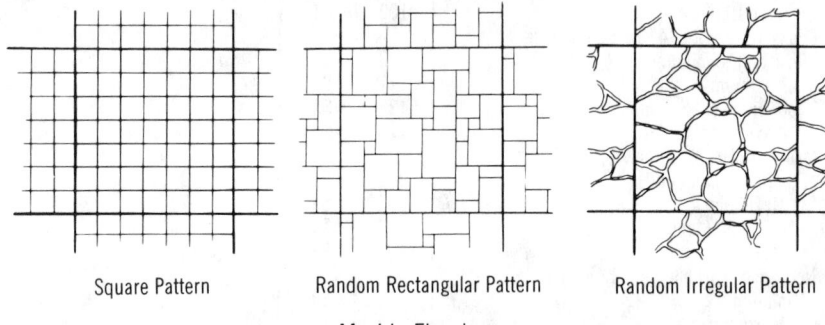

Square Pattern   Random Rectangular Pattern   Random Irregular Pattern

Marble Flooring

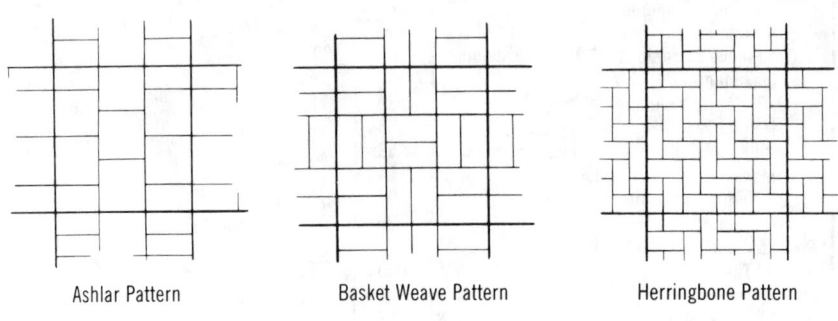

Ashlar Pattern   Basket Weave Pattern   Herringbone Pattern

Slate Flooring

Figure 9.23 Installation Time in Man-Hours for Stone Flooring (continued)

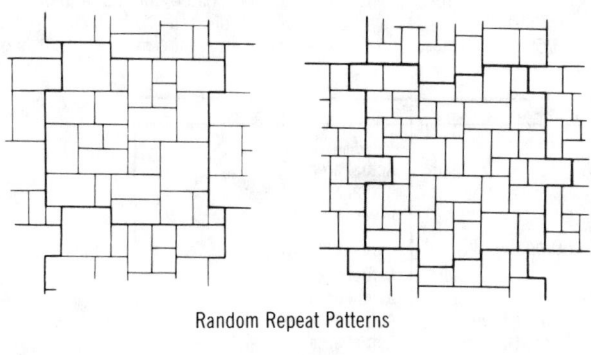

Random Repeat Patterns

Bluestone Flooring

Figure 9.24   Painting Quantity Factors

The *Painting and Decorating Contractors of America Estimating Guide* recommends that the following factors be used when estimating painting. For each type of area, the actual measured flat surface area of the item should be increased by the factors as indicated. These factors are for adjusting painting rates and outputs of flat wall painting to account for the extra amount of detail and associated extra time needed for trim work.

Area		Factor
Balustrades:		1 side x 4
Blinds:	Plain	Actual area x 2
	Slotted	Actual area x 4
Cabinets:		Front area x 5
Clapboards and Drop Siding:		Actual area x 1.1
Cornices:	1 Story	Actual area x 2
	2 Story	Actual area x 3
	1 Story Ornamental	Actual area x 4
	2 Story Ornamental	Actual area x 6
Doors:	Flush	150% per side
	Two Panel	175% per side
	Four Panel	200% per side
	Six Panel	225% per side
Door Trim:		L.F. + 50% per side
Fences:	Chain Link	1 side x 3 for both sides
	Picket	1 side x 4 for both sides
Gratings:		1 side x 4/6
Grilles:	Plain	1 side x 200%
	Lattice	Area x 2 per side
Moldings under 12" Wide		1 S.F./L.F.
Open Trusses:		Length x Depth x 2.5
Pipes:	Up to 4"	1 S.F. per L.F.
	4" to 8"	2 S.F. per L.F.
	8" to 12"	3 S.F. per L.F.
	12" x 16"	4 S.F. per L.F.
	Hangers Extra	
Radiator:		Face area x 7
Shingle Siding:		Area x 1.5
Stairs:		Number of risers x 8 widths
Tie Rods:		2 S.F. per L.F.
Wainscoting, Paneled:		Actual area x 2
Walls and Ceilings:		Length x Width, no deducts for less than 100 S.F.
Sanding and	Quality Work	Actual area x 2
Puttying:	Average Work	Actual area x 50%
	Industrial	Actual area x 25%
Downspouts and Gutters:		Actual area x 2
Window Sash:		1 L.F. of part = 1 S.F.

Figure 9.25   Paint Coverage

This table lists the average expected coverages for various painting tasks using various methods of application.

Item	Coat	S.F. Covered by One Gallon			Daily Coverage (in S.F.) per Worker (8 Hrs.)			Man Hours per 100 S.F.		
		Brush	Roller	Spray	Brush	Roller	Spray	Brush	Roller	Spray
Paint wood siding	Prime	275	250	325	1150	1400	4000	.695	.571	.200
	Others	300	275	325	1600	2200	4000	.500	.364	.200
Paint exterior trim	Prime	450	—	—	650	—	—	1.230	—	—
	1st	525	—	—	700	—	—	1.143	—	—
	2nd	575	—	—	750	—	—	1.067	—	—
Paint shingle siding	Prime	300	285	335	1050	1700	2800	.763	.470	.286
	Others	400	375	425	1200	2000	3200	.667	.400	.250
Stain shingle siding	1st	200	190	220	1200	1400	3200	.667	.571	.250
	2nd	300	275	325	1300	1700	4000	.615	.471	.200
Paint brick masonry	Prime	200	150	175	850	1700	4000	.941	.471	.200
	1st	300	250	320	1200	2200	4400	.364	.364	.182
	2nd	375	340	400	1300	2400	4400	.615	.333	.182
Paint interior plaster or drywall	Prime	450	425	550	1600	2500	4000	.500	.320	.200
	Others	500	475	550	1400	3000	4000	.571	.267	.200
Paint interior doors and windows	Prime	450	—	—	1300	—	—	.333	—	—
	1st	475	—	—	1150	—	—	.696	—	—
	2nd	500	—	—	1000	—	—	.800	—	—

Figure 9.26   Application Time in Man-Hours for Painting

Description	Man-Hours	Unit
Cabinets and Casework		
Labor Cost Includes Protection of Adjacent Items Not Painted		
Primer Coat, Oil Base, Brushwork	.020	S.F.
Paint, Oil Base, Brushwork		
1 Coat	.021	S.F.
2 Coats	.040	S.F.
Stain, Brushwork, Wipe Off	.022	S.F.
Shellac, 1 Coat, Brushwork	.021	S.F.
Varnish, 3 Coats, Brushwork	.034	S.F.
Doors and Windows		
Labor Cost Includes Protection of Adjacent Items Not Painted		
Flush Door and Frame, per Side, Oil Base,		
Primer Coat, Brushwork	.571	Ea.
Paint		
1 Coat	.320	Ea.
2 Coats	.640	Ea.
3 Coats	.889	Ea.
Stain, Brushwork, Wipe Off	.800	Ea.
Shellac, 1 Coat, Brushwork	.667	Ea.
Varnish, 3 Coats, Brushwork	1.600	Ea.
Panel Door and Frame, per Side, Oil Base,		
Primer Coat, Brushwork	.615	Ea.
Paint		
1 Coat	.667	Ea.
2 Coats	1.330	Ea.
3 Coats	2.000	Ea.
Stain, Brushwork, Wipeoff	1.330	Ea.
Shellac, 1 Coat, Brushwork	1.000	Ea.
Varnish, 3 Coats, Brushwork	2.670	Ea.
Windows, Including Frame and Trim, per Side		
Colonial Type, 2' x 3', Oil Base, Primer Coat		
Brushwork	.333	Ea.
Paint		
1 Coat	.364	Ea.
2 Coats	.615	Ea.
3 Coats	.800	Ea.
3' x 5' Opening, Primer Coat, Brushwork	.533	Ea.
Paint		
1 Coat	.615	Ea.
2 Coats	1.000	Ea.
3 Coats	1.380	
4' x 8' Opening, Primer Coat, Brushwork	.667	Ea.
Paint		
1 Coat	.800	Ea.
2 Coats	1.330	Ea.
3 Coats	1.860	Ea.

Figure 9.26 Application Time in
Man-Hours for Painting (continued)

Description	Man-Hours	Unit
Standard, 6 to 8 Lites, 2' x 3', Primer	.308	Ea.
Paint		
1 Coat	.333	Ea.
2 Coats	.615	Ea.
3 Coats	.800	Ea.
3' x 5', Primer	.471	Ea.
Paint		
1 Coat	.533	Ea.
2 Coats	.800	Ea.
3 Coats	1.000	Ea.
4' x 8', Primer	.571	Ea.
Paint		
1 Coat	.667	Ea.
2 Coats	1.000	Ea.
3 Coats	1.330	Ea.
Single Lite Type, 2' x 3', Oil Base, Primer Coat,		
Brushwork	.200	Ea.
Paint		
1 Coat	.216	Ea.
2 Coats	.400	Ea.
3 Coats	.571	Ea.
3' x 5' Opening, Primer Coat, Brushwork	.296	Ea.
Paint		
1 Coat	.320	Ea.
2 Coats	.571	Ea.
3 Coats	.800	Ea.
4' x 8' Opening, Primer Coat, Brushwork	.571	Ea.
Paint		
1 Coat	.615	Ea.
2 Coats	1.000	Ea.
3 Coats	1.230	Ea.
Miscellaneous		
Fence, Chain Link, Per Side, Oil Base,		
Primer Coat, Brushwork	.013	S.F.
Spray	.010	S.F.
Paint 1 Coat, Brushwork	.014	S.F.
Spray	.010	S.F.
Picket, Wood, Primer Coat, Brushwork	.039	S.F.
Spray	.031	S.F.
Paint 1 Coat, Brushwork	.040	S.F.
Spray	.031	S.F.
Floors, Concrete or Wood, Oil Base, Primer or		
Sealer Coat, Brushwork	.007	S.F.
Roller	.006	S.F.
Spray	.006	S.F.
Paint 1 Coat, Brushwork	.008	S.F.
Roller	.007	S.F.
Spray	.006	S.F.

(continued on next page)

Figure 9.26 Application Time in
Man-Hours for Painting (continued)

Description	Man-Hours	Unit
Stain, Wood Floor, Brushwork	.007	S.F.
Roller	.006	S.F.
Spray	.006	S.F.
Varnish, Wood Floor, Brushwork	.008	S.F.
Roller	.007	S.F.
Spray	.006	S.F.
Grilles, per Side, Oil Base, Primer Coat, Brushwork	.020	Ea.
Spray	.016	Ea.
Paint 1 Coat, Brushwork	.021	Ea.
Spray	.016	Ea.
Paint 2 Coats, Brushwork	.040	Ea.
Spray	.032	Ea.
Gutters and Downspouts, Oil Base, Primer Coat, Brushwork	.025	L.F.
Paint 1 Coat, Brushwork	.027	L.F.
Paint 2 Coats, Brushwork	.049	L.F.
Pipe, to 4" Diameter, Primer or Sealer Coat, Oil Base, Brushwork	.020	L.F.
Spray	.015	L.F.
Paint 1 Coat, Brushwork	.021	L.F.
Spray	.015	L.F.
Paint 2 Coats, Brushwork	.040	L.F.
Spray	.029	L.F.
To 8" Diameter, Primer or Sealer Coat, Brushwork	.040	L.F.
Spray	.025	L.F.
Paint 1 Coat, Brushwork	.046	L.F.
Spray	.025	L.F.
Paint 2 Coats, Brushwork	.080	L.F.
Spray	.043	L.F.
To 12" Diameter, Primer or Sealer Coat, Brushwork	.067	L.F.
Spray	.050	L.F.
Paint 1 Coat, Brushwork	.073	L.F.
Spray	.050	L.F.
Paint 2 Coats, Brushwork	.133	L.F.
Spray	.100	L.F.
To 16" Diameter, Primer or Sealer Coat, Brushwork	.083	L.F.
Spray	.067	L.F.
Paint 1 Coat, Brushwork	.089	L.F.
Spray	.067	L.F.
Paint 2 Coats, Brushwork	.160	L.F.
Spray	.123	L.F.
Trim, Wood, Including Puttying Under 6" Wide Primer Coat, Oil Base, Brushwork	.009	L.F.
Paint, Brushwork 1 Coat	.009	L.F.
2 Coats	.016	L.F.
3 Coats	.027	L.F.

Figure 9.26 Application Time in Man-Hours for Painting (continued)

Description	Man-Hours	Unit
Over 6" Wide, Primer Coat, Brushwork	.013	L.F.
Paint, Brushwork		
1 Coat	.018	L.F.
2 Coats	.027	L.F.
3 Coats	.042	L.F.
Cornice, Simple Design, Primer Coat,		
Oil Base, Brushwork	.029	S.F.
Paint, Brushwork		
1 Coat	.032	S.F.
2 Coats	.050	S.F.
Ornate Design, Primer Coat	.053	S.F.
Paint		
1 Coat	.057	S.F.
2 Coats	.089	S.F.
Balustrades, per Side, Primer Coat, Oil Base,		
Brushwork	.027	S.F.
Paint		
1 Coat	.028	S.F.
2 Coats	.047	S.F.
Trusses and Exposed Wood Frames		
Primer Coat Oil Base, Brushwork	.010	S.F.
Spray	.007	S.F.
Paint 1 Coat, Brushwork	.011	S.F.
Spray	.007	S.F.
Paint 2 Coats, Brushwork	.016	S.F.
Spray	.013	S.F.
Stain, Brushwork, Wipe Off	.013	S.F.
Varnish, 3 Coats, Brushwork	.029	S.F.
Siding Exterior		
Steel Siding, Oil Base, Primer or Sealer Coat,		
Brushwork	.009	S.F.
Spray	.005	S.F.
Paint 2 Coats, Brushwork	.012	S.F.
Spray	.006	S.F.
Stucco, Rough, Oil Base, Paint 2 Coats,		
Brushwork	.012	S.F.
Roller	.008	S.F.
Spray	.006	S.F.
Texture 1-11 or Clapboard, Oil Base		
Primer Coat, Brushwork	.006	S.F.
Spray	.004	S.F.
Paint 1 Coat, Brushwork	.006	S.F.
Spray	.004	S.F.
Paint 2 Coats, Brushwork	.013	S.F.
Spray	.008	S.F.
Stain 1 Coat, Brushwork	.006	S.F.
Spray	.004	S.F.
Stain 2 Coats, Brushwork	.013	S.F.
Spray	.008	S.F.

(continued on next page)

Figure 9.26   Application Time in
Man-Hours for Painting (continued)

Description	Man-Hours	Unit
Wood Shingles, Oil Base Primer Coat, Brushwork	.006	S.F.
Spray	.004	S.F.
Paint 1 Coat, Brushwork	.007	S.F.
Spray	.004	S.F.
Paint 2 Coats, Brushwork	.014	S.F.
Spray	.008	S.F.
Stain 1 Coat, Brushwork	.006	S.F.
Spray	.004	S.F.
Stain 2 Coats, Brushwork	.014	S.F.
Spray	.008	S.F.
Wall and Ceilings		
Concrete, Dry Wall or Plaster, Oil Base, Primer or Sealer Coat		
Smooth Finish, Brushwork	.004	S.F.
Roller	.004	S.F.
Spray	.002	S.F.
Sand Finish, Brushwork	.005	S.F.
Roller	.004	S.F.
Spray	.002	S.F.
Paint 1 Coat		
Smooth Finish, Brushwork	.004	S.F.
Roller	.004	S.F.
Spray	.002	S.F.
Sand Finish, Brushwork	.005	S.F.
Roller	.004	S.F.
Spray	.002	S.F.
Paint 2 Coats		
Smooth Finish, Brushwork	.008	S.F.
Roller	.007	S.F.
Spray	.004	S.F.
Sand Finish, Brushwork	.010	S.F.
Roller	.008	S.F.
Spray	.004	S.F.
Paint 3 Coats		
Smooth Finish, Brushwork	.012	S.F.
Roller	.010	S.F.
Spray	.005	S.F.
Sand Finish, Brushwork	.014	S.F.
Roller	.011	S.F.
Spray	.005	S.F.
Glaze Coating, 5 Coats, Spray		
Clear	.009	S.F.
Multicolor	.009	S.F.

Figure 9.26   Application Time in
Man-Hours for Painting (continued)

Description	Man-Hours	Unit
Masonry or Concrete Block, Oil Base, Primer or Sealer Coat		
Smooth Finish, Brushwork	.005	S.F.
Spray	.002	S.F.
Sand Finish, Brushwork	.006	S.F.
Spray	.002	S.F.
Paint 1 Coat		
Smooth Finish, Brushwork	.005	S.F.
Spray	.002	S.F.
Sand Finish, Brushwork	.006	S.F.
Spray	.002	S.F.
Paint 2 Coats		
Smooth Finish, Brushwork	.010	S.F.
Spray	.004	S.F.
Sand Finish, Brushwork	.011	S.F.
Spray	.004	S.F.
Paint 3 Coats		
Smooth Finish, Brushwork	.013	S.F.
Spray	.005	S.F.
Sand Finish, Brushwork	.023	S.F.
Spray	.005	S.F.
Glaze Coating, 5 Coats, Spray		
Clear	.009	S.F.
Multicolor	.009	S.F.
Block Filler, 1 Coat, Brushwork	.006	S.F.
Silicone, Water Repellent, 2 Coats, Spray	.009	S.F.
Varnish 1 Coat + Sealer, on Wood Trim, No Sanding Included	.009	S.F.
Hardwood Floors, 2 Coats, No Sanding Included	.010	S.F.
Wall Coatings Acrylic Glazed Coatings		
Minimum	.015	S.F.
Maximum	.026	S.F.
Epoxy Coatings		
Minimum	.015	S.F.
Maximum	.047	S.F.
Exposed Aggregate, Troweled on		
1/16" to 1/4"		
Minimum	.034	S.F.
Maximum (Epoxy or Polyacrylate)	.062	S.F.
1/2" to 5/8" Aggregate,		
Minimum	.062	S.F.
Maximum	.100	S.F.
1" Aggregate Size		
Minimum	.089	S.F.
Maximum	.145	S.F.
Exposed Aggregate, Sprayed on		
1/8" Aggregate		
Minimum	.027	S.F.
Maximum	.055	S.F.

(continued on next page)

Figure 9.26 Application Time in Man-Hours for Painting (continued)

Description	Man-Hours	Unit
High Build Epoxy, 50 Mil		
Minimum	.021	S.F.
Maximum	.084	S.F.
Laminated Epoxy with Fiberglass		
Minimum	.027	S.F.
Maximum	.055	S.F.
Sprayed Perlite or Vermiculite		
1/16" Thick		
Minimum	.003	S.F.
Maximum	.013	S.F.
Vinyl Plastic Wall Coating		
Minimum	.011	S.F.
Maximum	.033	S.F.
Urethane on Smooth Surface		
2 Coats		
Minimum	.007	S.F.
Maximum	.012	S.F.
3 Coats		
Minimum	.010	S.F.
Maximum	.017	S.F.
Ceramic-like Glazed Coating, Cementitious		
Minimum	.018	S.F.
Maximum	.023	S.F.
Resin Base		
Minimum	.013	S.F.
Maximum	.024	S.F.

Figure 9.27 Wall Covering Quantities

This table is helpful for estimating the number of rolls of wall, border, and/or ceiling covering required for various sized rooms.

Size of Room	Single Rolls of Side Wall Height of Ceiling			Yards of Border	Rolls of Ceiling
	8 Feet	9 Feet	10 Feet		
4 x 8	6	7	8	9	2
4 x 10	7	8	9	11	2
4 x 12	8	9	10	12	2
6 x 10	8	9	10	12	2
6 x 12	9	10	11	13	3
8 x 12	10	11	13	15	4
8 x 14	11	12	14	16	4
10 x 14	12	14	15	18	5
10 x 16	13	15	16	19	6
12 x 16	14	16	17	20	7
12 x 18	15	17	19	22	8
14 x 18	16	18	20	23	8
14 x 22	18	20	22	26	10
15 x 16	15	17	19	23	8
15 x 18	16	18	20	24	9
15 x 20	17	20	22	25	10
15 x 23	19	21	23	28	11
16 x 18	17	19	21	25	10
16 x 20	18	20	22	26	10
16 x 22	19	21	23	28	11
16 x 24	20	22	25	29	12
16 x 26	21	23	26	31	13
17 x 22	19	22	24	28	12
17 x 25	21	23	26	31	13
17 x 28	22	25	28	32	15
17 x 32	24	27	30	35	17
17 x 35	26	29	32	37	18
18 x 22	20	22	25	29	12
18 x 25	21	24	27	31	14
18 x 28	23	26	28	33	16
20 x 26	23	28	28	33	17
20 x 28	24	27	30	34	18
20 x 34	27	30	33	39	21

Deduct one single roll of side wall covering for every two ordinary sized doors or windows, or every 36 square feet of opening. For yard goods with no match such as wide vinyls, measure the area, deduct for openings, and allow 10" for waste.

(courtesy of Painting and Decorating Contractors of America, Falls Church, Va. 22046)

Figure 9.28   Installation Time in Man-Hours for Wall Coverings

Description	Man-Hours	Unit
Wall Covering		
Aluminum Foil	.029	S.F.
Copper Sheets, .025" Thick		
Vinyl Backing	.033	S.F.
Phenolic Backing	.033	S.F.
Cork Tiles, Light or Dark, 12" x 12"		
3/16" Thick	.033	S.F.
5/16" Thick	.034	S.F.
1/4" Basketweave	.033	S.F.
1/2" Natural, Non-directional Pattern	.033	S.F.
Granular Surface, 12" x 36"		
1/2" Thick	.021	S.F.
1" Thick	.022	S.F.
Polyurethane Coated, 12" x 12"		
3/16" Thick	.033	S.F.
5/16" Thick	.034	S.F.
Cork Wallpaper, Paperbacked		
Natural	.017	S.F.
Colors	.017	S.F.
Flexible Wood Veneer, 1/32" Thick		
Plain Woods	.080	S.F.
Exotic Woods	.084	S.F.
Gypsum-based, Fabric-backed, Fire Resistant		
for Masonry Walls		
Minimum	.020	S.F.
Average	.023	S.F.
Maximum	.027	S.F.
Acrylic, Modified, Semi-rigid PVC		
.028" Thick	.048	S.F.
.040" Thick	.050	S.F.
Vinyl Wall Covering, Fabric-backed		
Lightweight	.013	S.F.
Medium Weight	.017	S.F.
Heavy Weight	.018	S.F.
Grass Cloths with Lining Paper		
Minimum	.020	S.F.
Maximum	.023	S.F.

Checklist

For an estimate to be reliable, all items must be accounted for.
A complete estimate can also eliminate the need to include
contingencies. The following checklist can be used to help ensure
that all items are properly accounted for.

Bases
- ☐ Cove
- ☐ Sanitary

Ceilings
- ☐ Acoustical
- ☐ Dropped
- ☐ Drywall
- ☐ Insulation
    - ____ Thermal
    - ____ Acoustic
- ☐ Plaster
- ☐ Suspension system

Drywall _____

Flooring
- ☐ Brick
- ☐ Carpet
- ☐ Carpet tile
- ☐ Ceramic tile
- ☐ Composition
- ☐ Concrete topping
- ☐ Metal tile
- ☐ Paint
- ☐ Epoxy
- ☐ Urethane
- ☐ Mosaic tile
- ☐ Plastic tile
- ☐ Quarry tile
- ☐ Raised access
- ☐ Resilient
    - ____ Asphalt
    - ____ Conductive
    - ____ Cork
    - ____ Linoleum
    - ____ Polyurethane

_____ Rubber tile
_____ Sheet vinyl
_____ Vinyl tile
☐ Stone
☐ Terrazzo
☐ Wood

Wall Finishes
☐ Ceramic tile
☐ Planking
☐ Plastic tile
☐ Mosaic tile
☐ Quarry tile
☐ Metal tile
☐ Paint
☐ Panelling
☐ Plaster
☐ Stucco
☐ Wainscotting
☐ Wall coverings
_____ Cloth
_____ Paper
_____ Vinyl

Tips

General Note on Finishes

If, during the course of a project, the owner decides to cut back on costs, this is not a good area in which to do it. When a project is finished and the owner and prospective tenants walk through, they will not be able to see a complicated foundation, nor the marvelously engineered subsystems. What they will see are the finishes. If the project is "cheapened" at this point, it will show and dull the luster of an otherwise shining project.

Room Finish Schedule

A complete set of plans should contain a room finish schedule. If one is not available, it would be well worth the time and effort to put one together. A room finish schedule should contain the room number, room name (for clarity), floor materials, base materials, wainscot materials, wainscot height, wall materials (for each wall), ceiling materials, and special instructions. It is handy and easier to work out a room finish schedule on a grid system.

Studs for Lathing and Plaster

If not specified, the studs for plaster partitions are usually spaced at no greater than 16" on center.

Lathing Waste Factor

Allow a 5% waste factor when figuring the amount of lathing.

Plaster Allowances

Deductions for openings in plaster partitions vary by the preference and experience of the estimator, from 0% for openings of less than 2 S.F. to 50% for openings greater than 2 S.F.

For curved walls with a radius less than 6' allow twice as much plaster.

Pricing a plaster and lath job depends on the quality of work desired. A first class project will have variations in the wall level of less than 1/16" in 10 feet. An ordinary plaster job (or commercial grade) may have waves of 1/8" to 3/16" in 10 feet. Overall, labor for first class work is approximately 20% higher than ordinary or commercial grade work.

Drywall Deductions

A rule of thumb when estimating materials for drywall is to not deduct for door or window openings of less than 32 square feet.

Wallcoverings

A single roll of wallpaper will cover approximately 36 S.F. Allow approximately 6 S.F. per roll for waste. This means take the total square foot area to be covered and divide by 30 to determine the number of rolls needed.

For vinyls and grass cloth coverings with no patterns to match, allow approximately 10% (3 square feet per roll) for waste. For patterns requiring matching, allow up to 25% to 30%.

Waste can run as high as 50% to 60% for coverings with large, bold, or intricate patterns.

Always specify wallcoverings from the same batch with identical batch numbers. This will help to ensure that the colors and patterns will match.

Surplus Finishes

Review the specifications to determine if there is any requirement to provide certain amounts of extra floor tile, ceiling pads, paint, wallcoverings, etc., for the owner's maintenance department. In some cases, the owner may require a substantial amount of materials, especially if it is a special order or long lead time item.

Division Ten
Specialties

Introduction

Specialties include pre-finished, manufactured items that are usually installed at or near the end of a project—the finishing touches. Division 10 could be considered a catch-all type of division, containing miscellaneous items that do not quite fit into any other division. There is, of course the potential for overlap; e.g., do toilet partitions belong in Division 8—Doors, Division 9—Finishes, or Division 10—Specialties? (It has become accepted that they belong in Division 10, as they are a specialty item).

The first step in estimating Division 10 is to check out the available specifications. Often, a list of items is provided, along with the approved or accepted manufacturers of these items. The next step is to go through all drawings and to make a list of items to be priced.

Many of the items covered in Division 10 are priced on a delivered material only basis. It is important to keep this in mind, as the labor for the installation of these items will have to be figured separately.

Estimating Data

The following tables show the expected installation time for various specialty items. Each table lists the specialty item, the typical installation crew mix, the expected average daily output of that crew (the number of units the crew can be expected to install in one eight-hour day), and the man-hours required for the installation of one unit.

Table of Contents

Figure 10.1   Installation Time for Chalkboards

Chalkboards	Crew Makeup	Daily Output	Man-Hours	Unit
CHALKBOARDS Cement asbestos, no frame, economy	2 Carpenters	270	.059	S.F.
Deluxe		260	.062	S.F.
Hardboard, tempered, no frame, 1/4" thick		270	.059	S.F.
1/2" thick		260	.062	S.F.
Hardboard, not tempered, no frame, 1/4" thick		270	.059	S.F.
1/2" thick	↓	260	.062	S.F.
Porcelain enamel, 24 ga. steel, 4' high, with aluminum trim, alum. foil backing, with core materials as follows:				
3/8" gypsum core	2 Carpenters	260	.062	S.F.
1/4" hardboard core		270	.059	S.F.
7/16" hardboard core		255	.063	S.F.
3/8" honeycomb core		275	.058	S.F.
3/8" particleboard core		260	.062	S.F.
1/4" plywood core		270	.059	S.F.
3/8" plywood core		260	.062	S.F.
Slate, 3/8", frame not included, to 4' wide		240	.067	S.F.
To 4'-6" wide		230	.070	S.F.
Over 4'-6" to 5'-0"		220	.073	S.F.
Treated plastic on plywood, no frame, 1/4" thick		240	.067	S.F.
1/2" thick		230	.070	S.F.
Frame for chalkboards, aluminum, chalk tray		290	.055	L.F.
Trim		385	.042	L.F.
Map and display rail, economy		385	.042	L.F.
Deluxe		350	.046	L.F.
Factory fabricated, tempered hardbd., w/wood frame, 3'-6" high		265	.060	S.F.
4' high		260	.062	S.F.
Aluminum frame, 3'-6" high		270	.059	S.F.
4' high		265	.060	S.F.
Porcelain steel, aluminum frame, 3' high		260	.062	S.F.
4' high		255	.063	S.F.
Magnetic swing leaf panels, 36" x 24", 4 panels		3	5.330	Total
5 panels		3	5.330	Total
6 panels		3	5.330	Total
8 panels		3	5.330	Total
Vertical sliding with tempered hardboard, manual		150	.107	S.F.
Electric		115	.139	S.F.
Horizontal sliding, 4' high, 10' long, 2 track		40	.400	S.F.
3 track		40	.400	S.F.
4 track		40	.400	S.F.
Horizontal sliding, 4' high, 20' long, 2 track		80	.200	S.F.
3 track		80	.200	S.F.
4 track	↓	80	.200	S.F.

Figure 10.2 Installation Time for Toilet Compartments

Toilet Compartments	Crew Makeup	Daily Output	Man-Hours	Unit
PARTITIONS, TOILET Cubicles, ceiling hung, marble	2 Marble Setters	2	8.000	Ea.
Painted metal	2 Carpenters	4	4.000	Ea.
Plastic laminate on particle board		4	4.000	Ea.
Porcelain enamel		4	4.000	Ea.
Stainless steel	↓	4	4.000	Ea.
Floor & ceiling anchored, marble	2 Marble Setters	2.50	6.400	Ea.
Painted metal	2 Carpenters	5	3.200	Ea.
Painted laminate on particle board		5	3.200	Ea.
Porcelain enamal		5	3.200	Ea.
Stainless steel	↓	5	3.200	Ea.
Floor mounted, marble	2 Marble Setters	3	5.330	Ea.
Painted metal	2 Carpenters	7	2.290	Ea.
Plastic laminate on particle board		7	2.290	Ea.
Porcelain enamel		7	2.290	Ea.
Stainless steel	↓	7	2.290	Ea.
Headrail braced, marble	2 Marble Setters	3	5.330	Ea.
Painted metal	2 Carpenters	6	2.670	Ea.
Plastic laminate on particle board		6	2.670	Ea.
Porcelain enamel		6	2.670	Ea.
Stainless steel		6	2.670	Ea.
Wall hung partitions, painted metal		7	2.290	Ea.
Porcelain enamel		7	2.290	Ea.
Stainless steel	↓	7	2.290	Ea.
Screens, entrance, floor mounted, 54" high				
Marble	1 Bricklayer	35	.457	L.F.
	1 Bricklayer Helper			
Painted metal	2 Carpenters	60	.267	L.F.
Plastic laminate on particle board		60	.267	L.F.
Porcelain enamel		60	.267	L.F.
Stainless steel	↓	60	.267	L.F.
Urinal screen, 18" wide, ceiling braced, marble	1 Bricklayer	6	2.670	Ea.
	1 Bricklayer Helper			
Painted metal	2 Carpenters	8	2.000	Ea.
Plastic laminate on particle board		8	2.000	Ea.
Porcelain enamel		8	2.000	Ea.
Stainless steel	↓	8	2.000	Ea.
Floor mounted, headrail braced				
Marble	1 Bricklayer	6	2.670	Ea.
	1 Bricklayer Helper			
Painted metal	2 Carpenters	8	2.000	Ea.
Plastic laminate on particle board		8	2.000	Ea.
Porcelain enamel		8	2.000	Ea.
Stainless steel	↓	8	2.000	Ea.
Pilaster, flush, marble	1 Bricklayer	9	1.780	Ea.
	1 Bricklayer Helper			
Painted metal	2 Carpenters	10	1.600	Ea.
Plastic laminate on particle board		10	1.600	Ea.
Porcelain enamel		10	1.600	Ea.
Stainless steel	↓	10	1.600	Ea.

Figure 10.2   Installation Time for
Toilet Compartments (continued)

Toilet Compartments	Crew Makeup	Daily Output	Man-Hours	Unit
PARTITIONS, TOILET Post braced, marble	1 Bricklayer 1 Bricklayer Helper	9	1.780	Ea.
Painted metal	2 Carpenters	10	1.600	Ea.
Plastic laminate on particle board		10	1.600	Ea.
Porcelain enamel		10	1.600	Ea.
Stainless steel	▼	10	1.600	Ea.
Wall hung, bracket supported				
Painted metal	2 Carpenters	10	1.600	Ea.
Plastic laminate on particle board		10	1.600	Ea.
Porcelain enamel		10	1.600	Ea.
Stainless steel		10	1.600	Ea.
Flange supported, painted metal		10	1.600	Ea.
Plastic laminate on particle board		10	1.600	Ea.
Porcelain enamel		10	1.600	Ea.
Stainless steel		10	1.600	Ea.
Wedge type, painted metal		10	1.600	Ea.
Porcelain enamel		10	1.600	Ea.
Stainless steel	▼	10	1.600	Ea.

Figure 10.3   Installation Time for Shower Compartments

Shower Compartments	Crew Makeup	Daily Output	Man-Hours	Unit
PARTITIONS, SHOWER Economy, painted steel, steel base,				
no door or plumbing included	2 Sheet Metal Workers	5	3.200	Ea.
Square, 32" x 32", stock, with receptor & door, fiberglass		4.50	3.560	Ea.
Galvanized and painted steel	↓	5	3.200	Ea.
Shower stall, double wall, incl. receptor but not including				
door or plumbing, enameled steel	2 Sheet Metal Workers	5	3.200	Ea.
Porcelain enamel steel		5	3.200	Ea.
Stainless steel		5	3.200	Ea.
Circular fiberglass, 36" diameter, no plumbing included		4	4.000	Ea.
One piece, 36" diameter, less door		4	4.000	Ea.
With door		3.50	4.570	Ea.
Curved shell shower, no door needed		3	5.330	Ea.
Glass stalls, with doors, no receptors, chrome on brass		3	5.330	Ea.
Anodized aluminum	↓	4	4.000	Ea.
Marble shower stall, stock design with shower door	2 Marble Setters	1.20	13.330	Ea.
With curtain		1.30	12.310	Ea.
Receptors, precast terrazzo, 32" x 32"		14	1.140	Ea.
48" x 34"		12	1.330	Ea.
Plastic, simulated terrazzo receptor, 32" x 32"		14	1.140	Ea.
32" x 48"		12	1.330	Ea.
Precast concrete, colors, 32" x 32"		14	1.140	Ea.
48" x 48"	↓	12	1.330	Ea.
Shower doors, economy plastic, 24" wide	1 Sheet Metal Worker	9	.889	Ea.
Tempered glass door, economy		8	1.000	Ea.
Folding, tempered glass, aluminum frame		6	1.330	Ea.
Sliding, tempered glass, 48" opening		6	1.330	Ea.
Deluxe, tempered glass, chrome on brass frame, min.		5	1.600	Ea.
Maximum		5	1.600	Ea.
On anodized aluminum frame, minimum		5	1.600	Ea.
Maximum		5	1.600	Ea.
Tub enclosure, plastic panels, economy, sliding panel		4	2.000	Ea.
Folding panel		4	2.000	Ea.
Deluxe, tempered glass, anodized alum. frame, min.		2	4.000	Ea.
Maximum		1.50	5.330	Ea.
On chrome-plated brass frame, minimum		2	4.000	Ea.
Maximum	↓	1.50	5.330	Ea.

Figure 10.4 Installation Time for Metal Wall Louvers

Metal Wall Louvers	Crew Makeup	Daily Output	Man-Hours	Unit
LOUVERS Aluminum with screen, residential, 8" x 8"	1 Carpenter	38	.211	Ea.
12" x 12"		38	.211	Ea.
12" x 18"		35	.229	Ea.
14" x 24"		30	.267	Ea.
18" x 24"		27	.296	Ea.
30" x 24"		24	.333	Ea.
Triangle, adjustable, small		20	.400	Ea.
Large		15	.533	Ea.
Midget, aluminum, 3/4" deep, 1" diameter		85	.094	Ea.
3" diameter		60	.133	Ea.
4" diameter		50	.160	Ea.
6" diameter	↓	30	.267	Ea.
Louvers, ridge vent strip, mill finish	1 Sheet Metal Worker	155	.052	L.F.
Soffit vent, continuous, 3" wide, aluminum, mill finish	1 Carpenter	200	.040	L.F.
Baked enamel finish		200	.040	L.F.
Under eaves vent, aluminum, mill finish, 16" x 4"		75	.107	Ea.
16" x 8"		75	.107	Ea.
Vinyl wall louvers, 1-1/2" deep, 8" x 8"		38	.211	Ea.
12" x 12"		38	.211	Ea.
12" x 18"		35	.229	Ea.
14" x 24"	↓	30	.267	Ea.

Figure 10.5   Installation Time for Wall & Corner Guards

Wall & Corner Guards	Crew Makeup	Daily Output	Man-Hours	Unit
CORNER GUARDS Steel angle w/anchors, 1" x 1", 1.5#/L.F.	1 Struc. Steel Foreman 3 Struc. Steel Workers 1 Gas Welding Machine	320	.100	L.F.
2" x 2" angles, 3.5#/L.F.		300	.107	L.F.
3" x 3" angles, 6#/L.F.		275	.116	L.F.
4" x 4" angles, 8#/L.F.		240	.133	L.F.
Cast iron wheel guards, 3'-0" high		24	1.330	Ea.
5'-0" high		19	1.680	Ea.
Pipe bumper for truck doors, 8' long, 6" diameter		20	1.600	Ea.
8" diameter	▼	20	1.600	Ea.
CORNER PROTECTION				
Acrylic and vinyl, high impact type, stainless frame	1 Struc. Steel Worker	45	.178	L.F.
Shock-absorbing water filled bumpers, corner, 26" long		14	.571	Ea.
41" long		14	.571	Ea.
Half round, 26" long		14	.571	Ea.
40" long		14	.571	Ea.
Stainless steel, 16 ga., with adhesive		80	.100	L.F.
12 ga. stainless, with adhesive	▼	80	.100	L.F.
Vinyl adhesive type, 3-3/8" wide	1 Carpenter	128	.063	L.F.
WALLGUARD				
Neoprene with aluminum fastening strip, 1-1/2" x 2"	1 Carpenter	110	.073	L.F.
Trolley rail, PVC, clipped to wall, 5" high		185	.043	L.F.
8" high		180	.044	L.F.
Vinyl acrylic bed aligner and bumper, 37" long		10	.800	Ea.
43" long	▼	9	.889	Ea.

Figure 10.6   Installation Time for Access Flooring

Access Flooring	Crew Makeup	Daily Output	Man-Hours	Unit
PEDESTAL ACCESS FLOORS Computer room application, metal				
Particle board or steel panels, no covering, under 6000 S.F.	2 Carpenters	1,000	.016	S.F.
Metal covered, over 6000 S.F.		1,100	.015	S.F.
Aluminum, 24" panels		500	.032	S.F.
For snap on stringer system, add	▼	1,000	.016	S.F.
Office applications, to 8" high, steel panels,				
no covering, over 6000 S.F.	2 Carpenters	400	.040	S.F.
Machine cutouts after initial installation	1 Carpenter	10	.800	Ea.
Air conditioning grilles, 4" x 12"		17	.471	Ea.
6" x 18"	▼	14	.571	Ea.
Approach ramps, minimum	2 Carpenters	85	.188	S.F.
Maximum	"	60	.267	S.F.
Handrail, 2 rail aluminum	1 Carpenter	15	.533	L.F.

Figure 10.7 Installation Time for Prefabricated Fireplaces

Prefabricated Fireplaces	Crew Makeup	Daily Output	Man-Hours	Unit
FIREPLACE, PREFABRICATED Free standing or wall hung				
with hood & screen, minimum	1 Carpenter Power Tools	1.30	6.150	Ea.
Average		1	8.000	Ea.
Maximum		.90	8.890	Ea.
Double wall for chimney heights over 8'-6"				
7" diameter, add		33	.242	V.L.F.
10" diameter, add		32	.250	V.L.F.
12" diameter, add		31	.258	V.L.F.
14" diameter, add		30	.267	V.L.F.
Simulated brick chimney top, 4' high, 16" x 16"		10	.800	Ea.
24" x 24"		7	1.140	Ea.
Simulated logs, gas fired, 40,000 BTU, 2' long, mini.		7	1.140	Set
Maximum		6	1.330	Set
Electric, 1500 BTU, 1'-6" long, minimum		7	1.140	Set
11,500 BTU, maximum		6	1.330	Set

Figure 10.8   Installation Time for Fireplace Accessories

Fireplace Accessories	Crew Makeup	Daily Output	Man-Hours	Unit
FIREPLACE ACCESSORIES Chimney screens				
Galv., 13" x 13" flue	1 Bricklayer	8	1.000	Ea.
Galv., 24" x 24" flue		5	1.600	Ea.
Stainless steel, 13" x 13" flue		8	1.000	Ea.
20" x 20" flue		5	1.600	Ea.
Cleanout doors and frames, cast iron, 8" x 8"		12	.667	Ea.
12" x 12"		10	.800	Ea.
18" x 24"		8	1.000	Ea.
Cast iron frame, steel door, 24" x 30"		5	1.600	Ea.
Damper, rotary control, steel, 30" opening		6	1.330	Ea.
Cast iron, 30" opening		6	1.330	Ea.
36" opening		6	1.330	Ea.
48" opening		6	1.330	Ea.
60" opening		6	1.330	Ea.
72" opening		5	1.600	Ea.
84" opening		5	1.600	Ea.
96" opening		4	2.000	Ea.
Steel plate, poker control, 60" opening		8	1.000	Ea.
84" opening		5	1.600	Ea.
"Universal" type, chain operated, 32" x20" opening		8	1.000	Ea.
48" x 24" opening		5	1.600	Ea.
Dutch Oven door and frame, cast iron, 12" x 15" opening		13	.615	Ea.
Copper plated, 12" x 15" opening		13	.615	Ea.
Fireplace forms with registers, 25" opening		3	2.670	Ea.
34" opening		2.50	3.200	Ea.
48" opening		2	4.000	Ea.
72" opening		1.50	5.330	Ea.
Squirrel and bird screens, galvanized, 8" x 8" flue		16	.500	Ea.
13" x 13" flue	▼	12	.667	Ea.

Figure 10.9   Installation Time for Stoves

Stoves	Crew Makeup	Daily Output	Man-Hours	Unit
WOODBURNING STOVES Cast iron, minimum	2 Carpenters Power Tools	1.30	12.310	Ea.
Average		1	16.000	Ea.
Maximum	▼	.80	20.000	Ea.

Figure 10.10 Installation Time for Ground Set Flagpoles

Ground Set Flagpoles	Crew Makeup	Daily Output	Man-Hours	Unit
FLAGPOLE Not including base or foundation				
Aluminum, tapered, ground set 20' high	1 Carpenter	2	8.000	Ea.
	1 Truck Driver (light)			
	1 Truck w/Power Equip.			
25' high		1.70	9.410	Ea.
30' high		1.50	10.670	Ea.
35' high		1.40	11.430	Ea.
40' high		1.20	13.330	Ea.
50' high		1	16.000	Ea.
60' high		.90	17.780	Ea.
70' high		.80	20.000	Ea.
80' high		.70	22.860	Ea.
Counterbalanced, tapered, aluminum, 20' high		1.80	8.890	Ea.
30' high		1.50	10.670	Ea.
40' high		1.30	12.310	Ea.
50' high		1	16.000	Ea.
Aluminum, electronically operated, 30' high		1.40	11.430	Ea.
35' high		1.30	12.310	Ea.
40' high		1.10	14.550	Ea.
45' high		1	16.000	Ea.
50' high		.90	17.780	Ea.
Fiberglass, tapered, ground set, 25' high		2	8.000	Ea.
30' high		1.50	10.670	Ea.
35' high		1.40	11.430	Ea.
40' high		1.20	13.330	Ea.
50' high		1	16.000	Ea.
60' high		.90	17.780	Ea.
Yardarms and rigging for poles, 6' total length		1.90	8.420	Ea.
12' total length		1.80	8.890	Ea.
Steel, sectional, lightweight, ground set, 20' high		1.30	12.310	Ea.
25' high		1.20	13.330	Ea.
30' high		1.10	14.550	Ea.
35' high		1	16.000	Ea.
40' high		.90	17.780	Ea.
50' high		.80	20.000	Ea.
Tapered, heavyweight steel, ground set, 35' high		.80	20.000	Ea.
50' high		.70	22.860	Ea.
60' high		.70	22.860	Ea.
75' high		.60	26.670	Ea.
Bases, ornamental, minimum	1 Carpenter	6	1.330	Ea.
Average		4	2.000	Ea.
Maximum		2	4.000	Ea.
Wood poles, tapered, clear vertical grain fir with				
tilting base, not incl. foundation, 4" butt, 25' high	1 Carpenter	1.90	8.420	Ea.
	1 Truck Driver (light)			
	1 Truck w/Power Equip.			
6" butt, 30' high	" "	1.30	12.310	Ea.
Foundations for flagpoles, including				
excavation and concrete, to 35' high poles	3 Carpenters	10	3.200	Ea.
	1 Building Laborer			
	Power Tools			
40' to 50' high		3.50	9.140	Ea.
Over 60' high		2	16.000	Ea.

Figure 10.11 Installation Time for Wall-Mounted Flagpoles

Wall-Mounted Flagpoles	Crew Makeup	Daily Output	Man-Hours	Unit
FLAGPOLE Not including base or foundation, wall-mounted Aluminum, outrigger wall poles, including base				
10' long, minimum	1 Carpenter 1 Truck Driver (light) 1 Truck w/Power Equip.	1.70	9.410	Ea.
Maximum		1.40	11.430	Ea.
20' long outrigger, minimum		1.30	12.310	Ea.
Average		1.20	13.330	Ea.
Maximum		1	16.000	Ea.
Aluminum, vertical wall set, tapered, with base, 23' high		1.29	13.330	Ea.
28' high		1	16.000	Ea.
Outrigger poles with base, 12' long		1.30	12.310	Ea.
14' long		1	16.000	Ea.

Figure 10.12 Installation Time for Directories

Directories	Crew Makeup	Daily Output	Man-Hours	Unit
DIRECTORY BOARDS Plastic, glass covered, 30" x 20"	2 Carpenters	3	5.330	Ea.
36" x 48"		2	8.000	Ea.
Grooved cork, 30" x 20"		3	5.330	Ea.
36" x 48"		2	8.000	Ea.
Black felt, 30" x 20"		3	5.330	Ea.
36" x 48"		2	8.000	Ea.
Outdoor, weatherproof, black plastic, 36" x 24"		2	8.000	Ea.
36"x 36"		1.50	10.670	Ea.
Grooved cork, 36" x 24"		2	8.000	Ea.
36"x 36"		1.50	10.670	Ea.
Vinyl plastic, 36" x 24"		2	8.000	Ea.
36"x 36"		1.50	10.670	Ea.
Indoor, economy, open face 20" x 15"		7	2.290	Ea.
24" x 18"		7	2.290	Ea.
30" x 20"		6	2.670	Ea.
39" x 27"		6	2.670	Ea.
Building directory boards, alum., black felt panels				
24" x 18"		4	4.000	Ea.
39" x 22"		3.50	4.570	Ea.
48" x 32"		3	5.330	Ea.
36" x 48"		2.50	6.400	Ea.
48" x 60"		2	8.000	Ea.
48" x 72"		1	16.000	Ea.

Figure 10.13   Installation Time for Bulletin Boards

Bulletin Boards	Crew Makeup	Daily Output	Man-Hours	Unit
BULLETIN BOARDS Cork sheets, unbacked, no frame				
1/8" thick	2 Carpenters	290	.055	S.F.
1/4" thick		290	.055	S.F.
Burlap-faced cork, no frame, 1/8" thick,		290	.055	S.F.
With 1/8" cork backing		290	.055	S.F.
With 1/4" cork backing		290	.055	S.F.
With 3/8" backboard		290	.055	S.F.
1/16" vinyl cork, on 3/8" fiber board, no frame		280	.057	S.F.
1/4" vinyl cork, on 7/16" backboard, no frame		280	.057	S.F.
1/4" vinyl cork, on 1/4" hardboard, no frame		280	.057	S.F.
No backing, no frame		280	.057	S.F.
For map and display rail, economy, add		385	.042	L.F.
Bulletin boards, for map & display rail, deluxe, add	2 Carpenters	350	.046	L.F.
Prefabricated, 1/4" cork, 4' x 4' with aluminum frame		14	1.140	Ea.
Aluminum frame with glass door		13	1.230	Ea.
8' x 4' with aluminum frame		8	2.000	Ea.
Aluminum frame with glass door		7	2.290	Ea.
3' x 5' with wood frame		14	1.140	Ea.
Wood frame with glass door		13	1.230	Ea.
4' x 4' with wood frame		10	1.600	Ea.
6' x 4' with wood frame		8	2.000	Ea.
Prefabricated with vinyl covered cork, aluminum frame				
4' x 3'		12	1.330	Ea.
Vinyl on 1/8" vinyl cork plus 3/8" fiberboard, 4' x 3'		11	1.450	Ea.
Vinyl on 1/4" vinyl cork plus 1/4" hardboard, 8' x 4'		8	2.000	Ea.
Vinyl on 3/8" fiberboard, aluminum frame, 8' x 4'		7	2.290	Ea.
Prefabricated, sliding glass, enclosed, 3' x 4'		16	1.000	Ea.
5' x 3'		12	1.330	Ea.
5' x 4'		11	1.450	Ea.
6' x 4'		7	2.290	Ea.
For lights, add per cabinet	1 Electrician	13	.615	Ea.
Horizontal sliding units with 2 sliders, 8' x 4'	2 Carpenters	3	5.330	Ea.
12' x 4'		2	8.000	Ea.
4 sliding units, 16' x 4'		2	8.000	Ea.
24' x 4'		1.50	10.670	Ea.
CONTROL BOARDS Magnetic, porcelain finish, framed				
24" x 18"		8	2.000	Ea.
36" x 24"		7.50	2.130	Ea.
48" x 36"		7	2.290	Ea.
72" x 48"		6	2.670	Ea.
96" x 48"		5	3.200	Ea.
Roll type				
49" x 74" case, with 2 rollers, 8 to 60 S.F. sleeves		125	.128	S.F.
49" x 95" case, with 4 rollers, 37 to 120 S.F. sleeeves		130	.123	S.F.

Figure 10.14　Installation Time for Signs

Signs	Crew Makeup	Daily Output	Man-Hours	Unit
SIGNS Letters, individual, 2" high, cast aluminum	1 Carpenter	32	.250	Ea.
Cast bronze		32	.250	Ea.
4" high, 5/8" deep, cast aluminum		24	.333	Ea.
Cast bronze		24	.333	Ea.
6" high, 1" deep, cast aluminum		20	.400	Ea.
Cast bronze		20	.400	Ea.
12" high, 1-1/4" deep, cast aluminum		18	.444	Ea.
Cast bronze		18	.444	Ea.
18" high, 1-1/4" deep, cast aluminum		12	.667	Ea.
Cast bronze		12	.667	Ea.
Fabricated aluminum, 12" high, 3" deep		18	.444	Ea.
18" high, 3" deep		12	.667	Ea.
Fabricated stainless steel, 6" high, 3" deep		20	.400	Ea.
12" high, 3" deep		18	.444	Ea.
18" high, 3" deep		12	.667	Ea.
24" high, 4" deep		10	.800	Ea.
Painted sheet steel, 12" high, 2" deep		18	.444	Ea.
18" high, 3" deep		12	.667	Ea.
Plastic, 6" high, 1" deep		20	.400	Ea.
12" high, 2" deep		18	.444	Ea.
Plastic face, alum. frame, 20" high, 15" deep		5	1.600	Ea.
36" high, 24" deep		4	2.000	Ea.
Stainless steel frame, 12" high, 6" deep		5	1.600	Ea.
24" high, 8" deep	↓	4	2.000	Ea.
Plaques, 20" x 30", for up to 450 letters, cast alum.	2 Carpenters	4	4.000	Ea.
Cast bronze		4	4.000	Ea.
30" x 40", up to 900 letters cast aluminum		3	5.330	Ea.
Plaques, 30" x 40", for up to 900 letters, cast bronze		3	5.330	Ea.
36" x 48", for up to 1300 letters, cast bronze		2	8.000	Ea.
Signs, cast aluminum steel signs, 2-way		30	.533	Ea.
4-way		30	.533	Ea.
Acrylic exit signs, 15" x 6", surface mounted				
Minimum		30	.533	Ea.
Maximum		20	.800	Ea.
Bracket mounted, double face, minimum		30	.533	Ea.
Maximum		20	.800	Ea.
Plexiglass, exterior, illuminated, single face		100	.160	S.F.
Double face		75	.213	S.F.
Interior, illuminated, single face		100	.160	S.F.
Double face		75	.213	S.F.
Painted plywood (MDO), over 4' x 8'		120	.133	S.F.
Under 4' x 8'	↓	100	.160	S.F.

Figure 10.15 Installation Time for Turnstiles

Turnstiles	Crew Makeup	Daily Output	Man-Hours	Unit
TURNSTILES One way, 4 arm, 46" diameter, economy				
Manual	2 Carpenters	5	3.200	Ea.
Electric		1.20	13.330	Ea.
High security, galv., 5'-5" diameter, 7' high, manual		1	16.000	Ea.
Electric		.60	26.670	Ea.
Three arm, 24" opening, light duty, manual		2	8.000	Ea.
Heavy duty		1.50	10.670	Ea.
Manual, with registering & controls, light duty		2	8.000	Ea.
Heavy duty		1.50	10.670	Ea.
Electric, heavy duty	▼	1.10	14.550	Ea.
One way gate with horizontal bars, 5'-5" diameter				
7" high, recreation or transit type	2 Carpenters	.80	20.000	Ea.

Figure 10.16 Installation Time for Metal Lockers

Metal Lockers	Crew Makeup	Daily Output	Man-Hours	Unit
LOCKERS Steel, baked enamel, 60" or 72", single tier				
Minimum	1 Sheet Metal Worker	14	.571	Opng.
Maximum		12	.667	Opng.
2 tier, 60" or 72" total height, minimum		26	.308	Opng.
Maximum		20	.400	Opng.
5 tier box lockers, minimum		30	.267	Opng.
Maximum		24	.333	Opng.
6 tier box lockers, minimum		36	.222	Opng.
Maximum		30	.267	Opng.
Basket rack with 32 baskets, 9" x 13" x 8" basket		50	.160	Basket
24 baskets, 12" x 13" x 8" basket		50	.160	Basket
Athletic, wire mesh, no lock, 18" x 18" x 72" basket	▼	12	.667	Ea.
Overhead locker baskets on chains, 14" x 14" baskets	3 Sheet Metal Workers	96	.250	Basket
Overhead locker framing system, add		600	.040	Basket
Locking rail and bench units, add	▼	120	.200	Basket
Locker bench, laminated maple, top only	1 Sheet Metal Worker	100	.080	L.F.
Pedestals, steel pipe	"	25	.320	Ea.
Teacher and pupil wardrobes, enameled				
22" x 15" x 61" high, minimum	1 Sheet Metal Worker	10	.800	Ea.
Average		9	.889	Ea.
Maximum		8	1.000	Ea.
Duplex lockers with 2 doors, 72" high, 15" x 15"		10	.800	Ea.
15" x 21"	▼	10	.800	Ea.

Figure 10.17 Installation Time for Canopies

Canopies	Crew Makeup	Daily Output	Man-Hours	Unit
CANOPIES Wall hung, aluminum, prefinished, 8′ x 10′	1 Struc. Steel Foreman 1 Struc. Steel Worker 1 Truck Driver (light) 1 Truck w/Power Equip.	1.30	18.460	Ea.
8′ x 20′		1.10	21.820	Ea.
10′ x 10′		1.30	18.460	Ea.
10′ x 20′		1.10	21.820	Ea.
12′ x 20′		1	24.000	Ea.
12′ x 30′		.80	30.000	Ea.
12′ x 40′	↓	.60	40.000	Ea.
Aluminum entrance canopies, flat soffit				
3′-6″ x 4′-0″, clear anodized	2 Carpenters	4	4.000	Ea.
Bronze anodized		4	4.000	Ea.
Polyurethane painted		4	4.000	Ea.
4′-6″ x 10′-0″, clear anodized		2	8.000	Ea.
Bronze anodized		2	8.000	Ea.
Polyurethane painted	↓	2	8.000	Ea.
Wall downspout, 10 L.F., clear anodized	1 Carpenter	7	1.140	Ea.
Bronze anodized		7	1.140	Ea.
Polyurethane painted	↓	7	1.140	Ea.
Canvas awnings, including canvas, frame & lettering				
Minimum	2 Carpenters	100	.160	S.F.
Average		90	.178	S.F.
Maximum	↓	80	.200	S.F.
Carport, baked vinyl finish, 20′ x 10′, no foundations				
Minimum	1 Struc. Steel Forman 1 Struc. Steel Worker 1 Truck Driver (light) 1 Truck w/Power Equip.	4	6.000	Car
Maximum		2	12.000	Car
Walkway cover, to 12′ wide, steel, vinyl finish, no foundations				
Minimum		250	.096	S.F.
Maximum	↓	200	.120	S.F.

Figure 10.18 Installation Time for Mail Chutes

Mail Chutes	Crew Makeup	Daily Output	Man-Hours	Unit
MAIL CHUTES Aluminum & glass, 14-1/4″ wide				
4-5/8″ deep	2 Sheet Metal Workers	4	4.000	Floor
8-5/8″ deep		3.80	4.210	Floor
8-3/4″ x 3-1/2″, aluminum		5	3.200	Floor
Bronze or stainless		4.50	3.560	Floor
Lobby collection boxes, aluminum		5	3.200	Ea.
Bronze or stainless	↓	4.50	3.560	Ea.

Figure 10.19 Installation Time for Mail Boxes

Mail Boxes	Crew Makeup	Daily Output	Man-Hours	Unit
MAIL BOXES Horiz. key lock, 5"H x 6" W x 15"D, alum.				
Rear loading	1 Carpenter	34	.235	Ea.
Front loading		34	.235	Ea.
Double, 5"H x 12"W x 15"D, rear loading		26	.308	Ea.
Front loading		26	.308	Ea.
Quadruple, 10"H x 12"W x 15"D, rear loading		20	.400	Ea.
Front loading		20	.400	Ea.
Vertical, front loading, 15"H x 5" W x 6"D, aluminum		34	.235	Ea.
Bronze, duranodic finish		34	.235	Ea.
Steel, enameled		34	.235	Ea.
Alphabetical directories, 35 names		10	.800	Ea.
Letter collection box		6	1.330	Ea.
Letter slot, residential		20	.400	Ea.
Post office type		8	1.000	Ea.
Post office counter window, with grille		2	4.000	Ea.
Key keeper, single key, aluminum		26	.308	Ea.
Steel enameled		26	.308	Ea.

Figure 10.20 Installation Time for Partition Supports

Partition Supports	Crew Makeup	Daily Output	Man-Hours	Unit
TOILET PARTITION SUPPORTS Overhead framing for ceiling hung partions, 3' wide	1 Struc. Steel Foreman 3 Struc. Steel Workers 1 Gas Welding Machine	16	2.000	Stall

Figure 10.21   Installation Time for Wire Mesh Partitions

Wire Mesh Partitions	Crew Makeup	Daily Output	Man-Hours	Unit
PARTITIONS, WOVEN WIRE For tool or stockroom enclosures				
Channel frame, 1-1/2" diamond mesh, 10 ga. wire painted				
Wall panels, 4' wide, 7' high	2 Carpenters	25	.640	Ea.
8' high		23	.696	Ea.
10' high		18	.889	Ea.
Ceiling panels, 10' long, 2' wide		25	.640	Ea.
4' wide		15	1.070	Ea.
Panel with service window & shelf, 5' long, 7' high		20	.800	Ea.
10' high		15	1.070	Ea.
Sliding doors, 3' wide, 7' full height		6	2.670	Ea.
10' full height		5	3.200	Ea.
6' wide, 7' full height		5	3.200	Ea.
10'full height		4	4.000	Ea.
Swinging doors, 3' wide, 7' high, no transom		6	2.670	Ea.
7' high, 3' transom		5	3.200	Ea.

Figure 10.22   Installation Time for Folding Gates

Folding Gates	Crew Makeup	Daily Output	Man-Hours	Unit
SECURITY GATES				
Scissors type folding gate, painted steel				
Single, 6' high, 51" wide	2 Struc. Steel Workers	2.20	7.270	Opng.
75" wide		2	8.000	Opng.
99" wide		1.80	8.890	Opng.
Double gate, 92" wide		1.80	8.890	Opng.
124" wide		1.50	10.670	Opng.
156" wide		1.20	13.330	Opng.
172" wide		1	16.000	Opng.

Figure 10.23 Installation Time for Demountable Partitions

Demountable Partitions	Crew Makeup	Daily Output	Man-Hours	Unit
PARTITIONS, MOVABLE OFFICE Demountable, add for doors				
Do not deduct door openings from total L.F.				
Asbestos cement, 1-3/4" thick, prefinished, low walls	2 Carpenters	80	.200	L.F.
Full height		40	.400	L.F.
Gypsum, laminated 2-1/4" thick, 9' high, unpainted		40	.400	L.F.
Painted		40	.400	L.F.
3" thick, acoustical, unpainted		40	.400	L.F.
Painted		40	.400	L.F.
Vinyl clad drywall on 2-1/2" metal studs, to 9' high		60	.267	L.F.
42" high, plus 10" glass		60	.267	L.F.
Vinyl clad gypsum with air space, 3" thick		60	.267	L.F.
Steel clad gypsum, as above		60	.267	L.F.
Hardboard, vinyl faced, 7' high, 1-9/16" thick		60	.267	L.F.
2-1/4" thick, painted		50	.320	L.F.
With 40" glass		40	.400	L.F.
10' high, vinyl faced, 1-9/16" thick		40	.400	L.F.
Enameled, 2-3/4" thick		30	.533	L.F.
Metal, to 9'-6" high, enameled steel, no glass		40	.400	L.F.
Steel frame, all glass		40	.400	L.F.
Vinyl covered, no glass		40	.400	L.F.
Steel frame with 52% glass		40	.400	L.F.
Free standing, 4'-6" high, steel with glass		100	.160	L.F.
Acoustical		100	.160	L.F.
Low rails, 3'-3" high, enameled steel		100	.160	L.F.
Vinyl covered		100	.160	L.F.
Plywood, prefin, 1-3/4" thick, rotary cut veneers				
Minimum		80	.200	L.F.
Maximum		80	.200	L.F.
Sliced veneers, book matched, minimum		80	.200	L.F.
Maximum		80	.200	L.F.
Sliced veneers, random matched, minimum		80	.200	L.F.
Maximum		80	.200	L.F.
Trackless wall, cork finish, semi-acoustic, 1-5/8" thick				
Minimum		325	.049	S.F.
Maximum		190	.084	S.F.
Acoustic, 2" thick, minimum		305	.052	S.F.
Maximum		225	.071	S.F.
For doors, not incl. hardware, hollow metal door, add		4.30	3.720	Ea.
Hardwood door, add		3.40	4.710	Ea.
Hardware for doors, not incl. closers, keyed		29	.552	Ea.
Non-keyed	▼	29	.552	Ea.

Figure 10.24   Installation Time for Portable Partitions

Portable Partitions	Crew Makeup	Daily Output	Man-Hours	Unit
PARTITIONS, HOSPITAL Curtain track, box channel				
Ceiling mounted	1 Carpenter	135	.059	L.F.
Suspended	"	100	.080	L.F.
Curtains, 8' to 9', nylon mesh tops, fire resistant				
Cotton, 2.85 lbs. per S.Y.	1 Carpenter	425	.019	L.F.
Anti-bacterial, thermosplastic		425	.019	L.F.
Fiberglass, 7 oz. per S.Y.	↓	425	.019	L.F.
I.V. track systems				
I.V. track, 4' x 7' oval	1 Carpenter	135	.059	L.F.
I.V. trolley		32	.250	Ea.
I.V. pendent, (tree, 5 hook)	↓	32	.250	Ea.
PARTITIONS, PORTABLE Divider walls, free standing				
Fiber core panels, 5' long, 4'-6" high				
Burlap face, straight	2 Carpenters	160	.100	L.F.
Curved		158	.101	L.F.
Carpeted face, straight		160	.100	L.F.
Curved		158	.101	L.F.
Plastic laminated face, straight		160	.100	L.F.
Curved		158	.101	L.F.
5' high, burlap face, straight		150	.107	L.F.
Curved		148	.108	L.F.
Carpeted face, straight		150	.107	L.F.
Curved		148	.108	L.F.
Plastic laminated face, straight		150	.107	L.F.
Curved		148	.108	L.F.
6' high, burlap face, straight		125	.128	L.F.
Curved		120	.133	L.F.
Carpeted face, straight		125	.128	L.F.
Curved		120	.133	L.F.
Plastic laminated face, straight		125	.128	L.F.
Curved		120	.133	L.F.
Metal chalkboard, 6'-6" high, chalkboard 1 side		125	.128	L.F.
2 sides		120	.133	L.F.
Hardboard chalkboard, 1 side		125	.128	L.F.
Tackboard, both sides	↓	123	.130	L.F.
PARTITIONS, WORK STATIONS Incl. top cabinet & desk top				
Multi-station units, 1 to 3 person seating capacity				
Minimum per person	1 Carpenter	6	1.330	Ea.
Average per person		5	1.600	Ea.
Maximum per person	↓	4	2.000	Ea.
4 to 6 person seating capacity				
Minimum per person	1 Carpenter	4	2.000	Ea.
Average per person		3	2.670	Ea.
Maximum per person	↓	2	4.000	Ea.

Figure 10.25 Installation Time for Panel Partitions

Panel Partitions	Crew Makeup	Daily Output	Man-Hours	Unit
PARTITIONS, FOLDING LEAF Acoustic, wood				
Vinyl faced, to 18' high, 6 psf, minimum	2 Carpenters	60	.267	S.F.
Average		45	.356	S.F.
Maximum		30	.533	S.F.
Formica or hardwood finish, minimum		60	.267	S.F.
Maximum		30	.533	S.F.
Wood, low acoustical type, 4.5 psf, to 14' high		50	.320	S.F.
Steel, acoustical, 9 to 12 lb. per S.F., vinyl faced				
Minimum		60	.267	S.F.
Maximum		30	.533	S.F.
Aluminum framed, acoutical, to 12' high, 5.5 psf				
Minimum		60	.267	S.F.
Maximum		30	.533	S.F.
6.5 lb. per S.F., minimum		60	.267	S.F.
Maximum		30	.533	S.F.
PARTITIONS, OPERABLE Acoustic air wall, 1-5/8" thick				
Minimum		375	.043	S.F.
Maximum		365	.044	S.F.
2-1/4" thick, minimum		360	.044	S.F.
Maximum		330	.048	S.F.
Overhead track type, acoustical, 3" thick, 11 psf				
Minimum		350	.046	S.F.
Maximum	↓	300	.053	S.F.

Figure 10.26 Installation Time for Accordian Partitions

Accordian Partitions	Crew Makeup	Daily Output	Man-Hours	Unit
PARTITIONS, FOLDING ACCORDIAN				
Vinyl covered, over 150 S.F., frame not included				
Residential, 1.25 lb. per S.F., 8' maximum height	2 Carpenters	300	.053	S.F.
Commercial, 1.75 lb. per S.F., 8' maximum height		225	.071	S.F.
2 lb. per S.F., 17' maximum height		150	.107	S.F.
Industrial, 4 lb. per S.F., 27' maximum height		75	.213	S.F.
Acoustical, 3 lb. per S.F., 17' maximum height		100	.160	S.F.
5 lb. per S.F., 27' maximum height		95	.168	S.F.
5.5 lb. per S.F., 17' maximum height		90	.178	S.F.
Fire rated, 4.5 psf, 20' maximum height		160	.100	S.F.
Vinyl clad wood or steel, electric operation, 5.0 psf		160	.100	S.F.
Wood, non-acoustic, birch or mahogany, to 10' high	↓	300	.053	S.F.

Figure 10.27   Installation Time for Storage & Shelving

Storage & Shelving	Crew Makeup	Daily Output	Man-Hours	Unit
PARTS BINS Metal, gray baked enamel finish				
6'-3" high, 3' wide				
12 bins, 18" wide x 12" high, 12" deep	2 Building Laborers	10	1.600	Ea.
24" deep		10	1.600	Ea.
72 bins, 6" wide x 6" high, 12" deep		8	2.000	Ea.
24" deep	↓	8	2.000	Ea.
7'-3" high, 3' wide				
14 bins, 18" wide x 12" high, 12" deep	2 Building Laborers	10	1.600	Ea.
24" deep		10	1.600	Ea.
84 bins, 6" wide x 6" high, 12" deep		8	2.000	Ea.
24" deep	↓	8	2.000	Ea.
SHELVING Metal, industrial, cross-braced, 3' wide				
12" deep	1 Struc. Steel Worker	175	.046	S.F. Shlf.
24" deep		330	.024	S.F. Shlf.
4' wide, 12" deep		185	.043	S.F. Shlf.
24" deep		380	.021	S.F. Shlf.
Enclosed sides, cross-braced back, 3' wide				
12" deep		175	.046	S.F. Shlf.
24" deep		290	.028	S.F. Shlf.
Fully enclosed, sides and back, 3' wide, 12" deep		150	.053	S.F. Shlf.
24" deep		255	.031	S.F. Shlf.
4' wide, 12" deep		150	.053	S.F. Shlf.
24" deep		290	.028	S.F. Shlf.
Wide span, 1600 lb. capacity per shelf, 7' wide				
24" deep		380	.021	S.F. Shlf.
36" deep		440	.018	S.F. Shlf.
8' wide, 24" deep		440	.018	S.F. Shlf.
36" deep	↓	520	.015	S.F. Shlf.
Pallet racks, steel frame 2500 lb. capacity, 7' long	2 Struc. Steel Workers	450	.036	S.F. Shlf.
30" deep.		500	.032	S.F. Shlf.
36" deep		520	.031	S.F. Shlf.
42" deep	↓			

Figure 10.28 Installation Time for Bath Accessories

Bath Accessories	Crew Makeup	Daily Output	Man-Hours	Unit
BATHROOM ACCESSORIES				
Curtain rod, stainless steel, 5' long, 1" diameter	1 Carpenter	13	.615	Ea.
1-1/4" diameter	"	13	.615	Ea.
Dispenser units, combined soap & towel dispensers,				
mirror and shelf, flush mounted	1 Carpenter	10	.800	Ea.
Towel dispenser and waste receptacle,				
flush mounted	1 Carpenter	10	.800	Ea.
Grab bar, straight, 1" diameter, stainless steel, 12" long		24	.333	Ea.
18" long		23	.348	Ea.
24" long		22	.364	Ea.
36" long		20	.400	Ea.
1-1/2" diameter, 18" long		23	.348	Ea.
36" long		20	.400	Ea.
Tub bar, 1" diameter, horizontal		14	.571	Ea.
Plus vertical arm		12	.667	Ea.
End tub bar, 1" diameter, 90° angle		12	.667	Ea.
Hand dryer, surface mounted, electric, 110 volt		4	2.000	Ea.
220 volt		4	2.000	Ea.
Hat and coat strip, stainless steel, 4 hook, 36" long		24	.333	Ea.
6 hook, 60" long		20	.400	Ea.
Mirror with stainless steel, 3/4" square frame				
18" x 24"		20	.400	Ea.
36" x 24"		15	.533	Ea.
48" x 24"		10	.800	Ea.
72" x 24"		6	1.330	Ea.
Mirror with 5" stainless steel shelf, 3/4" sq. frame				
18" x 24"		20	.400	Ea.
36" x 24"		15	.533	Ea.
48" x 24"		10	.800	Ea.
72" x 24"		6	1.330	Ea.
Mop holder strip, stainless steel, 6 holders, 60" long		20	.400	Ea.
Napkin/tampon dispenser, surface mounted		15	.533	Ea.
Robe hook, single, regular		36	.222	Ea.
Heavy duty, concealed mounting		36	.222	Ea.
Soap dispenser, chrome, surface mounted, liquid		20	.400	Ea.
Powder		20	.400	Ea.
Recessed stainless steel, liquid		10	.800	Ea.
Powder		10	.800	Ea.
Soap tank, stainless steel, 1 gallon		10	.800	Ea.
5 gallon		5	1.600	Ea.
Shelf, stainless steel, 5" wide, 18 ga., 24" long		24	.333	Ea.
72" long		16	.500	Ea.
8" wide shelf, 18 ga., 24" long		22	.364	Ea.
72" long		14	.571	Ea.
Toilet seat cover dispenser, stainless steel, recessed		20	.400	Ea.
Surface mounted		15	.533	Ea.
Toilet tissue dispenser, surface mounted, stainless steel				
Single roll		30	.267	Ea.
Double roll		24	.333	Ea.
Towel bar, stainless steel, 18" long		23	.348	Ea.
30" long	▼	21	.381	Ea.

(continued on next page)

Figure 10.28 Installation Time
for Bath Accessories (continued)

Bath Accessories	Crew Makeup	Daily Output	Man-Hours	Unit
BATHROOM ACCESSORIES (cont.)				
Towel dispenser, stainless steel, surface mounted	1 Carpenter	16	.500	Ea.
Flush mounted, recessed		10	.800	Ea.
Towel holder, hotel type, 2 guest size		20	.400	Ea.
Towel shelf, stainless steel, 24" long, 8" wide		20	.400	Ea.
Tumbler holder, tumbler only		30	.267	Ea.
Soap, tumbler & toothbrush		30	.267	Ea.
Wall urn ash receiver, recessed, 14" long		12	.667	Ea.
Surface, 8" long		18	.444	Ea.
Waste receptacles, stainless steel, with top, 13 gallon		10	.800	Ea.
36 Gallon	↓	8	1.000	Ea.
MEDICINE CABINETS With mirror, stock, 16" x 22"				
Unlighted	1 Carpenter	14	.571	Ea.
Lighted		6	1.330	Ea.
Sliding mirror doors, 36" x 22", unlighted		7	1.140	Ea.
Lighted		5	1.600	Ea.
Center mirror, 2 end cabinets, unlighted, 48" long		7	1.140	Ea.
72" long	↓	5	1.600	Ea.
For lighting, 48" long, add	1 Electrician	3.50	2.290	Ea.
72" long, add	"	3	2.670	Ea.
Hotel cabinets, stainless, with lower shelf, unlighted	1 Carpenter	10	.800	Ea.
Lighted	"	5	1.600	Ea.

Figure 10.29 Installation Time for Telephone Enclosures

Telephone Enclosures	Crew Makeup	Daily Output	Man-Hours	Unit
TELEPHONE ENCLOSURE Desk-top type	2 Carpenters	8	2.000	Ea.
Shelf type, wall hung, minimum		5	3.200	Ea.
Maximum		5	3.200	Ea.
Booth type, painted steel, indoor or outdoor, minimum		1.50	10.670	Ea.
Maximum (stainless steel)		1.50	10.670	Ea.
Outdoor, acoustical, on post		3	5.330	Ea.
Phone carousel, pedestal mounted with dividers		.60	26.670	Ea.
Outdoor, drive-up type, wall mounted		4	4.000	Ea.
Post mounted, stainless steel posts	↓	3	5.330	Ea.
Directory shelf, wall mounted, stainless steel				
3 binders	2 Carpenters	8	2.000	Ea.
4 binders		7	2.290	Ea.
7 binders		6	2.670	Ea.
Table type, stainless steel, 4 binders		8	2.000	Ea.
7 binders	↓	7	2.290	Ea.

Figure 10.30 Installation Time for Scales

Scales	Crew Makeup	Daily Output	Man-Hours	Unit
SCALES Built-in floor scale, not incl. foundations				
Dial type, 5 ton capacity, 8' x 6' platform	3 Carpenters	.50	48.000	Ea.
9' x 7' platform		.40	60.000	Ea.
10 ton capacity, steel platform, 8' x 6' platform		.40	60.000	Ea.
9' x 7' platform	↓	.35	68.570	Ea.
Truck scales, incl. steel weigh bridge, not including foundation				
Dial type, mech., 24' x 10' platform, 20 ton cap.	3 Carpenters	.30	80.000	Ea.
30 ton capacity		.20	120.000	Ea.
50 ton capacity, 50' x 10' platform		.14	171.000	Ea.
70' x 10' platform		.12	200.000	Ea.
60 ton capacity, 60' x 10' platform		.13	185.000	Ea.
70' x 10' platform		.10	240.000	Ea.
Digital, electronic, 30 ton capacity, 12' x 10' platform		.20	120.000	Ea.
50 ton capacity, 40' x 10' platform		.14	171.000	Ea.
60 ton capacity, 60' x 10' platform		.13	185.000	Ea.
75' x 10' platform	↓	.12	200.000	Ea.
Concrete foundation pits for above, 8' x 6'				
5 C.Y. required	3 Carpenters 1 Building Laborer Power Tools	.50	64.000	Ea.
14' x 6' platform, 10 C.Y. required		.35	91.430	Ea.
50' x 10' platform, 30 C.Y. required		.25	128.000	Ea.
70' x 10' platform, 40 C.Y. required	↓	.15	213.000	Ea.
Low profile electronic warehouse scale, not incl. printer, 4' x 6' platform, 6000 lb. capacity	2 Carpenters	.30	53.330	Ea.
5' x 7' platform, 10,000 lb. capacity		.25	64.000	Ea.
20,000 lb. capacity	↓	.20	80.000	Ea.

Figure 10.31 Installation Time for Coat Racks/Wardrobes

Coat Racks/Wardrobes	Crew Makeup	Daily Output	Man-Hours	Unit
COAT RACKS & WARDROBES Dormitory units, wood or metal				
Stock units, 84" high, incl. door, minimum	1 Carpenter	10	.800	L.F.
Average		7	1.140	L.F.
Maximum		5	1.600	L.F.
Hospital type, 84" high, plastic faced wood		5	1.600	L.F.
Enameled steel		5	1.600	L.F.
Stainless steel		5	1.600	L.F.
Coat and hat rack, wall mounted tubular steel, 1 shelf		70	.114	L.F.
3 shelves	↓	50	.160	L.F.

Checklist

For an estimate to be reliable, all items must be accounted for. A complete estimate can also eliminate the need to include contingencies. The following checklist can be used to help ensure that all items are properly accounted for.

☐ Access flooring
☐ Ash trays/urns/units
☐ Bulletin boards
☐ Canopies
☐ Chalkboards
_____ Blackboards
_____ Whiteboards
_____ Writing screens
☐ Coat racks
☐ Control boards
☐ Chair rails
☐ Chutes
_____ Laundry
_____ Trash
_____ Mail
☐ Corner guards
☐ Directory boards
☐ Disappearing stairs
☐ Display cases
☐ Fireplace accessories
☐ Flagpoles
☐ Folding gates
☐ Grilles and screens — decorative
☐ Hat rack
☐ Hospital equipment
_____ Curtain track
_____ Equipment mounting racks
_____ I.V. tracks
_____ Other _____
☐ Key cabinets
☐ Locker room equipment
_____ Basket rack
_____ Benches
_____ Lockers

- ☐ Louvers
  - ____ Wall
  - ____ Soffit
- ☐ Mail items
  - ____ Mail boxes
  - ____ Mail slots
  - ____ Mail chute system
- ☐ Partitions
  - ____ Demountable
  - ____ Folding
  - ____ Operable
  - ____ Portable
  - ____ Toilet
    - ____ Handicapped _____
  - ____ Shower
  - ____ Work station
  - ____ Woven wire
- ☐ Protection screens
- ☐ Scales
- ☐ Screens
  - ____ Entrance
  - ____ Urinal
- ☐ Security devices
  - ____ Gates
  - ____ Miscellaneous _____
- ☐ Shelving
  - ____ Parts bins
  - ____ Storage
  - ____ Pallet racks
- ☐ Shower doors
  - ____ Screens
- ☐ Signage
  - ____ Building sign
  - ____ Interior door signage
  - ____ Plaques
- ☐ Storage cubicles
- ☐ Telephone enclosures
- ☐ Toilet accessories
  - ____ Ash urns
  - ____ Curtain rods
  - ____ Dispenser units

_____ Paper towel
_____ Napkin
_____ Soap
_____ Toilet paper
_____ Toilet seat covers
_____ Grab bars
_____ Hand dryers
_____ Handicapped bars
_____ Hat/coat rack
_____ Medicine chests
_____ Mirrors
_____ Mop holder
_____ Robe hook
_____ Shelving
_____ Towel bars
_____ Towel shelf
_____ Tooth brush holder
_____ Tumbler holder
_____ Waste receptacles
☐ Tub enclosures
☐ Turnstiles
☐ Vending machines
☐ Wall guards
☐ Wardrobes
_____ Coat rack
_____ Dormitory style
_____ Hat rack
_____ Hospital type
☐ Waste receptacles

Tips

Item Takeoff

When taking off Division 10 items, list each item called for, along with the specified manufacturer. Often no substitutions are allowed for specified Division 10 items.

Labor Cost for Installation

When receiving bids for Division 10 items, make sure that the costs of installation are included. In many cases these items are sold per unit without installation. If installation is not included, inquire if delivery is included. Items sold by the piece without installation usually do not include any shipping costs.

Support Systems

Note that many Division 10 items require some type of support system not usually supplied with the item — such as support brackets, plates or angles. These must be accounted for and may need to be added to the appropriate division. Phrases in the specifications that gloss over this subject and thus make it difficult to recover any added costs include "The contractor shall install all products in accordance with the manufacturers' recommendations" and the like. Remember, it is much more costly to install a behind-the-wall support system after the wall is in place.

Preparation of Items

In some cases, Division 10 items may require some assembly before installation. This assembly time can often exceed the time required for installation.

Shop Smart

There can be a wide variance in costs for the exact same item. Smart shopping for these items can help lower your overall bid.

Notes

Division Eleven
Equipment

Introduction

Division 11 — Equipment generally refers to the total package of equipment that is required for that project's type of construction. For example, Division 11 for a bank would include teller windows, package transfer units, automatic banking equipment, vault doors, safe deposit boxes, etc.

Division 11 items are usually counted individually, as most are sold individually. To estimate Division 11, begin with a review of the specifications. Items to be included may be listed singly, or grouped. Approved manufacturers may also be listed. Note that some items may not be listed in the specifications, but may appear on the drawings only. Review each and every drawing for items to be included or hinted at, (e.g., on the concrete drawings, an item such as "concrete base for book depository" may be included). Such items should be investigated to determine who is supplying them. The drawings may also designate items that are "by others" or not in the contractor's scope of work.

Quite often the equipment listed in Division 11 is purchased directly by the owner of the facility. In some cases, the contractor installs the equipment, while in others the owner arranges for the installation. Where it is determined that the installation of the owner-supplied equipment is the contractor's responsibility, the labor should be figured separately and on a unit basis.

Estimating Data

The following tables show the expected installation time for various specialty items. Each table lists the specialty item, the typical installation crew mix, the expected average daily output of that crew (the number of units the crew can be expected to install in one eight-hour day), and the man-hours required for the installation of one unit.

Table of Contents

Figure 11.1   Installation Time for Maintenance Equipment

Maintenance Equipment	Crew Makeup	Daily Output	Man-Hours	Unit
VACUUM CLEANING Central, 3 inlet, residential	1 Skilled Worker	.90	8.890	Total
Commercial		.70	11.430	Total
5 inlet system, residential		.50	16.000	Total
7 inlet system		.40	20.000	Total
9 inlet system	↓	.30	26.670	Total

Figure 11.2   Installation Time for Teller and Bank Service Equipment

Teller and Service Equipment	Crew Makeup	Daily Output	Man-Hours	Unit
BANK EQUIPMENT Alarm system, police	2 Electricians	1.60	10.000	Ea.
With vault alarm	"	.40	40.000	Ea.
Bullet-resistant teller window, 44" x 60"	1 Glazier	.60	13.330	Ea.
48" x 60"	"	.60	13.330	Ea.
Counters for banks, frontal only	2 Carpenters	1	16.000	Station
Complete with steel undercounter	"	.50	32.000	Station
Door and frame, bullet-resistant, with vision panel				
Minimum	2 Struc. Steel Workers	1.10	14.550	Ea.
Maximum		1.10	14.550	Ea.
Drive-up window, drawer & mike, not incl. glass				
Minimum		1	16.000	Ea.
Maximum		.50	32.000	Ea.
Night depository, with chest, minimum		1	16.000	Ea.
Maximum		.50	32.000	Ea.
Package receiver, painted		3.20	5.000	Ea.
Stainless steel	↓	3.20	5.000	Ea.
Partitions, bullet-resistant, 1-3/16" glass, 8' high	2 Carpenters	10	1.600	L.F.
Acrylic	"	10	1.600	L.F.
Pneumatic tube systems, 2 lane drive-up, complete	1 Carpenter .5 Electrician .5 Sheet Metal Worker	.25	64.000	Total
With T.V. viewer	"	.20	80.000	Total
Safety deposit boxes, minimum	1 Struc. Steel Worker	44	.182	Opng.
Maximum, 10" x 15" opening		19	.421	Opng.
Teller locker, average	↓	15	.533	Opng.
Pass thru, bullet-resistant window, painted steel				
24" x 36"	2 Struc. Steel Workers	1.60	10.000	Ea.
48" x 48"		1.20	13.330	Ea.
72" x 40"	↓	.80	20.000	Ea.
Surveillance system, 16mm film camera, complete	2 Electricians	1	16.000	Ea.
Surveillance system, video camera, complete	"	1	16.000	Ea.
Twenty-four hour teller, single unit, automated deposit, cash and memo	1 Carpenter .5 Electrician .5 Sheet Metal Worker	.25	64.000	Ea.

Figure 11.3 Installation Time for Ecclesiastical Equipment

Ecclesiastical Equipment	Crew Makeup	Daily Output	Man-Hours	Unit
CHURCH EQUIPMENT Altar, wood, custom design, plain	1 Carpenter	1.40	5.710	Ea.
Deluxe	"	.20	40.000	Ea.
Granite or marble, average	2 Marble Setters	.50	32.000	Ea.
Deluxe	"	.20	80.000	Ea.
Arks, prefabricated, plain	2 Carpenters	.80	20.000	Ea.
Deluxe, maximum	"	.20	80.000	Ea.
Bapistry, fiberglass, 3'-6" deep, x 13'-7" long, steps at both ends, incl. plumbing, minimum	2 Carpenters .5 Plumber	1	20.000	Ea.
Maximum	"	.70	28.570	Ea.
Confessional, wood, prefabricated, single, plain	1 Carpenter	.60	13.330	Ea.
Deluxe		.40	20.000	Ea.
Double, plain		.40	20.000	Ea.
Deluxe		.20	40.000	Ea.
Lecterns, wood, plain		5	1.600	Ea.
Deluxe		2	4.000	Ea.
Pews, bench type, hardwood, minimum	↓	20	.400	L.F.
Church equipment, pews, bench type, hardwood Maximum	1 Carpenter	15	.533	L.F.
Pulpits, hardwood, prefabricated, plain		2	4.000	Ea.
Deluxe		1.60	5.000	Ea.
Railing, hardwood, average		25	.320	L.F.
Seating, individual, oak, contour, laminated		21	.381	Person
Cushion seat		21	.381	Person
Fully upholstered		21	.381	Person
Combination, self-rising	↓	21	.381	Person
Steeples, translucent fiberglass, 30" square 15' high	4 Carpenters 1 Equip. Oper. (crane) 1 Hyd. Crane, 12 Ton Power Tools	2	20.000	Ea.
25' high		1.80	22.220	Ea.
Painted fiberglass, 24" square, 14' high		2	20.000	Ea.
28' high		1.80	22.220	Ea.
Porcelain enamel steeples, custom, 40' high		.50	80.000	Ea.
60' high	↓	.30	133.000	Ea.
Wall cross, aluminum, extruded, 2" x 2" section	1 Carpenter	34	.235	L.F.
4" x 4" section		29	.276	L.F.
Bronze, extruded, 1" x 2" section		31	.258	L.F.
2-1/2" x 2-1/2" section		34	.235	L.F.
Solid bar stock, 1/2" x 3" section		29	.276	L.F.
Fiberglass, stock		34	.235	L.F.
Stainless steel, 4" deep, channel section		29	.276	L.F.
4" deep box section	↓	29	.276	L.F.

Figure 11.4   Installation Time for Library Equipment

Library Equipment	Crew Makeup	Daily Output	Man-Hours	Unit
LIBRARY EQUIPMENT Bookshelf, mtl., 90" high, 10" shelf				
Single face	1 Carpenter	12	.667	L.F.
Double face		12	.667	L.F.
Carrels, hardwood, 36" x 24", minimum		5	1.600	Ea.
Maximum		4	2.000	Ea.
Charging desk, built-in, with counter,				
plastic laminated top	▼	7	1.140	L.F.

Figure 11.5   Installation Time for Theater/Stage Equipment

Theater/ Stage Equipment	Crew Makeup	Daily Output	Man-Hours	Unit
STAGE EQUIPMENT Control boards				
with dimmers & breakers, minimum	1 Electrician	1	8.000	Ea.
Average		.50	16.000	Ea.
Maximum	▼	.20	40.000	Ea.
Curtain track, straight, light duty	2 Carpenters	20	.800	L.F.
Heavy duty		18	.889	L.F.
Curved sections		12	1.330	L.F.
Curtains, velour, medium weight		600	.027	S.F.
Asbestos		50	.320	S.F.
Silica based yarn, fireproof	▼	50	.320	S.F.
Lights, border, quartz, reflector, vented,				
colored or white	1 Electrician	20	.400	L.F.
Spotlight, follow spot, with transformer, 2100 watt		4	2.000	Ea.
Stationary spot, fresnel quartz, 6" lens		4	2.000	Ea.
8" lens		4	2.000	Ea.
Ellipsoidal quartz, 1000W, 6" lens		4	2.000	Ea.
12" lens		4	2.000	Ea.
Strobe light, 1 to 15 flashes per second, quartz		3	2.670	Ea.
Color wheel, portable, five hole, motorized	▼	4	2.000	Ea.
Telescoping platforms, extruded alum., straight				
Minimum	4 Carpenters	157	.204	S.F. Stg.
Maximum		77	.416	S.F. Stg.
Pie-shaped, minimum		150	.213	S.F. Stg.
Maximum		70	.457	S.F. Stg.
Band risers, steel frame, plywood deck, minimum		275	.116	S.F. Stg.
Maximum	▼	138	.232	S.F. Stg.
Chairs for above, self-storing, minimum	2 Carpenters	43	.372	Ea.
Maximum	"	40	.400	Ea.
Rule of thumb: total stage equipment, minimum	4 Carpenters	100	.320	S.F. Stg.
Maximum	"	25	1.280	S.F. Stg.
MOVIE EQUIPMENT				
Lamphouses, incl. rectifiers, xenon, 1000 watt	1 Electrician	2	4.000	Ea.
1600 watt		2	4.000	Ea.
2000 watt		1.50	5.330	Ea.
4000 watt	▼	1.50	5.330	Ea.
Projection screens, rigid, in wall, acrylic, 1/4" thick	2 Glaziers	195	.082	S.F.
1/2" thick	"	130	.123	S.F.
Electric operated, heavy duty, 400 S.F.	2 Carpenters	1	16.000	Ea.
Sound systems, incl. amplifier, single system				
Minimum	1 Electrician	.90	8.890	Ea.
Dolby/Super Sound, maximum		.40	20.000	Ea.
Dual system, minimum		.70	11.430	Ea.
Dolby/Super Sound, maximum		.40	20.000	Ea.
Speakers, recessed behind screen, minimum		2	4.000	Ea.
Maximum	▼	1	8.000	Ea.
Seating, painted steel, upholstered, minimum	2 Carpenters	35	.457	Ea.
Maximum	"	28	.571	Ea.

Figure 11.6 Installation Time for Barber Shop Equipment

Barber Shop Equipment	Crew Makeup	Daily Output	Man-Hours	Unit
BARBER EQUIPMENT Chair, hydraulic, movable				
Minimum	1 Carpenter	24	.333	Ea.
Maximum	"	16	.500	Ea.
Wall hung styling station with mirrors, minimum	1 Carpenter	8	2.000	Ea.
	1 Helper			
Maximum	"	4	4.000	Ea.
Sink, hair washing basin, rough plumbing not incl.	1 Plumber	8	1.000	Ea.
Total equipment, rule of thumb, per chair				
Minimum	2 Carpenters	1	20.000	Ea.
	.5 Plumber			
Maximum	"	1	20.000	Ea.

Figure 11.7 Installation Time for Cash Register/Checkout

Cash Register/Checkout	Crew Makeup	Daily Output	Man-Hours	Unit
CHECKOUT COUNTER Supermarket conveyor				
Single belt	2 Building Laborers	10	1.600	Ea.
Double belt		9	1.780	Ea.
Power take-away		7	2.290	Ea.
Warehouse or bulk type	↓	6	2.670	Ea.

Figure 11.8 Installation Time for Display Cases

Display Cases	Crew Makeup	Daily Output	Man-Hours	Unit
REFRIGERATED FOOD CASES Dairy, multi-deck				
12' long	1 Steamfitter	3	5.330	Ea.
	1 Steamfitter Apprentice			
Delicatessen case, service deli, 12' long				
Single deck		3.90	4.100	Ea.
Multi-deck, 18 S.F. shelf display		3	5.330	Ea.
Freezer, self-contained, chest-type, 30 C.F.		3.90	4.100	Ea.
Glass door, upright, 78 C.F.		3.30	4.850	Ea.
Frozen food, chest type, 12' long		3.30	4.850	Ea.
Glass door, reach-in, 5 door		3	5.330	Ea.
Island case, 12' long, single deck		3.30	4.850	Ea.
Multi-deck		3	5.330	Ea.
Meat case, 12' long, single deck		3.30	4.850	Ea.
Multi-deck		3.10	5.160	Ea.
Produce, 12' long, single deck		3.30	4.850	Ea.
Multi-deck	↓	3.10	5.160	Ea.

Figure 11.9 Installation Time for Laundry/Dry Cleaning Equipment

Laundry/Dry Cleaning	Crew Makeup	Daily Output	Man-Hours	Unit
LAUNDRY EQUIPMENT Not including rough-in				
Residential, 16 lb. capacity, average	1 Plumber	3	2.670	Ea.
Commercial, 30 lb. capacity, coin operated				
Single		3	2.670	Ea.
Double stacked		2	4.000	Ea.
Industrial, 30 lb. capacity		2	4.000	Ea.
50 lb. capacity	↓	1.70	4.710	Ea.
Dry cleaners, electric, 20 lb. capacity	1 Electrician	.20	80.000	Ea.
	1 Plumber			
25 lb. capacity		.17	94.120	Ea.
30 lb. capacity		.15	107.000	Ea.
60 lb. capacity	↓	.09	178.000	Ea.
Folders, blankets & sheets, minimum	1 Electrician	.17	47.060	Ea.
King size with automatic stacker		.10	80.000	Ea.
For conveyor delivery, add		.45	17.780	Ea.
Ironers, institutional, 110", single roll	↓	.20	40.000	Ea.
Lint collector, ductwork not included				
8000 to 10,000 C.F.M	2 Sheet Metal Workers	.30	80.000	Ea.
	1 Sheet Metal Apprentice			
Washers, residential, 4 cycle, average	1 Plumber	3	2.670	Ea.
Commercial, coin operated, average	"	3	2.670	Ea.
Combination washer/extractor				
20 lb. capacity	1 Plumber	1.50	8.000	Ea.
	.5 Electrician			
30 lb. capacity		.80	15.000	Ea.
50 lb. capacity		.68	17.650	Ea.
75 lb. capacity		.30	40.000	Ea.
125 lb. capacity	↓	.16	75.000	Ea.

Figure 11.10 Installation Time for Projection Screens

Projection Screens	Crew Makeup	Daily Output	Man-Hours	Unit
PROJECTION SCREENS Wall or ceiling hung,				
glass beaded, manually operated, economy	2 Carpenters	500	.032	S.F.
Intermediate		450	.036	S.F.
Deluxe		400	.040	S.F.
Electric operated, glass beaded, 25 S.F., economy		5	3.200	Ea.
Deluxe		4	4.000	Ea.
50 S.F., economy		3	5.330	Ea.
Deluxe		2	8.000	Ea.
Heavy duty, electric operated, 200 S.F.		1.50	10.670	Ea.
400 S.F.	↓	1	16.000	Ea.
Rigid acrylic in wall, for rear projection, 1/4" thick	2 Glaziers	195	.082	S.F.
1/2" thick (maximum size 10' x 20')	"	130	.123	S.F.

Figure 11.11 Installation Time for Service Station Equipment

Service Station Equipment	Crew Makeup	Daily Output	Man-Hours	Unit
AUTOMOTIVE Compressors, electric, 1-1/2 H.P.,				
Standard controls	2 Skilled Workers	1.50	16.000	Ea.
Dual controls	1 Helper	1.50	16.000	Ea.
5 H.P., 115/230 volt, standard controls		1	24.000	Ea.
Dual controls		1	24.000	Ea.
Gasoline pumps, conventional, lighted, single		2.50	9.600	Ea.
Double		2	12.000	Ea.
Hoists, single post, 8000# capacity, swivel arms		.40	60.000	Ea.
Two posts, adjustable frames, 11,000# capacity		.25	96.000	Ea.
24,000# capacity		.15	160.000	Ea.
7500# capacity, frame supports		.50	48.000	Ea.
Four post, roll on ramp		.50	48.000	Ea.
Lube equipment, 3 reel type, with pumps,				
not including piping		.50	48.000	Set
Spray painting booth, 26' long, complete		.40	60.000	Ea.

Figure 11.12 Installation Time for Parking Control Equipment

Parking Control Equipment	Crew Makeup	Daily Output	Man-Hours	Unit
PARKING EQUIPMENT Traffic, detectors, magnetic	2 Electricians	2.70	5.930	Ea.
Single treadle		2.40	6.670	Ea.
Automatic gates, 8' arm, one way		1.10	14.550	Ea.
Two way		1.10	14.550	Ea.
Fee indicator, 1" display		4.10	3.900	Ea.
Ticket printer and dispenser, standard		1.40	11.430	Ea.
Rate computing		1.40	11.430	Ea.
Card control station, single period		4.10	3.900	Ea.
4 period		4.10	3.900	Ea.
Key station on pedestal		4.10	3.900	Ea.
Coin station, multiple coins		4.10	3.900	Ea.

Figure 11.13 Installation Time for Loading Dock Equipment

Loading Dock Equipment	Crew Makeup	Daily Output	Man-Hours	Unit
LOADING DOCK Bumpers, rubber blocks 4-1/2" thick				
10" high, 14" long	1 Carpenter	26	.308	Ea.
24" long		22	.364	Ea.
36" long		17	.471	Ea.
12" high, 14" long		25	.320	Ea.
24" long		20	.400	Ea.
36" long		15	.533	Ea.
Rubber blocks 6" thick, 10" high, 14" long		22	.364	Ea.
24" long		18	.444	Ea.
36" long		13	.615	Ea.
20" high, 11" long		13	.615	Ea.
Extruded rubber bumpers, T section				
22" x 22" x 3" thick		41	.195	Ea.
Molded rubber bumpers, 24" x 12" x 3" thick		20	.400	Ea.
Welded installation of above bumpers	1 Welder Foreman 1 Gas Welding Machine	8	1.000	Ea.
For drilled anchors, add per anchor	1 Carpenter	36	.222	Ea.
Door seal for door perimeter				
12" x 12", vinyl covered	"	26	.308	L.F.
Levelers, hinged for trucks, 10 ton capacity, 6' x 8'	2 Skilled Workers 1 Helper	1.90	12.630	Ea.
7' x 8'		1.90	12.630	Ea.
Hydraulic, 10 ton capacity, 6' x 8'		1.90	12.630	Ea.
7' x 8'		1.90	12.630	Ea.
Lights for loading docks, single arm, 24" long	1 Electrician	3.80	2.110	Ea.
Double arm, 60" long	"	3.80	2.110	Ea.
Shelters, fabric, for truck or train, scissor arms				
Minimum	1 Carpenter	1	8.000	Ea.
Maximum	"	.50	16.000	Ea.
DOCK BUMPERS Bolts not incl. 2" x 6" to 4" x 8"				
Average	1 Carpenter Power Tools	.30	26.670	M.B.F.

Figure 11.14   Installation Time for
Waste Handling Equipment

Waste Handling Equipment	Crew Makeup	Daily Output	Man-Hours	Unit
WASTE HANDLING Compactors, 115 volt, 250#/hour, Chute fed	2 Skilled Workers 1 Helper	1	24.000	Ea.
Hand fed		2.40	10.000	Ea.
Multi-bag, hand or chute fed, 230 volt, 600#/hour		1	24.000	Ea.
Containerized, hand fed, 2 to 6 C.Y. containers				
250#/hour		1	24.000	Ea.
For chute fed, add per floor		1	24.000	Ea.
Heavy duty industrial compactor, 0.5 C.Y. capacity		1	24.000	Ea.
1.0 C.Y. capacity		1	24.000	Ea.
2.5 C.Y. capacity		.50	48.000	Ea.
5.0 C.Y. capacity		.50	48.000	Ea.
Combination shredder/compactor (5000#/hour)		.50	48.000	Ea.
Crematory, not including building, 1 place	1 Plumber Foreman (ins.) 2 Plumbers 1 Plumber Apprentice	.20	160.000	Ea.
2 place	"	.10	320.000	Ea.
Incinerator, electric, 100#/hour, minimum	1 Labor Foreman (inside) 2 Building Laborers 1 Struc. Steel Worker .5 Electrician	.75	48.000	Ea.
Maximum		.70	51.430	Ea.
400#/hour, minimum		.60	60.000	Ea.
Maximum		.50	72.000	Ea.
100#/hour, minimum		.25	114.000	Ea.
Maximum		.20	180.000	Ea.
Gas, not incl. chimney, electric or pipe				
50#/hour, minimum	1 Plumber Foreman (ins.) 2 Plumbers 1 Plumber Apprentice	.80	40.000	Ea.
Maximum		.70	45.710	Ea.
200#/hour, minimum (batch type)		.60	53.330	Ea.
Maximum (with feeder)		.50	64.000	Ea.
400#/hour, minimum (batch type)		.30	107.000	Ea.
Maximum (with feeder)		.25	128.000	Ea.
800#/hour, with feeder, minimum		.20	160.000	Ea.
Maximum		.17	188.000	Ea.
1200#/hour, with feeder, minimum		.15	213.000	Ea.
Maximum		.11	291.000	Ea.
2000#/hour, with feeder, minimum		.10	320.000	Ea.
Maximum		.05	640.000	Ea.
For heat recovery system, add, minimum		.25	128.000	Ea.
Add, maximum		.11	291.000	Ea.
For automatic ash conveyer, add		.50	64.000	Ea.
Large municipal incinerators, incl. stack, minimum		.25	128.000	Ton/Day
Maximum		.10	320.000	Ton/Day

Figure 11.15   Installation Time for Detention Equipment

Detention Equipment	Crew Makeup	Daily Output	Man-Hours	Unit
DETENTION EQUIPMENT Bar front, rolling, 7/8" bars, 4" O.C., 7' high, 5' wide, with hardware	1 Struc. Steel Foreman 3 Struc. Steel Workers 1 Gas Welding Machine	2	16.000	Ea.
Doors & frames, 3' x 7', complete, with hardware				
Single plate	↓	4	8.000	Ea.
Double plate		4	8.000	Ea.
Cells, prefab., 5' to 6' wide, 7' to 8' high, 7' to 8' deep, bar front, cot, not incl. plumbing	1 Struc. Steel Foreman 3 Struc. Steel Workers 1 Gas Welding Machine	1.50	21.330	Ea.
Cot, bolted, single, painted steel	↓	20	1.600	Ea.
Stainless steel		20	1.600	Ea.
Toilet apparatus including wash basin, average	2 Carpenters .5 Plumber	1.50	13.330	Ea.
Visitor cubicle, vision panel, no intercom	1 Struc. Steel Foreman 3 Struc. Steel Workers 1 Gas Welding Machine	2	16.000	Ea.

Figure 11.16   Installation Time for Food Service Equipment

Food Service Equipment	Crew Makeup	Daily Output	Man-Hours	Unit
APPLIANCES Cooking range, 30" free standing, 1 oven				
Minimum	2 Building Laborers	10	1.600	Ea.
Maximum		4	4.000	Ea.
2 oven, minimum		10	1.600	Ea.
Maximum	↓	4	4.000	Ea.
Built-in, 30" wide, 1 oven, minimum	2 Carpenters	4	4.000	Ea.
Maximum		2	8.000	Ea.
2 oven, minimum		4	4.000	Ea.
Maximum	↓	2	8.000	Ea.
Free-standing, 21" wide range, 1 oven, minimum	2 Building Laborers	10	1.600	Ea.
Maximum	"	4	4.000	Ea.
Counter top cook tops, 4 burner, standard				
Minimum	1 Electrician	6	1.330	Ea.
Maximum		3	2.670	Ea.
As above, but with grille and griddle attachment				
Minimum		6	1.330	Ea.
Maximum		3	2.670	Ea.
Induction cooktop, 30" wide		3	2.670	Ea.
Microwave oven, minimum		4	2.000	Ea.
Maximum	↓	2	4.000	Ea.
Combination range, refrigerator and sink, 30" wide				
Minimum	1 Electrician 1 Plumber	2	8.000	Ea.
Maximum		1	16.000	Ea.
60" wide, average		1.40	11.430	Ea.
72" wide, average		1.20	13.330	Ea.
Office model, 48" wide		2	8.000	Ea.
Refrigerator and sink only	↓	2.40	6.670	Ea.
Combination range, refrigerator, sink, microwave oven and ice maker	1 Electrician 1 Plumber	.80	20.000	Ea.
Compactor, residential size, 4 to 1 compaction				
Minimum	1 Carpenter	5	1.600	Ea.
Maximum	"	3	2.670	Ea.
Deep freeze, 15 to 23 C.F., minimum	2 Building Laborers	10	1.600	Ea.
Maximum		5	3.200	Ea.
30 C.F., minimum	↓	8	2.000	Ea.
Maximum		3	5.330	Ea.
Dishwasher, built-in, 2 cycles, minimum	1 Electrician 1 Plumber	4	4.000	Ea.
Maximum		2	8.000	Ea.
4 or more cycles, minimum		4	4.000	Ea.
Maximum	↓	2	8.000	Ea.
Dryer, automatic, minimum	1 Carpenter 1 Helper	3	5.330	Ea.
Maximum	"	2	8.000	Ea.
Garbage disposer, sink type, minimum	1 Electrician 1 Plumber	5	3.200	Ea.
Maximum	"	3	5.330	Ea.

Figure 11.16 Installation Time for Food Service Equipment (continued)

Food Service Equipment	Crew Makeup	Daily Output	Man-Hours	Unit
Appliances, Heater, electric, built-in, 1250 watt				
Ceiling type, minimum	1 Electrician	4	2.000	Ea.
Maximum		3	2.670	Ea.
Wall type, minimum		4	2.000	Ea.
Maximum		3	2.670	Ea.
1500 watt wall type, with blower		4	2.000	Ea.
3000 watt	↓	3	2.670	Ea.
Hood for range, 2 speed, vented, 30" wide				
Minimum.	1 Carpenter .5 Electrician .5 Sheet Metal Worker	5	3.200	Ea.
Maximum		3	5.330	Ea.
42" wide, minimum	↓	5	3.200	Ea.
Maximum		3	5.330	Ea.
Icemaker, automatic, 13#/day	1 Plumber	7	1.140	Ea.
51#/day	"	2	4.000	Ea.
Refrigerator, no frost, 10 to 12 C.F., minimum	2 Building Laborers	10	1.600	Ea.
Maximum		6	2.670	Ea.
14 to 16 C.F., minimum		9	1.780	Ea.
Maximum		5	3.200	Ea.
18 to 20 C.F., minimum		8	2.000	Ea.
Maximum		4	4.000	Ea.
21 to 29 C.F., minimum		7	2.290	Ea.
Maximum	↓	3	5.330	Ea.
Sump pump cellar drainer, 1/3 H.P., minimum	1 Plumber	3	2.670	Ea.
Maximum		2	4.000	Ea.
Washing machine, automatic, minimum		3	2.670	Ea.
Maximum	↓	1	8.000	Ea.
Water heater, electric, glass lined, 30 gallon				
Minimum	1 Electrician 1 Plumber	5	3.200	Ea.
Maximum		3	5.330	Ea.
80 gallon	↓	2	8.000	Ea.
Maximum		1	16.999	Ea.
Water heater, gas, glass lined, 30 gallon, minimum	2 Plumbers	5	3.200	Ea.
Maximum		3	5.330	Ea.
50 gallon, minimum		2.50	6.400	Ea.
Maximum		1.50	10.670	Ea.
Water softener, automatic, to 30 grains/gallon	↓	5	3.200	Ea.
To 75 grains/gallon		4	4.000	Ea.
Vent kits for dryers	1 Carpenter	10	.800	Ea.
KITCHEN EQUIPMENT Bake oven, single deck	1 Plumber 1 Plumber Apprentice	8	2.000	Ea.
Double deck		7	2.290	Ea.
Triple deck	↓	6	2.670	Ea.
Electric convection, 40" x 45" x 57"	2 Carpenters 1 Building Laborer .5 Electrician	4	7.000	Ea.

(continued on next page)

Figure 11.16   Installation Time for Food Service Equipment (continued)

Food Service Equipment	Crew Makeup	Daily Output	Man-Hours	Unit
KITCHEN EQUIPMENT Broiler, without oven Standard	1 Plumber 1 Plumber Apprentice	8	2.000	Ea.
Infra-red	2 Carpenters 1 Building Laborer .5 Electrician	4	7.000	Ea.
Cooler, reach-in, beverage, 6' long	1 Plumber 1 Plumber Apprentice	6	2.670	Ea.
Dishwasher, commercial, rack type 10 to 12 racks/hour	1 Plumber 1 Plumber Apprentice	3.20	5.000	Ea.
Semi-automatic 38 to 50 racks/hour	"	1.30	12.310	Ea.
Automatic 190 to 230 racks/hour	2 Plumbers 1 Plumber Apprentice	.70	34.290	Ea.
235 to 275 racks/hour	↓	.50	48.000	Ea.
8750 to 12,500 dishes/hour		.20	120.000	Ea.
Fast food equipment, total package, minimum	6 Skilled Workers	.08	600.000	Ea.
Maximum	"	.07	686.000	Ea.
Food mixers, 20 quarts	2 Carpenters 1 Building Laborer .5 Electrician	7	4.000	Ea.
60 quarts	"	5	5.600	Ea.
Freezers, reach-in, 44 C.F.	1 Plumber 1 Plumber Apprentice	4	4.000	Ea.
68 C.F.		3	5.330	Ea.
Fryer, with submerger, single		7	2.290	Ea.
Double	↓	5	3.200	Ea.
Griddle, 3' long		7	2.290	Ea.
4' long		6	2.670	Ea.
Ice cube maker, 50#/day		6	2.670	Ea.
500#/day	↓	4	4.000	Ea.
Kettles, steam-jacketed, 20 gallons	2 Carpenters 1 Building Laborer .5 Electrician	7	4.000	Ea.
60 gallons	"	6	4.670	Ea.
Range, restaurant type, 6 burners and 1 oven, 36"	1 Plumber 1 Plumber Apprentice	7	2.290	Ea.
2 ovens, 60"		6	2.670	Ea.
Heavy duty, single 34" oven, open top		5	3.200	Ea.
Fry top		6	2.670	Ea.
Hood fire protection system, minimum		3	5.330	Ea.
Maximum		1	16.000	Ea.
Refrigerators, reach-in type, 44 C.F.		5	3.200	Ea.
With glass doors, 68 C.F.	↓	4	4.000	Ea.

Figure 11.16   Installation Time for Food Service Equipment (continued)

Food Service Equipment	Crew Makeup	Daily Output	Man-Hours	Unit
KITCHEN EQUIPMENT Steamer, electric, 27 KW	2 Carpenters 1 Building Laborer .5 Electrician	7	4.000	Ea.
Electric, 10 KW or gas 100,000	"	5	5.600	Ea.
Rule of thumb: Equipment cost based on kitchen work area				
Office buildings, minimum	2 Carpenters 1 Building Laborer .5 Electrician	77	.364	S.F.
Maximum		58	.483	S.F.
Public eating facilities, minimum		77	.364	S.F.
Maximum		46	.609	S.F.
Hospitals, minimum		58	.483	S.F.
Maximum	▼	39	.718	S.F.
WINE VAULT Redwood, air conditioned, walk-in type				
6'-8" high, incl. racks, 2' x 4' for 156 bottles	2 Carpenters	2	8.000	Ea.
4' x 6' for 614 bottles		1.50	10.670	Ea.
6' x 12' for 1940 bottles	▼	1	16.000	Ea.

Figure 11.17   Installation Time for Disappearing Stairs

Disappearing Stairs	Crew Makeup	Daily Output	Man-Hours	Unit
DISAPPEARING STAIRWAY No trim included				
Custom grade, pine, 8'-6" ceiling, minimum	1 Carpenter	4	2.000	Ea.
Average		3.50	2.290	Ea.
Maximum		3	2.670	Ea.
Heavy duty, pivoted, 8'-6" ceiling		3	2.670	Ea.
16' ceiling		2	4.000	Ea.
Economy folding, pine, 8'-6" ceiling		4	2.000	Ea.
9'-6" ceiling	▼	4	2.000	Ea.
Fire escape, galvanized steel, 8' to 10'-4" ceiling	2 Carpenters	1	16.000	Ea.
10'-6" to 13'-6" ceiling		1	16.000	Ea.
Automatic electric, aluminum, floor to floor height				
8' to 9'		1	16.000	Ea.
11' to 12'		.90	17.780	Ea.
14' to 15'	▼	.70	22.860	Ea.

Figure 11.18 Installation Time for Darkroom Equipment

Darkroom Processing	Crew Makeup	Daily Output	Man-Hours	Unit
DARKROOM EQUIPMENT Developing tanks				
5" deep, 24" x 48"	1 Plumber	2	8.000	Ea.
	1 Plumber Apprentice			
48" x 52"		1.70	9.410	Ea.
10" deep, 24" x 48"		1.70	9.410	Ea.
24" x 108"	↓	1.50	10.670	Ea.
Dryers, dehumidified filtered air				
36" x 25" x 68" high	2 Carpenters	6	4.670	Ea.
	1 Building Laborer			
	.5 Electrician			
48" x 25" x 68" high		5	5.600	Ea.
Processors, automatic, color print, minimum		4	7.000	Ea.
Maximum		1	28.000	Ea.
Black and white print, minimum		4	7.000	Ea.
Maximum		1	28.000	Ea.
Manual processor				
16" x 20" maximum print size		4	7.000	Ea.
20" x 24" maximum print size		1	28.000	Ea.
Viewing lites, 20" x 24"		6	4.670	Ea.
20" x 24" with color correction	↓	6	4.670	Ea.
Washers, round, minimum sheet 11" x 14"	1 Plumber	2	8.000	Ea.
	1 Plumber Apprentice			
Maximum sheet		1	16.000	Ea.
Square, minimum sheet		1	16.000	Ea.
Maximum sheet	↓	.80	20.000	Ea.
Combination tank sink, tray sink, washers,				
with dry side tables, average	1 Plumber	1	16.000	Ea.
	1 Plumber Apprentice			

Figure 11.19 Installation Time for Revolving Darkroom Doors

Revolving Darkroom Doors	Crew Makeup	Daily Output	Man-Hours	Unit
DARKROOM DOORS Revolving, standard				
2 way, 40" diameter	2 Carpenters	3.50	4.570	Opng.
3 way, 50" diameter		3	5.330	Opng.
4 way, 50" diameter		2	8.000	Opng.
Hinged safety, 2 way, 40" diameter		3.20	5.000	Opng.
3 way, 50" diameter		2.90	5.520	Opng.
Pop out safety, 2 way, 40" diameter		3.10	5.160	Opng.
3 way, 50" diameter	↓	2.80	5.710	Opng.

Figure 11.20 Installation Time for Athletic/Recreational Equipment

Athletic/Recreational	Crew Makeup	Daily Output	Man-Hours	Unit
HEALTH CLUB EQUIPMENT				
Circuit training apparatus, 12 machines, minimum	2 Building Laborers	1.25	12.800	Set
Average		1	16.000	Set
Maximum		.75	21.330	Set
Squat racks		5	3.200	Ea.
SCHOOL EQUIPMENT				
Basketball backstops, wall mounted, 6' extended, fixed				
Minimum	1 Carpenter 1 Helper	1	16.000	Ea.
Maximum		1	16.000	Ea.
Swing up, minimum		1	16.000	Ea.
Maximum		1	16.000	Ea.
Portable, manual, heavy duty, hydraulic		1.90	8.420	Ea.
Ceiling suspended, stationary, minimum		.78	20.510	Ea.
Fold up, with accessories, maximum		1	16.000	Ea.
For electrically operated, add	1 Electrician	1	8.000	Ea.
Benches, folding, in wall, 14' table, 2 benches	2 Skilled Workers 1 Helper	2	12.000	Set
Bleachers, telescoping, manual to 15 tier, minimum	1 Carpenter Foreman 3 Carpenters Power Tools	65	.492	Seat
Maximum		60	.533	Seat
16 to 20 tier, minimum		60	.533	Seat
Maximum		55	.582	Seat
21 to 30 tier, minimum		50	.640	Seat
Maximum		40	.800	Seat
For integral power operation, add, minimum	2 Electricians	300	.053	Seat
Maximum	"	250	.064	Seat
Exercise equipment				
Chinning bar, adjustable, wall mounted	1 Carpenter	5	1.600	Ea.
Exercise ladder, 16' x 1'-7", suspended	1 Carpenter 1 Helper	3	5.330	Ea.
High bar, floor plate attached	1 Carpenter	4	2.000	Ea.
Parellel bars, adjustable		4	2.000	Ea.
Uneven parallel bars, adjustable		4	2.000	Ea.
Wall mounted, adjustable	1 Carpenter 1 Helper	1.50	10.670	Set
Rope, ceiling mounted, 18' long	1 Carpenter	10	.800	Ea.
Side horse, vaulting		5	1.600	Ea.
Treadmill, motorized, deluxe, training type		5	1.600	Ea.
Weight lifting multi-station, minimum	2 Building Laborers	1	16.000	Ea.
Maximum	"	.50	32.000	Ea.
Scoreboards, baseball, minimum	1 Electrician Foreman 1 Electrician .5 Equip. Oper. (crane) .5 S.P. Crane, 5 Ton	2.40	8.330	Ea.
Maximum		.05	400.000	Ea.
Football, minimum		1.20	16.670	Ea.
Maximum		.20	100.000	Ea.
Basketball (one side), minimum		2.40	8.330	Ea.
Maximum		.30	66.670	Ea.
Hockey-basketball (four sides), minimum		.25	80.000	Ea.
Maximum		.15	133.000	Ea.

Figure 11.21   Installation Time for Bowling Alleys

Bowling Alleys	Crew Makeup	Daily Output	Man-Hours	Unit
BOWLING ALLEYS Including alley, pinsetter, scorer, counters and misc. supplies, minimum	4 Carpenters	.20	160.000	Lane
Average	↓	.19	168.000	Lane
Maximum		.18	178.000	Lane

Figure 11.22   Installation Time for Shooting Ranges

Shooting Ranges	Crew Makeup	Daily Output	Man-Hours	Unit
SHOOTING RANGE Including bullet traps, target provisions, controls, separators, ceiling system, etc. Not including structural shell Commercial	1 Labor Foreman (inside) 2 Building Laborers 1 Struc. Steel Worker .5 Electrician	.85	42.350	Point
Law enforcement		.65	55.380	Point
National Guard armories		.71	50.700	Point
Reserve training centers		.85	42.350	Point
Schools and colleges		.75	48.000	Point
Major academies	↓	.48	75.000	Point

Figure 11.23   Installation Time for Industrial Equipment

Industrial Equipment	Crew Makeup	Daily Output	Man-Hours	Unit
EQUIPMENT INSTALLATION Industrial equipment Minimum	1 Struc. Steel Foreman 4 Struc. Steel Workers 1 Equip. Oper. (crane) 1 Equip. Oper. Oiler 1 Crane, 90 Ton	12	4.670	Ton
Maximum	"	2	28.000	Ton

Figure 11.24 Installation Time for Specialized Equipment

Specialized Equipment	Crew Makeup	Daily Output	Man-Hours	Unit
VOCATIONAL SHOP EQUIPMENT Benches, work,				
wood, average	2 Carpenters	5	3.200	Ea.
Metal, average		5	3.200	Ea.
Combination belt & disc sander, 6"	↓	4	4.000	Ea.
Drill press, floor mounted, 12", 1/2 H.P.		4	4.000	Ea.
Dust collector, not incl. ductwork, 6' diameter	1 Sheet Metal Worker	1.10	7.270	Ea.
Grinders, double wheel, 1/2 H.P.	2 Carpenters	5	3.200	Ea.
Jointer, 4", 3/4 H.P.		4	4.000	Ea.
Kilns, 16 C.F., to 2000°		4	4.000	Ea.
Lathe, woodworking, 10", 1/2 H.P.		4	4.000	Ea.
Planer, 13" x 6"		4	4.000	Ea.
Potter's wheel, motorized		4	4.000	Ea.
Saws, band, 14", 3/4 H.P.		4	4.000	Ea.
Metal cutting band saw, 14"		4	4.000	Ea.
Radial arm saw, 10", 2 H.P.		4	4.000	Ea.
Scroll saw, 24"		4	4.000	Ea.
Table saw, 10", 3 H.P.		4	4.000	Ea.
Welder AC arc, 30 amp capacity	↓	4	4.000	Ea.

Figure 11.25 Installation Time for Laboratory Equipment

Laboratory Equipment	Crew Makeup	Daily Output	Man-Hours	Unit
LABORATORY EQUIPMENT Cabinets, base, door units,				
metal	2 Carpenters	18	.889	L.F.
Drawer units		18	.889	L.F.
Tall storage cabinets, open, 7' high		20	.800	L.F.
With glazed doors		20	.800	L.F.
Wall cabinets, metal, 12-1/2" deep, open		20	.800	L.F.
With doors		20	.800	L.F.
Counter tops, not incl. base cabinets, acidproof,				
Minimum		82	.195	S.F.
Maximum		70	.229	S.F.
Stainless steel	↓	82	.195	S.F.
Fume hood, with contertop & base, not including HVAC				
Simple, minimum	2 Carpenters	5.40	2.960	L.F.
Complex, including fixtures		2.40	6.670	L.F.
Special, maximum	↓	1.70	9.410	L.F.
Ductwork, minimum	2 Sheet Metal Workers	1	16.000	Hood
Maximum	"	.50	32.000	Hood
For sink assembly with hot and cold water, add	1 Plumber	1.40	5.710	Ea.
Glassware washer, distilled water rinse, minimum	1 Electrician	1.80	8.890	Ea.
	1 Plumber			
Maximum	"	1	16.000	Ea.
Sink, one piece plastic, flask wash, hose, free standing	1 Plumber	1.60	5.000	Ea.
Epoxy resin sink, 25" x 16" x 10"	"	2	4.000	Ea.
Utility table, acid resistant top with drawers	2 Carpenters	30	.533	L.F.

Figure 11.26 Installation Time for Medical Equipment

Medical Equipment	Crew Makeup	Daily Output	Man-Hours	Unit
MEDICAL EQUIPMENT Autopsy table, standard	1 Plumber	1	8.000	Ea.
Deluxe		.60	13.330	Ea.
Distiller, water, steam heated, 50 gal. capacity		1.40	5.710	Ea.
Automatic washer/sterilizer	↓	2	4.000	Ea.
Steam generators, electric 10 KW to 180 KW				
Minimum	1 Electrician	3	2.670	Ea.
Maximum	"	.70	11.430	Ea.
Surgery table, minor minimum	1 Struc. Steel Worker	.70	11.430	Ea.
Maximum		.50	16.000	Ea.
Major surgery table, minimum		.50	16.000	Ea.
Maximum	↓	.50	16.000	Ea.
Surgical lights, single arm	2 Electricians	.90	17.780	Ea.
Dual arm	"	.30	53.330	Ea.
Tables, physical therapy, walk off, electric	2 Carpenters	3	5.330	Ea.
Standard, vinyl top with base cabinets, minimum		3	5.330	Ea.
Maximum	↓	2	8.000	Ea.
Utensil washer-sanitizer	1 Plumber	2	4.000	Ea.

Figure 11.27 Installation Time for Dental Equipment

Dental Equipment	Crew Makeup	Daily Output	Man-Hours	Unit
DENTAL EQUIPMENT Central suction system, minimum	1 Plumber	1.20	6.670	Ea.
Maximum	"	.90	8.890	Ea.
Air compressor, minimum	1 Skilled Worker	.80	10.000	Ea.
Maximum		.50	16.000	Ea.
Chair, electric or hydraulic, minimum		.50	16.000	Ea.
Maximum		.25	32.000	Ea.
Drill console with accessories, minimum		.50	16.000	Ea.
Maximum		.33	24.240	Ea.
Light, floor or ceiling mounted, minimum		3.60	2.220	Ea.
Maximum		1.20	6.670	Ea.
X-ray unit, wall		1.90	4.210	Ea.
Panoramic unit		.60	13.330	Ea.
Developers, X-ray, minimum		1	8.000	Ea.
Maximum	↓	1	8.000	Ea.

Checklist

For an estimate to be reliable, all items must be accounted for.
A complete estimate can also eliminate the need to include
contingencies. The following checklist can be used to help ensure
that all items are properly accounted for.

Appliances
- ☐ Cook tops
- ☐ Compactors
- ☐ Dehumidifiers — unit
- ☐ Dishwasher
- ☐ Dryer
- ☐ Exhaust hoods
- ☐ Garbage disposer
- ☐ Heaters — electric unit
- ☐ Humidifiers — unit
- ☐ Ice makers
- ☐ Ovens
- ☐ Refrigerators
- ☐ Sump pump — unit
- ☐ Washing machine
- ☐ Water heater — unit
- ☐ Water softener — unit

Automotive Equipment
- ☐ Lifts
- ☐ Hoists
- ☐ Lube

Bank Equipment
- ☐ Counters
- ☐ Safes
- ☐ Safe deposit boxes
- ☐ Vaults
- ☐ Teller stations/windows

Barber Shop Equipment _____

Bowling Alley Equipment _____

Check Room Equipment _____

Church Equipment
- ☐ Altar
- ☐ Baptistries
- ☐ Bells

- ☐ Carillons
- ☐ Confessionals
- ☐ Crosses
- ☐ Organ
- ☐ Pews
- ☐ Pulpit
- ☐ Spires

Coin Stations ──────────────────────────────────

Commercial Equipment
- ☐ Cash register systems
- ☐ Check-out counters
- ☐ Display cases
- ☐ Refrigerated cases

Darkroom Equipment ──────────────────────────────

Data Processing Equipment ───────────────────────

Dental Equipment
- ☐ Chairs
- ☐ Drill/equipment stations
- ☐ Thruway equipment

Detention Equipment ─────────────────────────────

Dock Equipment
- ☐ Bumpers
- ☐ Boards
- ☐ Door seals
- ☐ Levellers
- ☐ Lights
- ☐ Shelters

Food Service Equipment
- ☐ Bar units
- ☐ Cooking equipment
- ☐ Dishwashing equipment
- ☐ Food preparation equipment
- ☐ Food serving equipment
- ☐ Refrigerated cases
- ☐ Tables

Gymnasium Equipment ─────────────────────────────

Industrial Equipment ────────────────────────────

Laboratory Equipment
- ☐ Casework
- ☐ Countertops

- ☐ Hoods
- ☐ Sinks
- ☐ Tables

Laundry Equipment _____

Library Equipment
- ☐ Bookshelves
- ☐ Book stacks
- ☐ Card files
- ☐ Carrels
- ☐ Charging desks
- ☐ Racks

Medical Equipment _____

Mortuary Equipment _____

Musical Equipment _____

Observatory Equipment _____

Parking Equipment
- ☐ Automatic gates
- ☐ Booths
- ☐ Card readers
- ☐ Control stations
- ☐ Ticket dispensers
- ☐ Traffic detectors

Playground Equipment _____

Projection Equipment _____

Prison Equipment _____

Residential Equipment
- ☐ Kitchen cabinets/appliances
- ☐ Lavatory cabinets
- ☐ Kitchen equipment
- ☐ Laundry equipment
- ☐ Unit kitchens
- ☐ Vacuum equipment

Safes _____

Saunas _____

School Equipment
- ☐ Arts and crafts
- ☐ Audiovisual
- ☐ Language labs
- ☐ Vocational/shop equipment
- ☐ Wall benches
- ☐ Wall tables

Service Station Equipment _____

Shop Equipment _____

Shooting Ranges _____

Stage Equipment _____

Steam Baths _____

Swimming Pool Equipment
☐ Diving board
☐ Diving platform
☐ Filtration systems
☐ Heaters
☐ Ladders
☐ Lifeguard chair
☐ Lights
☐ Pool covers
☐ Slides

Unit Kitchens _____

Vacuum Systems _____

Waste Disposal
☐ Compactors
☐ Incinerators

Special Equipment _____

_____

_____

_____

_____

Tips

Faulty Assumptions

Do not assume that items covered in Division 11 will be purchased and installed by others outside of your contract. This can be a costly error. Check all drawings for these items and for terms such as "NIC" (Not In Contract) or "By Others." If these or similar terms are not in evidence, then it is safe to conclude that these are in your scope of work.

Installation of Items

In many cases Division 11 items are purchased by others, but their installation is the contractor's responsibility. Check all drawings and specifications carefully for these items.

Handling Charges

In cases where Division 11 items are purchased by others but installed by the contractor, contractors often add a handling charge. (10% of the estimated material costs is common practice). This charge covers the receiving, handling, storage, protection, and final delivery of these items.

Support Systems

Note that Division 11 items may require some type of support system not usually supplied with the item — such as support brackets, plates, or angles. These need to be accounted for and may need to be added to the appropriate division. Phrases in the specifications that gloss over this subject and thus make it difficult to recover any added costs include "The contractor shall install all products in accordance with the manufacturers' recommendations" and the like. Remember, it is much more costly to install a behind-the-wall support system after the wall is in place.

Preparation of Items

In some cases, Division 11 items may require assembly before installation. Be aware that the assembly time can often exceed the time required for installation.

Notes

Division Twelve
Furnishings

Introduction

Division 12 — Furnishings generally refers to the items brought into the building for the use and/or comfort of the occupants. Items in this category include furniture, artwork, window treatments (blinds, curtains, etc.), mats, rugs, plants and the like. In the majority of cases, these items are supplied by the owner or are arranged for outside the scope of most construction-related contracts.

Items included within Division 12 are usually counted individually, as most items are bought/sold in this fashion. The procedures used to estimate this division are usually as follows.

Review all contract documents to determine what is and what is not included in the scope of responsibility. Review the specifications (if applicable), all drawings, sketches, owner-supplied lists, and/or any other relevant documentation for the items required, specific models, and/or for acceptable manufacturers. All items should be listed, accounted for, and costs attached.

Many of the items can be delivered in place or have installation costs already included in the price of the item. Others, however, may need some attention or at least cleaning prior to installation. Costs must be estimated for these items.

Estimating Data

The following tables show the expected installation time for various specialty items. Each table lists the specialty item, the typical installation crew mix, the expected average daily output of that crew (the number of units the crew can be expected to install in one eight-hour day), and the man-hours required for the installation of one unit.

Table of Contents

Figure 12.1   Installation Time for Artwork

Artwork	Crew Makeup	Daily Output	Man-Hours	Unit
ARTWORK Framed	1 Carpenter	36	.222	Ea.
Photography, minimum		30	.267	Ea.
Maximum		36	.222	Ea.
Posters, minimum		30	.267	Ea.
Maximum		36	.222	Ea.
Reproductions, minimum		30	.267	Ea.
Maximum	↓			

Figure 12.2   Installation Time for Metal Casework

Metal Casework	Crew Makeup	Daily Output	Man-Hours	Unit
KEY CABINETS Wall mounted, 60 key capacity	1 Carpenter	20	.400	Ea.
1200 key capacity	"	10	.800	Ea.
Drawer type, 600 key capacity	1 Building Laborer	15	.533	Ea.
2400 key capacity		20	.400	Ea.
Tray type, 20 key capacity		50	.160	Ea.
50 key capacity	↓	40	.200	Ea.

Figure 12.3   Installation Time for Hospital Casework

Hospital Casework	Crew Makeup	Daily Output	Man-Hours	Unit
CABINETS				
Hospital, base cabinets, laminated plastic	2 Carpenters	10	1.600	L.F.
Enameled steel		10	1.600	L.F.
Stainless steel		10	1.600	L.F.
Cabinet base trim, 4" high, enameled steel		200	.080	L.F.
Stainless steel		200	.080	L.F.
Counter top, laminated plastic, no backsplash		40	.400	L.F.
With backsplash		40	.400	L.F.
For sink cutout, add		12.20	1.310	Ea.
Stainless steel counter top		40	.400	L.F.
Nurses station, door type, laminated plastic		10	1.600	L.F.
Enameled steel		10	1.600	L.F.
Stainless steel		10	1.600	L.F.
Wall cabinets, laminated plastic		15	1.070	L.F.
Enameled steel		15	1.070	L.F.
Stainless steel		15	1.070	L.F.
Kitchen, base cabinets, metal, minimum		30	.533	L.F.
Maximum		25	.640	L.F.
Wall cabinets, metal, minimum		30	.533	L.F.
Maximum		25	.640	L.F.
School, 24" deep		15	1.070	L.F.
Counter height units		20	.800	L.F.
Wood, custom fabricated, 32" high counter		20	.800	L.F.
Add for counter top		56	.286	L.F.
84" high wall units	▼	15	1.070	L.F.

Figure 12.4   Installation Time for Display Casework

Display Casework	Crew Makeup	Daily Output	Man-Hours	Unit
DISPLAY CASES Free standing, all glass				
Aluminum frame, 42" high x 36" x 12" deep	2 Carpenters	8	2.000	Ea.
70" high x 48" x 18" deep	"	6	2.670	Ea.
Wall mounted, glass front, aluminum frame				
Non-illuminated, one section 3' x 4' x 1'-4"	2 Carpenters	5	3.200	Ea.
5' x 4' x 1'-4"		5	3.200	Ea.
6' x 4' x 1'-4"		4	4.000	Ea.
Two sections, 8' x 4' x 1'-4"		2	8.000	Ea.
10' x 4' x 1'-4"		2	8.000	Ea.
Three sections, 16' x 4' x 1'-4"		1.50	10.670	Ea.
Table exhibit cases, 2' wide, 3' high, 4' long, flat top		5	3.200	Ea.
3' wide, 3' high, 4' long, sloping top	▼	3	5.330	Ea.

Figure 12.5   Installation Time for Residential Casework

Residential Casework	Crew Makeup	Daily Output	Man-Hours	Unit

IRONING CENTER Including cabinet, board & light
    Minimum
    Maximum
	1 Carpenter
	"
		2
		1.50
			4.000
			5.330
				Ea.
				Ea.

Figure 12.6 Installation Time for Window Treatments — Blinds

Blinds	Crew Makeup	Daily Output	Man-Hours	Unit
BLINDS, INTERIOR Solid colors				
Horizontal, 1" aluminum slats, custom, minimum	1 Carpenter	590	.014	S.F.
Maximum		440	.018	S.F.
2" aluminum slats, custom, minimum		590	.014	S.F.
Maximum		440	.018	S.F.
Stock, minimum		590	.014	S.F.
Maximim		440	.018	S.F.
2" steel slats, stock, minimum		590	.014	S.F.
Maximum		440	.018	S.F.
Custom, minimum		590	.014	S.F.
Maximum		400	.020	S.F.
Alternate method of figuring:				
1" aluminum slats, 48" wide, 48" high	1 Carpenter	30	.267	Ea.
72" high		29	.276	Ea.
96" high		28	.286	Ea.
72" wide, 72" high		25	.320	Ea.
96" high		23	.348	Ea.
96" wide, 96" high		20	.400	Ea.
Vertical, 3" to 5" PVC or cloth strips, minimum		460	.017	S.F.
Maximum		400	.020	S.F.
4" aluminum slats, minimum		460	.017	S.F.
Maximum		400	.020	S.F.
Mylar mirror-finish strips, to 8" wide, minimum		460	.017	S.F.
Maximum		400	.020	S.F.
Alternate method of figuring:				
2" aluminum slats, 48" wide, 48" high	1 Carpenter	30	.267	Ea.
72" high		29	.276	Ea.
96" high		28	.286	Ea.
72" wide, 72" high		25	.320	Ea.
96" high		23	.348	Ea.
96" wide, 96" high		20	.400	Ea.
Mirror finish, 48" wide, 48" high		30	.267	Ea.
72" high		29	.276	Ea.
96" high		28	.286	Ea.
72" wide, 72" high		25	.320	Ea.
96" high		23	.348	Ea.
96" wide, 96" high		20	.400	Ea.
Decorative printed finish, 48" wide, 48" high		30	.267	Ea.
72" high		29	.276	Ea.
96" high		28	.286	Ea.
72" wide, 72" high		25	.320	Ea.
96" high		23	.348	Ea.
96" wide, 96" high		20	.400	Ea.

(continued on next page)

Figure 12.6　Installation Time for Window Treatments — Blinds (continued)

Blinds	Crew Makeup	Daily Output	Man-Hours	Unit
BLINDS, INTERIOR Wood folding panels with movable louvers	1 Carpenter			
7" x 20" each		17	.471	Pr.
8" x 28" each		17	.471	Pr.
9" x 36" each		17	.471	Pr.
10" x 40" each		17	.471	Pr.
Fixed louver type, stock units, 8" x 20" each		17	.471	Pr.
10" x 28" each		17	.471	Pr.
12" x 36" each		17	.471	Pr.
18" x 40" each		17	.471	Pr.
Insert panel type, stock, 7" x 20" each		17	.471	Pr.
8" x 28" each		17	.471	Pr.
9" x 36" each		17	.471	Pr.
10" x 40" each		17	.471	Pr.
Raised panel type, stock, 10" x 24" each		17	.471	Pr.
12" x 26" each		17	.471	Pr.
14" x 30" each		17	.471	Pr.
16" x 36" each		17	.471	Pr.

Figure 12.7　Installation Time for Window Treatments — Shades

Shades	Crew Makeup	Daily Output	Man-Hours	Unit
SHADES Basswood, roll-up, stain finish, 3/8" slats	1 Carpenter	300	.027	S.F.
7/8" slats		300	.027	S.F.
Vertical side slide, stain finish, 3/8" slats		300	.027	S.F.
7/8" slats		300	.027	S.F.
For fire retardant finishes, add		300		S.F.
For "B" rated finishes, add		300		S.F.
Mylar, single layer, non-heat reflective		685	.012	S.F.
Double layered, heat reflective		685	.012	S.F.
Triple layered, heat reflective		685	.012	S.F.
Vinyl coated cotton, standard		685	.012	S.F.
Lightproof decorator shades		685	.012	S.F.
Vinyl, lightweight, 4 gauge		685	.012	S.F.
Heavyweight, 6 gauge		685	.012	S.F.
Vinyl laminated fiberglass, 6 gauge, translucent		685	.012	S.F.
Lightproof		685	.012	S.F.
Woven aluminum, 3/8" thick, lightproof and fireproof		350	.023	S.F.
Insulative shades		125	.064	S.F.
Solar screening, fiberglass		85	.094	S.F.
Interior insulative shutter, stock unit, 15" x 60"		17	.471	Pr.

Figure 12.8 Installation Time for Window Treatments — Draperies and Curtains

Draperies and Curtains	Crew Makeup	Daily Output	Man-Hours	Unit
DRAPERIES				
Drapery installation, hardware and drapes,				
Labor cost only, minimum	1 Building Laborer	75	.107	L.F.
Maximum	"	20	.400	L.F.

Figure 12.9 Installation Time for Window Treatments — Drape/Curtain Hardware

Drape/Curtain Hardware	Crew Makeup	Daily Output	Man-Hours	Unit
DRAPERY HARDWARE				
Standard traverse, per foot, minimum	1 Carpenter	59	.136	L.F.
Maximum		51	.157	L.F.
Decorative traverse, 28" to 48", minimum		22	.364	Ea.
Maximum		21	.381	Ea.
48" to 84", minimum		20	.400	Ea.
Maximum		19	.421	Ea.
66" to 120", minimum		18	.444	Ea.
Maximum		17	.471	Ea.
84" to 156", minimum		16	.500	Ea.
Maximum		15	.533	Ea.
130" to 240", minimum		14	.571	Ea.
Maximum		13	.615	Ea.
Ripplefold, snap-a-pleat system, 3' or less				
Minimum		15	.533	Ea.
Maximum		14	.571	Ea.
Traverse rods, adjustable, 28" to 48"		22	.364	Ea.
48" to 84"		20	.400	Ea.
66" to 120"		18	.444	Ea.
84" to 156"		16	.500	Ea.
120" to 220"		14	.571	Ea.
228" to 312"		13	.615	Ea.

Figure 12.10 Installation Time for Interior Landscape Partitions

Landscape Partitions	Crew Makeup	Daily Output	Man-Hours	Unit
PANELS AND DIVIDERS Free standing				
Fabric panel, class A fire rated				
Minimum NRC .95, STC 28, fabric edged				
40" high, 24" wide	2 Building Laborers	30	.533	Ea.
36" wide		28	.571	Ea.
48" wide		26	.615	Ea.
60" wide		25	.640	Ea.
50" high, 24" wide		30	.533	Ea.
36" wide		28	.571	Ea.
48" wide		26	.615	Ea.
60" wide		25	.640	Ea.
60" high, 24" wide		29	.552	Ea.
36" wide		27	.593	Ea.
48" wide		25	.640	Ea.
60" wide		24	.667	Ea.
72" high, 24" wide		29	.552	Ea.
36" wide		27	.593	Ea.
48" wide		25	.640	Ea.
60" wide		24	.667	Ea.
Hardwood edged, 40" high, 24" wide		30	.533	Ea.
36" wide		28	.571	Ea.
48" wide		26	.615	Ea.
60" wide		25	.640	Ea.
50" high, 24" wide		30	.533	Ea.
36" wide		28	.571	Ea.
48" wide		26	.615	Ea.
60" wide		25	.640	Ea.
60" high, 24" wide		29	.552	Ea.
36" wide		27	.593	Ea.
48" wide		25	.640	Ea.
60" wide		24	.667	Ea.
72" high, 24" wide		29	.552	Ea.
36" wide		27	.593	Ea.
48" wide		25	.640	Ea.
60" wide		24	.667	Ea.

Figure 12.11   Installation Time for Furniture

Furniture	Crew Makeup	Daily Output	Man-Hours	Unit
DORMITORY				
Desk top, built-in, laminated plastic, 24" deep				
Minimum	2 Carpenters	50	.320	L.F.
Maximum		40	.400	L.F.
30" deep, minimum		50	.320	L.F.
Maximum		40	.400	L.F.
Dressing unit, built-in, minimum		12	1.330	L.F.
Maximum	↓	8	2.000	L.F.
LIBRARY				
Attendent desk, 36" x 62" x 29" high	1 Carpenter	16	.500	Ea.
Book display, "A" frame display, both sides		16	.500	Ea.
Table with bulletin board	↓	16	.500	Ea.
Book trucks, descending platform,				
Small, 14" x 30" x 35" high	1 Carpenter	16	.500	Ea.
Large, 14" x 40" x 42" high		16	.500	Ea.
Card catalogue, 30 tray unit		16	.500	Ea.
60 tray unit	↓	16	.500	Ea.
72 tray unit	2 Carpenters	16	1.000	Ea.
120 tray unit	"	16	1.000	Ea.
Carrels, single face, initial unit	1 Carpenter	16	.500	Ea.
Additional unit	"	16	.500	Ea.
Double face, initial unit	2 Carpenters	16	1.000	Ea.
Additional unit		16	1.000	Ea.
Cloverleaf	↓	11	1.450	Ea.
Chairs, sled base, arms, minimum	1 Carpenter	24	.333	Ea.
Maximum		16	.500	Ea.
No arms, minimum		24	.333	Ea.
Maximum		16	.500	Ea.
Standard leg base, arms, minimum		24	.333	Ea.
Maximum		16	.500	Ea.
No arms, minimum		24	.333	Ea.
Maximum	↓	16	.500	Ea.
Charge desk, modular unit, 35" x 27" x 39" high				
Wood front and edges, plastic laminate tops				
Book return	1 Carpenter	16	.500	Ea.
Book truck port		16	.500	Ea.
Card file drawer, 5 drawers		16	.500	Ea.
10 drawers		16	.500	Ea.
15 drawers		16	.500	Ea.
Card and legal file		16	.500	Ea.
Charging machine		16	.500	Ea.
Corner		16	.500	Ea.
Cupboard		16	.500	Ea.
Detachable end panel		16	.500	Ea.
Gate		16	.500	Ea.
Knee space		16	.500	Ea.
Open storage		16	.500	Ea.
Station charge		16	.500	Ea.
Work station		16	.500	Ea.
Dictionary stand, stationary		16	.500	Ea.
Revolving	↓	16	.500	Ea.

(continued on next page)

Figure 12.11   Installation Time for Furniture (continued)

Furniture	Crew Makeup	Daily Output	Man-Hours	Unit
LIBRARY (cont.)				
Exhibit case, table style, 60″ x 28″ x 36″	1 Carpenter	11	.727	Ea.
Globe stand		16	.500	Ea.
Magazine rack		16	.500	Ea.
Newspaper rack		16	.500	Ea.
Tables, card catalog reference, 24″ x 60″ x 42″		16	.500	Ea.
24″ x 60″ x 72″		16	.500	Ea.
Index, single tier, 48″ x 72″		16	.500	Ea.
Double tier, 48″ x 72″	↓	16	.500	Ea.
Study, panel ends, plastic lam. surfaces 29″ high				
36″ x 60″	2 Carpenters	16	1.000	Ea.
36″ x 72″		16	1.000	Ea.
36″ x 90″		16	1.000	Ea.
48″ x 72″	↓	16	1.000	Ea.
Parsons, 29″ high, plastic laminate top,				
wood legs and edges				
36″ x 36″	2 Carpenters	16	1.000	Ea.
36″ x 60″		16	1.000	Ea.
36″ x 72″		16	1.000	Ea.
36″ x 84″		16	1.000	Ea.
42″ x 90″		16	1.000	Ea.
48″ x 72″		16	1.000	Ea.
48″ x 120″		16	1.000	Ea.
Round, leg or pedestal base, 36″ diameter		16	1.000	Ea.
42″ diameter		16	1.000	Ea.
48″ diameter		16	1.000	Ea.
60″ diameter	↓	16	1.000	Ea.
RESTAURANT Bars, built-in, front bar	1 Carpenter	5	1.600	L.F.
Back bar	″	5	1.600	L.F.

Figure 12.12 Installation Time for Furniture Accessories

Furniture Accessories	Crew Makeup	Daily Output	Man-Hours	Unit
ASH/TRASH RECEIVERS				
Ash urn, cylindrical metal				
8" diameter, 20" high	1 Building Laborer	60	.133	Ea.
8" diameter, 25" high		60	.133	Ea.
10" diameter, 26" high		60	.133	Ea.
12" diameter, 30" high	↓	60	.133	Ea.
Combination ash/trash urn, metal				
8" diameter, 20" high	1 Building Laborer	60	.133	Ea.
8" diameter, 25" high		60	.133	Ea.
10" diameter, 26" high		60	.133	Ea.
12" diameter, 30" high	↓	60	.133	Ea.
Trash receptacle, metal				
8" diameter, 15" high	1 Building Laborer	60	.133	Ea.
10" diameter, 18" high		60	.133	Ea.
16" diameter, 16" high		60	.133	Ea.
18" diameter, 32" high	↓	60	.133	Ea.
Trash receptacle, plastic, fire resistant				
Rectangular 11" x 8" x 12" high	1 Building Laborer	60	.133	Ea.
16" x 8" x 14" high	"	60	.133	Ea.
Trash receptacle, plastic, with lid				
35 gallon	1 Building Laborer	60	.133	Ea.
45 gallon	"	60	.133	Ea.

Figure 12.13 Installation Time for Floor Mats and Frames

Floor Mats and Frames	Crew Makeup	Daily Output	Man-Hours	Unit
FLOOR MATS Recessed, in-laid black rubber,				
3/8" thick, solid	1 Building Laborer	155	.052	S.F.
Perforated		155	.052	S.F.
1/2" thick, solid		155	.052	S.F.
Perforated		155	.052	S.F.
In colors, 3/8" thick, solid		155	.052	S.F.
Perforated		155	.052	S.F.
1/2" thick, solid		155	.052	S.F.
Perforated		155	.052	S.F.
Link mats, including nosings, aluminum, 3/8" thick		155	.052	S.F.
Black rubber with galvanized tie rods		155	.052	S.F.
Steel, galvanized, 3/8" thick		155	.052	S.F.
Vinyl, in colors		155	.052	S.F.
Recess frames for above mats, aluminum	1 Carpenter	100	.080	L.F.
Bronze	"	100	.080	L.F.
Skate lock tile, 24" x 24" x 1/2" thick, rubber, black	1 Building Laborer	125	.064	S.F.
Color		125	.064	S.F.
12" x 24" border, black		75	.107	L.F.
Color		75	.107	L.F.
12" x 12" outside corner, black		100	.080	S.F.
Color		100	.080	S.F.
Duckboard, aluminum slats		155	.052	S.F.
Hardwood strips on rubber base, to 54" wide		155	.052	S.F.
Assembled with brass rods & vinyl spacers,				
to 48" wide		155	.052	S.F.
Tire fabric, 3/4" thick		155	.052	S.F.
Vinyl, 36" wide, in colors, hollow top and bottoms		155	.052	S.F.
Solid top and bottom members		155	.052	S.F.

Figure 12.14 Installation Time for Miscellaneous Seating

Miscellaneous Seating	Crew Makeup	Daily Output	Man-Hours	Unit
SEATING				
Lecture hall, pedestal type, minimum	2 Carpenters	35	.457	Ea.
Maximum		20	.800	Ea.
Auditorium chair, all veneer construction		35	.457	Ea.
Veneer back, padded seat		35	.457	Ea.
Fully upholstered, spring seat		35	.457	Ea.

Figure 12.15   Installation Time for Booths and Tables

Booths and Tables	Crew Makeup	Daily Output	Man-Hours	Unit
BOOTHS				
Banquet, upholstered seat and back, custom				
Straight, minimum	2 Carpenters	40	.400	L.F.
Maximum		36	.444	L.F.
"L" or "U" shape, minimum		35	.457	L.F.
Maximum	↓	30	.533	L.F.
Upholstered outside finished backs for				
single booths and custom banquets				
Minimum	2 Carpenters	44	.364	L.F.
Maximum	"	40	.400	L.F.
Fixed seating, one piece plastic chair and				
plastic laminate table top				
Two seat, 24" x 24" table, minimum	2 Carpenters 2 Building Laborers Power Tools	30	1.070	Ea.
Maximum		26	1.230	Ea.
Four seat, 24" x 48" table, minimum		28	1.140	Ea.
Maximum		24	1.330	Ea.
Six seat, 24" x 76" table, minimum		26	1.230	Ea.
Maximum		22	1.450	Ea.
Eight seat, 24" x 102" table, minimum		20	1.600	Ea.
Maximum	↓	18	1.780	Ea.
Free standing, wood fiber core with				
plastic laminate face, single booth				
24" wide	2 Carpenters	38	.421	Ea.
48" wide		34	.471	Ea.
60" wide		30	.533	Ea.
Double booth, 24" wide		32	.500	Ea.
48" wide		28	.571	Ea.
60" wide	↓	26	.615	Ea.
Upholstered seat and back				
Foursome, single booth, minimum	2 Carpenters	38	.421	Ea.
Maximum		30	.533	Ea.
Double booth, minimum		32	.500	Ea.
Maximum	↓	26	.615	Ea.
Mount in floor, wood fiber core with				
plastic laminate face, single booth				
24" wide	2 Carpenters 2 Building Laborers Power Tools	30	1.070	Ea.
48" wide		28	1.140	Ea.
60" wide		26	1.230	Ea.
Double booth, 24" wide		26	1.230	Ea.
48" wide		24	1.330	Ea.
60" wide	↓	22	1.450	Ea.

Checklist

For an estimate to be reliable, all items must be accounted for.
A complete estimate can also eliminate the need to include
contingencies. The following checklist can be used to help ensure
that all items are properly accounted for.

Artwork _____

Interior Landscaping
☐ Guarantee

Cabinets _____

Floor Mats _____

Furniture
☐ Beds
☐ Dressers
☐ Chairs
☐ Chests
☐ Desks
☐ Sofas
☐ Tables
☐ Wardrobes

Seating
☐ Auditorium
☐ Classroom
☐ Stadium
☐ Theatre

Window Treatments
☐ Blinds
☐ Shades
☐ Other _____

_____

_____

_____

Tips

Faulty Assumptions

Do not assume that items covered in Division 12 will be purchased and installed outside of your contract, especially if any furnishings or typical arrangements are shown on the drawings. This could be a costly assumption. Check all drawings and specifications for these items and if any are found, check for the terms "NIC" (Not In Contract) or "By Others." If these or similar terms are not present, it is safe to conclude that these items are in your scope of work.

Installation of Items

In many cases, Division 12 items are purchased by others, but the contractor is responsible for them. Check all drawings and specifications for these items.

Handling Charges

In cases where Division 12 items are purchased by others but are to be installed by the contractor, many contractors add a handling charge. (10% of the estimated material cost is common practice.) This charge covers the receiving, handling, storage, protection, and final delivery of these items.

Preparation of Items

In some cases, Division 12 items may require some assembly before installation. The assembly time can often exceed the time required for installation.

Cleaning Time

When installing materials purchased by others, be sure to allow for cleaning time. Invariably, these items will need some cleaning and this responsibility will fall upon the installer.

Notes

Division Thirteen
Special Construction

Introduction

Division 13 — Special Construction deals with specialty subsystems. Items usually associated with this division can range from air-supported and pre-engineered structures to darkrooms.

The procedure for estimating this division starts with reading and reviewing the specifications. Usually if a Special Construction item is required, there will be some text about it in the specifications. The plans, as well as any other related construction document, should be thoroughly studied when Special Construction items are involved; they can contain important information.

Usually the items included in this division are purchased as a whole, complete, installed assembly, rather than installed in parts by one or more subcontractors. In many cases Special Construction systems are bought and installed by the owner. In some cases, the owner may purchase the unit and will have the contractor be responsible for the installation. In this case, the labor must be added to the overall estimate.

Estimating Data

The following tables show the expected installation time for various specialty items. Each table lists the specialty item, the typical installation crew mix, the expected average daily output of that crew (the number of units the crew can be expected to install in one eight-hour day), and the man-hours required for the installation of one unit.

Table of Contents

Figure 13.1 Installation Time for Air-Supported Structures

Air-Supported Structures	Makeup	Daily Output	Man-Hours	Unit
AIR-SUPPORTED STRUCTURES				
Site preparation, incl. anchor placement and utilities	1 Equipment Oper. (med.)	2,000	.008	S.F. Flr.
	1 Building Laborer			
	1 Dozer, 200 H.P.			
	1 Air Powered Tamper			
	1 Air Compr. 365 C.F.M.			
	2-50' Air Hoses, 1-1/2" Dia.			
Warehouse, polyester/vinyl fabric, 24 oz. over 10 yr. life, welded seams, tension cables, primary & auxiliary inflation system airlock, personnel doors and liner				
5000 S.F.	4 Building Laborers	5,000	.006	S.F. Flr.
12,000 S.F.	"	6,000	.005	S.F. Flr.
24,000 S.F.	8 Building Laborers	12,000	.005	S.F. Flr.
50,000 S.F.	"	12,500	.005	S.F. Flr.
12 oz. reinforced vinyl fabric, 5 yr. life, sewn seams, accordian door including liner, 3000 S.F.	4 Building Laborers	3,000	.011	S.F. Flr.
12,000 S.F.	"	6,000	.005	S.F. Flr.
24,000 S.F.	8 Building Laborers	12,000	.005	S.F. Flr.
Tedlar/vinyl fabric, 17 oz., with liner, over 10 yr. life, incl. overhead and personnel doors, 3000 S.F.	4 Building Laborers	3,000	.011	S.F. Flr.
12,000 S.F.	"	6,000	.005	S.F. Flr.
24,000 S.F.	8 Building Laborers	12,000	.005	S.F. Flr.
Greenhouse/shelter, woven polyethylene with liner, 2 yr. life, sewn seams, including doors				
3000 S.F.	4 Building Laborers	3,000	.011	S.F. Flr.
12,000 S.F.	"	6,000	.005	S.F. Flr.
24,000 S.F.	8 Building Laborers	12,000	.005	S.F. Flr.
Tennis/gymnasium, polyester/vinyl fabric 24 oz., over 10 yr. life, including thermal liner, heat and lights				
7200 S.F.	4 Building Laborers	6,000	.005	S.F. Flr.
13,000 S.F.	"	6,500	.005	S.F. Flr.
Over 24,000 S.F.	8 Building Laborers	12,000	.005	S.F. Flr.
Stadium/convention center, teflon coated fiberglass, heavy weight, over 20 yr. life incl. thermal liner and heating system				
Minimum	9 Building Laborers	26,000	.003	S.F. Flr.
Maximum	"	19,000	.004	S.F. Flr.
Doors, air lock, 15' long, 10' x 10'	2 Carpenters	.80	20.000	Ea.
15' x 15'		.50	32.000	Ea.
Revolving personnel door, 6' diameter, 6'- 6" high	↓	.80	20.000	Ea.

Figure 13.1 Installation Time for Air-Supported Structures (continued)

Air-Supported Structures	Makeup	Daily Output	Man-Hours	Unit
AIR-SUPPORTED STORAGE TANK COVERS Vinyl polyester scrim, double layer with hardware, blower, standby & controls Round, 75' diameter	1 Labor Foreman (outside) 4 Building Laborers	5,000	.008	S.F.
100' diameter		6,000	.007	S.F.
150' diameter		6,000	.007	S.F.
Rectangular, 20' x 20'		6,000	.007	S.F.
30' x 40'		6,000	.007	S.F.
50' x 60'	▼	6,000	.007	S.F.

Figure 13.2 Installation Time for Integrated Ceilings

Integrated Ceilings	Crew Makeup	Daily Output	Man-Hours	Unit
INTEGRATED CEILINGS Lighting, ventilating & acoustical Luminaire, incl. HVAC & light., 5' x 5' modules 50% lighted	1 Carpenter .5 Electrician .5 Sheet Metal Worker	90	.178	S.F.
100% lighted	▼	50	.320	S.F.
Dimensionaire, 2' x 4' board system		50	.320	L.F.
Integrated ceilings, dimensionaire, 2' x 4' tile system	1 Carpenter	250	.032	S.F.
Grid system & ceiling tile only, flat profile 5' x 5' modules, 50% lighted, mineral fiber panels	1 Carpenter	700	.011	S.F.
Glass fiber panels		700	.011	S.F.
For vaulted coffer, 5' x 5' modules, 50% lighted	▼	500	.016	S.F.
Radiant electric ceiling board, strapped between joists	1 Electrician	250	.032	S.F.
2' x 4' heating panel for grid system, manila finish		25	.320	Ea.
Textured epoxy finish		22	.364	Ea.
Vinyl finish		19	.421	Ea.
Hair cell, ABS plastic finish	▼	13	.615	Ea.
2' x 4' alternate blank panel, for use with above Manila finish	1 Electrician	50	.160	Ea.
Textured epoxy finish		45	.178	Ea.
Vinyl finish		40	.200	Ea.
Hair cell, ABS plastic finish	▼	25	.320	Ea.

Figure 13.3   Installation Time for Special Purpose Rooms

Special Purpose Rooms	Crew Makeup	Daily Output	Man-Hours	Unit
DARKROOMS Shell, complete except for door, 64 S.F.	2 Carpenters			
8' high		128	.125	S.F. Flr.
12' high		64	.250	S.F. Flr.
120 S.F., floor, 8' high		120	.133	S.F. Flr.
12' high		60	.267	S.F. Flr.
240 S.F. floor, 8' high		120	.133	S.F. Flr.
12' high		60	.267	S.F. Flr.
Mini-cylindrical, revolving, unlined, 4' diameter		3.50	4.570	Ea.
5'-6" diameter		2.50	6.400	Ea.

Figure 13.4   Installation Time for Athletic Rooms

Athletic Rooms	Crew Makeup	Daily Output	Man-Hours	Unit
SPORT COURT				
Rule of thumb for components:				
Walls	3 Carpenters	.15	160.000	Court
Floor	"	.25	96.000	Court
Lighting	2 Electricians	.60	26.670	Court
Handball, racquetball court in existing building				
Minimum	3 Carpenters	.20	160.000	Court
	1 Building Laborer			
	Power Tools			
Maximum	"	.10	320.000	Court
Rule of thumb for components: walls	3 Carpenters	.12	200.000	Court
Floor		.25	96.000	Court
Ceiling		.33	72.730	Court
Lighting	2 Electricians	.60	26.670	Court

Figure 13.5   Installation Time for Audiometric Rooms

Audiometric Rooms	Crew Makeup	Daily Output	Man-Hours	Unit
AUDIOMETRIC ROOMS Under 500 S.F. surface	4 Carpenters	98	.327	S.F. Surf.
Over 500 S.F. surface	"	120	.267	S.F. Surf.

Figure 13.6 Installation Time for Cold Storage Rooms

Cold Storage Rooms	Crew Makeup	Daily Output	Man-Hours	Unit
REFRIGERATORS Curbs, 12" high, 4" thick, concrete	2 Carpenters	58	.276	L.F.
Finishes, 2 coat Portland cement plaster, 1/2" thick	1 Plasterer	48	.167	S.F.
For galvanized reinforcing mesh, add	1 Lather	335	.024	S.F.
3/16" thick latex cement	1 Plasterer	88	.091	S.F.
For glass cloth reinforced ceilings, add	"	450	.018	S.F.
Fiberglass panels, 1/8" thick	1 Carpenter	149.45	.054	S.F.
Polystyrene, plastic finish ceiling, 1" thick		274	.029	S.F.
2" thick		274	.029	S.F.
4" thick	↓	219	.037	S.F.
Floors, concrete, 4" thick	1 Cement Finisher	93	.086	S.F.
6" thick	"	85	.094	S.F.
Partitions, galv. sandwich panels, 4" thick, stock	2 Carpenters	219.20	.073	S.F.
Aluminum or fiberglass, 4" thick		219.20	.073	S.F.
Prefab walk-in, 7'-6" high, aluminum, incl. door & floors, not incl. partitions or refrigeration				
6' x 6' O.D. nominal		54.80	.292	S.F. Flr.
10' x 10' O.D. nominal		82.20	.195	S.F. Flr.
12' x 14' O.D. nominal		109.60	.146	S.F. Flr.
12' x 20' O.D. nominal		109.60	.146	S.F. Flr.
Rule of thumb for complete units, not incl. doors				
Cooler		146	.110	S.F. Flr.
Freezer		109.60	.146	S.F. Flr.
Shelving, plated or galvanized, steel wire type		360	.044	S.F. Hor.
Slat shelf type		375	.043	S.F. Hor.
Vapor barrier, on wood walls		1,644	.010	S.F.
On masonry walls	↓	1,315	.012	S.F.

Figure 13.7 Installation Time for Saunas

Saunas	Crew Makeup	Daily Output	Man-Hours	Unit
SAUNA Prefabricated, incl. heater & controls, 7' high				
6' x 4'	2 Carpenters 1 Building Laborer .5 Electrician	2.20	12.730	Ea.
6' x 5'		2	14.000	Ea.
6' x 6'		1.80	15.560	Ea.
6' x 9'		1.60	17.500	Ea.
8' x 12'		1.10	25.450	Ea.
8' x 8'		1.40	20.000	Ea.
8' x 10'		1.20	23.330	Ea.
10' x 12'	↓	1	28.000	Ea.
Door only, with tempered insulated glass window	2 Carpenters	3.40	4.710	Ea.
Prehung, incl. jambs, pulls & hardware	"	12	1.330	Ea.

Figure 13.8   Installation Time for Steam Baths

Steam Baths	Crew Makeup	Daily Output	Man-Hours	Unit
STEAM BATH Heater, timer & head, single, to 140 C.F.	1 Plumber	1.20	6.670	Ea.
To 300 C.F.		1.10	7.270	Ea.
Commercial size, to 800 C.F.		.90	8.890	Ea.
To 2500 C.F.	↓	.80	10.000	Ea.
Multiple baths, motels, apartment, 2 baths	1 Plumber	1.30	12.310	Ea.
	1 Plumber Apprentice			
4 baths	"	.70	22.860	Ea.

Figure 13.9   Installation Time for Acoustical Enclosures

Acoustical Enclosures	Crew Makeup	Daily Output	Man-Hours	Unit
ACOUSTICAL Enclosure, 4" thick wall and ceiling panels				
8# per S.F., up to 12' span	3 Carpenters	72	.333	S.F. Surf.
Better quality panels, 10.5# per S.F.		64	.375	S.F. Surf.
Reverb-chamber, 4" thick, parallel walls		60	.400	S.F. Surf.
Skewed walls, parallel roof, 4" thick panels		55	.436	S.F. Surf.
Skewed walls, skewed roof, 4" layers, 4" air space		48	.500	S.F. Surf.
Sound-absorbing panels, painted metal, 2'-6" x 8'				
Under 1000 S.F.		215	.112	S.F. Surf.
Over 2400 S.F.		240	.100	S.F. Surf.
Fabric faced	↓	240	.100	S.F. Surf.
Flexible transparent curtain, clear	3 Sheet Metal Workers	215	.112	S.F. Surf.
50% foam		215	.112	S.F. Surf.
75% foam		215	.112	S.F. Surf.
100% foam	↓	215	.112	S.F. Surf.
Audio masking system, including speakers,				
amplification and signal generator				
Ceiling mounted, 5000 S.F.	2 Electricians	2,400	.007	S.F.
10,000 S.F.		2,800	.006	S.F.
Plenum mounted, 5000 S.F.		3,800	.004	S.F.
10,000 S.F.	↓	4,400	.004	S.F.
MUSIC Practice room, modular, perforated steel				
Under 500 S.F.	2 Carpenters	70	.229	S.F. Surf.
Over 500 S.F.	"	80	.200	S.F. Surf.

Figure 13.10 Installation Time for Radiation Protection

Radiation Protection	Crew Makeup	Daily Output	Man-Hours	Unit
SHIELDING LEAD				
Lead lined door frame, not incl. steel frame				
or hardware, 1/16" thick	1 Lather	2.40	3.330	Ea.
Lead lath or sheets, 1/16" thick	2 Lathers	135	.119	S.F.
1/8" thick		120	.133	S.F.
Lead glass, 1/4" thick, 12" x 16"		13	1.230	Ea.
24" x 36"		8	2.000	Ea.
36" x 60"		2	8.000	Ea.
Frame with 1/16" lead and voice passage				
36" x 60"		2	8.000	Ea.
24" x 36" frame		8	2.000	Ea.
Lead gypsum board, 5/8" thick with 1/16" lead		160	.100	S.F.
1/8" lead		140	.114	S.F.
1/32" lead		200	.080	S.F.
Butt joints in 1/8" lead or thicker, lead strip, add		240	.067	S.F.
X-ray protection, average radiography or				
fluoroscopy room, up to 300 S.F. floor, 1/16" lead				
Minimum	2 Lathers	.25	64.000	Total
Maximum, 7' walls	"	.15	107.000	Total
Deep therapy X-ray room, 250 KV capacity,				
up to 300 S.F. floor, 1/4" lead, minimum	2 Lathers	.08	200.000	Total
Maximum, 7' walls	"	.06	267.000	Total
SHIELDING, RADIO FREQUENCY				
Prefabricated or screen-type copper or steel				
Minimum	2 Carpenters	180	.089	S.F. Surf.
Average		155	.103	S.F. Surf.
Maximum		145	.110	S.F. Surf.

Figure 13.11 Installation Time for Pre-Engineered Structures

Pre-Engineered Structures	Crew Makeup	Daily Output	Man-Hours	Unit
DOMES Revolving aluminum, electric drive, for astronomy observation, shell only,				
Stock units, 10′ diameter, 800#, dome	2 Carpenters	.25	64.000	Ea.
Base		.67	23.880	Ea.
18′ diameter, 2500#, dome		.17	94.120	Ea.
Base		.33	48.480	Ea.
24′ diameter, 4500#, dome		.08	200.000	Ea.
Base		.25	64.000	Ea.
Bulk storage, shell only, dual radius hemispher. arch, steel framing, corrugated steel covering, 150′ diameter	1 Struc. Steel Foreman 4 Struc. Steel Workers 1 Equip. Oper. (crane) 1 Equip. Oper. Oiler 1 Crane, 90 Ton	550	.102	S.F. Flr.
400′ diameter	"	720	.078	S.F. Flr.
Wood framing, wood decking, to 400′ diameter	4 Carpenters 1 Equip. Oper. (crane) 1 Equip. Oper. Oiler 1 Hyd. Crane, 55 Ton Power Tools	400	.120	S.F. Flr.
Radial framed wood (2″ x 6″), 1/2″ thick plywood, asphalt shingles				
50′ diameter	4 Carpenters 1 Equip. Oper. (crane) 1 Hyd. Crane, 12 Ton Power Tools	2,000	.020	S.F. Flr.
60′ diameter		1,900	.021	S.F. Flr.
72′ diameter		1,800	.022	S.F. Flr.
116′ diameter		1,730	.023	S.F. Flr.
150′ diameter		1,500	.027	S.F. Flr.

Figure 13.11 Installation Time for Pre-Engineered Structures (continued)

Pre-Engineered Structures	Crew Makeup	Daily Output	Man-Hours	Unit
GEODESIC DOME Shell only, interlocking plywood panels				
30' diameter	1 Carpenter Foreman 3 Carpenters Power Tools	1.60	20.000	Ea.
35' diameter	↓	1.14	28.070	Ea.
39' diameter		1	32.000	Ea.
45' diameter	↓	.90	35.560	Ea.
60' diameter	4 Carpenters 1 Equip. Oper. (crane) 1 Hyd. Crane, 12 Ton Power Tools	.40	100.000	Ea.
Aluminum panel, stressed skin, with 1-1/2" insulation				
82' diameter	1 Struc. Steel Foreman 5 Struc. Steel Workers 1 Equip. Oper. (crane) 1 Hyd. Crane, 25 Ton	900	.062	S.F. Flr.
232' diameter	"	1,300	.043	S.F. Flr.
Aluminum framed, plexiglass closure panels, 40' diameter	1 Struc. Steel Foreman 1 Struc. Steel Worker 1 Truck Driver (light) 1 Truck w/Power Equip.	250	.096	S.F. Flr.
200' diameter	1 Struc. Steel Foreman 5 Struc. Steel Workers 1 Equip. Oper. (crane) 1 Hyd. Crane, 25 Ton	1,000	.056	S.F. Flr.
Aluminum framed, aluminum closure panels, 40' diameter	1 Struc. Steel Foreman 1 Struc. Steel Worker 1 Truck Driver (light) 1 Truck w/Power Equip.	500	.048	S.F. Flr.
320' diameter	2 Skilled Worker Foremen 8 Skilled Workers .375 Equip. Oper. (crane) .375 Crane, 80 Ton & Tools .375 Hand Held Power Tools .375 Walk Behind Power Tools	1,900	.044	S.F. Flr.
415' diameter		2,300	.036	S.F. Flr.
Aluminum framed, fiberglass sandwich panel closure 6' diameter	2 Carpenters Power Tools	150	.107	S.F. Flr.
28' diameter	"	350	.046	S.F. Flr.

(continued on next page)

Figure 13.11 Installation Time for Pre-Engineered Structures (continued)

Pre-Engineered Structures	Crew Makeup	Daily Output	Man-Hours	Unit
GARAGES Residential, prefab shell, stock, wood, single car				
Minimum	2 Carpenters Power Tools	1	16.000	Total
Maximum	|	.67	23.880	Total
Two car, minimum	|	.67	23.880	Total
Maximum	↓	.50	32.000	Total
SILOS Concrete stave industrial, not incl. foundations, conical or sloping bottoms				
12' diameter, 35' high	3 Bricklayers 2 Bricklayer Helpers	.11	364.000	Ea.
16' diameter, 45' high	|	.08	500.000	Ea.
25' diameter, 75' high	↓	.05	800.000	Ea.
Steel, factory fab., 30,000 gallon capacity, painted, minimum	1 Struc. Steel Foreman 5 Struc. Steel Workers 1 Equip. Oper. (crane) 1 Hyd. Crane, 25 Ton	1	56.000	Ea.
Maximum	|	.50	112.000	Ea.
Epoxy lined, minimum	|	1	56.000	Ea.
Maximum	↓	.50	112.000	Ea.
TENSION STRUCTURES Rigid steel frame, vinyl coated polyester fabric shell, 72' clear span, not incl. foundations or floors				
4800 S.F.	1 Labor Foreman (outside) 4 Building Laborers .25 Equip. Oper. (crane) .25 Equip. Oper. Oiler .25 Crawler Crane, 40 Ton	1,000	.044	S.F. Flr.
12,000 S.F.	|	1,100	.040	S.F. Flr.
20,600 S.F.	↓	1,220	.036	S.F. Flr.
124' clear span, 11,000 S.F.	1 Struc. Steel Forman 5 Struc. Steel Workers 1 Equip. Oper. (crane) 1 Hyd. Crane, 25 Ton	2,175	.026	S.F. Flr.
25,750 S.F.	|	2,300	.024	S.F. Flr.
36,900 S.F.	↓	2,500	.022	S.F. Flr.
For roll-up door, 12' x 14' add	1 Carpenter 1 Helper	1	16.000	Ea.

Figure 13.12   Installation Time for Pre-Engineered Buildings

Pre-Engineered Buildings	Crew Makeup	Daily Output	Man-Hours	Unit
HANGARS Prefabricated steel T hangars Galv. steel roof & walls, incl. electric bi-folding doors, 4 or more units, not including floors or foundations, minimum	1 Struc. Steel Foreman 4 Struc. Steel Workers 1 Equip. Oper. (crane) 1 Equip. Oper. Oiler 1 Crane, 90 Ton	1,275	.044	S.F. Flr.
Maximum		1,063	.053	S.F. Flr.
With bottom rolling doors, minimum		1,386	.040	S.F. Flr.
Maximum	↓	966	.058	S.F. Flr.
Alternate pricing method: Galv. roof and walls, electric bi-folding doors, minimum	1 Struc. Steel Foreman 4 Struc. Steel Workers 1 Equip. Oper. (crane) 1 Equip. Oper. Oiler 1 Crane, 90 Ton	1.06	52.830	Plane
Maximum		.91	61.540	Plane
With bottom rolling doors, minimum		1.25	44.800	Plane
Maximum	↓	.97	57.730	Plane
Circular type, prefab., steel frame, plastic skin, electric door, including foundations, 80′ diameter, for up to 5 light planes Minimum	1 Struc. Steel Foreman 4 Struc. Steel Workers 1 Equip. Oper. (crane) 1 Equip. Oper. Oiler 1 Crane, 90 Ton	.50	112.000	Total
Maximum	"	.25	224.000	Total

Figure 13.13   Installation Time for Metal Building Systems

Metal Building Systems	Crew Makeup	Daily Output	Man-Hours	Unit
SHELTERS Aluminum frame, acrylic glazing 3′ x 9′ x 8′ high	2 Struc. Steel Workers	2	8.000	Ea.
9′ x 12′ x 8′ high	"	1	16.000	Ea.

Figure 13.14 Installation Time for Greenhouses

Greenhouses	Crew Makeup	Daily Output	Man-Hours	Unit

GREENHOUSE Shell only, stock units
    not incl. 2' stud walls, foundation, floors
    heat or compartments
Residential type, free standing, 8'-6" long				

7'-6" wide	2 Carpenters	59	.271	S.F. Flr.
10'-6" wide		85	.188	S.F. Flr.
13'-6" wide		108	.148	S.F. Flr.
17'-0" wide		160	.100	S.F. Flr.
Lean-to type, 3'-10" wide		34	.471	S.F. Flr.
6'-10" wide	↓	58	.276	S.F. Flr.
Geodesic hemisphere, 1/8" plexiglass glazing				
8' diameter	2 Carpenters	2	8.000	Ea.
24' diameter		.35	45.710	Ea.
48' diameter	↓	.20	80.000	Ea.

Figure 13.15 Installation Time for Portable Buildings

Portable Buildings	Crew Makeup	Daily Output	Man-Hours	Unit

COMFORT STATIONS Prefabricated, stock,
    with doors, windows & fixtures
Not incl. interior finish or electrical
    Permanent, including concrete slab				

Minimum	1 Equip. Oper. (crane) 1 Equip. Oper. Oiler 1 Gradall, 3 Ton, .5 C.Y.	50	.320	S.F.
Maximum	"	43	.372	S.F.
GARDEN HOUSE Prefabricated wood, no floors or foundations, 48 to 200 S.F.				
Minimum	2 Carpenters Power Tools	200	.080	S.F. Flr.
Maximum	"	48	.333	S.F. Flr.

Figure 13.16 Installation Time for Swimming Pools

Swimming Pools	Crew Makeup	Daily Output	Man-Hours	Unit
SWIMMING POOL ENCLOSURE Translucent, free standing, not incl. foundations, heat or light				
Economy, minimum	2 Carpenters	200	.080	S.F. Hor.
Maximum		100	.160	S.F. Hor.
Deluxe, minimum		100	.160	S.F. Hor.
Maximum	↓	70	.229	S.F. Hor.
SWIMMING POOL EQUIPMENT Diving stand, stainless steel, 3 meter		.40	40.000	Ea.
1 meter		2.70	5.930	Ea.
Diving boards, 16' long, aluminum		2.70	5.930	Ea.
Fiberglass	↓	2.70	5.930	Ea.
Filter system, sand or diatomite type, incl. pump				
6000 gallons per hour	2 Plumbers	1.80	8.890	Total
Add for chlorination system, 800 S.F. pool		3	5.330	Ea.
5000 S.F. pool	↓	3	5.330	Ea.
Gutter system, stainless steel, with grating, stock,				
contains supply and drainage system	1 Welder Foreman 1 Welder 1 Equip. Oper. (light) 1 Gas Welding Mach.	20	1.200	L.F.
Integral gutter and 5' high wall system, stainless steel	"	10	2.400	L.F.
Ladders, heavy duty, stainless steel, 2 tread	2 Carpenters	7	2.290	Ea.
4 tread		6	2.670	Ea.
Lifeguard chair, stainless steel, fixed	↓	2.70	5.930	Ea.
Lights, underwater, 12 volts, with transformer				
300 watt	1 Electrician	.40	20.000	Ea.
110 volt, 500 watt, standard		.40	20.000	Ea.
Low water cutoff type	↓	.40	20.000	Ea.
Slides, fiberglass, aluminum handrails & ladder				
6'-0", straight	2 Carpenters	1.60	10.000	Ea.
7'-6", curved		3	5.330	Ea.
10'-6", curved		1	16.000	Ea.
12'-0", straight with platform	↓	1.20	13.330	Ea.
Hydraulic lift, movable pool bottom, single ram				
Under 1000 S.F. area	1 Labor Foreman (ins.) 2 Building Laborers 1 Struc. Steel Worker .5 Electrician	.03	200.000	Ea.
Four ram lift, over 1000 S.F.	"	.02	800.000	Ea.
Removable access ramp, stainless steel	2 Building Laborers	2	8.000	Ea.
Removable stairs, stainless steel, collapsible	"	2	8.000	Ea.

Figure 13.17   Installation Time for Ice Rinks

Ice Rinks	Crew Makeup	Daily Output	Man-Hours	Unit
ICE SKATING Dasher boards, polyethylene coated plywood 3' acrylic screen at sides, 5' acrylic ends, 85' x 200'	1 Carpenter Foreman 3 Carpenters Power Tools	.07	457.000	Ea.
Fiberglass & aluminum construction, same sides and ends	"	.07	457.000	Ea.
Subsoil heating system (recycled from compressor) 85' x 200'	1 Steamfitter Foreman (ins.) 2 Steamfitters 1 Steamfitter Apprentice	.30	107.000	Ea.
Subsoil insulation, 2 lb. polystyrene with vapor barrier 85' x 200'	2 Carpenters Power Tools	.16	100.000	Ea.

Figure 13.18   Installation Time for Ground Storage Tanks

Ground Storage Tanks	Crew Makeup	Daily Output	Man-Hours	Unit
TANKS Not incl. pipe or pumps Wood tanks, ground level, 2" cypress, 3000 gallons	3 Carpenters 1 Building Laborer Power Tools	.19	168.000	Ea.
2-1/2" cypress, 10,000 gallons		.12	267.000	Ea.
3" redwood or 3" fir, 20,000 gallons		.10	320.000	Ea.
30,000 gallons		.08	400.000	Ea.
45,000 gallons	▼	.07	457.000	Ea.
Vinyl coated fabric pillow tanks, freestanding 5000 gallons	4 Building Laborers	4	8.000	Ea.
Supporting embankment not included 25,000 gallons	6 Building Laborers	2	24.000	Ea.
50,000 gallons	8 Building Laborers	1.50	42.670	Ea.
100,000 gallons	9 Building Laborers	.90	80.000	Ea.
150,000 gallons		.50	144.000	Ea.
200,000 gallons		.40	180.000	Ea.
250,000 gallons	▼	.30	240.000	Ea.

Figure 13.19   Installation Time for Power Control Systems

Power Control Systems	Crew Makeup	Daily Output	Man-Hours	Unit
RADIO TOWERS Guyed, 50' high, 40 lb. section,				
wind load 30 psf	2 Struc. Steel Workers	1	16.000	Ea.
wind load 50 psf	"	1	16.000	Ea.
200' high, 40 lb. section, wind load 30 psf	1 Struc. Steel Foreman	.33	72.730	Ea.
	1 Struc. Steel Worker			
	1 Truck Driver (light)			
	1 Truck w/Power Equip.			
70 lb. section, wind load 50 psf		.33	72.730	Ea.
300' high, 70 lb. section, wind load 30 psf		.20	120.000	Ea.
90 lb. section, wind load 50 psf		.20	120.000	Ea.
400' high, 90 lb. section, 30 psf wind load		.14	171.000	Ea.
Self-supporting, 30 psf wind load, 60' high		.80	30.000	Ea.
120' high		.40	60.000	Ea.
200' high		.20	120.000	Ea.

Checklist

For an estimate to be reliable, all items must be accounted for. A complete estimate can also eliminate the need to include contingencies. The following checklist can be used to help ensure that all items are properly accounted for.

- ☐ Acoustical Chambers
  - ____ Enclosures
  - ____ Panels
- ☐ Air curtains
- ☐ Air supported structures
- ☐ Bath houses
- ☐ Bowling alleys
- ☐ Broadcast studio
- ☐ Changing houses/cabanas
- ☐ Chimneys
- ☐ Clean rooms
- ☐ Comfort stations
- ☐ Darkrooms
- ☐ Domes
- ☐ Garages
- ☐ Garden house, shed
- ☐ Grandstand
- ☐ Greenhouses
- ☐ Hangars
- ☐ Hyperbaric rooms
- ☐ Incinerators
- ☐ Insulated rooms
  - ____ Coolers
  - ____ Freezers
- ☐ Integrated ceilings
- ☐ Jacuzzis
- ☐ Music practice rooms
- ☐ Pedestal floors
- ☐ Pre-engineered structures
- ☐ Pre-fabricated structures
- ☐ Radiation protection/containment
  - ____ Nuclear materials room
  - ____ Radiological room
  - ____ X-ray room

☐ Radio frequency shielding
☐ Radio towers
☐ Saunas/steam rooms
☐ Silos
☐ Squash/handball/racquetball rooms
☐ Storage vaults
☐ Swimming pool enclosure
☐ Swimming pools
☐ Tanks
☐ Therapeutic pools
☐ Vault fronts
☐ Zoo structures
☐ Other _____

_____

_____

_____

_____

Tips

Covering All Bases

If you take outside quotes as an aid to estimating Division 13, review the scope of work covered in those quotes. It may be that the outside agent has wrongfully assumed that you will provide traditionally supplied items such as excavation or an unloading crane. Check to make sure that it is covered somewhere in your estimate, and that it is not carried by both parties.

Interfaces

Review all documentation to ensure that all interfaces, such as electrical connections and control wiring, are accounted for. These items have a habit of falling through the cracks of an estimate.

Division Fourteen

Conveying
Systems

Introduction

Division 14 — Conveying Systems is concerned with inter/intra-building transportation devices, such as elevators, handicap lifts, escalators, dumbwaiters, correspondence lifts, pneumatic tube systems, moving walkways and ramps, material handling systems, hoists and cranes, parcel lifts, and vertical conveyors.

Passenger Elevators

Electric elevators are the most common, but hydraulic elevators can also be used for lifts up to 70' and where large capacities are required. Hydraulic speeds are limited to 150 F.P.M., but cars are self-leveling at the stops. On low rises, hydraulic installation runs about 15% less than standard electric elevators, but on higher rises this installation cost advantage is reduced. Maintenance of hydraulic elevators is about the same as for electric but the underground portion is not included in the maintenance contract.

In standard electric elevators, there are two basic control systems: rheostatic systems for speeds up to 150 F.P.M. and variable voltage systems for speeds over 150 F.P.M. The two types of drives are geared for low speeds and gearless for 450 F.P.M. and over. As a rule of thumb, each added 100 F.P.M. adds about 20% to the total cost.

Freight Elevators

Freight elevator capacities run from 1,500 lbs. to over 100,000 lbs., with 3,000 to 10,000 lb. capacities the most common. Travel speeds are generally slower and control systems less intricate than on passenger elevators.

Escalators, Moving Stairs

Escalators are often used in buildings where 600 or more people will be traveling to the second floor or beyond on a daily basis. Freight cannot be carried on escalators; therefore, at least one elevator must be available for this function. The carrying capacity of an escalator is 5,000 to 8,000 people per hour. The power requirement is 2 to 3 KW per hour, and the required incline angle is 30°.

Estimating Data

The following tables and illustrations provide selection and installation data for conveying systems. Also included is a worksheet that can be used to help size and price elevators, and organize data for the estimate.

Table of Contents

Figure 14.1   Elevator Nomenclature and Comparison of Elevator Types

This chart illustrates the types of elevators available and the common terminology used with elevators.

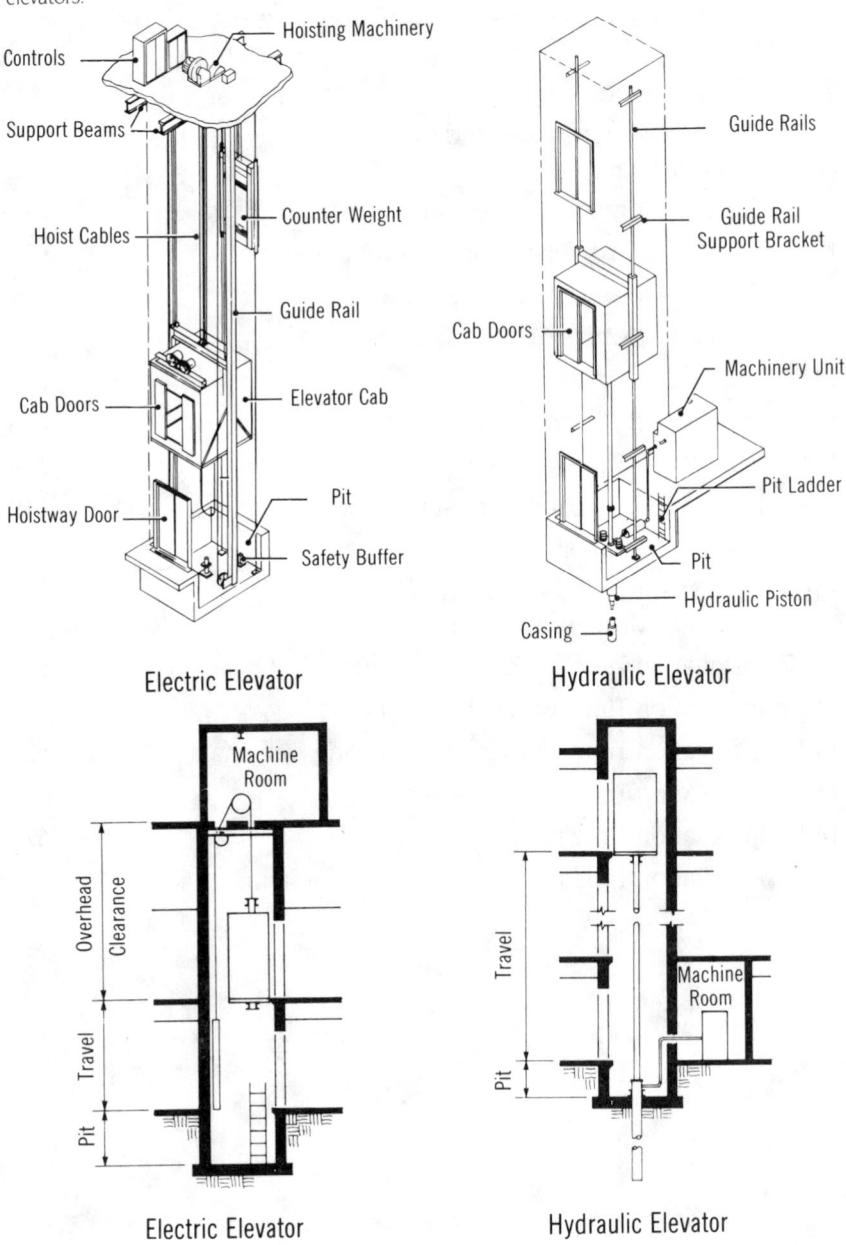

Controls

Hoisting Machinery

Support Beams

Hoist Cables

Counter Weight

Guide Rail

Cab Doors

Elevator Cab

Hoistway Door

Pit

Safety Buffer

Guide Rails

Guide Rail Support Bracket

Cab Doors

Machinery Unit

Pit Ladder

Pit

Hydraulic Piston

Casing

Electric Elevator

Hydraulic Elevator

Machine Room

Overhead Clearance

Travel

Pit

Electric Elevator

Travel

Pit

Machine Room

Hydraulic Elevator

Figure 14.2   Cab Size Comparisons for Various Types of Elevators

This chart shows size comparisons for commonly encountered elevator cabs.

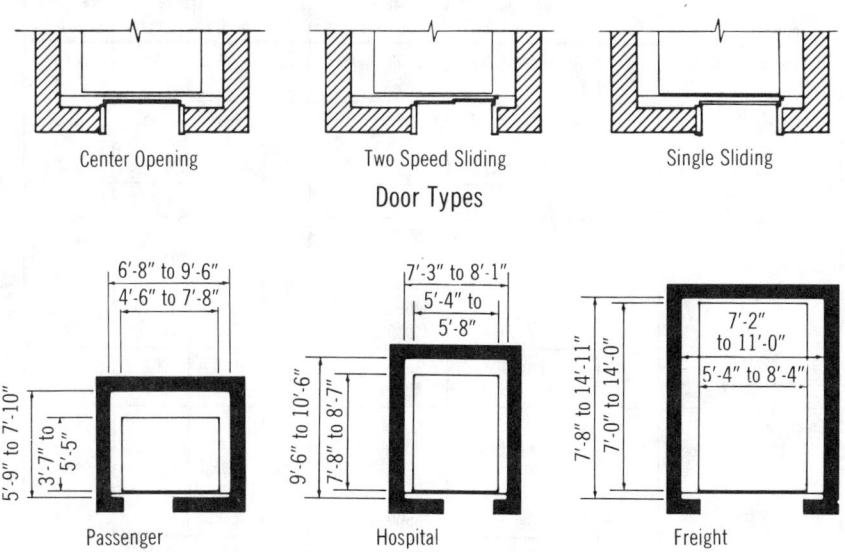

Center Opening   Two Speed Sliding   Single Sliding

Door Types

Passenger   Hospital   Freight

Figure 14.3   Elevator Speed Vs. Height Requirements

Building Type and Elevator Capacities	Travel Speeds in Feet per Minute								
	100 fpm	200 fpm	250 fpm	350 fpm	400 fpm	500 fpm	700 fpm	800 fpm	1000 fpm
Apartments 1200 lb. to 2500 lb.	to 70	to 100	to 125	to 150	to 175	to 250	to 350		
Department Stores 1200 lb. to 2500 lb.		100		125	175	250	350		
Hospitals 3500 lb. to 4000 lb.	to 70	100	125	150	175	250	350		
Office Buildings 2000 lb. to 4000 lb.		100	125	150		175	250	to 350	over 350

Note: Vertical transportation capacity may be determined by code occupancy requirements of the number of square feet per person divided into the total square feet of building type. If we are contemplating an office building, we find that the Occupancy Code Requirement is 100 S.F. per person. For a 20,000 S.F. building, we would have a legal capacity of two hundred people. Elevator handling capacity is subject to the five minute evacuation recommendation, but it may vary from 11 to 18%. Speed required is a function of the travel height, number of stops and capacity of the elevator.

Figure 14.4   Elevator Hoistway Size Requirements

This table lists and compares the common types of elevators, including their limitations, common usage, capacities, door/entry type, standard dimensions, and area per floor required.

Elevator Type	Floors	Building Type	Capacity Lbs.	Capacity Passengers	Entry *	Hoistway Width	Hoistway Depth	S.F. Area per Floor
Hydraulic	5	Apt./Small Office	1500	10	S	6'-7"	4'-6"	29.6
			2000	13	S	7'-8"	4'-10"	37.4
	7	Av. Office/Hotel	2500	16	S	8'-4"	5'-5"	45.1
			3000	20	S	8'-4"	5'-11"	49.3
		Lg. Office/Store	3500	23	S	8'-4"	6'-11"	57.6
		Freight Light Duty	2500		D	7'-2"	7'-10"	56.1
			5000		D	10'-2"	10'-10"	110.1
			7500		D	10'-2"	12'-10"	131
		Heavy Duty	5000		D	10'-2"	10'-10"	110.1
			7500		D	10'-2"	12'-10"	131
			10,000		D	10'-4"	14'-10"	153.3
		Hospital	3500		D	6'-10"	9'-2"	62.7
			4000		D	7'-4"	9'-6"	69.6
Electric Traction, High Speed	High Rise	Apt./Small Office	2000	13	S	7'-8"	5'-10"	44.8
			2500	16	S	8'-4"	6'-5"	54.5
			3000	20	S	8'-4"	6'-11"	57.6
			3500	23	S	8'-4"	7'-7"	63.1
		Store	3500	23	S	9'-5"	6'-10"	64.4
		Large Office	4000	26	S	9'-4"	7'-6"	70
		Hospital	3500		D	7'-6"	9'-2"	69.4
			4000		D	7'-10"	9'-6"	58.8
Geared, Low Speed		Apartment	1200	8	S	6'-4"	5'-3"	33.2
			2000	13	S	7'-8"	5'-8"	43.5
			2500	16	S	8'-4"	6'-3"	52
		Office	3000	20	S	8'-4"	6'-9"	56
		Store	3500	23	S	9'-5"	6'-10"	64.4
			Add 4" width for multiple units					

* S = Single Door   * D = Double Door

Elevator Passenger Capacity During Peak Periods

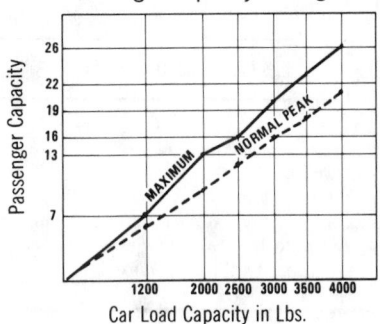

Car Load Capacity in Lbs.

Speed & Travel Chart

All speeds are available with any capacity car

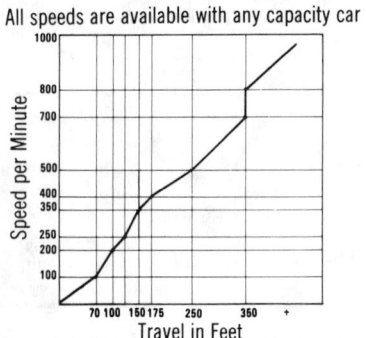

Travel in Feet

Figure 14.5   Elevator Selective Cost Worksheet

This worksheet can be used as a checklist when talking to elevator contractors or in conjunction with the Means *Building Construction Cost Data* Book section on elevator costs. As you discuss costs for the various requirements for your project, this form can be filled in for quick tabulation and solid documentation.

	Passenger		Freight		Hospital	
A. Base Unit	Hydraulic	Electric	Hydraulic	Electric	Hydraulic	Electric
Capacity	1500 Lb.	2000 Lb.	2000 Lb.	4000 Lb.	3500 Lb.	4000 Lb.
Speed	50 F.P.M.	200 F.P.M.	25 F.P.M.	50 F.P.M.	50 F.P.M.	200 F.P.M.
#Stops/Travel Ft.	2/12	4/40	2/20	4/40	2/20	4/40
Push Button Oper.	Yes	Yes	Yes	Yes	Yes	Yes
Telephone Box & Wire	"	"	"	"	"	"
Emergency Lighting	"	"	No	No	"	"
Cab	Plastic Lam. Walls	Plastic Lam. Walls	Painted Steel	Painted Steel	Plastic Lam. Walls	Plastic Lam. Walls
Cove Lighting	Yes	Yes	No	No	Yes	Yes
Floor	V.C.T.	V.C.T.	Wood w/Safety Treads	Wood w/Safety Treads	V.C.T.	V.C.T.
Doors, & Speedside Slide	Yes	Yes	No	No	Yes	Yes
Gates, Manual	No	No	Yes	Yes	No	No
Signals, Lighted Buttons	Car and Hall	Car and Hall	In Use Light	In Use Light	Car and Hall	Car and Hall
O.H. Geared Machine	N.A.	Yes	N.A.	Yes	N.A.	Yes
Variable Voltage Contr.	"	"	"	"	"	"
Emergency Alarm	Yes	"	Yes	"	Yes	"
Class "A" Loading	N.A.	N.A.	"	"	N.A.	N.A.
Base Cost	$	$	$	$	$	$
B. Capacity Adjustment						
2,000 Lb.	$	—	—	—	—	—
2,500	$	$	$	—	—	—
3,000	$	$	$	—	—	—
3,500	$	$	$	—	—	—
4,000	$	$	$	—	$	—
4,500	$	$	$	—	$	$
5,000	$	$	$	$	$	$
6,000	—	—	$	$	—	—
7,000	—	—	$	$	—	—
8,000	—	—	$	$	—	—
10,000	—	—	$	$	—	—
12,000	—	—	$	$	—	—
16,000	—	—	$	$	—	—
20,000	—	—	$	$	—	—
C. Travel Over Base	$ V.L.F.	$ V.L.F.	$ V.L.F.	$ V.L.F.	$ V.L.F.	$ V.L.F.
D. Additional Stops	$ Ea.	$ Ea.	$ Ea.	$ Ea.	$ Ea.	$ Ea.
E. Speed Adjustment						
50 F.P.M.	—	—	$	—	—	—
75	$	—	$	$	$	—
100	$	—	$	$	$	—
125	$	—	$	$	$	—
150	$	—	$	$	$	—
175	$	—	$	$	$	—
200	$	—	$	$	—	—
Geared 250	—	$	—	$	—	$
4 Flrs. 300	—	$	—	$	—	$
Min. 350	—	$	—	$	—	$
400	—	$	—	$	—	$
500	—	$	—	$	—	$
Gearless 600	—	$	—	$	—	$
10 Flrs. 700	—	$	—	$	—	$
Min. 800	—	$	—	$	—	$
1,000	—	Spec. Applic.	—	Spec. Applic.	—	Spec. Applic.
1,200	—	"	—	"	—	"
F. Other Than Class						
"A" Loading						
"B"	—	—	$	$	—	—
"C-1"	—	—	$	$	—	—
"C-2"	—	—	$	$	—	—
"C-3"	—	—	$	$	—	—

(continued on next page)

Figure 14.5 Elevator Selective Cost Worksheet (continued)

	Passenger	Freight	Hospital
G. Options			
1. Controls			
Automatic, 2 car group	$	—	$
3 car group	$	—	$
4 car group	$	—	$
5 car group	$	—	$
6 car group	$	—	$
Emergency, fireman service	$	—	$
Intercom service	$	—	$
Selective collective, single car	—	—	—
Duplex car	$		$
2. Doors			
Center opening, 1 speed	$	—	$
2 speed	$	—	$
Rear opening-opposite front	$	—	$
Side opening, 2 speed	$	—	$
Freight, bi-parting	—	$	—
Power operated door and gate	—	$	—
3. Emergency power switching, automatic	$	—	$
Manual	$	—	$
4. Finishes based on 3500# cab			
Ceilings, acrylic panel	$	—	
Aluminum egg crate	$	—	$
Doors, stainless steel	$	—	$
Floors, carpet, class "A"	$	—	—
Epoxy	$	—	$
Quarry tile	$	—	$
Slate	$	—	—
Steel plate	—	$	—
Textured rubber	$	—	$
Walls, plastic laminate	Std.	—	Std.
Stainless steel	$	—	$
Return at door	$	—	$
Steel plate, 1/4" x 4' high, 14 ga. above	—	$	—
Entrance, doors, baked enamel	Std.	—	Std.
Stainless steel	$	—	$
Frames, baked enamel	Std.	—	Std.
Stainless steel	$	—	$
5. Maintenance contract - 12 months	$	$	$
6. Signal devices			
Hall lantern, each	$	$	$
Position indicator, car or lobby	$	$	$
Add for over three each, per floor	$	$	$
7. Specialties			
High speed, heavy duty door opener	$	—	$
Variable voltage, O.H. gearless machine	$	—	$
Basement installed geared machine	$	$	$

Figure 14.6   Installation Time for Elevators

Elevators	Crew Makeup	Daily Output	Man-Hours	Unit
ELEVATORS				
2 story, hydraulic, 4,000 lb. capacity, minimum	3 Elevator Constructors 1 Elevator Apprentice Hand Tools	.09	356.000	Ea.
Maximum		.09	356.000	Ea.
10,000 lb. capacity, minimum		.07	457.000	Ea.
Maximum		.07	457.000	Ea.
6 story hydraulic, 4000 lb. capacity		.03	67.000	Ea.
10,000 lb. capacity		.03	67.000	Ea.
6 story geared electric 4000 lb. capacity		.04	800.000	Ea.
10,000 lb. capacity		.03	67.000	Ea.
12 story gearless electric 4000 lb. capacity		.03	67.000	Ea.
10,000 lb. capacity		.03	67.000	Ea.
20 story gearless electric 4000 lb. capacity		.02	600.000	Ea.
10,000 lb. capacity		.02	600.000	Ea.
Passenger, 2 story hydraulic, 2000 lb. capacity		.07	457.000	Ea.
5000 lb. capacity		.07	457.000	Ea.
6 story hydraulic, 2000 lb. capacity		.03	67.000	Ea.
5000 lb. capacity		.03	67.000	Ea.
6 story geared electric, 2000 lb. capacity		.04	800.000	Ea.
5000 lb. capacity		.04	800.000	Ea.
12 story gearless electric, 2000 lb. capacity		.03	67.000	Ea.
5000 lb. capacity		.03	67.000	Ea.
20 story gearless electric, 2000 lb. capacity		.02	600.000	Ea.
5000 lb. capacity		.02	600.000	Ea.
Passenger, pre-engineered, 5 story, hydraulic				
2500 lb. capacity		.04	800.000	Ea.
For less than 5 stops, deduct		.29	110.000	Stop
10 story, geared traction, 200 FPM				
2500 lb. capacity		.02	600.000	Ea.
For less than 10 stops, deduct		.34	94.120	Stop
For 4500 lb. capacity, general purpose		.02	600.000	Ea.
For hospital		.02	600.000	Ea.
Residential, cab type, 1 floor, 2 stop, minimum	2 Elevator Constructors	.20	80.000	Ea.
Maximum		.10	160.000	Ea.
2 floor, 3 stop, minimum		.12	133.000	Ea.
Maximum		.06	267.000	Ea.
Stair climber (chair lift) single seat, minimum		1	16.000	Ea.
Maximum		.20	80.000	Ea.
Wheelchair, porch lift, minimum		1	16.000	Ea.
Maximum		.50	32.000	Ea.
Stair lift, minimum		1	16.000	Ea.
Maximum		.20	80.000	Ea.

Figure 14.7   Dumbwaiter Illustration and Nomenclature

This chart shows the workings of a common dumbwaiter and some of the terms used for its components.

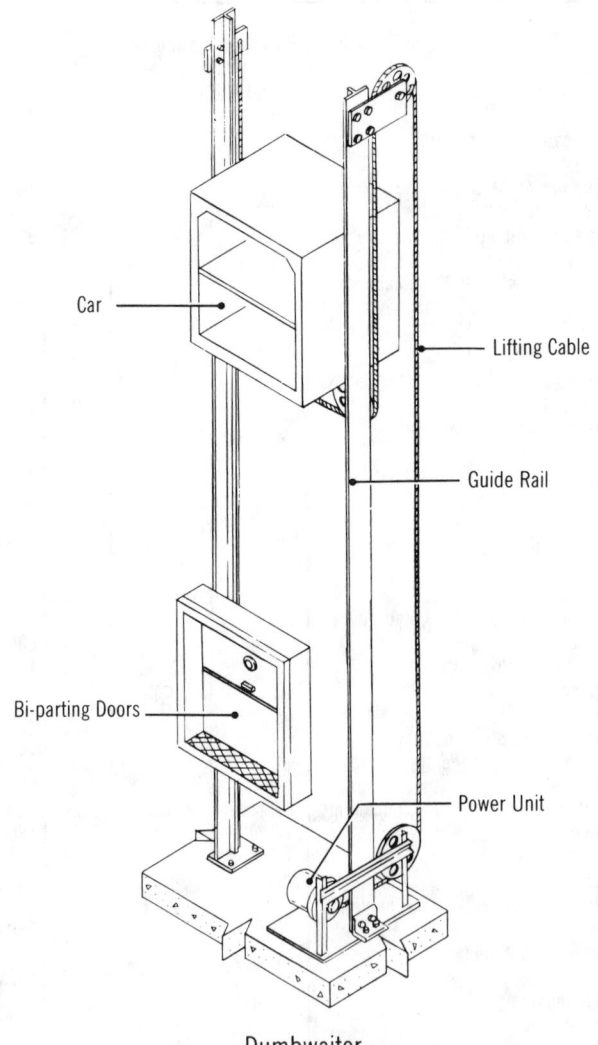

Dumbwaiter

Figure 14.8 Installation Time for Electric Dumbwaiters

Electric Dumbwaiters	Crew Makeup	Daily Output	Man-Hours	Unit
DUMBWAITERS 2 stop, electric, minimum	2 Elevator Constructors	.13	123.000	Ea.
Maximum		.11	145.000	Ea.
For each additional stop, add	↓	.54	29.630	Stop

Figure 14.9 Installation Time for Manual Dumbwaiters

Manual Dumbwaiters	Crew Makeup	Daily Output	Man-Hours	Unit
DUMBWAITERS 2 stop, hand, minimum	2 Elevator Constructors	.23	69.570	Ea.
Maximum		.19	84.210	Ea.
For each additional stop, add	↓	.60	26.670	Stop

Figure 14.10 Installation Time for Lifts

Lifts	Crew Makeup	Daily Output	Man-Hours	Unit
CORRESPONDENCE LIFT 1 floor, 2 stop,				
Electric, 25 lb. capacity	2 Elevator Constructors	.20	80.000	Ea.
Hand, 5 lb. capacity	"	.20	80.000	Ea.
PARCEL LIFT 20" x 20", 100 lb. capacity				
Electric, per floor	2 Millwrights	.25	64.000	Ea.

Figure 14.11   Section Through an Escalator and Common Sizing Proportions

This illustration shows the workings of an escalator, common terms for its components, and some standard dimensional information.

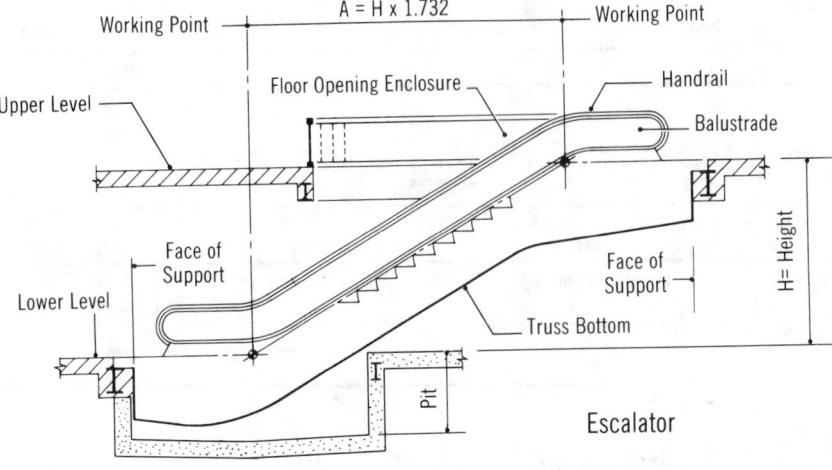

Escalator

Figure 14.12   Installation Time for Escalators

Escalators	Crew Makeup	Daily Output	Man-Hours	Unit
ESCALATORS Per single unit, minimum	3 Elevator Constructors 1 Elevator Apprentice Hand Tools	.06	533.000	Ea.
Maximum	"	.04	800.000	Ea.

Figure 14.13 Installation Time for Moving Walkways

Moving Walks	Crew Makeup	Daily Output	Man-Hours	Unit
MOVING RAMPS AND WALKS Walk, 24" tread width				
Minimum	3 Elevator Constructors 1 Elevator Apprentice Hand Tools	6.50	4.920	L.F.
Maximum		4.43	7.220	L.F.
40" tread width walk, minimum		4.43	7.220	L.F.
Maximum		3.82	8.380	L.F.
Ramp, 12° incline, 32" tread width, minimum		5.27	6.070	L.F.
Maximum		3.82	8.380	L.F.
40" tread width, minimum		3.57	8.960	L.F.
Maximum		2.91	11.000	L.F.

Figure 14.14 Installation Time for Material Handling Systems

Material Handling Systems	Crew Makeup	Daily Output	Man-Hours	Unit
MOTORIZED CAR Distribution systems, single track 20 lb. per car capacity, material handling				
Minimum	4 Millwrights	.19	168.000	Station
Maximum	"	.15	213.000	Station
Larger system, incl. hospital transport, track type, fully automated material handling system				
Minimum	4 Millwrights	.05	640.000	Station
Maximum	3 Struc. Steel Foremen 9 Struc. Steel Workers 1 Equip. Oper. (crane) 1 Welder 1 Equip. Oper. Oiler 1 Equip. Oper. (light) 1 Crane, 90 Ton 1 Gas Welding Machine 1 Torch, Gas & Air 1 Air Compr., 160 C.F.M. 2 Impact Wrenches	.05	560.000	Station

Figure 14.15 Installation Time for Conveyors

Conveyors	Crew Makeup	Daily Output	Man-Hours	Unit
MATERIAL HANDLING Conveyers, gravity type 2" rollers, 3" O.C., horizontal belt, center drive and takeup, 45 F.P.M., 16" belt, 26.5' length	2 Millwrights Power Tools	.50	32.000	Ea.
24" belt, 41.5' length	↓	.40	40.000	Ea.
61.5' length		.30	53.330	Ea.
Inclined belt, 25° incline with horizontal loader and end idler assembly, 34' length, 12" belt	2 Millwrights Power Tools	.30	53.330	Ea.
16" belt	↓	.20	80.000	Ea.
24" belt		.15	107.000	Ea.
Conveyer, overhead, automatic powered chain conveyer, 130 lb./L.F. Capacity	2 Millwrights Power Tools	17	.941	L.F.
Monorail, overhead, manual, channel type 125 lb. per L.F.	1 Millwright	26	.308	L.F.
500 lb. per L.F.	"	21	.381	L.F.
VERTICAL CONVEYER Automatic selective central control, to 10 floors, base price	2 Millwrights	.04	400.000	Total
For automatic service, any floor to any floor, add	"	1.15	13.910	Floor

Figure 14.16 Installation Time for Chutes

Chutes	Crew Makeup	Daily Output	Man-Hours	Unit
CHUTES Linen or refuse, incl. sprinklers Aluminized steel, 16 gauge, 18" diameter	2 Sheet Metal Workers	3.50	4.570	Floor
24" diameter		3.20	5.000	Floor
30" diameter		3	5.330	Floor
36" diameter		2.80	5.710	Floor
Galvanized steel, 16 gauge, 18" diameter		3.50	4.570	Floor
24" diameter		3.20	5.000	Floor
30" diameter		3	5.330	Floor
36" diameter		2.80	5.710	Floor
Stainless steel, 18" diameter		3.50	4.570	Floor
24" diameter		3.20	5.000	Floor
30" diameter		3	5.330	Floor
36" diameter		2.80	5.710	Floor
Linen bottom collector, aluminized steel		4	4.000	Ea.
Stainless steel		4	4.000	Ea.
Refuse bottom hopper, aluminized steel 18" diameter		3	5.330	Ea.
24" diameter		3	5.330	Ea.
36" diameter		3	5.330	Ea.
Package chutes, spiral type, minimum		4.50	3.560	Floor
Maximum	↓	1.50	10.670	Floor

Figure 14.17 Installation Time for Pneumatic Tube Systems

Tube Systems	Crew Makeup	Daily Output	Man-Hours	Unit
PNEUMATIC TUBE SYSTEM Single tube, 2 stations, 100' long, stock, economy,				
3" diameter	2 Steamfitters	.12	133.000	Total
4" diameter	"	.09	178.000	Total
Twin tube, two stations or more, conventional system				
2-1/2" round	2 Steamfitters	62.50	.256	L.F.
3" round		46	.348	L.F.
4" round		49.60	.323	L.F.
4" x 7" oval		37.60	.426	L.F.
Add for blower		2	8.000	System
Plus for each round station, add		7.50	2.130	Ea.
Plus for each oval station, add		7.50	2.130	Ea.
Alternate pricing method: base cost, minimum		.75	21.330	Total
Maximum		.25	64.000	Total
Plus total system length, add, minimum		93.40	.171	L.F.
Maximum		37.60	.426	L.F.
Completely automatic system, 4" round,				
15 to 50 stations		.29	55.170	Station
51 to 144 stations		.32	50.000	Station
6" round or 4" x 7" oval, 15 to 50 stations		.24	66.670	Station
51 to 144 stations		.23	69.570	Station

Checklist

For an estimate to be reliable, all items must be accounted for.
A complete estimate can also eliminate the need to include
contingencies. The following checklist can be used to help ensure
that all items are properly accounted for.

Ash Hoist _____

Conveyor Systems _____

Correspondence Lifts _____

Dumbwaiters
- [ ] Capacity
- [ ] Floors
- [ ] Size
- [ ] Speed
- [ ] Stops
- [ ] Finish

Elevators
- [ ] Hydraulic
- [ ] Electric
- [ ] Capacity
- [ ] Floors
- [ ] Stops
- [ ] Finish
- [ ] Door type
- [ ] Geared
- [ ] Gearless
- [ ] Size
- [ ] Number required
- [ ] Speed
- [ ] Machinery location
- [ ] Signals
- [ ] Special requirements _____

_____

_____

_____

Escalators _____

Hoists and Cranes
- [ ] Capacity
- [ ] Floors
- [ ] Story height

- ☐ Finish
- ☐ Incline angle
- ☐ Size
- ☐ Number required
- ☐ Speed
- ☐ Machinery location
- ☐ Special requirements _____

_____

_____

_____

Lifts _____

Material Handling Systems

- ☐ Automated
- ☐ Non-automated

Moving Stairs, Ramps, and Walks

- ☐ Capacity
- ☐ Floors
- ☐ Story height
- ☐ Finish
- ☐ Incline angle
- ☐ Size
- ☐ Number required
- ☐ Speed
- ☐ Machinery location
- ☐ Special requirements _____

_____

_____

_____

Pneumatic Tube Systems

- ☐ Automatic
- ☐ Size
- ☐ Length
- ☐ Manual
- ☐ Stations
- ☐ Special requirements _____

_____

_____

_____

_____

Vertical Conveyors
☐ Automatic
☐ Manual
Other Conveying _____

_____
_____
_____
_____

Tips

Elevator Doors

When preparing an estimate for elevators, check to make sure that doors have been included, not only for the elevator itself, but for each floor where the elevator stops.

Hydraulic Piston

When figuring the cost for a hydraulic elevator, make sure that the excavation for the elevator piston is included in the estimate. For each floor above grade that the elevator is to travel, there should be an equal length of piston below grade.

Fire Stops

When estimating any vertical conveying system, make sure all openings for doors, dumbwaiter access panels, etc., are fire-rated per the applicable code. In case of a fire emergency, their shaftways will otherwise act as a conduit for smoke and heat.

Help

When in doubt about price, what is standard on elevators, material selection, availability, etc., do not hesitate to call on any of the elevator manufacturing companies directly. The competition for your business can be fierce, and they will try to assist you in any way they can.

Tie-in

Make sure that either the conveying system estimate, the electrical estimate, or mechanical estimate include the tie-in to the rest of the electrical and control systems. This is another item that people often assume someone else has covered.

Notes

Division Fifteen
Mechanical

Introduction

Setting HVAC units, running waste piping, and the many other functions of the mechanical contractor may involve the services of a multitude of other trades. For this reason, the full service mechanical contractor, very much like the general contractor, may employ or subcontract for all of the individual mechanical trades, as well as temperature control specialists, balancing contractors, excavators, and even electricians. Although there are thousands of full service mechanical contractors, this number is far outweighed by the number of individual plumbing shops, HVAC firms, sprinkler contractors, and sheet metal contractors that accomplish the bulk of mechanical work. This division is designed to be used and understood by not only the total mechanical contractor, but also by the individual trades.

The trade divisions included in this division are:

- Pipe and fittings
- Pipe supports
- Valves
- Fixtures
- Water heaters
- Pumps
- Fire protection systems
- Heating systems
- Air conditioning
- Ventilation
- Cable trays
- Conduits
- Ductwork

Estimating Data

The following tables present effective estimating guidelines for items found in Division 15 — Mechanical. Please note that these guidelines can be used as indicators of what may be expected, and that each project must be evaluated individually.

Table of Contents

Figure 15.1   Conversion Factors and Equivalents

Conversion Factors and Equivalents		
Multiply	**By**	**To Obtain**
Acres	43.560	Square Feet
B.T.U.'s	0.2530	Kilogram-calories
B.T.U.'s	778.2	Foot-pounds
B.T.U.'s	.0002928	Kilowatt-hours
B.T.U.'s per min.	12.97	Foot-pounds/sec.
B.T.U.'s per min.	0.02356	Horsepower
B.T.U.'s per min.	0.01757	Kilowatts
B.T.U.'s per min.	17.57	Watts
Centimeters	0.3937	Inches
Cubic Feet	7.481	Gallons
Cubic feet/min.	0.1247	Gallons/sec.
Cubic feet/min.	62.43	Lbs. of water/min.
Cubic feet water	62.43	Lbs. of water
EDR	240	B.T.U.'s
Ft. head	2.31	Lbs.
Ft. of Water	62.43	Lbs./Sq. Ft.
Ft. of Water	0.8826	Inches of mercury
Ft. of Water	0.4335	Lbs./Sq. In.
Ft./min.	0.01136	Miles/Hr.
Foot-pounds	.001285	B.T.U.'s
Gallons	231	Cubic inches
Gallons/min.	.002228	Cu. Ft./sec.
Gallons water	8.345	Lbs. of water
Grains (troy)	0.06480	Grams
Grams	0.03527	Ounces
Horsepower	42.44	B.T.U.'s/min.
Horsepower	33,000	Ft.-Lbs./min.
Horsepower	550	Ft.-Lbs./sec.
Horsepower	745.7	Watts
Horsepower (Boiler)	33,479	B.T.U.'s/Hr.
Horsepower (Boiler)	9.804	Kilowatts
Horsepower-hours	2547	B.T.U.'s
Inches	2.540	Centimeters
Inches of mercury	1.133	Ft. of water
Inches of mercury	0.4912	Lbs./Sq. In.
Inches of water	0.03613	Lbs./Sq. In.
Kilograms	2.2046	Pounds
Kilowatts	56.92	B.T.U.'s/min.
Kilowatts	1.341	Horsepower
Kilowatt-hours	3415	B.T.U.'s
Liters	0.2642	Gallons
Miles	5280	Feet
Ounces	437.5	Grains
Ounces	28.35	Grams
Pounds	7000	Grains
Pounds	453.6	Grams
Pounds of water	27.68	Cubic inches
Pounds of water	0.1198	Gallons
Pounds of water	7000	Grains
Pounds of water/min.	0.2669	Cu. Ft./sec.
Square feet	144	Square inches
Square inches	1.273×10^{-6}	Circular mils
Square miles	640	Acres
Temp. (degs. C) + 273	1	Abs. temp. (degs. K)
Temp. (degs. C) + 17.8	1.8	Temp. (degs. Fahr.)
Temp. (degs. F) –32	5/9	Temp. (degs. Cent.)
Ton (Refrigeration)	200	B.T.U.'s/min.
Watts	.001341	Horsepower
Watt-hours	3.415	B.T.U.'s

Figure 15.1a   Air Requirements for Air Cooled Condenser

Refrigeration:
750 CFM per HP, 1000 CFM per ton
Air Conditioning:
1000 CFM per HP, 400 CFM per ton

Figure 15.1b   Water Requirements for Condensing Units

City Water — Tons refrigeration x 1.5 = GPM
Cooling Tower — Tons refrigeration x 3.0 = GPM

Figure 15.2   Units of Measure and Their Equivalents

Length
1 in. = 25.4 mm.
1 mm. = .03937 in.
1 ft. = 30.48 cm.
1 meter = 3.28083 ft.
1 micron = .001 mm.

Area
1 sq. in. = 6.4516 sq. cm.
1 sq. ft. = 929.03 sq. cm.
1 sq. cm. = 0.155 sq. in.
1 sq. cm. = 0.0010764 sq. ft.

Volume
1 cu. in. = 16.387 cu. cm.
1 cu. ft. = 1728 cu. in.
1 cu. ft. = 7.4805 U.S. gal.
1 cu. ft. = 6.229 British gal.
1 cu. ft. = 28.317 liters
1 U.S. gal. = 0.1337 cu. ft.
1 U.S. gal. = 231 cu. in.
1 U.S. gal. = 3.785 liters
1 British gal. = 1.20094 U.S. gal.
1 British gal. = 277.3 cu. in.
1 British gal. = 4.546 liters
1 liter = 61.023 cu. in.
1 liter = 0.03531 cu. ft.
1 liter = 0.2642 U.S. gal.

Weight
1 ounce av. = 28.35 g.
1 lb. av. = 453.59 g.
1 gram = 0.03527 oz. av.
1 kg. = 2.205 lb. av.
1 cu. ft. of water = 62.425 lb.
1 U.S. gal. of water = 8.33 lb.
1 cu. in. of water = 0.0361 lb.
1 British gal. of water = 10.04 lb.
1 cu. ft. of air at 32° F. & 1 atm = 0.080728 lb.

Velocity
1 ft. per sec. = 30.48 cm. per sec.
1 cm. per sec. = .032808 ft. per sec.

Flow
1 cu. ft. per sec. = 448.83 gal. per min.
1 cu. ft. per sec. = 1699.3 liters per min.
1 U.S. gal. per min. = 0.002228 cu. ft. per sec.
1 U.S. gal. per min. = 0.06308 liters per sec.
1 cu. cm. per sec. = 0.0021186 cu. ft. per min.

Density
1 lb. per cu. ft. = 16.018 kg. per cu. meter
1 lb. per cu. ft. = .0005787 lb. per cu. in.
1 kg. per cu. meter = 0.06243 lb. per cu. ft.
1 g. per cu. cm. = 0.03613 lb. per cu. in.

Viscosity
1 Centipoise = .000672 lb. per ft. sec.
1 Centistoke = .00001076 sq. ft. per sec.

Pressure
1 in. of water = 0.03613 lb. per sq. in.
1 in. of water = 0.07355 in. of Hg.
1 ft. of water = 0.4335 lb. per sq. in.
1 ft. of water = 0.88265 in. of Hg.
1 in. of mercury = 0.49116 lb. per sq. in.
1 in. of mercury = 13.596 in. of water
1 in. of mercury = 1.13299 ft. of water
1 atmosphere = 14.696 lb. per sq. in.
1 atmosphere = 760 mm. of Hg.
1 atmosphere = 29.921 in. of Hg.
1 atmosphere = 33.899 ft. of water
1 lb. per sq. in. = 27.70 in. of water
1 lb. per sq. in. = 2.036 in. of Hg.
1 lb. per sq. in. = .0703066 kg. per sq. cm.
1 kg. per sq. cm. = 14.223 lb. per sq. in.
1 dyne per sq. cm. = .0000145 lb. per sq. in.
1 micron = .00001943 lb. per sq. in.

Energy
1 B.T.U.* = 777.97 ft. lbs.
1 erg = 9.4805 x 10⁻¹¹ B.T.U.
1 erg = 7.3756 x 10⁻⁸ ft. lbs.
1 kilowatt hour = 2.655 x 10⁶ ft. lbs.
1 kilowatt hour = 1.3410 h.p. hr.
1 kg. calorie = 3.968 B.T.U.

Power
1 horsepower = 33,000 ft. lb. per min.
1 horsepower = 550 ft. lb. per sec.
1 horsepower = 2,546.5 B.T.U. per hr.
1 horsepower = 745.7 watts
1 watt = 0.00134 horsepower
1 watt = 44.26 ft. lbs. per min.

Temperature
Temperature Fahrenheit (F) = 9/5 Centigrade (C) + 32 =
   9/4 R + 32
Temperature Centigrade (C) = 5/9 Fahrenheit (F) – 32 =
   5/4 R
Temperature Reaumur (R) = 4/9 Fahrenheit (F) – 32 =
   4/5 C
Absolute Temperature Centigrade or Kelvin (K) =
Degrees C + 273.16
Absolute Temperature Fahrenheit or Rankine (R) =
Degrees F + 459.69

Heat Transfer
1 B.T.U. per sq. ft. = .2712g. cal. per sq. cm.
1 g. calorie per sq. cm. = 3.687 B.T.U. per sq. ft.
1 B.T.U. per hr. per sq. ft. per °F = 4.88 kg. cal. per hr. per
   sq. m. per °C:
1 Kg. cal. per hr. per sq. m. per °C = .205 B.T.U. per hr. per
   sq. ft. per °F:
1 Boiler Horsepower = 33,479 B.T.U. per hr.

*B.T.U. Formula (energy required to heat any substance):
B.T.U./Hr. = Matl. Wt. (lbs.) x Temp. Rise (°F) x Spec. Heat.

Figure 15.3   Mechanical Equipment Service Life*

Service life is a time value established by ASHRAE that reflects the expected life of a specific component. Service life should not be confused with useful life or depreciation period used for income tax purposes. It is the life expectancy of system components. Equipment life is highly variable because of the diverse equipment applications, the preventive maintenance given, the environment, technical advancements of new equipment and personal opinions. The values in this table are a median listing of replacement time of the components. Service life can be used to establish an amortization period; or, if an amortization period is given, service life can give insight into adjusting the maintenance or replacement costs of components.

Equipment Item	Median Years	Equipment Item	Median Years
Air conditioners		Coils	
Window unit	10	DX, water, or steam	20
Residential single or split package	15	Electric	15
Commercial through-the-wall	15	Heat exchangers	
Water-cooled package	15	Shell-and-tube	24
Computer room	15	Reciprocating compressors	20
Heat pumps		Package chillers	
Residential air-to-air	**	Reciprocating	20
Commercial air-to-air	15	Centrifugal	23
Commercial water-to-air	19	Absorption	23
Roof-top air conditioners		Cooling towers	
Single-zone	15	Galvanized metal	20
Multizone	15	Wood	20
Boilers, hot water (steam)		Ceramic	34
Steel water-tube	24 (30)	Air-cooled condensers	20
Steel fire-tube	25 (25)	Evaporative condensers	20
Cast iron	35 (30)	Insulation	
Electric	15	Molded	20
Burners	21	Blanket	24
Furnaces		Pumps	
Gas- or oil-fired	18	Base-mounted	20
Unit heaters		Pipe-mounted	10
Gas or electric	13	Sump and well	10
Hot water or steam	20	Condensate	15
Radiant heaters		Reciprocating engines	20
Electric	10	Steam turbines	30
Hot water or steam	25	Electric motors	18
Air terminals		Motor starters	17
Diffusers, grilles, and registers	27	Electric transformers	30
Induction and fan-coil units	20	Controls	
VAV and double-duct boxes	20	Pneumatic	20
Air washers	17	Electric	16
Duct work	30	Electronic	15
Dampers	20	Valve actuators	
Fans		Hydraulic	15
Centrifugal	25	Pneumatic	20
Axial	20	Self-contained	10
Propeller	15		
Ventilating roof-mounted	20		

*Obtained from a nation-wide survey conducted in 1977 by ASHRAE TC 1.8 (RP 186).
**Data removed by TC 1.8 because of changing technology.

Figure 15.4 Storage Tank Capacities

To calculate the capacity (in gallons) of rectangular tanks, reduce all dimensions to inches, then multiply length by width by height and divide the resulting figure by 231.

Length in Feet	12"	18"	24"	30"	36"	42"	48"	54"	60"	66"	72"	78"	84"	90"	96"	102"	108"
1	6	13	24	37	53	72	94	120	145	180	210	250	290	330	375	425	475
2	12	26	48	74	106	144	188	240	290	360	420	500	580	660	750	850	950
3	18	39	72	111	159	216	282	360	435	540	630	750	870	990	1125	1275	1425
4	24	52	96	148	212	288	376	480	580	720	840	1000	1160	1320	1500	1700	1900
5	30	65	120	185	265	360	470	600	725	900	1050	1250	1450	1650	1875	2125	2375
6	36	78	144	222	318	432	564	720	870	1080	1260	1500	1740	1980	2250	2550	2850
7	42	91	168	259	371	504	658	840	1015	1260	1470	1750	2030	2310	2625	2975	3325
8	48	104	192	296	424	576	752	960	1160	1440	1680	2000	2320	2640	3000	3400	3800
9	54	117	216	333	477	648	846	1080	1305	1620	1890	2250	2610	2970	3375	3825	4275
10	60	130	240	370	530	720	940	1200	1450	1800	2100	2500	2900	3300	3750	4250	4750
11	66	143	264	407	583	792	1034	1320	1595	1980	2310	2750	3190	3630	4125	4675	5225
12	72	156	288	444	636	864	1128	1440	1740	2160	2520	3000	3480	3960	4500	5100	5700
13	78	169	312	481	689	936	1222	1560	1885	2340	2730	3250	3770	4290	4875	5525	6175
14	84	182	336	518	742	1008	1316	1680	2030	2520	2940	3500	4060	4620	5250	5950	6650
15	90	195	360	555	795	1080	1410	1800	2175	2700	3150	3750	4350	4950	5625	6375	7125
16	96	208	384	592	848	1152	1504	1920	2320	2880	3360	4000	4640	5280	6000	6800	7600

Diameter in Inches

Figure 15.5 Capacity in Cubic Feet and U.S. Gallons of Pipes and Cylinders

Diameter in Inches	For 1 Foot Length		Diameter in Inches	For 1 Foot Length		Diameter in Inches	For 1 Foot Length	
	Cubic Feet	U.S. Gallons		Cubic Feet	U.S. Gallons		Cubic Feet	U.S. Gallons
1/4	0.0003	0.0025	6-3/4	0.2485	1.859	19	1.969	14.73
5/16	0.0005	0.0040	7	0.2673	1.999	19-1/2	2.074	15.51
3/8	0.0008	0.0057	7-1/4	0.2867	2.145	20	2.182	16.32
7/16	0.0010	0.0078	7-1/2	0.3068	2.295	20-1/2	2.292	17.15
1/2	0.0014	0.0102	7-3/4	0.3276	2.450	21	2.405	17.99
9/16	0.0017	0.0129	8	0.3491	2.611	21-1/2	2.521	18.86
5/8	0.0021	0.0159	8-1/4	0.3712	2.777	22	2.640	19.75
11/16	0.0026	0.0193	8-1/2	0.3941	2.948	22-1/2	2.761	20.66
3/4	0.0031	0.0230	8-3/4	0.4176	3.125	23	2.885	21.58
13/16	0.0036	0.0269	9	0.4418	3.305	23-1/2	3.012	22.53
7/8	0.0042	0.0312	9-1/4	0.4667	3.491	24	3.142	23.50
15/16	0.0048	0.0359	9-1/2	0.4922	3.682	25	3.409	25.50
1	0.0055	0.0408	9-3/4	0.5185	3.879	26	3.687	27.58
1-1/4	0.0085	0.0638	10	0.5454	4.080	27	3.976	29.74
1-1/2	0.0123	0.0918	10-1/4	0.5730	4.286	28	4.276	31.99
1-3/4	0.0167	0.1249	10-1/2	0.6013	4.498	29	4.587	34.31
2	0.0218	0.1632	10-3/4	0.6303	4.715	30	4.909	36.72
2-1/4	0.0276	0.2066	11	0.6600	4.937	31	5.241	39.21
2-1/2	0.0341	0.2550	11-1/4	0.6903	5.164	32	5.585	41.78
2-3/4	0.0412	0.3085	11-1/2	0.7213	5.396	33	5.940	44.43
3	0.0491	0.3672	11-3/4	0.7530	5.633	34	6.305	47.16
3-1/4	0.0576	0.4309	12	0.7854	5.875	35	6.681	49.98
3-1/2	0.0668	0.4998	12-1/2	0.8522	6.375	36	7.069	52.88
3-3/4	0.0767	0.5738	13	0.9218	6.895	37	7.467	55.86
4	0.0873	0.6528	13-1/2	0.9940	7.436	38	7.876	58.92
4-1/4	0.0985	0.7369	14	1.069	7.997	39	8.296	62.06
4-1/2	0.1104	0.8263	14-1/2	1.147	8.578	40	8.727	65.28
4-3/4	0.1231	0.9206	15	1.227	9.180	41	9.168	68.58
5	0.1364	1.020	15-1/2	1.310	9.801	42	9.621	71.97
5-1/4	0.1503	1.125	16	1.396	10.44	43	10.085	75.44
5-1/2	0.1650	1.234	16-1/2	1.485	11.11	44	10.559	78.99
5-3/4	0.1803	1.349	17	1.576	11.79	45	11.045	82.62
6	0.1963	1.469	17-1/2	1.670	12.49	46	11.541	86.33
6-1/4	0.2131	1.594	18	1.767	13.22	47	12.048	90.13
6-1/2	0.2304	1.724	18-1/2	1.867	13.96	48	12.566	94.00

Figure 15.6   Gas Pipe Capacities

*(Gas–ft.^{3}/hr.) Maximum capacity of pipe in cubic feet of gas per hour for gas pressures of 0.5 Psig or less and a pressure drop of 0.3″ water column, based on a 0.60 specific gravity natural gas. If 1.5 L.P. gas is used, multiply capacity by 0.633.

Nominal Iron Pipe Size, Inches	Internal Diameter, Inches	Length of Pipe in Feet													
		10	20	30	40	50	60	70	80	90	100	125	150	175	200
1/4	.364	32	22	18	15	14	12	11	11	10	9	8	8	7	6
3/8	.493	72	49	40	34	30	27	25	23	22	21	18	17	15	14
1/2	.622	132	92	73	63	56	50	46	43	40	38	34	31	28	26
3/4	.824	278	190	152	130	115	105	96	90	84	79	72	64	59	55
1	1.049	520	350	285	245	215	195	180	170	160	150	130	120	110	100
1-1/4	1.380	1,050	730	590	500	440	400	370	350	320	305	275	250	225	210
1-1/2	1.610	1,600	1,100	890	760	670	610	560	530	490	460	410	380	350	320
2	2.067	3,050	2,100	1,650	1,450	1,270	1,150	1,050	990	930	870	780	710	650	610
2-1/2	2.469	4,800	3,300	2,700	2,300	2,000	1,850	1,700	1,600	1,500	1,400	1,250	1,130	1,050	980
3	3.068	8,500	5,900	4,700	4,100	3,600	3,250	3,000	2,800	2,600	2,500	2,200	2,000	1,850	1,700
4	4.026	17,500	12,000	9,700	8,300	7,400	6,800	6,200	5,800	5,400	5,100	4,500	4,100	3,800	3,500

* Per AGA and NFPA

Figure 15.7   Linear Expansion of Copper Tubing

Boiler Water Temp. °F	Total expansion in inches, based on initial tubing temperature of 70°F. Note that amount of expansion is same for all tubing sizes.									
	10 Ft.	20 Ft.	30 Ft.	40 Ft.	50 Ft.	60 Ft.	70 Ft.	80 Ft.	90 Ft.	100 Ft.
80°F	.01	.02	.04	.05	.06	.07	.08	.09	.11	.12
100°F	.04	.07	.11	.14	.18	.22	.25	.28	.32	.35
120°F	.06	.12	.18	.24	.30	.35	.41	.47	.53	.59
140°F	.08	.16	.25	.33	.41	.50	.58	.66	.74	.82
160°F	.11	.21	.32	.42	.53	.64	.74	.85	.95	1.06
180°F	.13	.26	.39	.52	.65	.78	.91	1.04	1.17	1.29
200°F	.15	.31	.46	.61	.77	.92	1.07	1.22	1.38	1.53
220°F	.18	.35	.53	.71	.88	1.06	1.24	1.41	1.59	1.77
240°F	.20	.40	.60	.80	1.00	1.20	1.40	1.60	1.80	2.00
260°F	.22	.45	.67	.89	1.12	1.34	1.56	1.79	2.01	2.23
280°F	.25	.50	.74	.99	1.24	1.49	1.73	1.98	2.23	2.47

Note: Calculations based on average expansion coefficient of 0.0000098 per °F for copper.

Figure 15.8   Linear Expansion of Steel Pipe

Boiler Water Temp. °F	Total expansion in inches, based on initial pipe temperature of 70°F. Note that amount of expansion is same for all pipe sizes.									
	10 Ft.	20 Ft.	30 Ft.	40 Ft.	50 Ft.	60 Ft.	70 Ft.	80 Ft.	90 Ft.	100 Ft.
80°F	.01	.02	.03	.04	.04	.05	.06	.07	.07	.08
100°F	.03	.05	.07	.10	.12	.14	.17	.19	.22	.24
120°F	.04	.08	.12	.16	.20	.24	.28	.32	.36	.40
140°F	.06	.11	.17	.23	.28	.33	.39	.45	.50	.56
160°F	.07	.14	.22	.29	.36	.43	.50	.57	.65	.71
180°F	.09	.18	.27	.35	.44	.53	.61	.70	.79	.87
200°F	.10	.21	.31	.42	.52	.62	.72	.83	.93	1.03
220°F	.12	.24	.36	.48	.60	.72	.83	.95	1.07	1.19
240°F	.14	.27	.40	.57	.67	.81	.94	1.08	1.21	1.35
260°F	.15	.30	.45	.60	.75	.90	1.05	1.20	1.35	1.50
280°F	.17	.33	.50	.67	.83	1.00	1.16	1.33	1.50	1.67

Note: Calculations based on expansion coefficient 0.0000066 per °F for steel.

Figure 15.9   General Pipe Material Considerations and Background Data

1. Malleable fittings should be used for gas service.

2. Malleable fittings are used where there are stresses/strains due to expansion and vibration.

3. Cast fittings may be broken as an aid to disassembling of heating lines frozen by long use, temperature and minerals.

4. Cast iron pipe is extensively used for underground and submerged service.

5. Type M (light wall) copper tubing is available in hard temper only and is used for nonpressure and less severe applications than K and L.

6. Type L (medium wall) copper tubing, available hard or soft for interior service.

7. Type K (heavy wall) copper tubing, available in hard or soft temper for use where conditions are severe. For underground and interior service.

8. Hard drawn tubing requires fewer hangers or supports but should not be bent. Silver brazed fittings are recommended, however, soft solder is usually used.

Figure 15.10   Estimated Quantity of Soft Solder Required to Make 100 Joints

Size	Quantity in Pounds
3/8"	.5
1/2"	.75
3/4"	1.0
2"	1.5
1-1/4"	1.75
1-1/2"	2.0
2"	2.5
2-1/2"	3.4
3"	4.0
3-1/2"	4.8
4"	6.8
5"	8.0
6"	15.0
8"	32.0
10"	42.0

1. The quantity of hard solder used is dependent on the skill of the operator, but for estimating purposes, 75% of the above figures may be used.

2. Two oz. of solder flux will be required for each pound of solder.

3. Includes an allowance for waste.

4. Drainage fittings consume 20% less.

Figure 15.11 Pipe Sizing for Heating

Heating Load, BTU/HR.	GPM Circulated (20°T.D.)	Recommended Connecting Tubing Size (Type M) for Various Heating Loads and Connecting Tubing Lengths. (Figures based on 10,000 BTU per GPM, or on temperature drop of 20° thru the circuit.)				
		Total Length Ft. Connecting Tubing				
		0–50	50–100	100–150	150–200	200–300
		Tubing, Nominal O.D., Type M				
5,000	0.5	3/8	3/8	1/2	1/2	1/2
10,000	1.0	3/8	3/8	1/2	1/2	1/2
15,000	1.5	1/2	1/2	1/2	1/2	3/4
20,000	2.0	1/2	1/2	1/2	3/4	3/4
30,000	3.0	1/2	1/2	3/4	3/4	3/4
40,000	4.0	1/2	3/4	3/4	3/4	3/4
50,000	5.0	3/4	3/4	3/4	1	1
60,000	6.0	3/4	3/4	1	1	1
75,000	7.5	3/4	1	1	1	1
100,000	10.0	1	1	1	1-1/4	1-1/4
125,000	12.5	1	1	1-1/4	1-1/4	1-1/4
150,000	15.0	1	1-1/4	1-1/4	1-1/4	1-1/2
200,000	20.0	1-1/4	1-1/4	1-1/2	1-1/2	1-1/2
250,000	25.0	1-1/4	1-1/2	1-1/2	2	2
300,000	30.0	1-1/2	1-1/2	2	2	2
400,000	40.0	2	2	2	2	2
500,000	50.0	2	2	2	2-1/2	2-1/2
600,000	60.0	2	2	2-1/2	2-1/2	2-1/2
800,000	80.0	2-1/2	2-1/2	2-1/2	2-1/2	3
1,000,000	100.0	2-1/2	2-1/2	3	3	3
1,250,000	125.0	2-1/2	3	3	3	3-1/2
1,500,000	150.0	3	3	3	3-1/2	3-1/2
2,000,000	200.0	3	3-1/2	3-1/2	4	4
2,500,000	250.0	3-1/2	3-1/2	4	4	5
3,000,000	300.0	3-1/2	4	4	4	5
4,000,000	400.0	4	4	5	5	5
5,000,000	500.0	5	5	5	6	6
6,000,000	600.0	5	6	6	6	8
8,000,000	800.0	8	6	8	8	8
10,000,000	1,000.0	8	8	8	8	10

Figure 15.12   Pipe Sizing for Cooling

Cooling Load, BTU/HR.	Cooling Load, Tons	GPM Circulated 3 GPM/Ton 8°Rise	Recommended Connecting Tubing Size (Type M) for Various Cooling Loads and Connecting Tubing Lengths. (Figures based on 3 GPM/Ton, or on temperature rise of 8° thru the circuit.)				
			Total Length Ft. Connecting Tubing				
			0–50	50–100	100–150	150–200	200–300
			Tubing, Nominal O.D., Type M				
6,000	0.5	1.5	1/2	1/2	1/2	1/2	3/4
9,000	0.75	2.25	1/2	1/2	1/2	3/4	3/4
12,000	1.0	3.0	1/2	1/2	3/4	3/4	3/4
18,000	1.5	4.5	3/4	3/4	3/4	1	1
24,000	2.0	6.0	3/4	3/4	1	1	1
30,000	2.5	7.5	3/4	1	1	1	1
36,000	3.0	9.0	1	1	1	1-1/4	1-1/4
48,000	4.0	12.0	1	1	1-1/4	1-1/4	1-1/4
60,000	5.0	15.0	1	1-1/4	1-1/4	1-1/4	1-1/2
72,000	6.0	18.0	1-1/4	1-1/4	1-1/2	1-1/2	1-1/2
96,000	8.0	24.0	1-1/4	1-1/2	1-1/2	2	2
120,000	10.0	30.0	1-1/2	1-1/2	2	2	2
144,000	12.0	36.0	1-1/2	2	2	2	2
180,000	15.0	45.0	2	2	2	2	2-1/2
240,000	20.0	60.0	2	2	2-1/2	2-1/2	2-1/2
300,000	25.0	75.0	2-1/2	2-1/2	2-1/2	2-1/2	3
360,000	30.0	90.0	2-1/2	2-1/2	2-1/2	3	3
480,000	40.0	120.0	2-1/2	3	3	3	3-1/2
600,000	50.0	150.0	3	3	3	3-1/2	3-1/2
720,000	60.0	180.0	3	3-1/2	3-1/2	4	4
900,000	75.0	225.0	3-1/2	3-1/2	4	4	5
1,200,000	100.0	300.0	3-1/2	4	4	4	5
1,500,000	125.0	375.0	4	4	5	5	5
1,800,000	150.0	450.0	4	5	5	5	6
2,400,000	200.0	600.0	5	5	6	6	6
3,000,000	250.0	750.0	5	6	6	6	8
3,600,000	300.0	900.0	6	6	8	8	8
4,800,000	400.0	1200.0	8	8	8	8	10
6,000,000	500.0	1500.0	8	8	10	10	10
7,200,000	600.0	1800.0	10	10	10	10	12

Figure 15.13   Piping Labor Factors
Due to Elevated Installations

For installations higher than an average of 15', labor costs may be increased by the following suggested percentages.

Ceiling Height	Labor Increase
15' to 20'	10%
20' to 25'	20%
25' to 30'	30%
30' to 35'	40%
35' to 40'	50%
Over 40'	60%

Figure 15.14 Installation Time in Man-Hours for Cast Iron Soil Piping

Description	Man-Hours	Unit
Cast Iron Soil Pipe Service Weight, Single Hub with Hangers Every Five Feet, Lead and Oakum Joints Every Ten Feet		
2" Pipe Size	.254	L.F.
3" Pipe Size	.267	L.F.
4" Pipe Size	.291	L.F.
5" Pipe Size	.316	L.F.
6" Pipe Size	.329	L.F.
8" Pipe Size	.542	L.F.
10" Pipe Size	.593	L.F.
12" Pipe Size	.667	L.F.
Push on Gasket Joints Every Ten Feet		
2" Pipe Size	.242	L.F.
3" Pipe Size	.254	L.F.
4" Pipe Size	.281	L.F.
5" Pipe Size	.304	L.F.
6" Pipe Size	.320	L.F.
8" Pipe Size	.516	L.F.
10" Pipe Size	.571	L.F.
12" Pipe Size	.653	L.F.
Cast Iron Soil Pipe Fittings Hub and Spigot Service Weight Bends or Elbows		
2" Pipe Size	1.000	Ea.
3" Pipe Size	1.140	Ea.
4" Pipe Size	1.230	Ea.
5" Pipe Size	1.330	Ea.
6" Pipe Size	1.410	Ea.
8" Pipe Size	2.910	Ea.
10" Pipe Size	3.200	Ea.
12" Pipe Size	3.560	Ea.
Tees or Wyes		
2" Pipe Size	1.600	Ea.
3" Pipe Size	1.780	Ea.
4" Pipe Size	2.000	Ea.
5" Pipe Size	2.000	Ea.
6" Pipe Size	2.180	Ea.
8" Pipe Size	4.570	Ea.
10" Pipe Size	4.870	Ea.
12" Pipe Size	5.330	Ea.

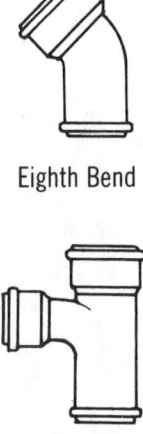

Eighth Bend

Sanitary Tee

Single Hub Soil Pipe

Figure 15.14 Installation Time in Man-Hours for Cast Iron Soil Piping (continued)

Description	Man-Hours	Unit
Push on Gasket Joints Every Ten Feet		
Bends or Elbows		
2" Pipe Size	.800	Ea.
3" Pipe Size	.941	Ea.
4" Pipe Size	1.070	Ea.
5" Pipe Size	1.140	Ea.
6" Pipe Size	1.260	Ea.
8" Pipe Size	2.670	Ea.
10" Pipe Size	2.910	Ea.
12" Pipe Size	3.200	Ea.
Tees or Wyes		
2" Pipe Size	1.330	Ea.
3" Pipe Size	1.600	Ea.
4" Pipe Size	1.780	Ea.
5" Pipe Size	1.850	Ea.
6" Pipe Size	2.180	Ea.
8" Pipe Size	4.000	Ea.
10" Pipe Size	4.870	Ea.
12" Pipe Size	5.330	Ea.
Cleanouts		
Floor Type		
2" Pipe Size	.800	Ea.
3" Pipe Size	1.000	Ea.
4" Pipe Size	1.333	Ea.
5" Pipe Size	2.000	Ea.
6" Pipe Size	2.667	Ea.
8" Pipe Size	4.000	Ea.
Cleanout Tee		
2" Pipe Size	2.000	Ea.
3" Pipe Size	2.222	Ea.
4" Pipe Size	2.424	Ea.
5" Pipe Size	2.909	Ea.
6" Pipe Size	3.200	Ea.
8" Pipe Size	6.400	Ea.
Drains		
Heelproof Floor Drain		
2" to 4" Pipe Size	1.600	Ea.
5" and 6" Pipe Size	1.778	Ea.
8" Pipe Size	2.000	Ea.
Shower Drain		
1-1/2" to 3" Pipe Size	2.000	Ea.
4" Pipe Size	2.286	Ea.
Cast Iron Service Weight Traps		
Deep Seal		
2" Pipe Size	1.143	Ea.
3" Pipe Size	1.333	Ea.
4" Pipe Size	1.455	Ea.

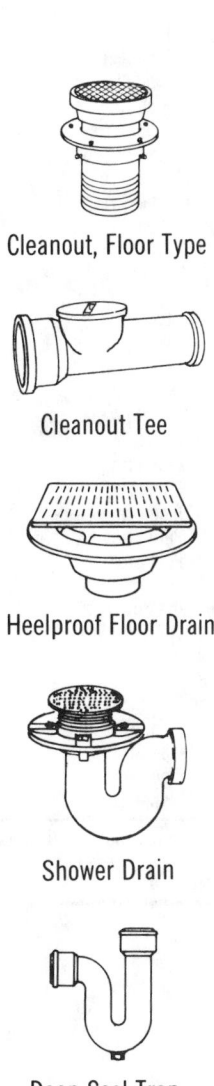

Cleanout, Floor Type

Cleanout Tee

Heelproof Floor Drain

Shower Drain

Deep Seal Trap

(continued on next page)

Figure 15.14   Installation Time in Man-Hours for Cast Iron Soil Piping (continued)

Description	Man-Hours	Unit
P Trap		
2" Pipe Size	1.000	Ea.
3" Pipe Size	1.143	Ea.
4" Pipe Size	1.231	Ea.
5" Pipe Size	1.333	Ea.
6" Pipe Size	1.412	Ea.
8" Pipe Size	2.909	Ea.
10" Pipe Size	3.200	Ea.
Running Trap with Vent		
3" Pipe Size	1.143	Ea.
4" Pipe Size	1.231	Ea.
5" Pipe Size	2.182	Ea.
6" Pipe Size	3.000	Ea.
8" Pipe Size	3.200	Ea.
S Trap		
2" Pipe Size	1.067	Ea.
3" Pipe Size	1.143	Ea.
4" Pipe Size	1.231	Ea.
No Hub with Couplings Every Ten Feet OC		
1-1/2" Pipe Size	.225	L.F.
2" Pipe Size	.239	L.F.
3" Pipe Size	.250	L.F.
4" Pipe Size	.276	L.F.
5" Pipe Size	.289	L.F.
6" Pipe Size	.304	L.F.
8" Pipe Size	.464	L.F.
10" Pipe Size	.525	L.F.
No Hub Couplings*		
1-1/2" Pipe Size	.333	Ea.
2" Pipe Size	.364	Ea.
3" Pipe Size	.421	Ea.
4" Pipe Size	.485	Ea.
5" Pipe Size	.545	Ea.
6" Pipe Size	.600	Ea.
8" Pipe Size	.970	Ea.
10" Pipe Size	1.230	Ea.

P Trap

Running Trap with Vent

S Trap

No Hub Coupling

*Note: In estimating labor for no hub fittings, all the labor is included in the no hub couplings. One coupling per joint.

Figure 15.15 Steel Pipe/Size and Weight Data

Nom. Size	O.D.	Schedule 10		Schedule 20		Schedule 30		Standard (P.E.)	
		Wall	Lbs./Ft.	Wall	Lbs./Ft.	Wall	Lbs./Ft.	Wall	Lbs./Ft.
1/8"	.405							.068	.24
1/4"	.540							.088	.42
3/8"	.675							.091	.57
1/2"	.840							.109	.85
3/4"	1.050							.113	1.13
1"	1.315							.133	1.68
1-1/4"	1.660							.140	2.27
1-1/2"	1.900							.145	2.72
2"	2.375							.154	3.65
2-1/2"	2.875							.203	5.79
3"	3.500							.216	7.58
3-1/2"	4.000							.226	9.11
4"	4.500							.237	10.79
5"	5.563							.258	14.62
6"	6.625							.280	18.97
8"	8.625			.250	22.36	.277	24.70	.322	28.55
10"	10.750			.250	28.04	.307	34.24	.365	40.48
12"	12.750			.250	33.38	.330	43.77	.375	49.56
14"	14.000	.250	36.71	.312	45.61	.375	54.57	.375	54.57
16"	16.000	.250	42.05	.312	52.27	.375	62.58	.375	62.58
18"	18.000	.250	47.39	.312	58.94	.438	82.15	.375	70.59
20"	20.000	.250	52.73	.375	78.60	.500	104.13	.375	78.60
22"	22.000	.250	58.07	.375	86.61	.500	114.81	.375	86.61
24"	24.000	.250	63.41	.375	94.62	.562	140.68	.375	94.62
26"	26.000	.312	85.60	.500	136.17			.375	102.63
28"	28.000	.312	92.41	.500	146.85	.625	182.73	.375	110.64
30"	30.000	.312	98.93	.500	157.53	.625	196.08	.375	118.65
32"	32.000	.312	105.59	.500	168.21	.625	209.43	.375	126.66
34"	34.000	.312	112.25	.500	178.89	.625	222.78	.375	134.67
36"	36.000	.312	118.92	.500	189.57	.625	236.13	.375	142.68

(continued on next page)

Figure 15.15 Steel Pipe/Size and Weight Data (continued)

Nom. Size	O.D.	Schedule 40		Schedule 60		Extra Strong		Schedule 80	
		Wall	Lbs./Ft.	Wall	Lbs./Ft.	Wall	Lbs./Ft.	Wall	Lbs./Ft.
1/8"	.405	.068	.24			.095	.31	.095	.31
1/4"	.540	.088	.42			.119	.54	.119	.54
3/8"	.675	.091	.57			.126	.74	.126	.74
1/2"	.840	.109	.85			.147	1.09	.147	1.09
3/4"	1.050	.113	1.13			.154	1.47	.154	1.47
1"	1.315	.133	1.68			.179	2.17	.179	2.17
1-1/4"	1.660	.140	2.27			.191	3.00	.191	3.00
1-1/2"	1.900	.145	2.72			.200	3.63	.200	3.63
2"	2.375	.154	3.65			.218	5.02	.218	5.02
2-1/2"	2.875	.203	5.79			.276	7.66	.276	7.66
3"	3.500	.216	7.58			.300	10.25	.300	10.25
3-1/2"	4.000	.226	9.11			.318	.12.50	.318	12.50
4"	4.500	.237	10.79			.337	14.98	.337	14.98
5"	5.563	.258	14.62			.375	20.78	.375	20.78
6"	6.625	.280	18.97			.432	28.57	.432	28.57
8"	8.625	.322	28.55	.406	35.64	.500	43.39	.500	43.39
10"	10.750	.365	40.48	.500	54.74	.500	54.74	.594	64.43
12"	12.750	.406	53.52	.562	73.15	.500	65.42	.688	88.63
14"	14.000	.438	63.44	.594	85.05	.500	72.09	.750	106.13
16"	16.000	.500	82.77	.656	107.50	.500	82.77	.844	136.61
18"	18.000	.562	104.67	.750	138.17	.500	93.45	.938	170.92
20"	20.000	.594	123.11	.812	166.40	.500	104.13	1.031	208.87
22"	22.000			.875	197.41	.500	114.81	1.125	250.81
24"	24.000	.688	171.29	.969	238.35	.500	125.49	1.219	296.53
26"	26.000					.500	136.17		
28"	28.000					.500	146.85		
30"	30.000					.500	157.53		
32"	32.000	.688	229.92			.500	168.21		
34"	34.000	.688	244.77			.500	178.89		
36"	36.000	.750	282.35			.500	189.57		

Figure 15.15   Steel Pipe/Size and Weight Data (continued)

Nom. Size	O.D.	Schedule 100		Schedule 120		Schedule 140		Schedule 160		XX Strong	
		Wall	Lbs./Ft.	Wall	Lbs./Ft.	Wall	Lbs./Ft.	Wall	Lbs./Ft.	Wall	Lbs./Ft.
1/8"	.405										
1/4"	.540										
3/8"	.675										
1/2"	.840							.188	1.31	.294	1.71
3/4"	1.050							.219	1.94	.308	2.44
1"	1.315							.250	2.84	.358	3.66
1-1/4"	1.660							.250	3.76	.382	5.21
1-1/2	1.900							.281	4.86	.400	6.41
2"	2.375							.344	7.46	.436	9.03
2-1/2"	2.875							.375	10.01	.552	13.69
3"	3.500							.438	14.32	.600	18.58
3-1/2"	4.000									.636	22.85
4"	4.500			.438	19.00			.531	22.51	.674	27.54
5"	5.563			.500	27.04			.625	32.96	.750	38.55
6"	6.625			.562	36.39			.719	45.35	.864	53.16
8"	8.625	.594	50.95	.719	60.71	.812	67.76	.906	74.69	.875	72.42
10"	10.750	.719	77.03	.844	89.29	1.000	104.13	1.125	115.64	1.000	104.13
12"	12.750	.844	107.32	1.000	125.49	1.125	139.67	1.312	160.27	1.000	125.49
14"	14.000	.938	130.85	1.094	150.79	1.250	170.21	1.406	189.11		
16"	16.000	1.031	164.82	1.219	192.43	1.438	223.64	1.594	245.25		
18"	18.000	1.156	207.96	1.375	244.14	1.562	274.22	1.781	308.55		
20"	20.000	1.281	256.19	1.500	296.37	1.750	341.10	1.969	379.14		
22"	22.000	1.375	302.88	1.625	353.61	1.875	403.01	2.125	451.07		
24"	24.000	1.531	367.45	1.812	429.50	2.062	483.24	2.344	542.09		
26"	26.000										
28"	28.000										
30"	30.000										
32"	32.000										
34"	34.000										
36"	36.000										

Figure 15.16 Bolting Information for Standard Flanges

150 Lb. Steel Flanges				
Pipe Size	Diam. of Bolt Circle	Diam. of Bolts	No. of Bolts	Bolt Length
1/2	2-3/8	1/2	4	1-3/4
3/4	2-3/4	1/2	4	2
1	3-1/8	1/2	4	2
1-1/4	3-1/2	1/2	4	2-1/4
1-1/2	3-7/8	1/2	4	2-1/4
2	4-3/4	5/8	4	2-3/4
2-1/2	5-1/2	5/8	4	3
3	6	5/8	4	3
3-1/2	7	5/8	8	3
4	7-1/2	5/8	8	3
5	8-1/2	3/4	8	3-1/4
6	9-1/2	3/4	8	3-1/4
8	11-3/4	3/4	8	3-1/2
10	14-1/4	7/8	12	3-3/4
12	17	7/8	12	4
14	18-3/4	1	12	4-1/4
16	21-1/4	1	16	4-1/2
18	22-3/4	1-1/8	16	4-3/4
20	25	1-1/8	20	5-1/4
22	27-1/4	1-1/4	20	5-1/2
24	29-1/2	1-1/4	20	5-3/4
26	31-3/4	1-1/4	24	6
30	36	1-1/4	28	6-1/4
34	40-1/2	1-1/2	32	7
36	42-3/4	1-1/2	32	7
42	49-1/2	1-1/2	36	7-1/2

Steel, cast iron, bronze, stainless, etc. Flanges have identical bolting requirements per pipe size and pressure rating.

Figure 15.17   Support Spacing for Metal Pipe

Nominal Pipe Size Inches	Span Water Feet	Steam, Gas, Air Feet	Rod Size
1	7	9	
1-1/2	9	12	3/8"
2	10	13	
2-1/2	11	14	
3	12	15	1/2"
3-1/2	13	16	
4	14	17	5/8"
5	16	19	
6	17	21	3/4"
8	19	24	
10	20	26	7/8"
12	23	30	
14	25	32	
16	27	35	1"
18	28	37	
20	30	39	
24	32	42	1-1/4"
30	33	44	

Figure 15.18   Installation Time in Man-Hours for Steel Pipe

Description	Man-Hours	Unit
Man-hours to Install Black Schedule #40 Flanged with a Pair of Weld Neck Flanges and a Roll Type Hanger Every Ten Feet. The Pipe Hanger Is Oversized to Allow for Insulation.		
1-1/4" Pipe Size	.250	L.F.
1-1/2" Pipe Size	.276	L.F.
2" Pipe Size	.356	L.F.
2-1/2" Pipe Size	.444	L.F.
3" Pipe Size	.500	L.F.
3-1/2" Pipe Size	.552	L.F.
4" Pipe Size	.615	L.F.
5" Pipe Size	.762	L.F.
6" Pipe Size	.960	L.F.
8" Pipe Size	1.263	L.F.
10" Pipe Size	1.500	L.F.
12" Pipe Size	1.714	L.F.
Fittings for Use with Steel Pipe Flanged Elbows, Cast Iron 90° or 45°, 125 lb.		
1-1/2" Pipe Size	1.140	Ea.
2" Pipe Size	1.231	Ea.
2-1/2" Pipe Size	1.333	Ea.
3" Pipe Size	1.455	Ea.
3-1/2" Pipe Size	2.000	Ea.
4" Pipe Size	2.000	Ea.
5" Pipe Size	2.286	Ea.
6" Pipe Size	2.667	Ea.
8" Pipe Size	3.000	Ea.
10" Pipe Size	3.429	Ea.
12" Pipe Size	4.000	Ea.
Flanged Tees, Cast Iron 125 lb.		
1-1/2" Pipe Size	1.778	Ea.
2" Pipe Size	1.778	Ea.
2-1/2" Pipe Size	2.000	Ea.
3" Pipe Size	2.286	Ea.
3-1/2" Pipe Size	3.200	Ea.
4" Pipe Size	3.200	Ea.
5" Pipe Size	4.000	Ea.
6" Pipe Size	4.000	Ea.
8" Pipe Size	4.800	Ea.
10" Pipe Size	6.000	Ea.
12" Pipe Size	8.000	Ea.
Gasket and Bolt Sets Required at Each Flanged Joint		
1-1/2" Pipe Size	.267	Ea.
2" Pipe Size	.267	Ea.
2-1/2" Pipe Size	.267	Ea.
3" Pipe Size	.267	Ea.
3-1/2" Pipe Size	.286	Ea.
4" Pipe Size	.296	Ea.
5" Pipe Size	.308	Ea.
6" Pipe Size	.333	Ea.
8" Pipe Size	.400	Ea.
10" Pipe Size	.444	Ea.
12" Pipe Size	.500	Ea.

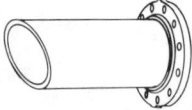

Flanged Steel Pipe

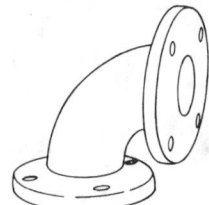

90° Elbow — Flanged

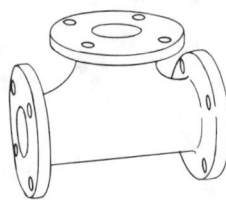

Tee — Flanged

Figure 15.18 Installation Time in Man-Hours for Steel Pipe (continued)

Description	Man-Hours	Unit
Steel Pipe Man-hours to Install Black, Schedule #10, Grooved Joint or Plain End with a Mechanical Joint Coupling and Pipe Hanger Every Ten Feet		
2" Pipe Size	.186	L.F.
2-1/2" Pipe Size	.262	L.F.
3" Pipe Size	.291	L.F.
3-1/2" Pipe Size	.302	L.F.
4" Pipe Size	.327	L.F.
5" Pipe Size	.400	L.F.
6" Pipe Size	.522	L.F.
8" Pipe Size	.585	L.F.
10" Pipe Size	.706	L.F.
12" Pipe Size	.800	L.F.
Black or Galvanized Schedule #40 Grooved Joint or Plain End with a Mechanical Joint Coupling and Pipe Hanger Every Ten Feet		
3/4" Pipe Size	.113	L.F.
1" Pipe Size	.127	L.F.
1-1/4" Pipe Size	.138	L.F.
1-1/2" Pipe Size	.157	L.F.
2" Pipe Size	.200	L.F.
2-1/2" Pipe Size	.281	L.F.
3" Pipe Size	.320	L.F.
3-1/2" Pipe Size	.340	L.F.
4" Pipe Size	.356	L.F.
5" Pipe Size	.432	L.F.
6" Pipe Size	.571	L.F.
8" Pipe Size	.649	L.F.
10" Pipe Size	.774	L.F.
12" Pipe Size	.889	L.F.
Fittings for Use with Grooved Joint or Plain End Steel Pipe Elbows 90° or 45°		
3/4" Pipe Size	.160	Ea.
1" Pipe Size	.160	Ea.
1-1/4" Pipe Size	.200	Ea.
1-1/2" Pipe Size	.242	Ea.
2" Pipe Size	.320	Ea.
2-1/2" Pipe Size	.400	Ea.
3" Pipe Size	.485	Ea.
4" Pipe Size	.640	Ea.
5" Pipe Size	.800	Ea.
6" Pipe Size	.960	Ea.
8" Pipe Size	1.143	Ea.
10" Pipe Size	1.333	Ea.
12" Pipe Size	1.600	Ea.

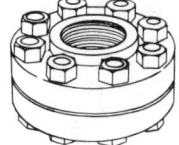

Flanged Joint

Grooved Joint Steel Pipe

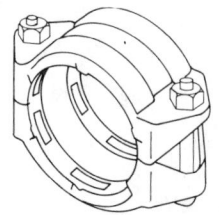

Grooved Joint Coupling

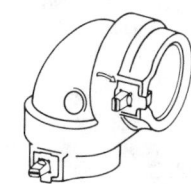

Mechanical Joint Elbow
90° — Plain End Pipe

(continued on next page)

Figure 15.18 Installation Time in Man-Hours for Steel Pipe (continued)

Description	Man-Hours	Unit
Tees		
3/4" Pipe Size	.211	Ea.
1" Pipe Size	.242	Ea.
1-1/4" Pipe Size	.296	Ea.
1-1/2" Pipe Size	.364	Ea.
2" Pipe Size	.471	Ea.
2-1/2" Pipe Size	.593	Ea.
3" Pipe Size	.727	Ea.
4" Pipe Size	.941	Ea.
5" Pipe Size	1.231	Ea.
6" Pipe Size	1.412	Ea.
8" Pipe Size	1.714	Ea.
10" Pipe Size	2.000	Ea.
12" Pipe Size	2.400	Ea.
Man-hours to Install Black or Galvanized Schedule #40 Threaded with a Coupling and Pipe Hanger Every Ten Feet. The Pipe Hanger is Oversized to Allow for Insulation.		
1/2" Pipe Size	.127	L.F.
3/4" Pipe Size	.131	L.F.
1" Pipe Size	.151	L.F.
1-1/4" Pipe Size	.180	L.F.
1-1/2" Pipe Size	.200	L.F.
2" Pipe Size	.250	L.F.
2-1/2" Pipe Size	.320	L.F.
3" Pipe Size	.372	L.F.
3-1/2" Pipe Size	.400	L.F.
4" Pipe Size	.444	L.F.
5" Pipe Size	.615	L.F.
6" Pipe Size	.774	L.F.
8" Pipe Size	.889	L.F.
10" Pipe Size	1.043	L.F.
12" Pipe Size	1.333	L.F.
Fittings for Use with Steel Pipe. Threaded Fittings, Cast Iron, 125 lb. or Malleable Iron Rated at 150 lb. Elbows, 90° or 45°		
1/2" Pipe Size	.533	Ea.
3/4" Pipe Size	.571	Ea.
1" Pipe Size	.615	Ea.
1-1/4" Pipe Size	.727	Ea.
1-1/2" Pipe Size	.800	Ea.
2" Pipe Size	.889	Ea.
2-1/2" Pipe Size	1.143	Ea.
3" Pipe Size	1.600	Ea.
3-1/2" Pipe Size	2.000	Ea.
4" Pipe Size	2.667	Ea.
5" Pipe Size	3.200	Ea.
6" Pipe Size	3.429	Ea.
8" Pipe Size	4.000	Ea.

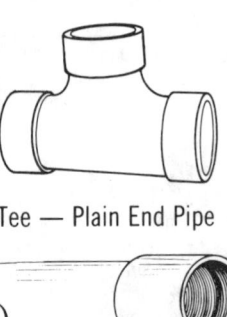

Tee — Plain End Pipe

Threaded and Coupled Steel Pipe

45° Elbow — Malleable Iron

Tee — Cast Iron

Figure 15.18   Installation Time in Man-Hours for Steel Pipe (continued)

Description	Man-Hours	Unit
Tees		
1/2" Pipe Size	.889	Ea.
3/4" Pipe Size	.889	Ea.
1" Pipe Size	1.000	Ea.
1-1/4" Pipe Size	1.143	Ea.
1-1/2" Pipe Size	1.231	Ea.
2" Pipe Size	1.455	Ea.
2-1/2" Pipe Size	1.778	Ea.
3" Pipe Size	2.667	Ea.
3-1/2" Pipe Size	3.200	Ea.
4" Pipe Size	4.000	Ea.
5" Pipe Size	5.333	Ea.
6" Pipe Size	6.000	Ea.
8" Pipe Size	8.000	Ea.
Unions Malleable Iron 150 lb.*		
1/2" Pipe Size	.571	Ea.
3/4" Pipe Size	.615	Ea.
1" Pipe Size	.667	Ea.
1-1/4" Pipe Size	.762	Ea.
1-1/2" Pipe Size	.842	Ea.
2" Pipe Size	.941	Ea.
2-1/2" Pipe Size	1.231	Ea.
3" Pipe Size	1.778	Ea.
3-1/2" Pipe Size	3.200	Ea.
4" Pipe Size	3.200	Ea.
Steel Pipe Man-hours to Install Black, Schedule #40 Welded with a Butt Joint and a Roll Type Hanger Every Ten Feet. The Pipe Hanger is Oversized to Allow for Insulation.		
1-1/4" Pipe Size	.190	L.F.
1-1/2" Pipe Size	.211	L.F.
2" Pipe Size	.262	L.F.
2-1/2" Pipe Size	.340	L.F.
3" Pipe Size	.372	L.F.
3-1/2" Pipe Size	.410	L.F.
4" Pipe Size	.432	L.F.
5" Pipe Size	.500	L.F.
6" Pipe Size	.667	L.F.
8" Pipe Size	.828	L.F.
10" Pipe Size	1.000	L.F.
12" Pipe Size	1.263	L.F.
Fittings for Use with Steel Pipe Butt Weld, Long Radius 90° Elbows 150 lb.		
1-1/4" Pipe Size	1.143	Ea.
1-1/2" Pipe Size	1.231	Ea.
2" Pipe Size	1.600	Ea.
2-1/2" Pipe Size	2.000	Ea.
3" Pipe Size	2.286	Ea.
3-1/2" Pipe Size	3.200	Ea.
4" Pipe Size	3.200	Ea.
5" Pipe Size	4.800	Ea.
6" Pipe Size	4.800	Ea.
8" Pipe Size	6.000	Ea.
10" Pipe Size	8.000	Ea.
12" Pipe Size	9.600	Ea.

*For unions larger than 4", flanges must be used.

Union — Malleable Iron

Bevel End Steel Pipe

90° Elbow, Butt Weld, Long Radius — Steel Pipe

(continued on next page)

Figure 15.18 Installation Time in Man-Hours for Steel Pipe (continued)

Description	Man-Hours	Unit
Forty-five Degree Elbows Require The Same Man-hours for Installation as Ninety Degree Elbows. Since Forty-five Degree Ells are Often Double the Price of Ninety Degree Ells, a Cost-conscious Pipe Fitter Will Cut a Ninety in Half to Make Two Forty-fives.		
Butt Weld Tees 150 lb.		
1-1/4" Pipe Size	1.778	Ea.
1-1/2" Pipe Size	2.000	Ea.
2" Pipe Size	2.667	Ea.
2-1/2" Pipe Size	3.200	Ea.
3" Pipe Size	4.000	Ea.
3-1/2" Pipe Size	5.333	Ea.
4" Pipe Size	5.333	Ea.
5" Pipe Size	8.000	Ea.
6" Pipe Size	8.000	Ea.
8" Pipe Size	9.600	Ea.
10" Pipe Size	12.000	Ea.
12" Pipe Size	15.000	Ea.
Weld Neck Flanges 150 lb.		
1-1/4" Pipe Size	.552	Ea.
1-1/2" Pipe Size	.615	Ea.
2" Pipe Size	.800	Ea.
2-1/2" Pipe Size	1.000	Ea.
3" Pipe Size	1.143	Ea.
3-1/2" Pipe Size	1.333	Ea.
4" Pipe Size	1.600	Ea.
5" Pipe Size	2.000	Ea.
6" Pipe Size	2.400	Ea.
8" Pipe Size	3.429	Ea.
10" Pipe Size	4.000	Ea.
12" Pipe Size	4.800	Ea.
Slip-on Weld Flanges 150 lb.		
1-1/4" Pipe Size	1.000	Ea.
1-1/2" Pipe Size	1.067	Ea.
2" Pipe Size	1.333	Ea.
2-1/2" Pipe Size	1.600	Ea.
3" Pipe Size	1.778	Ea.
3-1/2" Pipe Size	2.286	Ea.
4" Pipe Size	2.667	Ea.
5" Pipe Size	3.200	Ea.
6" Pipe Size	4.000	Ea.
8" Pipe Size	4.800	Ea.
10" Pipe Size	6.000	Ea.
12" Pipe Size	8.000	Ea.

Tee, Butt Weld —
Steel Pipe

Weld Neck Flange —
Steel Pipe

Slip-on Weld Flange —
Steel Pipe

Figure 15.19 Copper Tubing Dimensions and Weights

Nominal Size of Tube/In.	Actual Outside Diam. of Tube/In.	Type K Tube		Type L Tube		Type M Tube	
		Wall Thick-ness/In.	Weight Per Foot/Lbs.	Wall Thick-ness/In.	Weight Per Foot/Lbs.	Wall Thick-ness/In.	Weight Per Foot/Lbs.
1/4	3/8			.030	.126		
3/8	1/2	.049	.269	.035	.198		
1/2	5/8	.049	.344	.040	.285		
3/4	7/8	.065	.641	.045	.455		
1	1-1/8	.065	.839	.050	.655		
1-1/4	1-3/8	.065	1.04	.055	.884	.042	.68
1-1/2	1-5/8	.072	1.36	.060	1.14	.049	.94
2	2-1/8	.083	2.06	.070	1.75	.058	1.46
2-1/2	2-5/8	.095	2.93	.080	2.48	.065	2.03
3	3-1/8	.109	4.00	.090	3.33	.072	2.68
3-1/2	3-5/8	.120	5.12	.100	4.29	.083	3.58
4	4-1/8	.130	6.51	.110	5.38	.095	4.66
5	5-1/8	.160	9.67	.125	7.61	.109	6.66
6	6-1/8	.192	13.90	.140	10.20	.122	8.92
8	8-1/8	.271	25.90	.200	19.30	.170	16.50
10	10-1/8	.338	40.30	.250	30.10	.212	25.60
12	12-1/8	.405	57.80	.280	40.40	.254	36.70

Figure 15.20   Copper Tubing: Commercially Available Lengths

		Hard Drawn		Annealed	
Type K	Straight lengths up to: 8-inch diameter 10-inch diameter 12-inch diameter		20 ft. 18 ft. 12 ft.	Straight lengths up to: 8-inch diameter 10-inch diameter 12-inch diameter	20 ft. 18 ft. 12 ft.
				Coils up to: 1-inch diameter 1-1/4 and 1-1/2 inch diameter 2-inch diameter	60 ft. 100 ft. 60 ft. 40 ft. 45 ft.
Type L	Straight lengths up to: 10-inch diameter 12-inch diameter		20 ft. 18 ft.	Straight Lengths up to: 10-inch diameter 12-inch diameter	20 ft. 18 ft.
				Coils up to: 1-inch diameter 1-1/4 and 1-1/2 inch diameter 2-inch diameter	60 ft. 100 ft. 60 ft. 40 ft. 45 ft.
Type M	Straight lengths: All diameters		20 ft.	Straight lengths up to: 12-inch diameter	20 ft.
				Coils up to: 1-inch diameter 1-1/4 and 1-1/2-inch diameter 2-inch diameter	60 ft. 100 ft. 60 ft. 40 ft. 45 ft.
DWV	Straight lengths: All diameters		20 ft.	Not available	—
ACR	Straight lengths:		20 ft.	Coils:	50 ft.

Figure 15.21 Installation Time in Man-Hours for Copper Tubing

Description	Man-Hours	Unit
Man-hours to Install the Several Types of Copper Tubing Based on a Soft Soldered Coupling and a Clevis Hanger Every Ten Feet		
Type K Tubing		
1/2" Pipe Size	.103	L.F.
3/4" Pipe Size	.108	L.F.
1" Pipe Size	.121	L.F.
1-1/4" Pipe Size	.143	L.F.
1-1/2" Pipe Size	.160	L.F.
2" Pipe Size	.200	L.F.
2-1/2" Pipe Size	.267	L.F.
3" Pipe Size	.296	L.F.
3-1/2" Pipe Size	.381	L.F.
4" Pipe Size	.421	L.F.
5" Pipe Size	.500	L.F.
6" Pipe Size	.632	L.F.
8" Pipe Size	.706	L.F.
Type L Tubing		
1/2" Pipe Size	.099	L.F.
3/4" Pipe Size	.105	L.F.
1" Pipe Size	.118	L.F.
1-1/4" Pipe Size	.138	L.F.
1-1/2" Pipe Size	.154	L.F.
2" Pipe Size	.190	L.F.
2-1/2" Pipe Size	.258	L.F.
3" Pipe Size	.286	L.F.
3-1/2" Pipe Size	.372	L.F.
4" Pipe Size	.410	L.F.
5" Pipe Size	.471	L.F.
6" Pipe Size	.600	L.F.
8" Pipe Size	.667	L.F.
Type M Tubing		
1/2" Pipe Size	.095	L.F.
3/4" Pipe Size	.103	L.F.
1" Pipe Size	.114	L.F.
1-1/4" Pipe Size	.133	L.F.
1-1/2" Pipe Size	.148	L.F.
2" Pipe Size	.182	L.F.
2-1/2" Pipe Size	.250	L.F.
3" Pipe Size	.276	L.F.
3-1/2" Pipe Size	.356	L.F.
4" Pipe Size	.400	L.F.
5" Pipe Size	.444	L.F.
6" Pipe Size	.571	L.F.
8" Pipe Size	.632	L.F.
Type DWV Tubing		
1-1/4" Pipe Size	.133	L.F.
1-1/2" Pipe Size	.148	L.F.
2" Pipe Size	.182	L.F.
2-1/2" Pipe Size	.225	L.F.
3" Pipe Size	.276	L.F.
4" Pipe Size	.400	L.F.
5" Pipe Size	.444	L.F.
6" Pipe Size	.571	L.F.
8" Pipe Size	.632	L.F.

(continued on next page)

Figure 15.21   Installation Time in Man-Hours for Copper Tubing (continued)

Description	Man-Hours	Unit
Fittings for Copper Tubing, Pressurized Systems, Solder Joint		
Elbows 90° or 45°		
1/2" Tubing Size	.400	Ea.
3/4" Tubing Size	.421	Ea.
1" Tubing Size	.500	Ea.
1-1/4" Tubing Size	.533	Ea.
1-1/2" Tubing Size	.615	Ea.
2" Tubing Size	.727	Ea.
2-1/2" Tubing Size	1.231	Ea.
3" Tubing Size	1.455	Ea.
3-1/2" Tubing Size	1.600	Ea.
4" Tubing Size	1.778	Ea.
5" Tubing Size	2.667	Ea.
6" Tubing Size	2.667	Ea.
8" Tubing Size	3.000	Ea.
Tees		
1/2" Tubing Size	.615	Ea.
3/4" Tubing Size	.667	Ea.
1" Tubing Size	.800	Ea.
1-1/4" Tubing Size	.889	Ea.
1-1/2" Tubing Size	1.000	Ea.
2" Tubing Size	1.143	Ea.
2-1/2" Tubing Size	2.000	Ea.
3" Tubing Size	2.286	Ea.
3-1/2" Tubing Size	2.667	Ea.
4" Tubing Size	3.200	Ea.
5" Tubing Size	4.000	Ea.
6" Tubing Size	4.000	Ea.
8" Tubing Size	4.800	Ea.
Drainage Waste and Vent Systems, Solder Joint		
Elbows 90° or 45°		
1-1/4" Tubing Size	.615	Ea.
1-1/2" Tubing Size	.667	Ea.
2" Tubing Size	.800	Ea.
3" Tubing Size	1.600	Ea.
4" Tubing Size	1.778	Ea.
5" Tubing Size	2.667	Ea.
6" Tubing Size	2.667	Ea.
8" Tubing Size	3.000	Ea.

90° Elbow — Solder Joint

45° Elbow — Solder Joint

Tee — Solder Joint

Figure 15.22   Installation Time in Man-Hours for Flexible Plastic Tubing

Description	Man-Hours	Unit
Man-hours to Install Flexible Plastic Piping. Pipe Hangers and Couplings Are Not Included Since They Are Not Always Necessary Such as When This Piping is Laid in a Trench or Strung Through Joists or Clipped to Structural Members.		
1/2" Pipe Size	.020	L.F.
3/4" Pipe Size	.030	L.F.
1" Pipe Size	.040	L.F.
1-1/4" Pipe Size	.060	L.F.
1-1/2' Pipe Size	.080	L.F.
2" Pipe Size	.100	L.F.
3" Pipe Size	.130	L.F.
Fittings for Polybutylene or Polyethylene Pipe		
Acetal Insert Type with Crimp Rings (Hot and Cold Water)		
Elbows 90°		
1/2" Pipe Size	.350	Ea.
3/4" Pipe Size	.360	Ea.
Couplings		
1/2" Pipe Size	.350	Ea.
3/4" Pipe Size	.350	Ea.
Tees		
1/2" Pipe Size	.530	Ea.
3/4" Pipe Size	.570	Ea.
Nylon Insert Type with Stainless Steel Clamps (Cold Water Only)		
Elbows 90°		
3/4" Pipe Size	.360	Ea.
1" Pipe Size	.420	Ea.
1-1/4" Pipe Size	.440	Ea.
1-1/2" Pipe Size	.470	Ea.
2" Pipe Size	.500	Ea.
Couplings		
3/4" Pipe Size	.360	Ea.
1" Pipe Size	.420	Ea.
1-1/4" Pipe Size	.440	Ea.
1-1/2" Pipe Size	.470	Ea.
2" Pipe Size	.500	Ea.
Tees		
3/4" Pipe Size	.570	Ea.
1" Pipe Size	.620	Ea.
1-1/4" Pipe Size	.670	Ea.
1-1/2" Pipe Size	.730	Ea.
2" Pipe Size	.800	Ea.
Stainless Steel Clamps		
3/4" Pipe Size	.070	Ea.
1" Pipe Size	.070	Ea.
1-1/4" Pipe Size	.080	Ea.
1-1/2" Pipe Size	.080	Ea.
2" Pipe Size	.090	Ea.

90° Elbow — Acetal Insert

Coupling — Acetal Insert

Tee — Acetal Insert

90° Elbow — Nylon Insert

Coupling — Nylon Insert

Tee — Nylon Insert

Stainless Steel Clamp Ring

(continued on next page)

Figure 15.22   Installation Time in Man-Hours for Flexible Plastic Tubing (continued)

Description	Man-Hours	Unit	
Acetal Flare Type (Hot and Cold Water)			
Elbows 90°			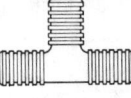
1/2" Pipe Size	.360	Ea.	
3/4" Pipe Size	.380	Ea.	90° Elbow —
1" Pipe Size	.440	Ea.	Acetal Flare Type
Couplings			
1/2" Pipe Size	.360	Ea.	
3/4" Pipe Size	.380	Ea.	
1" Pipe Size	.440	Ea.	
Tees			Coupling —
1/2" Pipe Size	.570	Ea.	Acetal Flare Type
3/4" Pipe Size	.620	Ea.	
1" Pipe Size	.670	Ea.	
Fusion Type (Hot and Cold Water)			
Elbows 90° or 45°			
1" Pipe Size	.420	Ea.	
1-1/4" Pipe Size	.440	Ea.	
1-1/2" Pipe Size	.470	Ea.	Tee — Acetal Flare Type
2" Pipe Size	.500	Ea.	
3" Pipe Size	.800	Ea.	
Couplings			
1" Pipe Size	.420	Ea.	
1-1/4" Pipe Size	.440	Ea.	
1-1/2" Pipe Size	.470	Ea.	
2" Pipe Size	.500	Ea.	90° Elbow —
3" Pipe Size	.800	Ea.	Fusion Type
Tees			
1" Pipe Size	.620	Ea.	
1-1/4" Pipe Size	.670	Ea.	
1-1/2" Pipe Size	.730	Ea.	
2" Pipe Size	.800	Ea.	
3" Pipe Size	1.140	Ea.	
Brass Insert Type Fittings (Hot and Cold Water)			Coupling —
Elbows 90°			Fusion Type
1/2" Pipe Size	.350	Ea.	
3/4" Pipe Size	.360	Ea.	
Couplings			
1/2" Pipe Size	.350	Ea.	
3/4" Pipe Size	.360	Ea.	
Tees			
1/2" Pipe Size	.530	Ea.	Tee — Fusion Type
3/4" Pipe Size	.570	Ea.	

Tee —
Brass Insert Type

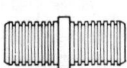

Coupling —
Brass Insert Type

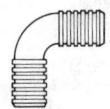

90° Elbow —
Brass Insert Type

Figure 15.23 Installation Time in Man-Hours for Plastic Piping

Description	Man-Hours	Unit
Man-hours to Install DWV Piping, ABS, or PVC with a Coupling and 3 Pipe Hangers Every Ten Feet		
1-1/4" Pipe Size	.190	L.F.
1-1/2" Pipe Size	.222	L.F.
2" Pipe Size	.271	L.F.
3" Pipe Size	.302	L.F.
4" Pipe Size	.333	L.F.
6" Pipe Size	.410	L.F.
Fittings for Use with Plastic Pipe DWV, ABS or PVC, with Socket Joints		
Bends - Quarter or Eighth		
1-1/4" Pipe Size	.471	Ea.
1-1/2" Pipe Size	.500	Ea.
2" Pipe Size	.571	Ea.
3" Pipe Size	.941	Ea.
4" Pipe Size	1.143	Ea.
6" Pipe Size	2.000	Ea.
Couplings		
1-1/4" Pipe Size	.471	Ea.
1-1/2" Pipe Size	.500	Ea.
2" Pipe Size	.571	Ea.
3" Pipe Size	.727	Ea.
4" Pipe Size	.941	Ea.
6" Pipe Size	1.333	Ea.
Tees and Wyes - Sanitary		
1-1/4" Pipe Size	.727	Ea.
1-1/2" Pipe Size	.800	Ea.
2" Pipe Size	.941	Ea.
3" Pipe Size	1.455	Ea.
4" Pipe Size	1.778	Ea.
6" Pipe Size	3.200	Ea.
Man-hours to Install PVC Schedule 40 Piping with a Coupling and 3 Hangers Every Ten Feet. PVC Piping is Used for Water Service, Gas Service, Irrigation and Air Conditioning.		
1/2" Pipe Size	.148	L.F.
3/4" Pipe Size	.157	L.F.
1" Pipe Size	.174	L.F.
1-1/4" Pipe Size	.190	L.F.
1-1/2" Pipe Size	.222	L.F.
2" Pipe Size	.271	L.F.
2-1/2" Pipe Size	.286	L.F.
3" Pipe Size	.302	L.F.
4" Pipe Size	.333	L.F.
5" Pipe Size	.372	L.F.
6" Pipe Size	.410	L.F.
8" Pipe Size	.500	L.F.

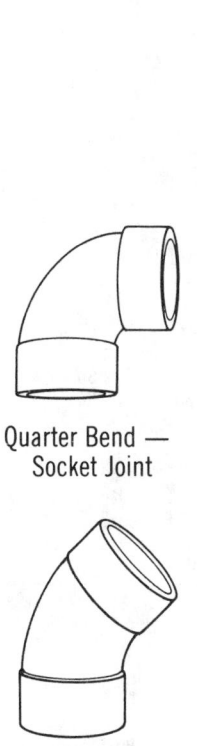

Quarter Bend — Socket Joint

Eighth Bend — Socket Joint

(continued on next page)

Figure 15.23 Installation Time in Man-Hours for Plastic Piping (continued)

Description	Man-Hours	Unit
Fittings for Use with PVC Schedule 40, with Socket Joints, Elbows, 90° or 45°		
1/2" Pipe Size	.364	Ea.
3/4" Pipe Size	.381	Ea.
1" Pipe Size	.444	Ea.
1-1/4" Pipe Size	.471	Ea.
1-1/2" Pipe Size	.500	Ea.
2" Pipe Size	.571	Ea.
2-1/2" Pipe Size	.727	Ea.
3" Pipe Size	.941	Ea.
4" Pipe Size	1.143	Ea.
5" Pipe Size	1.333	Ea.
6" Pipe Size	2.000	Ea.
8" Pipe Size	2.400	Ea.
Couplings		
1/2" Pipe Size	.364	Ea.
3/4" Pipe Size	.381	Ea.
1" Pipe Size	.444	Ea.
1-1/4" Pipe Size	.471	Ea.
1-1/2" Pipe Size	.500	Ea.
2" Pipe Size	.571	Ea.
2-1/2" Pipe Size	.700	Ea.
3" Pipe Size	.842	Ea.
4" Pipe Size	1.000	Ea.
5" Pipe Size	1.143	Ea.
6" Pipe Size	1.333	Ea.
8" Pipe Size	1.714	Ea.
Tees		
1/2" Pipe Size	.571	Ea.
3/4" Pipe Size	.615	Ea.
1" Pipe Size	.667	Ea.
1-1/4" Pipe Size	.727	Ea.
1-1/2" Pipe Size	.800	Ea.
2" Pipe Size	.941	Ea.
2-1/2" Pipe Size	1.143	Ea.
3" Pipe Size	1.455	Ea.
4" Pipe Size	1.778	Ea.
5" Pipe Size	2.000	Ea.
6" Pipe Size	3.200	Ea.
8" Pipe Size	4.000	Ea.
Man-hours to Install PVC Schedule 80 Piping with a Coupling and 3 Hangers Every Ten Feet		
1/2" Pipe Size	.160	L.F.
3/4" Pipe Size	.170	L.F.
1" Pipe Size	.186	L.F.
1-1/4" Pipe Size	.205	L.F.
1-1/2" Pipe Size	.235	L.F.
2" Pipe Size	.291	L.F.

Vent Tee — Socket Joint

90° Elbow — Socket Joint

Coupling — Socket Joint

Tee — Socket Joint

Figure 15.23 Installation Time in Man-Hours for Plastic Piping (continued)

Description	Man-Hours	Unit
2-1/2" Pipe Size	.308	L.F.
3" Pipe Size	.320	L.F.
4" Pipe Size	.348	L.F.
5" Pipe Size	.381	L.F.
6" Pipe Size	.421	L.F.
8" Pipe Size	.511	L.F.
Fittings for Use with PVC Schedule 80, Socket or Threaded Joints, Elbows, 90° or 45°		
1/2" Pipe Size	.444	Ea.
3/4" Pipe Size	.471	Ea.
1" Pipe Size	.533	Ea.
1-1/4" Pipe Size	.571	Ea.
1-1/2" Pipe Size	.615	Ea.
2" Pipe Size	.727	Ea.
2-1/2" Pipe Size	.941	Ea.
3" Pipe Size	1.143	Ea.
4" Pipe Size	1.333	Ea.
5" Pipe Size	2.000	Ea.
6" Pipe Size	2.286	Ea.
8" Pipe Size	3.000	Ea.
Couplings		
1/2" Pipe Size	.444	Ea.
3/4" Pipe Size	.471	Ea.
1" Pipe Size	.533	Ea.
1-1/4" Pipe Size	.571	Ea.
1-1/2" Pipe Size	.615	Ea.
2" Pipe Size	.727	Ea.
2-1/2" Pipe Size	.800	Ea.
3" Pipe Size	.842	Ea.
4" Pipe Size	1.000	Ea.
5" Pipe Size	1.143	Ea.
6" Pipe Size	1.333	Ea.
8" Pipe Size	1.714	Ea.
Tees		
1/2" Pipe Size	.667	Ea.
3/4" Pipe Size	.727	Ea.
1" Pipe Size	.800	Ea.
1-1/4" Pipe Size	.889	Ea.
1-1/2" Pipe Size	1.000	Ea.
2" Pipe Size	1.143	Ea.
2-1/2" Pipe Size	1.455	Ea.
3" Pipe Size	1.778	Ea.
4" Pipe Size	2.000	Ea.
5" Pipe Size	3.200	Ea.
6" Pipe Size	3.200	Ea.
8" Pipe Size	4.000	Ea.

45° Elbow — Threaded Joint

Coupling — Threaded Joint

Tee — Threaded Joint

Note: The man-hours are the same to install CPVC schedule 80 piping (chlorinated polyvinyl chloride) with a coupling and 3 hangers every ten feet. CPVC piping was developed for hot water use and is used for cold water also.

Figure 15.24 Installation Time in Man-Hours for Piping Supports

Description	Man-Hours	Unit
Side Beam Brackets	.167	Ea.
Wall Brackets	.235	Ea.
C Clamps	.050	Ea.
I Beam Clamps		
2" Flange Size	.083	Ea.
3" Flange Size	.084	Ea.
4" Flange Size	.086	Ea.
5" Flange Size	.087	Ea.
6" Flange Size	.089	Ea.
7" Flange Size	.091	Ea.
8" Flange Size	.093	Ea.
Riser Clamps		
3/4" Pipe Size	.167	Ea.
1" Pipe Size	.170	Ea.
1-1/4" Pipe Size	.174	Ea.
1-1/2" Pipe Size	.178	Ea.
2" Pipe Size	.186	Ea.
2-1/2" Pipe Size	.195	Ea.
3" Pipe Size	.200	Ea.
3-1/2" Pipe Size	.205	Ea.
4" Pipe Size	.211	Ea.
5" Pipe Size	.216	Ea.
6" Pipe Size	.222	Ea.
8" Pipe Size	.235	Ea.
10" Pipe Size	.250	Ea.
12" Pipe Size	.286	Ea.
Split Ring Clamps		
1/2" Pipe Size	.117	Ea.
3/4" Pipe Size	.119	Ea.
1" Pipe Size	.121	Ea.
1-1/4" Pipe Size	.124	Ea.
1-1/2" Pipe Size	.127	Ea.
2" Pipe Size	.129	Ea.
2-1/2" Pipe Size	.133	Ea.
3" Pipe Size	.137	Ea.
3-1/2" Pipe Size	.140	Ea.
4" Pipe Size	.145	Ea.
5" Pipe Size	.151	Ea.
6" Pipe Size	.154	Ea.
8" Pipe Size	.160	Ea.
High Temperature Clamps		
4" Pipe Size	.151	Ea.
6" Pipe Size	.151	Ea.
8" Pipe Size	.165	Ea.
10" Pipe Size	.190	Ea.
12" Pipe Size	.222	Ea.

Side Beam Bracket

Medium Welded Steel Bracket

Brackets

C-Clamp

I-Beam Clamp

Extension Pipe or Riser Clamp

Alloy Steel Pipe Clamp

Medium Pipe Clamp

Split Ring Pipe Clamp

Clamp Types

Figure 15.24   Installation Time in Man-Hours for Piping Supports (continued)

Description	Man-Hours	Unit
Two Bolt Clamps		
1/2" Pipe Size	.117	Ea.
3/4" Pipe Size	.119	Ea.
1" Pipe Size	.121	Ea.
1-1/4" Pipe Size	.123	Ea.
1-1/2" Pipe Size	.127	Ea.
2" Pipe Size	.129	Ea.
2-1/2" Pipe Size	.133	Ea.
3" Pipe Size	.137	Ea.
3-1/2" Pipe Size	.140	Ea.
4" Pipe Size	.145	Ea.
5" Pipe Size	.151	Ea.
6" Pipe Size	.154	Ea.
8" Pipe Size	.160	Ea.
10" Pipe Size	.167	Ea.
12" Pipe Size	.180	Ea.
Pipe Alignment Guides		
1" Pipe Size	.308	Ea.
1-1/4" to 2" Pipe Size	.381	Ea.
2-1/2" to 3-1/2" Pipe Size	.444	Ea.
4" to 5" Pipe Size	.500	Ea.
6" Pipe Size	.762	Ea.
8" Pipe Size	1.000	Ea.
10" Pipe Size	1.333	Ea.
12" Pipe Size	1.412	Ea.
Band Hangers		
1/2" Pipe Size	.113	Ea.
3/4" Pipe Size	.114	Ea.
1" Pipe Size	.117	Ea.
1-1/4" Pipe Size	.119	Ea.
1-1/2" Pipe Size	.122	Ea.
2" Pipe Size	.124	Ea.
2-1/2" Pipe Size	.128	Ea.
3" Pipe Size	.131	Ea.
3-1/2" Pipe Size	.134	Ea.
4" Pipe Size	.140	Ea.
5" Pipe Size	.145	Ea.
6" Pipe Size	.148	Ea.
8" Pipe Size	.154	Ea.
Adjustable Swivel Rings		
1/2" Pipe Size	.117	Ea.
3/4" Pipe Size	.118	Ea.
1" Pipe Size	.121	Ea.
1-1/4" Pipe Size	.124	Ea.
1-1/2" Pipe Size	.127	Ea.
2" Pipe Size	.129	Ea.
2-1/2" Pipe Size	.133	Ea.
3" Pipe Size	.137	Ea.
3-1/2" Pipe Size	.140	Ea.
4" Pipe Size	.145	Ea.
5" Pipe Size	.151	Ea.
6" Pipe Size	.154	Ea.
8" Pipe Size	.160	Ea.

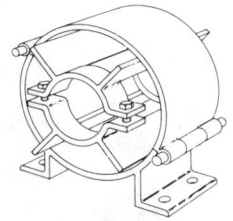

Pipe Alignment Guide

Adjustable Band Hanger

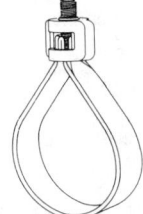

Adjustable Pipe Ring

(continued on next page)

Figure 15.24 Installation Time in Man-Hours for Piping Supports (continued)

Description	Man-Hours	Unit
Clevis		
1/2" Pipe Size	.117	Ea.
3/4" Pipe Size	.119	Ea.
1" Pipe Size	.121	Ea.
1-1/4" Pipe Size	.124	Ea.
1-1/2" Pipe Size	.127	Ea.
2" Pipe Size	.129	Ea.
2-1/2" Pipe Size	.133	Ea.
3" Pipe Size	.137	Ea.
3-1/2" Pipe Size	.140	Ea.
4" Pipe Size	.145	Ea.
5" Pipe Size	.151	Ea.
6" Pipe Size	.154	Ea.
8" Pipe Size	.160	Ea.
10" Pipe Size	.167	Ea.
12" Pipe Size	.180	Ea.
Insulation Protection Saddles		
3/4" Pipe Size	.118	Ea.
1" Pipe Size	.118	Ea.
1-1/4" Pipe Size	.118	Ea.
1-1/2" Pipe Size	.121	Ea.
2" Pipe Size	.121	Ea.
2-1/2" Pipe Size	.125	Ea.
3" Pipe Size	.125	Ea.
3-1/2" Pipe Size	.129	Ea.
4" Pipe Size	.129	Ea.
5" Pipe Size	.133	Ea.
6" Pipe Size	.133	Ea.
Adjustable Yoke with Roll		
2-1/2" Pipe Size	.117	Ea.
3" Pipe Size	.122	Ea.
3-1/2" Pipe Size	.129	Ea.
4" Pipe Size	.137	Ea.
5" Pipe Size	.145	Ea.
6" Pipe Size	.154	Ea.
8" Pipe Size	.167	Ea.
10" Pipe Size	.200	Ea.
12" Pipe Size	.235	Ea.
Two Rod Rolls		
2-1/2" Pipe Size	.136	Ea.
3" Pipe Size	.139	Ea.
3-1/2" Pipe Size	.142	Ea.
4" Pipe Size	.143	Ea.
5" Pipe Size	.145	Ea.
6" Pipe Size	.158	Ea.
8" Pipe Size	.178	Ea.
10" Pipe Size	.200	Ea.
12" Pipe Size	.235	Ea.

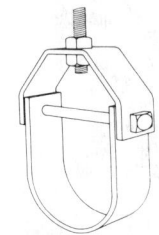

Adjustable Clevis

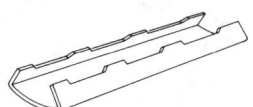

Pipe Covering
Protection Saddle

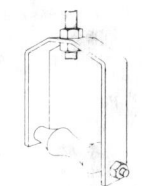

Adjustable Steel Yoke
Pipe Roll

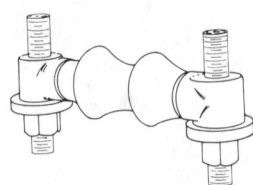

Adjustable Two-Rod
Roller Hanger

Figure 15.24   Installation Time in Man-Hours for Piping Supports (continued)

Description	Man-Hours	Unit
Chair Roll		
2″ Pipe Size	.118	Ea.
2-1/2″ Pipe Size	.123	Ea.
3″ Pipe Size	.129	Ea.
3-1/2″ Pipe Size	.133	Ea.
4″ Pipe Size	.138	Ea.
5″ Pipe Size	.143	Ea.
6″ Pipe Size	.151	Ea.
8″ Pipe Size	.160	Ea.
10″ Pipe Size	.167	Ea.
12″ Pipe Size	.174	Ea.
Pipe Straps		
1/2″ Pipe Size	.113	Ea.
3/4″ Pipe Size	.114	Ea.
1″ Pipe Size	.117	Ea.
1-1/4″ Pipe Size	.119	Ea.
1-1/2″ Pipe Size	.122	Ea.
2″ Pipe Size	.124	Ea.
2-1/2″ Pipe Size	.128	Ea.
3″ Pipe Size	.131	Ea.
3-1/2″ Pipe Size	.134	Ea.
4″ Pipe Size	.140	Ea.
U-Bolts		
1/2″ Pipe Size	.050	Ea.
3/4″ Pipe Size	.051	Ea.
1″ Pipe Size	.053	Ea.
1-1/4″ Pipe Size	.054	Ea.
1-1/2″ Pipe Size	.056	Ea.
2″ Pipe Size	.058	Ea.
2-1/2″ Pipe Size	.060	Ea.
3″ Pipe Size	.063	Ea.
3-1/2″ Pipe Size	.066	Ea.
4″ Pipe Size	.068	Ea.
5″ Pipe Size	.070	Ea.
6″ Pipe Size	.072	Ea.
8″ Pipe Size	.073	Ea.
10″ Pipe Size	.075	Ea.
12″ Pipe Size	.077	Ea.
U-Hooks		
3/4″ to 2″ Pipe Size	.083	Ea.

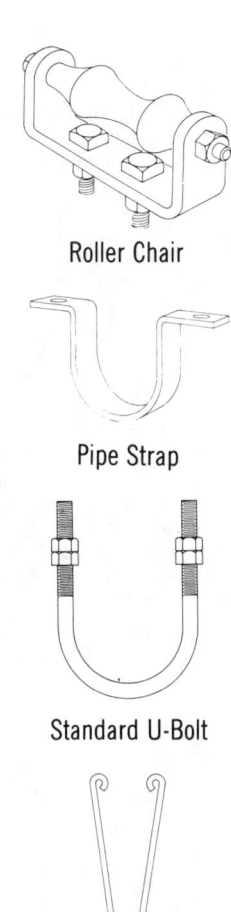

Roller Chair

Pipe Strap

Standard U-Bolt

U-Hook

Figure 15.25 Valve Selection Considerations

In any piping application, valve performance is critical. Valves should be selected to give the best performance at the lowest cost.

The following is a list of performance characteristics generally expected of valves.

1. Stopping flow or starting it.
2. Throttling flow (modulation).
3. Changing the direction of flow.
4. Checking backflow (permitting flow in only one direction).
5. Relieving or regulating pressure.

In order to properly select the right valve, some facts must be determined.

A. What liquid or gas will flow through the valve?
B. Does the fluid contain suspended particles?
C. Does the fluid remain in liquid form at all times?
D. Which metals does the fluid corrode?
E. What are the pressure and temperature limits? (As temperature and pressure rise, so will the price of the valve.)
F. Is there constant line pressure?
G. Is the valve merely an on–off valve?
H. Will checking of backflow be required?
I. Will the valve operate frequently or infrequently?

Valves are classified by design type into such classifications as Gate, Globe, Angle, Check, Ball, Butterfly and Plug. They are also classified by end connection, stem, pressure restrictions and material such as bronze, cast iron, etc. Each valve has a specific use. A quality valve used correctly will provide a lifetime of trouble-free service, but even a high quality valve installed in the wrong service may require frequent attention.

Figure 15.26   Valve Materials, Service Pressures, Definitions

Valve Materials

Bronze:
Bronze is one of the oldest materials used to make valves. It is most commonly used in hot and cold water systems and other non-corrosive services. It is often used as a seating surface in larger iron body valves to ensure tight closure.

Carbon Steel:
Carbon steel is a high strength material. Therefore, valves made from this metal are used in higher pressure services, such as steam lines up to 600 psi at 850°F. Many steel valves are available with butt-weld ends for economy and are generally used in high pressure steam service as well as other higher pressure non-corrosive services.

Forged Steel:
Valves from tough carbon steel are used in service up to 2000 psi and temperatures up to 1000°F in Gate, Globe and Check valves.

Iron:
Valves are normally used in medium to large pipe lines to control non-corrosive fluid and gases, where pressures do not exceed 250 psi at 450° or 500 psi cold water, oil or gas.

Stainless Steel:
Developed steel alloys can be used in over 90% corrosive services.

Plastic PVC:
This is used in a great variety of valves, generally in high corrosive service with lower temperatures and pressures.

Valve Service Pressures
Pressure ratings on valves provide an indication of the safe operating pressure for a valve at some elevated temperature. This temperature is dependent upon the materials used and the fabrication of the valve. When specific data is not available, a good "rule-of-thumb" to follow is the temperature of saturated steam on the primary rating indicated on the valve body. Example: The valve has the number 150S printed on the side indicating 150 psi and hence, a maximum operating temperature of 367°F (temperature of saturated steam and 150 psi).

Valve Definitions
1. "WOG" – Water, oil, gas (cold working pressures).
2. "SWP" – Steam working pressure.
3. 100% area (full port) – means the area through the valve is equal to or greater than the area of standard pipe.
4. "Standard Opening" – means that the area through the valve is less than the area of standard pipe and therefore these valves should be used only where restriction of flow is unimportant.
5. "Round Port" – means the valve has a full round opening through the plug and body, of the same size and area as standard pipe.
6. "Rectangular Port" – valves have rectangular shaped ports through the plug body. The area of the port is either equal to 100% of the area of standard pipe, or restricted (standard opening). In either case it is clearly marked.
7. "ANSI" – American National Standards Institute.

Figure 15.27a   Stem Types

(O.S. & Y) - Rising Stem — Outside Screw and Yoke
Offers a visual indication of whether the valve is
open or closed. The stem threads are engaged
by the yoke bushing so the stem rises through
the hand wheel as it is turned.

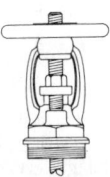

(R.S.) – Rising Stem – Inside Screw
Adequate clearance for operation must be
provided because both the hand wheel and the
stem rise.
The valve wedge position is indicated by the
position of the stem and hand wheel.

(N.R.S.) – Non-Rising Stem – Inside Screw
A minimum clearance is required for operating
this type of valve. Excessive wear or damage to
stem threads inside the valve may be caused by
heat, corrosion, and solids. Because the hand
wheel and stem do not rise, wedge position
cannot be visually determined.

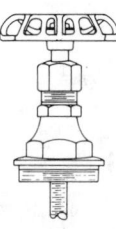

Figure 15.27b   Valve Types

Gate Valves
Provide full flow, minute pressure drop, minimum turbulence and minimum fluid trapped in the line.
They are normally used where operation is infrequent.

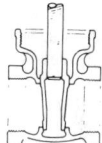

Globe Valves
Globe valves are designed for throttling and/or frequent operation with positive shut-off. Particular attention must be paid to the several types of seating materials available to avoid unnecessary wear. The seats must be compatible with the fluid in service and may be composition or metal. The configuration of the Globe valve opening causes turbulence which results in increased resistance. Most bronze Globe valves are rising stem-inside screw, but they are also available on O.S. & Y.

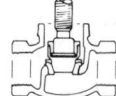

Angle Valves
The fundamental difference between the Angle valve and the Globe valve is the fluid flow through the Angle valve. It makes a 90° turn and offers less resistance to flow than the Globe valve while replacing an elbow. An Angle valve thus reduces the number of joints and installation time.

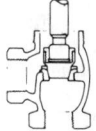

Check Valves
Check valves are designed to prevent backflow by automatically seating when the direction of fluid is reversed.
Swing Check valves are generally installed with Gate valves, as they provide comparable full flow.
Usually recommended for lines where flow velocities are low and should not be used on lines with pulsating flow. Recommended for horizontal installation, or in vertical lines only where flow is upward.

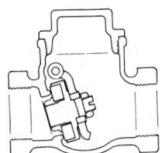

(continued on next page)

Figure 15.27b   Valve Types (continued)

Lift Check Valves

These are commonly used with Globe and Angle valves since they have similar diaphragm seating arrangements and are recommended for preventing backflow of steam, air, gas and water, and on vapor lines with high flow velocities. For horizontal lines, horizontal lift checks should be used and vertical lift checks for vertical lines.

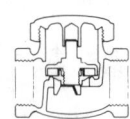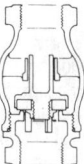

Ball Valves

Ball valves are light and easily installed, yet because of modern elastomeric seats, provide tight closure. Flow is controlled by rotating up to 90° a drilled ball which fits tightly against resilient seals. This ball seats with flow in either direction, and valve handle indicates the degree of opening. Recommended for frequent operation, readily adaptable to automation, ideal for installation where space is limited.

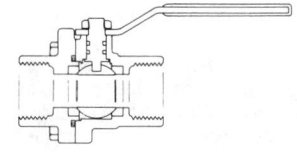

Butterfly Valves

Butterfly valves provide bubble-tight closure with excellent throttling characteristics. They can be used for full-open, closed and for throttling applications.

The Butterfly valve consists of a disc within the valve body which is controlled by a shaft.

In its closed position, the valve disc seals against a resilient seat. The disc position throughout the full 90° rotation is visually indicated by the position of the operator.

A Butterfly valve is only a fraction of the weight of a Gate valve and requires no gaskets between flanges in most cases. Recommended for frequent operation and adaptable to automation where space is limited.

Wafer and Lug type bodies, when installed between two pipe flanges, can be easily removed from the line. The pressure of the bolted flanges holds the valve in place.

Locating lugs makes installation easier.

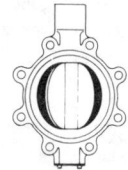

Plug Valves

Lubricated plug valves, because of the wide range of service to which they are adapted, may be classified as all purpose valves. They can be safely used at all pressures and vacuums, and at all temperatures up to the limits of available lubricants. They are the most satisfactory valves for the handling of gritty suspensions and many other destructive, erosive, corrosive and chemical solutions.

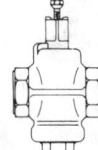

Figure 15.28 Installation Time in Man-Hours for Valves

Description	Man-Hours	Unit
Valves		
Threaded or Solder End		
1/8" Diameter	.333	Ea.
1/4" Diameter	.333	Ea.
3/8" Diameter	.333	Ea.
1/2" Diameter	.333	Ea.
3/4" Diameter	.400	Ea.
1" Diameter	.421	Ea.
1-1/4" Diameter	.533	Ea.
1-1/2" Diameter	.615	Ea.
2" Diameter	.727	Ea.
2-1/2" Diameter	1.067	Ea.
3" Diameter	1.231	Ea.
Flanged End		
2" Diameter	1.600	Ea.
2-1/2" Diameter	3.200	Ea.
3" Diameter	3.556	Ea.
3-1/2" Diameter	5.333	Ea.
4" Diameter	5.333	Ea.
5" Diameter	7.059	Ea.
6" Diameter	8.000	Ea.
8" Diameter	9.600	Ea.
10" Diameter	10.909	Ea.
12" Diameter	14.118	Ea.
Plastic Threaded or Socket End		
1/4" Diameter	.308	Ea.
1/2" Diameter	.308	Ea.
3/4" Diameter	.320	Ea.
1" Diameter	.348	Ea.
1-1/4" Diameter	.381	Ea.
1-1/2" Diameter	.400	Ea.
2" Diameter	.471	Ea.
2-1/2" Diameter	.615	Ea.
3" Diameter	.667	Ea.
Backflow Preventers		
Threaded End		
3/4" Diameter	.500	Ea.
1" Diameter	.571	Ea.
1-1/2" Diameter	.800	Ea.
2" Diameter	1.143	Ea.
Flanged End		
2-1/2" Diameter	3.200	Ea.
3" Diameter	3.556	Ea.
4" Diameter	5.333	Ea.
6" Diameter	8.000	Ea.
8" Diameter	12.000	Ea.
10" Diameter	24.000	Ea.

Double Check

Reduced Pressure — Flanged

Reduced Pressure — Threaded

Backflow Preventers

Figure 15.29 Roof Drain Sizing Chart

Pipe Diameter	Max. S.F. Roof Area	Gal./Min.
2″	544	23
3″	1,610	67
4″	3,460	144
5″	6,280	261
6″	10,200	424
8″	22,000	913

Figure 15.30 Installation Time in Man-Hours for Roof Storm Drains

Description	Man-Hours						Units
	2″	3″	4″	5″	6″	8″	
Roof Drain	1.143	1.231	1.333	1.600	2.000	2.290	Ea.
Cleanout Tee	2.000	2.222	2.424	2.909	3.200	6.400	Ea
C.I. Pipe (no hub)	.262	.276	.302	.324	.343	.533	L.F.
Elbow	1.230	1.334	1.454	1.714	2.000	3.368	Ea.
Tee	1.845	2.001	2.181	2.571	3.000	5.052	Ea.
Copper Tube	.348	.533	.696	.842	1.000	1.143	L.F.
Elbow	.727	1.455	1.778	2.667	2.667	3.000	Ea.
Tee	1.143	2.286	3.200	3.700	4.000	4.800	Ea.
Galvanized Steel Pipe	.250	.372	.444	.615	.774	.889	L.F.
Elbow	.889	1.600	2.667	3.200	3.429	4.000	Ea.
Tee	1.455	2.667	4.000	5.333	6.000	8.000	Ea.
DWV Pipe	.271	.302	.333	.370	.410	.500	L.F.
(ABS/PVC) Elbow	.571	.941	1.143	1.333	2.000	2.400	Ea.
Tee	.941	1.455	1.778	2.000	3.200	4.000	Ea.

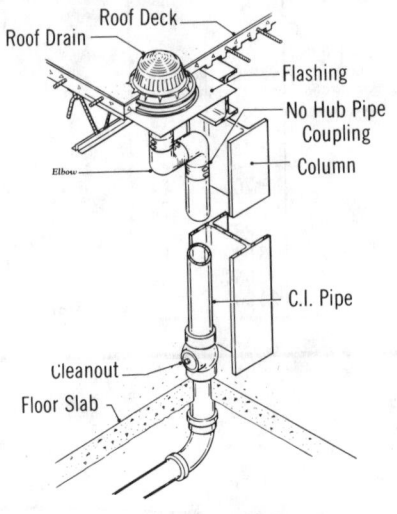

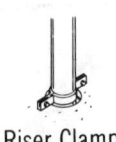

Riser Clamp

Figure 15.31 Minimum Plumbing Fixture Requirements

Type of Building/Use	Water Closets		Urinals		Lavatories	
	Number of Persons	Required Fixtures	Number of Persons	Required Fixtures	Number of Persons	Required Fixtures
Assembly Halls Auditoriums Theater Public Assembly	1-100 101-200 201-400	1 2 3	1-200 201-400 401-600	1 2 3	1-200 201-400 401-750	1 2 3
	Over 400 add 1 fixture for ea. 500 men; 1 fixture for ea. 300 women		Over 600 add 1 fixture for each 300 men		Over 750 add 1 fixture for each 500 persons.	

Bathtubs/Showers: No minimum required
Drinking Fountains: 1 for each 1000 persons
Service Sink: 1 minimum required

Type of Building/Use	Water Closets		Urinals		Lavatories	
	Number of Persons	Required Fixtures	Number of Persons	Required Fixtures	Number of Persons	Required Fixtures
Assembly Public Worship	300 men 150 women	1 1	300 men	1	300 men 150 women	1 1

Type of Building/Use	Water Closets	Urinals	Lavatories
Dormitories	Men: 1 for each 10 persons; Women: 1 for each 8 persons	1 for each 25 men, over 150 and 1 fixture for each 50 men	1 for each 12 persons 1 separate dental lavatory for each 50 persons recommended.

Bathtubs/Showers: 1 for each 8 persons; for over 150 persons add 1 for each 20 persons; for women add 1 additional for each 30 persons
Drinking Fountains: 1 for each 75 persons
Service Sink: 1 for each 100 persons
Laundry Trays: 1 for each 50 persons

Type of Building/Use	Water Closets	Urinals	Lavatories
Dwellings Apartments and homes	1 fixture for each unit		1 fixture for each unit

Bathtubs/Showers: 1 for each unit

(continued on next page)

Figure 15.31   Minimum Plumbing Fixture Requirements (continued)

Type of Building/Use	Water Closets		Urinals	Lavatories	
	Persons	Required		Persons	Required
Hospitals					
Individual Room		1			1
Ward	8 persons	1		10 persons	1
Waiting room		1			1

Bathtubs/Showers: 1 for each individual room;
                    1 for each 20 persons in a ward
Drinking Fountains: 1 for each 100 persons in a ward
Service Sink: 1 for each floor

Type of Building/Use	Water Closets		Urinals		Lavatories	
	Number of Persons	Required Fixtures	Number of Persons	Required Fixtures	Number of Persons	Required Fixtures
Industrial						
Mfg. plants	1-10	1	0-30	1	1-100	1 for ea. 10
Warehouses	11-25	2	31-80	2		
	26-50	3	81-160	3		
	51-75	4	161-240	4	over 100	1 for ea. 15
	76-100	5				
	1 fixture for each additional 30 persons					

Bathtubs/Showers: 1 shower for each 15 persons subject to excessive heat or
                    occupational hazard
Drinking Fountains: 1 for each 75 persons

Type of Building/Use	Water Closets		Urinals	Lavatories	
	Number of Persons	Required Fixtures		Number of Persons	Required Fixtures
Public Buildings					
Businesses	1-15	1	Urinals may be	1-15	1
Offices	16-35	2	provided in place of	16-35	2
	36-55	3	water closets but may	36-60	3
	56-80	4	not replace more than	61-90	4
	81-110	5	1/3 required number of	91-125	5
	111-150	6	men's water closets		
	1 fixture for each additional 40 persons			1 fixture for each additional 45 persons	

Drinking Fountains: 1 for each 75 persons
Service Sinks: 1 for each floor

Figure 15.31 Minimum Plumbing Fixture Requirements (continued)

Type of Building/Use	Water Closets	Urinals	Lavatories
Schools Elementary	1 for ea. 30 boys 1 for ea. 25 girls	1 for ea. 25 boys	1 for ea. 35 boys 1 for ea. 35 girls

Showers: (For gym or pool) 1 for each 5 students
Drinking Fountains: 1 for each 40 students

Type of Building/Use	Water Closets	Urinals	Lavatories
Schools Secondary	1 for ea. 40 boys 1 for ea. 30 girls	1 for ea. 25 boys	1 for ea. 40 boys 1 for ea. 40 girls

Showers: (For gym or pool) 1 for each 5 students
Drinking Fountains: 1 for each 50 students

Figure 15.32 Fixture Demands in Gallons per Hour per Fixture

Table below is based on 140°F final temperature except for dishwashers in public places (*) where 180°F water is mandatory.

Fixture	Apt. House	Club	Gym	Hos-pital	Hotel	Indust. Plant	Office	Private Home	School
Bathtubs	20	20	30	20	20			20	
Dishwashers, automatic*	15	50-150		50-150	50-200	20-100		15	20-100
Kitchen sink	10	20		20	30	20	20	10	20
Laundry, stationary tubs	20	28		28	28			20	
Laundry, automatic wash	75	75		100	150			75	
Private lavatory	2	2	2	2	2	2	2	2	2
Public lavatory	4	6	8	6	8	12	6		15
Showers	30	150	225	75	75	225	30	30	225
Service sink	20	20		20	30	20	20	15	20
Demand factor	0.30	0.30	0.40	0.25	0.25	0.40	0.30	0.30	0.40
Storage capacity factor	1.25	0.90	1.00	0.60	0.80	1.00	2.00	0.70	1.00

*To obtain the probable maximum demand multiply the total demands for the fixtures (gal./fixture/hour) by the demand factor. The heater should have a heating capacity in gallons per hour equal to this maximum. The storage tank should have a capacity in gallons equal to the probable maximum demand multiplied by the storage capacity factor.

Figure 15.33 Water Cooler Application/Capacities

Type of Service	Requirement
Office, School or Hospital	12 persons per gallon per hour
Office Lobby or Department Store	4 or 5 gallons per hour per fountain
Light manufacturing	7 persons per gallon per hour
Heavy manufacturing	5 persons per gallon per hour
Hot heavy manufacturing	4 persons per gallon per hour
Hotels	.08 gallons per hour per room
Theatre	1 gallon per hour per 100 seats

Figure 15.34 Drainage Requirements for Plumbing Fixtures

Drainage lines must have a slope to maintain flow for proper operation. This slope should not be less than 1/4" per foot for 3" diameter or smaller pipe and not less than 1/8" per foot for 4" diameter or larger pipe. The capacity of building drainage systems is calculated on a basis of "drainage fixture units" (d.f.u.) as per the following chart.

Type of Fixture	d.f.u. Value	Type of Fixture	d.f.u. Value
Automatic clothes washer (2" standpipe)	3	Service sink (trap standard)	3
Bathroom group (water closet, lavatory and		Service sink (P trap)	2
bathtub or shower) tank type closet	6	Urinal, pedestal, syphon jet blowout	6
Bathtub (with or without overhead shower)	2	Urinal, wall hung	4
Clinic sink	6	Urinal, stall washout	4
Combination sink & tray with food disposal	4	Wash sink (circ. or mult.) per faucet set	2
Dental unit or cuspidor	1	Water closet, tank operated	4
Dental lavatory	1	Water closet, valve operated	6
Drinking fountain	1/2	Fixtures not listed above	
Dishwasher, domestic	2	Trap size 1-1/4" or smaller	1
Floor drains with 2" waste	3	Trap size 1-1/2"	2
Kitchen sink, domestic with one 1-1/2" trap	2	Trap size 2"	3
Kitchen sink, domestic with food disposal	2	Trap size 2-1/2"	4
Lavatory with 1-1/4" waste	1	Trap size 3"	5
Laundry tray (1 or 2 compartment)	2	Trap size 4"	6
Shower stall, domestic	2		

For continuous or nearly continuous flow into the system from a pump, air conditioning equipment or other item, allow 2 fixture units for each gallon per minute of flow.

When the "drainage fixture units" (d.f.u.) for each horizontal branch or vertical stack is computed from the table above, the appropriate pipe size for each branch or stack is determined from the table below.

Figure 15.35   Allowable Fixture Units (d.f.u.) for Branches and Stacks

Pipe Diam.	Horiz. Branch (not incl. drains)	Stack Size for 3 Stories or 3 Levels	Stack Size for Over 3 Levels	Maximum for 1 Story Building Stack
1-1/2"	3	4	8	2
2"	6	10	24	6
2-1/2"	12	20	42	9
3"	20*	48*	72*	20*
4"	160	240	500	90
5"	360	540	1100	200
6"	620	960	1900	350
8"	1400	2200	3600	600
10"	2500	3800	5600	1000
12"	3900	6000	8400	1500
15"	7000			

*Not more than two water closets or bathroom groups within each branch interval nor more than six water closets or bathroom groups on the stack. Stacks sized for the total may be reduced as load decreases at each story to a minimum diameter of 1/2 the maximum diameter.

Figure 15.36 Installation Time in Man-Hours
for Rough-in of Plumbing Fixtures

Description	Man-Hours	Unit
For Roughing-In		
Bath Tub	7.730	Ea.
Bidet	8.990	Ea.
Dental Fountain	6.900	Ea.
Drinking Fountain	4.370	Ea.
Lavatory		
Vanity Top	6.960	Ea.
Wall Hung	9.640	Ea.
Laundry Sinks	7.480	Ea.
Prison/Institution Fixtures		
Lavatory	10.670	Ea.
Service Sink	17.978	Ea.
Urinal	10.738	Ea.
Water Closet	13.445	Ea.
Combination Water Closet and Lavatory	16.000	Ea.
Shower Stall	7.800	Ea.
Sinks		
Corrosion Resistant	7.920	Ea.
Kitchen, Countertop	7.480	Ea.
Kitchen, Raised Deck	8.650	Ea.
Service, Floor	9.760	Ea.
Service, Wall	12.310	Ea.
Urinals		
Wall Hung	5.650	Ea.
Stall Type	8.040	Ea.
Wash Fountain, Group	8.790	Ea.
Water Closets		
Tank Type, Wall Hung	6.150	Ea.
Floor Mount, One Piece	8.250	Ea.
Bowl Only, Wall Hung	7.800	Ea.
Bowl Only, Floor Mount	8.890	Ea.
Gang, Side by Side, First	8.120	Ea.
Each Additional	7.480	Ea.
Gang, Back to Back, First Pair	9.090	Pair
Each Additional Pair	8.840	Pair
Water Conserving Type	8.250	Ea.
Water Cooler	3.620	Ea.

Figure 15.37 Installation Time in Man-Hours for Plumbing Fixtures

Description	Man-Hours	Unit
For Setting Fixture and Trim		
Bath Tub	3.636	Ea.
Bidet	3.200	Ea.
Dental Fountain	2.000	Ea.
Drinking Fountain	2.500	Ea.
Lavatory		
Vanity Top	2.500	Ea.
Wall Hung	2.000	Ea.
Laundry Sinks	2.667	Ea.
Prison/Institution Fixtures		
Lavatory	2.000	Ea.
Service Sink	5.333	Ea.
Urinal	4.000	Ea.
Water Closet	2.759	Ea.
Combination Water Closet and Lavatory	3.200	Ea.
Shower Stall	8.000	Ea.
Sinks		
Corrosion Resistant	5.333	Ea.
Kitchen, Countertop	3.330	Ea.
Kitchen, Raised Deck	7.270	Ea.
Service, Floor	3.640	Ea.
Service, Wall	4.000	Ea.
Urinals		
Wall Hung	5.333	Ea.
Stall Type	6.400	Ea.
Wash Fountain, Group	9.600	Ea.
Water Closets		
Tank Type, Wall Hung	3.019	Ea.
Floor Mount, One Piece	3.019	Ea.
Bowl Only, Wall Hung	2.759	Ea.
Bowl Only, Floor Mount	2.759	Ea.
Gang, Side by Side, First	2.759	Ea.
Each Additional	2.759	Ea.
Gang, Back to Back, First Pair	5.520	Pair
Each Additional Pair	5.520	Pair
Water Conserving Type	2.963	Ea.
Water Cooler	4.000	Ea.

Figure 15.38 Hot Water Consumption Rates for Commercial Applications

Type of Building	Size Factor	Maximum Hourly Demand	Average Day Demand
Apartment Dwellings	No. of apartments: Up to 20 21 to 50 51 to 75 76 to 100 101 to 200 201 up	12.0 gal. per apt. 10.0 gal. per apt. 8.5 gal. per apt. 7.0 gal. per apt. 6.0 gal. per apt. 5.0 gal. per apt.	42.0 gal. per apt. 40.0 gal. per apt. 38.0 gal. per apt. 37.0 gal. per apt. 36.0 gal. per apt. 35.0 gal. per apt.
Dormitories	Men Women	3.8 gal. per man 5.0 gal. per woman	13.1 gal. per man 12.3 gal. per woman
Hospitals	Per bed	23.0 gal. per Patient	90.0 gal. per Patient
Hotels	Single room with bath Double room with bath	17.0 gal. per unit 27.0 gal. per unit	50.0 gal. per unit 80.0 gal. per unit
Motels	No. of units: Up to 20 21 to 100 101 Up	6.0 gal. per unit 5.0 gal. per unit 4.0 gal. per unit	20.0 gal. per unit 14.0 gal. per unit 10.0 gal. per unit
Nursing Homes		4.5 gal. per bed	18.4 gal. per bed
Office Buildings		0.4 gal. per person	1.0 gal. per person
Restaurants	Full meal type Drive-in snack type	1.5 gal./max. meals/hr. 0.7 gal./max. meals/hr.	2.4 gal. per meal 0.7 gal. per meal
Schools	Elementary Secondary and high	0.6 gal. per student 1.0 gal. per student	0.6 gal. per student 1.8 gal. per student

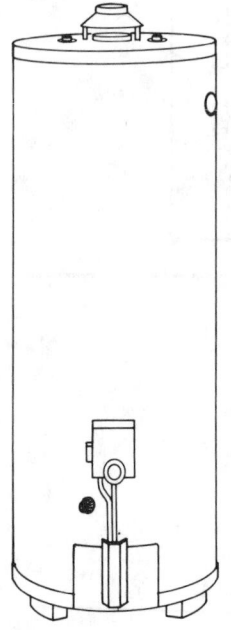

Residential Gas Fired Water Heater

Figure 15.39 Installation Time in Man-Hours for Commercial Gas Hot Water Heaters

Description	Man-Hours	Unit
Gas Fired Water, Commercial, 100°F Rise		
75.5 MBH Input, 63 Gal./Hr.	5.714	Ea.
100 MBH Input, 91 Gal./Hr.	5.714	Ea.
155 MBH Input, 150 Gal./Hr.	10.000	Ea.
200 MBH Input, 192 Gal./Hr.	13.333	Ea.
300 MBH Input, 278 Gal./Hr.	20.000	Ea.
390 MBH Input, 374 Gal./Hr.	20.000	Ea.
500 MBH Input, 480 Gal./Hr.	22.857	Ea.
600 MBH Input, 576 Gal./Hr.	26.667	Ea.
800 MBH Input, 768 Gal./Hr.	32.000	Ea.
1000 MBH Input, 960 Gal./Hr.	32.000	Ea.
1500 MBH Input, 1440 Gal./Hr.	40.000	Ea.
1800 MBH Input, 1730 Gal./Hr.	48.000	Ea.
2450 MBH Input, 2350 Gal./Hr.	60.000	Ea.
3000 MBH Input, 2880 Gal./Hr.	80.000	Ea.
3750 MBH Input, 3600 Gal./Hr.	80.000	Ea.

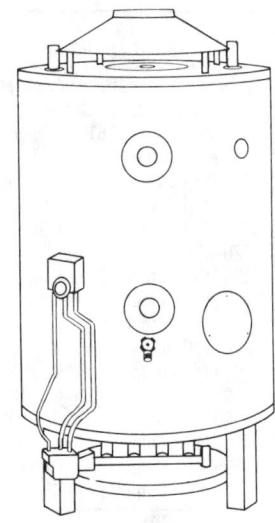

Gas Fired
Water Heater —
Commercial

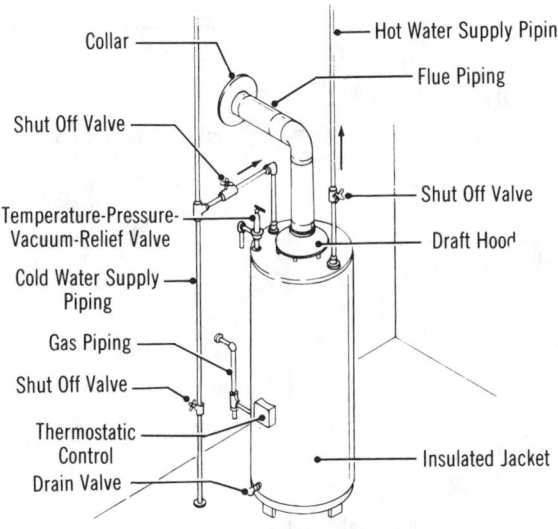

Collar

Hot Water Supply Piping

Flue Piping

Shut Off Valve

Temperature-Pressure-
Vacuum-Relief Valve

Shut Off Valve

Draft Hood

Cold Water Supply
Piping

Gas Piping

Shut Off Valve

Thermostatic
Control

Drain Valve

Insulated Jacket

Gas Fired Hot Water System

Figure 15.40 Installation Time in Man-Hours for Commercial Oil Fired Hot Water Heaters

Description	Man-Hours	Unit
Oil Fired Water Heater, Commercial, 100°F Rise		
97 MBH Output, 116 Gal./Hr.	7.273	Ea.
134 MBH Output, 161 Gal./Hr.	10.000	Ea.
161 MBH Output, 192 Gal./Hr.	13.333	Ea.
187 MBH Output, 224 Gal./Hr.	16.000	Ea.
262 MBH Output, 315 Gal./Hr.	22.857	Ea.
341 MBH Output, 409 Gal./Hr.	22.857	Ea.
420 MBH Output, 504 Gal./Hr.	26.667	Ea.
525 MBH Output, 630 Gal./Hr.	32.000	Ea.
630 MBH Output, 756 Gal./Hr.	32.000	Ea.
735 MBH Output, 880 Gal./Hr.	40.000	Ea.
840 MBH Output, 1000 Gal./Hr.	40.000	Ea.
1050 MBH Output, 1260 Gal./Hr.	40.000	Ea.
1365 MBH Output, 1640 Gal./Hr.	48.000	Ea.
1680 MBH Output, 2000 Gal./Hr.	60.000	Ea.
2310 MBH Output, 2780 Gal./Hr.	80.000	Ea.
2835 MBH Output, 3400 Gal./Hr.	80.000	Ea.
3150 MBH Output, 3780 Gal./Hr.	120.000	Ea.

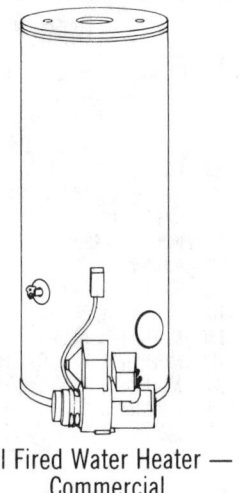

Oil Fired Water Heater — Commercial

Figure 15.41 Installation Time in Man-Hours for Commercial Electric Hot Water Heaters

Description	Man-Hours	Unit
Electric Water Heater, Commercial, 100°F Rise		
50 Gal. Tank, 9 KW 37 Gal./Hr.	4.444	Ea.
80 Gal., 12 KW 49 Gal./Hr.	5.333	Ea.
36 KW 147 Gal./Hr.	5.333	Ea.
120 Gal., 36 KW 147 Gal./Hr.	6.667	Ea.
150 Gal., 120 KW 490 Gal./Hr.	8.000	Ea.
200 Gal., 120 KW 490 Gal./Hr.	9.412	Ea.
250 Gal., 150 KW 615 Gal./Hr.	10.667	Ea.
300 Gal., 180 KW 738 Gal./Hr.	12.308	Ea.
350 Gal., 30 KW 123 Gal./Hr.	14.545	Ea.
180 KW 738 Gal./Hr.	14.545	Ea.
500 Gal., 30 KW 123 Gal./Hr.	20.000	Ea.
240 KW 984 Gal./Hr.	20.000	Ea.
700 Gal., 30 KW 123 Gal./Hr.	24.000	Ea.
300 KW 1230 Gal./Hr.	24.000	Ea.
1000 Gal., 60 KW 245 Gal./Hr.	34.286	Ea.
480 KW 1970 Gal./Hr.	34.286	Ea.
1500 Gal., 60 KW 245 Gal./Hr.	48.000	Ea.
480 KW 1970 Gal./Hr.	48.000	Ea.
2000 Gal., 60 KW 245 Gal./Hr.	80.000	Ea.
480 KW 1970 Gal./Hr.	80.000	Ea.

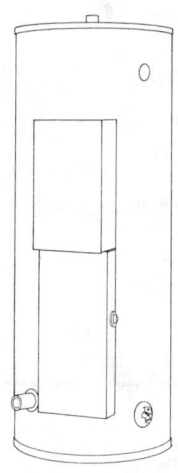

Electric Water Heater — Commercial

Figure 15.42 Installation Time in Man-Hours for Commercial Hot Water Heater Components

Description	Man-Hours	Unit
Water Heater, Commercial, Gas, 75.5 MBH, 63 gal./hr	5.714	Ea.
Copper Tubing, Type L, Solder Joint, Hanger 10' On Center, 1" Diameter	.118	L.F.
Wrought Copper 90° Elbow for Solder Joints 1" Diameter	.500	Ea.
Wrought Copper Tee for Solder Joints 1" Diameter	.800	Ea.
Wrought Copper Union for Solder Joints 1" Diameter	.533	Ea.
Valve, Gate, Bronze, 125 lb., NRS, Soldered 1" Diameter	.421	Ea.
Relief Valve, Bronze, Press and Temp, Self Close, 3/4" IPS	.286	Ea.
Copper Tubing, Type L, Solder Joints, 3/4" Diameter	.105	L.F.
Wrought Copper 90° Elbow for Solder Joints 3/4" Diameter	.421	Ea.
Wrought Copper, Adapter, CTS to MPT, 3/4" IPS	.381	Ea.
Pipe Steel Black, Schedule 40, Threaded, 3/4" Diameter	.131	L.F.
Pipe, 90° Elbow, Malleable Iron Black, 150 lb. Threaded, 3/4" Diameter	.571	Ea.
Pipe, Union with Brass Seat, Malleable Iron Black, 3/4" Diameter	.615	Ea.
Valve, Gas Stop w/o Check, Brass, 3/4" IPS	.364	Ea.

Figure 15.43 Installation Time in Man-Hours for Residential Hot Water Heaters

Number of Bedrooms	H.W. Storage Capacity	Man-Hours	Unit
1	20 gal.	3.810	Ea.
2	30 gal.	4.000	Ea.
3	40 gal.	4.211	Ea.
4	50 gal.	4.444	Ea.

(For each additional bedroom go to the next larger size 70, 85, and 100 gallon.)

Figure 15.44   Installation Time in Man-Hours for Pumps

Description	Man-Hours	Unit
Circulating Pumps, Heating, Cooling, Potable Water		
In Line Type		
3/4" Pipe Size 1/40 H.P.	1.000	Ea.
3/4" Through 1-1/2" Pipe Size 1/3 H.P.	2.667	Ea.
2" Pipe Size 1/6 H.P.	3.200	Ea.
2-1/2" Pipe Size 1/4 H.P.	3.200	Ea.
3" Pipe Size 1/3 H.P. through 1 H.P.	4.000	Ea.
Close Coupled Type		
1-1/2" Pipe Size 1-1/2 H.P. 50 GPM	5.333	Ea.
2" Pipe Size 3 H.P. 90 GPM	6.957	Ea.
2-1/2" Pipe Size 3 H.P. 150 GPM	8.000	Ea.
3" Pipe Size 5 H.P. 225 GPM	8.889	Ea.
4" Pipe Size 7-1/2 H.P. 350 GPM	10.000	Ea.
5" Pipe Size 15 H.P. 1000 GPM	14.118	Ea.
6" Pipe Size 25 H.P. 1550 GPM	16.000	Ea.
Base Mounted Type		
2-1/2" Pipe Size 3 H.P. 150 GPM	8.889	Ea.
3" Pipe Size 5 H.P. 225 GPM	10.000	Ea.
4" Pipe Size 7-1/2 H.P. 350 GPM	10.667	Ea.
5" Pipe Size 15 H.P. 1000 GPM	15.000	Ea.
6" Pipe Size 25 H.P. 1550 GPM	17.143	Ea.
8" Pipe Size 30 H.P. 1660 GPM	20.000	Ea.
10" Pipe Size 40 H.P. 1800 GPM	20.000	Ea.
12" Pipe Size 50 H.P. 2200 GPM	24.000	Ea.
Pumps		
Sewage Ejector with Basin and Cover		
Single		
110 GPM 1/2 H.P.	6.400	Ea.
173 GPM 3/4 H.P.	8.000	Ea.
218 GPM 1 H.P.	10.000	Ea.
285 GPM 2 H.P.	12.308	Ea.
325 GPM 3 H.P.	17.143	Ea.
370 GPM 5 H.P.	24.000	Ea.
Duplex		
110 GPM 1/2 H.P.	8.000	Ea.
173 GPM 3/4 H.P.	10.000	Ea.
218 GPM 1 H.P.	20.000	Ea.
285 GPM 2 H.P.	24.000	Ea.
325 GPM 3 H.P.	30.000	Ea.
370 GPM 5 H.P.	48.000	Ea.
Sump Pumps, Submersible		
Pedestal Type Cellar Drainer with External Float		
42 GPM 1/3 H.P.	1.600	Ea.
Submersible with Built in Float		
22 GPM 1/4 H.P.	1.330	Ea.
68 GPM 1/2 H.P.	1.600	Ea.
94 GPM 1/2 H.P.	1.600	Ea.
105 GPM 1/2 H.P.	2.000	Ea.

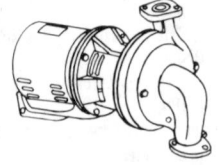

In-Line
Centrifugal Pump

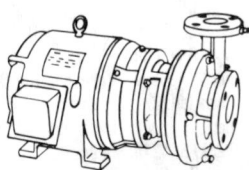

Close Coupled,
Centrifugal Pump

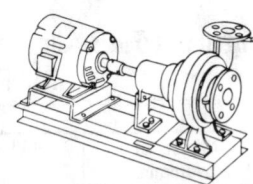

Base Mounted
Centrifugal Pump

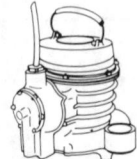

Submersible
Sump Pump

Figure 15.44 Installation Time in Man-Hours for Pumps (continued)

Description	Man-Hours	Unit
Well Pumps		
Deep Well Multi-Stage	10.000	Ea.
Shallow Well Jet Type	4.000	Ea.
Fire Pumps with Controller and Fittings		
Electric		
4″ Pipe Size 100 H.P. 500 GPM	51.619	Ea.
6″ Pipe Size 250 H.P. 1000 GPM	88.889	Ea.
8″ Pipe Size 300 H.P. 2000 GPM	114.286	Ea.
10″ Pipe Size 450 H.P. 3500 GPM	133.333	Ea.
Diesel		
4″ Pipe Size 111 H.P. 500 GPM	53.333	Ea.
6″ Pipe Size 255 H.P. 1000 GPM	80.000	Ea.
8″ Pipe Size 255 H.P. 2000 GPM	100.000	Ea.
10″ Pipe Size 525 H.P. 3500 GPM	160.000	Ea.

Figure 15.45   Types of Automatic Sprinkler Systems

Sprinkler systems may be classified by type as follows:

1. **Wet Pipe System.** A system employing automatic sprinklers attached to a piping system containing water and connected to a water supply so that water discharges immediately from sprinklers opened by a fire.

2. **Dry Pipe System.** A system employing automatic sprinklers attached to a piping system containing air under pressure. When the pressure is released from the opening of sprinklers, the water pressure opens a valve known as a "dry pipe valve." The water then flows into the piping system and out the opened sprinklers.

3. **Pre-Action System.** A system employing automatic sprinklers attached to a piping system containing air that may or may not be under pressure. Pre-action systems have heat-activated devices that are more sensitive than the automatic sprinklers themselves, installed in the same areas as the sprinklers. Actuation of the heat responsive system, as from a fire, opens a valve which permits water to flow into the sprinkler piping system and to be discharged from any sprinklers which may be open.

4. **Deluge System.** A dry system connected to a water supply through a valve which is opened by the operation of a heat responsive system (installed in the same areas as the sprinklers). When this valve opens, water flows into the piping system and discharges from all attached sprinklers.

5. **Combined Dry Pipe and Pre-Action Sprinkler System.** A system employing automatic sprinklers attached to a piping system containing air under pressure. A supplemental heat responsive system of generally more sensitive characteristics than the automatic sprinklers themselves is installed in the same areas as the sprinklers. Operation of the heat responsive system, as from a fire, actuates tripping devices which open dry pipe valves simultaneously and without loss of air pressure in the system. Operation of the heat responsive system also opens approved air exhaust valves at the end of the feed main which facilitates the filling of the system with water (which usually precedes the opening of sprinklers). The heat responsive system also serves as an automatic fire alarm system.

6. **Limited Water Supply System.** A system employing automatic sprinklers and conforming to automatic sprinkler standards, but supplied by a pressure tank of limited capacity.

7. **Chemical Systems.** Systems using halon, carbon dioxide, dry chemical or high expansion foam as selected for special requirements. The agent may extinguish flames by chemically inhibiting flame propagation, suffocate flames by excluding oxygen, interrupting chemical action of oxygen uniting with fuel or sealing and cooling the combustion center.

8. **Firecycle System.** Firecycle is a fixed fire protection sprinkler system utilizing water as its extinguishing agent. It is a time delayed, recycling, pre-action type which automatically shuts the water off when the heat is reduced below the detector operating temperature. Water is turned on again when that temperature is exceeded. The system senses a fire condition through a closed circuit electrical detector which controls water flow to the fire automatically. Batteries supply up to 90 hour emergency power supply for system operation. The piping system is dry (until water is required) and is monitored with pressurized air. Should any leak in the system piping occur, an alarm will sound, but water will not enter the system until heat is sensed by a Firecycle detector.

Area coverage sprinkler systems may be laid out and fed from the supply in any one of several patterns as shown in the illustration below. It is desirable, if possible, to utilize a central feed and achieve a shorter flow path from the riser to the furthest sprinkler. This permits use of the smallest sizes of pipe possible, with resulting savings.

Figure 15.45 Types of Automatic Sprinkler Systems (continued)

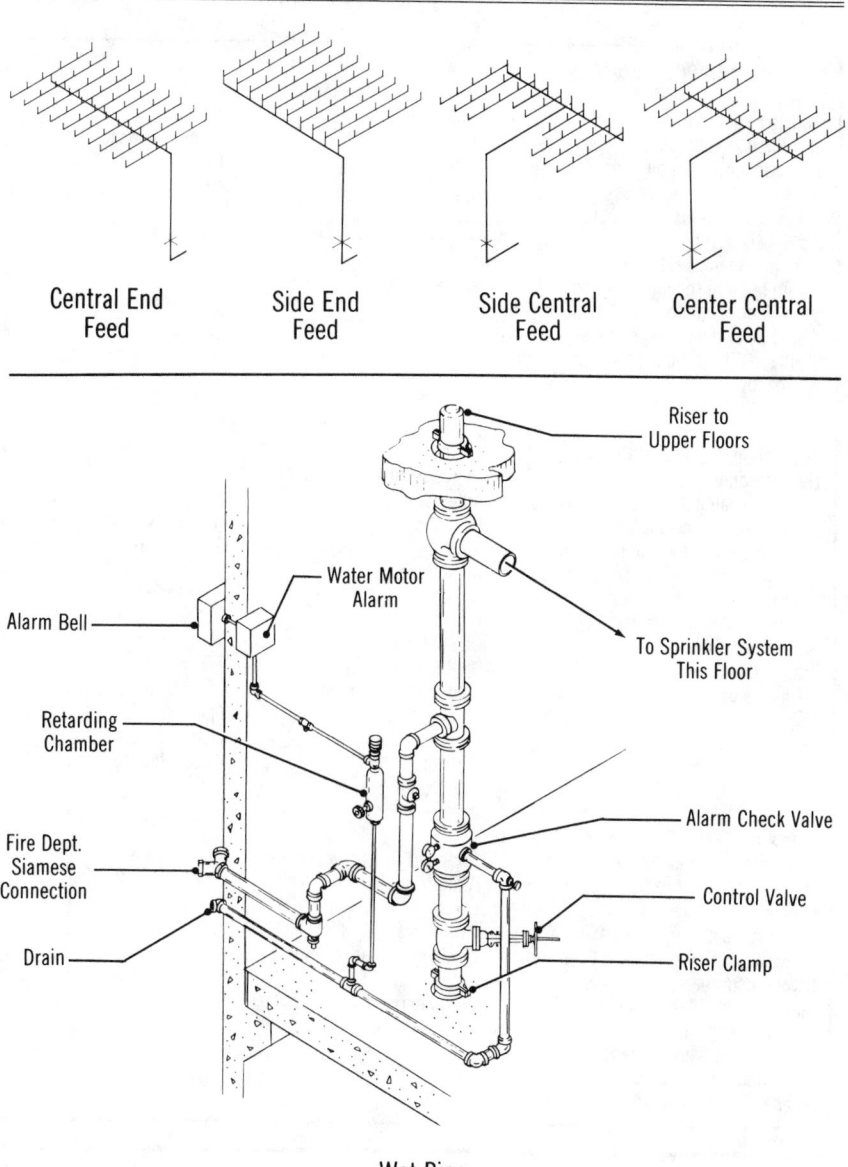

Central End Feed

Side End Feed

Side Central Feed

Center Central Feed

Riser to Upper Floors

Water Motor Alarm

Alarm Bell

To Sprinkler System This Floor

Retarding Chamber

Alarm Check Valve

Fire Dept. Siamese Connection

Control Valve

Drain

Riser Clamp

Wet Pipe

Figure 15.46   Classification of Sprinkler Systems

Rules for installation of sprinkler systems vary depending on the classification of occupancy falling into one of three categories as follows:

LIGHT HAZARD OCCUPANCY

The protection area allotted per sprinkler should not exceed 200 S.F. with the maximum distance between lines and sprinklers on lines being 15'. The sprinklers do not need to be staggered. Branch lines should not exceed eight sprinklers on either side of a cross main. Each large area requiring more than 100 sprinklers and without a sub-dividing partition should be supplied by feed mains or risers sized for ordinary hazard occupancy.

Included in this group are:

Auditoriums	Museums
Churches	Nursing Homes
Clubs	Offices
Educational	Residential
Hospitals	Restaurants
Institutional	Schools
Libraries	Theaters
(except large stack rooms)	

ORDINARY HAZARD OCCUPANCY

The protection area allotted per sprinkler shall not exceed 130 S.F. of noncombustible ceiling and 120 S.F. of combustible ceiling. The maximum allowable distance between sprinkler lines and sprinklers on lines is 15'. Sprinklers shall be staggered if the distance between heads exceeds 12'. Branch lines should not exceed eight sprinklers on either side of a cross main.

Included in this group are:

Automotive garages	Electric generating stations
Bakeries	Feed mills
Beverage manufacturing	Grain elevators
Bleacheries	Ice manufacturing
Boiler houses	Laundries
Canneries	Machine shops
Cement plants	Mercantiles
Clothing factories	Paper mills
Cold storage warehouses	Printing and Publishing
Dairy products	Shoe factories
manufacturing	Warehouses
Distilleries	Wood product assembly
Dry cleaning	

EXTRA HAZARD OCCUPANCY

The protection area allotted per sprinkler shall not exceed 90 S.F. of noncombustible ceiling and 80 S.F. of combustible ceiling. The maximum allowable distance between lines and between sprinklers on lines is 12'. Sprinklers on alternate lines shall be staggered if the distance between sprinklers on lines exceeds 8'. Branch lines should not exceed six sprinklers on either side of a cross main.

Included in this group are:

Aircraft hangars	Paint shops
Chemical works	Shade cloth manufacturing
Explosives manufacturing	Solvent extracting
Linoleum manufacturing	Varnish works
Linseed oil mills	Volatile, flammable, liquid
Oil refineries	manufacturing and use.

Figure 15.47   Sprinkler Head Quantities for Various Sizes and Types of Pipe

Sprinkler Quantities: The table below lists the usual maximum number of sprinkler heads for each size of copper and steel pipe for both wet and dry systems. These quantities may be adjusted to meet individual structural needs or local code requirements. Maximum area on any one floor for one system is: light hazard and ordinary hazard 52,000 S.F., extra hazardous 25,000 S.F.

Pipe Size Diameter	Light Hazard Occupancy		Ordinary Hazard Occupancy		Extra Hazard Occupancy	
	Steel Pipe Sprinklers	Copper Pipe Sprinklers	Steel Pipe Sprinklers	Copper Pipe Sprinklers	Steel Pipe Sprinklers	Copper Pipe Sprinklers
1"	2	2	2	2	1	1
1-1/4"	3	3	3	3	2	2
1-1/2"	5	5	5	5	5	5
2"	10	12	10	12	8	8
2-1/2"	30	40	20	25	15	20
3"	60	65	40	45	27	30
3-1/2"	100	115	65	75	40	45
4"			100	115	55	65
5"			160	180	90	100
6"			275	300	150	170

Dry Pipe Systems: A dry pipe system should be installed where a wet pipe system is impractical, as in rooms or buildings which cannot be properly heated.

The use of an approved dry pipe system is more desirable than shutting off the water supply during cold weather.

Not more than 750 gallons of system capacity should be controlled by one dry pipe valve. Where two or more dry pipe valves are used, systems should preferably be divided horizontally.

Figure 15.48 Installation Time in Man-Hours for Automatic Sprinkler Systems

Description	Man-Hours	Unit
Valve, Gate, Iron Body, Flanged OS and Y, 125 lb. 4" Diameter	5.333	Ea.
Valve, Swing Check, w/Ball Drip, Flanged, 4" Diameter	5.333	Ea.
Valve, Swing Check, Bronze, Thread End, 2-1/2" Diameter	1.067	Ea.
Valve, Angle, Bronze, Thread End, 2" Diameter	.727	Ea.
Valve, Gate, Bronze, Thread End, 1" Diameter	.421	Ea.
Alarm Valve, 2-1/2" Diameter	5.333	Ea.
Alarm, Water Motor, with Gong	2.000	Ea.
Fire Alarm Horn, Electric	.308	Ea.
Pipe, Black Steel, Threaded, Schedule 40, 4" Diameter	.444	L.F.
2-1/2" Diameter	.320	L.F.
2" Diameter	.250	L.F.
1-1/4" Diameter	.180	L.F.
1" Diameter	.151	L.F.
Pipe Tee, 150 lb. Black Malleable 4" Diameter	4.000	Ea.
2-1/2" Diameter	1.778	Ea.
2" Diameter	1.455	Ea.
1-1/4" Diameter	1.143	Ea.
1" Diameter	1.000	Ea.
Pipe Elbow, 150 lb. Black Malleable 1" Diameter	.615	Ea.
Sprinkler Head, 135° to 286°, 1/2" Diameter	.500	Ea.
Sprinkler Head, Dry Pendent 1" Diameter	.571	Ea.
Dry Pipe Valve, w/Trim and Gauges, 4" Diameter	16.000	Ea.
Deluge Valve, w/Trim and Gauges, 4" Diameter	16.000	Ea.
Deluge System Monitoring Panel 120 Volt	.444	Ea.
Thermostatic Release	.400	Ea.
Heat Detector	.500	Ea.
Firecycle Controls w/Panel, Batteries, Valves and Switches	32.000	Ea.
Firecycle Package, Check and Flow Control Valves, Trim, 4" Diameter	16.000	Ea.
Air Compressor, Automatic, 200 gal. Sprinkler System 1/3 H.P.	6.154	Ea.

Figure 15.48 Installation Time in Man-Hours for Automatic Sprinkler Systems (continued)

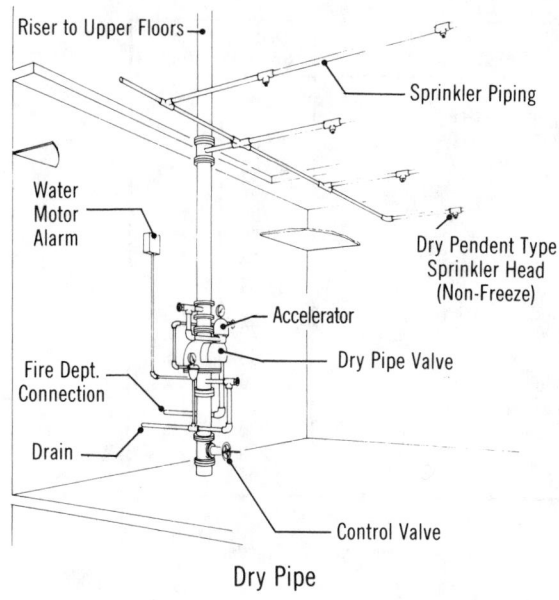

Riser to Upper Floors

Sprinkler Piping

Water Motor Alarm

Dry Pendent Type Sprinkler Head (Non-Freeze)

Accelerator

Dry Pipe Valve

Fire Dept. Connection

Drain

Control Valve

Dry Pipe

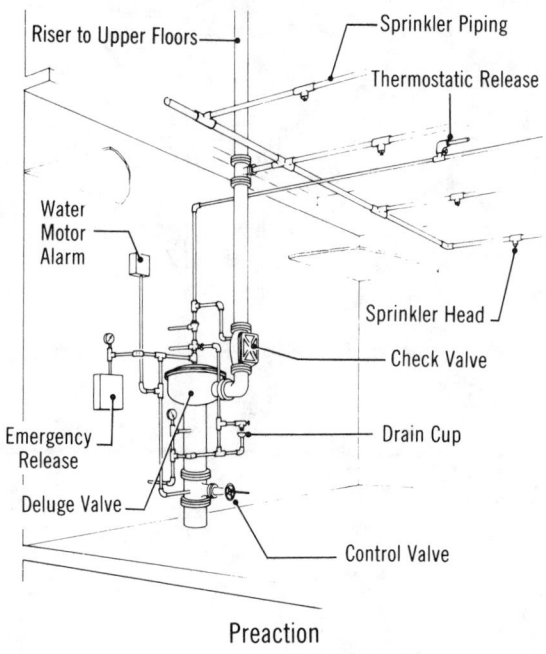

Riser to Upper Floors

Sprinkler Piping

Thermostatic Release

Water Motor Alarm

Sprinkler Head

Check Valve

Emergency Release

Drain Cup

Deluge Valve

Control Valve

Preaction

Figure 15.49 Installation Time in Man-Hours for Halon Fire Suppression Systems

Description	Man-Hours	Unit
Halon System, Filled, Including Mounting Bracket		
26 lb. Cylinder	2.000	Ea.
44 lb. Cylinder	2.286	Ea.
63 lb. Cylinder	2.667	Ea.
101 lb. Cylinder	3.200	Ea.
196 lb. Cylinder	4.000	Ea.
Electro/Mechanical Release	4.000	Ea.
Manual Pull Station	1.333	Ea.
Pneumatic Damper Release	1.000	Ea.
Discharge Nozzle	.570	Ea.
Control Panel Single Zone	8.000	Ea.
Control Panel Multi-zone (4)	16.000	Ea.
Battery Standby Power	2.810	Ea.
Heat Detector	1.000	Ea.
Smoke Detector	1.290	Ea.
Audio Alarm	1.194	Ea.

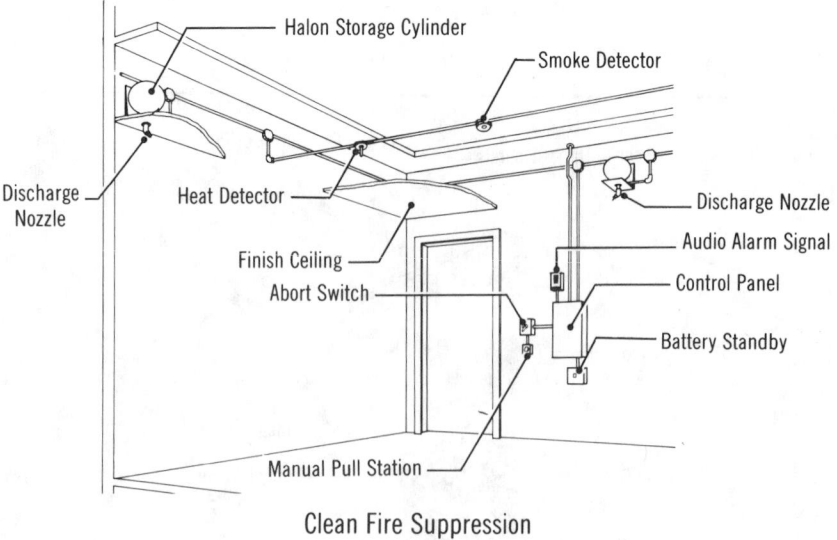

Clean Fire Suppression

Figure 15.50   Standpipe Systems

The basis for standpipe system design is National Fire Protection Association NFPA 14. However, the authority having jurisdiction should be consulted for special conditions, local requirements and approval.

Standpipe systems, properly designed and maintained, are an effective and valuable time-saving aid for extinguishing fires, especially in the upper stories of tall buildings, the interior of large commercial or industrial malls, or other areas where construction features or access make the laying of temporary hose lines time-consuming and/or hazardous. Standpipes are frequently installed with automatic sprinkler systems for maximum protection.

There are three general classes of service for standpipe systems:

Class I for use by fire departments and personnel with special training for heavy streams (2-1/2" hose connections).

Class II for use by building occupants until the arrival of the fire department (1-1/2" hose connector with hose).

Class III for use by either fire departments and trained personnel or by the building occupants (both 2-1/2" and 1-1/2" hose connections or one 2-1/2" hose valve with an easily removable 2-1/2" by 1-1/2" adapter).

Standpipe systems are also classified by the way water is supplied to the system. The four basic types are:

Type 1: Wet standpipe system having supply valve open and water pressure maintained at all times.

Type 2: Standpipe system so arranged through the use of approved devices as to admit water to the system automatically by opening a hose valve.

Type 3: Standpipe system arranged to admit water to the system through manual operation of approved remote control devices located at each hose station.

Type 4: Dry standpipe having no permanent water supply.

Class	Design-Use	Pipe Size Minimums	Water Supply Minimums
Class I	2-1/2" hose connection on each floor All areas within 30' of nozzle with 100' of hose Fire Department trained personnel	Height to 100', 4" diam. Heights above 100', 6" diam. (275' max. except with pressure regulators 400' max.)	For each standpipe riser 500 GPM flow For common supply pipe allow 500 GPM for first standpipe plus 250 GPM for each additional standpipe (2500 GPM max. total), 30 min. duration, 65 PSI at 500 GPM
Class II	1-1/2" hose connection with hose on each floor All areas within 30' of nozzle with 100' of hose Occupant personnel	Height to 50', 2" diam. Height above 50', 2-1/2" diam.	For each standpipe riser 100 GPM flow For multiple riser common supply pipe 100 GPM 30 min. duration, 65 PSI at 100 GPM,
Class III	Both of above. Class I valved connections will meet Class III with addition of 2-1/2" by 1-1/2" adapter and 1-1/2" hose	Same as Class I	Same as Class I

(continued on next page)

Figure 15.50   Standpipe Systems (continued)

Combined Systems (or two separate valves)

Combined systems are systems where the risers supply both automatic sprinklers and 2-1/2" hose connection outlets for fire department use. In such a system the sprinkler spacing pattern shall be in accordance with NFPA 13 while the risers and supply piping will be sized in accordance with NFPA 14. When the building is completely sprinklered the risers may be sized by hydraulic calculation. The minimum size riser for buildings not completely sprinklered is 6".

The minimum water supply of a completely sprinklered, light hazard, high-rise occupancy building will be 500 GPM while the supply required for other types of completely sprinklered high-rise buildings is 1000 GPM.

General System Requirements

1. Approved valves will be provided at the riser for controlling branch lines to hose outlets.
2. A hose valve will be provided at each outlet for attachment of hose.
3. Where pressure at any standpipe outlet exceeds 100 PSI, a pressure reducer must be installed to limit the pressure to 100 PSI. Note that the pressure head due to gravity in 100' of riser is 43.4 PSI. This must be overcome by city pressure, fire pumps, or gravity tanks to provide adequate pressure at the top of the riser.
4. Each hose valve on a wet system having linen hose shall have an automatic drip connection to prevent valve leakage from entering the hose.
5. Each riser will have a valve to isolate it from the rest of the system.
6. One or more fire department connections as an auxiliary supply shall be provided for each Class I or Class III standpipe system. In buildings having two or more zones, a connection will be provided for each zone.
7. There will be no shutoff valve in the fire department connection, but a check valve will be located in the line before it joins the system.
8. All hose connections street side will be identified on a cast plate or fitting as to purpose.

Figure 15.51 Installation Time in Man-Hours for Fire Standpipes

Description	Man-Hours 4"	6"	8"	Unit
Black Steel Pipe	.444	.774	.889	L.F.
Pipe Tee	4.000	6.000	8.000	Ea.
Pipe Elbow	2.667	3.429	4.000	Ea.
Pipe Nipple, 2-1/2"	1.000	1.000	1.000	Ea.
Hose Valve, 2-1/2"	1.140	1.140	1.140	Ea.
Pressure Restricting Valve, 2-1/2"	1.140	1.140	1.140	Ea.
Check Valve with Ball Drip	5.333	8.000	10.667	Ea.
Siamese Inlet	3.200	3.478	3.478	Ea.
Roof Manifold with Valves	3.333	3.478	3.478	Ea.

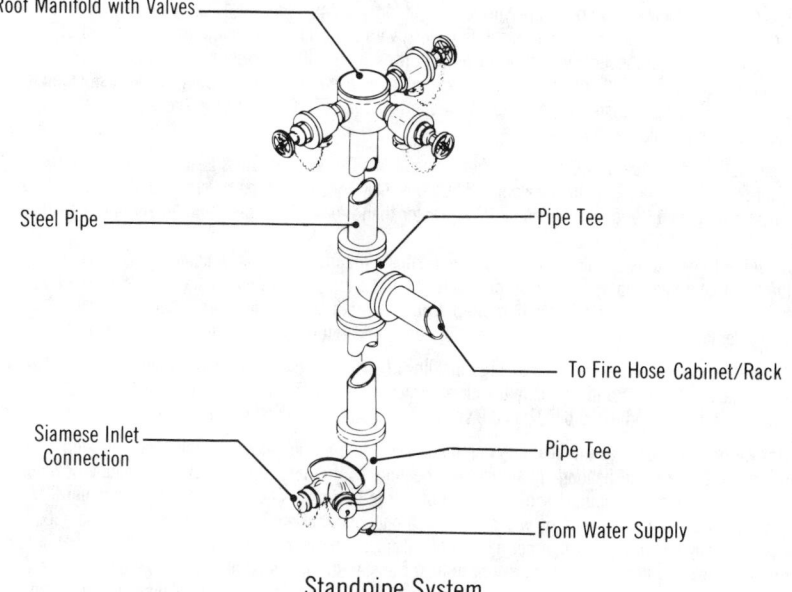

Roof Manifold with Valves

Steel Pipe

Pipe Tee

To Fire Hose Cabinet/Rack

Siamese Inlet Connection

Pipe Tee

From Water Supply

Standpipe System

Figure 15.52   Basics of a Heating System

The function of a heating system is to achieve and maintain a desired temperature in a room or building by replacing the amount of heat being dissipated. There are four kinds of heating systems: hot-water, steam, warm-air and electric resistance. Each has certain essential and similar elements, with the exception of electric resistance heating.

The basic elements of a heating system are:

A. A **combustion chamber** in which fuel is burned and heat transferred to a conveying medium.

B. The **"fluid"** used for conveying the heat (water, steam or air).

C. **Conductors** or pipes for transporting the fluid to specific desired locations.

D. A means of disseminating the heat, sometimes called **terminal units.**

A. The **combustion chamber** in a furnace heats air which is then distributed. This is called a *warm air system.*

The combustion chamber in a boiler heats water which is either distributed as hot water or steam. This is termed a *hydronic system.*

The maximum allowable working pressures are limited by ASME "Code for Heating Boilers" to 15 PSI for steam and 160 PSI for hot water heating boilers, with a maximum temperature limitation of 250°F. Hot water boilers are generally rated for a working pressure of 30 PSI. High pressure boilers are governed by the ASME "Code for Power Boilers" which is used almost universally for boilers operating over 15 PSIG. High pressure boilers used for a combination of heating/process loads are usually designed for 150 PSIG.

Boiler ratings are usually indicated as either Gross or Net Output. The Gross Load is equal to the Net Load plus a piping and pickup allowance. When this allowance cannot be determined, divide the gross output rating by 1.25 for a value equal to or greater than the net heat loss requirement of the building.

B. Of the three **fluids** used, steam carries the greatest amount of heat per unit volume. This is due to the fact that it gives up its latent heat of vaporization at a temperature considerably above room temperature. Another advantage is that the pressure to produce a positive circulation is readily available. Piping conducts the steam to terminal units and returns condensate to the boiler.

The **steam system** is well adapted to large buildings because of its positive circulation, its comparatively economical installation and its ability to deliver large quantities of heat. Nearly all large office buildings, stores, hotels, and industrial buildings are so heated, in addition to many residences.

Hot water, when used as the heat carrying fluid, gives up a portion of its sensible heat and then returns to the boiler or heating apparatus for reheating. As the heat conveyed by each pound of water is about one-fiftieth of the heat conveyed by a pound of steam, it is necessary to circulate about fifty times as much water as steam by weight (although only one-thirtieth as much by volume). The hot water system is usually, although not necessarily, designed to operate at temperatures below that of the ordinary steam system and so the amount of heat transfer surface must be correspondingly greater. A temperature of 190°F to 200°F is normally the maximum. Circulation in small buildings may depend on the difference in density between hot water and the cool water returning to the boiler. Circulating pumps are normally used to maintain a desired rate of flow. Pumps permit a greater degree of flexibility and better control.

In **warm-air** furnace systems, cool air is taken from one or more points in the building, passed over the combustion chamber and flue gas passages, and then distributed through a duct system. A disadvantage of this system is that the ducts take up much more building volume than steam or hot water pipes. Advantages of this system are the relative ease with which humidification can be accomplished by the evaporation of water as the air circulates through the heater, and the lack of need for expensive disseminating units as the warm air simply becomes part of the interior atmosphere of the building.

Figure 15.52 Basics of a Heating System (continued)

C. Conductors (pipes and ducts) have been lightly treated in the discussion of conveying fluids. For more detailed information such as sizing and distribution methods, the reader is referred to technical publications such as the American Society of Heating, Refrigerating and Air-Conditioning Engineers "Handbook of Fundamentals."

D. Terminal units come in an almost infinite variety of sizes and styles, but the basic principles of operation are very limited. As previously mentioned, warm-air systems require only a simple register or diffuser to mix heated air with that present in the room. Special application items such as radiant coils and infra-red heaters are available to meet particular conditions but are not usually considered for general heating needs. Most heating is accomplished by having air flow over coils or pipes containing the heat transporting medium (steam, hot-water, electricity). These units, while varied, may be separated into two general types: (1) radiator/convectors and (2) unit heaters.

Radiator/convectors may be cast, fin-tube or pipe assemblies. They may be direct, indirect, exposed, concealed or mounted within a cabinet enclosure, upright or baseboard style. These units are often collectively referred to as "radiators" or "radiation," although none gives off heat either entirely by radiation or by convection, but rather by a combination of both. The air flows over the units as a gravity "current." It is necessary to have one or more heat-emitting units in each room. The most efficient placement is low along an outside wall or under a window to counteract the cold coming into the room and achieve an even distribution.

In contrast to radiator/convectors which operate most effectively against the walls of smaller rooms, **unit heaters** utilize a fan to move air over heating coils and are very effective in locations of relatively large volume. Unit heaters, while usually suspended overhead, may also be floor-mounted. They also may take in fresh outside air for ventilation. The heat distributed by unit heaters may be from a remote source and conveyed by a fluid, or it may be from the combustion of fuel in each individual heater. In the latter case, the only piping required would be for fuel. However, a vent for the products of combustion would be necessary. The following list gives many of the advantages of unit heaters for applications other than office or residential:

a. large capacity, so smaller number of units are required, **b.** piping system simplified, **c.** space saved, where they are located overhead out of the way, **d.** rapid heating directed where needed with effective wide distribution, **e.** difference between floor and ceiling temperature reduced, **f.** circulation of air obtained, and ventilation with introduction of fresh air possible, and **g.** heat output flexible and easily controlled.

Figure 15.53 Determining Heat Loss for Various Types of Buildings

General: While the most accurate estimates of heating requirements would naturally be based on detailed information about the building being considered, it is possible to arrive at a reasonable approximation using the following procedure:

1. Calculate the cubic volume of the room or building.

2. Select the appropriate factor from the table below. Note that the factors apply only to inside temperatures listed in the first column and to 0°F outside temperature.

3. If the building has bad north and west exposures, multiply the heat loss factor by 1.1.

4. If the outside design temperature is other than 0°F, multiply the factor from the table below by the factor from Figure 15.54.

5. Multiply the cubic volume by the factor selected from the table below. This will give the estimated BTUH heat loss which must be made up to maintain inside temperature.

Building Type	Conditions	Qualifications	Loss Factor*
Factories & Industrial Plants General Office Areas 70°F	One Story	Skylight in Roof No Skylight in Roof	6.2 5.7
	Multiple Story	Two Story Three Story Four Story Five Story Six Story	4.6 4.3 4.1 3.9 3.6
	All Walls Exposed	Flat Roof Heated Space Above	6.9 5.2
	One Long Warm Common Wall	Flat Roof Heated Space Above	6.3 4.7
	Warm Common Walls on Both Long Sides	Flat Roof Heated Space Above	5.8 4.1
Warehouses 60°F	All Walls Exposed	Skylights in Roof No Skylights in Roof Heated Space Above	5.5 5.1 4.0
	One Long Warm Common Wall	Skylight in Roof No Skylight in Roof Heated Space Above	5.0 4.9 3.4
	Warm Common Walls on Both Long Sides	Skylight in Roof No Skylight in Roof Heated Space Above	4.7 4.4 3.0

*Note: This table tends to be conservative, particularly for new buildings designed for minimum energy consumption.

Figure 15.54 Outside Design Temperature Correction Factor (for Degrees Fahrenheit)

Outside Design Temperature	50	40	30	20	10	0	–10	–20	–30
Correction Factor	0.29	0.43	0.57	0.72	0.86	1.00	1.14	1.28	1.43

Figure 15.55   Basic Properties of Fuels

Fuel Type	Unit	Heating Value BTU/Unit	Overall System Efficiency	Remarks
Electricity	Kilowatt	3,412	100	
Steam	Pounds (at atmospheric pressure)	1,000	—	
Oil #2	Gallon	138,500	60–88	
Oil #4	Gallon	145,000	60–88	
Oil #6	Gallon	152,000	—	Preheat
Natural Gas	CCF (100 C.F.)*	103,000	65–92	
Propane	Gallon	95,500	65–90	
Coal	Pound	13,000	45–75	
Solar	S.F. of collector—varies with location and collector			

*Note: 1 therm = 1.013 CCF

Figure 15.56   Boiler Selection Chart

Several types of boilers are available to meet the hot water and heating needs of both residential and commercial buildings. Some different boiler types are shown in this table.

Most hot water boilers operate at less than 30 psig (low-pressure boilers). Low-pressure steam boilers operate at less than 15 psig. Above 30 psig, high-pressure boilers must conform to stricter requirements. Water is lost to a heating system through minor leaks and evaporation; therefore, a makeup water line to feed the boiler with fresh makeup water must be provided.

Boiler Type	Output Capacity Range – MBH Efficiency Range	Fuel Types	Uses
Cast Iron Sectional	80–14,500 80–92%	Oil, Gas, Coal, Wood/Fossil	Steam/Hot Water
Steel	1,200–18,000 80–92%	Oil, Gas, Coal, Wood/Fossil, Electric	Steam/Hot Water
Scotch Marine	3,400–24,000 80–92%	Oil, Gas	Steam/Hot Water
Pulse Condensing	40–150 90–95%	Gas	Hot Water
Residential/Wall Hung	15–60 90–95%	Gas, Electric	Hot Water

Efficiencies shown are averages and will vary with specific manufacturers. For existing equipment, efficiencies may be 60-75%.

Figure 15.57 Installation Time in Man-Hours for Boilers

Description	Man-Hours	Unit
Electric Fired, Steel Output		
60 MBH	20.000	Ea.
500 MBH	36.293	Ea.
1000 MBH	60.000	Ea.
2000 MBH	94.118	Ea.
3000 MBH	114.286	Ea.
7000 MBH	177.778	Ea.
Gas Fired, Cast Iron Output		
80 MBH	21.918	Ea.
500 MBH	53.333	Ea.
1000 MBH	64.000	Ea.
2000 MBH	99.250	Ea.
3000 MBH	133.333	Ea.
7000 MBH	400.000	Ea.
Oil Fired, Cast Iron Output		
100 MBH	26.667	Ea.
500 MBH	64.000	Ea.
1000 MBH	76.000	Ea.
2000 MBH	114.286	Ea.
3000 MBH	139.130	Ea.
7000 MBH	400.000	Ea.
Scotch Marine Packaged Units, Gas or Oil Fired Output		
1300 MBH	80.000	Ea.
3350 MBH	80.000	Ea.
4200 MBH	85.000	Ea.
5000 MBH	90.000	Ea.
6700 MBH	90.000	Ea.
8370 MBH	110.000	Ea.
10,000 MBH	120.000	Ea.
20,000 MBH	130.000	Ea.
23,400 MBH	140.000	Ea.

Figure 15.57   Installation Time in Man-Hours for Boilers (continued)

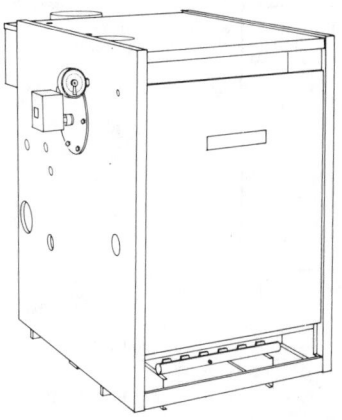

Packaged, Cast Iron Sectional,
Gas Fired Boiler-Residential

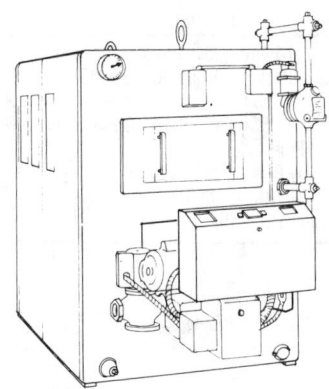

Packaged, Cast Iron Sectional,
Gas/Oil Fired Boiler-Commercial

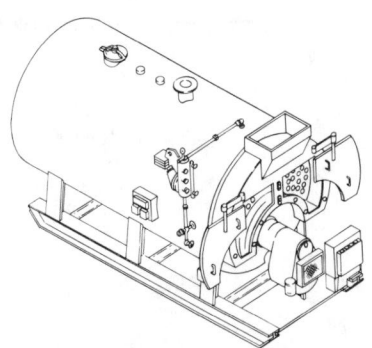

Packaged, Oil Fired, Modified
Scotch Marine Boiler-
Commercial

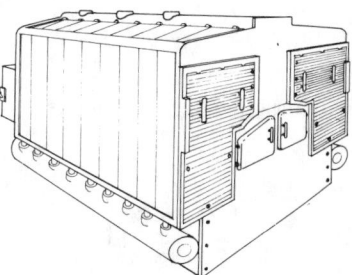

Cast Iron Sectional Boiler-
Commercial

Figure 15.58 Installation Time in Man-Hours for Forced Hot Water Heating Systems

Description	Man-Hours	Unit
Hot Water Heating System, Area to 2400 S.F.		
Boiler Package, Oil Fired, 225 MBH	17.143	Ea.
Oil Piping System	4.584	Ea.
Oil Tank, 550 Gallon, with Black Steel		
Fill Pipe	4.000	Ea.
Supply Piping, 3/4" Copper Tubing	.182	L.F.
Supply Fittings, Copper	.421	L.F.
Baseboard Radiation	.667	L.F.

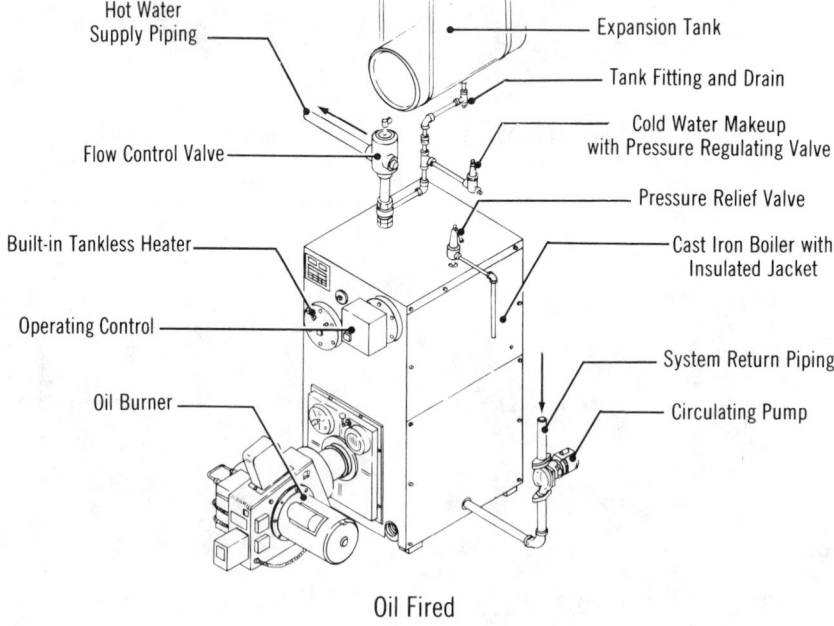

Hot Water Supply Piping — Flow Control Valve — Built-in Tankless Heater — Operating Control — Oil Burner — Expansion Tank — Tank Fitting and Drain — Cold Water Makeup with Pressure Regulating Valve — Pressure Relief Valve — Cast Iron Boiler with Insulated Jacket — System Return Piping — Circulating Pump

Oil Fired

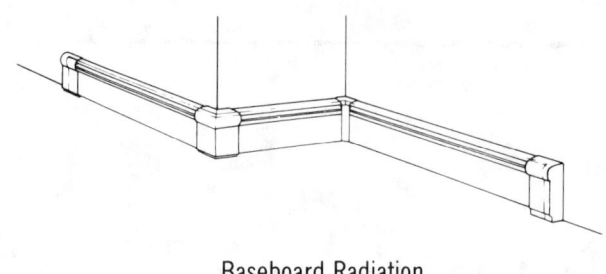

Baseboard Radiation

Figure 15.59 Installation Time in Man-Hours for Forced Air Heating Systems

Description	Man-Hours	Unit
Heating Only, Gas Fired Hot Air, One Zone, 1200 S.F. Building		
Furnace, Gas, Up Flow	4.710	Ea.
Intermittent Pilot	4.710	Ea.
Supply Duct, Rigid Fiberglass	.007	L.F.
Return Duct, Sheet Metal, Galvanized	.102	lb.
Lateral Ducts, 6″ Flexible Fiberglass	.062	L.F.
Register, Elbows	.267	Ea.
Floor Registers, Enameled Steel	.250	Ea.
Floor Grille, Return Air	.364	Ea.
Thermostat	1.000	Ea.
Plenum	1.000	Ea.
Ductwork Fabricated Rectangular, Includes Fittings, Joists, Supports, Allowance for Flexible Connections, No Insulation.		
Aluminum, Alloy 3003-H14,		
Under 300 lbs.	.320	lb.
300 to 500 lbs.	.300	lb.
500 to 1000 lbs.	.253	lb.
1000 to 2000 lbs.	.200	lb.
2000 to 10,000 lbs.	.185	lb.
Over 10,000 lbs.	.166	lb.
Galvanized Steel, Under 400 lbs..		
400 to 1000 lbs.	.094	lb.
1000 to 2000 lbs.	.091	lb.
2000 to 5000 lbs.	.087	lb.
5000 to 10,000 lbs.	.084	lb.
Over 10,000 lbs.	.080	lb.

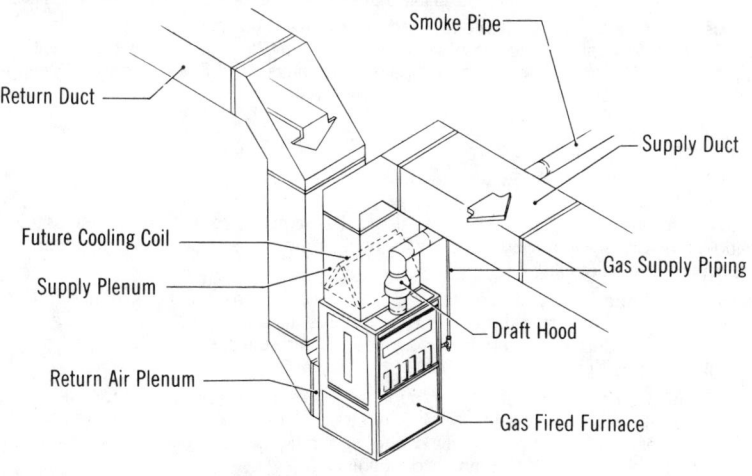

Gas Fired, Warm Air System

(continued on next page)

Figure 15.59   Installation Time in Man-Hours for Forced Air Heating Systems (continued)

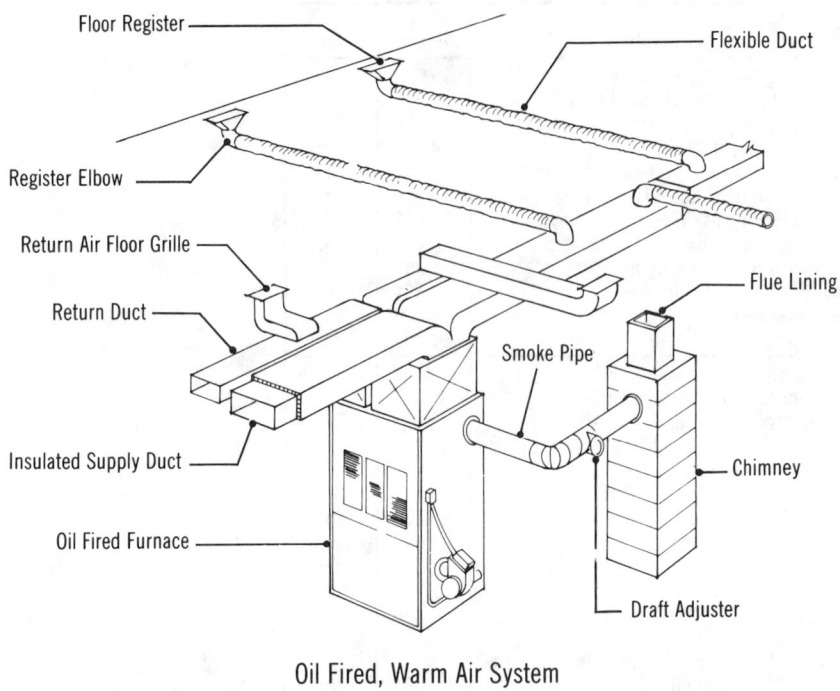

Floor Register

Flexible Duct

Register Elbow

Return Air Floor Grille

Flue Lining

Return Duct

Smoke Pipe

Insulated Supply Duct

Chimney

Oil Fired Furnace

Draft Adjuster

Oil Fired, Warm Air System

Figure 15.60   Air-Conditioning Basics

The purpose of air-conditioning is to control the environment of a space so that comfort is provided for the occupants, and/or conditions are suitable for the processes or equipment contained therein. The several items which should be evaluated to define system objectives are:

Temperature control
Humidity control
Cleanliness
Odor, smoke and fumes
Ventilation

Efforts to control the above parameters must also include consideration of the degree or tolerance of variation, the noise level introduced, the velocity of air motion, and the energy requirements to accomplish the desired results.

The variation in **temperature** and **humidity** is a function of the sensor and the controller. The controller reacts to a signal from the sensor and produces the appropriate response in either the terminal unit, the conductor of the transporting medium (air, steam, chilled water, etc.), or the source (boiler, evaporating coils, etc.).

The **noise level** is a byproduct of the energy supplied to moving components of the system. Those items which usually contribute the most noise are pumps, blowers, fans, compressors and diffusers. The level of noise can be partially controlled through use of vibration pads, isolators, proper sizing, shields, baffles and sound absorbing liners.

Figure 15.60 Air-Conditioning Basics (continued)

Some **air motion** is necessary to prevent stagnation and stratification. The maximum acceptable velocity varies with the degree of heating or cooling which is taking place. Most people feel air moving past them at velocities in excess of 25 FPM as an annoying draft. However, velocities up to 45 FPM may be acceptable in certain cases. Ventilation, expressed as air changes per hour and percentage of fresh air, is usually an item regulated by local codes.

Selection of the system to be used for a particular application is usually a trade-off. In some cases, the building size, style, or room available for mechanical use limits the range of possibilities. Prime factors influencing the decision are first cost and total life (operating, maintenance, and replacement costs). The accuracy with which each parameter is determined will be an important measure of the reliability of the decision and subsequent satisfactory operation of the installed system.

Heat delivery may be desired from an air conditioning system. Heating capability usually is added as follows: A gas fired burner or hot water/steam/electric coils may be added to the air handling unit directly and heat all air equally. For limited or localized heat requirements the water/steam/electric coils may be inserted into the duct branch supplying the cold areas. Gas fired duct furnaces are also available. Note: when water or steam coils are used, the cost of the piping and fuel must also be added. For a rough estimate, use the cost per square foot of the appropriate sized hydronic system with unit heater. This will provide a cost for the boiler and piping, and the unit heaters of the system would equate to the approximate cost of the heating coils.

Air-Conditioning Requirements

BTU's per Hour per S.F. of Floor Area and S.F. per Ton of Air Conditioning								
Type Building	BTU per S.F.	S.F. per Ton	Type Building	BTU per S.F.	S.F. per Ton	Type Building	BTU per S.F.	S.F. per Ton
Apartments, Individual	26	450	Dormitory, Rooms	40	300	Libraries	50	240
Corridors	22	550	Corridors	30	400	Low Rise Office, Exterior	38	320
Auditoriums & Theaters	40	300/18*	Dress Shops	43	280	Interior	33	360
Banks	50	240	Drug Stores	80	150	Medical Centers	28	425
Barber Shops	48	250	Factories	40	300	Motels	28	425
Bars & Taverns	133	90	High Rise Office —			Office (small suite)	43	280
			Exterior Rooms	46	263			
Beauty Parlors	66	180	Interior Rooms	37	325	P. O., Individual Office	42	285
Bowling Alleys	68	175	Hospitals, Core	43	280	Central Area	46	260
Churches	36	330/20*	Perimeter	46	260	Residences	20	600
Cocktail Lounges	68	175	Hotel, Guest Rooms	44	275	Restaurants	60	200
Computer Rooms	141	85	Corridors	30	400	Schools & Colleges	46	260
Dental Offices	52	230	Public Spaces	55	220	Shoe Stores	55	220
Department Stores,			Industrial Plants,			Shopping Centers,		
Basement	34	350	Offices	38	320	Super Markets	34	350
Main Floor	40	300	General Offices	34	350	Retail Stores	48	250
Upper Floor	30	400	Plant Areas	40	300	Specialty	60	200

*Persons per ton
12,000 BTU = 1 ton of air conditioning

(continued on next page)

Figure 15.60   Air-Conditioning Basics (continued)

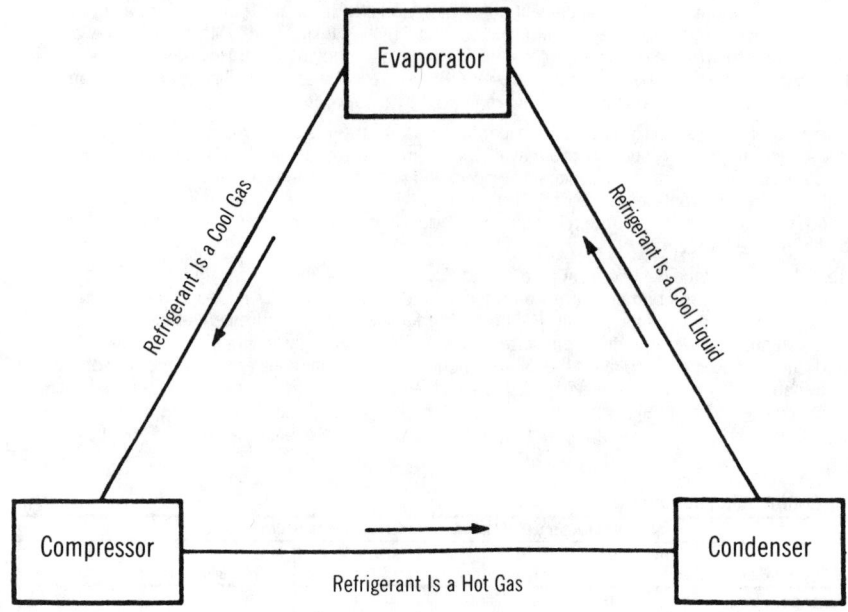

Basic Refrigeration Cycle
(Compression System)

All compression systems are composed of three basic components plus controls, regulators etc.

1. When the compressor and condenser are combined, the assembly is called a *condensing unit*.

2. The condenser may be air cooled or water cooled. (Water cooled systems require a water source, usually a water tower or pond.)

3. Evaporators may also be called *air handlers, water chillers*, or *cooling coils*.

Figure 15.61 Recommended Ventilation Air Changes

Range of Time in Minutes per Change for Various Types of Facilities					
Assembly Halls	2–10	Dance Halls	2–10	Laundries	1– 3
Auditoriums	2–10	Dining Rooms	3–10	Markets	2–10
Bakeries	2– 3	Dry Cleaners	1– 5	Offices	2–10
Banks	3–10	Factories	2– 5	Pool Rooms	2– 5
Bars	2– 5	Garages	2–10	Recreation Rooms	2–10
Beauty Parlors	2– 5	Generator Rooms	2– 5	Sales Rooms	2–10
Boiler Rooms	1– 5	Gymnasiums	2–10	Theaters	2– 8
Bowling Alleys	2–10	Kitchens – Hospitals	2– 5	Toilets	2– 5
Churches	5–10	Kitchens – Restaurants	1– 3	Transformer Rooms	1– 5

CFM air required for changes = Volume of room in cubic feet ÷ Minutes per change.

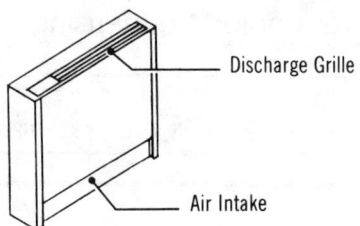

Discharge Grille

Air Intake

Room Size Fan Coil Unit

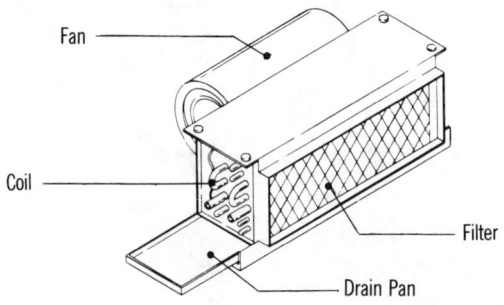

Fan

Coil

Filter

Drain Pan

Concealed Fan Coil Element

Figure 15.62 Air Requirements for Air Cooled Condensing Units

Refrigeration:
750 CFM per HP, 1000 CFM per ton
Air Conditioning:
1000 CFM per HP, 1000 CFM per ton

Figure 15.63 Water Requirements for Water Cooled Condensing Units

City Water—Tons refrigeration × 1.5 = GPM
Cooling Tower—Tons refrigeration × 3.0 = GPM

Figure 15.64 Installation Time in Man-Hours for Air Handling Units

Description	Man-Hours	Unit
Fan Coil Unit, Free Standing Finished Cabinet, 3 Row Cooling or Heating Coil, Filter		
200 CFM	2.000	Ea.
400 CFM	2.667	Ea.
600 CFM	2.909	Ea.
1000 CFM	3.200	Ea.
1200 CFM	4.000	Ea.
4000 CFM	8.571	Ea.
6000 CFM	16.000	Ea.
8000 CFM	30.000	Ea.
12,000 CFM	40.000	Ea.
Direct Expansion Cooling Coil, Filter		
2000 CFM	5.333	Ea.
3000 CFM	5.333	Ea.
4000 CFM	9.231	Ea.
8000 CFM	34.286	Ea.
12,000 CFM	40.000	Ea.
16,000 CFM	53.333	Ea.
20,000 CFM	63.158	Ea.
Central Station Unit, Factory Assembled, Modular, 4, 6 or 8 Row Coils, Filter and Mixing Box		
1500 CFM	13.333	Ea.
2200 CFM	14.545	Ea.
3800 CFM	20.000	Ea.
5400 CFM	30.000	Ea.
8000 CFM	40.000	Ea.
12,100 CFM	52.174	Ea.
18,400 CFM	72.727	Ea.
22,300 CFM	82.759	Ea.
33,700 CFM	126.316	Ea.
52,500 CFM	200.000	Ea.
63,000 CFM	246.154	Ea.

(continued on next page)

Figure 15.64   Installation Time in
Man-Hours for Air Handling Units (continued)

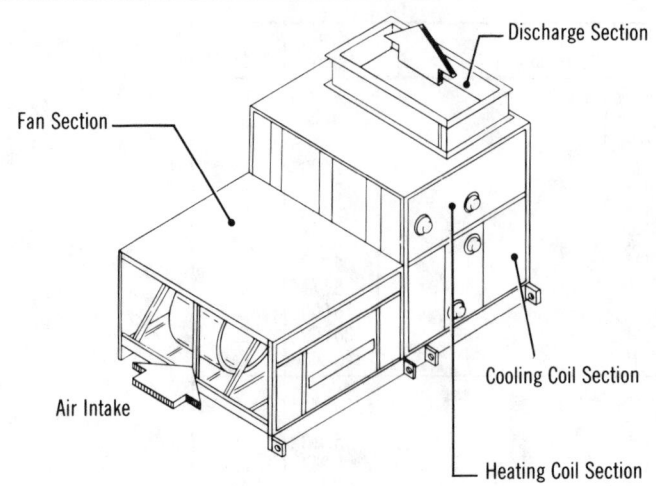

Discharge Section

Fan Section

Cooling Coil Section

Air Intake

Heating Coil Section

Central Station Air Handling Unit

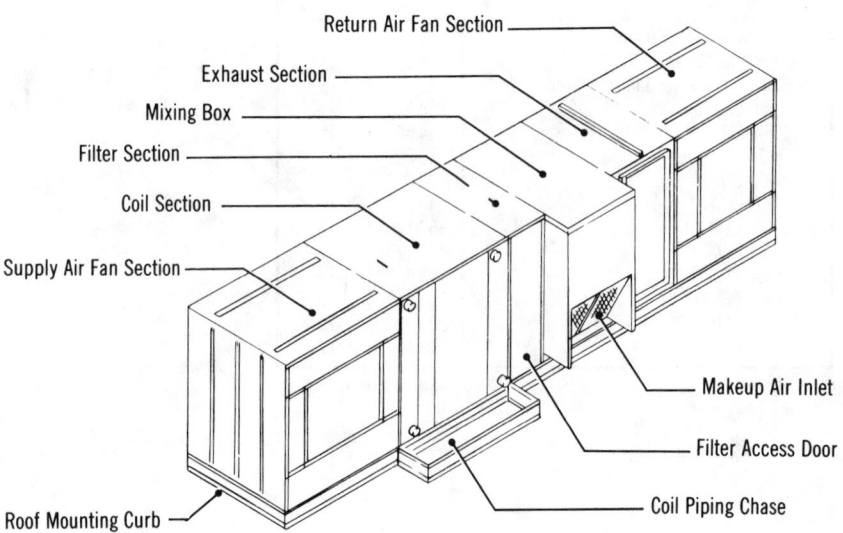

Return Air Fan Section

Exhaust Section

Mixing Box

Filter Section

Coil Section

Supply Air Fan Section

Makeup Air Inlet

Filter Access Door

Coil Piping Chase

Roof Mounting Curb

Central Station Air Handling Unit for Rooftop Location

Figure 15.65 Installation Time in Man-Hours for Packaged Rooftop Air-Conditioner Units

Description	Man-Hours	Unit
Single Zone, Electric Cool, Gas Heat		
2 Ton Cooling, 60 M BTU/Hr. Heating	10.667	Ea.
4 Ton Cooling, 95 M BTU/Hr. Heating	14.545	Ea.
5 Ton Cooling, 112 M BTU/Hr. Heating	28.571	Ea.
10 Ton Cooling, 200 M BTU/Hr. Heating	52.174	Ea.
15 Ton Cooling, 270 M BTU/Hr. Heating	77.419	Ea.
20 Ton Cooling, 360 M BTU/Hr. Heating	100.000	Ea.
30 Ton Cooling, 540 M BTU/Hr. Heating	145.455	Ea.
40 Ton Cooling, 675 M BTU/Hr. Heating	200.000	Ea.
50 Ton Cooling, 810 M BTU/Hr. Heating	246.154	Ea.
Multi-Zone, Electric Cool, Gas Heat, Economizer		
15 Ton Cooling, 360 M BTU/Hr. Heating	145.455	Ea.
20 Ton Cooling, 360 M BTU/Hr. Heating	152.381	Ea.
25 Ton Cooling, 450 M BTU/Hr. Heating	177.778	Ea.
28 Ton Cooling, 450 M BTU/Hr. Heating	200.000	Ea.
30 Ton Cooling, 540 M BTU/Hr. Heating	213.333	Ea.
37 Ton Cooling, 540 M BTU/Hr. Heating	246.154	Ea.
70 Ton Cooling, 1500 M BTU/Hr. Heating	355.556	Ea.
80 Ton Cooling, 1500 M BTU/Hr. Heating	400.000	Ea.
90 Ton Cooling, 1500 M BTU/Hr. Heating	457.143	Ea.
105 Ton Cooling, 1500 M BTU/Hr. Heating	533.333	Ea.

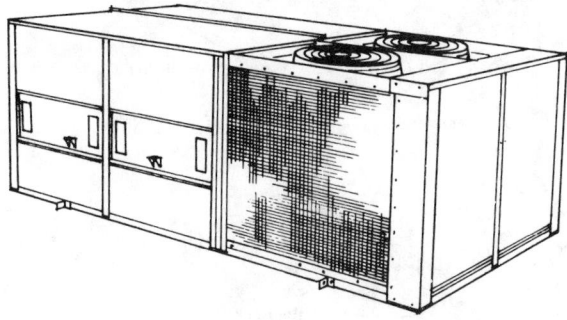

Packaged Rooftop Air Conditioner

(continued on next page)

Figure 15.65 Installation Time in Man-Hours for Packaged Rooftop Air-Conditioner Units (continued)

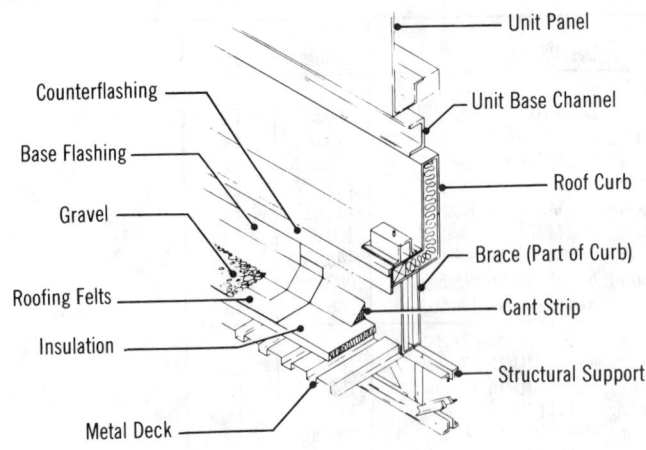

Curb Mounted Directly on Roof Supports

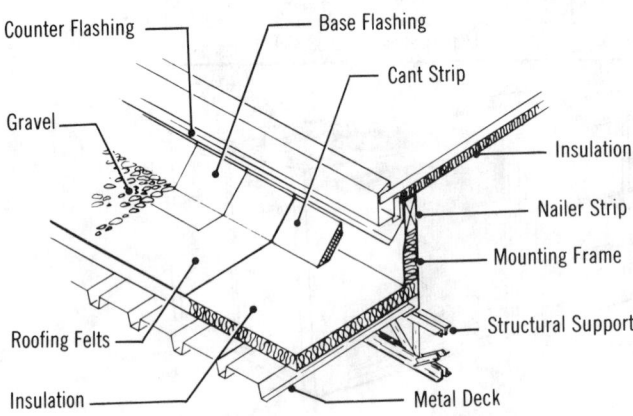

Flashing for Roof Mounting Frame

Figure 15.66 Installation Time in Man-Hours for Packaged Water Chillers

Description	Man-Hours	Unit
Packaged Water Chillers		
Centrifugal, Water Cooled, Hermetic		
200 Ton	80.000	Ea.
400 Ton	96.000	Ea.
1000 Ton	174.000	Ea.
1300 Ton	294.000	Ea.
1500 Ton	324.000	Ea.
Reciprocating, Air Cooled		
20 Ton	91.429	Ea.
40 Ton	133.333	Ea.
65 Ton	228.571	Ea.
100 Ton	320.000	Ea.
110 Ton	355.556	Ea.
125 Ton	400.000	Ea.
Water Cooled Multiple Compressor		
Semi Hermetic		
15 Ton	80.000	Ea.
20 Ton	88.889	Ea.
25 Ton	100.000	Ea.
30 Ton	118.519	Ea.
40 Ton	133.333	Ea.
50 Ton	152.381	Ea.
60 Ton	168.421	Ea.
80 Ton	213.333	Ea.
100 Ton	266.667	Ea.
120 Ton	320.000	Ea.
140 Ton	355.556	Ea.
Air Cooled for Remote Condenser		
40 Ton	118.519	Ea.
60 Ton	152.381	Ea.
80 Ton	200.000	Ea.
100 Ton	290.909	Ea.
120 Ton	320.000	Ea.
160 Ton	355.556	Ea.
Absorption		
Gas Fired, Air Cooled		
5 Ton	26.667	Ea.
10 Ton	40.000	Ea.
Steam or Hot Water, Water Cooled		
100 Ton	108.000	Ea.
600 Ton	126.000	Ea.
1100 Ton	165.000	Ea.
1500 Ton	210.000	Ea.

(continued on next page)

Figure 15.66   Installation Time in Man-Hours
for Packaged Water Chillers (continued)

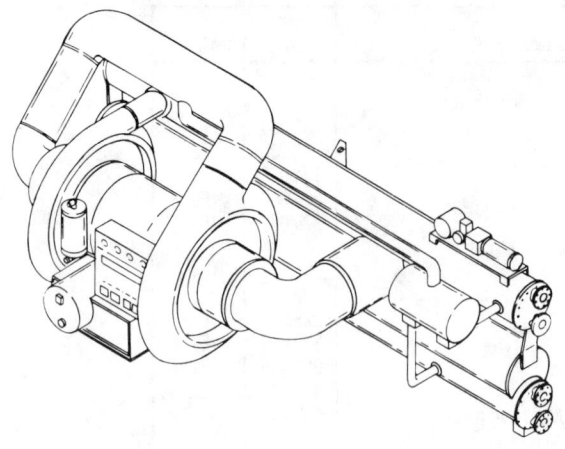

Centrifugal, Water Cooled, Hermetic

Figure 15.67   Installation Time in
Man-Hours for Cooling Towers

Description	Man-Hours	Unit
Labor to Set in Place - Rigging not Included		
60 Ton Single Flow	8	Ea.
90 Ton Single Flow	16	Ea.
100 Ton Single Flow	16	Ea.
125 Ton Double Flow	24	Ea.
150 Ton Double Flow	30	Ea.
300 Ton Double Flow	40	Ea.
600 Ton Double Flow	80	Ea.
840 Ton Double Flow	136	Ea.
1000 Ton Double Flow	152	Ea.

Figure 15.67   Installation Time in Man-Hours for Cooling Towers (continued)

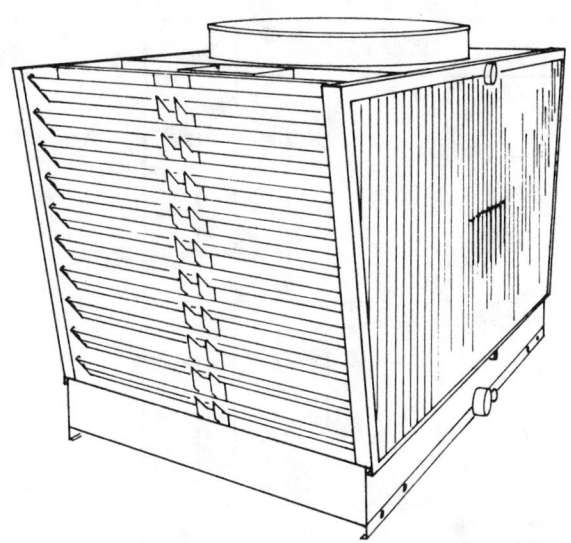

Induced Air, Double Flow, Cooling Tower

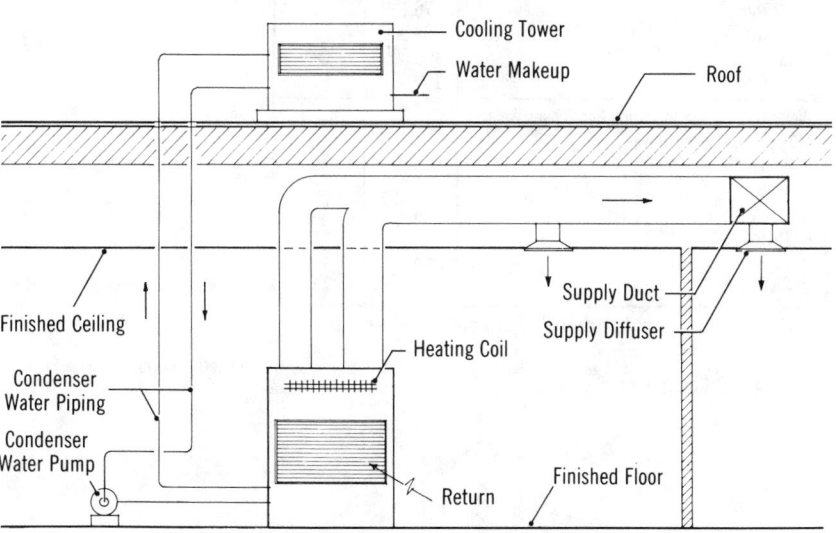

Cooling Tower and Condenser Water System

Figure 15.68 Installation Time in Man-Hours for Computer Room Cooling Systems

Description	Man-Hours	Unit
Computer Room Unit, Air Cooled, Includes Remote Condenser		
3 Ton	32.000	Ea.
5 Ton	35.556	Ea.
8 Ton	59.259	Ea.
10 Ton	64.000	Ea.
15 Ton	72.727	Ea.
20 Ton	82.759	Ea.
23 Ton	85.714	Ea.
Chilled Water, for Connection to Existing Chiller System		
5 Ton	21.622	Ea.
8 Ton	32.000	Ea.
10 Ton	32.653	Ea.
15 Ton	33.333	Ea.
20 Ton	34.783	Ea.
23 Ton	38.095	Ea.
Glycol System, Complete Except for Interconnecting Tubing		
3 Ton	40.000	Ea.
5 Ton	42.105	Ea.
8 Ton	69.565	Ea.
10 Ton	76.190	Ea.
15 Ton	92.308	Ea.
20 Ton	100.000	Ea.
23 Ton	109.091	Ea.
Water Cooled, Not Including Condenser Water Supply or Cooling Tower		
3 Ton	25.806	Ea.
5 Ton	29.630	Ea.
8 Ton	44.485	Ea.
15 Ton	59.259	Ea.
20 Ton	63.158	Ea.
23 Ton	70.588	Ea.

Note: The man-hours do not include ductwork or piping.

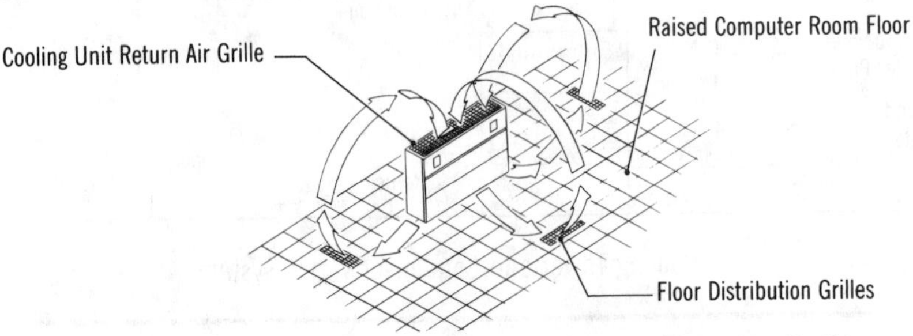

Self-Contained Computer Room Cooling Unit — Under Floor Distribution

Figure 15.69 Installation Time in Man-Hours for Fans

Description	Man-Hours	Unit
Fans		
Belt Drive, In-Line Centrifugal		
3800 CFM	5.882	Ea.
6400 CFM	7.143	Ea.
10,500 CFM	8.333	Ea.
15,600 CFM	12.500	Ea.
23,000 CFM	28.571	Ea.
28,000 CFM	50.000	Ea.
Direct Drive Ceiling Fan		
95 CFM	1.000	Ea.
210 CFM	1.053	Ea.
385 CFM	1.111	Ea.
885 CFM	1.250	Ea.
1650 CFM	1.538	Ea.
2960 CFM	1.818	Ea.
Direct Drive Paddle Blade Fan		
36", 4000 CFM	3.333	Ea.
52", 7000 CFM	5.000	Ea.
Direct Drive Roof Fan		
420 CFM	2.857	Ea.
675 CFM	3.333	Ea.
770 CFM	4.000	Ea.
1870 CFM	4.762	Ea.
2150 CFM	5.000	Ea.
Belt Drive Roof Fan		
1660 CFM	3.333	Ea.
2830 CFM	4.000	Ea.
4600 CFM	5.000	Ea.
8750 CFM	6.667	Ea.
12,500 CFM	10.000	Ea.
21,600 CFM	20.000	Ea.
Direct Drive Utility Set		
150 CFM	3.125	Ea.
485 CFM	3.448	Ea.
1950 CFM	4.167	Ea.
2410 CFM	4.545	Ea.
3328 CFM	6.667	Ea.
Belt Drive Utility Set		
800 CFM	3.333	Ea.
1300 CFM	4.000	Ea.
2000 CFM	4.348	Ea.
2900 CFM	4.762	Ea.
3600 CFM	5.000	Ea.
4800 CFM	5.714	Ea.
6700 CFM	6.667	Ea.
11,000 CFM	10.000	Ea.
13,000 CFM	12.500	Ea.
15,000 CFM	20.000	Ea.
17,000 CFM	25.000	Ea.
20,000 CFM	25.000	Ea.

(continued on next page)

Figure 15.69 Installation Time in Man-Hours for Fans (continued)

Description	Man-Hours	Unit
Belt Drive Propeller Fan		
12", 1000 CFM	.571	Ea.
14", 1500 CFM	.667	Ea.
16", 2000 CFM	.889	Ea.
30", 4800 CFM	1.143	Ea.
36", 7000 CFM	1.333	Ea.
42", 10,000 CFM	1.600	Ea.
48", 16,000 CFM	2.000	Ea.
Belt Drive Airfoil Centrifugal		
12,420 CFM	7.273	Ea.
18,620 CFM	8.000	Ea.
27,580 CFM	8.889	Ea.
40,980 CFM	10.667	Ea.
60,920 CFM	16.000	Ea.
74,520 CFM	20.000	Ea.
90,160 CFM	22.857	Ea.
110,300 CFM	32.000	Ea.
134,960 CFM	40.000	Ea.

Figure 15.70 Installation Time in Man-Hours for Roof Ventilators

Description	Man-Hours	Unit
Rotary Syphons		
6" Diameter 185 CFM	1.000	Ea.
8" Diameter 215 CFM	1.143	Ea.
10" Diameter 260 CFM	1.333	Ea.
12" Diameter 310 CFM	1.600	Ea.
14" Diameter 500 CFM	1.600	Ea.
16" Diameter 635 CFM	1.778	Ea.
18" Diameter 835 CFM	1.778	Ea.
20" Diameter 1080 CFM	2.000	Ea.
24" Diameter 1530 CFM	2.000	Ea.
30" Diameter 2500 CFM	2.286	Ea.
36" Diameter 3800 CFM	2.667	Ea.
42" Diameter 4500 CFM	4.000	Ea.
Spinner Ventilators		
4" Diameter 180 CFM	.800	Ea.
5" Diameter 210 CFM	.889	Ea.
6" Diameter 250 CFM	1.000	Ea.
8" Diameter 360 CFM	1.143	Ea.
10" Diameter 540 CFM	1.333	Ea.
12" Diameter 770 CFM	1.600	Ea.
14" Diameter 830 CFM	1.600	Ea.
16" Diameter 1200 CFM	1.778	Ea.
18" Diameter 1700 CFM	1.778	Ea.
20" Diameter 2100 CFM	2.000	Ea.
24" Diameter 3100 CFM	2.000	Ea.
30" Diameter 4500 CFM	2.286	Ea.
36" Diameter 5500 CFM	2.667	Ea.
Stationary Gravity Syphons		
3" Diameter 40 CFM	.667	Ea.
4" Diameter 50 CFM	.800	Ea.
5" Diameter 58 CFM	.889	Ea.
6" Diameter 66 CFM	1.000	Ea.
7" Diameter 86 CFM	1.067	Ea.
8" Diameter 110 CFM	1.143	Ea.
10" Diameter 140 CFM	1.333	Ea.
12" Diameter 160 CFM	1.600	Ea.
14" Diameter 250 CFM	1.600	Ea.
16" Diameter 380 CFM	1.778	Ea.
18" Diameter 500 CFM	1.778	Ea.
20" Diameter 625 CFM	2.000	Ea.
24" Diameter 900 CFM	2.000	Ea.
30" Diameter 1375 CFM	2.286	Ea.
36" Diameter 2000 CFM	2.667	Ea.
42" Diameter 3000 CFM	4.000	Ea.

Rotary Syphon

Spinner Ventilator

Stationary
Gravity Syphon

(continued on next page)

Figure 15.70 Installation Time in Man-Hours for Roof Ventilators (continued)

Description	Man-Hours	Unit
Rotating Chimney Caps		
4" Diameter	.800	Ea.
5" Diameter	.889	Ea.
6" Diameter	1.000	Ea.
7" Diameter	1.067	Ea.
8" Diameter	1.143	Ea.
10" Diameter	1.333	Ea.
Relief Hoods, Intake/Exhaust		
500 CFM 12" x 16"	2.000	Ea.
750 CFM 12" x 20"	2.222	Ea.
1000 CFM 12" x 24"	2.424	Ea.
1500 CFM 12" x 36"	2.759	Ea.
3000 CFM 20" x 42"	4.000	Ea.
6000 CFM 20" x 84"	6.154	Ea.
8000 CFM 24" x 96"	6.957	Ea.
10,000 CFM 48" x 60"	8.889	Ea.
12,500 CFM 48" x 72"	10.000	Ea.
15,000 CFM 48" x 96"	12.308	Ea.
20,000 CFM 48" x 120"	13.333	Ea.
25,000 CFM 60" x 120"	17.778	Ea.
30,000 CFM 72" x 120"	22.857	Ea.
40,000 CFM 96" x 120"	26.667	Ea.
50,000 CFM 96" x 144"	32.000	Ea.

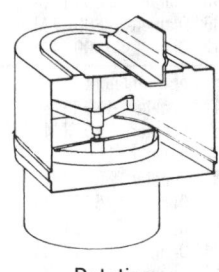

Rotating
Chimney Cap

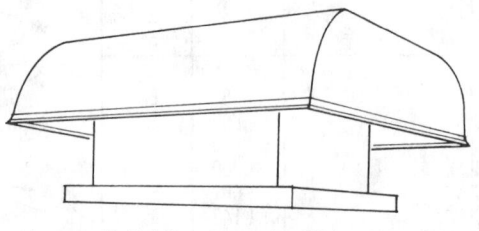

Relief Hood, Intake/Exhaust

Figure 15.71 Sheet Metal Calculator —
Weight in Pounds per Linear Foot

Example: If duct is 34" x 20" x 15' long, 34" is greater than 30" maximum for 24 ga. so 22 ga. must be used. 34" + 20" = 54". Going across from 54" find 13.5 lbs. per foot in 22 ga. column. 13.5 x 15' = 202.5 lbs. For S.F. of surface area 202.5 ÷ 1.406 = 144 S.F.
Note: Figures include an allowance for scrap.

Gauge	26	24	22	20	18	16	Gauge	26	24	22	20	18	16
Wt.–Lb./S.F.	.906	1.156	1.406	1.656	2.156	2.656	Wt.–Lb./S.F.	.906	1.156	1.406	1.656	2.156	2.656
SMACNA Max. Dimension Long Side		30"	54"	84"	85" Up		SMACNA Max. Dimension Long Side		30"	54"	84"	85" Up	
Sum–2 Sides							Sum–2 Sides						
2	.30	.40	.50	.60	.80	.90	42	7.0	9.0	10.5	12.2	16.0	18.9
3	.50	.65	.80	.90	1.1	1.4	43	7.2	9.2	10.8	12.5	16.3	19.4
4	.70	.85	1.0	1.2	1.5	1.8	44	7.3	9.5	11.0	12.8	16.7	19.8
5	.80	1.1	1.3	1.5	1.9	2.3	45	7.5	9.7	11.3	13.1	17.1	20.3
6	1.0	1.3	1.5	1.7	2.3	2.7	46	7.7	9.9	11.5	13.3	17.5	20.7
7	1.2	1.5	1.8	2.0	2.7	3.2	47	7.8	10.1	11.8	13.6	17.9	21.2
8	1.3	1.7	2.0	2.3	3.0	3.6	48	8.0	10.3	12.0	13.9	18.2	21.6
9	1.5	1.9	2.3	2.6	3.4	4.1	49	8.2	10.5	12.3	14.2	18.6	22.1
10	1.7	2.2	2.5	2.9	3.8	4.5	50	8.3	10.7	12.5	14.5	19.0	22.5
11	1.8	2.4	2.8	3.2	4.2	5.0	51	8.5	11.0	12.8	14.8	19.4	23.0
12	2.0	2.6	3.0	3.5	4.6	5.4	52	8.7	11.2	13.0	15.1	19.8	23.4
13	2.2	2.8	3.3	3.8	4.9	5.9	53	8.8	11.4	13.3	15.4	20.1	23.9
14	2.3	3.0	3.5	4.1	5.3	6.3	54	9.0	11.6	13.5	15.7	20.5	24.3
15	2.5	3.2	3.8	4.4	5.7	6.8	55	9.2	11.8	13.8	16.0	20.9	24.8
16	2.7	3.4	4.0	4.6	6.1	7.2	56	9.3	12.0	14.0	16.2	21.3	25.2
17	2.8	3.7	4.3	4.9	6.5	7.7	57	9.5	12.3	14.3	16.5	21.7	25.7
18	3.0	3.9	4.5	5.2	6.8	8.1	58	9.7	12.5	14.5	16.8	22.0	26.1
19	3.2	4.1	4.8	5.5	7.2	8.6	59	9.8	12.7	14.8	17.1	22.4	26.6
20	3.3	4.3	5.0	5.8	7.6	9.0	60	10.0	12.9	15.0	17.4	22.8	27.0
21	3.5	4.5	5.3	6.1	8.0	9.5	61	10.2	13.1	15.3	17.7	23.2	27.5
22	3.7	4.7	5.5	6.4	8.4	9.9	62	10.3	13.3	15.5	18.0	23.6	27.9
23	3.8	5.0	5.8	6.7	8.7	10.4	63	10.5	13.5	15.8	18.3	24.0	28.4
24	4.0	5.2	6.0	7.0	9.1	10.8	64	10.7	13.7	16.0	18.6	24.3	28.8
25	4.2	5.4	6.3	7.3	9.5	11.3	65	10.8	13.9	16.3	18.9	24.7	29.3
26	4.3	5.6	6.5	7.5	9.9	11.7	66	11.0	14.1	16.5	19.1	25.1	29.7
27	4.5	5.8	6.8	7.8	10.3	12.2	67	11.2	14.3	16.8	19.4	25.5	30.2
28	4.7	6.0	7.0	8.1	10.6	12.6	68	11.3	14.6	17.0	19.7	25.8	30.6
29	4.8	6.2	7.3	8.4	11.0	13.1	69	11.5	14.8	17.3	20.0	26.2	31.1
30	5.0	6.5	7.5	8.7	11.4	13.5	70	11.7	15.0	17.5	20.3	26.6	31.5
31	5.2	6.7	7.8	9.0	11.8	14.0	71	11.8	15.2	17.8	20.6	27.0	32.0
32	5.3	6.9	8.0	9.3	12.2	14.4	72	12.0	15.4	18.0	20.9	27.4	32.4
33	5.5	7.1	8.3	9.6	12.5	14.9	73	12.2	15.6	18.3	21.2	27.7	32.9
34	5.7	7.3	8.5	9.9	12.9	15.3	74	12.3	15.8	18.5	21.5	28.1	33.3
35	5.8	7.5	8.8	10.2	13.3	15.8	75	12.5	16.1	18.8	21.8	28.5	33.8
36	6.0	7.8	9.0	10.4	13.7	16.2	76	12.7	16.3	19.0	22.0	28.9	34.2
37	6.2	8.0	9.3	10.7	14.1	16.7	77	12.8	16.5	19.3	22.3	29.3	34.7
38	6.3	8.2	9.5	11.0	14.4	17.1	78	13.0	16.7	19.5	22.6	29.6	35.1
39	6.5	8.4	9.8	11.3	14.8	17.6	79	13.2	16.9	19.8	22.9	30.0	35.6
40	6.7	8.6	10.0	11.6	15.2	18.0	80	13.3	17.1	20.0	23.2	30.4	36.0
41	6.8	8.8	10.3	11.9	15.6	18.5	81	13.5	17.3	20.3	23.5	30.8	36.5

(continued on next page)

Figure 15.71   Sheet Metal Calculator — Weight in Pounds per Linear Foot (continued)

Gauge	26	24	22	20	18	16	Gauge	26	24	22	20	18	16
Wt.-Lb./S.F.	.906	1.156	1.406	1.656	2.156	2.656	Wt.-Lb./S.F.	.906	1.156	1.406	1.656	2.156	2.656
SMACNA Max. Dimension Long Side		30"	54"	84"	85" Up		SMACNA Max. Dimension Long Side		30"	54"	84"	85" Up	
Sum–2 Sides							Sum–2 Sides						
82	13.7	17.5	20.5	23.8	31.2	36.9	97	16.2	20.8	24.3	28.1	36.9	43.7
83	13.8	17.8	20.8	24.1	31.5	37.4	98	16.3	21.0	24.5	28.4	37.2	44.1
84	14.0	18.0	21.0	24.4	31.9	37.8	99	16.5	21.2	24.8	28.7	37.6	44.6
85	14.2	18.2	21.3	24.7	32.3	38.3	100	16.7	21.4	25.0	29.0	38.0	45.0
86	14.3	18.4	21.5	24.9	32.7	38.7	101	16.8	21.6	25.3	29.3	38.4	45.5
87	14.5	18.6	21.8	25.2	33.1	39.2	102	17.0	21.8	25.5	29.6	38.8	45.9
88	14.7	18.8	22.0	25.5	33.4	39.6	103	17.2	22.0	25.8	29.9	39.1	46.4
89	14.8	19.0	22.3	25.8	33.8	40.1	104	17.3	22.3	26.0	30.2	39.5	46.8
90	15.0	19.3	22.5	26.1	34.2	40.5	105	17.5	22.5	26.3	30.5	39.9	47.3
91	15.2	19.5	22.8	26.4	34.6	41.0	106	17.7	22.7	26.5	30.7	40.3	47.7
92	15.3	19.7	23.0	26.7	35.0	41.4	107	17.8	22.9	26.8	31.0	40.7	48.2
93	15.5	19.9	23.3	27.0	35.3	41.9	108	18.0	23.1	27.0	31.3	41.0	48.6
94	15.7	20.1	23.5	27.3	35.7	42.3	109	18.2	23.3	27.3	31.6	41.4	49.1
95	15.8	20.3	23.8	27.6	36.1	42.8	110	18.3	23.5	27.5	31.9	41.8	49.5
96	16.0	20.5	24.0	27.8	36.5	43.2							

Figure 15.72   Ductwork Packages (per ton of cooling)

System	Sheet Metal	Insulation	Diffusers	Return Register
Rooftop Unit Single Zone	120 lbs.	52 S.F.	1	1
Rooftop Unit Multizone	240 lbs.	104 S.F.	2	1
Self-contained Air or Water Cooled	108 lbs.	—	2	—
Split System Air Cooled	102 lbs.	—	2	—

Systems reflect most common usage

Figure 15.73 Recommended Thickness of Insulation for Fiberglass and Rock Wool on Piping & Ductwork

		Process Temperature (°F)									
Nominal Pipe Size (in inches)		150°	250°	350°	450°	550°	650°	750°	850°	950°	1050°
1/2	Thickness	1	1-1/2	2	2-1/2	3	3-1/2	4	4	4-1/2	5-1/2
	Heat Loss	8	16	24	33	43	54	66	84	100	114
	Surf. Temp.	72	75	76	78	79	81	82	86	87	87
1	Thickness	1	1-1/2	2	2-1/2	3-1/2	4	4	4-1/2	5	5-1/2
	Heat Loss	11	21	30	41	49	61	79	96	114	135
	Surf. Temp.	73	76	78	80	79	81	84	86	88	89
1-1/2	Thickness	1	2	2-1/2	3	4	4	4	5-1/2	5-1/2	6
	Heat Loss	14	22	33	45	54	73	94	103	128	152
	Surf. Temp.	73	74	77	79	79	82	86	84	88	90
2	Thickness	1-1/2	2	3	3-1/2	4	4	4	5-1/2	6	6
	Heat Loss	13	25	24	47	61	81	105	114	137	168
	Surf. Temp.	71	75	75	77	79	83	87	85	87	91
3	Thickness	1-1/2	2-1/2	3-1/2	4	4	4-1/2	4-1/2	6	6-1/2	7
	Heat Loss	16	28	39	54	75	94	122	133	154	184
	Surf. Temp.	72	74	75	77	81	83	87	86	87	90
4	Thickness	1-1/2	3	4	4	4	5	5-1/2	6	7	7-1/2
	Heat Loss	19	29	42	63	88	102	126	152	174	206
	Surf. Temp.	72	73	74	78	82	86	85	87	88	90
6	Thickness	2	3	4	4	4-1/2	5	5-1/2	6-1/2	7-1/2	8
	Heat Loss	21	38	54	81	104	130	159	181	208	246
	Surf. Temp.	71	74	75	79	82	84	87	88	89	91
8	Thickness	2	3-1/2	4	4	5	5	5-1/2	7	8	8-1/2
	Heat Loss	26	42	65	97	116	155	189	204	234	277
	Surf. Temp.	71	73	76	80	81	86	89	88	89	92
10	Thickness	2	3-1/2	4	4	5	5-1/2	5-1/2	7-1/2	8-1/2	9
	Heat Loss	32	50	77	115	136	170	220	226	259	307
	Surf. Temp.	72	74	77	81	82	85	90	87	89	91

Heat Loss = BTU/per ft./per hr.
Based on 65°F ambient temperature
Users are advised to consult manufacturer's literature for specific product temperature limitations.

(continued on next page)

Figure 15.73   Recommended Thickness of
Insulation for Fiberglass and
Rock Wool on Piping & Ductwork (continued)

Process Temperature (°F)										
Nominal Pipe Size (in inches)	**150°**	**250°**	**350°**	**450°**	**550°**	**650°**	**750°**	**850°**	**950°**	**1050°**
12 Thickness	2	3-1/2	4	4	5	5-1/2	5-1/2	7-1/2	8-1/2	9-1/2
Heat Loss	36	57	87	131	154	192	249	253	290	331
Surf. Temp.	72	74	77	82	82	86	91	88	89	91
14 Thickness	2	3-1/2	4	4	5	5-1/2	6-1/2	7-1/2	9	9-1/2
Heat Loss	40	61	94	141	165	206	236	271	297	352
Surf. Temp.	72	74	77	82	83	86	87	89	89	91
16 Thickness	2-1/2	3-1/2	4	4	5-1/2	5-1/2	7	8	9	10
Heat Loss	37	68	105	157	171	228	247	284	326	372
Surf. Temp.	71	74	78	83	82	87	86	88	89	91
18 Thickness	2-1/2	3-1/2	4	4	5-1/2	5-1/2	7	8	9	10
Heat Loss	41	75	115	173	187	250	270	310	354	404
Surf. Temp.	71	74	78	83	83	87	87	88	90	91
20 Thickness	2-1/2	3-1/2	4	4	5-1/2	5-1/2	7	8	9	10
Heat Loss	45	82	126	189	204	272	292	335	383	436
Surf. Temp.	71	75	78	83	83	87	87	89	90	92
24 Thickness	2-1/2	4	4	4	5-1/2	6	7-1/2	8	9	10
Heat Loss	53	86	147	221	237	295	320	386	439	498
Surf. Temp.	71	74	78	83	83	86	86	89	91	93
30 Thickness	2-1/2	4	4	4	5-1/2	6-1/2	7-1/2	8-1/2	10	10
Heat Loss	65	105	179	268	286	332	383	439	481	591
Surf. Temp.	71	74	79	84	84	85	87	89	89	94
36 Thickness	2-1/2	4	4	4	5-1/2	7	8	9	10	10
Heat Loss	77	123	211	316	335	364	422	486	556	683
Surf. Temp.	71	74	79	84	84	84	86	88	90	94
Flat Thickness	2	3-1/2	4	4-1/2	5-1/2	8-1/2	9-1/2	10	10	10
Heat Loss	10	14	20	27	31	27	31	38	47	58
Surf. Temp.	72	74	77	80	82	80	82	85	89	93

Heat Loss = BTU/per ft./per hr.
Based on 65°F ambient temperature
Users are advised to consult manufacturer's literature for specific product temperature limitations.

Figure 15.74 Ductwork Fabrication and Installation Guidelines

The labor cost for sheet metal duct includes both the cost of fabrication and installation of the duct. The split is approximately 60% for fabrication, 40% for installation. It is for this reason that the percentage add for elevated installation is less than the percentage add for prefabricated duct.

Example: assume a piece of duct costs $100 installed (labor only)

Sheet Metal Fabrication	=	60%	= $60
Installation	=	40%	= $40

The add for elevated installation is:
Based on total labor = $100 × 4% = $4.00
(fabrication & installation)

Based on installation cost only
(material purchased prefabricated) = $40 × 10% = $4.00

The $4.00 markup (10' to 15' high) is the same.

Figure 15.75 Installation Time in Man-Hours for Ductwork

Description	Man-Hours	Unit
Ductwork		
Fabricated Rectangular, Includes Fittings,		
Joints, Supports		
Allowance for Flexible Connections,		
No Insulation		
Aluminum, Alloy 3003-H14, Under 300 lbs.	.320	lb.
300 to 500 lbs.	.300	lb.
500 to 1000 lbs.	.253	lb.
1000 to 2000 lbs.	.200	lb.
2000 to 10,000 lbs.	.185	lb.
Over 10,000 lbs.	.166	lb.
Galvanized Steel, Under 400 lbs.	.102	lb.
400 to 1000 lbs.	.094	lb.
1000 to 2000 lbs.	.091	lb.
2000 to 5000 lbs.	.087	lb.
5000 to 10,000 lbs.	.084	lb.
Over 10,000 lbs.	.080	lb.
Stainless Steel, Type 304, Under 400 lbs.	.145	lb.
400 to 1000 lbs.	.130	lb.
1000 to 2000 lbs.	.120	lb.
2000 to 10,000 lbs.	.107	lb.
Over 10,000 lbs.	.102	lb.
Flexible, Vinyl Coated Spring Steel or		
Aluminum, Pressure to 10" (WG) UL-181		
Non-Insulated, 3" Diameter	.040	L.F.
4" Diameter	.044	L.F.
5" Diameter	.050	L.F.
Ductwork, Flexible, Non-Insulated		
6" Diameter	.057	L.F.
7" Diameter	.067	L.F.
8" Diameter	.080	L.F.
9" Diameter	.089	L.F.
10" Diameter	.100	L.F.
12" Diameter	.133	L.F.
14" Diameter	.200	L.F.
16" Diameter	.267	L.F.
Insulated, 4" Diameter	.047	L.F.
5" Diameter	.053	L.F.
6" Diameter	.062	L.F.
7" Diameter	.073	L.F.
8" Diameter	.089	L.F.
9" Diameter	.100	L.F.
10" Diameter	.114	L.F.
12" Diameter	.160	L.F.
14" Diameter	.200	L.F.
16" Diameter	.267	L.F.
18" Diameter	.356	L.F.
20" Diameter	.369	L.F.

Figure 15.75   Installation Time in Man-Hours for Ductwork (continued)

Description	Man-Hours	Unit
Fiberglass, Aluminized Jacket, 1-1/2" Blanket		
4" Diameter	.047	L.F.
5" Diameter	.053	L.F.
6" Diameter	.062	L.F.
7" Diameter	.073	L.F.
8" Diameter	.089	L.F.
9" Diameter	.100	L.F.
10" Diameter	.114	L.F.
12" Diameter	.160	L.F.
14" Diameter	.200	L.F.
16" Diameter	.267	L.F.
18" Diameter	.356	L.F.
Rigid Fiberglass, Round, .003" Foil Scrim Jacket		
4" Diameter	.052	L.F.
5" Diameter	.058	L.F.
6" Diameter	.067	L.F.
7" Diameter	.073	L.F.
8" Diameter	.089	L.F.
9" Diameter	.100	L.F.
10" Diameter	.114	L.F.
12" Diameter	.160	L.F.
14" Diameter	.200	L.F.
16" Diameter	.267	L.F.
18" Diameter	.356	L.F.
20" Diameter	.369	L.F.
22" Diameter	.400	L.F.
24" Diameter	.436	L.F.
26" Diameter	.480	L.F.
28" Diameter	.533	L.F.
30" Diameter	.600	L.F.
Rectangular, 1" Thick, Aluminum Faced, No Additional Insulation Required	.069	S.F. surf.

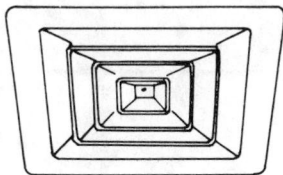

Supply Diffuser

(continued on next page)

Figure 15.75 Installation Time in Man-Hours for Ductwork (continued)

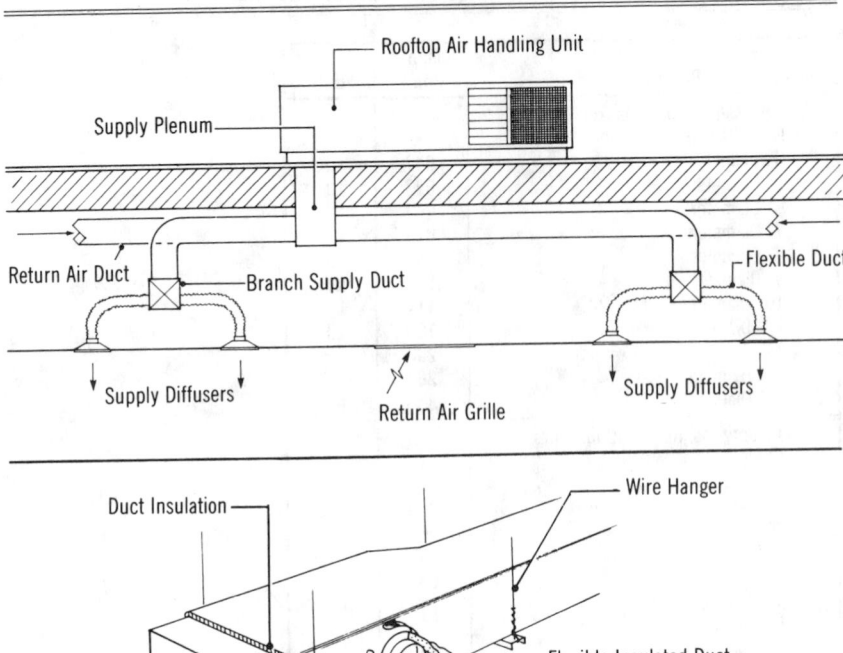

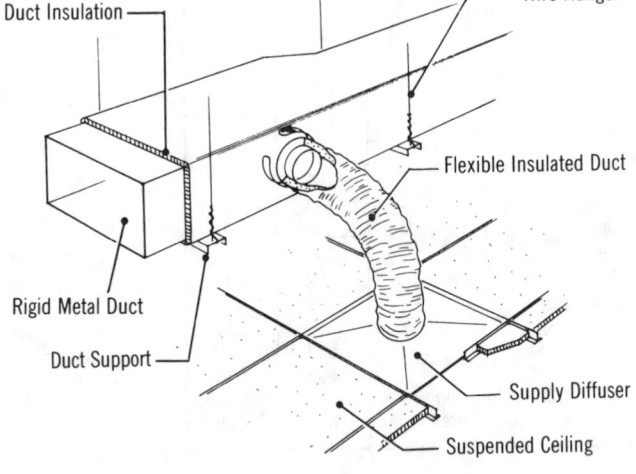

Ductwork System

Checklist

For an estimate to be reliable, all items must be accounted for.
A complete estimate can also eliminate the need to include
contingencies. The following checklist can be used to help ensure
that all items are properly accounted for.

Air Handling Units
☐ Balance
☐ Cooling
☐ Heating
☐ Filtration
☐ Supply fans
☐ Supply system
☐ Other _____

Balance Air Systems _____

Building Drainage
☐ Area drains
☐ Balcony drains
☐ Floor drains
☐ Roof drains
☐ Shower drains
☐ Sump drains

Coils
☐ Cooling
☐ Preheat
☐ Reheat

Cooling Plant
☐ Chillers
☐ Compressors
☐ Condensers
☐ Cooling towers
☐ Pumps
☐ Other _____

Domestic Water — Cold
☐ Boiler feed system
☐ Booster pump
☐ Hose bibs
☐ Pressure tank
☐ Reducing valve

- [ ] Standpipe system
- [ ] Water conditioning
- [ ] Water filtration
- [ ] Water softening
- [ ] Other _____

Domestic Water—Hot
- [ ] Boiler
- [ ] Conditioner
- [ ] Hose bibs
- [ ] Pumps
- [ ] Storage tanks
- [ ] Other _____

Ductwork System
- _____ Double duct
- _____ High velocity
- _____ With coils
- [ ] Diffusers
- [ ] Grilles
- [ ] Hoods
- [ ] Registers
- [ ] Terminal boxes
  - _____ Double duct
  - _____ High velocity
  - _____ With coils
  - [ ] Volume dampers

Fans
- [ ] Exhaust
- [ ] Return
- [ ] Other _____

Fire Protection Systems
- [ ] Building alarms
- [ ] Carbon dioxide system
- [ ] Extinguishers
- [ ] Fire department connection
  - _____ Standpipe
  - _____ Siamese connection
- [ ] Halon system
- [ ] Hydrants
- [ ] Roof manifold

☐ Sprinkler systems
 _____ Dry system
  _____ Compressed air supply
 _____ Wet system
 _____ Pre-action system
 _____ Deluge
 _____ Fire-cycle system
 _____ High temperature heads
☐ Standpipe system
 _____ Hose cabinets
 _____ Hose racks
☐ Special systems _____

_____

_____

Fixtures
☐ Bathtubs
☐ Drinking fountains
☐ Hose bibs
☐ Lavatories
☐ Showers
☐ Sinks
 _____ Bar
 _____ Janitor
 _____ Kitchen
 _____ Laundry
 _____ Slop
☐ Toilets
☐ Urinals
☐ Wash centers
☐ Wash fountains
☐ Water closets
☐ Water coolers
☐ Other _____

_____

_____

_____

Gas Supply System
☐ Bottled gas
☐ Natural gas
☐ Laboratory gas systems
☐ Oxygen

- ☐ Industrial gasses
- ☐ Control valves
- ☐ Fittings
- ☐ Meters
- ☐ Shutoffs
- ☐ Other _____

Heating
- ☐ Boilers
- ☐ Forced air
- ☐ Fuel storage tanks
- ☐ Heat pumps
- ☐ Rooftop units
- ☐ Solar
- ☐ Temperature control systems
- ☐ Unit heaters
- ☐ Water storage tanks
- ☐ Systems
  - ____ Multi-zone
  - ____ Perimeter radiant
  - ____ Radiant panels
  - ____ Single zone
  - ____ Terminal reheat
  - ____ VAV box system
- ☐ Other _____
  _____
  _____

Insulation
- ☐ Duct
- ☐ Piping
  - ____ Cold water
  - ____ Hot water
  - ____ Process
  - ____ Steam
  - ____ Tanks
- ☐ Sound
- ☐ Water heaters

Oil Supply System
- ☐ Filters
- ☐ Meters
- ☐ Pumps

- ☐ Safety valves
- ☐ Tanks

Piping

- ☐ Air chambers
- ☐ Concrete encasement
- ☐ Escutcheons
- ☐ Expansion joints
- ☐ Excavation for piping
- ☐ Fittings
- ☐ Hangers
- ☐ Materials
- ☐ Meters
- ☐ Shock absorbers
- ☐ Specialty
  - ____ Carbon dioxide
  - ____ Compressed air
  - ____ Industrial gasses
  - ____ Nitrous oxide
  - ____ Oxygen
  - ____ Process
  - ____ Sanitary
  - ____ Vacuum
- ☐ Other _____

Sanitary System

- ☐ Sump pumps
- ☐ Bilge pumps
- ☐ Trash pumps
- ☐ Soils piping
- ☐ Stacks

Temperature Control Systems

- ☐ Devices
- ☐ Wiring
- ☐ Other _____

Special Systems _____

_____

_____

_____

_____

_____

_____

Tips

Plans and Specifications

Review all construction documentation before proceeding with the estimate for Division 15. It is not unusual to find items on the plans, but not in the specifications; or they may be in the specifications but not on the drawings. Also check any and all communications; the owner may have requested a change or expressed a concern or a specific need, and not have had it included in your drawings or specifications.

Review All Drawing Divisions

It is important to review all of the drawings available to ensure that the systems called for will physically fit where they are supposed to. For example, sometimes the architect subcontracts out the HVAC design function as well as the structural design. If the drawings are not carefully coordinated, a conflict may result, such as inadequate space under deep trusses where main ductwork was supposed to be installed and a required finished ceiling height maintained.

Check Drawing Scales

When utilizing drawings from different sections of the plans, always check to see if the drawing scales are the same. It has happened that an estimator has ended up with twice the length of piping needed (or worse, half) because he/she was not aware that the drawing scales varied from structural to mechanical.

Document All Inconsistencies

All inconsistencies with standard practice, conflicts, omissions and other concerns should be addressed before the final estimate is resolved. Do not rely on memory which can become quickly overloaded in a rush situation.

Temperature Control System

When estimating the cost of an HVAC system, check to see who is responsible for providing and installing the temperature control wiring system. In many cases, this item is overlooked, as it was assumed that it would be included in the electrical estimate. This can be a costly assumption.

Connecting to an Existing System

When tapping into an existing system for water, drainage or waste, check with the local authorities (or building owner, in the case of interior work) to see if it is required that they or their designated contractor make the tap. Such surprises could lead to an unexpected expense or stoppage.

Non-Listed Items

It is important to include items that are not shown on the plans, but must be priced. These items include, but are not limited to, roof penetrations and pitchpots, dust protection, coring floors and walls, temporary water supply, testing and balancing HVAC systems, cleaning piping, purifying potable water systems, clean-up, and final adjustments.

Cutting and Patching

Not all projects will be done under ideal conditions, that is, wide open spaces with no interferences. Many times the mechanical contractor will have to run duct or piping through pre-existing partitions or partitions that have been recently installed. Allowances must be made for cutting into the partition and the ensuing patching. Otherwise, a statement must be included with the estimate that all cutting and patching will be the responsibility of others.

Electrical Connections

Clarify who is responsible for connecting the electrical wiring to mechanical items. It has been argued that the electrician is responsible only for bringing the wiring to the unit, while others maintain that it is the electrician's job to connect the wiring.

Notes

Division Sixteen
Electrical

Introduction

The electrical systems in building construction can be grouped under four major headings: Power, Lighting, Motor Control Centers, and Special Systems. Like mechanical systems, electrical systems have three distinct portions: Source, Conductor/Connection, and Terminal Unit. For example, take lighting: The source is a wall switch, the conductor/connector is wiring in conduit, and the terminal unit is the fixture. The first two components, the source and the conductor/connector, represent the rough-in phase of the project, and the terminal unit represents the finish portion of the electrical work.

Each of the major categories of electrical systems share common components: conduit or raceways, conductors, devices, switches, panelboards and fixtures. Raceways and conduits are channels that carry and protect conductors from the source to the terminal unit. Raceways consist of conduit, ducts, cable trays, and surface raceways.

A conductor is a wire or metal bar with a low level of resistance to the flow of electric current. Conductors are generally made of copper or aluminum, and are enclosed in an insulating jacket. Cables are conductors made up of heavy or multiple wires contained in a common jacket. Cable contained in a metallic sheathing (armored) is used where physical protection of the cable is necessary, but the use of conduit is not practical. Other types of conductors include high voltage shielded cable, flat wire or cable for undercarpet installations, low voltage wiring for telephones, and coaxial cable for data transmission.

Boxes are used in electrical wiring at each junction point or at each electrical device to provide easy and safe access for connections. Simply stated, devices control or conduct the flow of electricity without consuming any. Devices include wall switches, outlets, and light dimmers.

Switches are devices that are used to open, close, or change the condition of an electric circuit.

Boards serve as a means of grouping and mounting electric components, switches, safety devices, or controls. A load center is a specialized type of board used principally for containing circuit breakers.

Fixtures are by far the most elemental portion of the electrical work. Lighting, in building construction, is still the largest single electrical cost center.

The following topics are covered by the estimating charts in this section:

- Cable trays
- Conduits
- Conductors
- Boxes
- Wiring devices
- Starters and controls
- Boards
- Switchgear
- Transformers
- Lighting
- Utilities
- Special systems

Estimating Data

The following tables present effective estimating guidelines for items found in Division 16 — Electrical. Please note that these guidelines can be used as indicators of what may be expected, and that each project must be evaluated individually.

Table of Contents

Figure 16.1 Electrical Symbols

This table contains an extensive list of legend symbols as an aid to the electrical estimator. These designations are commonly used and accepted on electrical plan drawings, schematic diagrams, wiring diagrams, and takeoff forms.

Frequently, drawings do not contain complete, detailed information for the estimator. The legend symbol must, therefore, be used to interpret sizes, capacities, and/or equipment requirements in order to identify and select unit prices for the estimate.

Lighting Outlets

○	Ceiling Surface Incandescent Fixture
⊢○	Wall Surface Incandescent Fixture
ⓡ	Ceiling Recess Incandescent Fixture
⊢ⓡ	Wall Recess Incandescent Fixture
ᴬ○₃ᵦ	Standard Designation for All Lighting Fixtures — A = Fixture Type, 3 = Circuit Number, b = Switch Control
ⓑ	Ceiling Blanked Outlet
⊢ⓑ	Wall Blanked Outlet
ⓔ	Ceiling Electrical Outlet
⊢ⓔ	Wall Electrical Outlet
ⓙ	Ceiling Junction Box
⊢ⓙ	Wall Junction Box
ⓛ₍ₚₛ₎	Ceiling Lamp Holder with Pull Switch
⊢ⓛ₍ₚₛ₎	Wall Lamp Holder with Pull Switch
ⓛ	Ceiling Outlet Controlled by Low Voltage Switching When Relay Is Installed in Outlet Box
⊢ⓛ	Wall Outlet — Same as Above
◇	Outlet Box with Extension Ring
EX →	Exit Sign with Arrow as Indicated
▭○▭	Surface Fluorescent Fixture
▭○▭ P	Pendant Fluorescent Fixture
OR	Recessed Fluorescent Fixture
▭○▭	Wall Surface Fluorescent Fixture
⊣▭⊢	Channel Mounted Fluorescent Fixture

▭○▭▭	Surface or Pendant Continuous Row Fluorescent Fixtures
OR▭▭	Recessed Continuous Row Fluorescent Fixtures
●	Incandescent Fixture on Emergency Circuit
▭•▭	Fluorescent Fixture on Emergency Circuit

Receptacle Outlets

⊢⊖	Single Receptacle Outlet
⊢⊜	Duplex Receptacle Outlet
⊢⊜ˣ	Duplex Receptacle Outlet "X" Indicates Above Counter Max. Height = 42" or Above Counter
⊜_G.F.I.	Receptacle, Equipped with Ground Fault Circuit Interrupter
⊢⊜_WP	Weatherproof Receptacle Outlet
⊢⊕	Triplex Receptacle Outlet
⊢⊕	Quadruplex Receptacle Outlet
⊢⊜	Duplex Receptacle Outlet — Split Wired
⊢⊕	Triplex Receptacle Outlet — Split Wired
⊢⊘	Single Special Purpose Receptacle Outlet
⊢⊘	Duplex Special Purpose Receptacle Outlet
⊢⊜_R	Range Outlet
⊢⬤_DW	Special Purpose Connection — Dishwasher

(continued on next page)

Figure 16.1   Electrical Symbols (continued)

Receptacle Outlets (continued)

⊖xp   Explosion-Proof Receptacle Outlet
Max. Height = 36" to ℄

⊖,x   Multi-Outlet Assembly

Ⓒ   Clock Hanger Receptacle

Ⓕ   Fan Hanger Receptacle

⊖   Floor Single Receptacle Outlet

⊖   Floor Duplex Receptacle Outlet

△   Floor Special Purpose Outlet

◀   Floor Telephone Outlet — Public

▷   Floor Telephone Outlet — Private

Underfloor Duct and Junction Box for Triple, Double, or Single Duct System as Indicated by Number of Parallel Lines

Cellular Floor Header Duct

Switch Outlets

S   Single Pole Switch
Max. = 42" to ℄

S_2 Double Pole Switch

S_3 Three-Way Switch

S_4 Four-Way Switch

S_D Automatic Door Switch

S_K Key Operated Switch

S_P Switch & Pilot Lamp

S_{CB} Circuit Breaker

S_{WCB} Weatherproof Circuit Breaker

S_{MC} Momentary Contact Switch

S_{RC} Remote Control Switch (Receiver)

S_{WP} Weatherproof Switch

S_F Fused Switch

S_L Switch for Low Voltage Switching System

S_{LM} Master Switch for Low Voltage Switching System

S_T Time Switch

S_{TH} Thermal Rated Motor Switch

S_{DM} Incandescent Dimmer Switch

S_{FDM} Fluorescent Dimmer Switch

⊖s   Switch & Single Receptacle

⊖s   Switch & Double Receptacle

⊖A
S_A Special Outlet Circuits

Institutional, Commercial & Industrial System Outlets

⬠   Nurses Call System Devices — Any Type

◇   Paging System Devices — Any Type

▭   Fire Alarm System Devices — Any Type

F   Fire Alarm Manual Station — Max. Height = 48" to ℄

◁F   Fire Alarm Horn with Integral Warning Light

⊗   Fire Alarm Thermodetector, Fixed Temperature

Ⓢ   Smoke Detector

⊗   Fire Alarm Thermodetector, Rate of Rise

Figure 16.1   Electrical Symbols (continued)

Institutional, Commercial & Industrial System Outlets (continued)

⊞ F   Fire Alarm Master Box — Max. Height per Fire Department

H   Magnetic Door Holder

ANN   Fire Alarm Annunciator

◇   Staff Register System — Any Type

🕐   Electrical Clock System Devices — Any Type

◀   Public Telephone System Devices

◁   Private Telephone system Devices — Any Type

⌂   Watchman System Devices

L   Sound System, L = Speaker, V = Volume Control

⊕   Other Signal System Devices — CTV = Television Antenna, DP = Data Processing

SC   Signal Central Station

Telephone Interconnection Box

PE   Pneumatic/Electric Switch

EP   Electric/Pneumatic Switch

GP   Operating Room Grounding Plate

P 6   Patient Ground Point — 6 = Number of Jacks

Panelboards

Flush Mounted Panelboard & Cabinet

Surface Mounted Panelboard & Cabinet

Lighting Panel

Power Panel

Heating Panel

Controller (Starter)

Externally Operated Disconnect Switch

Busducts & Wireways

Trolley Duct

Busway (Service, Feeder, or Plug-In)

Cable Through Ladder or Channel

Wireway

J   Bus Duct Junction Box

Electrical Distribution or Lighting System, Aerial, Lightning Protection

○   Pole

Street Light & Bracket

△   Transformer

———   Primary Circuit

– – –   Secondary Circuit

— – —   Auxiliary System Circuits

→   Down Guy

•   Head Guy

→   Sidewalk Guy

Service Weather Head

Lightning Rod

— L —   Lightning Protection System Conductor

Residential Signaling System Outlets

•   Push Button

Buzzer

Bell

(continued on next page)

Figure 16.1   Electrical Symbols (continued)

Residential Signaling System Outlets (continued)

⊄̶	Bell and Buzzer Combination
◇	Annunciator
◀	Outside Telephone
◁	Interconnecting Telephone
▶◀	Telephone Switchboard
BT	Bell Ringing Transformer
D	Electric Door Opener
M	Maid's Signal Plug
R	Radio Antenna Outlet
CH	Chime
TV	Television Antenna Outlet
T	Thermostat

Underground Electrical Distribution or Lighting System

M	Manhole
H	Handhole
TM	Transformer — Manhole or Vault
TP	Transformer Pad
– – – –	Underground Direct Burial Cable
⊢⊦ - –	Underground Duct Line
⋈	Street Light Standard Fed from Underground Circuit

Panel Circuits & Miscellaneous

– – – –	Conduit Concealed in Floor or Walls
– – – – – –	Wiring Exposed

▸—▸	Home Run to Panelboard — Number of Arrows Indicates Number of Circuits
——	Home Run to Panelboard — Two-Wire Circuit
—///—	Home Run to Panelboard — Number of Slashes Indicates Number of Wires (When more than two)
LS–L1,3,5	Home Run to Panelboard — 'LS' Indicates Panel Designation; LI, 3, 5, Indicates Circuit Breaker No.
⏚	Ground Connection
Y≟	Grounded Wye
ᴖ 30 A OR ᴖ 30	Fuse, 30 A Type
225/125 3 P	Circuit Breaker, Molded Case, 3 Pole, 225 A Frame/125 A Trip
3 P 100/60 A OR 3 P 100 A 60 A	Fused Disconnect Switch, 3 Pole, 100 A, 60 A Fuse
—C—	Clock Circuit, Conduit and Wire
—E—	Emergency Conduit and Wiring
—T—	Telephone Conduit and Wiring
——	Feeders
—o	Conduit Turned Up
—●	Conduit Turned Down
Ⓖ	Generator
Ⓜ	Motor
Ⓢ̲	Motor — Numeral Indicates Horsepower
Ⓘ	Instrument (Specify)

Figure 16.1   Electrical Symbols (continued)

Panel Circuits & Miscellaneous (continued)

| T | Transformer |

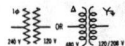 Transformer, 2 Winding, 1 ϕ Indicates
Single Phase; Δ-Y Indicates 3 Phase
and Type of Connection; Numbers
Indicate Voltages.

Remote Start-Stop Push Button
Station

Remote Start-Stop Push Button
Station w/Pilot Light

HTR   Electric Heater Wall Unit (Plan View)

Figure 16.2   Electrical Formulas

The information in this section can be used as a rule or method for performing estimates, either before or after the preliminary design stage.

OHM'S LAW

Ohm's Law is a method of explaining the relation existing between voltage, current, and resistance in an electrical circuit. It is practically the basis of all electrical calculations. The term "electromotive force" is often used to designate pressure in volts. This formula can be expressed in various forms.

To find the current in amperes:

$$\text{Current} = \frac{\text{Voltage}}{\text{Resistance}} \quad \text{or} \quad \text{Amperes} = \frac{\text{Volts}}{\text{Ohms}} \quad \text{or} \quad I = \frac{E}{R}$$

The flow of current in amperes through any circuit is equal to the voltage or electromotive force divided by the resistance of that circuit.

To find the pressure or voltage:

$$\text{Voltage} = \text{Current} \times \text{Resistance} \quad \text{or} \quad \begin{array}{c} \text{Volts} = \text{Amperes} \times \text{Ohms} \\ \text{or} \quad E = I \times R \end{array}$$

The voltage required to force a current through a circuit is equal to the resistance of the circuit multiplied by the current.

To find the resistance:

$$\text{Resistance} = \frac{\text{Voltage}}{\text{Current}} \quad \text{or} \quad \text{Ohms} = \frac{\text{Volts}}{\text{Amperes}} \quad \text{or} \quad R = \frac{E}{I}$$

The resistance of a circuit is equal to the voltage divided by the current flowing through that circuit.

(continued on next page)

Figure 16.2   Electrical Formulas (continued)

POWER FORMULAS

One horsepower = 746 watts   One kilowatt = 1000 watts

The power factor of electric motors varies from 80% to 90% in the larger size motors.

SINGLE-PHASE ALTERNATING CURRENT CIRCUITS
Power in Watts = Volts x Amperes x Power Factor

To find current in amperes:

$$\text{Current} = \frac{\text{Watts}}{\text{Volts x Power Factor}} \qquad \text{or}$$

$$\text{Amperes} = \frac{\text{Watts}}{\text{Volts x Power Factor}} \qquad \text{or} \qquad I = \frac{W}{E \times PF}$$

To find current of a motor, single phase:

$$\text{Current} = \frac{\text{Horsepower x 746}}{\text{Volts x Power Factor x Efficiency}} \qquad \text{or}$$

$$I = \frac{HP \times 746}{E \times PF \times \text{Eff.}}$$

To find horsepower of a motor, single phase:

$$\text{Horsepower} = \frac{\text{Volts x Current x Power Factor x Efficiency}}{746 \text{ Watts}} \qquad \text{or}$$

$$HP = \frac{E \times I \times PF \times \text{Eff.}}{746}$$

To find power in watts of a motor, single phase:

Watts = Volts x Current x Power Factor x Efficiency   or

Watts = E x I x PF x Eff.

To find single phase KVA:

$$1 \text{ Phase KVA} = \frac{\text{Volts x Amps}}{1000}$$

Figure 16.2   Electrical Formulas (continued)

POWER FORMULAS (continued)

THREE-PHASE ALTERNATING CURRENT CIRCUITS
Power in Watts = Volts x Amperes x Power Factor x 1.73

To find **current in amperes in each wire:**

$$\text{Current} = \frac{\text{Watts}}{\text{Voltage x Power Factor x 1.73}} \quad \text{or}$$

$$\text{Amperes} = \frac{\text{Watts}}{\text{Volts x Power Factor x 1.73}} \quad \text{or} \quad I = \frac{W}{E \text{ x PF x } 1.73}$$

To find **current of a motor, 3 phase:**

$$\text{Current} = \frac{\text{Horsepower x 746}}{\text{Volts x Power Factor x Efficiency x 1.73}} \quad \text{or}$$

$$I = \frac{\text{HP x 746}}{E \text{ x PF x Eff. x } 1.73}$$

To find **horsepower of a motor, 3 phase:**

$$\text{Horsepower} = \frac{\text{Volts x Current x 1.73 x Power Factor}}{746 \text{ Watts}} \quad \text{or}$$

$$\text{HP} = \frac{E \text{ x I x } 1.73 \text{ x PF}}{746}$$

To find **power in watts of a motor, 3 phase:**

Watts = Volts x Current x 1.73 x Power Factor x Efficiency   or

Watts = E x I x 1.73 x PF x Eff.

To find **3 phase KVA:**

$$3 \text{ phase KVA} = \frac{\text{Volts x Amps x 1.73}}{1000} \quad \text{or}$$

$$\text{KVA} = \frac{V \text{ x A x } 1.73}{1000}$$

Power Factor (PF) is the percentage ratio of the measured watts (effective power) to the volt-amperes (apparent watts).

$$\text{Power Factor} = \frac{\text{Watts}}{\text{Volts x Amperes}} \text{ x } 100\%$$

Figure 16.3   Typical Commercial Service Entrance

This figure illustrates the basic components of a typical commercial service entrance.

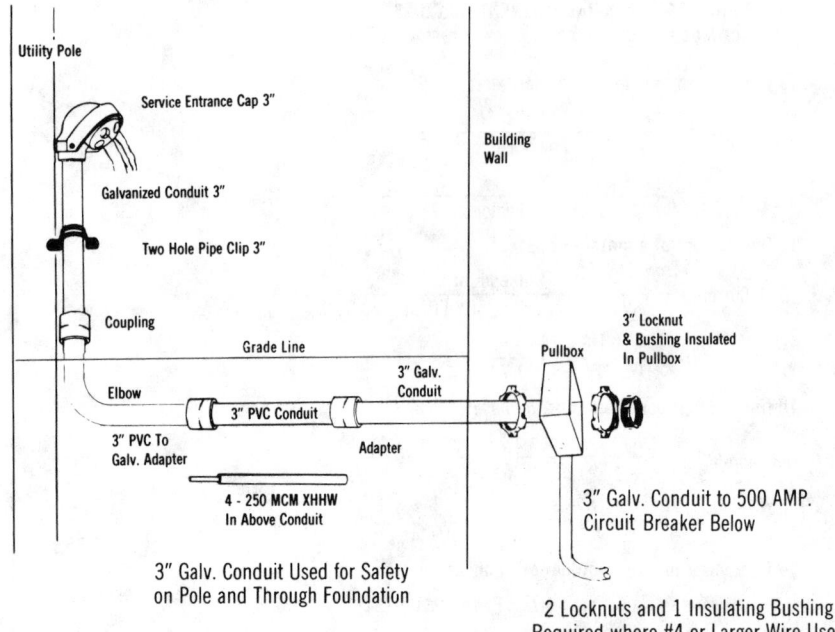

Utility Pole

Service Entrance Cap 3"

Building Wall

Galvanized Conduit 3"

Two Hole Pipe Clip 3"

Coupling

Grade Line

3" Locknut
& Bushing Insulated
In Pullbox

Pullbox

3" Galv.
Conduit

Elbow

3" PVC Conduit

3" PVC To
Galv. Adapter

Adapter

4 - 250 MCM XHHW
In Above Conduit

3" Galv. Conduit to 500 AMP.
Circuit Breaker Below

3" Galv. Conduit Used for Safety
on Pole and Through Foundation

2 Locknuts and 1 Insulating Bushing
Required where #4 or Larger Wire Used

Figure 16.4 Typical Commercial Electric System

This figure shows the basic lighting and power components used for the interior of a typical commercial project.

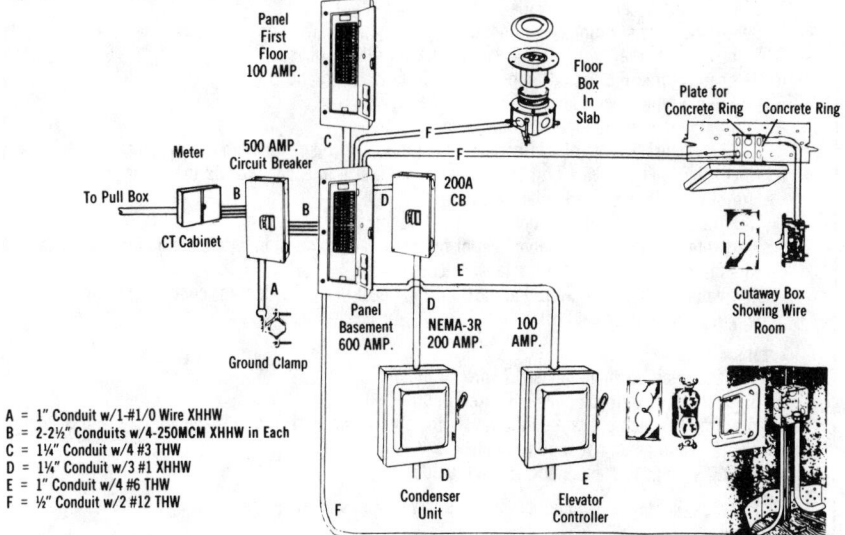

A = 1" Conduit w/1-#1/0 Wire XHHW
B = 2-2½" Conduits w/4-250MCM XHHW in Each
C = 1¼" Conduit w/4 #3 THW
D = 1¼" Conduit w/3 #1 XHHW
E = 1" Conduit w/4 #6 THW
F = ½" Conduit w/2 #12 THW

Figure 16.5   KW Value/Cost Determination

General: Lighting and electric heating loads are expressed in watts and kilowatts.

Cost Determination: The proper ampere values can be obtained as follows:
1. Convert watts to kilowatts (watts ÷ 1000 = kilowatts)
2. Determine voltage rating of equipment.
3. Determine whether equipment is single phase or three phase.
4. Refer to Figure 16.6 to find ampere value from KW, Ton and BTU/hr. values.
5. Determine type of wire insulation — TW, THW, THWN.
6. Determine if wire is copper or aluminum.
7. Refer to Figure 16.9 to obtain copper or aluminum wire size from ampere values.
8. Next refer to Figure 16.10 for the proper conduit size to accommodate the number and size of wires in each particular case.
9. Next refer to the per linear foot cost of the conduit.
10. Next refer to the per linear foot cost of the wire. Mulitply cost of wire L.F. x number of wires in the circuits to obtain total wire cost per L.F.
11. Add values obtained in Step 9 and 10 for total cost per linear foot for conduit and wire x length of circuit = Total Cost.

Notes:
1. 1 Phase refers to single phase, 2 wire circuits.
2. 3 Phase refers to three phase, 3 wire circuits.
3. For circuits which operate continuously for 3 hours or more, multiply the ampere values by 1.25 for a given KW requirement.
4. For KW ratings not listed, add ampere values.

   For example: find the ampere value of 9 KW at 208 volt, single phase.

$$\begin{array}{r} 4KW = 19.2A \\ 5KW = 24.0A \\ \hline 9KW = 43.2A \end{array}$$

5. "Length of Circuit" refers to the one-way distance of the run, not to the total sum of wire lengths.

Figure 16.6 Ampere Values as Determined by
KW Requirements, BTU/Hr.
or Ton, Voltage and Phase Values

KW	Ton	BTU/Hr.	Ampere Values						
			120V	208V		240V		277V	480V
			1 Phase	1 Phase	3 Phase	1 Phase	3 Phase	1 Phase	3 Phase
0.5	.1422	1,707	4.2A	2.4A	1.4A	2.1A	1.2A	1.8A	0.6A
0.75	.2133	2,560	6.2	3.6	2.1	3.1	1.9	2.7	0.9
1.0	.2844	3,413	8.3	4.9	2.8	4.2	2.4	3.6	1.2
1.25	.3555	4,266	10.4	6.0	3.5	5.2	3.0	4.5	1.5
1.5	.4266	5,120	12.5	7.2	4.2	6.3	3.1	5.4	1.8
2.0	.5688	6,826	16.6	9.7	5.6	8.3	4.8	7.2	2.4
2.5	.7110	8,533	20.8	12.0	7.0	10.4	6.1	9.1	3.1
3.0	.8532	10,239	25.0	14.4	8.4	12.5	7.2	10.8	3.6
4.0	1.1376	13,652	33.4	19.2	11.1	16.7	9.6	14.4	4.8
5.0	1.4220	17,065	41.6	24.0	13.9	20.8	12.1	18.1	6.1
7.5	2.1331	25,598	62.4	36.0	20.8	31.2	18.8	27.0	9.0
10.0	2.8441	34,130	83.2	48.0	27.7	41.6	24.0	36.5	12.0
12.5	3.5552	42,663	104.2	60.1	35.0	52.1	30.0	45.1	15.0
15.0	4.2662	51,195	124.8	72.0	41.6	62.4	37.6	54.0	18.0
20.0	5.6883	68,260	166.4	96.0	55.4	83.2	48.0	73.0	24.0
25.0	7.1104	85,325	208.4	120.2	70.0	104.2	60.0	90.2	30.0
30.0	8.5325	102,390		144.0	83.2	124.8	75.2	108.0	36.0
35.0	9.9545	119,455		168.0	97.1	145.6	87.3	126.0	42.1
40.0	11.3766	136,520		192.0	110.8	166.4	96.0	146.0	48.0
45.0	12.7987	153,585			124.8	187.5	112.8	162.0	54.0
50.0	14.2208	170,650			140.0	208.4	120.0	180.4	60.0
60.0	17.0650	204,780			166.4		150.4	216.0	72.0
70.0	19.9091	238,910			194.2		174.6		84.2
80.0	22.7533	273,040			221.6		192.0		96.0
90.0	25.5975	307,170					225.6		108.0
100.0	28.4416	341,300							120.0

Figure 16.7   Electric Circuit Voltages

The following method provides the user with a simple non-technical means of obtaining comparative costs of wiring circuits. The circuits considered serve the electrical loads of motors, electric heating, lighting and transformers, for example, those that require low voltage 60 Hertz alternating current.

The method used here is suitable only for obtaining estimated costs. It is not intended to be used as a substitute for electrical engineering design applications.

Conduit and wire circuits can represent from twenty to thirty percent of the total building electrical cost. By following the described steps and using the tables, the user can translate the various types of electric circuits into estimated costs.

Wire Size: Wire size is a function of the electric load which is usually listed in one of the following units:

1. Amperes (A)
2. Watts (W)
3. Kilowatts (KW)
4. Volt amperes (VA)
5. Kilovolt amperes (KVA)
6. Horsepower (HP)

The units of electric load must be converted to amperes in order to obtain the size of wire necessary to carry the load. To convert electric load units to amperes one must have an understanding of the voltage classification of the power source and the voltage characteristics of the electrical equipment or load to be energized. The seven A.C. circuits commonly used are illustrated in Figures A through G showing the transformer load voltage and the point of use voltage at the point on the circuit where the load is connected. The difference between the source and point of use voltages is attributed to the circuit voltage drop and is considered to be approximately 4%.

Motor Voltages: Motor voltages are listed by their point of use voltage and not the power source voltage.

For example: 460 volts instead of 480 volts
200 volts instead of 208 volts
115 volts instead of 120 volts

Lighting and Heating Voltages: Lighting and heating equipment voltages are listed by the power source voltage and not the point of wire voltage.

For example: 480, 277, 120 volt lighting
480 volt heating or air conditioning unit
208 volt heating unit

Transformer Voltages: Transformer primary (input) and secondary (output) voltages are listed by the power source voltage.

For example: Single phase 10 KVA
Primary 240/480 volts
Secondary 120/240 volts

In this case, the primary voltage may be 240 volts with a 120 volts secondary or may be 480 volts with either a 120v or a 240v secondary.

For example: Three phase 10 KVA
Primary 480 volts
Secondary 208Y/120 volts

In this case the transformer is suitable for connection to a circuit with a 3 phase 3 wire or 3 phase 4 wire circuit with a 480 voltage. This application will provide a secondary circuit of 3 phase 4 wire with 208 volts between phase wires and 120 volts between any phase wire and the neutral (white) wire.

Figure 16.7 Electric Circuit Voltages (continued)

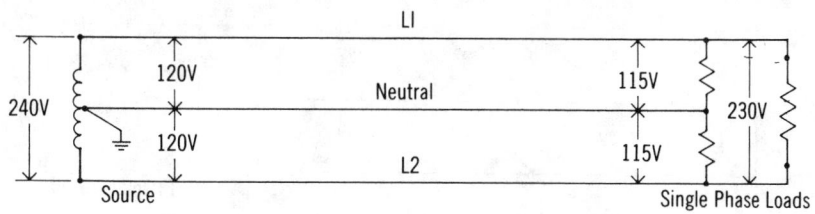

3 Wire, 1 Phase, 120/240 Volt System

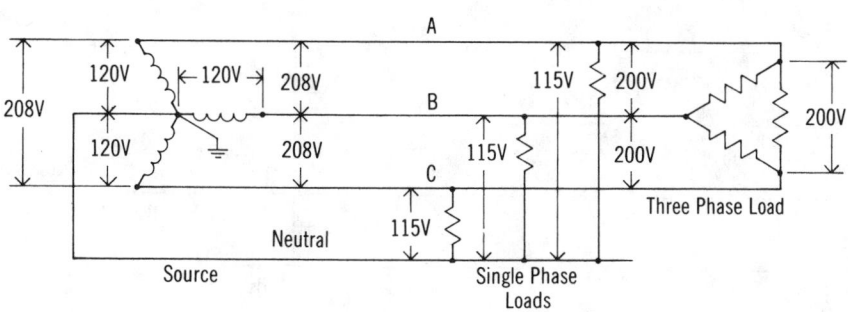

4 Wire, 3 Phase, 208Y/120 Volt System

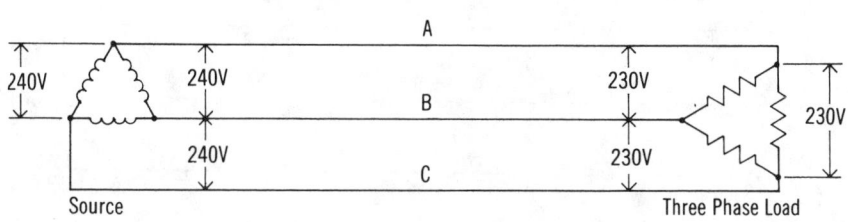

3 Wire, 3 Phase 240 Volt System

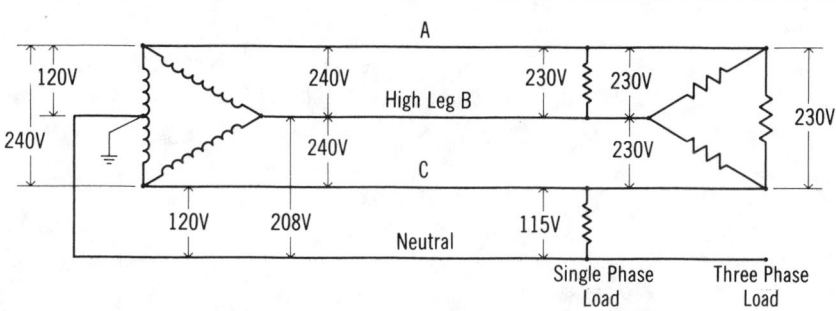

4 Wire, 3 Phase, 240/120 Volt System

(continued on next page)

Figure 16.7 Electric Circuit Voltages (continued)

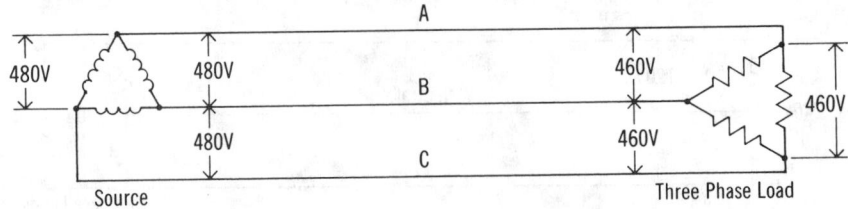

3 Wire, 3 Phase 480 Volt System

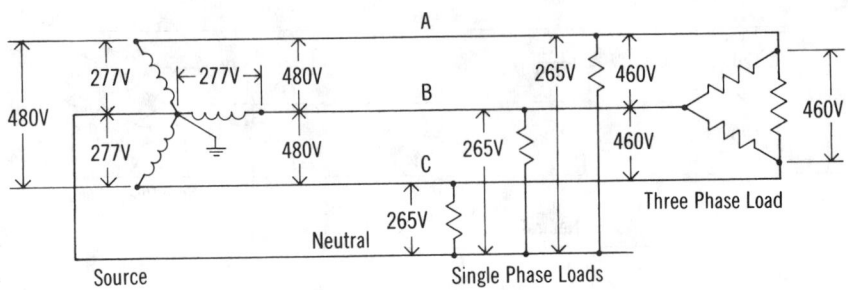

4 Wire, 3 Phase, 480Y/277 Volt System

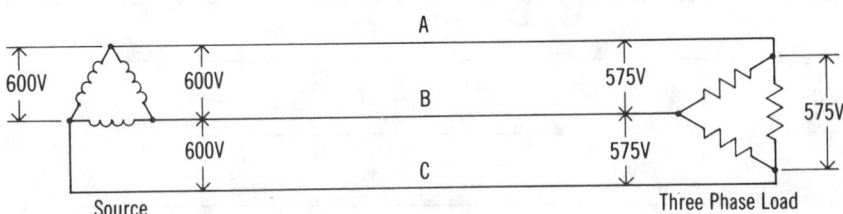

3 Wire, 3 Phase, 600 Volt System

Figure 16.8 Nominal Watts per S.F. for Electrical Systems by Various Building Types

This table is used in the preliminary design stage to develop the total load of the building in watts, based upon nominal watts per S.F. for electric systems for various building types.

The "Rule of Thumb" formula (see below) can be used to determine the service current required, based on the available voltage.

Type Construction	1. Lighting	2. Devices	3. HVAC	4. Misc.	5. Elevator	Total Watts
Apartment, luxury high-rise	2	2.2	3	1		
Apartment, low-rise	2	2	3	1		
Auditorium	2.5	1	3.3	.8		
Bank, branch office	3	2.1	5.7	1.4		
Bank, main office	2.5	1.5	5.7	1.4		
Church	1.8	.8	3.3	.8		
College, science building	3	3	5.3	1.3		
College, library	2.5	.8	5.7	1.4		
College, physical education center	2	1	4.5	1.1		
Department store	2.5	.9	4	1		
Dormitory, college	1.5	1.2	4	1		
Drive-in donut shop	3	4	6.8	1.7		
Garage, commercial	.5	.5	0	.5		
Hospital, general	2	4.5	5	1.3		
Hospital, pediatric	3	3.8	5	1.3		
Hotel, airport	2	1	5	1.3		
Housing for the elderly	2	1.2	4	1		
Manufacturing, food processing	3	1	4.5	1.1		
Manufacturing, apparel	2	1	4.5	1.1		
Manufacturing, tools	4	1	4.5	1.1		
Medical clinic	2.5	1.5	3.2	1		
Nursing home	2	1.6	4	1		
Office building, hi-rise	3	2	4.7	1.2		
Radio-TV studio	3.8	2.2	7.6	1.9		
Restaurant	2.5	2	6.8	1.7		
Retail store	2.5	.9	5.5	1.4		
School, elementary	3	1.9	5.3	1.3		
School, junior high	3	1.5	5.3	1.3		
School, senior high	2.3	1.7	5.3	1.3		
Supermarket	3	1	4	1		
Telephone exchange	1	.6	4.5	1.1		
Theater	2.5	1	3.3	.8		
Town Hall	2	1.9	5.3	1.3		
U.S. Post Office	3	2	5	1.3		
Warehouse, grocery	1	.6	0	.5		

Rule of Thumb: 1 KVA = 1 HP (Single Phase)

Three Phase:
Watts = 1.73 x Volts x Current x Power Factor x Efficiency

$$\text{Horsepower} = \frac{\text{Volts x Current x 1.73 x Power Factor}}{746 \text{ Watts}}$$

Figure 16.9   Economy of Scale for Electrical Installations

For large installations, the economy of scale may have a definite impact on the electrical costs. If large quantities of a particular item are installed in the same general area, certain deductions can be made for labor. This table lists some suggested deductions for larger scale electrical installations:

Description	Quantity	Labor Deduction
Under floor ducts, bus ducts, conduit, cable systems:	150 to 250 L.F. 250 to 350 L.F. 350 to 500 L.F. Over 500 L.F.	-10% -15% -20% -25%
Outlet boxes:	25 to 50 Ea. 50 to 75 Ea. 75 to 100 Ea. Over 100 Ea.	-15% -20% -25% -30%
Wiring devices:	10 to 25 Ea. 25 to 50 Ea. 50 to 100 Ea. Over 100 Ea.	-20% -25% -30% -35%
Lighting fixtures:	25 to 50 Ea. 50 to 75 Ea. 75 to 100 Ea. Over 100 Ea.	-15% -20% -25% -30%

Figure 16.10   Maximum Number of Wires (insulations noted) for Various Conduit Sizes

The table below lists the maximum number of conductors for various sized conduit using THW, TW, or THWN insulations.

Copper Wire Size	1/2"			3/4"			1"			1-1/4"			1-1/2"			2"			2-1/2"			3"			3-1/2"			4"	
	TW	THW	THWN	TW	THW	THWN	TW	THW	THWN	TW	THW	THWN	TW	THW	THWN	TW	THW	THWN	TW	THW	THWN	THW	THWN	THW	THWN	THW	THWN
#14	9	6	13	15	10	24	25	16	39	44	29	69	60	40	94	99	65	154	142	93		143		192			
#12	7	4	10	12	8	18	19	13	29	35	24	51	47	32	70	78	53	114	111	76	164	117		157			
#10	5	4	6	9	6	11	15	11	18	26	19	32	36	26	44	60	43	73	85	61	104	95	160	127		163	
#8	2	1	3	4	3	5	7	5	9	12	10	16	17	13	22	28	22	36	40	32	51	49	79	66	106	85	136
#6		1	1		2	4		4	6		7	11		10	15		16	26		23	37	36	57	48	76	62	98
#4		1	1		1	2		3	4		5	7		7	9		12	16		17	22	27	35	36	47	47	60
#3		1	1		1	1		2	3		4	6		6	8		10	13		15	19	23	29	31	39	40	51
#2		1	1		1	1		2	3		4	5		5	7		9	11		13	16	20	25	27	33	34	43
#1					1	1		1	1		3	3		4	5		6	8		9	12	14	18	19	25	25	32
1/0					1	1		1	1		2	3		3	4		5	7		8	10	12	15	16	21	21	27
2/0					1	1		1	1		1	2		3	3		5	6		7	8	10	13	14	17	18	22
3/0					1	1		1	1		1	1		2	3		4	5		6	7	9	11	12	14	15	18
4/0						1		1	1		1	1		1	2		3	4		5	6	7	9	10	12	13	15
250MCM								1	1		1	1		1	1		2	3		4	4	6	7	8	10	10	12
300								1	1		1	1		1	1		2	3		3	4	5	6	7	8	9	11
350									1		1	1		1	1		1	2		3	3	4	5	6	7	8	9
400											1	1		1	1		1	1		2	3	4	5	5	6	7	8
500											1	1		1	1		1	1		1	2	3	4	4	5	6	7
600												1		1	1		1	1		1	1	3	3	4	4	5	5
700														1	1		1	1		1	1	2	3	3	4	4	5
750														1	1		1	1		1	1	2	2	3	3	4	4

Figure 16.11 Conduit Weight Comparisons (lbs. per 100 ft.) Empty

Type	1/2"	3/4"	1"	1-1/4"	1-1/2"	2"	2-1/2"	3"	3-1/2"	4"	5"	6"
Rigid Aluminum	28	37	55	72	89	119	188	246	296	350	479	630
Rigid Steel	79	105	153	201	249	332	527	683	831	972	1314	1745
Intermediate Steel (IMC)	60	82	116	150	182	242	401	493	573	638		
Electrical Metallic Tubing (EMT)	29	45	65	96	111	141	215	260	365	390		
Polyvinyl Chloride Schedule 40	16	22	32	43	52	69	109	142	170	202	271	350
Polyvinyl Chloride Encased Burial						38		67	88	105	149	202
Fibre Duct Encased Burial						127		164	180	206	400	511
Fibre Duct Direct Burial						150		251	300	354		
Transite Encased Burial						160		240	290	330	450	550
Transite Direct Burial						220		310		400	540	640

Figure 16.12 Conduit Weight Comparisons (lbs. per 100 ft.) Filled* with Maximum Conductors Allowed

Type	1/2"	3/4"	1"	1-1/4"	1-1/2"	2"	2-1/2"	3"	3-1/2"	4"	5"	6"
Rigid Galvanized Steel (RGS)	104	140	235	358	455	721	1022	1451	1749	2148	3083	4343
Intermediate Steel (IMC)	84	113	186	293	379	611	883	1263	1501	1830		
Electrical Metallic Tubing (EMT)	54	116	183	296	368	445	641	930	1215	1540		

*Conduit & Heaviest Conductor Combination

Figure 16.13   Conduit Metric Equivalents

U.S. vs. European Conduit — Approximate Equivalents			
United States		European	
Trade Size	Inside Diameter Inch/MM	Trade Size	Inside Diameter MM
1/2	.622/15.8	11	16.4
3/4	.824/20.9	16	19.9
1	1.049/26.6	21	25.5
1-1/4	1.380/35.0	29	34.2
1-1/2	1.610/40.9	36	44.0
2	2.067/52.5	42	51.0
2-1/2	2.469/62.7		
3	3.068/77.9		
3-1/2	3.548/90.12		
4	4.026/102.3		
5	5.047/128.2		
6	6.065/154.1		

Figure 16.14   Installation Time in Man-Hours for Conduit

Description	Man-Hours	Unit
Rigid Galvanized Steel 1/2" Diameter	.089	L.F.
1-1/2" Diameter	.145	L.F.
3" Diameter	.320	L.F.
6" Diameter	.800	L.F.
Aluminum 1/2" Diameter	.080	L.F.
1-1/2" Diameter	.123	L.F.
3" Diameter	.178	L.F.
6" Diameter	.400	L.F.
IMC 1/2" Diameter	.080	L.F.
1-1/2" Diameter	.133	L.F.
3" Diameter	.267	L.F.
4" Diameter	.320	L.F.
Plastic Coated Rigid Steel 1/2" Diameter	.100	L.F.
1-1/2" Diameter	.178	L.F.
3" Diameter	.364	L.F.
6" Diameter	.800	L.F.
EMT 1/2" Diameter	.047	L.F.
1-1/2" Diameter	.089	L.F.
3" Diameter	.160	L.F.
4" Diameter	.200	L.F.
PVC Nonmetallic 1/2" Diameter	.042	L.F.
1-1/2" Diameter	.080	L.F.
3" Diameter	.145	L.F.
6" Diameter	.267	L.F.

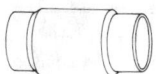

Rigid Steel,
Plastic Coated Coupling

PVC Conduit

PVC Elbow

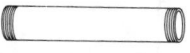

Aluminum Conduit

EMT Set Screw Connector

Aluminum Elbow

EMT Connector

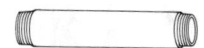

Rigid Steel, Plastic Coated Conduit

EMT to Conduit Adapter

Rigid Steel, Plastic Coated Elbow

EMT to Greenfield Adapter

Figure 16.15   Elevated Bus Ducts and Cable Tray Installation Adjustments

For the installation of bus ducts and cable trays elevated more than 15' above the floor, the labor costs should be adjusted to allow for the added complexity. This table lists suggested adjustment factors.

Installation Height Above Floor	Labor Adjustments
15' to 20'	+ 10%
20' to 25'	+ 20%
25' to 30'	+ 25%
30' to 35'	+ 30%
35' to 40'	+ 35%
Over 40'	+ 40%

Figure 16.16   Installation Time in Man-Hours for Cable Tray Systems

Description	Man-Hours	Unit
Cable Tray		
Ladder Type 36" Wide	.267	L.F.
Elbows Vertical 36"	3.810	Ea.
Elbows Horizontal 36"	3.810	Ea.
Tee Vertical 36"	4.440	Ea.
Tee Horizontal 36"	5.330	Ea.
Drop-Out 36"	1.000	Ea.
Reducer 36" to 12"	2.290	Ea.
Wall Bracket 12"	.364	Ea.
Cover Straight 36"	.100	L.F.
Cover Elbow 36"	.320	Ea.

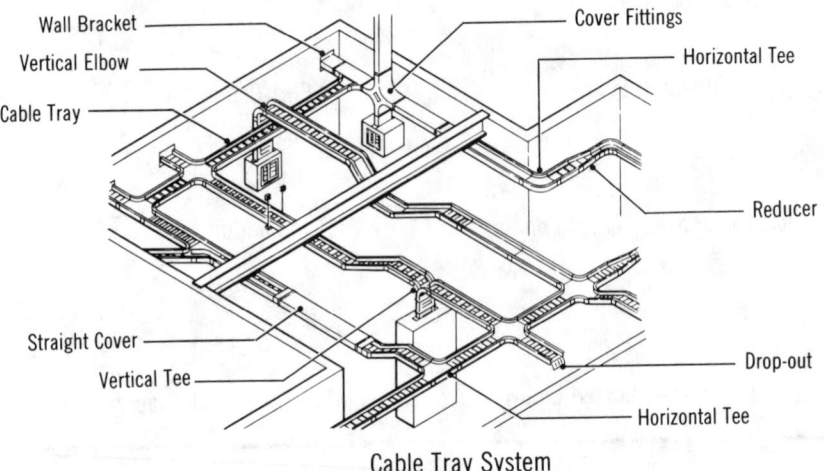

Cable Tray System

Figure 16.17 Installation Time in Man-Hours for Underfloor Raceway Systems

Description	Man-Hours	Unit
Blank Duct	.100	L.F.
Insert Duct	.110	L.F.
Elbow Vertical	.800	Ea.
Elbow Horizontal	.300	Ea.
Panel Connector	.250	Ea.
Junction Box, Single Duct	2.000	Ea.
Double Duct	2.500	Ea.
Triple Duct	2.950	Ea.
Saddle Support, Single Duct	.290	Ea.
Double Duct	.500	Ea.
Triple Duct	.720	Ea.
Insert to Conduit Adapter	.250	Ea.
Outlet, Low Tension (Telephone and Signal)	1.000	Ea.
Outlet, High Tension (Power)	1.000	Ea.
Offset Duct Type	.300	Ea.

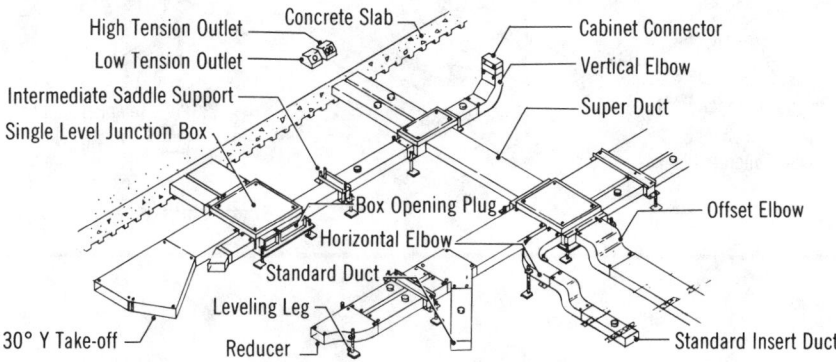

Figure 16.18   Installation Time in Man-Hours for Cellular Concrete Floor Raceway Systems

Description	Man-Hours	Unit
Underfloor Header Duct 3-1/8" Wide	.100	L.F.
7-1/4" Wide	.133	L.F.
Header Duct, Single Compartment 9" Wide	.400	L.F.
24" Wide	.727	L.F.
Double Compartment, 24" Wide	.800	L.F.
Triple Compartment, 36" Wide	1.330	L.F.
Outlet Drive Plug	.670	Ea.
Location Market Plug	.250	Ea.

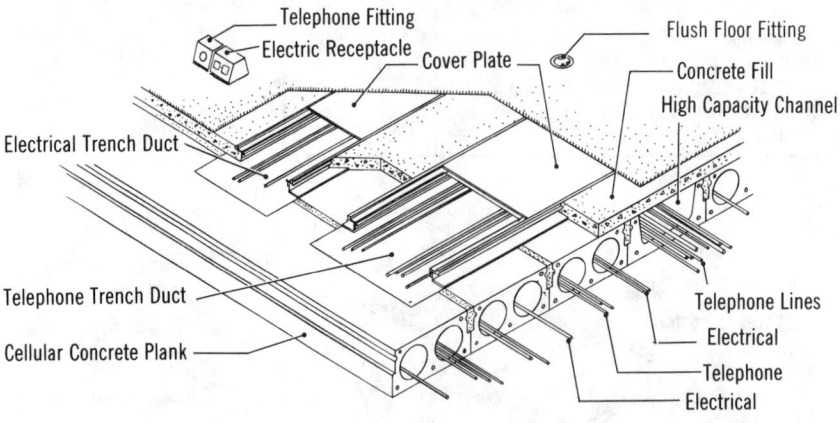

Figure 16.19 Minimum Wire Size Allowed for Various Powered Circuits

The table below lists the minimum copper and aluminim wire size allowed for various types of insulation and design.

Minimum Wire Sizes									
	Copper		Aluminum			Copper		Aluminum	
Amperes	THW THWN or XHHW	THHN XHHW *	THW XHHW	THHN XHHW *	Amperes	THW THWN or XHHW	THHN XHHW *	THW XHHW	THHN XHHW *
15A	#14	#14	#12	#12	195		2/0		
20	#12	#12			200	3/0			
25			#10	#10	205			250MCM	4/0
30	#10	#10			225		3/0		
40			#8		230	4/0		300MCM	250MCM
45				#8	250			350MCM	
50	#8		#6		255	250MCM			300MCM
55		#8			260		4/0		
60				#6	270			400MCM	
65	#6		#4		280				350MCM
75		#6	#3	#4	285	300MCM			
85	#4			#3	290		250MCM		
90			#2		305				400MCM
95		#4			310	350MCM		500MCM	
100	#3		#1	#2	320		300MCM		
110		#3			335	400MCM			
115	#2			#1	340			600MCM	
120			1/0		350		350MCM		500MCM
130	#1	#2			375			700MCM	
135			2/0	1/0	380	500MCM	400MCM		
150	1/0	#1		2/0	385			750MCM	600MCM
155			3/0		420	600MCM			700MCM
170		1/0			430		500MCM		
175	2/0			3/0	435				750MCM
180			4/0		475		600MCM		

*Dry Locations Only

Notes:
1. Size #14 to 4/0 is in AWG units (American Wire Gauge)
2. Size 250 to 750 is in MCM units (Thousand Circular Mils).
3. Use next higher ampere value if exact value is not listed in table.
4. For loads that operate continuously increase ampere value by 25% to obtain proper wire size.
5. Refer to other Figures for the maximum circuit length for the various size wires.

Figure 16.20   Maximum Circuit Length (Approximate) for Various Power Requirements

This figure assumes the use of THW, copper wire @ 75°C, and a 4% voltage drop.

Maximum Circuit Length: This Figure indicates the typical maximum installed length a circuit can have and still maintain an adequate voltage level at the point of use. The circuit length is similar to the conduit length.

If the circuit length for an ampere load and a copper wire size exceeds the length obtained from this table, use the next largest wire size to compensate for voltage drop.

Example: A 130 ampere load at 480 volts, 3 phase, 3 wire with No. 1 wire can be run a maximum of 555 L.F. and provide satisfactory operation. If the same load is to be wired at the end of a 625 L.F. circuit, then a larger wire must be used .

		Maximum Circuit Length in Feet				
	Wire	2 Wire, 1 Phase		3 Wire, 3 Phase		
Amperes	Size	120V	240V	240V	480V	600V
15	14*	50	105	120	240	300
	14	50	100	120	235	295
20	12*	60	125	145	290	360
	12	60	120	140	280	350
30	10*	65	130	155	305	380
	10	65	130	150	300	375
50	8	60	125	145	285	355
65	6	75	150	175	345	435
85	4	90	185	210	425	530
115	2	110	215	250	500	620
130	1	120	240	275	555	690
150	1/0	130	260	305	605	760
175	2/0	140	285	330	655	820
200	3/0	155	315	360	725	904
230	4/0	170	345	395	795	990
255	250	185	365	420	845	1055
285	300	195	395	455	910	1140
310	350	210	420	485	975	1220
380	500	245	490	565	1130	1415

*Solid Conductor.
Note: The circuit length is the one way distance between the origin and the load.

Figure 16.21   Metric Equivalents for Wire

U.S. vs. European Wire — Approximate Equivalents			
United States		European	
Size AWG or MCM	Area Cir. Mils. (CM) MM²	Size MM²	Area Cir. Mils.
18	1620/.82	.75	1480
16	2580/1.30	1.0	1974
14	4110/2.08	1.5	2961
12	6530/3.30	2.5	4935
10	10,380/5.25	4	7896
8	16,510/8.36	6	11,844
6	26,240/13.29	10	19,740
4	41,740/21.14	16	31,584
3	52,620/26.65	25	49,350
2	66,360/33.61	—	—
1	83,690/42.39	35	69,090
1/0	105,600/53.49	50	98,700
2/0	133,100/67.42	—	—
3/0	167,800/85.00	70	138,180
4/0	211,600/107.19	95	187,530
250	250,000/126.64	120	236,880
300	300,000/151.97	150	296,100
350	350,000/177.30	—	—
400	400,000/202.63	185	365,190
500	500,000/253.29	240	473,760
600	600,000/303.95	300	592,200
700	700,000/354.60	—	—
750	750,000/379.93	—	—

Figure 16.22   Installation Time in Man-Hours for Electrical Conductors: Wire and Cable

Description	Man-Hours	Unit
600V Copper #14 AWG	.610	CLF
#12 AWG	.720	CLF
#10 AWG	.800	CLF
#8 AWG	1.000	CLF
#6 AWG	1.230	CLF
#4 AWG	1.510	CLF
#3 AWG	1.600	CLF
#2 AWG	1.780	CLF
#1 AWG	2.000	CLF
#1/0	2.420	CLF
#2/0	2.760	CLF
#3/0	3.200	CLF
#4/0	3.640	CLF
250 MCM	4.000	CLF
500 MCM	5.000	CLF
1000 MCM	9.000	CLF

(continued on next page)

Figure 16.22 Installation Time in Man-Hours for Electrical Conductors: Wire and Cable (continued)

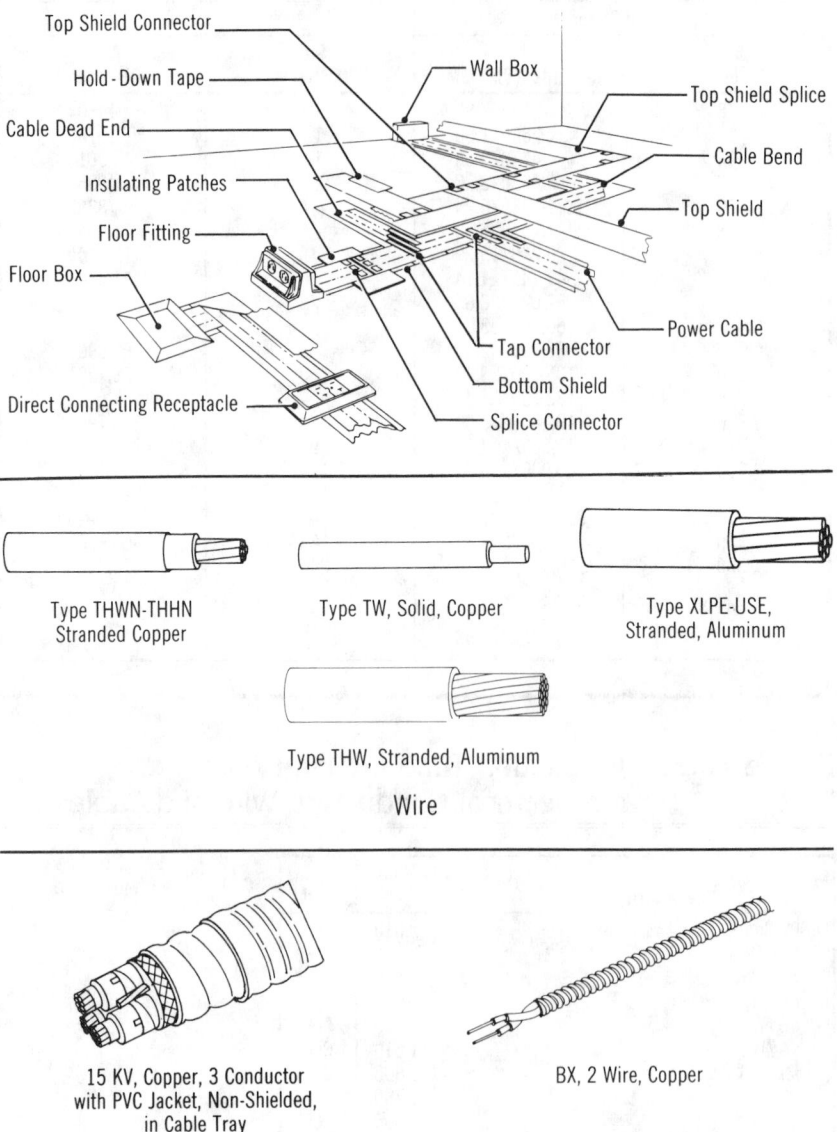

Top Shield Connector

Hold-Down Tape

Cable Dead End

Insulating Patches

Floor Fitting

Floor Box

Direct Connecting Receptacle

Wall Box

Top Shield Splice

Cable Bend

Top Shield

Power Cable

Tap Connector

Bottom Shield

Splice Connector

Type THWN-THHN
Stranded Copper

Type TW, Solid, Copper

Type XLPE-USE,
Stranded, Aluminum

Type THW, Stranded, Aluminum

Wire

15 KV, Copper, 3 Conductor
with PVC Jacket, Non-Shielded,
in Cable Tray

BX, 2 Wire, Copper

Armored Cable

Figure 16.22 Installation Time in Man-Hours for Electrical Conductors: Wire and Cable (continued)

Compression Adapter for
Aluminum Wire

Split Bolt Connector, Tapped

Crimp, 2-Way Connector

Crimp, 1-Hole Lug

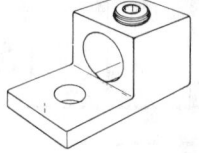

Terminal Lug, Solderless

Compression Hand Tool

Cable Terminations, Solderless

PVC Jacket Connector

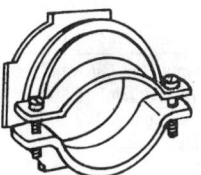

SER, Insulated, Aluminum

600 Volt, Armored

5 KV Armored

Cable Connectors

Figure 16.23 Installation Time in Man-Hours for Undercarpet Power Systems

Description	Man-Hours	Unit
Cable 3 Conductor #12 with Bottom Shield	.008	L.F.
Cable 5 Conductor #12 with Bottom Shield	.010	L.F.
Splice 3 Conductor with Insulating Patch	.334	Ea.
Splice 5 Conductor with Insulating Patch	.334	Ea.
Tap 3 Conductor with Insulating Patch	.367	Ea.
Tap 5 Conductor with Insulating Patch	.367	Ea.
Receptacle with Floor Box Pedestal Type	.500	Ea.
Receptacle Direct Connect	.320	Ea.
Top Shield	.005	L.F.
Transition Block	.104	Ea.
Transition Box, Flush Mount with Cover	.400	Ea.

Figure 16.24 Installation Time in Man-Hours for Undercarpet Telephone Systems

Description	Man-Hours	Unit
Cable Assembly 25 Pair with Connectors 50'	.670	Ea.
3 Pair with Connectors 50'	.340	Ea.
4 Pair with Connectors 50'	.350	Ea.
Cable (Bulk) 3 Pair	.006	L.F.
4 Pair	.007	L.F.
Bottom Shield for 25 Pair Cable	.005	L.F.
3-4 Pair Cable	.005	L.F.
Top Shield for all Cable	.005	L.F.
Transition Box, Flush Mount	.330	Ea.
In Floor Service Box	2.000	Ea.
Floor Fitting with Duplex Jack and Cover	.380	Ea.
Floor Fitting Miniature with Duplex Jack	.150	Ea.
Floor Fitting with 25 Pair Kit	.380	Ea.
Floor Fitting Call Director Kit	.420	Ea.

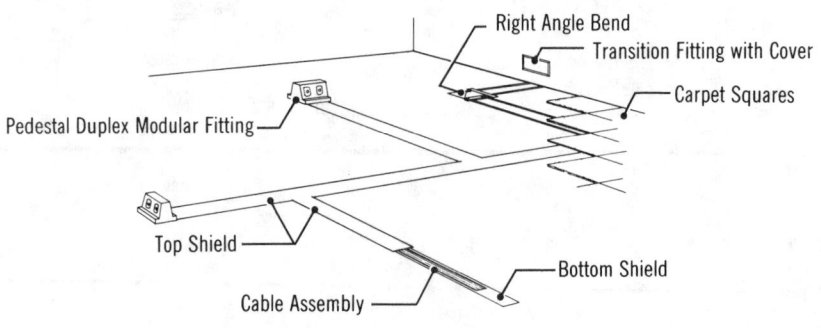

Figure 16.25 Installation Time in Man-Hours for Undercarpet Data Systems

Description	Man-Hours	Unit
Cable Assembly with Connectors 40'		
Single Lead	.360	Ea.
Dual Lead	.380	Ea.
Cable (Bulk) Single Lead	.010	L.F.
Dual Lead	.010	L.F.
Cable Notching 90 Degree	.080	Ea.
180 Degree	.130	Ea.
Connectors BNC Coax	.200	Ea.
Connectors TNC Coax	.200	Ea.
Transition Box, Flush Mount	.330	Ea.
In Floor Service Box	2.000	Ea.
Floor Fitting with Slotted Cover	.380	Ea.
With Blank Cover	.380	Ea.

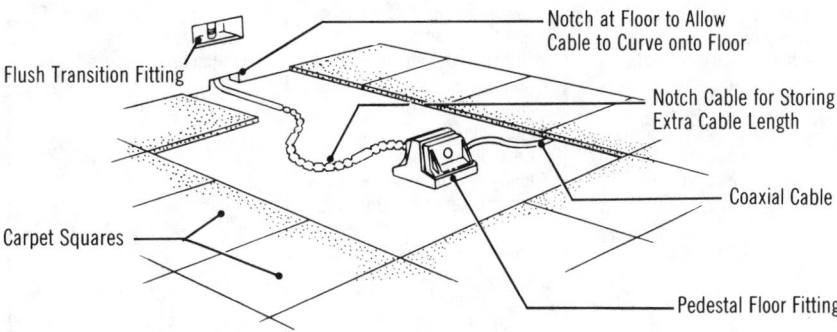

Flush Transition Fitting

Notch at Floor to Allow Cable to Curve onto Floor

Notch Cable for Storing Extra Cable Length

Coaxial Cable

Carpet Squares

Pedestal Floor Fitting

Figure 16.26 Hazardous Area Wiring

All hazardous locations require special material and wiring methods as determined by the classifications defined in the National Electrical Code. The descriptions of hazardous locations listed below have been abbreviated. The National Electrical Code should be consulted for complete descriptions.

Class 1 Locations	Class 2 Locations	Class 3 Locations
Locations in which flammable gases or vapors are or may be present in the air in quantities sufficient to produce explosive or ignitable mixtures.	Locations in which hazardous conditions exist because of the presence of combustible dust.	Locations in which the presence of easily flammable fibers or flyings are present in the air, but not in sufficient quantities to produce ignitable mixtures under normal conditions.
Division 1: This classification includes locations where a hazardous atmosphere is expected during normal operations; locations where a breakdown in operation of processing equipment results in release of hazardous vapors and the simultaneous failure of electrical equipment.	**Division 1:** This classification includes locations where combustible dust may be suspended in the air under normal conditions in sufficient quantities to produce explosive or ignitable mixtures; locations where a breakdown in operation of machinery or equipment might cause a hazardous condition to exist while creating a source of ignition with the simultaneous failure of electrical equipment; locations in which combustible dust of an electrically conductive nature may be present.	**Division 1:** This classification includes locations where easily ignitable fibers or materials producing combustible flyings are handled, manufactured, or used.
Division 2: This classification includes locations in which volatile flammable liquids or gases are handled, processed, or used, but will normally be confined to closed containers or systems from which they can escape only by accidental rupture or breakdown; locations where hazardous conditions will occur only under abnormal circumstances.	**Division 2:** This classification includes locations where air-suspended combustible dust is not at hazardous levels, but where an accumulation of dust may interfere with the safe dissipation of heat from electrical equipment, or may be ignited by arcs, sparks, or burning material located near electrical equipment.	**Division 2:** This classification includes locations where easily ignitable fibers are stored or handled.

Figure 16.27 Installation Time in Man-Hours for Hazardous Area Wiring

Description	Man-Hours	Unit
Sealing Fitting 1/2" Diameter	.660	Ea.
2" Diameter	1.600	Ea.
3" Diameter	2.000	Ea.
4" Diameter	2.670	Ea.
Flexible Coupling 3/4" Diameter x 12" Long	.800	Ea.
2" Diameter x 12" Long	1.740	Ea.
3" Diameter x 12" Long	2.670	Ea.
4" Diameter x 12" Long	3.330	Ea.
Pulling Elbow 3/4" Diameter	1.000	Ea.
2" Diameter	2.000	Ea.
3" Diameter	2.670	Ea.
Conduit LB 3/4" Diameter	1.000	Ea.
T 3/4" Diameter	1.330	Ea.
Cast Box NEMA 7		
6" Long x 6" Wide x 6" Deep	4.000	Ea.
12" Long x 12" Wide x 6" Deep	8.000	Ea.
18" Long x 18" Wide x 8" Deep	16.000	Ea.

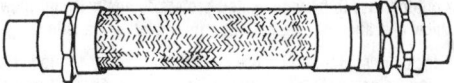

Explosionproof Flexible Coupling

Explosionproof
Sealing Fitting

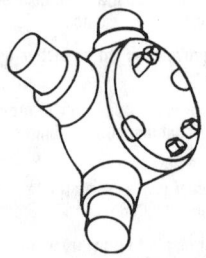

Explosionproof Round Box with Cover, 3 Threaded Hubs

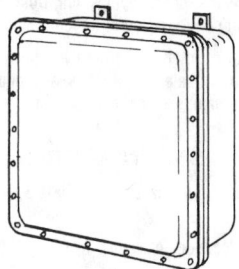

Explosionproof NEMA 7, Surface Mounted, Pull Box

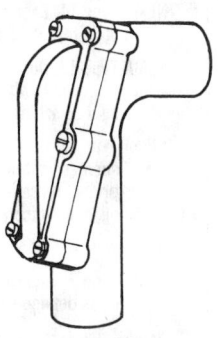

Explosionproof
Pulling Elbow

Figure 16.28   Standard Electrical Enclosure Types

NEMA Enclosures

Electrical enclosures serve two basic purposes: they protect people from accidental contact with enclosed electrical devices and connections, and they protect the enclosed devices and connections from specified external conditions. The National Electrical Manufacturers Association (NEMA) has established the following standards. These brief summaries are not intended to be complete representations of NEMA listings, and consultation of NEMA literature is advised for detailed information.

The following definitions and descriptions pertain to NONHAZARDOUS locations.

NEMA Type 1: General purpose enclosures intended for use indoors, primarily to prevent accidental contact of personnel with the enclosed equipment in areas that do not involve unusual conditions.

NEMA Type 2: Dripproof indoor enclosures intended to protect the enclosed equipment against dripping noncorrosive liquids and falling dirt.

NEMA Type 3: Dustproof, raintight and sleet-resistant (ice-resistant) enclosures intended for use outdoors to protect the enclosed equipment against wind-blown dust, rain, sleet, and external ice formation.

NEMA Type 3R: Rainproof and sleet-resistant (ice-resistant) enclosures which are intended for use outdoors to protect the enclosed equipment against rain. These enclosures are constructed so that the accumulation and melting of sleet (ice) will not damage the enclosure and its internal mechanisms.

NEMA Type 3S: Enclosures intended for outdoor use to provide limited protection against wind-blown dust, rain, and sleet (ice) and to allow operation of external mechanisms when ice-laden.

NEMA Type 4: Watertight and dust-tight enclosures intended for use indoors and out — to protect the enclosed equipment against splashing water, seepage of water, falling or hose-directed water, and severe external condensation.

NEMA Type 4X: Watertight, dust-tight, and corrosion-resistant indoor and outdoor enclosures featuring the same provisions as Type 4 enclosures, plus corrosion resistance.

NEMA Type 5: Indoor enclosures intended primarily to provide limited protection against dust and falling dirt.

NEMA Type 6: Enclosures intended for indoor and outdoor use — primarily to provide limited protection against the entry of water during occasional temporary submersion at a limited depth.

NEMA Type 6R: Enclosures intended for indoor and outdoor use — primarily to provide limited protection against the entry of water during prolonged submersion at a limited depth.

NEMA Type 11: Enclosures intended for indoor use —primarily to provide, by means of oil immersion, limited protection to enclosed equipment against the corrosive effects of liquids and gases.

NEMA Type 12: Dust-tight and driptight indoor enclosures intended for use indoors in industrial locations to protect the enclosed equipment against fibers, flyings, lint, dust, and dirt, as well as light splashing, seepage, dripping, and external condensation of noncorrosive liquids.

NEMA Type 13: Oil-tight and dust-tight indoor enclosures intended primarily to house pilot devices, such as limit switches, foot switches, push buttons, selector switches, and pilot lights, and to protect these devices against lint and dust, seepage, external condensation, and sprayed water, oil, and noncorrosive coolant.

The following definitions and descriptions pertain to HAZARDOUS, or CLASSIFIED, locations:

NEMA Type 7: Enclosures intended to use in indoor locations classified as Class 1, Groups A, B, C, or D, as defined in the National Electrical Code.

NEMA Type 9: Enclosures intended for use in indoor locations classified as Class 2, Groups E, F, or G, as defined in the National Electrical Code.

Figure 16.29   Installation Time in Man-Hours for NEMA Enclosures

Description	Man-Hours	Unit
NEMA 1		
12" Long x 12" Wide x 4" Deep	1.330	Ea.
NEMA 3R		
12" Long x 12" Wide x 6" Deep	1.600	Ea.
NEMA 4		
12" Long x 12" Wide x 6" Deep	4.000	Ea.
NEMA 7		
12" Long x 12" Wide x 6" Deep	8.000	Ea.
NEMA 9		
12" Long x 12" Wide x 6" Deep	5.000	Ea.
NEMA 12		
12" Long x 14" Wide x 6" Deep	1.510	Ea.

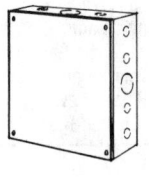

Screw Cover - NEMA 1

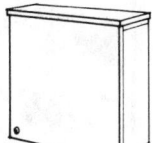

Rainproof and Weatherproof, Screw Cover - NEMA 3R

Hinged Cover - NEMA 1

Sheet Metal Pull Boxes

Figure 16.30   Installation Time in Man-Hours for Wiring Devices

Description	Man-Hours	Unit
Receptacle 20A 250V	.290	Ea.
Receptacle 30A 250V	.530	Ea.
Receptacle 50A 250V	.720	Ea.
Receptacle 60A 250V	1.000	Ea.
Box, 4" Square	.400	Ea.
Box, Single Gang	.290	Ea.
Box, Cast Single Gang	.660	Ea.
Cover, Weatherproof	.120	Ea.
Cover, Raised Device	.150	Ea.
Cover, Brushed Brass	.100	Ea.

30 Amp, 125 Volt, Nema 5

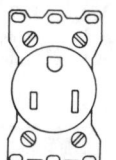

50 Amp, 125 Volt, Nema 5

20 Amp, 250 Volt, Nema 6

Receptacles

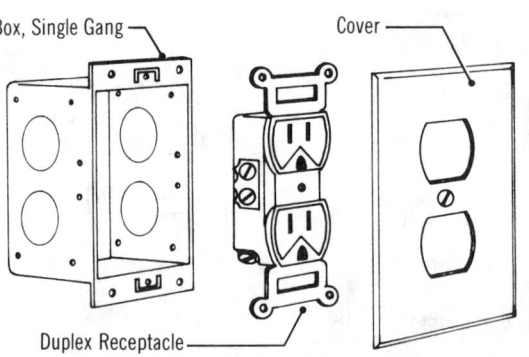

Box, Single Gang   Cover

Duplex Receptacle

Receptacle, Including Box and Cover

Figure 16.31   Elevated Equipment Installation Factors

For the installation of electrical equipment elevated more than 10' above the floor, the labor costs should be adjusted to allow for the added complexity. This table lists suggested adjustments.

Installation Height Above Floor	Labor Adjustments
10' to 15'	+ 15%
15' to 25'	+ 30%
Over 25'	+ 35%

Figure 16.32 Maximum Horsepower for Starter Size by Voltage

Starter Size	Maximum HP (3Φ)			
	208V	240V	480V	600V
00	1-1/2	1-1/2	2	2
0	3	3	5	5
1	7-1/2	7-1/2	10	10
2	10	15	25	25
3	25	30	50	50
4	40	50	100	100
5		100	200	200
6		200	300	300
7		300	600	600
8		450	900	900
8L		700	1500	1500

Figure 16.33 Wattages for Motors

The power factor of electric motors varies from 80% to 90% in larger size motors.

90% Power Factor & Efficiency @ 200 or 460V			
HP	Watts	HP	Watts
10	9024	30	25784
15	13537	40	33519
20	17404	50	41899
25	21916	60	49634

Figure 16.34 Power Requirements for Elevators with 3 Phase Motors

Elevator Type	Maximum Travel Height in Ft.	Travel Speeds in FPM	Capacity of Cars in Lbs.								
			1200	1500	1800	2000	2500	3000	3500	4000	4500
Hydraulic	70	70	10	15	15	15	20	20	20	25	30
		85	15	15	15	20	20	25	25	30	30
		100	15	15	20	20	25	30	30	40	40
		110	20	20	20	20	25	30	40	40	50
		125	20	20	20	25	30	40	40	50	50
		150	25	25	25	30	40	50	50	50	60
		175	25	30	30	40	50	50	60		
		200	30	30	40	40	50	60	60		
Geared Traction	300	200				10	10	15	15		23
		350				15	15	23	23		35

Figure 16.35 Installation Time in Man-Hours for Starters

Description	Man-Hours	Unit
Starter 3-Pole 2 HP Size 00	2.290	Ea.
5 HP Size 0	3.480	Ea.
10 HP Size 1	5.000	Ea.
25 HP Size 2	7.270	Ea.
50 HP Size 3	8.890	Ea.
100 HP Size 4	13.330	Ea.
200 HP Size 5	17.780	Ea.
400 HP Size 6	20.000	Ea.
Control Station Stop/Start	1.000	Ea.
Stop/Start, Pilot Light	1.290	Ea.
Hand/Off/Automatic	1.290	Ea.
Stop/Start/Reverse	1.510	Ea.

Figure 16.36 Installation Time in Man-Hours for Starters in Hazardous Areas

Description	Man-Hours	Unit
Circuit Breaker NEMA 7 600 Volts 3 Pole		
50 Amps	3.480	Ea.
150 Amps	8.000	Ea.
400 Amps	13.330	Ea.
Control Station Stop/Start	1.330	Ea.
Stop/Start Pilot Light	2.000	Ea.
Magnetic Starter FVNR 480 Volts 5 HP Size 0	5.000	Ea.
25 HP Size 2	8.890	Ea.
Combination 10 HP Size 1	8.000	Ea.
50 HP Size 3	20.000	Ea.
Panelboard 225 Amps M.L.O. 120/208 Volts		
24 Circuit	40.000	Ea.
Main Breaker	53.330	Ea.
Wall Switch, Single Pole 15 Amps	1.510	Ea.
Receptacle 15 Amps	1.510	Ea.

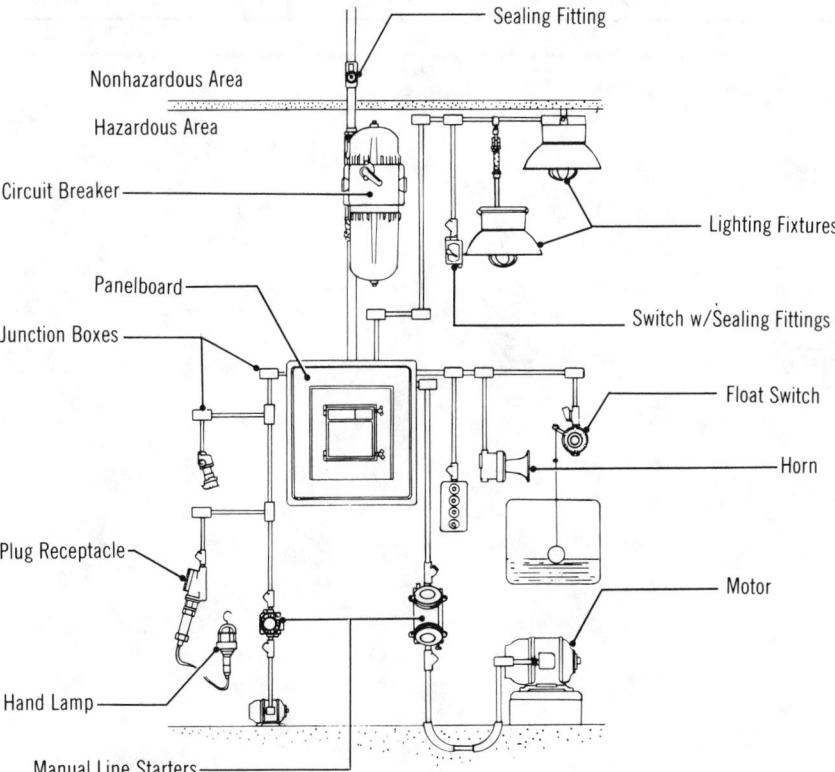

Starters for Class I, Division 2, Power and Lighting Installation

Figure 16.37   Motor Control Center
Standard Classifications

Motor control centers are available in two standard classes, with several types of units making up each class. These classifications and types are defined as follows:

Class 1

Type A: This type consists of a control unit with a circuit breaker or fusible disconnect wired to the line side of the starter only.

Type B: This type is the same as Type A, but the control circuit leads are wired to a fixed terminal block on the control unit.

Type C: This type is the same as Type B, but the leads are brought to the control unit terminal boards which are located at the top or bottom of the motor control center.

Interwiring and interlocking do not exist between starters or cubicles in any type in this class of unit.

Class 2

Type B: This type is the same as Class 1, Type B, but it contains wiring between control units in the same or adjacent cubicles.

Type C: This type is the same as Class 2, Type B, but it provides interwiring from the master terminal boards at the top and bottom of the control center.

Figure 16.38 Installation Time in Man-Hours for Motor Control Systems

Description	Man-Hours	Unit
Heavy Duty Fused Disconnect 30 Amps	2.500	Ea.
60 Amps	3.480	Ea.
100 Amps	4.210	Ea.
200 Amps	6.150	Ea.
600 Amps	13.330	Ea.
1200 Amps	20.000	Ea.
Starter 3-pole 2 HP Size 00	2.290	Ea.
5 HP Size 0	3.480	Ea.
10 HP Size 1	5.000	Ea.
25 HP Size 2	7.270	Ea.
50 HP Size 3	8.890	Ea.
100 HP Size 4	13.330	Ea.
200 HP Size 5	17.780	Ea.
400 HP Size 6	20.000	Ea.
Control Station Stop/Start	1.000	Ea.
Stop/Start, Pilot Light	1.290	Ea.
Hand/Off/Automatic	1.290	Ea.
Stop/Start/Reverse	1.510	Ea.

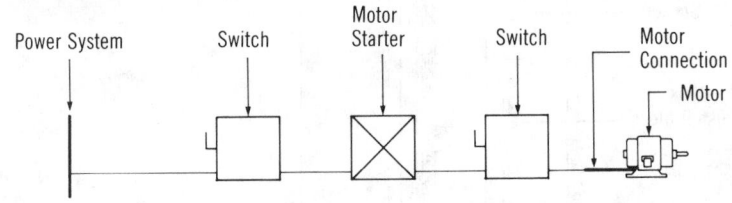

Power System   Switch   Motor Starter   Switch   Motor Connection   Motor

Figure 16.39 Installation Time in Man-Hours for Motor Control Centers

Description	Man-Hours	Unit
Structures 300 Amps 72" High	10.000	Ea.
Structures 300 Amps 72" High		
Back to Back Type	13.300	Ea.
Starters Class 1 Type B Size 1	3.000	Ea.
Size 2	4.000	Ea.
Size 3	8.000	Ea.
Size 4	11.400	Ea.
Size 5	16.000	Ea.
Size 6	20.000	Ea.
Pilot Light Wiring in Starter	.500	Ea.
Push Button Wiring in Starter	.500	Ea.
Auxilliary Contacts in Starter	.500	Ea.

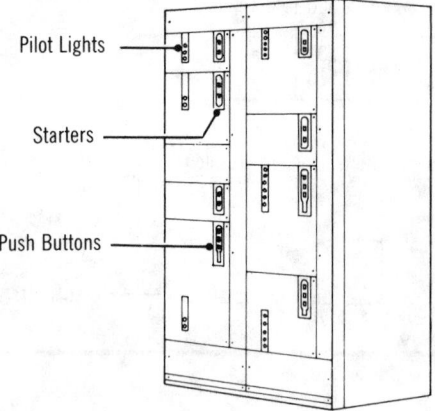

Pilot Lights

Starters

Push Buttons

Figure 16.40 Installation Time in Man-Hours for Panelboards

Description	Man-Hours	Unit
Panelboard 3-Wire 225 Amps Main Lugs		
38 Circuit	22.220	Ea.
4 Wire 225 Amps Main Lugs 42 Circuit	23.530	Ea.
3 Wire 400 Amps Main Circuit Breaker		
42 Circuit	32.000	Ea.
4 Wire 400 Amps Main Circuit Breaker		
42 Circuit	33.330	Ea.
3 Wire 100 Amps Main Lugs 20 Circuit	12.310	Ea.
4 Wire 100 Amps Main Circuit Breaker		
24 Circuit	17.020	Ea.

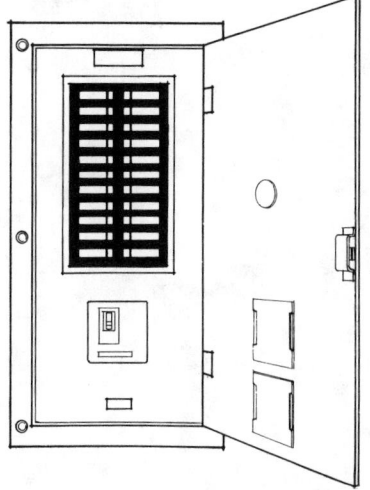 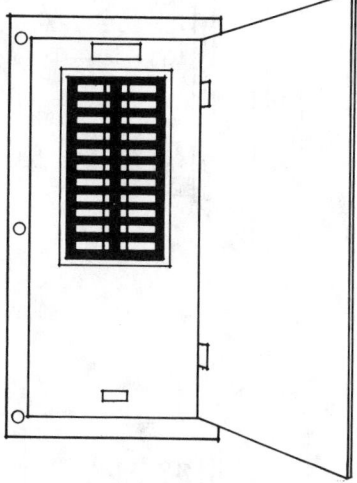

Main Circuit Breaker Panelboard   Main Lugs Only Panelboard

Figure 16.41   Installation Time in Man-Hours
for Safety Switches

Description	Man-Hours	Unit
Safety Switch NEMA 1 600V 3P 200 Amps	6.150	Ea.
NEMA 3R	6.670	Ea.
NEMA 7	10.000	Ea.
NEMA 12	6.670	Ea.

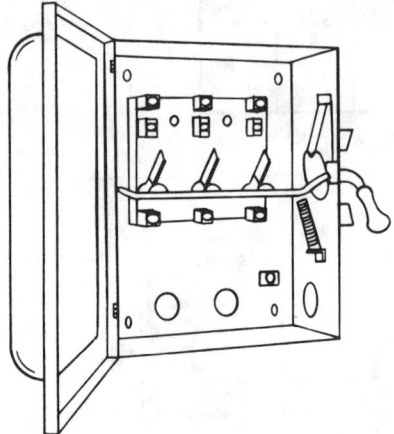

NEMA 1, Non-fusible, 600 Volt

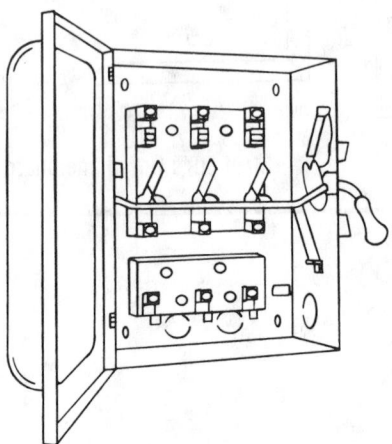

NEMA 1, Fusible, 600 Volt

Safety Swiches

Figure 16.42 Installation Time in Man-Hours for Metering Switchboards

Description	Man-Hours	Unit
Main Breaker Section 800 Amps	17.800	Ea.
1200 Amps	21.000	Ea.
1600 Amps	23.500	Ea.
Main Section 100 Amps 6 Meters	26.700	Ea.
8 Meters	30.800	Ea.
10 Meters	33.300	Ea.

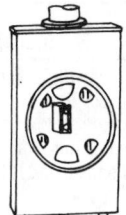

Meter Socket

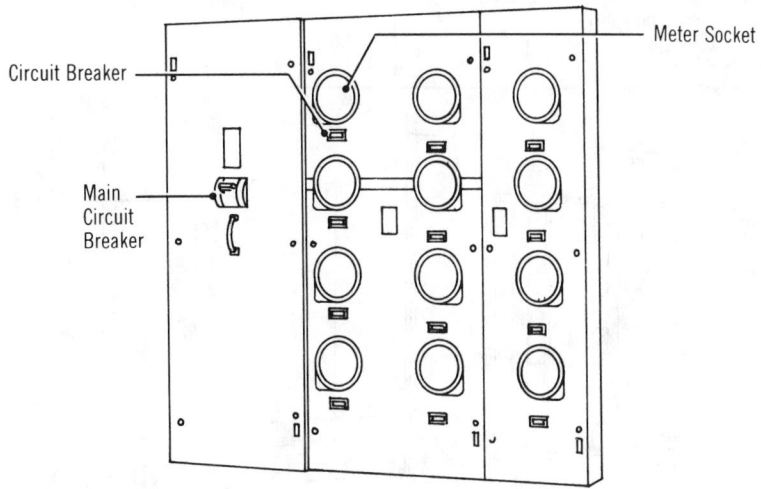

Meter Socket

Circuit Breaker

Main
Circuit
Breaker

Figure 16.43   Installation Time in Man-Hours for Multi-Section Switchboards

Description	Man-Hours	Unit
Main Switchboard Section 1200 Amps	18.000	Ea.
1600 Amps	19.000	Ea.
2000 Amps	20.000	Ea.
Main Ground Fault Protector 1200-2000 Amps	2.960	Ea.
Bus Way Connections 1200 Amps	6.150	Ea.
1600 Amps	6.670	Ea.
2000 Amps	8.000	Ea.
Auxilliary Pull Section	8.000	Ea.
Distribution Section 1200 Amps	22.220	Ea.
1600 Amps	24.240	Ea.
2000 Amps	25.810	Ea.
Breakers, 1 Pole 60 Amps	1.000	Ea.
2 Pole 60 Amps	1.150	Ea.
3 Pole 60 Amps	1.500	Ea.

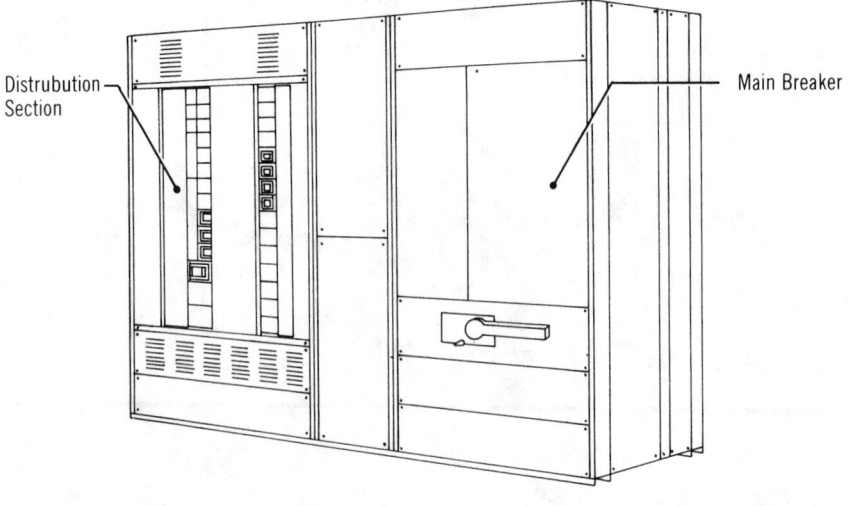

Distrubution Section

Main Breaker

Figure 16.44 Installation Time in Man-Hours for Metal-Clad Switchgear

Description	Man-Hours	Unit
Metal-Clad Structures 1200 Amps	17.500	Ea.
2000 Amps	19.500	Ea.
3000 Amps	24.000	Ea.
Breakers 1200 Amps	10.000	Ea.
2000 Amps	13.000	Ea.
3000 Amps	18.000	Ea.
Instrument Wiring	3.500	Ea.
Bus Bar Connections per Structure 1200 Amps	10.000	Ea.
2000 Amps	15.000	Ea.
3000 Amps	19.000	Ea.
Ground Bus Connection per Structure	6.000	Ea.

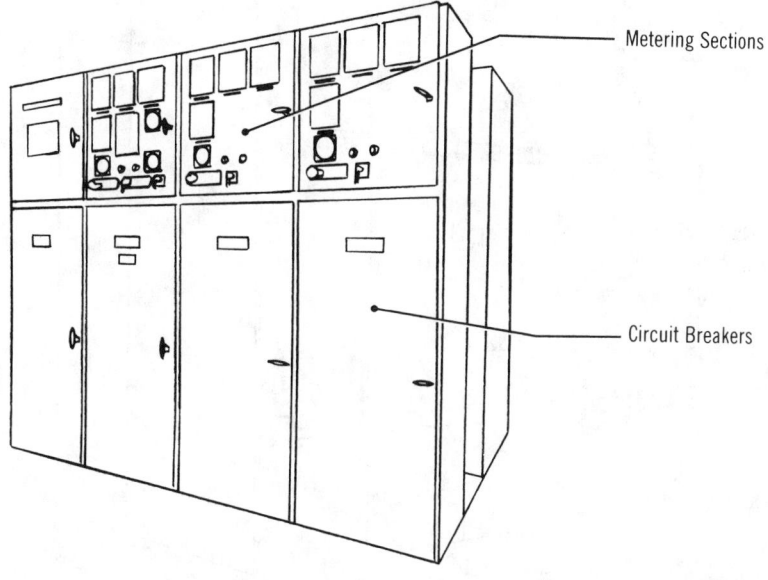

Metering Sections

Circuit Breakers

Figure 16.45 Installation Time in Man-Hours for Low Voltage Metal Enclosed Switchgear

Description	Man-Hours	Unit
Structures 1600 Amps	17.500	Ea.
4000 Amps	26.000	Ea.
Breakers Draw-Out 225 Amps	4.000	Ea.
600 Amps	8.000	Ea.
1600 Amps	11.000	Ea.
4000 Amps	23.000	Ea.
Bus Bar Connections 1600 Amps per Structure	12.000	Ea.
4000 Amps per Structure	21.000	Ea.

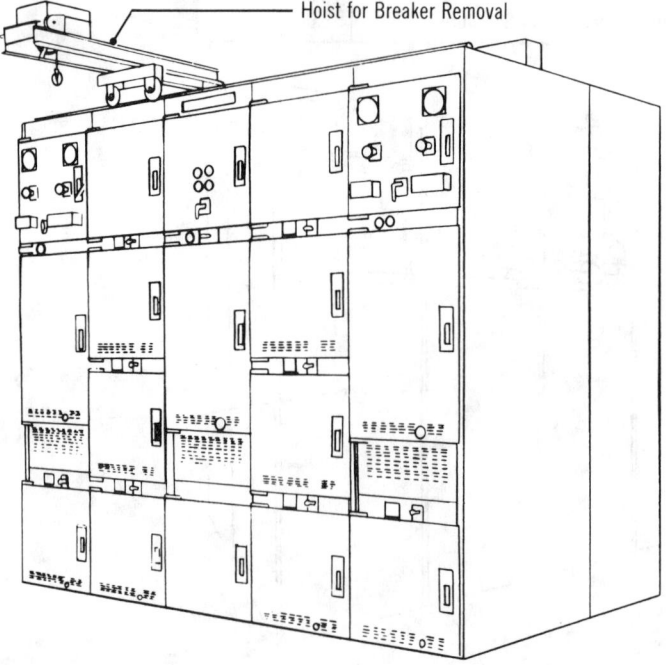

Hoist for Breaker Removal

Figure 16.46   Transformer Weight (lbs.) by KVA

Oil Filled 3 Phase 5/15 KV to 480/277			
KVA	Lbs.	KVA	Lbs.
150	1800	1000	6200
300	2900	1500	8400
500	4700	2000	9700
750	5300	3000	15000

Dry 240/480 to 120/240 Volt			
1 Phase		3 Phase	
KVA	Lbs.	KVA	Lbs.
1	23	3	90
2	36	6	135
3	59	9	170
5	73	15	220
7.5	131	30	310
10	149	45	400
15	205	75	600
25	255	112.5	950
37.5	295	150	1140
50	340	225	1575
75	550	300	1870
100	670	500	2850
167	900	750	4300

Figure 16.47 Installation Time in Man-Hours for Transformers

Description	Man-Hours	Unit
Oil Filled 5 KV Primary 277/480 Volt Secondary		
3 Phase 150 KVA	30.770	Ea.
1000 KVA	76.920	Ea.
3750 KVA	125.000	Ea.
Silicon Filled 5 KV Primary 277/480 Volt		
Secondary 3 Phase 225 KVA	36.360	Ea.
1000 KVA	76.920	Ea.
2500 KVA	105.000	Ea.
Dry 480 Volt Primary 120/208 Volt Secondary		
3 Phase 15 KVA	14.550	Ea.
112 KVA	23.530	Ea.
500 KVA	44.440	Ea.

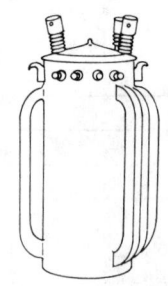

Oil Filled Transformer

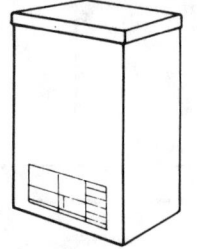
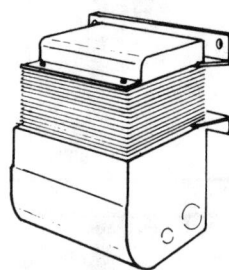

Dry Type Transformer, 3 Phase   Dry Type Transformer, Single Phase   Buck-Boost Transformer

Transformer Types

Figure 16.48 Installation Time in Man-Hours for Secondary Unit Substations

Description	Man-Hours	Unit
Load Interrupter Switch, 300 KVA and below	60.000	Ea.
400 KVA and above	63.000	Ea.
Transformer Section 112 KVA	37.000	Ea.
300 KVA	59.000	Ea.
500 KVA	67.000	Ea.
750 KVA	83.000	Ea.
Low Voltage Breakers		
2 Pole, 15 to 60 Amps, Type FA	1.430	Ea.
3 Pole, 15 to 60 Amps, Type FA	1.510	Ea.
2 Pole, 125 to 225 Amps, Type KA	2.350	Ea.
3 Pole, 125 to 225 Amps, Type KA	2.500	Ea.
2 Pole, 700 and 800 Amps, Type MA	5.330	Ea.
3 Pole, 700 and 800 Amps, Type MA	6.150	Ea.

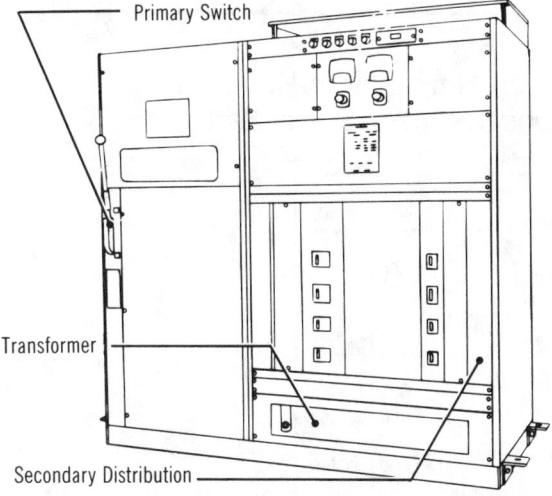

Primary Switch

Transformer

Secondary Distribution

Figure 16.49   Generator Weights (lbs.) by KW

3 Phase 4 Wire 277/480 Volt			
Gas		Diesel	
KW	Lbs.	KW	Lbs.
7.5	600	30	1800
10	630	50	2230
15	960	75	2250
30	1500	100	3840
65	2350	125	4030
85	2570	150	5500
115	4310	175	5650
170	6530	200	5930
		250	6320
		300	7840
		350	8220
		400	10750
		500	11900

Figure 16.50   Installation Time in Man-Hours for Emergency/Standby Power Systems

Description	Man-Hours	Unit
Battery Light Unit		
6 Volt Lead Battery and 2 Lights	2.000	Ea.
12 Volt Nickel Cadmium and 2 Lights	2.000	Ea.
Remote Mount Sealed Beam Light, 25W, 6 Volts	.300	Ea.
Self Contained Fluorescent Lamp Pack	.800	Ea.
Engine Generator 10KW Gas/Gasoline		
277/480 Volts Complete	34.000	System
170KW Complete	96.000	System
Engine Generator 500KW Diesel		
277/480 Volts Complete	133.000	System
1000KW Complete	180.000	System

Figure 16.50 Installation Time in Man-Hours for Emergency/ Standby Power Systems (continued)

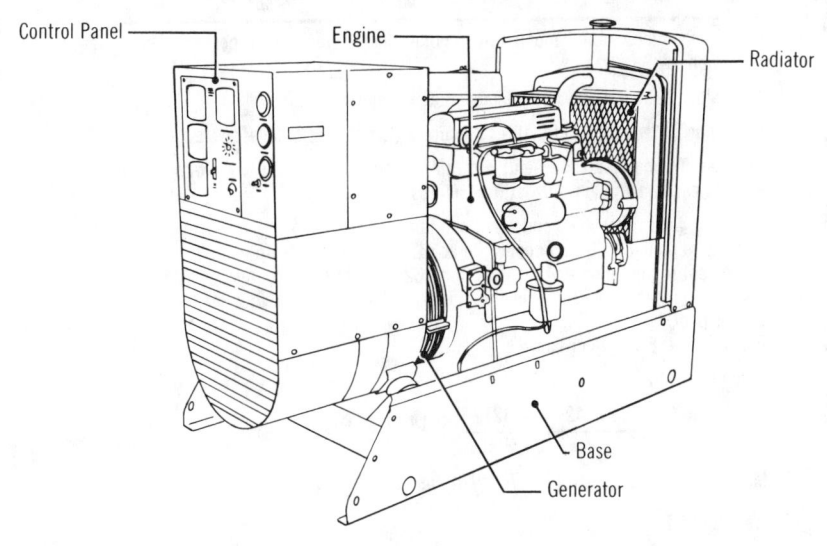

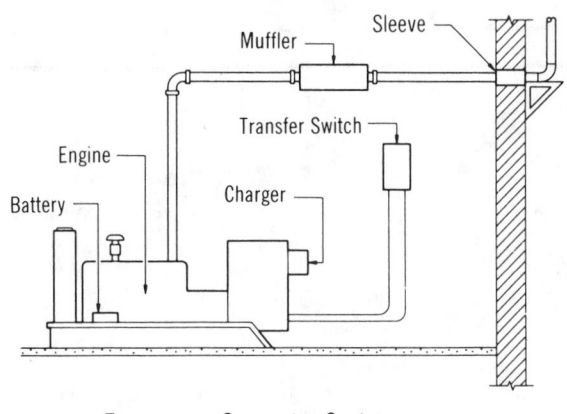

Emergency Generator System

Figure 16.51   Lighting: Calculating Footcandles and Watts per Square Foot

1. Initial footcandles = number of fixtures x lamps per fixture x lumens per lamp x coefficient of utilization ÷ S.F.

2. Maintained footcandles = initial footcandles x maintenance factor

3. Watts per S.F. = number of fixtures x lamps x (lamp watts + ballast watts) ÷ S.F.

Example: To find footcandles and watts per S.F. for an office 20' x 20' with 11 fluorescent fixtures each having four 40 watt C.W. lamps.

Based on good reflectance and clean conditions:
Lumens per lamp = 40 watt cool white at 3150 lumens per lamp
Coefficient of utilization = .42 (varies from .62 for light colored areas to .27 for dark)
Maintenance factor = .75 (varies from .80 for clean areas with good maintenance to .50 for poor)
Ballast loss = 8 watts per lamp.

1. Initial footcandles:

$$\frac{11 \times 4 \times 3150 \times .42}{400} = \frac{58212}{400} = 145 \text{ footcandles}$$

2. Maintained footcandles: 145 x .75 = 109 footcandles

3. Watts per S.F.

$$\frac{11 \times 4 \times (40 + 8)}{400} = \frac{2112}{400} = 5.3 \text{ watts per S.F.}$$

Figure 16.52 I.E.S.* Recommended Illumination Levels in Footcandles

Commercial Buildings			Industrial Buildings		
Type	Description	Foot-Candles	Type	Description	Foot-Candles
Banks	Lobby	50	Assembly Areas	Rough bench & machine work	50
	Customer Areas	70		Medium bench & machine work	100
	Teller Stations	150		Fine bench & machine work	500
	Accounting Areas	150	Inspection Areas	Ordinary	50
Offices	Routine Work	100		Difficult	100
	Accounting	150		Highly Difficult	200
	Drafting	200	Material Handling	Loading	20
	Corridors, Halls, Washrooms	30		Stock Picking	30
Schools	Reading or Writing	70		Packing, Wrapping	50
	Drafting, Labs, Shops	100	Stairways Washrooms	Service Areas	20
	Libraries	70		Service Areas	20
	Auditoriums, Assembly	15	Storage Areas	Inactive	5
	Auditoriums, Exhibition	30		Active, Rough, Bulky	10
Stores	Circulation Areas	30		Active, Medium	20
	Stock Rooms	30		Active, Fine	50
	Merchandise Areas, Service	100	Garages	Active Traffic Areas	20
	Self-Service Areas	200		Service & Repair	100

*Illuminating Engineering Society

Figure 16.53   General Lighting Loads by Occupancies

Type of Facility	Unit Load per S.F. (Watts)
Armories and Auditoriums	1
Banks	5
Barber Shops and Beauty Parlors	3
Churches	1
Clubs	2
Court Rooms	2
Dwelling Units (1)	3
Garages — Commercial (storage)	1/2
Hospitals	2
Hotels and Motels, including apartment houses without provisions for cooking by tenants (1)	2
Industrial Commercial (Loft) Buildings	2
Lodge Rooms	1-1/2
Office Buildings	5
Restaurants	2
Schools	3
Stores	3
Warehouses (storage)	1/4
(1) In any of the above occupancies except one-family dwellings and individual dwelling units of multi-family dwellings:	
Assembly Halls and Auditoriums	1
Halls, Corridors, Closets	1/2
Storage Spaces	1/4

Figure 16.54   Lighting Limit for Listed Occupancies

Type of Use	Maximum Watts per S.F.
INTERIOR	
Category A: Classrooms, office areas, automotive mechanical areas, museums, conference rooms, drafting rooms, clerical areas, laboratories, merchandising areas, kitchens, examining rooms, book stacks, athletic facilities.	3.00
Category B: Auditoriums, waiting areas, spectator areas, restrooms, dining areas, transportation terminals, working corridors in prisons and hospitals, book storage areas, active inventory storage, hospital bedrooms, hotel and motel bedrooms, enclosed shopping mall concourse areas, stairways.	1.00
Category C: Corridors, lobbies, elevators, inactive storage areas.	0.50
Category D: Indoor parking.	0.25
EXTERIOR	
Category E: Building perimeter: wall-wash, facade, canopy.	5.00 (per L.F.)
Category F: Outdoor parking.	0.10

Figure 16.55 Approximate Watts per S.F. for Popular Fixture Types

This figure provides the wattage required to maintain a given number of footcandles for a known type of lighting fixture.

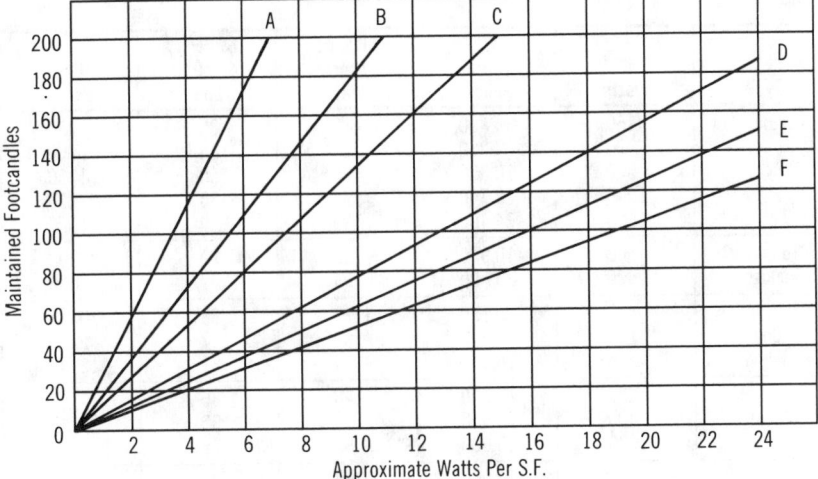

Due to the many variables involved, use for preliminary estimating only:
A. Fluorescent — industrial system
B. Fluorescent — lens unit system
C. Fluorescent — louvered unit system
D. Incandescent — open reflector system
E. Incandescent — lens unit system
F. Incandescent — down light system

Figure 16.56 Site Floodlighting Lamp Comparison (includes floodlight, ballast & lamp for pole mounting)

This chart compares typical types of lamps used for floodlighting, by required watts, intensity (measured in lumens, lumens per watt, and lumens at 40% life), and actual life of the lamp.

Type	Watts	Initial Lumens	Lumens per Watt	Lumens @ 40% Life	Life (Hours)
Incandescent	150	2880	19	85	750
	300	6360	21	84	750
	500	10,850	22	80	1000
	1000	23,740	24	80	1000
	1500	34,400	23	80	1000
Tungsten Halogen	500	10,950	22	97	2000
	1500	35,800	24	97	2000
Fluorescent Cool White	40	3150	79	88	20,000
	110	9200	84	87	12,000
	215	16,000	74	81	12,000
Deluxe Mercury	250	12,100	48	86	24,000
	400	22,500	56	85	24,000
	1000	63,000	63	75	24,000
Metal Halide	175	14,000	80	77	7500
	400	34,000	85	75	15,000
	1000	100,000	100	83	10,000
	1500	155,000	103	92	1500
High Pressure Sodium	70	5800	83	90	20,000
	100	9500	95	90	20,000
	150	16,000	107	90	24,000
	400	50,000	125	90	24,000
	1000	140,000	140	90	24,000
Low Pressure Sodium	55	4600	131	98	18,000
	90	12,750	142	98	18,000
	180	33,000	183	98	18,000

Color: High Pressure Sodium — Slightly Yellow   Mercury Vapor — Green-Blue   Note: Pole not
Low Pressure Sodium — Yellow   Metal Halide — Blue-White   included.

Figure 16.57 Installation Time in Man-Hours for Incandescent Lighting

Description	Man-Hours	Unit
Ceiling, Recess Mounted Alzak Reflector		
150W	1.000	Ea.
300W	1.190	Ea.
Surface Mounted Metal Cylinder		
150W	.800	Ea.
300W	1.000	Ea.
Opal Glass Drum 10" 2-60W	1.000	Ea.
Pendant Mounted Globe 150W	1.000	Ea.
Vaportight 200W	1.290	Ea.
Chandelier 24" Diameter x 42" High		
6 Candle	1.330	Ea.
Track Light Spotlight 150W PAR	.500	Ea.
Wall Washer Quartz 250W	.500	Ea.
Exterior Wall Mounted Quartz 500W	1.510	Ea.
1500W	1.900	Ea.
Ceiling, Surface Mounted Vaportight		
100W	2.650	Ea.
150W	2.950	Ea.
175W	2.950	Ea.
250W	2.950	Ea.
400W	3.350	Ea.
1000W	4.450	Ea.

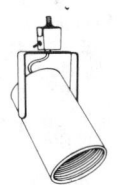

Track Lighting Spotlight

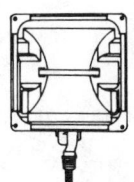

Exterior Fixture,
Wall Mounted, Quartz

Round Ceiling Fixture
with Concentric Louver

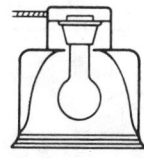

Round Ceiling Fixture
with Reflector, No Lens

Round Ceiling Fixture,
Recessed, with Alzak Reflector

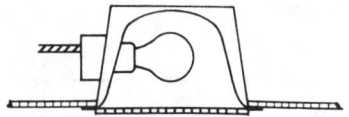

Square Ceiling Fixture, Recessed, with Glass Lens, Metal Trim

Fixtures

Figure 16.58　Installation Time in Man-Hours for Fluorescent Lighting

Description	Man-Hours	Unit
Troffer with Acrylic Lens 4-40W RS 2' x 4'	1.700	Ea.
2-40W URS 2' x 2'	1.400	Ea.
Surface Mounted Acrylic Wrap-around Lens		
4-40W RS 16" x 48"	1.500	Ea.
Industrial Pendant Mounted		
4' Long, 2-40W RS	1.400	Ea.
8' Long, 2-75W SL	1.820	Ea.
2-110W HO	2.000	Ea.
2-215W VHO	2.110	Ea.
Surface Mounted Strip, 4' Long, 1-40W RS	.940	Ea.
8' Long, 1-75W SL	1.190	Ea.

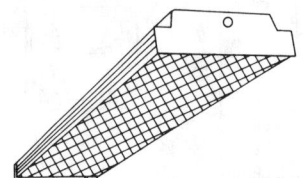

Surface or Pendant Mounted Fixture with Wrap Around Acrylic Lens, 4 Tube

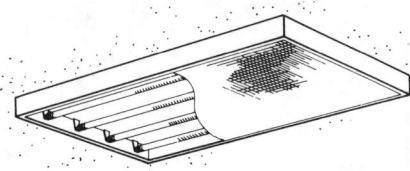

Surface Mounted Fixture with Acrylic Lens, 4 Tube

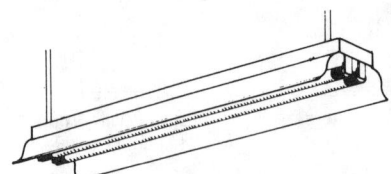

Pendant Mounted Industrial Fixture,

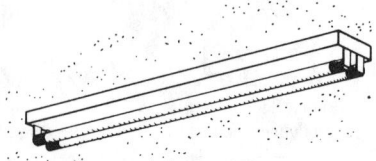

Surface Mounted Strip Fixture, 2 Tube

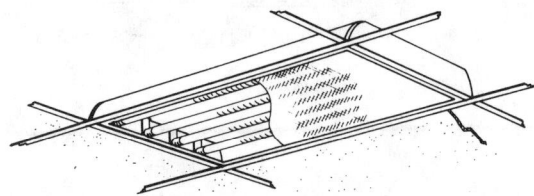

Troffer Mounted Fixture with Acrylic Lens, 4 Tube

Figure 16.59   Installation Time in Man-Hours for High Intensity Lighting

Description	Man-Hours	Unit
Ceiling Recessed Mounted Prismatic Lens		
Integral Ballast 2' x 2' HID 150W	2.500	Ea.
250W	2.500	Ea.
400W	2.760	Ea.
Surface Mounted 250W	2.960	Ea.
400W	3.330	Ea.
High Bay Aluminum Reflector 400W	3.480	Ea.
1000W	4.000	Ea.

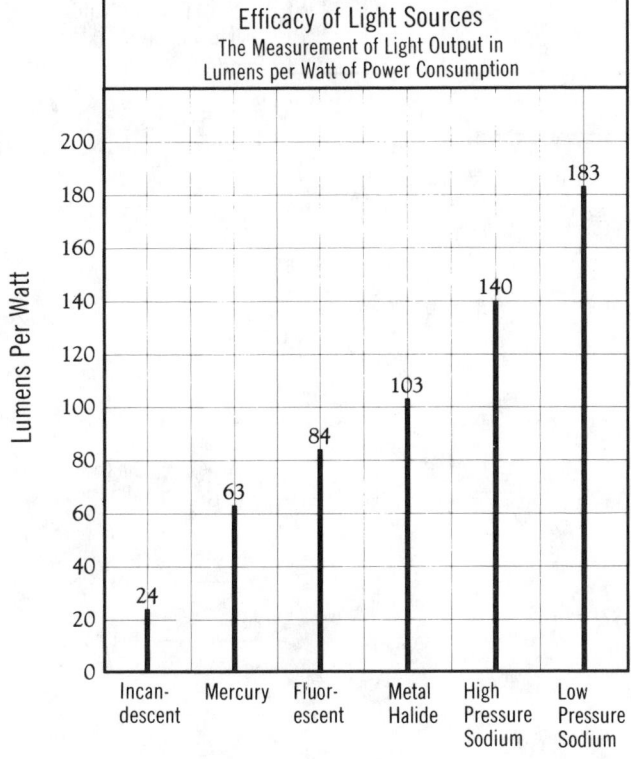

Efficacy of Light Sources
The Measurement of Light Output in
Lumens per Watt of Power Consumption

Figure 16.60   Installation Time in Man-Hours for Hazardous Area Lighting

Description	Man-Hours	Unit
Fixture, Pendant Mounted, Fluorescent 4' Long		
2-40W RS	3.480	Ea.
4-40W RS	4.710	Ea.
Incandescent 200W	2.290	Ea.
Ceiling Mounted, Incandescent 200W	2.000	Ea.
Ceiling, HID, Surface Mounted 100W	2.670	Ea.
150W	2.960	Ea.
250W	2.960	Ea.
400W	3.330	Ea.
Pendent Mounted 100W	2.960	Ea.
150W	3.330	Ea.
250W	3.330	Ea.
400W	3.810	Ea.

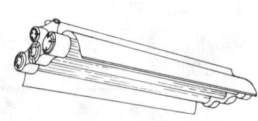

Explosionproof
Fluorescent Fixture,
Pendent Mounted. 3 Tube

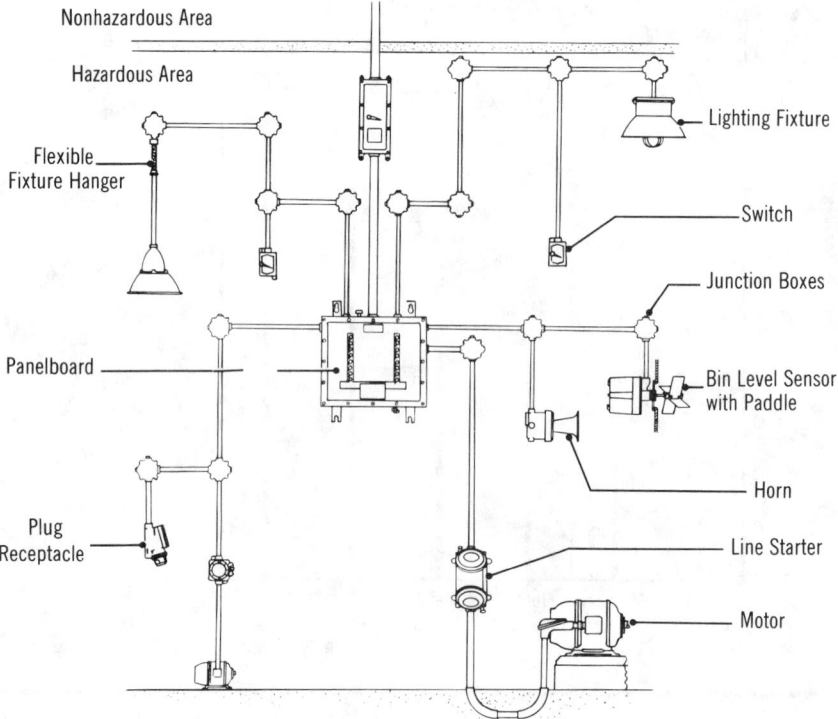

Class II Lighting Installation

Figure 16.61 Installation Time in Man-Hours for Parking Area Lighting

Description	Man-Hours	Unit
Luminaire HID 100 Watt	2.960	Ea.
150W	2.960	Ea.
175W	2.960	Ea.
250W	3.330	Ea.
400W	3.640	Ea.
1000W	4.000	Ea.
Bracket Arm 1 Arm	1.000	Ea.
2 Arm	1.000	Ea.
3 Arm	1.510	Ea.
4 Arm	1.510	Ea.
Aluminum Pole 20' High	8.300	Ea.
30' High	9.250	Ea.
40' High	12.000	Ea.
Steel Pole 20' High	9.250	Ea.
30' High	10.500	Ea.
40' High	14.200	Ea.
Fiberglass Pole 20' High	6.000	Ea.
30' High	6.700	Ea.
40' High	8.600	Ea.
Transformer Base	2.670	Ea.

Large Luminaire
Light Fixture, 1000 Watt

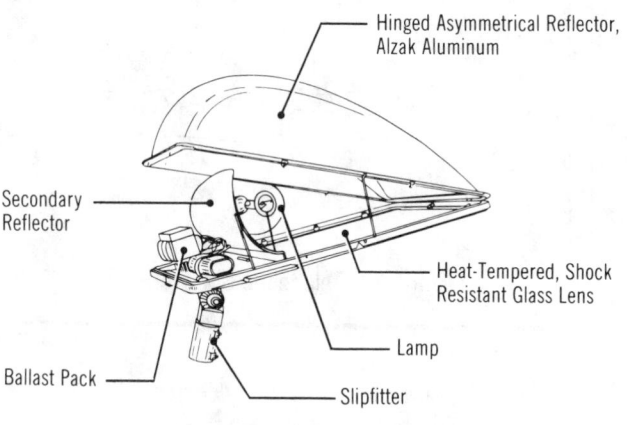

Hinged Asymmetrical Reflector, Alzak Aluminum

Secondary Reflector

Heat-Tempered, Shock Resistant Glass Lens

Lamp

Ballast Pack

Slipfitter

Luminaire Construction Features

Figure 16.62   Service Electric Utilities

Service includes the excavation and structures for bringing power to the building, and the actual installation of the primary cables. The figure below illustrates typical methods by which power is carried to the project.

Service covers the distribution methods used to route power, control, and communications cables to a facility's property and between its buildings and structures. There are three basic options: 1) direct burial cables, 2) underground in duct banks, and 3) overhead on poles.

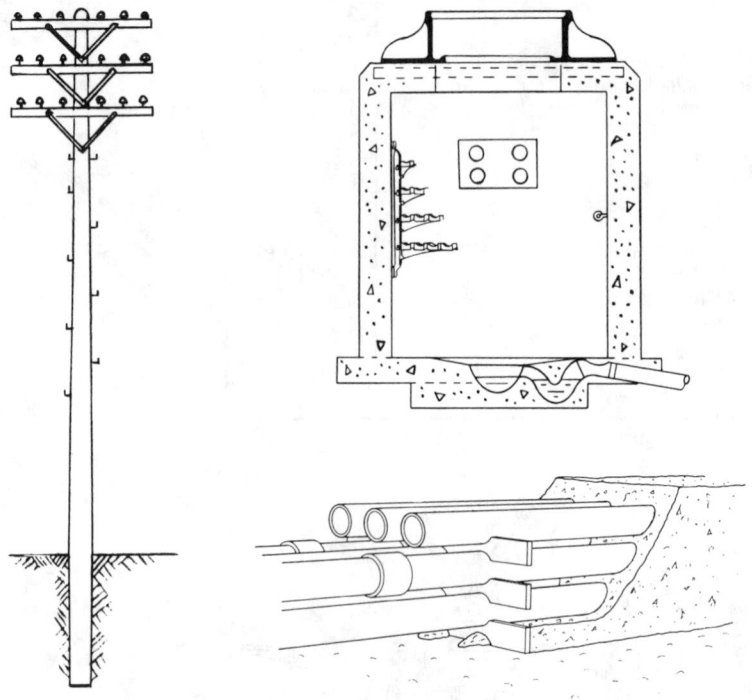

Electric and Telephone Site Work

Figure 16.63   Concrete for Conduit Encasement

The table below lists C.Y. of concrete for 100 L.F. of trench. Conduits separation center to center should meet 7.5″ (N.E.C.).

Number of Conduits	1	2	3	4	6	8	9	Number of Conduits
Trench Dimension	11.5″ x 11.5″	11.5″ x 19″	11.5″ x 27″	19″ x 19″	19″ x 27″	19″ x 38″	27″ x 27″	Trench Dimension
Conduit Diameter 2″	3.29	5.39	7.64	8.83	12.51	17.66	17.72	Conduit Diameter 2″
2.5″	3.23	5.29	7.49	8.62	12.19	17.23	17.25	2.5″
3.0″	3.15	5.13	7.24	8.29	11.71	16.59	16.52	3.0″
3.5″	3.08	4.97	7.02	7.99	11.26	15.98	15.84	3.5″
4.0″	2.99	4.80	6.76	7.65	10.74	15.30	15.07	4.0″
5″	2.78	4.37	6.11	6.78	9.44	13.57	13.12	5″
6.0″	2.52	3.84	5.33	5.74	7.87	11.48	10.77	6.0″

Figure 16.64   Fire Alarm Systems

Fire alarm systems consist of control panels, annunciator panels, battery with rack and charger, various sensing devices, such as smoke and heat detectors, and alarm horn and light signals. Some fire alarm systems are very sophisticated and include speakers, telephone lines, door closer controls, and other components. Some are connected directly to the fire station. Requirements for fire alarm systems are generally regulated by codes and by local authorities.

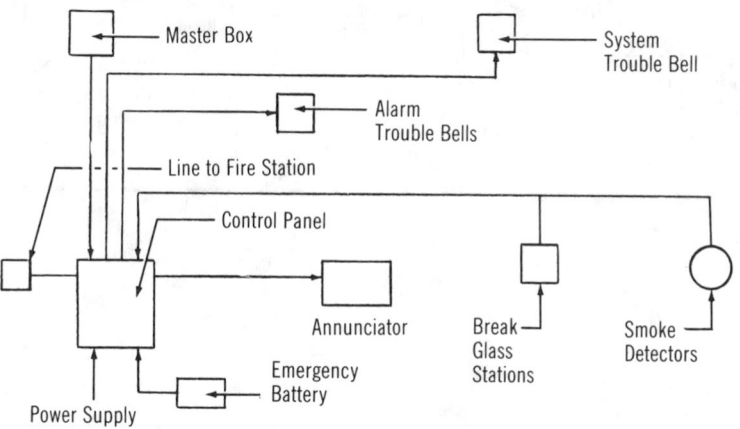

Fire Alarm

Figure 16.65 Burglar Alarm Systems

Burglar alarm systems consist of control panels, indicator panels, various types of alarm devices, and switches. The control panel is usually line powered with a battery backup supply. Some systems have a direct connection to the police or protection company, while others have auto-dial telephone capabilities. Most, however, simply have local control monitors and an annunciator. The sensing devices are various pressure switches, magnetic door switches, glass break sensors, infrared beams, infrared sensors, microwave detectors, and ultrasonic motion detectors. The alarms are sirens, horns, and/or flashing lights..

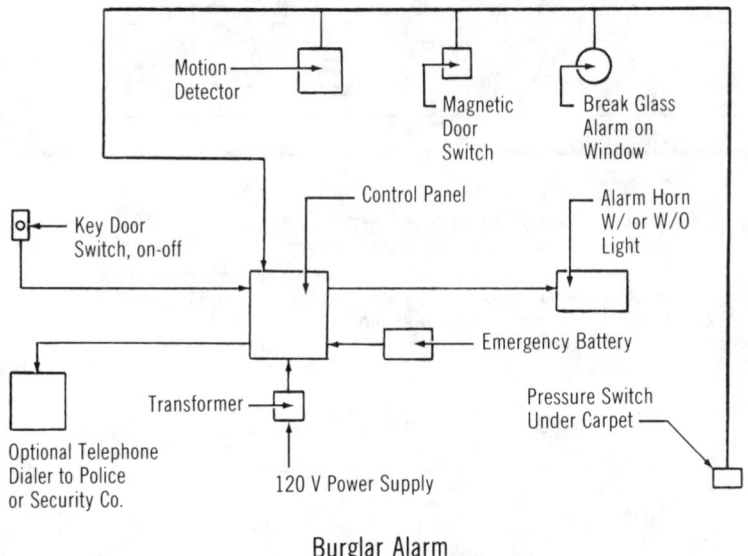

Burglar Alarm

Figure 16.66 Television Systems

Master TV antenna systems are used in schools, dormitories and apartment buildings. Each system consists of the antenna, lightning arrester, amplifier, splitters, and outlets. The signal is received by the antenna and increased by the amplifier. It then goes through the main cable to the splitter. Here, the signal is split among several branch circuit cables.

A closed-circuit surveillance system consists of a TV camera and monitor. These systems are used indoors or outdoors for security surveillance. Some applications require pan, tilt, and zoom (PTZ) mechanisms for remote control of the camera.

Educational TV studio equipment is comprised of 3 cameras and a monitor.

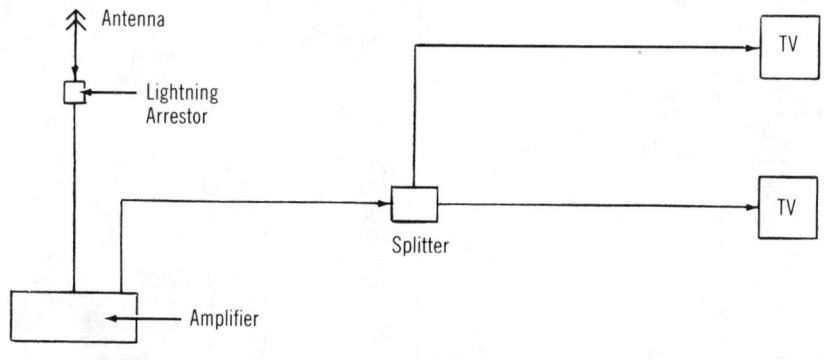

Figure 16.67 Doctor's In-Out Register

This is a system used in hospitals to facilitate communication between staff members. A recording register is located at the entrance. The doctors push their buttons, sending a signal to a register at the hospital control center. The attendant is then able to page or signal the doctor. Pocket paging devices are a material cost only.

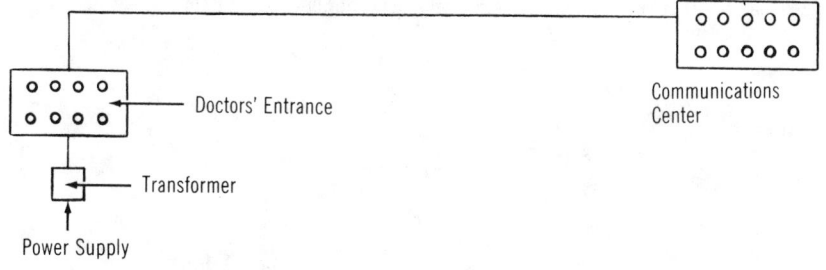

Figure 16.68   Installation Time in Man-Hours for Electric Heating

Description	Man-Hours	Unit
Baseboard heaters		
2' long, 375 watt	1	Ea.
3' long, 500 watt	1	Ea.
4' long, 750 watt	1.190	Ea.
5' long, 935 watt	1.400	Ea.
6' long, 1125 Wall	1.600	Ea.
7' long, 1310 watt	1.820	Ea.
8' long, 1500 watt	2	Ea.
9' long, 1680 watt	2.220	Ea.
10' long, 1875 watt	2.420	Ea.
Wall heaters with fan, 120 to 277 volt		
750 watt	1.140	Ea.
1000 watt	1.140	Ea.
1250 watt	1.330	Ea.
1500 watt	1.600	Ea.
2000 watt	1.600	Ea.
2500 watt	1.780	Ea.
3000 watt	2	Ea.
4000 watt	2.290	Ea.
Recessed heaters		
750 watt	1.330	Ea.
1000 watt	1.330	Ea.
1250 watt	1.600	Ea.
1500 watt	2	Ea.
2000 watt	2	Ea.
2500 watt	2.290	Ea.
3000 watt	2.670	Ea.
4000 watt	2.960	Ea.

Rule of thumb: For a quick estimate of installation time for baseboard units, including controls, allow 1.820 man-hours per KW. For duct heaters, including controls, allow 1.510 man-hours per KW.

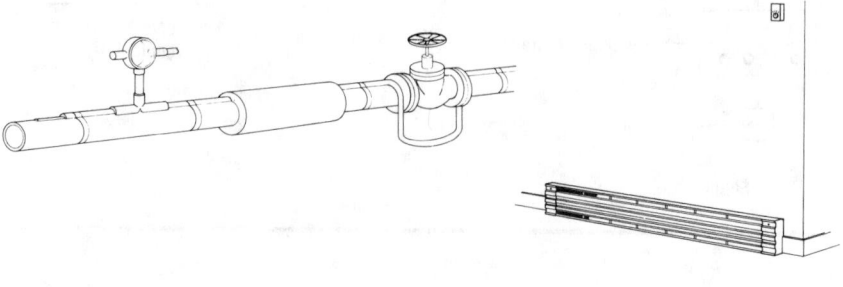

Electric Heating

Figure 16.69 Installation Time in Man-Hours for Clocks and Clock Systems

Description	Man-Hours	Unit
Clocks		
12″ diameter, single face	1	Ea.
Double face	1.290	Ea.
Clock Systems		
Time system components, master controller	24.240	Ea.
Program bell	1	Ea.
Combination clock & speaker	2.500	Ea.
Frequency generator	4	Ea.
Job time automatic stamp recorder, minimum	2	Ea.
Maximum	2	Ea.
Master time clock system, clocks & bells		
20 room	160	Ea.
50 room	400	Ea.
Time clock, 100 cards in & out, 1 color	2.500	Ea.
2 colors	2.500	Ea.
With 3 circuit program device, minimum	4	Ea.
Maximum	4	Ea.
Metal rack for 25 cards	1.140	Ea.
Watchman's tour station	1	Ea.
Annunciator with zone indication	8	Ea.

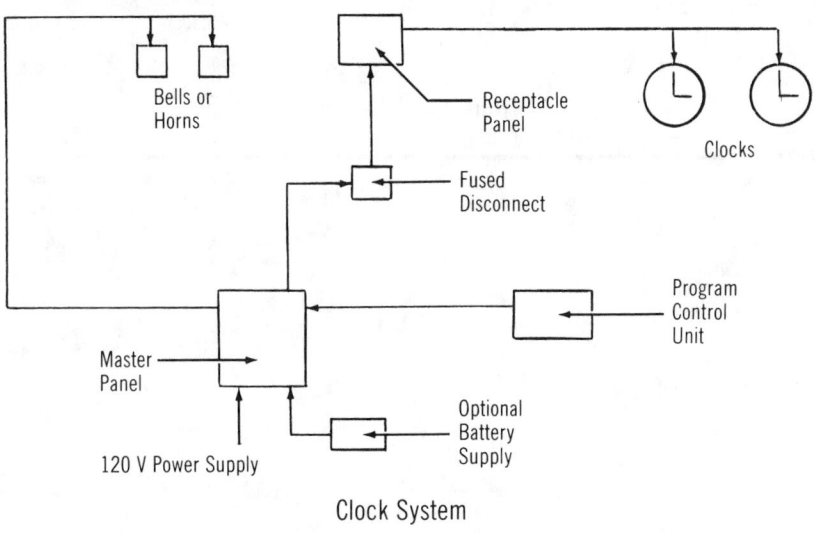

Clock System

Figure 16.70   Installation Time in Man-Hours for Nurses' Call Systems

Description	Man-Hours	Unit
Call station, single bedside	1	Ea.
Double bedside	2	Ea.
Ceiling speaker station	1	Ea.
Emergency call station	1	Ea.
Pillow speaker	1	Ea.
Duty station	2	Ea.
Standard call button	1	Ea.
Lights, corridor, dome or zone indicator	1	Ea.
Master control station for 20 stations	24.620	Total

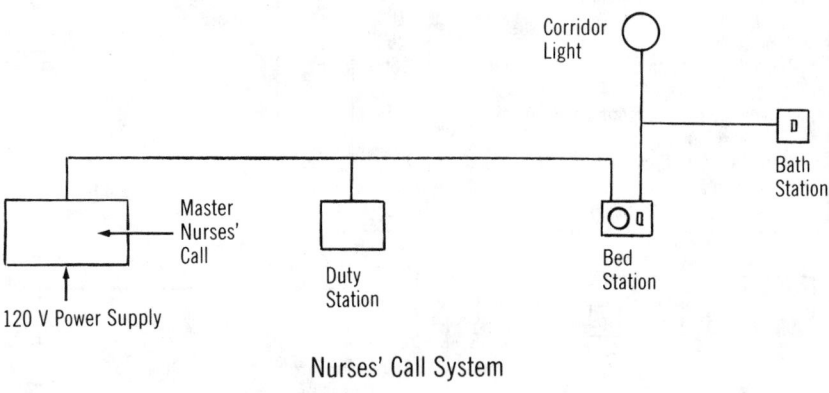

Nurses' Call System

Figure 16.71 Installation Time in Man-Hours for Doorbell Systems

Description	Man-Hours	Unit
Doorbell system including transformer, button & signal		
6" bell	2	Ea.
Buzzer	2	Ea.
Door chimes, 2 notes, minimum	.500	Ea.
Maximum	.666	Ea.
Tube type, 3 tube system	.666	Ea.
4 tube system	.800	Ea.
For transformer & button, minimum add	1.600	Ea.
Maximum, add	1.780	Ea.
For push button only, minimum	.333	Ea.
Maximum	.400	Ea.
Bell transformer	.500	Ea.

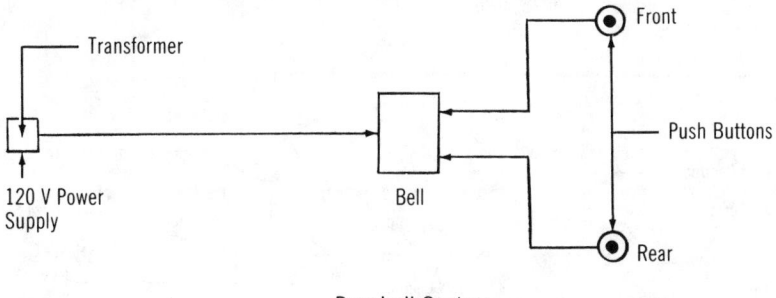

Doorbell System

Figure 16.72 Installation Time in Man-Hours for Lightning Protection

Lightning protection for the rooftops of buildings is achieved by a series of lightning rods or air terminals joined together by either copper or aluminum cable. The cable size is determined by the height of the building. The lightning cable system is connected through a downlead to a ground rod that is a minimum of 2' below grade and 1-1/2' to 3' out from the foundation wall.

Description	Man-Hours	Unit
Air Terminals, copper		
3/8" diameter x 10" (to 75' high)	1	Ea.
1/2" diameter x 12" (over 75' high)	1	Ea.
Aluminum, 1/2" diameter x 12" (to 75' high)	1	Ea.
5/8" diameter x 12" (over 75' high)	1	Ea.
Cable, copper		
220 lb. per thousand ft. (to 75' high)	.025	L.F.
375 lb. per thousand ft. (over 75' high)	.034	L.F.
Aluminum		
101 lb. per thousand ft. (to 75' high)	.028	L.F.
199 lb. per thousand ft. (over 75' high)	.033	L.F.
Arrestor, 175 volt AC to ground	1	Ea.
650 volt AC to ground	1.190	Ea.

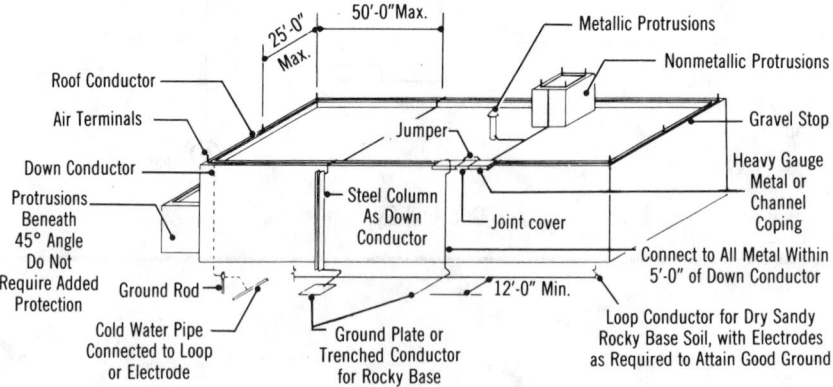

Checklist

For an estimate to be reliable, all items must be accounted for.
A complete estimate can also eliminate the need to include
contingencies. The following checklist can be used to help ensure
that all items are properly accounted for.

Bells, Bell Systems _____

Building Service
☐ Main switchboard
☐ Panels
   ___ Distribution
   ___ Lighting
   ___ Power

Bus Duct Systems _____

Cable _____

Cable Trays _____

Circuit Breakers _____

Clock Systems _____

Comfort Systems
☐ Electric heat
☐ Snow melting
☐ Other _____

Computer Power Supplies _____

Conduit _____

Controls _____

Devices
☐ Boxes for outlets, receptacles, light switches, dimmers
☐ Cover plates
☐ Dimmers
☐ Lighting control switches
☐ Outlets
☐ Receptacles

Disconnect Switches _____

Doorbells _____

Emergency Systems
☐ Area protection relay system
☐ Exit lighting
☐ Battery emergency lighting
☐ Generator

- [ ] Lighting
- [ ] Power
- [ ] Power connection to second service entrance
- [ ] Transfer switch

Equipment Connections _____

Fire Alarm System
- [ ] Alarm consoles
- [ ] Detectors
    - ____ Fixed temperature
    - ____ Rate of rise
    - ____ Smoke
- [ ] Horns
- [ ] Lights
- [ ] Separate cable trays
- [ ] Stations
- [ ] Zones

Fixtures
- [ ] Incandescent
- [ ] Fluorescent
- [ ] Mercury vapor
- [ ] Metal halide
- [ ] High pressure sodium
- [ ] Area lighting
- [ ] Bollards
- [ ] Exit lights
- [ ] Flood lighting
- [ ] Fountain lighting
- [ ] High Intensity Discharge (H.I.D.)
- [ ] Site lighting
- [ ] Step lighting

Grounding Systems _____

Heating Systems
- [ ] Area
- [ ] Perimeter
- [ ] Unit
- [ ] Wall

Incoming Service
- [ ] Connection to power company system
- [ ] Overhead
- [ ] Underground

- ☐ Primary
- ☐ Secondary
- ☐ Service feeds
- ☐ Main transformer
- ☐ Unit substation
  - ___ Manholes
  - ___ Feeder
  - ___ Primary
  - ___ Secondary
- ☐ Meters

Junction Boxes _____

Lightning Protection _____

Motors _____

Motor Control Centers _____

Paging Systems _____

Public Address Systems _____

Power Poles _____

Pull Boxes _____

Raceways _____

Safety Devices _____

Security Systems
- ☐ Card readers
- ☐ Closed circuit television
- ☐ Control panels
- ☐ Devices
- ☐ Door alarms
- ☐ Scanners
- ☐ Sensors

Sound Systems _____

Starters _____

Substations _____

Supports _____

Switches _____

Television Systems
- ☐ Antennae
- ☐ Cables
- ☐ Cameras
- ☐ Satellite dishes

Temperature Control Devices
- ☐ Thermostats

☐ Automated systems

Transformers _____

Transmission Line Poles _____

Transmission Line Towers _____

Under Floor Cabling Systems _____

Under Carpet Cabling/Wiring Systems _____

Utilities _____

Wiring

☐ Wire

☐ Cable

_____ Armored

_____ Shielded

_____ Insulated

_____ Metallic

_____ Nonmetallic

_____ Coaxial

☐ Connections

☐ Grounding

Tips

Conduit

Conduit should be taken off in three main categories: power distribution, branch power, and branch lighting. In this way, all conduit does not have to be taken off in one session. Instead the estimator can concentrate on systems and components, thereby making it easier to ensure that all items have been accounted for.

Aluminum Wiring

Remember that aluminum wiring of equal ampacity is larger in diameter than copper and may require larger conduit.

Switchgear

When estimating the costs for the installation of switchgear, especially large items, factors to review include access to the site, access and setting at the installation site, required connections, uncrating, pads, anchors, levelling, final assembly of the components, and temporary protection from physical damage, including from exposure to the environment.

Pads, Supports, and Panel Backing

While supports and pads may be shown on drawings for the larger equipment, in many cases nothing is shown for smaller pieces such as panelboards and area transformers. Whether a special floor-to-ceiling support system is required, or just a piece of plywood for the back of a panel, it must be included in the costs.

Installation Equipment

Do not overlook the costs for equipment used in the installation. If highlifts, scaffolding, or cherry pickers are available, the field will use them in lieu of the proposed ladders.

Material Weights

The estimator should take the weights of materials into consideration when completing a takeoff. Topics to consider include: How will the materials be supported? What methods of support are available? How high will the support structure have to reach? Will the final support structure be able to withstand the total burden? Is the support material included or separate from the fixture/equipment/material specified?

Non-Listed Items

It is important to include items that are not documented in the plans but must be priced. These items include, but are not limited to, testing, equipment hookups, motor controls, disconnect switches, special systems, dust protection, roof penetrations, pitch pots for the roof, coring concrete floors and walls, cleanup, and final adjustments.

Temporary Light and Power

Examine all contract documents to determine if providing temporary power and lighting is included in your scope of work. In many cases, temporary lighting and power requirements are not specified in contract documents. Either way, you will probably be required at some point to provide a cost for these items. Depending on your bidding strategy, you may or may not want to list this item as an alternate to your estimate. However, if it is stated in the contract documents, then you will be fully responsible for providing this service.

Cutting and Patching

Not all projects will be performed under ideal conditions — that is, wide open spaces with no interferences. Many times the electrical contractor will have to run conduit through pre-existing partitions or partitions that were recently installed. Allowances must be made for this cutting into the partition and the ensuing patching. Otherwise, a statement must be included with the estimate that all cutting and patching will be the responsibility of others.

Mechanical Connections

Clarify who is responsible for the connecting of the electrical wiring to mechanical items. It has been argued that the electrician is responsible only for bringing the wiring to the unit, while others say it is the electrician's job to connect the wiring.

Fixture and Device Counts

Performing a fixture takeoff is a good way to become familiar with a proposed project. Fixtures should be taken off room-by-room, using the fixture schedule (if provided), specifications, and the ceiling plan. While performing the count, it is a good idea to also take off the controlling devices and accessories (plaster rings, outlet boxes, cover plates, etc.). A spreadsheet works well for this purpose.

When finished, this takeoff can be used for purchasing, accounting and billing, and cost control.

Special Systems

When estimating material costs for special systems, it is always prudent to obtain manufacturers' quotes for equipment prices. Also, some installations will require special accessories. Often, the sales engineers are a good source of information on these requirements.

Appendix

List of Appendices

The Appendix is divided into nine sections, A — I. Each contains tables, illustrations, and/or listings of information that may be useful in creating an estimate. The specific contents of each Appendix section are listed in the table of contents below.

Standard Weights and Measures

Linear Measure		Square Measure	
1000 mils =	1 inch	144 square inches =	1 square foot
12 inches =	1 foot	9 square feet =	1 square yard
3 feet =	1 yard		
2 yards =	{ 1 fathom 6 feet	30-1/4 square yds. =	{ 1 square rod 272-1/4 square feet
5-1/2 yards =	{ 1 rod 16-1/2 feet	160 square rods =	{ 1 acre 43,560 square feet
40 rods =	{ 1 furlong 660 feet	640 acres =	{ 1 square mile 27,878,400 square feet
8 furlongs =	{ 1 mile 5280 feet	A circular mil is the area of a circle 1 mil, or 0.001 inch in diameter.	
1.15156 miles =	{ 1 nautical mile, or knot 6080.26 feet	1 square inch =	1,273,239 circular mils
3 nautical miles =	{ 1 league 18,240.78 feet	A circular inch is the area of a circle = 1 inch in diameter	0.7854 square inches.
		1 square inch =	1.2732 circular inches

Dry Measure		Weight—Avoirdupois or Commercial	
2 pints =	1 quart	437.5 grains =	1 ounce
8 quarts =	1 peck	16 ounces =	1 pound
4 pecks =	{ 1 bushel 2150.42 cubic in. 1.2445 cubic feet	112 pounds =	1 hundredweight
		20 hundredweight =	{ 1 gross, or long ton 2240 pounds
		2000 pounds =	1 net, or short ton
		2204.6 pounds =	1 metric ton
		1 lb. of water (39.1°F) = = = =	27.681217 cu. in. 0.016019 cu. ft. 0.119832 U.S. gallon 0.453617 liter

Decimals of a Foot for Each 32nd of an Inch

Inch	0	1	2	3	4	5	6	7	8	9	10	11
0	0	.0833	.1667	.2500	.3333	.4167	.5000	.5833	.6667	.7500	.8333	.9167
1/32	.0026	.0859	.1693	.2526	.3359	.4193	.5026	.5859	.6693	.7526	.8359	.9193
1/16	.0052	.0885	.1719	.2552	.3385	.4219	.5052	.5885	.6719	.7552	.8385	.9219
3/32	.0078	.0911	.1745	.2578	.3411	.4245	.5078	.5911	.6745	.7578	.8411	.9245
1/8	.0104	.0938	.1771	.2604	.3438	.4271	.5104	.5938	.6771	.7604	.8438	.9271
5/32	.0130	.0964	.1797	.2630	.3464	.4297	.5130	.5964	.6797	.7630	.8464	.9297
3/16	.0156	.0990	.1823	.2656	.3490	.4323	.5156	.5990	.6823	.7656	.8490	.9323
7/32	.0182	.1016	.1849	.2682	.3516	.4349	.5182	.6016	.6849	.7682	.8516	.9349
1/4	.0208	.1042	.1875	.2708	.3542	.4375	.5208	.6042	.6875	.7708	.8542	.9375
9/32	.0234	.1068	.1901	.2734	.3568	.4401	.5234	.6068	.6901	.7734	.8568	.9401
5/16	.0260	.1094	.1927	.2760	.3594	.4427	.5260	.6094	.6927	.7760	.8594	.9427
11/32	.0286	.1120	.1953	.2786	.3620	.4453	.5286	.6120	.6953	.7786	.8620	.9453
3/8	.0313	.1146	.1979	.2812	.3646	.4479	.5313	.6146	.6979	.7813	.8646	.9479
13/32	.0339	.1172	.2005	.2839	.3672	.4505	.5339	.6172	.7005	.7839	.8672	.9505
7/16	.0365	.1198	.2031	.2865	.3698	.4531	.5365	.6198	.7031	.7865	.8698	.9531
15/32	.0391	.1224	.2057	.2891	.3724	.4557	.5391	.6224	.7057	.7891	.8724	.9557
1/2	.0417	.1250	.2083	.2917	.3750	.4583	.5417	.6250	.7083	.7917	.8750	.9583
17/32	.0443	.1276	.2109	.2943	.3776	.4609	.5443	.6276	.7109	.7943	.8776	.9609
9/16	.0469	.1302	.2135	.2969	.3802	.4635	.5469	.6302	.7135	.7969	.8802	.9635
19/32	.0495	.1328	.2161	.2995	.3828	.4661	.5495	.6328	.7161	.7995	.8828	.9661
5/8	.0521	.1354	.2188	.3021	.3854	.4688	.5521	.6354	.7188	.8021	.8854	.9688
21/32	.0547	.1380	.2214	.3047	.3880	.4714	.5547	.6380	.7214	.8047	.8880	.9714
11/16	.0573	.1406	.2240	.3073	.3906	.4740	.5573	.6406	.7240	.8073	.8906	.9740
23/32	.0599	.1432	.2266	.3099	.3932	.4766	.5599	.6432	.7266	.8099	.8932	.9766
3/4	.0625	.1458	.2292	.3125	.3958	.4792	.5625	.6458	.7292	.8125	.8958	.9792
25/32	.0651	.1484	.2318	.3151	.3984	.4818	.5651	.6484	.7318	.8151	.8984	.9818
13/16	.0677	.1510	.2344	.3177	.4010	.4844	.5677	.6510	.7344	.8177	.9010	.9844
27/32	.0703	.1536	.2370	.3203	.4036	.4870	.5703	.6536	.7370	.8203	.9036	.9870
7/8	.0729	.1563	.2396	.3229	.4063	.4896	.5729	.6563	.7396	.8229	.9063	.9896
29/32	.0755	.1589	.2422	.3255	.4089	.4922	.5755	.6589	.7422	.8255	.9089	.9922
15/16	.0781	.1615	.2448	.3281	.4115	.4948	.5781	.6615	.7448	.8281	.9115	.9948
31/32	.0807	.1641	.2474	.3307	.4141	.4974	.5807	.6641	.7474	.8307	.9141	.9974

Water: Various Pressure and Flow Units

Comparison of Heads of Water in Feet with Pressures in Various Units
One foot of water at 39.1° F = 62.425 pounds per square foot (psf)
= 0.4335 pounds per square inch
= 0.0295 atmosphere
= 0.8826 inches of mercury at 30° F
= 773.3 feet of air at 32° F and atmospheric pressure
One foot of water at 62° F = 62.355 pounds per square foot
= 0.43302 pounds per square inch
One pound of water on the square inch at 62° F = 2.3094 feet of water
One ounce of water on the square inch at 62° F = 1.732 inches of water
1 atmosphere at sea level (32° F) = 14.697 lbs. per sq. in.
= 29.921 in. of mercury
1 inch of mercury (32° F) = 0.49119 lbs. per sq. in.
Flowing Water
cfs = cubic feet per second, or second feet
gpm = gallons per minute
1 cfs = 60 cu. ft. per min.
= 86,400 cu. ft. per 24 hrs.
= 448.83 U.S. gals. per min.
= 646,317 U.S. gals. per 24 hrs.
= 1.9835 acre-foot per 24 hrs. (usually taken as 2)
= 1 acre-inch per hour (approximate)
= .028317 cu. meter per second
1 U.S. gpm = 1440 U.S. gals per 24 hrs.
= 0.00442 acre-foot per 24 hrs.
= 0.0891 Miners inches, Ariz., Calif.
1 million U.S. gal. per day = 1.5472 cfs
= 3.07 acre-feet
= 2.629 cu. meters per min.

Volume and Capacity

Units and Equivalents	
1 cu. ft. of water at 39.1° F	= 62.425 lbs.
1 United States gallon	= 231 cu. in.
1 imperial gallon	= 277.274 cu. in.
1 cubic foot of water	= 1728 cu. in.
	= 7.480519 U.S. gallons
	= 6.232103 imperial gallons
1 cubic yard	= 27 cu. ft. = 46.656 cu. in.
1 quart	= 2 pints
1 gallon	= 4 quarts
1 U.S. gallon	= 231 cu. in.
	= 0.133681 cu. ft.
	= 0.83311 imperial gallons
	= 8.345 lbs.
1 barrel	= 31.5 gallons = 4.21 cu. ft.
1 U.S. bushel	= 1.2445 cu. ft.
1 fluid ounce	= 1.8047 cu. in.
1 acre-foot	= 43,560 cu. ft.
	= 1,613.3 cu. yds.
1 acre-inch	= 3,630 cu. ft.
1 million U.S. gallons	= 133,681 cu. ft.
	= 3.0689 acre-ft.
1 ft. depth on 1 sq. mi.	= 27,878,400 cu. ft.
	= 640 acre-ft.
1 cord	= 128 cu. ft.

Mechanical-Electrical Equivalents

	Power	
1 horsepower (hp)	=	550 foot-pounds (ft.-lbs.) per second (sec.)
	=	33,000 ft.-lbs. per minute (min.)
	=	1,980,000 ft.-lbs. per hour (hr.)
	=	.275 ft.-tons per sec.
	=	16.5 ft.-tons per min.
	=	990 ft.-tons per hr.
1 horsepower-second (hp-sec.)	=	550 ft.-lbs.
	=	.275 ft.-tons.
1 horsepower-minute (hp-min.)	=	33,000 ft.-lbs.
	=	16.5 ft.-tons.
1 horsepower-hour (hp-hr.)	=	1,980,000 ft.-lbs.
	=	990 ft.-tons
1 horsepower (hp)	=	746 watts (w)
	=	.746 kilowatts (kw)
	Energy	
1 horsepower-hour	=	2544 BTU
	=	.746 kw-hr.
1 kilowatt-hour	=	3413 BTU
	Pressure	
1 lb. per sq. in.	=	2.0360" of mercury at 32° F
	=	27.71" of water at 32° F
	=	2.3091 ft. of water at 60° F
	=	144 lbs. per sq. ft.
1 in. of mercury	=	.491 lbs. per sq. in.
1 in. of water	=	5.2 lbs. per sq. ft. = .0361 PSI

Average Weights for Various Materials

Substance	Weight Lbs. per C.F.	Substance	Weight Lbs. per C.F.
Ashlar Masonry		Excavations in Water	
Granite, syenite, gneiss	165	Sand or gravel	60
Limestone, marble	160	Sand or gravel and clay	65
Sandstone, bluestone	140	Clay	80
Mortar Rubble Masonry		River mud	90
Granite, syenite, gneiss	155	Soil	70
Limestone, marble	150	Stone riprap	65
Sandstone, bluestone	130	Minerals	
Dry Rubble Masonry		Asbestos	153
Granite, syenite, gneiss	130	Barytes	281
Limestone, marble	125	Basalt	184
Sandstone, bluestone	110	Bauxite	159
Brick Masonry		Borax	109
Pressed brick	140	Chalk	137
Common brick	120	Clay, marl	137
Soft brick	100	Dolomite	181
Concrete Masonry		Feldspar, orthoclase	159
Cement, stone, sand	144	Gneiss, serpentine	159
Cement, slag, etc.	130	Granite, syenite	175
Cement, cinder, etc.	100	Greenstone, trap	187
Expanded slag aggregate	100	Gypsum, alabaster	159
Haydite (burned clay agg.)	90	Hornblende	187
Vermiculite/perlite, load bearing	70–105	Limestone, marble	165
Vermiculite and perlite, non-load		Magnesite	187
bearing	25–50	Phosphate rock, apatite	200
Concrete Masonry Reinforced		Porphyry	172
Stone aggregate	150	Pumice, natural	40
Slag aggregate	138	Quartz, flint	165
Lightweight aggregates	30–106	Sandstone, bluestone	147
Various Building Materials		Shale, slate	175
Ashes, cinders	40–45	Soapstone, talc	169
Cement, Portland, loose	90	Stone, Quarried, Piled	
Cement, Portland, set	183	Basalt, granite, gneiss	96
Lime, gypsum, loose	53–64	Limestone, marble, quartz	95
Mortar, set	103	Sandstone	82
Slags, bank slag	67–72	Shale	92
Slags, bank screenings	98–117	Greenstone, hornblende	107
Slags, machine slag	96	Bituminous Substances	
Slags, slag sand	49–55	Asphaltum	81
Earth, Etc., Excavated		Coal, anthracite	97
Clay, dry	63	Coal, bituminous	84
Clay, damp, plastic	110	Coal, lignite	78
Clay and gravel, dry	100	Coal, peat, turf, dry	47
Earth, dry, loose	76	Coal, charcoal, pine	23
Earth, dry, packed	95	Coal, charcoal, oak	33
Earth, moist, loose	78	Coal, coke	75
Earth, moist, packed	96	Graphite	131
Earth, mud, flowing	108	Paraffine	56
Earth, mud, packed	115	Petroleum	54
Riprap, limestone	80–85	Petroleum, refined	50
Riprap, sandstone	90	Petroleum, benzine	46
Riprap, shale	105	Petroleum, gasoline	42
Sand, gravel, dry, loose	90–105	Pitch	69
Sand, gravel, dry, packed	100–120	Tar, bituminous	75
Sand, gravel, wet	118–120		

(continued on next page)

Average Weights for Various Materials (continued)

Substance	Weight Lbs. per C.F.	Substance	Weight Lbs. per C.F.
Coal and Coke, Piled		Rubber goods	94
Coal, anthracite	47–58	Salt, granulated, piled	48
Coal, bituminous, lignite	40–54	Saltpeter	67
Coal, peat, turf	20–26	Starch	96
Coal, charcoal	10–14	Sulphur	125
Coal, coke	23–32	Wool	82
Metals, Alloys, Ores		Timber, U.S. Seasoned	
Aluminum, cast, hammered	165	Moisture Content by Weight:	
Brass, cast, rolled	534	Seasoned timber 15 to 20%	
Bronze, 7.9 to 14% Sn	509	Green timber up to 50%	
Bronze, aluminum	481	Ash, white, red	40
Copper, cast, rolled	556	Cedar, white, red	22
Copper ore, pyrites	262	Chestnut	41
Gold, cast, hammered	1205	Cypress	30
Iron, cast, pig	450	Fir, Douglas spruce	32
Iron, wrought	485	Fir, eastern	25
Iron, spiegel-eisen	468	Elm, white	45
Iron, ferro-silicon	437	Hemlock	29
Iron ore, hematite	325	Hickory	49
Iron ore, hematite in bank	160–180	Locust	46
Iron ore, hematite loose	130–160	Maple, hard	43
Iron ore, limonite	237	Maple, white	33
Iron ore, magnetite	315	Oak, chestnut	54
Iron slag	172	Oak, live	59
Lead	710	Oak, red, black	41
Lead ore, galena	465	Oak, white	46
Magnesium, alloys	112	Pine, Oregon	32
Manganese	475	Pine, red	30
Manganese ore, pyrolusite	259	Pine, white	26
Mercury	849	Pine, yellow, long-leaf	44
Monel Metal	565	Pine, yellow, short-leaf	38
Nickel	565	Poplar	30
Platinum, cast, hammered	1330	Redwood, California	26
Silver, cast, hammered	565	Spruce, white, black	27
Steel, rolled	490	Walnut, black	38
Tin, cast, hammered	459	Walnut, white	26
Tin ore, cassiterite	418	Various Liquids	
Zinc, cast, rolled	440	Alcohol, 100%	49
Zinc ore, blende	253	Acids, muriatic 40%	75
Various Solids		Acids, nitric 91%	94
Cereals, oats, bulk	32	Acids, sulphuric 87%	112
Cereals, barley, bulk	39	Lye, soda 66%	106
Cereals, corn, rye, bulk	48	Oils, vegetable	58
Cereals, wheat, bulk	48	Oils, mineral, lubricants	57
Hay and straw, bales	20	Water, 4°C. max. density	62.428
Cotton, flax, hemp	93	Water, 100°C.	59.830
Fats	58	Water, ice	56
Flour, loose	28	Water, snow, fresh fallen	8
Flour, pressed	47	Water, sea water	64
Glass, common	156	Gases	
Glass, plate or crown	161	Air, 0°C. 760 mm	.08071
Glass, crystal	184	Ammonia	.0478
Leather	59	Carbon dioxide	.1234
Paper	58	Carbon monoxide	.0781
Potatoes, piled	42	Gas, illuminating	.028–.036
Rubber, caoutchouc	59	Gas, natural	.038–.039

Average Weights for Various Materials (continued)

Material	Weight Lbs. per S.F.	Material	Weight Lbs. per S.F.
Gases (cont.)		Partitions	
Hydrogen	.00559	Clay Tile	
Nitrogen	.0784	3 in.	17
Oxygen	.0892	4 in.	18
Ceilings		6 in.	28
Channel suspended system	1	8 in.	34
Lathing and plastering	See Partitions	10 in.	40
Acoustical fiber tile	1	Gypsum Block	
		2 in.	9-1/2
Floors		3 in.	10-1/2
Steel Deck	See Mfg.	4 in.	12-1/2
		5 in.	14
Concrete–Reinforced 1 in.		6 in.	18-1/2
Stone	12-1/2	Wood Studs 2 x 4	
Slag	11-1/2	12–16 in. o.c.	2
Lightweight	6 to 10	Steel partitions	4
Concrete–Plain 1 in.		Plaster 1 inch	
Stone	12	Cement	10
Slag	11	Gypsum	5
Lightweight	3 to 9	Lathing	
		Metal	1/2
Fills 1 in.		Gypsum Board 1/2 in.	2
Gypsum	6		
Sand	8	Walls	
Cinders	4	Brick	
		4 in.	40
Finishes		8 in.	80
Terrazzo 1 in.	13	12 in.	120
Ceramic or Quarry Tile 3/4 in.	10	Hollow Concrete Block	
Linoleum 1/4 in.	1	(Heavy Aggregate)	
Mastic 3/4 in.	9	4 in.	30
Hardwood 7/8 in.	4	6 in.	43
Softwood 3/4 in.	2-1/2	8 in.	55
Roofs		12-1/2 in.	80
Copper or tin	1	Hollow Concrete Block	
3-ply ready roofing	1	(Light Aggregate)	
3-ply felt and gravel	5-1/2	4 in.	21
5-ply felt and gravel	6	6 in.	30
		8 in.	38
Shingles		12 in.	55
Wood	2	Clay Tile	
Asphalt	3	(Load Bearing)	
Clay tile	9 to 14	4 in.	25
Slate 1/4 in.	10	6 in.	30
Sheathing		8 in.	33
Wood 3/4 in.	3	12 in.	45
Gypsum 1 in.	4	Stone 4 in.	55
		Glass Block 4 in.	18
Insulation 1 in.		Windows, Glass, Frame & Sash	8
Loose	1/2	Curtain Walls	See Mfg.
Poured in place	2	Structural Glass 1 in.	15
Rigid	1-1/2	Corrugated Cement	
		Asbestos 1/4 in.	3

Basic Metric Units and Prefixes

Basic Metric Units		Prefixes for Metric Units		
Quantity	Unit	Multiple and Submultiple	Prefix	Symbol
length	meter (m)	$1,000,000,000,000 = 10^{12}$	tera	T
mass	kilogram (kg)	$1,000,000,000 = 10^{9}$	giga	G
time	second (s)	$1,000,000 = 10^{6}$	mega	M
electric current	ampere (A)	$1,000 = 10^{3}$	kilo	k
temperature (thermodynamic)	kelvin (K)	$100 = 10^{2}$	hecto	h
amount of substance	mole (mol)	$10 = 10$	deka	da
luminous intensity	candela (cd)	$0.1 = 10^{-1}$	deci	d
		$0.01 = 10^{-2}$	centi	c
		$0.001 = 10^{-3}$	milli	m
		$0.000\ 001 = 10^{-6}$	micro	μ
		$0.000\ 000\ 001 = 10^{-9}$	nano	n
		$0.000\ 000\ 000\ 001 = 10^{-12}$	pico	p
		$0.000\ 000\ 000\ 000\ 001 = 10^{-15}$	femto	f
		$0.000\ 000\ 000\ 000\ 000\ 001 = 10^{-18}$	atto	a

Metric Equivalents

Length	
CM. = 0.3937 in.	In. = 2.5400 cm.
Meter = 3.2808 ft.	Ft. = 0.3048 m.
Meter = 1.0936 yd.	Yd. = 0.9144 m.
Km. = 0.6214 mile	Mile = 1.6093 km.

Area	
Sq. cm. = 0.1550 sq. in.	Sq. in. = 6.4516 sq. cm.
Sq. m. = 10.7639 sq. ft.	Sq. ft. = 0.0929 sq. m.
Sq. m. = 1.1960 sq. yd.	Sq. yd. = 0.8361 sq. m.
Hectare = 2.4710 acres	Acre = 0.4047 hectare
Sq. km. = 0.3861 sq. mile	Sq. mile = 2.5900 sq. km.

Volume	
Cu. cm. = 0.0610 cu. in.	Cu. in. = 16.3872 cu. cm.
Cu. m. = 35.3145 cu. ft.	Cu. ft. = 0.0283 cu. cm.
Cu. m. = 1.3079 cu. yd.	Cu. yd. = 0.7646 cu. m.

Capacity	
Liter = 61.0250 cu. in.	Cu. in. = 0.0164 liters
Liter = 0.0353 cu. ft.	Cu. ft. = 28.3162 liters
Liter = 0.2642 gal. (U.S.)	Gal. = 3.7853 liters
Liter = 0.0284 bu. (U.S.)	Bu. = 35.2383 liters

$$\text{Liter} = \begin{cases} 1000.027 \text{ cu. cm.} \\ 1.0567 \text{ qt. (liquid) or } 0.9081 \text{ qt. (dry)} \\ 2.2046 \text{ lbs. of pure water at } 4°C + 1 \text{ kg.} \end{cases}$$

Weight	
Gram = 15.4324 grains	Grain = 0.0648 g.
Gram = 0.0353 oz.	Oz. = 28.3495 g.
Kg. = 2.2046 lbs.	Lb. = 0.4536 kg.
Kg. = 0.0011 ton (short)	Ton (short) = 907.1848 kg.
Ton (met.) = 1.1023 ton (short)	Ton (short) = 0.9072 ton (met.)
Ton (met.) = 0.9842 ton (large)	Ton (large) = 1.0160 ton (met.)

Pressure
1 kg. per sq. cm. = 14.223 lbs. per sq. in.
1 lb. per sq. in. = 0.0703 kg. per sq. cm.
1 kg. per sq. m. = 0.2048 lbs. per sq. ft.
1 lb. per sq. ft. = 4.8824 kg. per sq. m.
1 kg. per sq. cm. = 0.9678 normal atmosphere

$$1 \text{ normal atmosphere} = \begin{cases} 1.0332 \text{ kg. per sq. cm.} \\ 1.0133 \text{ bars} \\ 14.696 \text{ lbs. per sq. in.} \end{cases}$$

Metric Conversion Table

Conversion Formulas: Inches x 2.54 = Centimeters
Feet x .3048 = Meters
Pounds x .4536 = Kilograms

	Inches to Centimeters	Feet to Meters	Pounds to Kilograms		Inches to Centimeters	Feet to Meters	Pounds to Kilograms
1	2.54	.3048	.4536	51	129.54	15.5448	23.1336
2	5.08	.6096	.9072	52	132.08	15.8496	23.5872
3	7.62	.9144	1.3608	53	134.62	16.1544	24.0408
4	10.16	1.2192	1.8144	54	137.16	16.4592	24.4944
5	12.7	1.524	2.268	55	139.7	16.764	24.948
6	15.24	1.8288	2.7216	56	142.24	17.0688	25.4016
7	17.78	2.1336	3.1752	57	144.78	17.3736	25.8552
8	20.32	2.4384	3.6288	58	147.32	17.6784	26.3088
9	22.86	2.7432	4.0824	59	149.86	17.9832	26.7624
10	25.4	3.048	4.536	60	152.4	18.288	27.216
11	27.94	3.3528	4.9896	61	154.94	18.5928	27.6696
12	30.48	3.6576	5.4432	62	157.48	18.8976	28.1232
13	33.02	3.9624	5.8968	63	160.02	19.2024	28.5768
14	35.56	4.2672	6.3504	64	162.56	19.5072	29.0304
15	38.1	4.572	6.804	65	165.1	19.812	29.488
16	40.64	4.8768	7.2576	66	167.64	20.1168	29.9376
17	43.18	5.1816	7.7112	67	170.18	20.4216	30.3912
18	45.72	5.4864	8.1648	68	172.72	20.7264	30.8448
19	48.26	5.7912	8.6184	69	175.26	21.0312	31.2984
20	50.8	6.096	9.072	70	177.8	21.336	31.752
21	53.34	6.4008	9.5256	71	180.34	21.6408	32.2056
22	55.88	6.7056	9.9792	72	182.88	21.9456	32.6592
23	58.42	7.0104	10.4328	73	185.42	22.2504	33.1128
24	60.96	7.3152	10.8864	74	187.96	22.5552	33.5664
25	63.5	7.62	11.34	75	190.5	22.86	34.02
26	66.04	7.9248	11.7936	76	193.04	23.1648	34.4736
27	68.58	8.2296	12.2472	77	195.58	23.4696	34.9272
28	71.12	8.5344	12.7008	78	198.12	23.7744	35.3808
29	73.66	8.8392	13.1544	79	200.66	24.0792	35.8344
30	76.2	9.144	13.608	80	203.2	24.384	36.288
31	78.74	9.4488	14.0616	81	205.74	24.6888	36.7416
32	81.28	9.7536	14.5152	82	208.28	24.9936	37.1952
33	83.82	10.0584	14.9680	83	210.82	25.2984	37.6488
34	86.36	10.3632	15.4224	84	213.36	25.6032	38.1024
35	88.9	10.668	15.876	85	215.9	25.908	38.556
36	91.44	10.9728	16.3296	86	218.44	26.2128	39.0096
37	93.98	11.2776	16.7832	87	220.98	26.5176	39.4632
38	96.52	11.5824	17.2368	88	223.52	26.8224	39.9168
39	99.06	11.8872	17.6904	89	226.06	27.1272	40.3704
40	101.6	12.192	18.144	90	228.6	27.432	40.824
41	104.14	12.4968	18.5976	91	231.14	27.7368	41.2776
42	106.68	12.8016	19.0512	92	233.68	28.0416	41.7312
43	109.22	13.1064	19.5048	93	236.22	28.3464	42.1848
44	111.76	13.4112	19.9584	94	238.76	28.6512	42.6384
45	114.3	13.716	20.412	95	241.3	28.956	43.092
46	116.84	14.0208	20.8656	96	243.84	29.2608	43.5456
47	119.38	14.3256	21.3192	97	246.38	29.5656	43.9992
48	121.92	14.6304	21.7728	98	248.92	29.8704	44.4528
49	124.46	14.9352	22.2264	99	252.46	30.1752	44.9064
50	127	15.24	22.68	100	254	30.48	45.36

Standard and Metric Linear and Area Conversion Tables

Linear Conversions							
Inches	Feet	Yards	Rods	Miles	Centi-meters	Meters	Kilo-meters
1	0.083	0.028	0.005	—	2.540	0.0254	—
12	1	0.333	0.061	0.0002	30.480	0.305	0.0003
36	3	1	0.182	0.0006	91.440	0.914	0.0009
0.3937	0.033	0.011	—	—	1	0.01	—
39.37	3.281	1.094	0.199	0.0006	100	1	0.001
					Furlongs		
198	16.5	5.5	1	0.003	0.025	5.029	0.005
	5,280	1,760	320	1	8	1,609,347	1.609
	660	220	40	0.125	1	201.168	0.201
	3,280.83	1,093.61	198.838	0.621	4.971	1,000	1

Area Conversions							
Square Inches	Square Feet	Square Yards	Acres	Square Centi-meters	Square Meters	Hectares	Square Kilo-meters
1	0.007	—	—	6.452	0.0006	—	—
144	1	0.111	0.00002	929.034	0.093	—	—
1,296	9	1	0.0002	8,361.31	0.836	—	—
0.155	0.001	—	—	1	0.0001	—	—
1,549.997	10.764	1.196	0.0002	10,000	1	0.0001	—
				Square Miles			
	43,560	4,840	1	0.002	4,046.87	0.405	0.004
	27,878,400	3,097,600	640	1	2,589,998	258.999	2.590
	107,638.7	11,959.9	2.471	0.004	10,000	1	0.01
	10,763,867	1,195,985	247.104	0.386	1,000,000	100	1

General Conversion Factors

Multiply	By	To Obtain
acres	43,560	square feet
acres	4047	square meters
acres	1.562×10^{-3}	square miles
acres	5645.8	square varas
acres	4840	square yards
amperes	1/10	abamperes
amperes	3×10^9	statamperes
atmospheres	76.0	cms. of mercury
atmospheres	29.92	inches of mercury
atmospheres	33.90	feet of water
atmospheres	10.333	kgs. per sq. meter
atmospheres	14.70	pounds per sq. inch
atmospheres	1.058	tons per sq. foot
British thermal units	0.2520	kilogram-calories
British thermal units	777.5	foot-pounds
British thermal units	3.927×10^{-4}	horse-power-hours
British thermal units	1054	joules
British thermal units	107.5	kilogram-meters
British thermal units	2.928×10^{-4}	kilowatt-hours
B.t.u. per min.	12.96	foot-pounds per sec.
B.t.u. per min.	0.02356	horse-power
B.t.u. per min.	0.01757	kilowatts
B.t.u. per min.	17.57	watts
B.t.u. per sq. ft. per min.	0.1220	watts per sq. inch
bushels	1.244	cubic feet
bushels	2150	cubic inches
bushels	0.03524	cubic meters
bushels	4	pecks
bushels	64	pints (dry)
bushels	32	quarts (dry)
centimeters	0.3397	inches
centimeters	0.01	meters
centimeters	393.7	mils
centimeters	10	millimeters
centimeter-grams	980.7	centimeter-dynes
centimeter-grams	10^{-5}	meter-kilograms
centimeter-grams	7.233×10^{-5}	pound-feet
centimeters of mercury	0.01316	atmospheres
centimeters of mercury	0.4461	feet of water
centimeters of mercury	136.0	kgs. per sq. meter
centimeters of mercury	27.85	pounds per sq. foot
centimeters of mercury	0.1934	pounds per sq. inch
centimeters per second	1.969	feet per minute
centimeters per second	0.03281	feet per second
centimeters per second	0.036	kilometers per hour
centimeters per second	0.6	meters per minute
centimeters per second	0.02237	miles per hour
centimeters per second	3.728×10^{-4}	miles per minute
cubic centimeters	3.531×10^{-5}	cubic feet
cubic centimeters	6.102×10^{-2}	cubic inches
cubic centimeters	10^{-6}	cubic meters
cubic centimeters	1.308×10^{-6}	cubic yards
cubic centimeters	2.642×10^{-4}	gallons
cubic centimeters	10^{-3}	liters

General Conversion Factors (continued)

Multiply	By	To Obtain
cubic centimeters	2.113×10^{-3}	pints (liquid)
cubic centimeters	1.057×10^{-3}	quarts (liquid)
cubic feet	62.43	pounds of water
cubic feet	2.832×10^4	cubic cms.
cubic feet	1728	cubic inches
cubic feet	0.02832	cubic meters
cubic feet	0.03704	cubic yards
cubic feet	7.481	gallons
cubic feet	28.32	liters
cubic feet	59.84	pints (liquid)
cubic feet	29.92	quarts (liquid)
cubic feet per minute	472.0	cubic cms. per sec.
cubic feet per minute	0.1247	gallons per sec.
cubic feet per minute	0.4720	liters per second
cubic feet per minute	62.4	lbs. of water per min.
cubic inches	16.39	cubic centimeters
cubic inches	5.787×10^{-4}	cubic feet
cubic inches	1.639×10^{-5}	cubic meters
cubic inches	2.143×10^{-5}	cubic yards
cubic inches	4.329×10^{-3}	gallons
cubic inches	1.639×10^{-2}	liters
cubic inches	0.03463	pints (liquid)
cubic inches	0.01732	quarts (liquid)
cubic yards	7.646×10^5	cubic centimeters
cubic yards	27	cubic feet
cubic yards	46,656	cubic inches
cubic yards	0.7646	cubic meters
cubic yards	202.0	gallons
cubic yards	764.6	liters
cubic yards	1616	pints (liquid)
cubic yards	807.9	quarts (liquid)
cubic yards per minute	0.45	cubic feet per sec.
cubic yards per minute	3.367	gallons per second
cubic yards per minute	12.74	liters per second
degrees (angle)	60	minutes
degrees (angle)	0.01745	radians
degrees (angle)	3600	seconds
dynes	1.020×10^{-3}	grams
dynes	7.233×10^{-5}	poundals
dynes	2.248×10^{-6}	pounds
ergs	9.486×10^{-11}	British thermal units
ergs	1	dyne-centimeters
ergs	7.376×10^{-8}	foot-pounds
ergs	1.020×10^{-3}	gram-centimeters
ergs	10^{-7}	joules
ergs	2.390×10^{-11}	kilogram-calories
ergs	1.020×10^{-8}	kilogram-meters
feet	30.48	centimeters
feet	12	inches
feet	0.3048	meters
feet	.36	varas
feet	1/3	yards
feet of water	0.02950	atmospheres
feet of water	0.8826	inches of mercury

(continued on next page)

General Conversion Factors (continued)

Multiply	By	To Obtain
feet of water	304.8	kgs. per sq. meter
feet of water	62.43	pounds per sq. ft.
feet of water	0.4335	pounds per sq. inch
feet per second	30.48	centimeters per second
feet per sec. per sec.	0.3048	meters per sec. per sec.
foot-pounds	1.286×10^{-3}	British thermal units
foot-pounds	1.356×10^{7}	ergs
foot-pounds	5.050×10^{-7}	horse-power hours
foot-pounds	1.356	joules
foot-pounds	3.241×10^{-4}	kilogram-calories
foot-pounds	0.1383	kilogram-meters
foot-pounds	3.766×10^{-7}	kilowatt-hours
foot-pounds per min.	1.286×10^{-3}	B.t. units per minute
foot-pounds per min.	0.01667	foot-pounds per sec.
foot-pounds per min.	3.030×10^{-5}	horse-power
foot-pounds per min.	3.241×10^{-4}	kg.-calories per min.
foot-pounds per min.	2.260×10^{-5}	kilowatts
foot-pounds per sec.	7.717×10^{-2}	B.t. units per minute
foot-pounds per sec.	1.818×10^{-3}	horse-power
foot-pounds per sec.	1.945×10^{-2}	kg.-calories per min.
foot-pounds per sec.	1.356×10^{-3}	kilowatts
gallons	8.345	pounds of water
gallons	3785	cubic centimeters
gallons	0.1337	cubic feet
gallons	231	cubic inches
gallons	3.785×10^{-3}	cubic meters
gallons	4.951×10^{-3}	cubic yards
gallons	3.785	liters
gallons	8	pints (liquid)
gallons	4	quarts (liquid)
gallons per minute	2.228×10^{-3}	cubic ft. per second
gallons per minute	0.06308	liters per second
grains (troy)	1	grains (av.)
grains (troy)	0.06480	grams
grains (troy)	0.04167	pennyweights (troy)
grams	980.7	dynes
grams	15.43	grains (troy)
grams	10^{-3}	kilograms
grams	10^{3}	milligrams
grams	0.03527	ounces
grams	0.03215	ounces (troy)
grams	0.07093	poundals
grams	2.205×10^{-3}	pounds
horse-power	42.44	B.t. units per min.
horse-power	33,000	foot-pounds per min.
horse-power	550	foot-pounds per sec.
horse-power	1.014	horse-power (metric)
horse-power	10.70	kg.-calories per min.
horse-power	0.7457	kilowatts
horse-power	745.7	watts
horse-power (boiler)	33,520	B.t.u. per hour
horse-power (boiler)	9.804	kilowatts
horse-power-hours	2547	British thermal units
horse-power-hours	1.98×10^{6}	foot-pounds

General Conversion Factors (continued)

Multiply	By	To Obtain
horse-power-hours	2.684×10^6	joules
horse-power-hours	641.7	kilogram-calories
horse-power-hours	2.737×10^5	kilogram-meters
horse-power-hours	0.7457	kilowatt-hours
inches	2.540	centimeters
inches	10^3	mils
inches	.03	varas
inches of mercury	0.03342	atmospheres
inches of mercury	1.133	feet of water
inches of mercury	345.3	kgs. per sq. meter
inches of mercury	70.73	pounds per sq. ft.
inches of mercury	0.4912	pounds per sq. in.
inches of water	0.002458	atmospheres
inches of water	0.07355	inches of mercury
inches of water	25.40	kgs. per sq. meter
inches of water	0.5781	ounces per sq. in.
inches of water	5.204	pounds per sq. ft.
inches of water	0.03613	pounds per sq. in.
kilograms	980,665	dynes
kilograms	10^3	grams
kilograms	70.93	poundals
kilograms	2.2046	pounds
kilograms	1.102×10^{-3}	tons (short)
kilogram-calories	3.968	British thermal units
kilogram-calories	3086	foot-pounds
kilogram-calories	1.558×10^{-3}	horse-power-hours
kilogram-calories	4183	joules
kilogram-calories	426.6	kilogram-meters
kilogram-calories	1.162×10^{-3}	kilowatt-hours
kg.-calories per min.	51.43	foot-pounds per sec.
kg.-calories per min.	0.09351	horse-power
kg.-calories per min.	0.06972	kilowatts
kilometers	10^5	centimeters
kilometers	3281	feet
kilometers	10^3	meters
kilometers	0.6214	miles
kilometers	1093.6	yards
kilowatts	56.92	B.t. units per min.
kilowatts	4.425×10^4	foot-pounds per min.
kilowatts	737.6	foot-pounds per sec.
kilowatts	1.341	horse-power
kilowatts	14.34	kg.-calories per min.
kilowatts	10^3	watts
kilowatt-hours	3415	British thermal units
kilowatt-hours	2.655×10^6	foot-pounds
kilowatt-hours	1.341	horse-power-hours
kilowatt-hours	3.6×10^6	joules
kilowatt-hours	860.5	kilogram-calories
kilowatt-hours	3.671×10^5	kilogram-meters
$\log^{10} N$	2.303	$\log_\epsilon N$ or $\ln N$
$\log^\epsilon N$ or $\ln N$	0.4343	$\log_{10} N$
meters	100	centimeters
meters	3.2808	feet
meters	39.37	inches

(continued on next page)

General Conversion Factors (continued)

Multiply	By	To Obtain
meters	10^{-3}	kilometers
meters	10^3	millimeters
meters	1.0936	yards
miles	1.609×10^5	centimeters
miles	5280	feet
miles	1.6093	kilometers
miles	1760	yards
miles	1900.8	varas
miles per hour	44.70	centimeters per sec.
miles per hour	88	feet per minute
miles per hour	1.467	feet per second
miles per hour	1.6093	kilometers per hour
miles per hour	0.8684	knots per hour
miles per hour	26.82	meters per minute
miles per hour per sec.	44.70	cms. per sec. per sec.
miles per hour per sec.	1.467	ft. per sec. per sec.
miles per hour per sec.	1.6093	kms. per hr. per sec.
miles per hour per sec.	0.4470	M. per sec. per sec.
months	30.42	days
months	730	hours
months	43,800	minutes
months	2.628×10^6	seconds
ounces	8	drams
ounces	437.5	grains
ounces	28.35	grams
ounces	.0625	pounds
ounces per square inch.	0.0625	pounds per sq. inch
pints (dry)	33.60	cubic inches
pints (liquid)	28.87	cubic inches
pounds	444,823	dynes
pounds	7000	grains
pounds	453.6	grams
pounds	16	ounces
pounds	32.17	poundals
pounds of water	0.01602	cubic feet
pounds of water	27.68	cubic inches
pounds of water	0.1198	gallons
pounds of water per min.	2.669×10^{-4}	cubic feet per sec.
pounds per cubic foot	0.01602	grams per cubic cm.
pounds per cubic foot	16.02	kgs. per cubic meter
pounds per cubic foot	5.787×10^{-4}	pounds per cubic in.
pounds per cubic foot	5.456×10^{-9}	pounds per mil foot
pounds per square foot	0.01602	feet of water
pounds per square foot	4.882	kgs. per sq. meter
pounds per square foot	6.944×10^{-3}	pounds per sq. inch
pounds per square inch	0.06804	atmospheres
pounds per square inch	2.307	feet of water
pounds per square inch	2.036	inches of mercury
pounds per square inch	703.1	kgs. per sq. meter
pounds per square inch	144	pounds per sq. foot
quarts	32	fluid ounces
quarts (dry)	67.20	cubic inches
quarts (liquid)	57.75	cubic inches
rods	16.5	feet

General Conversion Factors (continued)

Multiply	By	To Obtain
square centimeters	1.973×10^5	circular mils
square centimeters	1.076×10^{-3}	square feet
square centimeters	0.1550	square inches
square centimeters	10^{-6}	square meters
square centimeters	100	square millimeters
square feet	2.296×10^{-5}	acres
square feet	929.0	square centimeters
square feet	144	square inches
square feet	0.09290	square meters
square feet	3.587×10^{-8}	square miles
square feet	.1296	square varas
square feet	1/9	square yards
square inches	1.273×10^6	circular mils
square inches	6.452	square centimeters
square inches	6.944×10^{-3}	square feet
square inches	10^6	square mils
square inches	645.2	square millimeters
square miles	640	acres
square miles	27.88×10^6	square feet
square miles	2.590	square kilometers
square miles	3,613,040.45	square varas
square miles	3.098×10^6	square yards
square yards	2.066×10^{-4}	acres
square yards	9	square feet
square yards	0.8361	square meters
square yards	3.228×10^{-7}	square miles
square yards	1.1664	square varas
temp. (degs. C.) + 17.8	1.8	temp. (degs. Fahr.)
temp. (degs. F.) − 32	5/9	temp. (degs. Cent.)
tons (long)	2240	pounds
tons (short)	2000	pounds
yards	.9144	meters

Approximate Comparison of Drawing Scales

Metric			Approximate English Equivalent			Usage
1:1500 (means 1 cm	=	15 m)	1 in.	=	100 ft.	Location plans
1:1000 (means 1 cm	=	10 m)	1 in.	=	80 ft.	Site plans
1:500 (means 1 cm	=	5 m)	1 in.	=	40 ft.	Plot plans
1:200 (means 0.5 cm	=	1 m)	1/16 in.	=	1 ft.	Small scale plans
1:100 (means 1 cm	=	1 m)	1/8 in.	=	1 ft.	Normal scale plans
1:50 (means 2 cm	=	1 m)	1/4 in.	=	1 ft.	Large scale plans
1:20 (means 5 cm	=	1 m)	1/2 in.	=	1 ft.	Sections and details
1:10 (means 10 cm	=	1 m)	1 in.	=	1 ft.	Details
1:8 (means 12.5 cm	=	1 m)	1-1/2 in.	=	1 ft.	Details
1:1 (means 1 m	=	1 m)	Full size			Large details

Trigonometric Functions

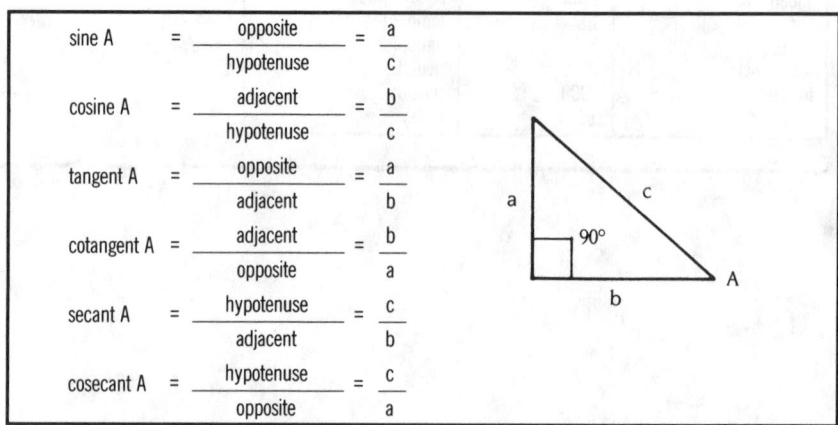

$$\text{sine } A = \frac{\text{opposite}}{\text{hypotenuse}} = \frac{a}{c}$$

$$\text{cosine } A = \frac{\text{adjacent}}{\text{hypotenuse}} = \frac{b}{c}$$

$$\text{tangent } A = \frac{\text{opposite}}{\text{adjacent}} = \frac{a}{b}$$

$$\text{cotangent } A = \frac{\text{adjacent}}{\text{opposite}} = \frac{b}{a}$$

$$\text{secant } A = \frac{\text{hypotenuse}}{\text{adjacent}} = \frac{c}{b}$$

$$\text{cosecant } A = \frac{\text{hypotenuse}}{\text{opposite}} = \frac{c}{a}$$

By using these relationships and the following Table of Natural Trigonometric Functions, unknown angles and sides of right triangles may be found.

Table of Natural Trigonometric Functions

Degrees	Sin	Cos	Tan	Cot	Sec	Csc	
0°00′	.0000	1.0000	.0000	—	1.000	—	90°00′
10	029	000	029	343.8	000	343.8	50
20	058	000	058	171.9	000	171.9	40
30	.0087	1.0000	.0087	114.6	1.000	114.6	30
40	116	9999	116	85.94	000	85.95	20
50	145	999	145	68.75	000	68.76	10
1°00′	.0175	.9998	.0175	57.29	1.000	57.30	89°00′
10	204	998	204	49.10	000	49.11	50
20	233	997	233	42.96	000	42.98	40
30	.0262	.9997	.0262	38.19	1.000	38.20	30
40	291	996	291	34.37	000	34.38	20
50	320	995	320	31.24	001	31.26	10
2°00′	.0349	.9994	.0349	28.64	1.001	28.65	88°00′
10	378	993	378	26.43	001	26.45	50
20	407	992	407	24.54	001	24.56	40
30	.0436	.9990	.0437	22.90	1.001	22.93	30
40	465	989	466	21.47	001	21.49	20
50	494	988	495	20.21	001	20.23	10
3°00′	.0523	.9986	.0524	19.08	1.001	19.11	87°00′
10	552	985	553	18.07	002	18.10	50
20	581	983	582	17.17	002	17.20	40
30	.0610	.9981	.0612	16.35	1.002	16.38	30
40	640	980	641	15.60	002	15.64	20
50	669	978	670	14.92	002	14.96	10
4°00′	.0698	.9976	.0699	14.30	1.002	14.34	86°00′
10	727	974	729	13.73	003	13.76	50
20	756	971	758	13.20	003	13.23	40
30	.0785	.9969	.0787	12.71	1.003	12.75	30
40	814	967	816	12.25	003	12.29	20
50	843	964	846	11.83	004	11.87	10
5°00′	.0872	.9962	.0875	11.43	1.004	11.47	85°00′
10	901	959	904	11.06	004	11.10	50
20	929	957	934	10.71	004	10.76	40
30	.0958	.9954	.0963	10.39	1.005	10.43	30
40	987	951	992	10.08	005	10.13	20
50	.1016	948	.1022	9.788	005	9.839	10
6°00′	.1045	.9945	.1051	9.514	1.006	9.567	84°00′
10	074	942	080	9.255	006	9.309	50
20	103	939	110	9.010	006	9.065	40
30	.1132	.9936	.1139	8.777	1.006	8.834	30
40	161	932	169	8.556	007	8.614	20
50	190	929	198	8.345	007	8.405	10
7°00′	.1219	.9925	.1228	8.144	1.008	8.206	83°00′
10	248	922	257	7.953	008	8.016	50
20	276	918	287	7.770	008	7.834	40
30	.1305	.9914	1317	7.596	1.009	7.661	30
40	334	911	346	7.429	009	7.496	20
50	363	907	376	7.269	009	7.337	10
8°00′	.1392	.9903	.1405	7.115	1.010	7.185	82°00′
10	421	899	435	6.968	010	7.040	50
20	449	894	465	6.827	011	6.900	40
30	.1478	.9890	.1495	6.691	1.011	6.765	30
40	507	886	524	6.561	012	6.636	20
50	536	881	554	6.435	012	6.512	10
9°00′	.1564	.9877	.1584	6.314	1.012	6.392	81°00′
	Cos	Sin	Cot	Tan	Csc	Sec	Degrees

(continued on next page)

Table of Natural Trigonometric Functions (continued)

Degrees	Sin	Cos	Tan	Cot	Sec	Csc	
9°00′	.1564	.9877	.1584	6.314	1.012	6.392	81°00′
10	593	872	614	197	013	227	50
20	622	868	644	084	013	166	40
30	.1650	.9863	.1673	5.976	1.014	6.059	30
40	679	858	703	871	014	5.955	20
50	708	853	733	769	015	855	10
10°00′	.1736	.9848	.1763	5.671	1.015	5.759	80°00′
10	765	843	793	576	016	665	50
20	794	838	823	485	016	575	40
30	.1822	.9833	.1853	5.396	1.017	5.487	30
40	851	827	883	309	018	403	20
50	880	822	914	226	018	320	10
11°00′	.1908	.9816	.1944	5.145	1.019	5.241	79°00′
10	937	811	974	066	019	164	50
20	965	805	.2004	4.989	020	089	40
30	.1994	.9799	.2035	4.915	1.020	5.016	30
40	.2022	793	065	843	021	4.945	20
50	051	787	095	773	022	876	10
12°00′	.2079	.9781	.2126	4.705	1.022	4.810	78°00′
10	108	775	156	638	023	745	50
20	136	769	186	574	024	682	40
30	.2164	.9763	.2217	4.511	1.024	4.620	30
40	193	757	247	449	025	560	20
50	221	750	278	390	026	502	10
13°00′	.2250	.9744	.2309	4.331	1.026	4.445	77°00′
10	278	737	339	275	027	390	50
20	306	730	370	219	028	336	40
30	.2334	.9724	.2401	4.165	1.028	4.284	30
40	363	717	432	113	029	232	20
50	391	710	462	061	030	182	10
14°00′	.2419	.9703	.2493	4.011	1.031	4.134	76°00′
10	447	696	524	3.962	031	086	50
20	476	689	555	914	032	039	40
30	.2504	.9681	.2586	3.868	1.033	3.994	30
40	532	674	617	821	034	950	20
50	560	667	648	776	034	906	10
15°00′	.2588	.9659	.2679	3.732	2.732	3.864	75°00′
10	616	652	711	689	036	822	50
20	644	644	742	647	037	782	40
30	.2672	.9636	.2773	3.606	1.038	3.742	30
40	700	628	805	566	039	703	20
50	728	621	836	526	039	665	10
16°00′	.2756	.9613	.2867	3.487	1.040	3.628	74°00′
10	784	605	899	450	041	592	50
20	812	596	931	412	042	556	40
30	.2840	.9588	.2962	3.376	1.043	3.521	30
40	868	580	994	340	044	487	20
50	896	572	.3026	305	045	453	10
17°00′	.2924	.9563	.3057	3.271	1.046	3.420	73°00′
10	952	555	089	237	047	388	50
20	979	546	121	204	048	356	40
30	.3007	.9537	.3153	3.172	1.049	3.326	30
40	035	528	185	140	049	295	20
50	062	520	217	108	050	265	10
18°00′	.3090	.9511	.3249	3.078	1.051	3.236	72°00′
	Cos	Sin	Cot	Tan	Csc	Sec	Degrees

Table of Natural Trigonometric Functions (continued)

Degrees	Sin	Cos	Tan	Cot	Sec	Csc	
18°00'	.3090	.9511	.3249	3.078	1.051	3.236	72°00'
10	118	502	281	047	052	207	50
20	145	492	314	018	053	179	40
30	.3173	.9843	.3346	2.989	1.054	3.152	30
40	201	474	378	960	056	124	20
50	228	465	411	932	057	098	10
19°00'	.3256	.9455	.3443	2.904	1.058	3.072	71°00'
10	283	446	476	877	059	046	50
20	311	436	508	850	060	021	40
30	.3388	.9426	.3541	2.824	1.061	2.996	30
40	365	417	574	798	062	971	20
50	393	407	607	773	063	947	10
20°00'	.3420	.9397	.3640	2.747	1.064	2.924	70°00'
10	448	387	673	723	065	901	50
20	475	377	706	699	066	878	40
30	.3502	.9367	.3739	2.675	1.068	2.855	30
40	529	356	772	651	069	833	20
50	557	346	805	628	070	812	10
21°00'	.3584	.9336	.3839	2.605	1.071	2.790	69°00'
10	611	325	872	583	072	769	50
20	638	315	906	560	074	749	40
30	.3665	.9304	.3939	2.539	1.075	2.729	30
40	692	293	973	517	076	709	20
50	719	283	.4006	496	077	689	10
22°00'	.3746	.0272	.4040	2.475	1.079	2.669	68°00'
10	773	261	074	455	080	650	50
20	800	250	108	434	081	632	40
30	.3827	.9239	.4142	2.414	1.082	2.613	30
40	854	228	176	394	084	595	20
50	881	216	210	375	085	577	10
23°00'	.3907	.9205	.4245	2.356	1.086	2.559	67°00'
10	934	194	279	337	088	542	50
20	961	182	314	318	089	525	40
30	.3987	.9171	.4348	2.300	1.090	2.508	30
40	.4014	159	383	282	092	491	20
50	041	147	417	264	093	475	10
24°00'	.4067	.9135	.4452	2.246	1.095	2.459	66°00'
10	094	124	487	229	096	443	50
20	120	112	522	211	097	427	40
30	.4147	.9100	.4557	2.194	1.099	2.411	30
40	173	088	592	177	100	396	20
50	200	075	628	161	102	381	10
25°00'	.4226	.9063	.4663	2.145	1.103	2.366	65°00'
10	253	051	699	128	105	352	50
20	279	038	734	112	106	337	40
30	.4305	.9026	.4770	2.097	1.108	2.323	30
40	331	013	806	081	109	309	20
50	358	001	841	066	111	295	10
26°00'	.4384	.8988	.4877	2.050	1.113	2.281	64°00'
10	410	975	913	035	114	268	50
20	436	962	950	020	116	254	40
30	.4462	.8949	.4986	2.006	1.117	2.241	30
40	488	936	.5022	1.991	119	228	20
50	514	923	059	977	121	215	10
27°00'	.4540	.8910	.5095	1.963	1.122	2.203	63°00'
	Cos	Sin	Cot	Tan	Csc	Sec	Degrees

(continued on next page)

Table of Natural Trigonometric Functions (continued)

Degrees	Sin	Cos	Tan	Cot	Sec	Csc	
27°00′	.4540	.8910	.5095	1.963	1.122	2.203	63°00′
10	566	897	132	949	124	190	50
20	592	884	169	935	126	178	40
30	.4617	.8870	.5206	1.921	1.127	2.166	30
40	643	857	243	907	129	154	20
50	669	843	280	894	131	142	10
28°00′	.4695	.8829	.5317	1.881	1.133	2.130	62°00′
10	720	816	354	868	143	118	50
20	746	802	392	855	136	107	40
30	.4772	.8788	.5430	1.842	1.138	2.096	30
40	797	774	467	829	140	085	20
50	823	760	505	816	142	074	10
29°00′	.4848	.8746	.5543	1.804	1.143	2.063	61°00′
10	874	732	581	792	145	052	50
20	899	718	619	780	147	041	40
30	.4924	.8704	.5658	1.767	1.149	2.031	30
40	950	689	696	756	151	020	20
50	975	675	735	744	153	010	10
30°00′	.500	.8660	.5774	1.732	1.155	2.000	60°00′
10	025	646	812	720	157	1.990	50
20	050	631	851	709	159	980	40
30	.5075	.8616	.5890	1.698	1.161	1.970	30
40	100	601	930	686	163	961	20
50	125	587	969	675	165	951	10
31°00′	.5150	.8572	.6009	1.664	1.167	1.942	59°00′
10	175	557	048	653	169	932	50
20	200	542	088	643	171	923	40
30	.5225	.8526	.6128	1.632	1.173	1.914	30
40	250	511	168	621	175	905	20
50	275	496	208	611	177	896	10
32°00′	.5299	.8480	.6249	1.600	1.179	1.887	58°00′
10	324	465	289	590	181	878	50
20	348	450	330	580	184	870	40
30	.5373	.8434	.6371	1.570	1.186	1.861	30
40	398	418	412	560	188	853	20
50	422	403	453	550	190	844	10
33°00′	.5446	.8387	.6494	1.540	1.192	1.836	57°00′
10	471	371	536	530	195	828	50
20	495	355	577	530	197	820	40
30	.5519	.8339	.6619	1.511	1.199	1.812	30
40	544	323	661	501	202	804	20
50	568	307	703	1.492	204	796	10
34°00′	.5592	.8290	.6745	1.483	1.206	1.788	56°00′
10	616	274	787	473	209	781	50
20	640	258	830	464	211	773	40
30	.5664	.8241	.6873	1.455	1.213	1.766	30
40	688	225	916	446	216	758	20
50	712	208	959	437	218	751	10
35°00′	.5736	.8192	.7002	1.428	1.221	1.743	55°00′
10	760	175	046	419	223	736	50
20	783	158	089	411	226	729	40
30	.5807	.8141	.7133	1.402	1.228	1.722	30
40	831	124	177	393	231	715	20
50	854	107	221	385	233	708	10
36°00′	.5878	.8090	.7265	1.376	1.236	1.701	54°00′
	Cos	Sin	Cot	Tan	Csc	Sec	Degrees

Table of Natural Trigonometric Functions (continued)

Degrees	Sin	Cos	Tan	Cot	Sec	Csc	
36°00′	.5878	.8090	.7265	1.376	1.236	1.701	54°00′
10	901	073	310	368	239	695	50
20	925	056	355	360	241	688	40
30	.5948	.8039	.7400	1.351	1.244	1.681	30
40	972	021	445	343	247	675	20
50	995	004	490	335	249	668	10
37°00′	.6018	.7986	.7536	1.327	1.252	1.662	53°00′
10	041	969	581	319	255	655	50
20	065	951	627	311	258	649	40
30	.6088	.7934	.7673	1.303	1.260	1.643	30
40	111	916	720	295	263	636	20
50	134	898	766	288	266	630	10
28°00′	.6157	.7880	.7813	1.280	1.269	1.624	52°00′
10	180	862	860	272	272	618	50
20	202	844	907	265	275	612	40
30	.6225	.7826	.7954	1.257	1.278	1.606	30
40	248	808	.8002	250	281	601	20
50	271	790	050	242	284	595	10
39°00′	.6293	.7771	.8098	1.235	1.287	1.589	51°00′
10	316	753	146	228	290	583	50
20	338	735	195	220	293	578	40
30	.6361	.7716	.8243	1.213	1.296	1.572	30
40	383	698	292	206	299	567	20
50	406	679	342	199	302	561	10
40°00′	.6428	.7660	.8391	1.192	1.305	1.556	50°00′
10	450	642	441	185	309	550	50
20	472	623	491	178	312	545	40
30	.6494	.7604	.8541	1.171	1.315	1.540	30
40	517	585	591	164	318	535	20
50	539	566	642	157	322	529	10
41°00′	.6561	.7547	.8693	1.150	1.325	1.524	49°00′
10	583	528	744	144	328	519	50
20	604	509	796	137	332	514	40
30	.6626	.7490	.8847	1.130	1.335	1.509	30
40	648	470	899	124	339	504	20
50	670	451	952	117	342	499	10
42°00′	.6691	.7431	.9004	1.111	1.346	1.494	48°00′
10	713	412	057	104	349	490	50
20	734	392	110	098	353	485	40
30	.6756	.7373	.9163	1.091	1.356	1.480	30
40	777	353	217	085	360	476	20
50	799	333	271	079	364	471	10
43°00′	.6820	.7314	.9325	1.072	1.367	1.466	47°00′
10	841	294	380	066	371	462	50
20	862	274	435	060	375	457	40
30	.6884	.7254	.9490	1.054	1.379	1.453	30
40	905	234	545	048	382	448	20
50	926	214	601	042	386	444	10
44°00′	.6947	.7193	.9657	1.036	1.390	1.440	46°00′
10	967	173	713	030	394	435	50
20	988	153	770	024	398	431	40
30	.7009	.7133	.9827	1.018	1.402	1.427	30
40	030	112	884	012	406	423	20
50	050	092	942	006	410	418	10
45°00′	.7071	.7071	1.0000	1.000	1.414	1.414	45°00′
	Cos	Sin	Cot	Tan	Csc	Sec	Degrees

Measure of Angles

Degrees	Rise in Inches per Ft.	Rise in Inches per Ft.	Degrees and Minutes	Percent Rise in Ft. per 100 Ft.	Degrees and Minutes	Percent Rise in Ft. per 100 Ft.	Degrees and Minutes
1	.210	1/4	1° 11′	1	34.4′	36	19° 48′
2	.419	1/2	2° 23′	2	1° 8.7′	37	20° 18′
3	.629	3/4	3° 35′	3	1° 43.1′	38	20° 48′
4	.839	1	4° 46′	4	2° 17.5′	39	21° 18′
5	1.050	1-1/4	5° 56′	5	2° 51.8′	40	21° 48′
6	1.261	1-1/2	7° 7′	6	3° 26.0′	41	22° 18′
7	1.473	1-3/4	8° 18′	7	4° 0.3′	42	22° 47′
8	1.686	2	9° 28′	8	4° 34.4′	43	23° 16′
9	1.901	2-1/4	10° 37′	9	5° 8.6′	44	23° 45′
10	2.116	2-1/2	11° 46′	10	5° 42.6′	45	24° 14′
11	2.333	2-3/4	12° 54′	11	6° 16.6′	46	24° 42′
12	2.551	3	14° 2′	12	6° 50.6′	47	25° 10′
13	2.770	3-1/4	15° 9′	13	7° 24.4′	48	25° 38′
14	2.992	3-1/2	16° 15′	14	7° 58.2′	49	26° 6′
15	3.215	3-3/4	17° 21′	15	8° 31.9′	50	26° 34′
16	3.441	4	18° 26′	16	9° 5.4′	51	27° 1′
17	3.669	4-1/4	19° 30′	17	9° 38.9′	52	27° 28′
18	3.900	4-1/2	20° 33′	18	10° 12.2′	53	27° 55′
19	4.132	4-3/4	21° 36′	19	10° 45.5′	54	28° 22′
20	4.368	5	22° 37′	20	11° 18.6′	55	28° 49′
21	4.606	5-1/4	23° 38′	21	11° 51.6′	56	29° 15′
22	4.843	5-1/2	24° 37′	22	12° 24.5′	57	29° 41′
23	5.094	5-3/4	25° 36′	23	12° 57.2′	58	30° 7′
24	5.313	6	26° 34′	24	13° 29.8′	59	30° 32′
25	5.596	6-1/4	27° 31′	25	14° 2.2′	60	30° 58′
26	5.853	6-1/2	28° 27′	26	14° 34.5′	61	31° 23′
27	6.114	6-3/4	29° 22′	27	15° 6.6′	62	31° 48′
28	6.381	7	30° 16′	28	15° 38.5′	63	32° 13′
29	6.652	7-1/4	31° 8′	29	16° 10.3′	64	32° 37′
30	6.928	7-1/2	32°	30	16° 42.0′	65	33° 1′
31	7.210	7-3/4	32° 51′	31	17° 13.4′	66	33° 25′
32	7.498	8	33° 41′	32	17° 44.7′	67	33° 49′
33	7.793	8-1/4	34° 30′	33	18° 15.8′	68	34° 13′
34	8.094	8-1/2	35° 19′	34	18° 46.7′	69	34° 36′
35	8.403	8-3/4	36° 5′	35	19° 17.0′	70	35″ 0′

Area Calculations

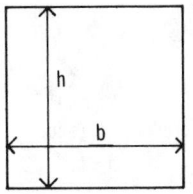

Square   Area = h x b

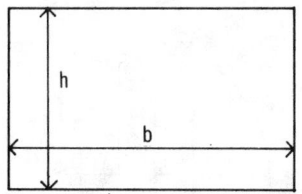

Rectangle   Area = h x b

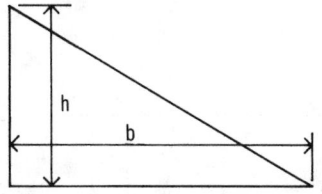

Triangle   Area = $\dfrac{h \times b}{2}$

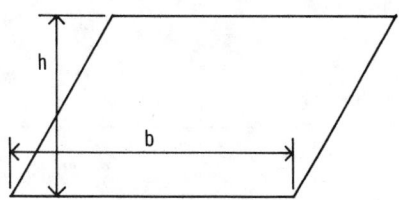

Parallelogram   Area = h x b

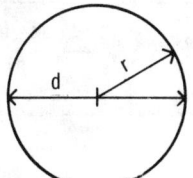

Circle   Area = $\dfrac{\pi d^2}{4} = \pi r^2$

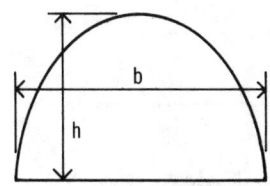

Parabola   Area = $\dfrac{2hb}{3}$

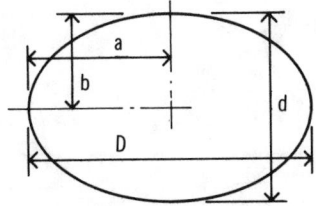

Ellipse   Area = 0.7854Dd

(continued on next page)

Area Calculations (continued)

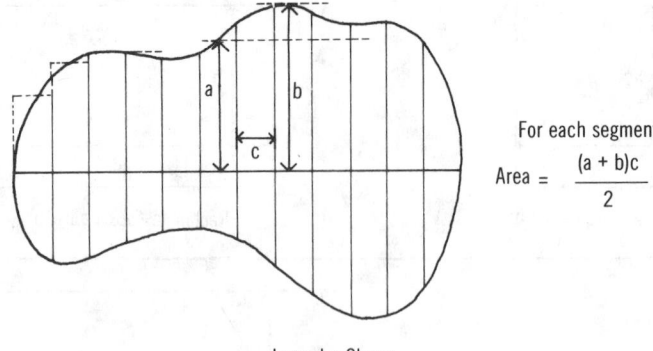

For each segment:

$$Area = \frac{(a + b)c}{2}$$

Irregular Shape

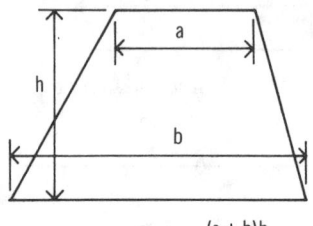

Trapezoid Area $= \dfrac{(a + b)h}{2}$

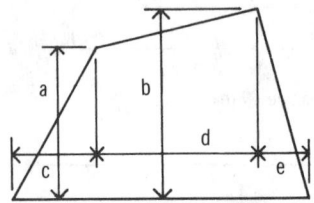

Trapezium Area $= \dfrac{bd - ad - be + ac}{2}$

Volume Calculations

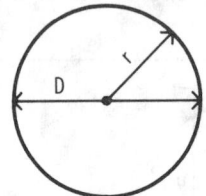

Sphere
Volume $= 0.5236\ D^3$
Surface Area $= 4\pi r^2$ ($\pi = 3.1416$)

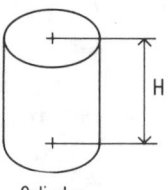

Cylinder

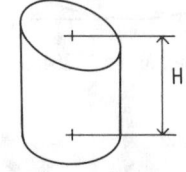

Slant Top Cylinder

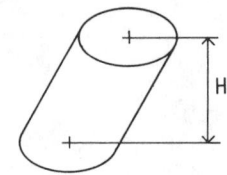

Oblique Cylinder

Volume = Area of the Base x Height (H)

Volume Calculations (continued)

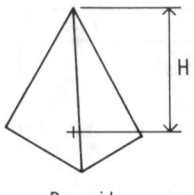
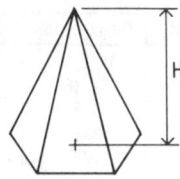
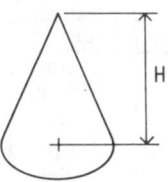

Pyramid   5 Sided Pyramid   Cone

Volume = Area of Base x 1/3 Height (H)

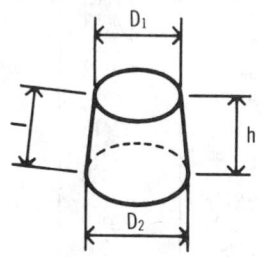

Frustum of a Cone

Volume =

$$\frac{D_1 + D_2 + (Area\ of\ Top + Area\ of\ Base) \times h \times 0.7854}{3}$$

$$*Surface\ Area = \frac{(D_1 + D_2)1}{2}$$

*Excludes Top & Base Areas

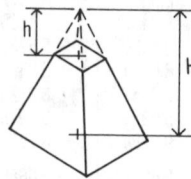
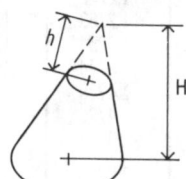

Truncated Pyramid   Truncated Oblique Cone

h = Height of Cut-Off
H = Height of Whole

Volume = Volume of the Whole Solid Less Volume of the Portion Cut Off

Circumferences and Areas of Circles

Diameter In.	Circum. In	Area Sq. In.	Diameter In.	Circum. In.	Area Sq. In.
1/64	.04909	.00019	1	3.1416	.7854
1/32	.09818	.00077	1/16	3.3379	.8866
3/64	.14726	.00173	1/8	3.5343	.9940
1/16	.19635	.00307	3/16	3.7306	1.1075
5/64	.24544	.00479	1/4	3.9270	1.2272
3/32	.29452	.00690	5/16	4.1233	1.3530
7/64	.34361	.00940	3/8	4.3197	1.4849
			7/16	4.5160	1.6230
1/8	.39270	.01227	1/2	4.7124	1.7671
9/64	.44179	.01553	9/16	4.9087	1.9175
5/32	.49087	.01917	5/8	5.1051	2.0739
11/64	.53996	.02320	11/16	5.3014	2.2365
3/16	.58905	.02761	3/4	5.4978	2.4053
13/64	.63814	.03241	13/16	5.6941	2.5802
7/32	.68722	.03758	7/8	5.8905	2.7612
15/64	.73631	.04314	15/16	6.0868	2.9483
1/4	.78540	.04909	2	6.2832	3.1416
17/64	.83449	.05542	1/16	6.4795	3.3410
9/32	.88357	.06213	1/8	6.6759	3.5466
19/64	.93266	.06922	3/16	6.8722	3.7583
5/16	.98175	.07670	1/4	7.0686	3.9761
21/64	1.03084	.08456	5/16	7.2649	4.2000
11/32	1.0799	.09281	3/8	7.4613	4.4301
23/64	1.1290	.10143	7/16	7.6576	4.6664
3/8	1.1781	.11045	1/2	7.8540	4.9087
25/64	1.2272	.11984	9/16	8.0503	5.1572
13/32	1.2763	.12962	5/8	8.2467	5.4119
27/64	1.3254	.13978	11/16	8.4430	5.6727
7/16	1.3744	.15033	3/4	8.6394	5.9396
29/64	1.4235	.16126	13/16	6.8357	6.2126
15/32	1.4726	.17257	7/8	9.0321	5.4918
31/64	1.5217	.18427	15/16	9.2284	6.7771
1/2	1.5708	.19635	3	9.4248	7.0686
17/32	1.6690	.22166	1/16	9.6211	7.3662
9/16	1.7671	.24850	1/8	9.8175	7.6699
19/32	1.8653	.27688	3/16	10.014	7.9798
5/8	1.9635	.30680	1/4	10.210	8.2958
21/32	2.0617	.33824	5/16	10.407	8.6179
11/16	2.1598	.37122	3/8	10.603	8.9462
23/32	2.2580	.40574	7/16	10.799	9.2806
3/4	2.3562	.44179	1/2	10.996	9.6211
25/32	2.4544	.47937	9/16	11.192	9.9678
13/16	2.5525	.51849	5/8	11.388	10.321
27/32	2.6507	.55914	11/16	11.585	10.680
7/8	2.7489	.60132	3/4	11.781	11.045
29/32	2.8471	.64504	13/16	11.977	11.416
15/16	2.9452	.69029	7/8	12.174	11.793
31/32	3.0434	.73708	15/16	12.370	12.177

Circumferences and Areas of Circles (continued)

Diameter In.	Circum. In	Area Sq. In.	Diameter In.	Circum. In.	Area Sq. In.
4	12.566	12.566	8	25.133	50.265
1/16	12.763	12.962	1/8	25.525	51.849
1/8	12.959	13.364	1/4	25.918	53.456
3/16	13.155	13.772	3/8	26.311	55.088
1/4	13.352	14.186	1/2	26.704	56.745
5/16	13.548	14.607	5/8	27.096	58.426
3/8	13.744	15.033	3/4	27.489	60.132
7/16	13.941	15.466	7/8	27.882	61.862
1/2	14.137	15.904	9	28.274	63.617
9/16	14.334	15.349	1/8	28.667	65.397
5/8	14.530	16.800	1/4	29.060	67.201
11/16	14.726	17.257	3/8	29.452	69.029
3/4	14.923	17.721	1/2	29.845	70.882
13/16	15.119	18.190	5/8	30.238	72.760
7/8	15.315	18.665	3/4	30.631	74.662
15/16	15.512	19.147	7/8	31.023	76.589
5	15.708	19.635	10	31.416	78.540
1/16	15.904	20.129	1/8	31.809	80.516
1/8	16.101	20.629	1/4	32.201	82.516
3/16	16.297	21.135	3/8	32.594	84.541
1/4	16.493	21.648	1/2	32.987	86.590
5/16	16.690	22.166	5/8	33.379	88.664
3/8	16.886	22.691	3/4	33.772	90.763
7/16	17.082	23.221	7/8	34.165	92.886
1/2	17.279	23.758	11	34.558	95.033
9/16	17.475	24.301	1/8	34.950	97.205
5/8	17.671	24.850	1/4	35.343	99.402
11/16	16.868	25.406	3/8	35.736	101.62
3/4	18.064	25,867	1/2	36.128	103.87
13/16	18.261	26.535	5/8	36.521	106.14
7/8	18.457	27.109	3/4	36.914	108.43
15/16	18.653	27.688	7/8	37.306	110.75
6	18.850	28.274	12	37.699	113.10
1/8	19.242	29.465	1/8	38.092	115.47
1/4	19.635	30.680	1/4	38.485	117.86
3/8	20.028	31.919	3/8	38.877	120.28
1/2	20.420	33.183	1/2	39.270	122.72
5/8	20.813	34.472	5/8	59.663	125.19
3/4	21.206	35.785	3/4	40.055	127.68
7/8	21.598	37.122	7/8	40.448	130.19
7	21.991	38.485	13	40.841	132.73
1/8	22.384	39.871	1/8	41.233	135.30
1/4	22.776	41.282	1/4	41.626	137.89
3/8	23.169	42.718	3/8	42.019	140.50
1/2	23.562	44.179	1/2	42.412	143.14
5/8	23.955	45.664	5/8	42.804	145.80
3/4	24.347	47.173	3/4	43.197	148.49
7/8	24.740	48.707	7/8	43.590	151.20

(continued on next page)

Circumferences and Areas of Circles (continued)

Diameter In.	Circum. In	Area Sq. In.	Diameter In.	Circum. In.	Area Sq. In.
14	43.982	153.94	17-1/4	54.192	233.71
1/8	44.375	156.70	3/8	54.585	237.10
1/4	44.768	159.48	1/2	54.978	240.53
3/8	45.160	162.30	5/8	55.371	243.98
1/2	45.553	165.13	3/4	55.763	247.45
5/8	45.946	167.99	7/8	56.156	250.95
3/4	46.338	170.87	18	56.549	254.47
7/8	46.731	173.78	1/8	56.941	258.02
15	47.124	176.71	1/4	57.334	261.59
1/8	47.517	179.67	3/8	57.727	265.18
1/4	47.909	182.65	1/2	58.119	268.80
3/8	48.302	185.66	5/8	58.512	272.45
1/2	48.695	188.69	3/4	58.905	276.12
5/8	49.087	191.75	7/8	59.298	279.81
3/4	49.480	194.83	19	59.690	283.53
7/8	49.873	197.93	1/8	60.083	287.27
16	50.265	201.06	1/4	60.476	291.04
1/8	50.658	204.22	3/8	60.868	294.83
1/4	51.051	207.39	1/2	61.261	298.65
3/8	51.444	210.60	5/8	61.654	302.49
1/2	51.836	213.82	3/4	62.046	306.35
5/8	52.229	217.08	7/8	62.439	310.24
3/4	52.622	220.35	20	62.832	314.16
7/8	53.014	223.65	1/8	63.225	318.10
17	53.407	226.98	1/4	63.617	322.06
1/8	53.800	230.33	3/8	64.010	326.05

Weather Data and Design Conditions
(winter design @ 97.5% – summer design @ 2.5%)

City	Latitude (1)		Winter Temperatures (1)			Winter Degree Days (2)	Summer (Design Dry Bulb) Temperatures and Relative Humidity		
			Med. of Annual						
	0	1′	Extremes	99%	97-1/2%		1%	2-1/2%	5%
UNITED STATES									
Albuquerque, NM	35	0	6	12	16	4,400	96/61	94/61	92/61
Atlanta, GA	33	4	14	17	22	3,000	95/74	92/74	90/73
Baltimore, MD	39	2	12	14	17	4,600	94/75	92/75	89/74
Birmingham, AL	33	3	17	17	21	2,600	97/74	94/75	93/74
Bismark, ND	46	5	-31	-23	-19	8,800	95/68	91/68	88/67
Boise, ID	43	3	0	3	10	5,800	96/65	93/64	91/64
Boston, MA	42	2	-1	6	9	5,600	91/73	88/71	85/70
Burlington, VT	44	3	-18	-12	-7	8,200	88/72	85/70	83/69
Charleston, WV	38	2	1	7	11	4,400	92/74	90/73	88/72
Charlotte, NC	35	1	13	18	22	3,200	96/74	94/74	92/74
Casper, WY	42	5	-20	-11	-5	7,400	92/58	90/57	87/57
Chicago, IL	41	5	-5	-3	2	6,600	94/75	91/74	88/73
Cincinnati, OH	39	1	2	1	6	4,400	94/73	92/72	90/72
Cleveland, OH	41	2	-2	1	5	6,400	91/73	89/72	86/71
Columbia, SC	34	0	16	20	24	2,400	98/76	96/75	94/75
Dallas, TX	32	5	14	18	22	2,400	101/75	99/75	97/75
Denver, CO	39	5	-9	-5	1	6,200	92/59	90/59	89/59
Des Moines, IA	41	3	-13	-10	-5	6,600	95/75	92/74	89/73
Detroit, MI	42	2	0	3	6	6,200	92/73	88/72	85/71
Great Falls, MT	47	3	-29	-21	-15	7,800	91/60	88/60	85/59
Hartford, CT	41	5	-4	3	7	6,200	90/74	88/73	85/72
Houston, TX	29	5	24	28	33	1,400	96/77	94/77	92/77
Indianapolis, IN	39	4	-2	-2	2	5,600	93/74	91/74	88/73
Jackson, MS	32	2	17	21	25	2,200	98/76	96/76	94/76
Kansas City, MO	39	1	-2	2	6	4,800	100/75	97/74	94/74
Las Vegas, NV	36	1	18	25	28	2,800	108/66	106/65	104/65
Lexington, KY	38	0	0	3	8	4,600	94/73	92/72	90/72
Little Rock, AR	34	4	13	15	20	3,200	99/76	96/77	94/77
Los Angeles, CA	34	0	38	41	43	2,000	94/70	90/70	87/69
Memphis, TN	35	0	11	13	18	3,200	98/77	96/76	94/76
Miami, FL	25	5	39	44	47	200	92/77	90/77	89/77
Milwaukee, WI	43	0	-11	-8	-4	7,600	90/74	87/73	84/71
Minneapolis, MN	44	5	-19	-16	-12	8,400	92/75	89/73	86/71
New Orleans, LA	30	0	29	29	33	1,400	93/78	91/78	90/77
New York, NY	40	5	6	11	15	5,000	94/74	91/73	88/72
Norfolk, VA	36	5	18	20	22	3,400	94/77	91/76	89/76
Oklahoma City, OK	35	2	4	9	13	3,200	100/74	97/74	95/73
Omaha, NE	41	2	-12	-8	-3	6,600	97/76	94/75	91/74
Philadelphia, PA	39	5	-7	10	14	4,400	93/75	90/74	87/72
Phoenix, AZ	33	3	25	31	34	1,800	108/71	106/71	104/71

(continued on next page)

Weather Data and Design Conditions (winter design @ 97.5% – summer design @ 2.5%) (continued)

City	Latitude (1)		Winter Temperatures (1)			Winter Degree Days (2)	Summer (Design Dry Bulb) Temperatures and Relative Humidity		
	0	1′	Med. of Annual Extremes	99%	97-1/2%		1%	2-1/2%	5%
UNITED STATES									
Pittsburgh, PA	40	3	1	3	7	6,000	90/72	88/71	85/70
Portland, ME	43	4	-14	-6	-1	7,600	88/72	85/71	81/69
Portland, OR	45	4	17	17	23	4,600	89/68	85/67	81/65
Portsmouth, NH	43	1	-8	-2	2	7,200	88/73	86/71	83/70
Providence, RI	41	4	0	5	9	6,000	89/73	86/72	83/70
Rochester, NY	43	1	-5	1	5	6,800	91/73	88/71	85/70
Salt Lake City, UT	40	5	-2	3	8	6,000	97/62	94/62	92/61
San Francisco, CA	37	5	38	38	40	3,000	80/63	77/62	83/61
Seattle, WA	47	4	22	22	27	5,200	81/68	79/66	76/65
Sioux Falls, SD	43	4	-21	-15	-11	7,800	95/73	92/72	89/71
St. Louis, MO	38	4	1	3	8	5,000	96/75	94/75	92/74
Tampa, FL	28	0	32	36	40	680	92/77	91/77	90/76
Trenton, NJ	40	1	7	11	14	5,000	92/75	90/74	87/73
Washington, DC	38	5	12	14	17	4,200	94/75	92/74	90/74
Wichita, KS	37	4	-1	3	7	4,600	102/72	99/73	96/73
Wilmington, DE	39	4	6	10	14	5,000	93/74	93/74	20/73
ALASKA									
Anchorage	61	1	-29	-23	-18	10,800	73/59	70/58	67/56
Fairbanks	64	5	-59	-51	-47	14,280	82/62	78/60	75/59
CANADA									
Edmonton, Alta.	53	3	-30	-29	-25	11,000	86/66	83/65	80/63
Halifax, N.S.	44	4	-4	1	5	8,000	83/66	80/65	77/64
Montreal, Que.	45	3	-20	-16	-10	9,000	88/73	86/72	84/71
Saskatoon, Sask.	52	1	-35	-35	-31	11,000	90/68	86/66	83/65
St. Johns, Nwf.	47	4	1	3	7	8,600	79/66	77/65	75/64
Saint John, N.B.	45	2	-15	-12	-8	8,200	81/67	79/65	77/64
Toronto, Ont.	43	4	-10	-5	-1	7,000	90/73	87/72	85/71
Vancouver, B.C.	49	1	13	15	19	6,000	80/67	78/66	76/65
Winnipeg, Man.	49	5	-31	-30	-27	10,800	90/73	87/71	84/70

(1) Handbook of Fundamentals, ASHRAE, Inc., NY 1972/1985
(2) Local Climatological Annual Survey, USDC Env. Science Services Administration, Ashville, NC

Snow Loads — 50 Year

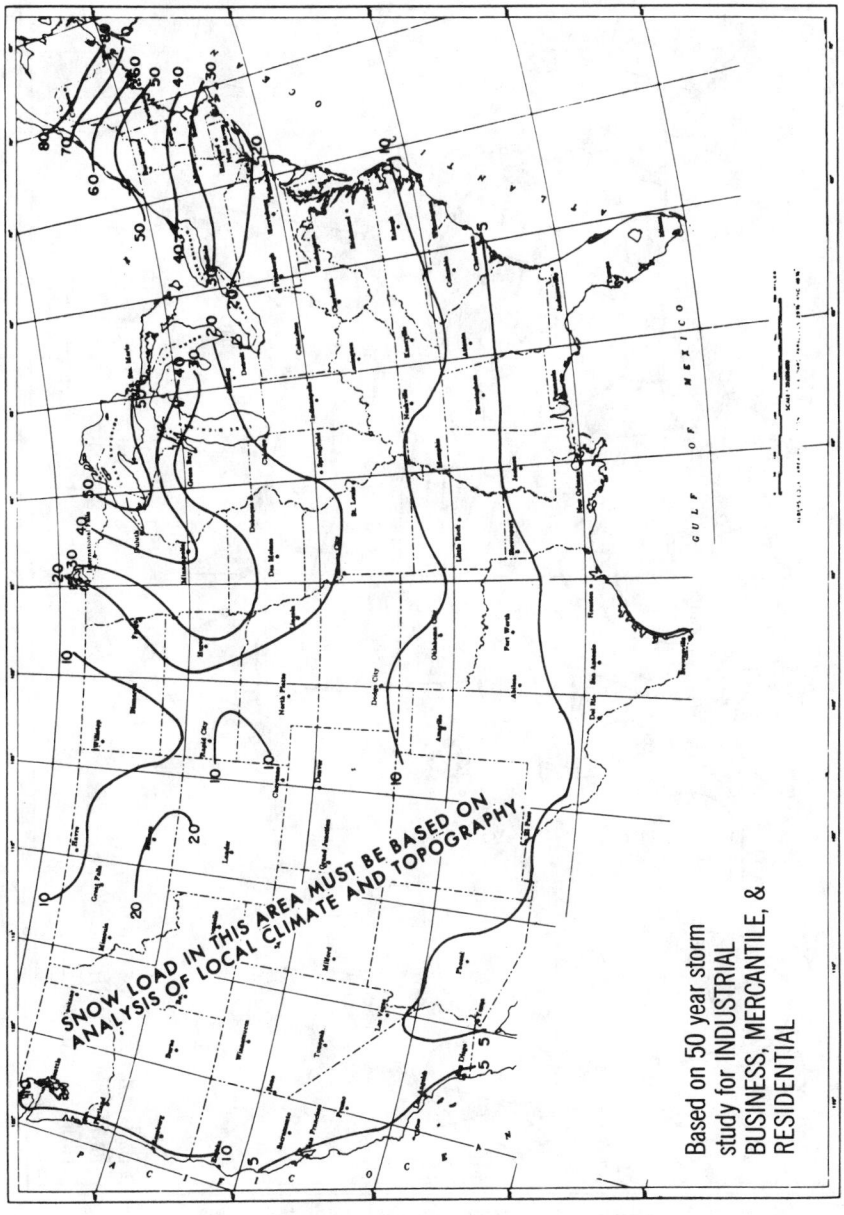

Based on 50 year storm study for INDUSTRIAL BUSINESS, MERCANTILE, & RESIDENTIAL

Snow Load in Pounds per Square Foot on the Ground

Snow Loads — 100 Year

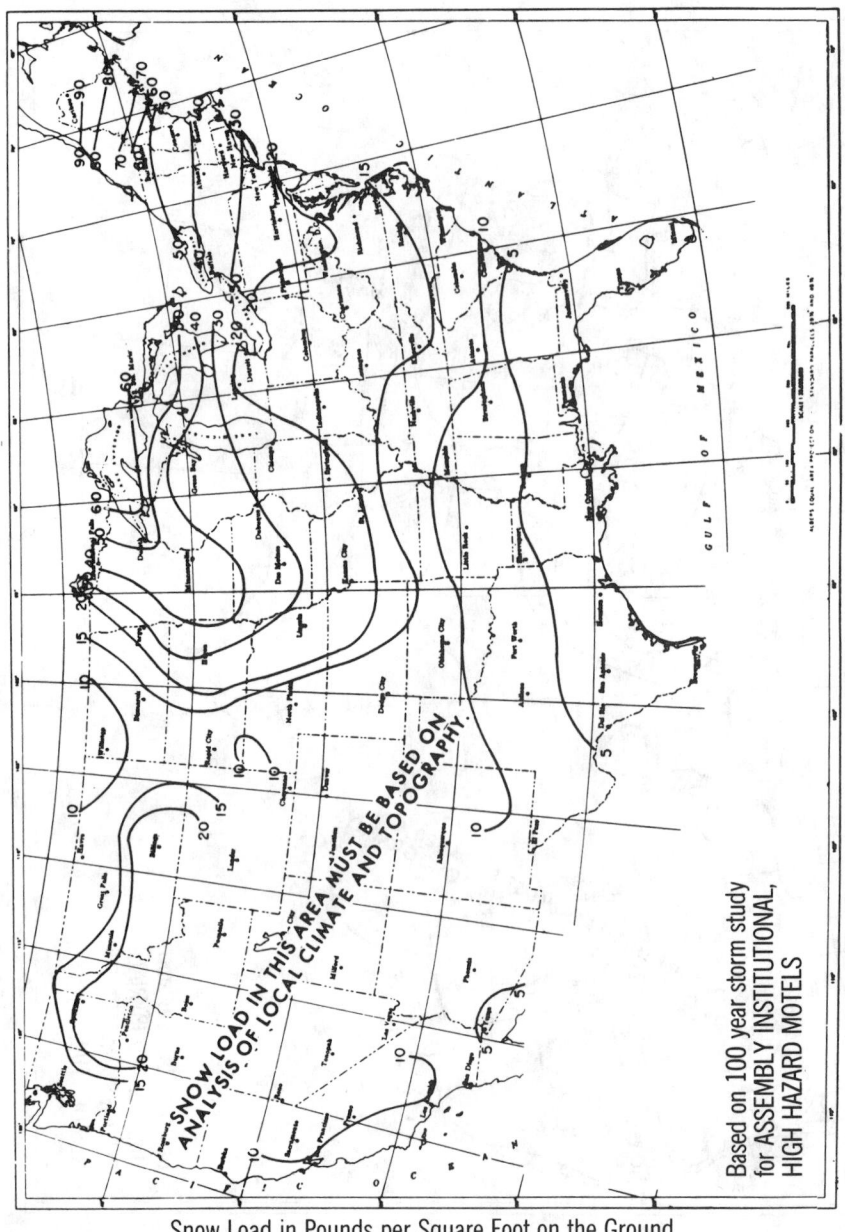

Snow Load in Pounds per Square Foot on the Ground

Frost Penetration

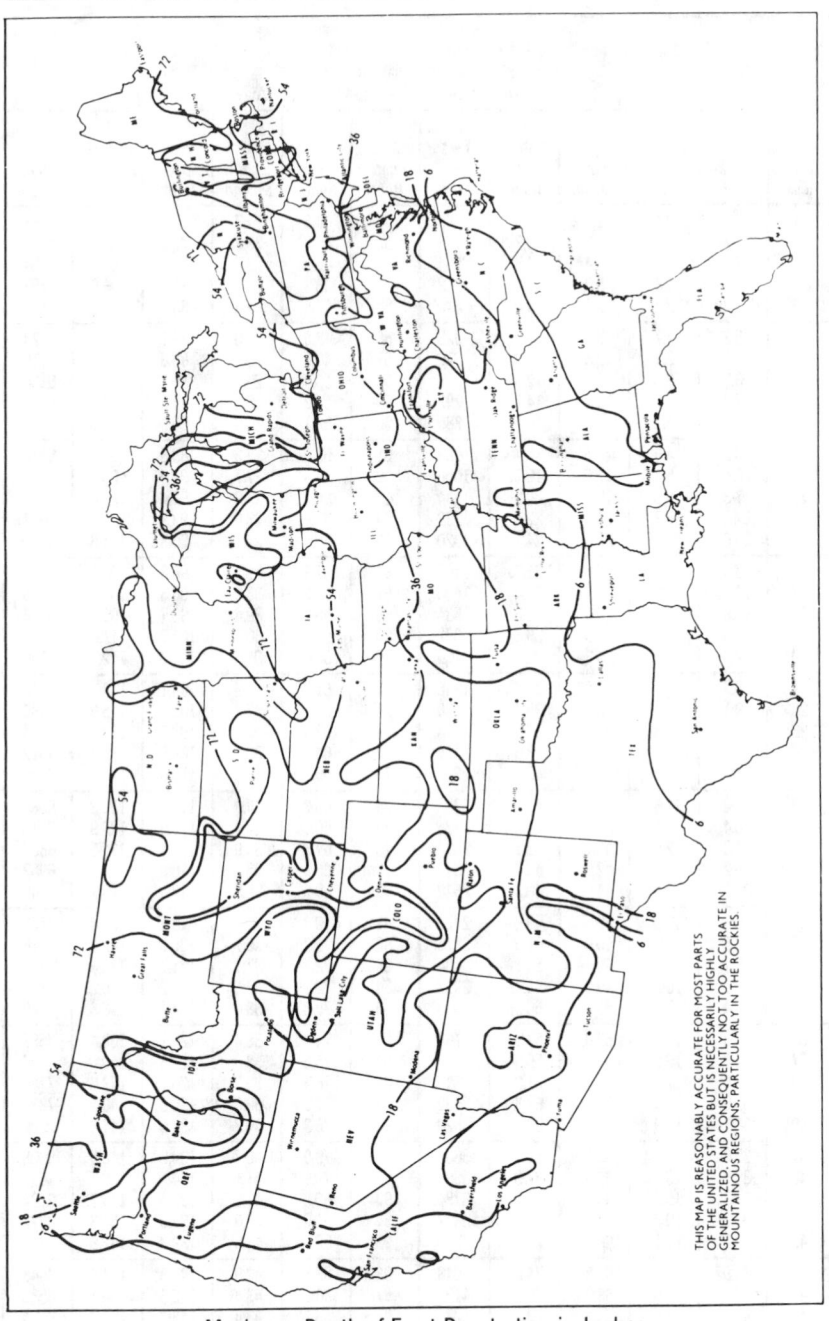

THIS MAP IS REASONABLY ACCURATE FOR MOST PARTS OF THE UNITED STATES BUT IS NECESSARILY HIGHLY GENERALIZED, AND CONSEQUENTLY NOT TOO ACCURATE IN MOUNTAINOUS REGIONS, PARTICULARLY IN THE ROCKIES.

Maximum Depth of Frost Penetration in Inches

Wind Conversion Table

					Wind Speed Units:						
	1 mile per hour	=	0.868391 knot		= 1.609344 km/hr.						
		=	1.4667 ft./sec		= 88 ft./min.						
		=	0.44704 m/sec.		= 0.34754° at lat./day						

Miles per Hour	Knots	Meters per Second	Feet per Second	Kilo-meters per Hour	Feet per Minute	Miles per Hour	Knots	Meters per Second	Feet per Second	Kilo-meters per Hour	Feet per Minute
1	0.9	0.4	1.5	1.6	88	51	44.3	22.8	74.8	82.1	4488
2	1.7	0.9	2.9	3.2	176	52	45.2	23.2	76.3	83.7	4576
3	2.6	1.3	4.4	4.8	264	53	46.0	23.7	77.7	85.3	4664
4	3.5	1.8	5.9	6.4	352	54	46.9	24.1	79.2	86.9	4752
5	4.3	2.2	7.3	8.0	440	55	47.8	24.6	80.7	88.5	4840
6	5.2	2.7	8.8	9.7	528	56	48.6	25.0	82.1	90.1	4928
7	6.1	3.1	10.3	11.3	616	57	49.5	25.5	83.6	91.7	5016
8	6.9	3.6	11.7	12.9	704	58	50.4	25.9	85.1	93.3	5104
9	7.8	4.0	13.2	14.5	792	59	51.2	26.4	86.5	95.0	5192
10	8.7	4.5	14.7	16.1	880	60	52.1	26.8	88.0	96.6	5280
11	9.6	4.9	16.1	17.7	968	61	53.0	27.3	89.5	98.2	5368
12	10.4	5.4	17.6	19.3	1056	62	53.8	27.7	90.0	99.8	5456
13	11.3	5.8	19.1	20.9	1144	63	54.7	28.2	92.4	101.4	5544
14	12.2	6.3	20.5	22.5	1232	64	55.6	28.6	93.9	103.0	5632
15	13.0	6.7	22.0	24.1	1320	65	56.4	29.1	95.3	104.6	5720
16	13.9	7.2	23.5	25.7	1408	66	57.3	29.5	96.8	106.2	5808
17	14.8	7.6	24.9	27.4	1496	67	58.2	30.0	98.3	107.8	5896
18	15.6	8.0	26.4	29.0	1584	68	59.1	30.4	99.7	109.4	5984
19	16.5	8.5	27.9	30.6	1672	69	59.9	30.8	101.2	111.0	6072
20	17.4	8.9	29.3	32.2	1760	70	60.8	31.3	102.7	112.7	6160
21	18.2	9.4	30.8	33.8	1848	71	61.7	31.7	104.1	114.3	6248
22	19.1	9.8	32.3	35.4	1936	72	62.5	32.2	105.6	115.9	6336
23	20.0	10.3	33.7	37.0	2024	73	63.4	32.6	107.1	117.5	6424
24	20.8	10.7	35.2	38.6	2112	74	64.3	33.1	108.5	119.1	6512
25	21.7	11.2	36.7	40.2	2200	75	65.1	33.5	110.0	120.7	6600
26	22.6	11.6	38.1	41.8	2288	76	66.0	34.0	111.5	122.3	6688
27	23.4	12.1	39.6	43.5	2376	77	66.9	34.4	112.9	123.9	6776
28	24.3	12.5	41.1	45.1	2464	78	67.7	34.9	114.4	125.5	6864
29	25.2	13.0	42.5	46.7	2552	79	68.6	35.3	115.9	127.1	6952
30	26.1	13.4	44.0	48.3	2640	80	69.5	35.8	117.3	128.7	7040
31	26.9	13.9	45.5	49.9	2728	81	70.3	36.2	118.8	130.4	7128
32	27.8	14.3	46.9	51.5	2816	82	71.2	36.7	120.3	132.0	7216
33	28.7	14.8	48.4	53.1	29.04	83	72.1	37.1	121.7	133.6	7304
34	29.5	15.2	49.9	54.7	2992	84	72.9	37.6	123.2	135.2	7392
35	30.4	15.6	51.3	56.3	3080	85	73.8	38.0	124.7	136.8	7480
36	31.3	16.1	52.8	57.9	3168	86	74.7	38.4	126.1	138.4	7568
37	32.1	16.5	54.3	59.5	3256	87	75.5	38.9	127.6	140.0	7656
38	33.0	17.0	55.7	61.2	3344	88	76.4	39.3	129.1	141.6	7744
39	33.9	17.4	57.2	62.8	3432	89	77.3	39.8	130.5	143.2	7832
40	34.7	17.9	58.7	64.4	3520	90	78.2	40.2	132.0	144.8	7920
41	35.6	18.3	60.1	66.0	3608	91	79.0	40.7	133.5	146.5	8008
42	36.5	18.8	61.6	67.6	3696	92	79.9	41.1	134.9	148.1	8096
43	37.3	19.2	63.1	69.2	3784	93	80.8	41.6	136.4	149.7	8184
44	38.2	19.7	64.5	70.8	3872	94	81.6	42.0	137.9	151.3	8272
45	39.1	20.1	66.0	72.4	3960	95	82.5	42.5	139.3	152.9	8360
46	39.9	20.6	67.5	74.0	4048	96	83.4	42.9	140.8	154.4	8448
47	40.8	21.0	68.9	75.6	4136	97	84.2	43.4	142.3	156.1	8536
48	41.7	21.5	70.4	77.2	4224	98	85.1	43.8	143.7	157.7	8624
49	42.6	21.9	71.9	78.9	4312	99	86.0	44.3	145.2	159.3	8712
50	43.4	22.4	73.3	80.5	4400	100	86.8	44.7	146.7	160.9	8800

Wind Chill Factors

Wind Speed M.P.H	Actual Thermometer Reading (°F)										
	50	40	30	20	10	0	−10	−20	−30	−40	−50
	Wind Chill Temperature (°F)										
0	50	40	30	20	10	0	−10	−20	−30	−40	−50
5	48	37	27	16	6	−5	−15	−26	−36	−47	−57
10	40	29	16	−4	−9	−21	−33	−46	−58	−70	−83
15	36	22	9	−5	−18	−36	−46	−58	−70	−85	−99
20	32	18	4	−10	−25	−39	−53	−67	−82	−96	−110
25	30	16	0	−15	−29	−44	−59	−74	−88	−104	−113
30	28	13	−2	−18	−33	−48	−63	−79	−94	−109	−123
35	27	11	−4	−20	−35	−49	−67	−82	−98	−113	−129
40	26	10	−6	−21	−37	−53	−69	−85	−100	−116	−132
	Very Cold				Bitter Cold			Extreme Cold			

Miscellaneous Temperature and Weather Data

Temperature

Freezing point of water = 32° Fahrenheit
= 0° Celsius

Boiling point of water (at normal air pressure) = 212° Fahrenheit
= 100° Celsius

1 degree Fahrenheit = 0.5556 degree (Celsius)
1 degree Celsius = 1.8 degrees Fahrenheit

Ice and Snow

1 cubic foot of ice at 32° F weighs 57.50 pounds; 1 pound of ice at 32° F has a volume of 0.0174 cubic foot = 30.067 cubic inches.

1 cubic foot of fresh snow, according to humidity of atmosphere, weighs 5 pounds to 12 pounds. 1 cubic foot of snow moistened and compacted by rain weighs 15 pounds to 50 pounds.

Preliminary Building Size Determination

The following tables are helpful when estimating conceptually.

With basic information such as type of occupancy and number of people intended to occupy the structure, and/or rough dimensions, it is possible to calculate other needed information.

Example:
A potential client wants to know how big a building he will need and how much it will cost. All he knows at this time is that there will be approximately 75 people working in the building. To begin a conceptual estimate, some size information is needed.

Step 1: Determine the net area. Consult the "Occupancy Determinations—Net Areas" table. The table states that for each occupant, 100 S.F. (net) of office space is required. (Note that in this case all codes have the same requirements. Check your area for the code used.)

Step 2: Determine the gross area. (Net area does not include common areas, stairwells, toilet area, mechanical areas, etc.) To obtain the gross area, consult the "Floor Area Requirements" table. The *gross to net ratio* for offices is 135%. Therefore, 100 S.F. (net) x 1.35 (gross/net), or 135 square feet per person, is needed for this office building. Since we expect an occupancy of 75 persons, the total office space will be 10,125 square feet.

(Note: In some cases, such as in apartment buildings, schools, hospitals, etc., the basic information will be in unit form, such as "I want an apartment building with 50 units," or "a hospital with 200 beds," etc. In these cases, use the "Unit Gross Area" table provided. To use this table, it is necessary to know a little about the quality of the structure. If the quality is to be "above average," then the units will be larger; if they are to be of "economy" construction, then you can assume the units will be smaller. If, in the case of an apartment building, the client wants 50 units of average quality, to obtain the overall size, multiply the number of units by the median gross square footage—in this case 50 units x 860 square feet per unit or 43,000 square feet *gross area.*)

Step 3: Determine the size. A 10,000+ square foot building lends itself nicely to a 100' x 100' economical square building. If the land available does not allow for this design, adjust the dimensions to fit. You now have the basic building size and shape to start the estimate.

Occupancy Determinations - Net Areas

	Description	S.F. Required Per Person*			
		BBC	BOCA	SBC	UBC
Assembly Areas	Fixed Seats	6	**	6	7
	Movable Seats	15		15	15
	Concentrated		7		
	Unconcentrated		15		
	Standing Space		3		
Educational	Unclassified	40			
	Classrooms		20	40	20
	Shop Areas		50	100	50
Institutional	Unclassified	150		125	
	In-Patient Areas		240		
	Sleeping Areas		120		
Mercantile	Basement	30	30	30	20
	Ground Floor	30	30	30	30
	Upper Floors	60	60	60	50
Office		100	100	100	100

* BBC=Basic Building Code
  BOCA=Building Officials & Code Administrators
  SBC=Southern Building Code
  UBC=Uniform Building Code
** The occupancy load for assembly area with fixed seats shall be determined by the number of fixed
  seats installed.

Unit Gross Area Requirements

The figures in the table below indicate typical ranges in square feet as a function of the "occupant" unit. This table is best used in the preliminary design stages to help determine the probable size requirement for the total project. The "1/4" means that of all buildings of this type built, 1/4 will be smaller in unit size and 3/4 will be larger; "median" means that 50% of the units will be smaller and 50% will be larger; the "3/4" indicates that 3/4 of the building units will be smaller and 1/4 will be larger.

Building Type	Unit	Gross Area in S.F.		
		1/4	Median	3/4
Apartments	Unit	660	860	1,100
Auditorium & Play Theaters	Seat	18	25	38
Bowling Alleys	Lane		940	
Churches & Synagogues	Seat	20	28	39
Dormitories	Bed	200	230	275
Fraternity & Sorority Houses	Bed	220	315	370
Garages, Parking	Car	325	355	385
Hospitals	Bed	685	850	1,075
Hotels	Rental Unit	475	600	710
Housing for the Elderly	Unit	515	635	755
Housing, Public	Unit	700	875	1,030
Ice Skating Rinks	Total	27,000	30,000	36,000
Motels	Rental Unit	360	465	620
Nursing Homes	Bed	290	350	450
Restaurants	Seat	23	29	39
Schools, Elementary	Pupil	65	77	90
Junior High & Middle		85	110	129
Senior High		102	130	145
Vocational		110	135	195
Shooting Ranges	Point		450	
Theaters & Movies	Seat		15	

Floor Area Ratios: Commonly Used
Gross-to-Net and Net-to-Gross Area Ratios
Expressed in % for Various Building Types

Building Type	Gross-to-Net Ratio	Net-to-Gross Ratio	Building Type	Gross-to-Net Ratio	Net-to-Gross Ratio
Apartment	156	64	School Buildings (campus type)		
Bank	140	72	Administrative	150	67
Church	142	70	Auditorium	142	70
Courthouse	162	61	Biology	161	62
Department Store	123	81	Chemistry	170	59
Garage	118	85	Classroom	152	66
Hospital	183	55	Dining Hall	138	72
Hotel	158	63	Dormitory	154	65
Laboratory	171	58	Engineering	164	61
Library	132	76	Fraternity	160	63
Office	135	75	Gymnasium	142	70
Restaurant	141	70	Science	167	60
Warehouse	108	93	Service	120	83
			Student Union	172	59

The gross area of a building is the total floor area based on outside dimensions.
The net area of a building is the usable floor area for the function intended and excludes such items as stairways, corridors and mechanical rooms. In the case of a commercial building, it might be considered as the "leasable area."

Definitions of Abbreviations Used in Formulas

F = Future amount; the amount of money you will end up with or want to end up with at the end of the time period specified.

P = Present value; the amount of money you have or need to have at the beginning of the time period specified.

n = Number of time periods being calculated (coordinate with i).

i = Interest rate per time period. (Note: if n, the number of time periods, is in days, weeks, months or years, i must also be in that time period.)

Formulas for Compound Interest

A) $F = P(1 + i)^n$

Use: To find the value of funds at the end of the number of time periods (n) if the interest rate is i per time period. A time period is the time the interest rate is compounded. For example, if you wish to know how much you will have in an account at the end of five years and the bank compounds their rate monthly at 5% interest, then F is what you are looking for, n would equal 5 years x 4 quarters per year = 20, and i = 5% per year divided by 4 quarters per year = 1.25.

B) $$P = F \left[\frac{1}{(1 + i)^n} \right] = F(1 + i)^{-n}$$

Use: To find the amount needed today (present value) to end up with the desired amount F (future value) at the end of time period n, at the interest rate of i per time period.

C) Annuity Compound Amount

$$F = A \left[\frac{(1 + i)^n - 1}{i} \right]$$

Use: To find the future value of an annuity if amount A is added (or paid) to an account at regular intervals for the time period of n intervals, at interest rate of i per interval. This formula can also be used to determine what the total of all payments (principle and interest) of a loan will be.

D) Sinking Fund

$$A = F \left[\frac{i}{(1 + i)^n - 1} \right]$$

Use: To find the amount A needed to add to an account at regular intervals, to end up with a future amount F after a time period of n intervals at an interest rate of i per interval.

E) Capital Recovery (Loan Repayment)

$$A = P \left[\frac{i (1 + i)^n}{(1 + i)^n - 1} \right] = P \left[\frac{i}{1 - (1 + i)^{-n}} \right]$$

Use: To find what payments are necessary per time interval to pay back a loan R over the number of time intervals n at an interest rate of i per time interval. For example, if you wanted to find your *monthly* payment for a loan of $100,000 spread over 30 years at a fixed rate of 10% per year, then P = $100,000, n = 30 years x 12 months per year = 360, and i = 10%/year/12 months/year = 0.833.

F) Present Value of a Loan

$$P = A \left[\frac{((1 + i)^n) - 1}{(i(1 + i)^n)} \right] = A \left[\frac{1 - (1 + i)^{-n}}{i} \right]$$

Use: To find the present value of payments, if you are making payments A over a number of time intervals n at the interest rate i per time interval

Professional Associations

ARI   Air-Conditioning and Refrigeration Institute
1501 Wilson Blvd., 6th Floor
Arlington, VA 22209

AMCA   Air Movement and Control Association
30 West University Drive
Arlington Heights, IL 60004

AA   Aluminum Association
900 19th Street, N.W., Suite 300
Washington, DC 20006

AACE   American Association of Cost Engineers
308 Monongahela Blvd.
Morgantown, WV 26505

AAN   American Association of Nurserymen, Inc.
1250 I St., N.W., Suite 500
Washington, DC 20005

ACI   American Concrete Institute
Box 19150
Reford Station
Detroit, MI 48219

ACEC   American Consulting Engineers Council
1015 15th Street, N.W.
Washington, D.C. 20005

AIA   American Institute of Architects
1735 New York Avenue, N.W.
Washington, D.C. 20006

AIC   American Institute of Constructors
20 South Front Street
Columbus, OH 43215

AISC   American Institute of Steel Construction
400 North Michigan Avenue
Eighth Floor
Chicago, IL 60611

AITC   American Institute of Timber Construction
333 W. Hampden Ave.
Englewood, CO 80110

AISI   American Iron and Steel Institute
1133 15th Street, N.W.
Washington, DC 20036

ANSI   American National Standards Institute
            1430 Broadway
            New York, NY 10018

APA   American Plywood Association
            Box 11700
            Tacoma, WA 98411

ASCE   American Society of Civil Engineers
            345 East 47th Street
            New York, NY 10017

ASHRAE   American Society of Heating,
            Refrigerating and Air Conditioning Engineers
            1791 Tullie Circle, N.E.
            Atlanta, GA 30329

ASLA   American Society of Landscape Architects
            1733 Connecticut Ave., N.W.
            Washington, DC 20009

ASME   American Society of Mechanical Engineers
            345 Each 47th Street
            New York, NY 10017

ASPE   American Society of Professional Estimators
            6911 Richmond, Hwy., Suite 230
            Alexandria, VA 22306

ASTM   American Society for Testing and Materials
            1916 Race Street
            Philadelphia, PA 19103

ASA   American Subcontractors Association
            1004 Duke Street
            Alexandria, VA 22314

AWWA   American Water Works Association
            6666 West Quincy Avenue
            Denver, CO 80235

AWS   American Welding Society
            550 N.W. 42nd Avenue
            Miami, FL 33136

AWC   American Wood Council
            1250 Connecticut Avenue, N.W.
            Suite 230
            Washington, DC 20036

AWPA   American Wood-Preservers' Association
            1945 Old Gallows
            Vienna, VA 22182

AWI   Architectural Woodwork Institute
2310 South Walter Reed Drive
Arlington, VA 22206

AI   Asphalt Institute
Asphalt Institute Building
College Park, MD 20740-1802

ABC   Associated Builders & Contractors, Inc.
729 15th Street, N.W.
Washington, DC 20005

AED   Associated Equipment Distributors
615 W. 22nd Street
Oak Brook, IL 60151

AGC   Associated General Contractors of America
1957 E. Street, N.W.
Washington, DC 20006

ALCA   Associated Landscape Contractors of America
405 Washington Street
North Falls Church, VA 20046

BIA   Brick Institute of America
1750 Old Meadow Road
McLean, VA 22102

BHMA   Builders Hardware Mfg. Assoc.
60 E. 42nd Street, Rm. 511
New York, NY 10165

CFMA   Construction Financial Management Association
40 Brunswick Ave., Suite 202
Edison, NJ 08818

CIMA   Construction Industry Mfg. Assoc.
Marine Plaza, Suite 1700
111 E. Wisconsin Ave.
Milwaukee, WI 53202

CMAA   Construction Management Association of America
12355 Sunrise Valley Drive, Suite 640
Reston, VA 22091

CSI   Construction Specifications Institute
601 Madison Street
Alexandria, VA 22314

CRSI   Concrete Reinforcing Steel Institute
933 N. Plum Grove Road
Schaumberg, IL 60195

CDA   Copper Development Association
           Greenwich Office Park II
           Box 1840
           Greenwich, CT 06836

DHI   Door & Hardware Institute
           7711 Old Springhouse Rd.
           McLean, VA 22102

FMRC   Factory Mutual Engineering & Research Group
           1151 Boston-Providence Turnpike
           Norwood, MA 02062

FGMA   Flat Glass Marketing Association
           White Lakes Professional Building
           3310 Harrison
           Topeka, KS 66611

GSA   Federal Specification
           General Services Administration
           Specifications and Consumer Information
           Distribution Section (WFSIS)
           7th & D Street, S.W., Room 6654
           Washington, DC 20407

GA   Gypsum Association
           1603 Orrington Avenue
           Evanston, IL 60201

IEEE   Institute of Electrical and Electronic Engineers
           345 East 47th Street
           New York, NY 10017

IMIAC   International Masonry Industry All-Weather Council
           International Masonry Institute
           815 15th Street, N.W.
           Washington, D.C. 20005

MCAA   Mechanical Contractors Association of America, Inc.
           1385 Piccard Drive
           Rockville, MD 20850

ML/SFA   Metal Lath/Steel Framing Association
           600 South Federal, Suite 400
           Chicago, IL 60605

NAAMM   National Association of Architectural Metal Manufacturers
           600 South Federal, Suite 400
           Chicago, IL 60605

NAHB   National Association of Home Builders
          15th & M Street, N.W.
          Washington, DC 20005

NAPHCC   National Association of Plumbing, Heating, and Cooling
              Contractors
          1016 20th Street, N.W.
          Washington, DC 20036

NECA   National Electrical Contractors Association
          7315 Wisconsin Avenue
          Bethesda, MD 20814

NEMA   National Electrical Manufacturers' Association
          2101 L. Street, N.W., Suite 300
          Washington, DC 20037

NIBS   National Institute of Building Science
          1015 15th St. N.W., Suite 700
          Washington, DC 20005

NFPA   National Fire Protection Association
          Batterymarch Park
          Quincy, MA 02269

NFPA   National Forest Products Association
          1250 Connecticut Avenue, Suite 200
          Washington, DC 20036

NPCA   National Precast Concrete Association
          825 E. 64th Street
          Indianapolis, IN 46220

NRCA   National Roofing Contractors Association
          6250 River Road
          Rosemont, IL 60018

NSPE   National Society of Professional Engineers
          1420 King Street
          Alexandria, VA 22314

NSWMA   National Solid Wastes Management Association
          1730 Rhode Island Avenue, N.W.
          Suite 1000
          Washington, D.C. 20036

NUCA   National Utility Contractors Association
          1235 Jefferson Davis Hwy., Suite 606
          Arlington, VA 22202

PCA   Portland Cement Association
5420 Old Orchard Road
Skokie, IL 60077

PCEAA   Professional Construction Estimators Association
P.O. Box 1107
Cornelius, NC 28031

PS   Product Standard
U.S. Department of Commerce
14th & Constitutional Avenue
Washington, DC 20230

SIGMA   Sealed Insulating Glass Manfuacturers Association
111 East Wacker Drive, Suite 600
Chicago, IL 60601

SMACNA   Sheet Metal and Air Conditioning Contractors' National
Association
8224 Old Court House Road
Vienna, VA 22180

SDI   Steel Door Institute
712 Lakewood Center North
14600 Detroit Avenue
Cleveland, OH 44107

SSPC   Steel Structures Painting Council
4400 Fifth Avenue
Pittsburgh, PA 15213

TAS   Technical Aid Series
Construction Specifications Institute
601 North Madison Street
Alexandria, VA 22314

TCA   Tile Council of America, Inc.
Box 326
Princeton, NJ 08542-0326

UL   Underwriters' Laboratories, Inc.
333 Pfingsten Road
Northbrook, IL 60062

USFS   U.S. Forest Service
Forest Products Laboratory
P.O. Box 5130
Madison, WI 53705

WCLIB   West Coast Lumber Inspection Bureau
Box 23145
Portland, OR 97223

WWPA   Western Wood Products Association
Yeon Bldg.
522 S.W. 5th Ave.
Portland, OR 97204

Abbreviations

A	Area Square Feet; Ampere	Brng.	Bearing
ABS	Acrylonitrile Butadiene Styrene;	Brs.	Brass
	Asbestos Bonded Steel	Brz.	Bronze
A.C.	Alternating Current;	Bsn.	Basin
	Air Conditioning;	Btr.	Better
	Asbestos Cement	BTU	British Thermal Unit
A.C.I.	American Concrete Institute	BTUH	BTU per Hour
Addit.	Additional	BX	Interlocked Armored Cable
Adj.	Adjustable	c	Conductivity
af	Audio-frequency	C	Hundred; Centigrade
A.G.A.	American Gas Association	C/C	Center to Center
Agg.	Aggregate	Cab.	Cabinet
A.H.	Ampere Hours	Cair.	Air Tool Laborer
A hr	Ampere-hour	Calc	Calculated
A.H.U.	Air Handling Unit	Cap.	Capacity
A.I.A.	American Institute of Architects	Carp.	Carpenter
AIC	Ampere Interrupting Capacity	C.B.	Circuit Breaker
Allow.	Allowance	C.C.A.	Chromate Copper Arsenate
alt.	Altitude	C.C.F.	Hundred Cubic Feet
Alum.	Aluminum	cd	Candela
a.m.	Ante Meridiem	cd/sf	Candela per Square Foot
Amp.	Ampere	CD	Grade of Plywood Face & Back
Anod.	Anodized	CDX	Plywood, grade C&D, exterior glue
Approx.	Approximate	Cefi.	Cement Finisher
Apt.	Apartment	Cem.	Cement
Asb.	Asbestos	CF	Hundred Feet
A.S.B.C.	American Standard Building Code	C.F.	Cubic Feet
Asbe.	Asbestos Worker	CFM	Cubic Feet per Minute
A.S.H.R.A.E.	American Society of Heating,	c.g.	Center of Gravity
	Refrig. & AC Engineers	CHW	Chilled Water
A.S.M.E.	American Society of	C.I.	Cast Iron
	Mechanical Engineers	C.I.P.	Cast in Place
A.S.T.M.	American Society for	Circ.	Circuit
	Testing and Materials	C.L.	Carload Lot
Attchmt.	Attachment	Clab.	Common Laborer
Avg.	Average	C.L.F.	Hundred Linear Feet
A.W.G.	American Wire Gauge	CLF	Current Limiting Fuse
Bbl.	Barrel	CLP	Cross Linked Polyethylene
B.&B.	Grade B and Better;	cm	Centimeter
	Balled & Burlapped	CMP	Corr. Metal Pipe
B.&S.	Bell and Spigot	C.M.U.	Concrete Masonry Unit
B.&W.	Black and White	Col.	Column
b.c.c.	Body-centered Cubic	CO_2	Carbon Dioxide
BE	Bevel End	Comb.	Combination
B.F.	Board Feet	Compr.	Compressor
Bg. Cem.	Bag of Cement	Conc.	Concrete
BHP	Boiler Horse Power	Cont.	Continuous; Continued
	Brake Horse Power	Corr.	Corrugated
B.I.	Black Iron	Cos	Cosine
Bit.; Bitum.	Bituminous	Cot	Cotangent
Bk.	Backed	Cov.	Cover
Bkrs.	Breakers	CPA	Control Point Adjustment
Bldg.	Building	Cplg.	Coupling
Blk.	Block	C.P.M.	Critical Path Method
Bm.	Beam	CPVC	Chlorinated Polyvinyl Chloride
Boil.	Boilermaker	C. Pr.	Hundred Pair
B.P.M.	Blows per Minute	CRC	Cold Rolled Channel
BR	Bedroom	Creos.	Creosote
Brg.	Bearing	Crpt.	Carpet & Linoleum Layer
Brhe.	Bricklayer Helper	CRT	Cathode-Ray Tube
Bric.	Bricklayer	CS	Carbon Steel
Brk.	Brick	Csc	Cosecant

Abbreviations (continued)

C.S.F.	Hundred Square Feet	Equip.	Equipment
CSI	Construction Specification	ERW	Electric Resistance Welded
	Institute	Est.	Estimated
C.T.	Current Transformer	esu	Electrostatic Units
CTS	Copper Tube Size	E.W.	Each Way
Cu	Cubic	EWT	Entering Water Temperature
Cu. Ft.	Cubic Foot	Excav.	Excavation
cw	Continuous Wave	Exp.	Expansion
C.W.	Cool White; Cold Water	Ext.	Exterior
Cwt.	100 Pounds	Extru.	Extrusion
C.W.X.	Cool White Deluxe	f.	Fiber stress
C.Y.	Cubic Yard (27 cubic feet)	F	Fahrenheit; Female; Fill
C.Y./Hr.	Cubic Yard per Hour	Fab.	Fabricated
Cyl.	Cylinder	FBGS	Fiberglass
d	Penny (nail size)	F.C.	Footcandles
D	Deep; Depth; Discharge	f.c.c.	Face-centered Cubic
Dis.; Disch.	Discharge	f'c.	Compressive Stress in Concrete;
Db.	Decibel		Extreme Compressive Stress
Dbl.	Double	F.E.	Front End
DC	Direct Current	FEP	Fluorinated Ethylene
Demob.	Demobilization		Propylene (Teflon)
d.f.u.	Drainage Fixture Units	F.G.	Flat Grain
D.H.	Double Hung	F.H.A.	Federal Housing Administration
DHW	Domestic Hot Water	Fig.	Figure
Diag.	Diagonal	Fin.	Finished
Diam.	Diameter	Fixt.	Fixture
Distrib.	Distribution	Fl. Oz.	Fluid Ounces
Dk.	Deck	Flr.	Floor
D.L.	Dead Load; Diesel	F.M.	Frequency Modulation;
Do.	Ditto		Factory Mutual
Dp.	Depth	Fmg.	Framing
D.P.S.T.	Double Pole, Single Throw	Fndtn.	Foundation
Dr.	Driver	Fori.	Foreman, inside
Drink.	Drinking	Fount.	Fountain
D.S.	Double Strength	FPM	Feet per Minute
D.S.A.	Double Strength A Grade	FPT	Female Pipe Thread
D.S.B.	Double Strength B Grade	Fr.	Frame
Dty.	Duty	F.R.	Fire Rating
DWV	Drain Waste Vent	FRK	Foil Reinforced Kraft
DX	Deluxe White, Direct Expansion	FRP	Fiberglass Reinforced Plastic
dyn	Dyne	FS	Forged Steel
e	Eccentricity	FSC	Cast Body; Cast Switch Box
E	Equipment Only; East	Ft.	Foot; Feet
Ea.	Each	Ftng.	Fitting
E.B.	Encased Burial	Ftg.	Footing
Econ.	Economy	Ft. Lb.	Foot Pound
EDP	Electronic Data Processing	Furn.	Furniture
E.D.R.	Equiv. Direct Radiation	FVNR	Full Voltage Non-Reversing
Eq.	Equation	FXM	Female by Male
Elec.	Electrician; Electrical	Fy.	Minimum Yield Stress of Steel
Elev.	Elevator; Elevating	g	Gram
EMT	Electrical Metallic Conduit;	G	Gauss
	Thin Wall Conduit	Ga.	Gauge
Eng.	Engine	Gal.	Gallon
EPDM	Ethylene Propylene	Gal./Min.	Gallon per Minute
	Diene Monomer	Galv.	Galvanized
Eqhv.	Equip. Oper., heavy	Gen.	General
Eqlt.	Equip. Oper., light	G.F.I.	Ground Fault Interrupter
Eqmd.	Equip. Oper., medium	Glaz.	Glazier
Eqmm.	Equip. Oper., Master Mechanic	GPD	Gallons per Day
Eqol.	Equip. Oper., oilers	GPH	Gallons per Hour

Abbreviations (continued)

GPM	Gallons per Minute	Km	Kilometer
GR	Grade	K.L.F.	Kips per Linear Foot
Gran.	Granular	K.S.F.	Kips per Square Foot
Grnd.	Ground	K.S.I.	Kips per Square Inch
H	High; High Strength Bar Joist;	K.V.	Kilovolt
	Henry	K.V.A.	Kilovolt Ampere
H.C.	High Capacity	K.V.A.R.	Kilovar (Reactance)
H.D.	Heavy Duty; High Density	KW	Kilowatt
H.D.O.	High Density Overlaid	KWh	Kilowatt-hour
Hdr.	Header	L	Labor Only; Length; Long;
Hdwe.	Hardware		Medium Wall Copper Tubing
Help.	Helper average	Lab.	Labor
HEPA	High Efficiency Particulate Air Filter	lat	Latitude
Hg	Mercury	Lath.	Lather
HIC	High Interrupting Capacity	Lav.	Lavatory
H.O.	High Output	lb.; #	Pound
Horiz.	Horizontal	L.B.	Load Bearing; L Conduit Body
H.P.	Horsepower; High Pressure	L. & E.	Labor & Equipment
H.P.F.	High Power Factor	lb./hr.	Pounds per Hour
Hr.	Hour	lb./L.F.	Pounds per Linear Foot
Hrs./Day	Hours per Day	lbf/sq in.	Pound-force per Square Inch
HSC	High Short Circuit	L.C.L.	Less than Carload Lot
Ht.	Height	Ld.	Load
Htg.	Heating	L.F.	Linear Foot
Htrs.	Heaters	Lg.	Long; Length; Large
HVAC	Heating, Ventilating &	L. & H.	Light and Heat
	Air Conditioning	L.H.	Long Span High Strength Bar Joist
Hvy.	Heavy	L.J.	Long Span Standard Strength
HW	Hot Water		Bar Joist
Hyd.; Hydr.	Hydraulic	L.L.	Live Load
Hz.	Hertz (cycles)	L.L.D.	Lamp Lumen Depreciation
I.	Moment of Inertia	lm	Lumen
I.C.	Interrupting Capacity	lm/sf	Lumen per Square Foot
ID	Inside Diameter	lm/W	Lumen per Watt
I.D.	Inside Dimension;	L.O.A.	Length Over All
	Identification	log	Logarithm
I.F.	Inside Frosted	L.P.	Liquefied Petroleum;
I.M.C.	Intermediate Metal Conduit		Low Pressure
In.	Inch	L.P.F.	Low Power Factor
Incan.	Incandescent	LR	Long Radius
Incl.	Included; Including	L.S.	Lump Sum
Int.	Interior	Lt.	Light
Inst.	Installation	Lt. Ga.	Light Gauge
Insul.	Insulation	L.T.L.	Less than Truckload Lot
I.P.	Iron Pipe	Lt. Wt.	Lightweight
I.P.S.	Iron Pipe Size	L.V.	Low Voltage
I.P.T.	Iron Pipe Threaded	M	Thousand; Material; Male;
I.W.	Indirect Waste		Light Wall Copper Tubing
J	Joule	m/hr	Man-hour
J.I.C.	Joint Industrial Council	mA	Milliampere
K	Thousand; Thousand Pounds;	Mach.	Machine
	Heavy Wall Copper Tubing	Mag. Str.	Magnetic Starter
K.A.H.	Thousand Amp. Hours	Maint.	Maintenance
K.D.A.T.	Kiln Dried After Treatment	Marb.	Marble Setter
kg	Kilogram	Mat; Mat'l.	Material
kG	Kilogauss	Max.	Maximum
kgf	Kilogram force	MBF	Thousand Board Feet
kHz	Kilohertz	MBH	Thousand BTU's per hr.
Kip	1000 Pounds	MC	Metal Clad Cable
KJ	Kiljoule	M.C.F.	Thousand Cubic Feet
K.L.	Effective Length Factor		

Abbreviations (continued)

M.C.F.M.	Thousand Cubic Feet per minute	N.R.C.	Noise Reduction Coefficient
M.C.M.	Thousand Circular Mils	N.R.S.	Non Rising Stem
M.C.P.	Motor Circuit Protector	ns	Nanosecond
MD	Medium Duty	nW	Nanowatt
M.D.O.	Medium Density Overlaid	OB	Opposing Blade
Med.	Medium	OC	On Center
MF	Thousand Feet	OD	Outside Diameter
M.F.B.M.	Thousand Feet Board Measure	O.D.	Outside Dimension
Mfg.	Manufacturing	ODS	Overhead Distribution System
Mfrs.	Manufacturers	O & P	Overhead and Profit
mg	Milligram	Oper.	Operator
MGD	Million Gallons per Day	Opng.	Opening
MGPH	Thousand Gallons per Hour	Orna.	Ornamental
MH	Manhole; Metal Halide; Man-Hour	O.S.&Y.	Outside Screw and Yoke
MHz	Megahertz	Ovhd.	Overhead
Mi.	Mile	OWG	Oil, Water or Gas
MI	Malleable Iron; Mineral Insulated	Oz.	Ounce
mm	Millimeter	P.	Pole; Applied Load; Projection
Mill.	Millwright	p.	Page
Min.	Minimum	Pape.	Paperhanger
Misc.	Miscellaneous	P.A.P.R.	Powered Air Purifying Respirator
ml	Milliliter	PAR	Weatherproof Reflector
M.L.F.	Thousand Linear Feet	Pc.	Piece
Mo.	Month	P.C.	Portland Cement; Power Connector
Mobil.	Mobilization	P.C.M.	Phase Contrast Microscopy
Mog.	Mogul Base	P.C.F.	Pounds per Cubic Foot
MPH	Miles per Hour	P.E.	Professional Engineer; Porcelain Enamel; Polyethylene; Plain End
MPT	Male Pipe Thread		
MRT	Mile Round Trip		
ms	Millisecond	Perf.	Perforated
M.S.F.	Thousand Square Feet	Ph.	Phase
Mstz.	Mosaic & Terrazzo Worker	P.I.	Pressure Injected
M.S.Y.	Thousand Square Yards	Pile.	Pile Driver
Mtd.	Mounted	Pkg.	Package
Mthe.	Mosaic & Terrazzo Helper	Pl.	Plate
Mtng.	Mounting	Plah.	Plasterer Helper
Mult.	Multi; Multiply	Plas.	Plasterer
M.V.A.	Million Volt Amperes	Pluh.	Plumbers Helper
M.V.A.R.	Million Volt Amperes Reactance	Plum.	Plumber
MV	Megavolt	Ply.	Plywood
MW	Megawatt	p.m.	Post Meridiem
MXM	Male by Male	Pord.	Painter, Ordinary
MYD	Thousand yards	pp	Pages
N	Natural; North	PP; PPL	Polypropylene
nA	Nanoampere	P.P.M.	Parts per Million
NA	Not Available; Not Applicable	Pr.	Pair
N.B.C.	National Building Code	Prefab.	Prefabricated
NC	Normally Closed	Prefin.	Prefinished
N.E.M.A.	National Electrical Manufacturers Association	Prop.	Propelled
		PSF; psf	Pounds per Square Foot
NEHB	Bolted Circuit Breaker to 600V.	PSI; psi	Pounds per Square Inch
N.L.B.	Non-Load-Bearing	PSIG	Pounds per Square Inch Gauge
NM	Non-Metallic Cable	PSP	Plastic Sewer Pipe
nm	Nanometer	Pspr.	Painter, Spray
No.	Number	Psst.	Painter, Structural Steel
NO	Normally Open	P.T.	Potential Transformer
N.O.C.	Not Otherwise Classified	P. & T.	Pressure & Temperature
Nose.	Nosing	Ptd.	Painted
N.P.T.	National Pipe Thread	Ptns.	Partitions
NQOB	Bolted Circuit Breaker to 240V.	Pu	Ultimate Load

Abbreviations (continued)

PVC	Polyvinyl Chloride	S.S.	Single Strength; Stainless Steel
Pvmt.	Pavement	S.S.B.	Single Strength B Grade
Pwr.	Power	Sswk.	Structural Steel Worker
Q	Quantity Heat Flow	Sswl.	Structural Steel Welder
Quan.; Qty.	Quantity	St.; Stl.	Steel
Q.C.	Quick Coupling	S.T.C.	Sound Transmission Coefficient
r	Radius of Gyration	Std.	Standard
R	Resistance	STP	Standard Temperature & Pressure
R.C.P.	Reinforced Concrete Pipe	Stpi.	Steamfitter, Pipefitter
Rect.	Rectangle	Str.	Strength; Starter; Straight
Reg.	Regular	Strd.	Stranded
Reinf.	Reinforced	Struct.	Structural
Req'd.	Required	Sty.	Story
Resi	Residential	Subj.	Subject
Rgh.	Rough	Subs.	Subcontractors
R.H.W.	Rubber, Heat & Water Resistant;	Surf.	Surface
	Residential Hot Water	Sw.	Switch
rms	Root Mean Square	Swbd.	Switchboard
Rnd.	Round	S.Y.	Square Yard
Rodm.	Rodman	Syn.	Synthetic
Rofc.	Roofer, Composition	Sys.	System
Rofp.	Roofer, Precast	t.	Thickness
Rohe.	Roofer Helpers (Composition)	T	Temperature; Ton
Rots.	Roofer, Tile & Slate	Tan	Tangent
R.O.W.	Right of Way	T.C.	Terra Cotta
RPM	Revolutions per Minute	T & C	Threaded and Coupled
R.R.	Direct Burial Feeder Conduit	T.D.	Temperature Difference
R.S.	Rapid Start	T.E.M.	Transmission Electron Microscopy
RT	Round Trip	TFE	Tetrafluoroethylene (Teflon)
S.	Suction; Single Entrance;	T. & G.	Tongue & Groove;
	South		Tar & Gravel
Scaf.	Scaffold	Th.; Thk.	Thick
Sch.; Sched.	Schedule	Thn.	Thin
S.C.R.	Modular Brick	Thrded	Threaded
S.D.	Sound Deadening	Tilf.	Tile Layer Floor
S.D.R.	Standard Dimension Ratio	Tilh.	Tile Layer Helper
S.E.	Surfaced Edge	THW	Insulated Strand Wire
S.E.R.; S.E.U.	Service Entrance Cable	THWN;	
S.F.	Square Foot	THHN	Nylon Jacketed Wire
S.F.C.A.	Square Foot Contact Area	T.L.	Truckload
S.F.G.	Square Foot of Ground	Tot.	Total
S.F. Hor.	Square Foot Horizontal	T.S.	Trigger Start
S.F.R.	Square Feet of Radiation	Tr.	Trade
S.F.Shlf.	Square Foot of Shelf	Transf.	Transformer
S4S	Surface 4 Sides	Trhv.	Truck Driver, Heavy
Shee.	Sheet Metal Worker	Trir.	Trailer
Sin.	Sine	Trlt.	Truck Driver, Light
Skwk.	Skilled Worker	TV	Television
SL	Saran Lined	T.W.	Thermoplastic Water
S.L.	Slimline		Resistant Wire
Sldr.	Solder	UCI	Uniform Construction Index
S.N.	Solid Neutral	UF	Underground Feeder
S.P.	Static Pressure; Single Pole;	U.H.F.	Ultra High Frequency
	Self Propelled	U.L.	Underwriters Laboratory
Spri.	Sprinkler Installer	Unfin.	Unfinished
Sq.	Square; 100 square feet	URD	Underground Residential
S.P.D.T.	Single Pole, Double Throw		Distribution
S.P.S.T.	Single Pole, Single Throw	V	Volt
SPT	Standard Pipe Thread	V.A.	Volt Amperes
Sq. Hd.	Square Head	V.A.C.	Vinyl Composition Tile
Sq. In.	Square Inch	VAV	Variable Air Volume

Abbreviations (continued)

Vent.	Ventilating	W.S.P.	Water, Steam, Petroleum
Vert.	Vertical	WT, Wt.	Weight
V.F.	Vinyl Faced	WWF	Welded Wire Fabric
V.G.	Vertical Grain	XFMR	Transformer
V.H.F.	Very High Frequency	XHD	Extra Heavy Duty
VHO	Very High Output	XHHW; XLPE	Cross-Linked Polyethylene
Vib.	Vibrating		Wire Insulation
V.L.F.	Vertical Linear Foot	Y	Wye
Vol.	Volume	yd	Yard
W	Wire; Watt; Wide; West	yr	Year
w/	With	Δ	Delta
W.C.	Water Column; Water Closet	%	Percent
W.F.	Wide Flange	~	Approximately
W.G.	Water Gauge	Ø	Phase
Wldg.	Welding	@	At
W. Mile	Wire Mile	#	Pound; Number
W.R.	Water Resistant	<	Less Than
Wrck.	Wrecker	>	Greater Than

Notes

Index

Index

Notes